RULES OF DIFFERENTIATION

Basic Formulas

1. $\dfrac{d}{dx}(c) = 0$

2. $\dfrac{d}{dx}(cu) = cu\dfrac{du}{dx}$

3. $\dfrac{d}{dx}(u \pm v) = \dfrac{du}{dx} \pm \dfrac{dv}{dx}$

4. $\dfrac{d}{dx}(uv) = u\dfrac{dv}{dx} + v\dfrac{du}{dx}$

5. $\dfrac{d}{dx}\left(\dfrac{u}{v}\right) = \dfrac{v\dfrac{du}{dx} - u\dfrac{dv}{du}}{v^2}$

6. $\dfrac{d}{dx}f(g(x)) = f'(g(x))g'(x)$

7. $\dfrac{d}{dx}(u^n) = nu^{n-1}\dfrac{du}{dx}$

Exponential and Logarithmic Functions

8. $\dfrac{d}{dx}(e^u) = e^u\dfrac{du}{dx}$

9. $\dfrac{d}{dx}(a^u) = (\ln a)a^u\dfrac{du}{dx}$

10. $\dfrac{d}{dx}\ln|u| = \dfrac{1}{u}\dfrac{du}{dx}$

11. $\dfrac{d}{dx}(\log_a u) = \dfrac{1}{u\ln a}\dfrac{du}{dx}$

Trigonometric Functions

12. $\dfrac{d}{dx}(\sin u) = \cos u\dfrac{du}{dx}$

13. $\dfrac{d}{dx}(\cos u) = -\sin u\dfrac{du}{dx}$

14. $\dfrac{d}{dx}(\tan u) = \sec^2 u\dfrac{du}{dx}$

15. $\dfrac{d}{dx}(\csc u) = -\csc u\cot u\dfrac{du}{dx}$

16. $\dfrac{d}{dx}(\sec u) = \sec u\tan u\dfrac{du}{dx}$

17. $\dfrac{d}{dx}(\cot u) = -\csc^2 u\dfrac{du}{dx}$

Inverse Trigonometric Functions

18. $\dfrac{d}{dx}(\sin^{-1} u) = \dfrac{1}{\sqrt{1 - u^2}}\dfrac{du}{dx}$

19. $\dfrac{d}{dx}(\cos^{-1} u) = -\dfrac{1}{\sqrt{1 - u^2}}\dfrac{du}{dx}$

20. $\dfrac{d}{dx}(\tan^{-1} u) = \dfrac{1}{1 + u^2}\dfrac{du}{dx}$

21. $\dfrac{d}{dx}(\csc^{-1} u) = -\dfrac{1}{|u|\sqrt{u^2 - 1}}\dfrac{du}{dx}$

22. $\dfrac{d}{dx}(\sec^{-1} u) = \dfrac{1}{|u|\sqrt{u^2 - 1}}\dfrac{du}{dx}$

23. $\dfrac{d}{dx}(\cot^{-1} u) = -\dfrac{1}{1 + u^2}\dfrac{du}{dx}$

Hyperbolic Functions

24. $\dfrac{d}{dx}(\sinh u) = \cosh u\dfrac{du}{dx}$

25. $\dfrac{d}{dx}(\cosh u) = \sinh u\dfrac{du}{dx}$

26. $\dfrac{d}{dx}(\tanh u) = \operatorname{sech}^2 u\dfrac{du}{dx}$

27. $\dfrac{d}{dx}(\operatorname{csch} u) = -\operatorname{csch} u\coth u\dfrac{du}{dx}$

28. $\dfrac{d}{dx}(\operatorname{sech} u) = -\operatorname{sech} u\tanh u\dfrac{du}{dx}$

29. $\dfrac{d}{dx}(\coth u) = -\operatorname{csch}^2 u\dfrac{du}{dx}$

Inverse Hyperbolic Functions

30. $\dfrac{d}{dx}(\sinh^{-1} u) = \dfrac{1}{\sqrt{1 + u^2}}\dfrac{du}{dx}$

31. $\dfrac{d}{dx}(\cosh^{-1} u) = \dfrac{1}{\sqrt{u^2 - 1}}\dfrac{du}{dx}$

32. $\dfrac{d}{dx}(\tanh^{-1} u) = \dfrac{1}{1 - u^2}\dfrac{du}{dx}$

33. $\dfrac{d}{dx}(\operatorname{csch}^{-1} u) = -\dfrac{1}{|u|\sqrt{u^2 + 1}}\dfrac{du}{dx}$

34. $\dfrac{d}{dx}(\operatorname{sech}^{-1} u) = -\dfrac{1}{u\sqrt{1 - u^2}}\dfrac{du}{dx}$

35. $\dfrac{d}{dx}(\coth^{-1} u) = \dfrac{1}{1 - u^2}\dfrac{du}{dx}$

TABLE OF INTEGRALS

Basic Forms

1. $\displaystyle\int u^n\,du = \frac{u^{n+1}}{n+1} + C, \quad n \neq -1$

2. $\displaystyle\int \frac{du}{u} = \ln|u| + C$

3. $\displaystyle\int \sin u\,du = -\cos u + C$

4. $\displaystyle\int \cos u\,du = \sin u + C$

5. $\displaystyle\int \tan u\,du = \ln|\sec u| + C$

6. $\displaystyle\int e^u\,du = e^u + C$

7. $\displaystyle\int a^u\,du = \frac{a^u}{\ln a} + C$

8. $\displaystyle\int \sec u\,du = \ln|\sec u + \tan u| + C$

9. $\displaystyle\int \csc u\,du = \ln|\csc u - \cot u| + C$

10. $\displaystyle\int \cot u\,du = \ln|\sin u| + C$

11. $\displaystyle\int \sec^2 u\,du = \tan u + C$

12. $\displaystyle\int \csc^2 u\,du = -\cot u + C$

13. $\displaystyle\int \sec u \tan u\,du = \sec u + C$

14. $\displaystyle\int \csc u \cot u\,du = -\csc u + C$

15. $\displaystyle\int \frac{du}{\sqrt{a^2 - u^2}} = \sin^{-1}\frac{u}{a} + C$

16. $\displaystyle\int \frac{du}{u\sqrt{u^2 - a^2}} = \frac{1}{a}\sec^{-1}\frac{u}{a} + C$

17. $\displaystyle\int \frac{du}{a^2 + u^2} = \frac{1}{a}\tan^{-1}\frac{u}{a} + C$

18. $\displaystyle\int \frac{du}{a^2 - u^2} = \frac{1}{2a}\ln\left|\frac{u + a}{u - a}\right| + C$

Forms Involving $a + bu$

19. $\displaystyle\int \frac{u\,du}{a + bu} = \frac{1}{b^2}\left(a + bu - a\ln|a + bu|\right) + C$

20. $\displaystyle\int \frac{u^2\,du}{a + bu}$

$\displaystyle\qquad = \frac{1}{2b^3}\left[(a + bu)^2 - 4a(a + bu) + 2a^2\ln|a + bu|\right] + C$

21. $\displaystyle\int \frac{u\,du}{(a + bu)^2} = \frac{a}{b^2(a + bu)} + \frac{1}{b^2}\ln|a + bu| + C$

22. $\displaystyle\int \frac{u^2\,du}{(a + bu)^2} = \frac{1}{b^3}\left(a + bu - \frac{a^2}{a + bu} - 2a\ln|a + bu|\right) + C$

23. $\displaystyle\int \frac{du}{u(a + bu)} = \frac{1}{a}\ln\left|\frac{u}{a + bu}\right| + C$

24. $\displaystyle\int \frac{du}{u^2(a + bu)} = -\frac{1}{au} + \frac{b}{a^2}\ln\left|\frac{a + bu}{u}\right| + C$

25. $\displaystyle\int \frac{du}{u(a + bu)^2} = \frac{1}{a(a + bu)} - \frac{1}{a^2}\ln\left|\frac{a + bu}{u}\right| + C$

26. $\displaystyle\int \frac{du}{u^2(a + bu)^2} = -\frac{1}{a^2}\left[\frac{a + 2bu}{u(a + bu)} + \frac{2b}{a}\ln\left|\frac{u}{a + bu}\right|\right] + C$

Forms Involving $\sqrt{a + bu}$

27. $\displaystyle\int u\sqrt{a + bu}\,du = \frac{2}{15b^2}(3bu - 2a)(a + bu)^{3/2} + C$

28. $\displaystyle\int \frac{u\,du}{\sqrt{a + bu}} = \frac{2}{3b^2}(bu - 2a)\sqrt{a + bu} + C$

29. $\displaystyle\int \frac{u^2\,du}{\sqrt{a + bu}} = \frac{2}{15b^3}(8a^2 + 3b^2u^2 - 4abu)\sqrt{a + bu} + C$

30. $\displaystyle\int \frac{du}{u\sqrt{a + bu}} = \begin{cases} \dfrac{1}{\sqrt{a}}\ln\left|\dfrac{\sqrt{a + bu} - \sqrt{a}}{\sqrt{a + bu} + \sqrt{a}}\right| + C & \text{if } a > 0 \\[3mm] \dfrac{2}{\sqrt{-a}}\tan^{-1}\sqrt{\dfrac{a + bu}{-a}} + C & \text{if } a < 0 \end{cases}$

31. $\displaystyle\int \frac{\sqrt{a + bu}}{u}\,du = 2\sqrt{a + bu} + a\int \frac{du}{u\sqrt{a + bu}}$

32. $\displaystyle\int \frac{\sqrt{a + bu}}{u^2}\,du = -\frac{\sqrt{a + bu}}{u} + \frac{b}{2}\int \frac{du}{u\sqrt{a + bu}}$

33. $\displaystyle\int u^n\sqrt{a + bu}\,du$

$\displaystyle\qquad = \frac{2}{b(2n + 3)}\left[u^n(a + bu)^{3/2} - na\int u^{n-1}\sqrt{a + bu}\,du\right]$

34. $\displaystyle\int \frac{u^n\,du}{\sqrt{a + bu}} = \frac{2u^n\sqrt{a + bu}}{b(2n + 1)} - \frac{2na}{b(2n + 1)}\int \frac{u^{n-1}\,du}{\sqrt{a + bu}}$

35. $\displaystyle \int \frac{du}{u^n \sqrt{a + bu}} = -\frac{\sqrt{a + bu}}{a(n-1)u^{n-1}} - \frac{b(2n-3)}{2a(n-1)} \int \frac{du}{u^{n-1}\sqrt{a + bu}}$

36. $\displaystyle \int \frac{\sqrt{a + bu}}{u^n} \, du = \frac{-1}{a(n-1)} \left[\frac{(a + bu)^{3/2}}{u^{n-1}} + \frac{(2n-5)b}{2} \int \frac{\sqrt{a + bu}}{u^{n-1}} \, du \right], \quad n \neq 1$

Forms Involving $\sqrt{a^2 + u^2}, a > 0$

37. $\displaystyle \int \sqrt{a^2 + u^2} \, du = \frac{u}{2}\sqrt{a^2 + u^2} + \frac{a^2}{2}\ln\left(u + \sqrt{a^2 + u^2}\right) + C$

38. $\displaystyle \int u^2 \sqrt{a^2 + u^2} \, du = \frac{u}{8}(a^2 + 2u^2)\sqrt{a^2 + u^2}$
$$- \frac{a^4}{8}\ln\left(u + \sqrt{a^2 + u^2}\right) + C$$

39. $\displaystyle \int \frac{\sqrt{a^2 + u^2}}{u} \, du = \sqrt{a^2 + u^2} - a\ln\left|\frac{a + \sqrt{a^2 + u^2}}{u}\right| + C$

40. $\displaystyle \int \frac{\sqrt{a^2 + u^2}}{u^2} \, du = -\frac{\sqrt{a^2 + u^2}}{u} + \ln\left(u + \sqrt{a^2 + u^2}\right) + C$

41. $\displaystyle \int \frac{du}{\sqrt{a^2 + u^2}} = \ln\left(u + \sqrt{a^2 + u^2}\right) + C$

42. $\displaystyle \int \frac{u^2 \, du}{\sqrt{a^2 + u^2}} = \frac{u}{2}\sqrt{a^2 + u^2} - \frac{a^2}{2}\ln\left(u + \sqrt{a^2 + u^2}\right) + C$

43. $\displaystyle \int \frac{du}{u\sqrt{a^2 + u^2}} = -\frac{1}{a}\ln\left|\frac{\sqrt{a^2 + u^2} + a}{u}\right| + C$

44. $\displaystyle \int \frac{du}{u^2\sqrt{a^2 + u^2}} = -\frac{\sqrt{a^2 + u^2}}{a^2 u} + C$

45. $\displaystyle \int \frac{du}{(a^2 + u^2)^{3/2}} = \frac{u}{a^2\sqrt{a^2 + u^2}} + C$

Forms Involving $\sqrt{a^2 - u^2}, a > 0$

46. $\displaystyle \int \sqrt{a^2 - u^2} \, du = \frac{u}{2}\sqrt{a^2 - u^2} + \frac{a^2}{2}\sin^{-1}\frac{u}{a} + C$

47. $\displaystyle \int u^2 \sqrt{a^2 - u^2} \, du = \frac{u}{8}(2u^2 - a^2)\sqrt{a^2 - u^2} + \frac{a^4}{8}\sin^{-1}\frac{u}{a} + C$

48. $\displaystyle \int \frac{\sqrt{a^2 - u^2}}{u} \, du = \sqrt{a^2 - u^2} - a\ln\left|\frac{a + \sqrt{a^2 - u^2}}{u}\right| + C$

49. $\displaystyle \int \frac{\sqrt{a^2 - u^2}}{u^2} \, du = -\frac{1}{u}\sqrt{a^2 - u^2} - \sin^{-1}\frac{u}{a} + C$

50. $\displaystyle \int \frac{u^2 \, du}{\sqrt{a^2 - u^2}} = -\frac{u}{2}\sqrt{a^2 - u^2} + \frac{a^2}{2}\sin^{-1}\frac{u}{a} + C$

51. $\displaystyle \int \frac{du}{u\sqrt{a^2 - u^2}} = -\frac{1}{a}\ln\left|\frac{a + \sqrt{a^2 - u^2}}{u}\right| + C$

52. $\displaystyle \int \frac{du}{u^2\sqrt{a^2 - u^2}} = -\frac{1}{a^2 u}\sqrt{a^2 - u^2} + C$

53. $\displaystyle \int (a^2 - u^2)^{3/2} \, du$
$$= -\frac{u}{8}(2u^2 - 5a^2)\sqrt{a^2 - u^2} + \frac{3a^4}{8}\sin^{-1}\frac{u}{a} + C$$

54. $\displaystyle \int \frac{du}{(a^2 - u^2)^{3/2}} = \frac{u}{a^2\sqrt{a^2 - u^2}} + C$

Forms Involving $\sqrt{u^2 - a^2}, a > 0$

55. $\displaystyle \int \sqrt{u^2 - a^2} \, du = \frac{u}{2}\sqrt{u^2 - a^2} - \frac{a^2}{2}\ln\left|u + \sqrt{u^2 - a^2}\right| + C$

56. $\displaystyle \int u^2 \sqrt{u^2 - a^2} \, du$
$$= \frac{u}{8}(2u^2 - a^2)\sqrt{u^2 - a^2} - \frac{a^4}{8}\ln\left|u + \sqrt{u^2 - a^2}\right| + C$$

57. $\displaystyle \int \frac{\sqrt{u^2 - a^2}}{u} \, du = \sqrt{u^2 - a^2} - a\cos^{-1}\frac{a}{|u|} + C$

58. $\displaystyle \int \frac{\sqrt{u^2 - a^2}}{u^2} \, du = -\frac{\sqrt{u^2 - a^2}}{u} + \ln\left|u + \sqrt{u^2 - a^2}\right| + C$

59. $\displaystyle \int \frac{du}{\sqrt{u^2 - a^2}} = \ln\left|u + \sqrt{u^2 - a^2}\right| + C$

60. $\displaystyle \int \frac{u^2 \, du}{\sqrt{u^2 - a^2}} = \frac{u}{2}\sqrt{u^2 - a^2} + \frac{a^2}{2}\ln\left|u + \sqrt{u^2 - a^2}\right| + C$

61. $\displaystyle \int \frac{du}{u^2\sqrt{u^2 - a^2}} = \frac{\sqrt{u^2 - a^2}}{a^2 u} + C$

62. $\displaystyle \int \frac{du}{(u^2 - a^2)^{3/2}} = -\frac{u}{a^2\sqrt{u^2 - a^2}} + C$

Forms Involving sin u, cos u, tan u

63. $\displaystyle\int \sin^2 u \, du = \frac{1}{2} u - \frac{1}{4} \sin 2u + C$

64. $\displaystyle\int \cos^2 u \, du = \frac{1}{2} u + \frac{1}{4} \sin 2u + C$

65. $\displaystyle\int \tan^2 u \, du = \tan u - u + C$

66. $\displaystyle\int \sin^3 u \, du = -\frac{1}{3}(2 + \sin^2 u) \cos u + C$

67. $\displaystyle\int \cos^3 u \, du = \frac{1}{3}(2 + \cos^2 u) \sin u + C$

68. $\displaystyle\int \tan^3 u \, du = \frac{1}{2} \tan^2 u + \ln|\cos u| + C$

69. $\displaystyle\int \sin^n u \, du = -\frac{1}{n} \sin^{n-1} u \cos u + \frac{n-1}{n} \int \sin^{n-2} u \, du$

70. $\displaystyle\int \cos^n u \, du = \frac{1}{n} \cos^{n-1} u \sin u + \frac{n-1}{n} \int \cos^{n-2} u \, du$

71. $\displaystyle\int \tan^n u \, du = \frac{1}{n-1} \tan^{n-1} u - \int \tan^{n-2} u \, du$

72. $\displaystyle\int \sin au \sin bu \, du = \frac{\sin(a-b)u}{2(a-b)} - \frac{\sin(a+b)u}{2(a+b)} + C$

73. $\displaystyle\int \cos au \cos bu \, du = \frac{\sin(a-b)u}{2(a-b)} + \frac{\sin(a+b)u}{2(a+b)} + C$

74. $\displaystyle\int \sin au \cos bu \, du = -\frac{\cos(a-b)u}{2(a-b)} - \frac{\cos(a+b)u}{2(a+b)} + C$

75. $\displaystyle\int u \sin u \, du = \sin u - u \cos u + C$

76. $\displaystyle\int u \cos u \, du = \cos u + u \sin u + C$

77. $\displaystyle\int u^n \sin u \, du = -u^n \cos u + n \int u^{n-1} \cos u \, du$

78. $\displaystyle\int u^n \cos u \, du = u^n \sin u - n \int u^{n-1} \sin u \, du$

79. $\displaystyle\int \sin^n u \cos^m u \, du$

$\displaystyle = -\frac{\sin^{n-1} u \cos^{m+1} u}{n+m} + \frac{n-1}{n+m} \int \sin^{n-2} u \cos^m u \, du$

$\displaystyle = \frac{\sin^{n+1} u \cos^{m-1} u}{n+m} + \frac{m-1}{n+m} \int \sin^n u \cos^{m-2} u \, du$

Forms Involving cot u, sec u, csc u

80. $\displaystyle\int \cot^2 u \, du = -\cot u - u + C$

81. $\displaystyle\int \cot^3 u \, du = -\frac{1}{2} \cot^2 u - \ln|\sin u| + C$

82. $\displaystyle\int \sec^3 u \, du = \frac{1}{2} \sec u \tan u + \frac{1}{2} \ln|\sec u + \tan u| + C$

83. $\displaystyle\int \csc^3 u \, du = -\frac{1}{2} \csc u \cot u + \frac{1}{2} \ln|\csc u - \cot u| + C$

84. $\displaystyle\int \cot^n u \, du = \frac{-1}{n-1} \cot^{n-1} u - \int \cot^{n-2} u \, du$

85. $\displaystyle\int \sec^n u \, du = \frac{1}{n-1} \tan u \sec^{n-2} u + \frac{n-2}{n-1} \int \sec^{n-2} u \, du$

86. $\displaystyle\int \csc^n u \, du = \frac{-1}{n-1} \cot u \csc^{n-2} u + \frac{n-2}{n-1} \int \csc^{n-2} u \, du$

Forms Involving Inverse Trigonometric Functions

87. $\displaystyle\int \sin^{-1} u \, du = u \sin^{-1} u + \sqrt{1-u^2} + C$

88. $\displaystyle\int \cos^{-1} u \, du = u \cos^{-1} u - \sqrt{1-u^2} + C$

89. $\displaystyle\int \tan^{-1} u \, du = u \tan^{-1} u - \frac{1}{2} \ln(1+u^2) + C$

90. $\displaystyle\int u \sin^{-1} u \, du = \frac{2u^2-1}{4} \sin^{-1} u + \frac{u\sqrt{1-u^2}}{4} + C$

91. $\displaystyle\int u \cos^{-1} u \, du = \frac{2u^2-1}{4} \cos^{-1} u - \frac{u\sqrt{1-u^2}}{4} + C$

92. $\displaystyle\int u \tan^{-1} u \, du = \frac{u^2+1}{2} \tan^{-1} u - \frac{u}{2} + C$

93. $\displaystyle\int u^n \sin^{-1} u \, du = \frac{1}{n+1}\left[u^{n+1} \sin^{-1} u - \int \frac{u^{n+1} \, du}{\sqrt{1-u^2}} \right],$
$\qquad n \neq -1$

94. $\displaystyle\int u^n \cos^{-1} u \, du = \frac{1}{n+1}\left[u^{n+1} \cos^{-1} u + \int \frac{u^{n+1} \, du}{\sqrt{1-u^2}} \right],$
$\qquad n \neq -1$

95. $\displaystyle\int u^n \tan^{-1} u \, du = \frac{1}{n+1}\left[u^{n+1} \tan^{-1} u - \int \frac{u^{n+1} \, du}{\sqrt{1+u^2}} \right],$
$\qquad n \neq -1$

Forms Involving Exponential and Logarithmic Functions

96. $\displaystyle\int ue^{au}\,du = \frac{1}{a^2}\,(au-1)e^{au} + C$

97. $\displaystyle\int u^n e^{au}\,du = \frac{1}{a}u^n e^{au} - \frac{n}{a}\int u^{n-1}e^{au}\,du$

98. $\displaystyle\int e^{au}\sin bu\,du = \frac{e^{au}}{a^2+b^2}\,(a\sin bu - b\cos bu) + C$

99. $\displaystyle\int e^{au}\cos bu\,du = \frac{e^{au}}{a^2+b^2}\,(a\cos bu + b\sin bu) + C$

100. $\displaystyle\int \frac{du}{1+be^{au}} = u - \frac{1}{a}\ln(1+be^{au}) + C$

101. $\displaystyle\int \ln u\,du = u\ln u - u + C$

102. $\displaystyle\int u^n \ln u\,du = \frac{u^{n+1}}{(n+1)^2}\,[(n+1)\ln u - 1] + C$

103. $\displaystyle\int \frac{1}{u\ln u}\,du = \ln|\ln u| + C$

Forms Involving Hyperbolic Functions

104. $\displaystyle\int \sinh u\,du = \cosh u + C$

105. $\displaystyle\int \cosh u\,du = \sinh u + C$

106. $\displaystyle\int \tanh u\,du = \ln\cosh u + C$

107. $\displaystyle\int \coth u\,du = \ln|\sinh u| + C$

108. $\displaystyle\int \operatorname{sech} u\,du = \tan^{-1}|\sinh u| + C$

109. $\displaystyle\int \operatorname{csch} u\,du = \ln\left|\tanh\tfrac{1}{2}u\right| + C$

110. $\displaystyle\int \operatorname{sech}^2 u\,du = \tanh u + C$

111. $\displaystyle\int \operatorname{csch}^2 u\,du = -\coth u + C$

112. $\displaystyle\int \operatorname{sech} u\tanh u\,du = -\operatorname{sech} u + C$

113. $\displaystyle\int \operatorname{csch} u\coth u\,du = -\operatorname{csch} u + C$

Forms Involving $\sqrt{2au^2 - u^2},\ a > 0$

114. $\displaystyle\int \sqrt{2au - u^2}\,du = \frac{u-a}{2}\sqrt{2au-u^2}$
$\displaystyle\qquad + \frac{a^2}{2}\cos^{-1}\!\left(\frac{a-u}{a}\right) + C$

115. $\displaystyle\int u\sqrt{2au - u^2}\,du = \frac{2u^2 - au - 3a^2}{6}\sqrt{2au-u^2}$
$\displaystyle\qquad + \frac{a^3}{2}\cos^{-1}\!\left(\frac{a-u}{a}\right) + C$

116. $\displaystyle\int \frac{\sqrt{2au-u^2}}{u}\,du = \sqrt{2au-u^2} + a\cos^{-1}\!\left(\frac{a-u}{a}\right) + C$

117. $\displaystyle\int \frac{\sqrt{2au-u^2}}{u^2}\,du = -\frac{2\sqrt{2au-u^2}}{u} - \cos^{-1}\!\left(\frac{a-u}{a}\right) + C$

118. $\displaystyle\int \frac{du}{\sqrt{2au-u^2}} = \cos^{-1}\!\left(\frac{a-u}{a}\right) + C$

119. $\displaystyle\int \frac{u\,du}{\sqrt{2au-u^2}} = -\sqrt{2au-u^2} + a\cos^{-1}\!\left(\frac{a-u}{a}\right) + C$

120. $\displaystyle\int \frac{u^2\,du}{\sqrt{2au-u^2}} = -\frac{(u+3a)}{2}\sqrt{2au-u^2}$
$\displaystyle\qquad + \frac{3a^2}{2}\cos^{-1}\!\left(\frac{a-u}{a}\right) + C$

121. $\displaystyle\int \frac{du}{u\sqrt{2au-u^2}} = -\frac{\sqrt{2au-u^2}}{au} + C$

MULTIVARIABLE

CALCULUS

MULTIVARIABLE
CALCULUS

SOO T. TAN
STONEHILL COLLEGE

BROOKS/COLE
CENGAGE Learning™

Australia • Brazil • Japan • Korea • Mexico • Singapore • Spain • United Kingdom • United States

Multivariable Calculus
Soo T. Tan

Senior Acquisitions Editor: Liz Covello

Publisher: Richard Stratton

Senior Developmental Editor: Danielle Derbenti

Developmental Editor: Ed Dodd

Developmental Project Editor: Terri Mynatt

Associate Editor: Jeannine Lawless

Editorial Assistant: Lauren Hamel

Media Editor: Peter Galuardi

Marketing Manager: Jennifer Jones

Marketing Assistant: Erica O'Connell

Marketing Communications Manager: Mary Anne Payumo

Senior Content Project Manager: Cheryll Linthicum

Creative Director: Rob Hugel

Senior Art Director: Vernon T. Boes

Print Buyer: Becky Cross

Rights Acquisitions Account Manager, Text: Roberta Broyer

Rights Acquisitions Account Manager, Image: Amanda Groszko

Production Service: Martha Emry

Text Designer: Diane Beasley

Art Editor: Leslie Lahr, Lisa Torri

Photo Researcher: Kathleen Olsen

Copy Editor: Barbara Willette

Illustrator: Precision Graphics, Matrix Art Services, Network Graphics

Cover Designer: Terri Wright

Cover Image: Nathan Fariss for *Popular Mechanics*

Compositor: Graphic World

For product information and technology assistance, contact us at
Cengage Learning Customer & Sales Support, 1-800-354-9706

For permission to use material from this text or product, submit all requests online at **www.cengage.com/permissions**
Further permissions questions can be e-mailed to
permissionrequest@cengage.com

Library of Congress Control Number: 2009921551

ISBN-13: 978-0-534-46575-9

ISBN-10: 0-534-46575-7

Brooks/Cole
10 Davis Drive
Belmont, CA 94002-3098
USA

Cengage Learning is a leading provider of customized learning solutions with office locations around the globe, including Singapore, the United Kingdom, Australia, Mexico, Brazil, and Japan. Locate your local office at **www.cengage.com/global**

Cengage Learning products are represented in Canada by Nelson Education, Ltd.

To learn more about Brooks/Cole, visit **www.cengage.com/brookscole**

Purchase any of our products at your local college store or at our preferred online store **www.ichapters.com**

Printed in Canada
1 2 3 4 5 6 7 13 12 11 10 09

To Olivia, Maxwell, Sasha, Isabella, and Ashley

About the Author

SOO T. TAN received his S.B. degree from the Massachusetts Institute of Technology, his M.S. degree from the University of Wisconsin–Madison, and his Ph.D. from the University of California at Los Angeles. He has published numerous papers on optimal control theory, numerical analysis, and the mathematics of finance. He is also the author of a series of textbooks on applied calculus and applied finite mathematics.

One of the most important lessons I have learned from my many years of teaching undergraduate mathematics courses is that most students, mathematics and non-mathematics majors alike, respond well when introduced to mathematical concepts and results using real-life illustrations.

This awareness led to the intuitive approach that I have adopted in all of my texts. As you will see, I try to introduce each abstract mathematical concept through an example drawn from a common, real-life experience. Once the idea has been conveyed, I then proceed to make it precise, thereby assuring that no mathematical rigor is lost in this intuitive treatment of the subject. Another lesson I learned from my students is that they have a much greater appreciation of the material if the applications are drawn from their fields of interest and from situations that occur in the real world. This is one reason you will see so many examples and exercises in my texts that are drawn from various and diverse fields such as physics, chemistry, engineering, biology, business, and economics. There are also many exercises of general and current interest that are modeled from data gathered from newspapers, magazines, journals, and other media. Whether it be global warming, brain growth and IQ, projected U.S. gasoline usage, or finding the surface area of the Jacqueline Kennedy Onassis Reservoir, I weave topics of current interest into my examples and exercises to keep the book relevant to all of my readers.

Contents

Author's Commitment to Accuracy

As with all of my projects, accuracy is of paramount importance. For this reason, I solved every problem myself and wrote the solutions for the solutions manual. In this accuracy checking process, I worked very closely with several professors who contributed in different ways and at different stages throughout the development of the text and manual: Jason Aubrey (*University of Missouri*), Kevin Charlwood (*Washburn University*), Jerrold Grossman (*Oakland University*), Tao Guo (*Rock Valley College*), James Handley (*Montanta Tech of the University of Montana*), Selwyn Hollis (*Armstrong Atlantic State University*), Diane Koenig (*Rock Valley College*), Michael Montano (*Riverside Community College*), John Samons (*Florida Community College*), Doug Shaw (*University of Northern Iowa*), and Richard West (*Francis Marion University*).

Accuracy Process

First Round
- The first draft of the manuscript was reviewed by numerous calculus instructors, all of whom either submitted written reviews, participated in a focus group discussion, or class-tested the manuscript.

Second Round
- The author provided revised manuscript to be reviewed by additional calculus instructors who went through the same steps as the first group and submitted their responses.
- Simultaneously, author Soo Tan was writing the solutions manual, which served as an additional check of his work on the text manuscript.

Third Round
- Two calculus instructors checked the revised manuscript for accuracy while simultaneously checking the solutions manual, sending their corrections back to the author for inclusion.
- Additional groups of calculus instructors participated in focus groups and class testing of the revised manuscript.
- First drafts of the art were produced and checked for accuracy.
- The manuscript was edited by a professional copyeditor.
- Biographies were written by a calculus instructor and submitted for copyedit.

Fourth Round
- Once the manuscript was declared final, a compositor created galley pages, whose accuracy was checked by several calculus instructors.
- Revisions were made to the art, and revised art proofs were checked for accuracy.
- Further class testing and live reviews were completed.
- Galley proofs were checked for consistency by the production team and carefully reviewed by the author.
- Biographies were checked and revised for accuracy by another calculus instructor.

Fifth Round
- First round page proofs were distributed, proofread, and checked for accuracy again. As with galley proofs, these pages were carefully reviewed by the author with art seen in place with the exposition for the first time.
- The revised art was again checked for accuracy by the author and the production service.

Sixth Round
- Revised page proofs were checked by a second proofreader and the author.

Seventh Round
- Final page proofs were checked for consistency by the production team and the author performed his final review of the pages.

Preface

Throughout my teaching career I have always enjoyed teaching calculus and helping students to see the elegance and beauty of calculus. So when I was approached by my editor to write this series, I welcomed the opportunity. Upon reflecting, I see that I started this project from a strong vantage point. I have written an *Applied Mathematics* series, and over the years I have gotten a lot of feedback from many professors and students using the books in the series. The wealth of suggestions that I gained from them coupled with my experience in the classroom served me well when I embarked upon this project.

In writing the *Calculus* series, I have constantly borne in mind two primary objectives: first, to provide the instructor with a book that is easy to teach from and yet has all the content and rigor of a traditional calculus text, and second, to provide students with a book that motivates their interest and at the same time is easy for them to read. In my experience, students coming to calculus for the first time respond best to an intuitive approach, and I try to use this approach by introducing abstract ideas with concrete, real-life examples that students can relate to, wherever appropriate. Often a simple real-life illustration can serve as motivation for a more complex mathematical concept or theorem. Also, I have tried to use a clear, precise, and concise writing style throughout the book and have taken special care to ensure that my intuitive approach does not compromise the mathematical rigor that is expected of an engineering calculus text.

In addition to the applications in mathematics, engineering, physics, and the other natural and social sciences, I have included many other examples and exercises drawn from diverse fields of current interest. The solutions to all the exercises in the book are provided in a separate manual. In keeping with the emphasis on conceptual understanding, I have included concept questions at the beginning of each exercise set. In each end-of-chapter review section I have also included fill-in-the-blank questions for a review of the concepts. I have found these questions to be an effective learning tool to help students master the definitions and theorems in each chapter. Furthermore, I have included many questions that ask for the interpretation of graphical, numerical, and algebraic results in both the examples and the exercise sets.

Finally, I have employed a unique approach to the introduction of the limit concept. Many calculus textbooks introduce this concept via the slope of a tangent line to a curve and then follow by relating the slope to the notion of the rate of change of one quantity with respect to another. In my text I do precisely the opposite: I introduce the limit concept by looking at the rate of change of the maglev (magnetic levitation train). This approach is more intuitive and captures the interest of the student from the very beginning—it shows immediately the relevance of calculus to the real world. I might add that this approach has worked very well for me not only in the classroom; it has also been received very well by the users of my applied calculus series. This intuitive approach (using the maglev as a vehicle) is carried into the introduction and explanation of some of the fundamental theorems in calculus, such as the Intermediate Value Theorem and the Mean Value Theorem. Consistently woven throughout the text, this idea permeates much of the text—from concepts in limits, to continuity, to integration, and even to inverse functions.

Soo T. Tan

■ Tan *Calculus* Series

The Tan *Calculus* series includes the following textbooks:

- *Calculus* © 2010 (ISBN 0-534-46579-X)
- *Single Variable Calculus* © 2010 (ISBN 0-534-46566-8)
- *Multivariable Calculus* © 2010 (ISBN 0-534-46575-7)
- *Calculus: Early Transcendentals* © 2011 (ISBN 0-534-46554-4)
- *Single Variable Calculus: Early Transcendentals* © 2011 (ISBN 0-534-46570-6)

■ Features

An Intuitive Approach . . . Without Loss of Rigor

Beginning with each chapter opening vignette and carrying through each chapter, Soo Tan's intuitive approach links the abstract ideas of calculus with concrete, real-life examples. This intuitive approach is used to advantage to introduce and explain many important concepts and theorems in calculus, such as tangent lines, Rolles's Theorem, absolute extrema, increasing and decreasing functions, limits at infinity, and parametric equations. In this example from Chapter 5 the discussion of the area between two curves is motivated with a real-life illustration that is followed by the precise discussion of the mathematical concepts involved.

■ A Real-Life Interpretation

Two cars are traveling in adjacent lanes along a straight stretch of a highway. The velocity functions for Car A and Car B are $v = f(t)$ and $v = g(t)$, respectively. The graphs of these functions are shown in Figure 1.

FIGURE 1
The shaded area S gives the distance that Car A is ahead of Car B at time $t = b$.

The area of the region under the graph of f from $t = 0$ to $t = b$ gives the total distance covered by Car A in b seconds over the time interval $[0, b]$. The distance covered by Car B over the same period of time is given by the area under the graph of g on the interval $[0, b]$. Intuitively, we see that the area of the (shaded) region S between the graphs of f and g on the interval $[0, b]$ gives the distance that Car A will be ahead of Car B at time $t = b$.

■ The Area Between Two Curves

Suppose f and g are continuous functions with $f(x) \geq g(x)$ for all x in $[a, b]$, so that the graph of f lies on or above that of g on $[a, b]$. Let's consider the region S bounded by the graphs of f and g between the vertical lines $x = a$ and $x = b$ as shown in Figure 2. To define the *area* of S, we take a regular partition of $[a, b]$,

$$a = x_0 < x_1 < x_2 < x_3 < \cdots < x_n = b$$

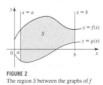

FIGURE 2
The region S between the graphs of f and g on $[a, b]$

and form the Riemann sum of the function $f - g$ over $[a, b]$ with respect to this partition:

$$\sum_{k=1}^{n} [f(c_k) - g(c_k)]\Delta x$$

where c_k is an evaluation point in the subinterval $[x_{k-1}, x_k]$ and $\Delta x = (b - a)/n$. The kth term of this sum gives the area of a rectangle with height $[f(c_k) - g(c_k)]$ and width Δx. As you can see in Figure 3, this area is an approximation of the area of the subregion of S that lies between the graphs of f and g on $[x_{k-1}, x_k]$.

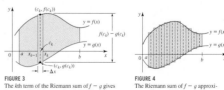

FIGURE 3
The kth term of the Riemann sum of $f - g$ gives the area of the kth rectangle of width Δx.

FIGURE 4
The Riemann sum of $f - g$ approximates the area of S.

Unique Applications in the Examples and Exercises

Our relevant, unique applications are designed to illustrate mathematical concepts and at the same time capture students' interest.

69. Constructing a New Road The following figures depict three possible roads connecting the point $A(-1000, 0)$ to the point $B(1000, 1000)$ via the origin. The functions describing the dashed center lines of the roads follow:

$$f(x) = \begin{cases} 0 & \text{if } -1000 \le x \le 0 \\ x & \text{if } 0 < x \le 1000 \end{cases}$$

$$g(x) = \begin{cases} 0 & \text{if } -1000 \le x \le 0 \\ 0.001x^2 & \text{if } 0 < x \le 1000 \end{cases}$$

$$h(x) = \begin{cases} 0 & \text{if } -1000 \le x \le 0 \\ 0.000001x^3 & \text{if } 0 < x \le 1000 \end{cases}$$

Show that f is not differentiable on the interval $(-1000, 1000)$, g is differentiable but not twice differentiable on $(-1000, 1000)$, and h is twice differentiable on $(-1000, 1000)$. Taking into consideration the dynamics of a moving vehicle, which proposal do you think is most suitable?

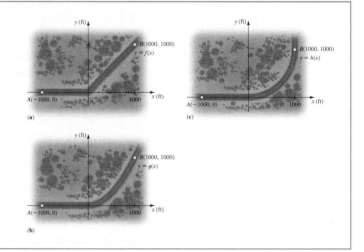

Connections

One particular example—the maglev (magnetic levitation) train—is used as a common thread throughout the development of calculus from limits through integration. The goal here is to show students the connection between the important theorems and concepts presented. Topics that are introduced through this example include the Intermediate Value Theorem, the Mean Value Theorem, the Mean Value Theorem for Definite Integrals, limits, continuity, derivatives, antiderivatives, initial value problems, inverse functions, and indeterminate forms.

■ A Real-Life Example

A prototype of a maglev (magnetic levitation train) moves along a straight monorail. To describe the motion of the maglev, we can think of the track as a coordinate line. From data obtained in a test run, engineers have determined that the maglev's displacement (directed distance) measured in feet from the origin at time t (in seconds) is given by

$$s = f(t) = 4t^2 \qquad 0 \le t \le 30 \qquad \text{(1)}$$

where f is called the position function of the maglev. The position of the maglev at time $t = 0, 1, 2, 3, \dots, 30$, measured in feet from its initial position, is

$$f(0) = 0, \qquad f(1) = 4, \qquad f(2) = 16, \qquad f(3) = 36, \qquad \dots, \qquad f(30) = 3600$$

(See Figure 1.)

FIGURE 1
A maglev moving along an elevated monorail track

Precise Figures That Help Students Visualize the Concepts

Carefully constructed art helps the student to visualize the mathematical ideas under discussion.

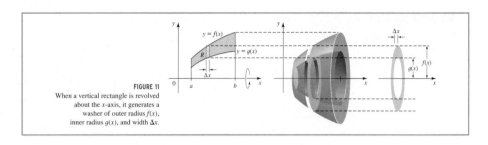

FIGURE 11
When a vertical rectangle is revolved about the *x*-axis, it generates a washer of outer radius $f(x)$, inner radius $g(x)$, and width Δx.

12.1 CONCEPT QUESTIONS

1. a. What is a vector-valued function?
 b. Give an example of a vector function. What is the parameter interval of the function that you picked?
2. Let $\mathbf{r}(t)$ be a vector function defined by $\mathbf{r}(t) = \langle f(t), g(t), h(t) \rangle$.
 a. Define $\lim_{t \to a} \mathbf{r}(t)$.
 b. Give an example of a vector function $\mathbf{r}(t)$ such that $\lim_{t \to 1} \mathbf{r}(t)$ does not exist.

3. a. What does it mean for a vector function $\mathbf{r}(t)$ to be continuous at a? Continuous on an interval I?
 b. Give an example of a function $\mathbf{r}(t)$ that is defined on the interval $(-1, 1)$ but fails to be continuous at 0.

12.1 EXERCISES

In Exercises 1–6, find the domain of the vector function.

1. $\mathbf{r}(t) = t\mathbf{i} + \dfrac{1}{t}\mathbf{j}$

2. $\mathbf{r}(t) = \cos t\mathbf{i} + 2\sin t\mathbf{j} + \sqrt{t+1}\,\mathbf{k}$

3. $\mathbf{r}(t) = \left\langle \sqrt{t}, \dfrac{1}{t-1}, \ln t \right\rangle$ **4.** $\mathbf{r}(t) = \left\langle \dfrac{1}{\sqrt{t-1}}, e^{-t} \right\rangle$

5. $\mathbf{r}(t) = \ln t\mathbf{i} + \cosh t\mathbf{j} + \tanh t\mathbf{k}$

6. $\mathbf{r}(t) = \sqrt[3]{t}\mathbf{i} + e^{1/t}\mathbf{j} + \dfrac{1}{t+2}\mathbf{k}$

In Exercises 7–12, match the vector functions with the curves labeled (a)–(f). Explain your choices.

7. $\mathbf{r}(t) = t^2\mathbf{i} + t^3\mathbf{j} + t^2\mathbf{k}$

8. $\mathbf{r}(t) = 2\cos 2t\mathbf{i} + t\mathbf{j} + 2\sin 2t\mathbf{k}$

9. $\mathbf{r}(t) = t\mathbf{i} + t\mathbf{j} + \left(\dfrac{1}{t^2+1}\right)\mathbf{k}$

10. $\mathbf{r}(t) = t\sin t\mathbf{i} + t\cos t\mathbf{j} + t\mathbf{k}$, $0 \le t \le 10\pi$

11. $\mathbf{r}(t) = 2\cos t\mathbf{i} + 3\sin t\mathbf{j} + e^{0.1t}\mathbf{k}$, $t \ge 0$

12. $\mathbf{r}(t) = \cos t\mathbf{i} + \sin t\mathbf{j} + \sin 3t\mathbf{k}$
 (a)   **(b)**

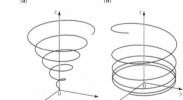

(c)   **(d)**

(e)   **(f)**

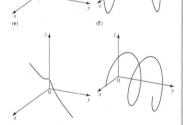

In Exercises 13–26, sketch the curve with the given vector function, and indicate the orientation of the curve.

13. $\mathbf{r}(t) = 2t\mathbf{i} + (3t+1)\mathbf{j}$, $-1 \le t \le 2$

14. $\mathbf{r}(t) = \sqrt{t}\mathbf{i} + (4-t)\mathbf{j}$, $t \ge 0$

15. $\mathbf{r}(t) = \langle t^2, t^3 \rangle$, $-1 \le t \le 2$

16. $\mathbf{r}(t) = 2\sin t\mathbf{i} + 3\cos t\mathbf{j}$, $0 \le t \le 2\pi$

17. $\mathbf{r}(t) = e^t\mathbf{i} + e^{2t}\mathbf{j}$, $-\infty < t < \infty$

18. $\mathbf{r}(t) = \langle 1 + 2\cos t, 3 + 2\sin t \rangle$, $0 \le t \le 2\pi$

19. $\mathbf{r}(t) = (1+t)\mathbf{i} + (2-t)\mathbf{j} + (3-2t)\mathbf{k}$, $-\infty < t < \infty$

20. $\mathbf{r}(t) = (2+t)\mathbf{i} + (3-2t)\mathbf{j} + (2+4t)\mathbf{k}$, $0 \le t \le 1$

21. $\mathbf{r}(t) = \langle t, t^2, t^3 \rangle$, $t \ge 0$

  Videos for selected exercises are available online at **www.academic.cengage.com/login**.

 $= 0$

all values
the form

 a. Use Rolle's Theorem to show that f has exactly two distinct zeros.
 b. Plot the graph of f using the viewing window $[-3, 3] \times [-5, 5]$.
32. Let
$$f(x) = \begin{cases} x\sin\dfrac{\pi}{x} & \text{if } x > 0 \\ 0 & \text{if } x = 0 \end{cases}$$

39. Complete the proof of Rolle's Theorem by considering the case in which $f(x) < d$ for some number x in (a, b).
40. Let f be continuous on $[a, b]$ and differentiable on (a, b). Put $h = b - a$.
 a. Use the Mean Value Theorem to show that there exists at least one number θ that satisfies $0 < \theta < 1$ such that
$$\frac{f(a+h) - f(a)}{h} = f'(a + \theta h)$$
 b. Find θ in the formula in part (a) for the function $f(x) = x^2$.
41. Let $f(x) = x^4 - 2x^3 + x - 2$.
 a. Show that f satisfies the hypotheses of Rolle's Theorem on the interval $[-1, 2]$.
 b. Use a calculator or a computer to estimate all values of c accurate to five decimal places that satisfy the conclusion of Rolle's Theorem.
 c. Plot the graph of f and the (horizontal) tangent lines to the graph of f at the point(s) $(c, f(c))$ for the values of c found in part (b).

Concept Questions

Designed to test student understanding of the basic concepts discussed in the section, these questions encourage students to explain learned concepts in their own words.

Exercises

Each exercise section contains an ample set of problems of a routine computational nature, followed by a set of application-oriented problems (many of them sourced) and true/false questions that ask students to explain their answer.

Graphing Utility and CAS Exercises

Indicated by and **cas** icons next to the corresponding exercises, these exercises offer practice in using technology to solve problems that might be difficult to solve by hand. Sourced problems using real-life data are often included.

CHAPTER 13 REVIEW

CONCEPT REVIEW

In Exercises 1–17, fill in the blanks.

1. **a.** A function f of two variables, x and y, is a _____ that assigns to each ordered pair _____ in the domain of f, exactly one real number $f(x, y)$.
   b. The number $z = f(x, y)$ is called a _____ variable, and x and y are _____ variables. The totality of the numbers z is called the _____ of the function f.
   c. The graph of f is the set $S =$ _____.

2. **a.** The curves with equation $f(x, y) = k$, where k is a constant in the range of f, are called the _____ of f.
   b. A level surface of a function f of three variables is the graph of the equation _____, where k is a constant in the range of f.

3. $\lim_{(x, y) \to (a, b)} f(x, y) = L$ means there exists a number _____ such that $f(x, y)$ can be made as close to _____ as we please by restricting (x, y) to be sufficiently close to _____.

4. If $f(x, y)$ approaches L_1 as (x, y) approaches (a, b) along one path, and $f(x, y)$ approaches L_2 as (x, y) approaches (a, b) along another path with $L_1 \neq L_2$, then $\lim_{(x, y) \to (a, b)} f(x, y)$ _____ exist.

5. **a.** $f(x, y)$ is continuous at (a, b) if $\lim_{(x, y) \to (a, b)} f(x, y) =$ _____.
   b. $f(x, y)$ is continuous on a region R if f is continuous at every point (x, y) in _____.

6. **a.** A polynomial function is continuous _____; a rational function is continuous at all points in its _____.
   b. If f is continuous at (a, b) and g is continuous at $f(a, b)$, then the composite function $h = g \circ f$ is continuous at _____.

7. **a.** The partial derivative of $f(x, y)$ with respect to x is _____ if the limit exists. The partial derivative $(\partial f / \partial x)(a, b)$ gives the slope of the tangent line to the curve obtained by the intersection of the plane _____ and the graph of $z = f(x, y)$ at _____; it also measures the rate of change of $f(x, y)$ in the _____-direction with y held _____ at _____.
   b. To compute $\partial f / \partial x$ where f is a function of x and y, treat _____ as a constant and differentiate with respect to _____ in the usual manner.

8. If $f(x, y)$ and its partial derivatives f_x, f_y, f_{xy}, and f_{yx} are continuous on an open region R, then $f_{xy}(x, y) =$ _____ for all (x, y) in R.

9. **a.** The total differential dz of $z = f(x, y)$ is $dz =$
   b. If $\Delta z = f(x + \Delta x, y + \Delta y) - f(x, y)$, then $\Delta z \approx$
   c. $\Delta z = f_x(x, y)\, \Delta x + f_y(x, y)\, \Delta y + \varepsilon_1\, \Delta x + \varepsilon_2\, \Delta y$, where ε_1 and ε_2 are functions of _____ and _____ such that $\lim_{(\Delta x, \Delta y) \to (0, 0)} \varepsilon_1 =$ _____ and $\lim_{(\Delta x, \Delta y) \to (0, 0)} \varepsilon_2 =$ _____.
   d. The function $z = f(x, y)$ is differentiable at (a, b) if Δz can be expressed in the form $\Delta z =$ _____, where _____ and _____ as $(\Delta x, \Delta y) \to$ _____.

10. **a.** If f is a function of x and y, and f_x and f_y are continuous on an open region R, then f is _____ in R.
    b. If f is differentiable at (a, b), then f is _____ at (a, b).

11. **a.** If $w = f(x, y)$, $x = g(t)$, and $y = h(t)$, then under suitable conditions the Chain Rule gives $dw/dt =$ _____.
    b. If $w = f(x, y)$, $x = g(u, v)$, and $y = h(u, v)$, then $\partial w / \partial u =$ _____.
    c. If $F(x, y) = 0$, _____
    d. If $F(x, y, z) = 0$ _____ prov _____ z implicitly as _____ and _____ and

12. **a.** If f is a functio_____ vector, then the _____ of $\mathbf{u}$ is $D_\mathbf{u} f(x$,_____
    b. The directional _____ change of f at _____
    c. If f is differenti_____
    d. The gradient of _____
    e. In terms of the _____

13. **a.** The maximum _____ occurs when $\mathbf{u}$ _____
    b. The minimum _____ occurs when $\mathbf{u}$ _____

14. **a.** ∇f is _____
    b. ∇F is _____
    c. The tangent pla_____ point $P(a, b, c)$ _____ through $P(a, b,$ _____

15. **a.** If $f(x, y) \geq f(a$_____ ing (a, b), then _____
    b. If $f(x, y) \geq f(a$_____ has an _____

REVIEW EXERCISES

In Exercises 1–4, sketch the curve with the given vector equation, and indicate the orientation of the curve.

1. $\mathbf{r}(t) = (2 + 3t)\mathbf{i} + (2t - 1)\mathbf{j}$
2. $\mathbf{r}(t) = t^3 \mathbf{i} + t^2 \mathbf{j}; \quad 0 \leq t \leq 2$
3. $\mathbf{r}(t) = (\cos t - 1)\mathbf{i} + (\sin t + 2)\mathbf{j} + 2\mathbf{k}$
4. $\mathbf{r}(t) = 2 \cos t\,\mathbf{i} + 3 \sin t\,\mathbf{j} + t^2\mathbf{k}; \quad 0 \leq t \leq 2\pi$

5. Find the domain of $\mathbf{r}(t) = \dfrac{1}{\sqrt{5 - t}}\mathbf{i} + \dfrac{\sin t}{t}\mathbf{j} + \ln(1 + t)\mathbf{k}$.

6. Find $\lim_{t \to 0^+} \mathbf{r}(t)$, where $\mathbf{r}(t) = \dfrac{\sqrt{t}}{1 + t^2}\mathbf{i} + \dfrac{t^2}{\sin t}\mathbf{j} + \dfrac{e^t - 1}{t}\mathbf{k}$.

7. Find the interval in which
$$\mathbf{r}(t) = \sqrt{t + 1}\,\mathbf{i} + \frac{e^t}{\sqrt{2 - t}}\mathbf{j} + \frac{t^2}{(t - 1)^2}\mathbf{k}$$
is continuous.

8. Find $\mathbf{r}'(t)$ if $\mathbf{r}(t) = \left[\displaystyle\int_0^t \cos^2 u\, du\right]\mathbf{i} + \left[\displaystyle\int_0^{t^2} \sin u\, du\right]\mathbf{j}$.

In Exercises 9–12, find $\mathbf{r}'(t)$ and $\mathbf{r}''(t)$.

9. $\mathbf{r}(t) = \sqrt{t}\,\mathbf{i} + t^2 \mathbf{j} + \dfrac{1}{t + 1}\mathbf{k}$
10. $\mathbf{r}(t) = e^{-t}\mathbf{i} + t \cos t\,\mathbf{j} + t \sin t\,\mathbf{k}$
11. $\mathbf{r}(t) = (t^2 + 1)\mathbf{i} + 2t\mathbf{j} + \ln t\,\mathbf{k}$
12. $\mathbf{r}(t) = \langle t \sin t, t \cos t, e^{2t}\rangle$

In Exercises 13 and 14, find parametric equations for the tangent line to the curve with the given parametric equations at the point with the given value of t.

13. $x = t^2 + 1, \quad y = 2t - 3, \quad z = t^3 + 1; \quad t = 0$
14. $x = t \cos t - \sin t, \quad y = t \sin t + \cos t, \quad z = t^2; \quad t = \dfrac{\pi}{2}$

In Exercises 15 and 16, evaluate the integral.

15. $\displaystyle\int \left(\sqrt{t}\,\mathbf{i} + e^{-2t}\mathbf{j} + \frac{1}{t + 1}\mathbf{k}\right) dt$
16. $\displaystyle\int_0^1 (2t\mathbf{i} + t^2\mathbf{j} + t^{3/2}\mathbf{k})\, dt$

In Exercises 17 and 18, find $\mathbf{r}(t)$ for the vector function $\mathbf{r}'(t)$ or $\mathbf{r}''(t)$ and the given initial condition(s).

17. $\mathbf{r}'(t) = 2\sqrt{t}\,\mathbf{i} + 3 \cos 2\pi t\,\mathbf{j} - e^{-t}\mathbf{k}; \quad \mathbf{r}(0) = \mathbf{i} + 2\mathbf{j}$

In Exercises 19 and 20, find the unit tangent and the unit normal vectors for the curve C defined by $\mathbf{r}(t)$ for the given value of t.

19. $\mathbf{r}(t) = t\mathbf{i} + t^2\mathbf{j} + t^3\mathbf{k}; \quad t = 1$
20. $\mathbf{r}(t) = 2 \cos t\,\mathbf{i} + 2 \sin t\,\mathbf{j} + e^t\mathbf{k}; \quad t = 0$

In Exercises 21 and 22, find the length of the curve.

21. $\mathbf{r}(t) = 2 \sin 2t\,\mathbf{i} + 2 \cos 2t\,\mathbf{j} + 3t\mathbf{k}; \quad 0 \leq t \leq 2$
22. $\mathbf{r}(t) = \sqrt{2}\,t\mathbf{i} + \dfrac{1}{2}t^2\mathbf{j} + \ln t\,\mathbf{k}; \quad 1 \leq t \leq 2$

In Exercises 23 and 24, find the curvature of the curve.

23. $\mathbf{r}(t) = t\mathbf{i} + t^2\mathbf{j} + t^3\mathbf{k}$
24. $\mathbf{r}(t) = t \sin t\,\mathbf{i} + t \cos t\,\mathbf{j} + t\mathbf{k}$

In Exercises 25 and 26, find the curvature of the plane curve, and determine the point on the curve at which the curvature is largest.

25. $y = x - \dfrac{1}{4}x^2$ 26. $y = e^{-x}$

In Exercises 27 and 28, find the velocity, acceleration, and speed of the object with the given position vector.

27. $\mathbf{r}(t) = 2t\mathbf{i} + e^{-2t}\mathbf{j} + \cos t\,\mathbf{k}$
28. $\mathbf{r}(t) = te^{-t}\mathbf{i} + \cos 2t\,\mathbf{j} + \sin 2t\,\mathbf{k}$

In Exercises 29 and 30, find the velocity and position vectors of an object with the given acceleration and the given initial velocity and position.

29. $\mathbf{a}(t) = t\mathbf{i} + \dfrac{1}{3}t^2\mathbf{j} + 3\mathbf{k}; \quad \mathbf{v}(0) = 2\mathbf{i} + 3\mathbf{j} + \mathbf{k}, \quad \mathbf{r}(0) = \mathbf{0}$
30. $\mathbf{a}(t) = e^t\mathbf{i} + e^{-t}\mathbf{j} + t\mathbf{k}; \quad \mathbf{v}(0) = 2\mathbf{i}, \quad \mathbf{r}(0) = \mathbf{i} + \mathbf{k}$

In Exercises 31–34, find the scalar tangential and normal components of acceleration of a particle with the given position vector.

31. $\mathbf{r}(t) = \mathbf{i} + t\mathbf{j} + t^2\mathbf{k}$
32. $\mathbf{r}(t) = 2 \cos t\,\mathbf{i} + 3 \sin t\,\mathbf{j} + t\mathbf{k}$
33. $\mathbf{r}(t) = \cos t\,\mathbf{i} + \sin 2t\,\mathbf{j}$
34. $\mathbf{r}(t) = \sqrt{2}\,t\mathbf{i} + e^t\mathbf{j} + e^{-t}\mathbf{k}$

35. **A Shot Put** In a track and field meet, a shot putter heaves a shot at an angle of $45°$ with the horizontal. As the shot leaves her hand, it is at a height of 7 ft and moving at a speed of 40 ft/sec. Set up a coordinate system so that the shot putter is at the origin.

PROBLEM-SOLVING TECHNIQUES

The following example shows that rewriting a_____ times pays dividends.

EXAMPLE Find $f^{(n)}(x)$ if $f(x) = \dfrac{x}{x^2 - 1}$.

Solution Our first instinct is to use the Quotient Rule to compute $f'(x)$, $f''(x)$, and so on. The expectation here is either that the rule for $f^{(n)}$ will become apparent or that at least a pattern will emerge that will enable us to guess at the form for $f^{(n)}(x)$. But the futility of this approach will be evident when you compute the first two derivatives of f.

Let's see whether we can transform the expression for $f(x)$ before we differentiate. You can verify that $f(x)$ can be written as

$$f(x) = \frac{x}{x^2 - 1} = \frac{\frac{1}{2}(x - 1) + \frac{1}{2}(x + 1)}{(x + 1)(x - 1)} = \frac{1}{2}\left[\frac{1}{x + 1} + \frac{1}{x - 1}\right]$$

Concept Review Questions

Beginning each end of chapter review, these questions give students a chance to check their knowledge of the basic definitions and concepts from the chapter.

Review Exercises

Offering a solid review of the chapter material, these exercises contain routine computational exercises as well as applied problems.

Problem-Solving Techniques

At the end of selected chapters the author discusses problem-solving techniques that provide students with the tools they need to make seemingly complex problems easier to solve.

CHALLENGE PROBLEMS

1. **a.** Use the definition of the double integral as a limit of a Riemann sum to compute $\iint_R (3x^2 + 2y)\, dA$, where $R = \{(x, y)\,|\, 0 \le x \le 2,\, 0 \le y \le 1\}$.
   Hint: Take $\Delta x = 2/m$, and $\Delta y = 1/n$, so that $x_i = 2i/m$, where $1 \le i \le m$, and $y_j = j/n$, where $1 \le j \le n$.
   b. Verify the result of part (a) by evaluating an appropriate iterated integral.

2. The following figure shows a triangular lamina. Its mass density at (x, y) is $f(x, y) = \cos(y^2)$. Find its mass.

4. Using the result of Problem 3, show that the area of the parallelogram determined by the vectors $\mathbf{a} = \langle a_1, a_2 \rangle$ and $\mathbf{b} = \langle b_1, b_2 \rangle$ is $|a_1 b_2 - a_2 b_1|$.

5. **Monte Carlo Integration** This is a method that is used to find the area of complicated bounded regions in the xy-plane. To describe the method, suppose that D is such a region completely enclosed by a rectangle $R = \{(x, y)\,|\, a \le x \le b, c \le y \le d\}$, as shown in the figure. Using a random number generator, we then pick points in R. If $A(D)$ denotes the area of D, then

$$\frac{A(D)}{A(R)} \approx \frac{N(D)}{n}$$

where $N(D)$ denotes the number of points landing in D, $A(R) = (b - a)(d - c)$, and n is the number of points picked. Then

$$A(D) \approx \frac{(b - a)(d - c)}{n} N(D)$$

Challenge Problems

Providing students with an opportunity to stretch themselves, the Challenge Problems develop their skills beyond the basics. These can be solved by using the techniques developed in the chapter but require more effort than the problems in the regular exercise sets do.

Guidance When Students Need It

The caution icon advises students how to avoid common mistakes and misunderstandings. This feature addresses both student misconceptions and situations in which students often follow unproductive paths.

> ⚠ Theorem 1 states that a relative extremum of f can occur only at a critical number of f. It is important to realize, however, that the converse of Theorem 1 is false. In other words, you may *not* conclude that if c is a critical number of f, then f must have a relative extremum at c. (See Example 3.)

Biographies to Provide Historical Context

Historical biographies provide brief looks at the people who contributed to the development of calculus, focusing not only on their discoveries and achievements, but on their human side as well.

Videos to Help Students Draw Complex Multivariable Calculus Artwork

Unique to this book, Tan's *Calculus* provides video lessons for the multivariable sections of the text that help students learn, step-by-step, how to draw the complex sketches required in multivariable calculus. Videos of these lessons will be available at the text's companion website.

Historical Biography

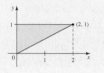

Sheila Terry/Photo Researchers, Inc.

BLAISE PASCAL
(1623-1662)

A great mathematician who was not acknowledged in his lifetime, Blaise Pascal came extremely close to discovering calculus before Leibniz (page 157) and Newton (page 179), the two people who are most commonly credited with the discovery. Pascal was something of a prodigy and published his first important mathematical discovery at the age of sixteen. The work consisted of only a single printed page, but it contained a vital step in the development of projective geometry and a proposition called *Pascal's mystic hexagram* that discussed a property of a hexagon inscribed in a conic section. Pascal's interests varied widely, and from 1642 to 1644 he worked on the first manufactured calculator, which he designed to help his father with his tax work. Pascal manufactured about 50 of the machines, but they proved too costly to continue production. The basic principle of Pascal's calculating machine was still used until the electronic age. Pascal and Pierre de Fermat (page 307) also worked on the mathematics in games of chance and laid the foundation for the modern theory of probability. Pascal's later work, *Treatise on the Arithmetical Triangle*, gave important results on the construction that would later bear his name, *Pascal's Triangle*.

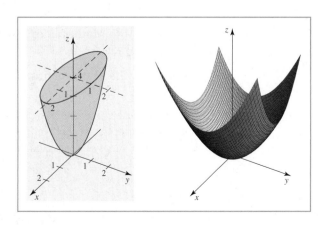

■ Instructor Resources

Instructor's Solutions Manual for Single Variable Calculus (ISBN 0-534-46569-2)
Instructor's Solutions Manual for Multivariable Calculus (ISBN 0-534-46578-1)
Prepared by Soo T. Tan

These manuals provide worked-out solutions to all problems in the text.

PowerLecture CD (ISBN 0-534-49443-9)
This comprehensive CD-ROM includes the *Instructor's Solutions Manual;* PowerPoint slides with art, tables, and key definitions from the text; and ExamView computerized testing, featuring algorithmically generated questions to create, deliver, and customize tests. A static version of the test bank will also be available online.

Solution Builder (ISBN 0-534-41829-5)
The online Solution Builder lets instructors easily build and save personal solution sets either for printing or for posting on password-protected class websites. Contact your local sales representative for more information on obtaining an account for this instructor-only resource.

Enhanced WebAssign (ISBN 0-534-41830-9)
Instant feedback and ease of use are just two reasons why WebAssign is the most widely used homework system in higher education. WebAssign allows instructors to assign, collect, grade, and record homework assignments via the Web. Now this proven homework system has been enhanced to include links to textbook sections, video examples, and problem-specific tutorials. Enhanced WebAssign is more than a homework system—it is a complete learning system for math students.

■ Student Resources

Student Solutions Manual for Single Variable Calculus (ISBN 0-534-46568-4)
Student Solutions Manual for Multivariable Calculus (ISBN 0-534-46577-3)
Prepared by Soo T. Tan

Providing more in-depth explanations, this insightful resource includes fully worked-out solutions for the answers to select exercises included at the back of the textbook, as well as problem-solving strategies, additional algebra steps, and review for selected problems.

CalcLabs with Maple: Single Variable Calculus, 4e by Phil Yasskin and Art Belmonte (ISBN 0-495-56062-6)

CalcLabs with Maple: Multivariable Calculus, 4e by Phil Yasskin and Art Belmonte (ISBN 0-495-56058-8)

CalcLabs with Mathematica: Single Variable Calculus, 4e by Selwyn Hollis (ISBN 0-495-56063-4)

CalcLabs with Mathematica: Multivariable Calculus, 4e by Selwyn Hollis (ISBN 0-495-82722-3)

Each of these comprehensive lab manuals helps students learn to effectively use the technology tools that are available to them. Each lab contains clearly explained exercises and a variety of labs and projects to accompany the text.

■ Acknowledgments

I want to express my heartfelt thanks to the reviewers for their many helpful comments and suggestions at various stages during the development of this text. I also want to thank Kevin Charlwood, Jerrold Grossman, Tao Guo, James Handley, Selwyn Hollis, Diane Koenig, and John Samons, who checked the manuscript and text for accuracy; Richard West, Richard Montano, and again Kevin Charlwood for class testing the manuscript; and Andrew Bulman-Fleming for his help with the production of the solutions manuals. A special thanks to Tao Guo for his contributions to the content and accuracy of the solutions manuals.

I feel fortunate to have worked with a wonderful team during the development and production of this text. I wish to thank the editorial, production, and marketing staffs of Cengage Learning: Richard Stratton, Liz Covello, Cheryll Linthicum, Danielle Derbenti, Ed Dodd, Terri Mynatt, Leslie Lahr, Jeannine Lawless, Peter Galuardi, Lauren Hamel, Jennifer Jones, Angela Kim, and Mary Ann Payumo. My editor, Liz Covello, who joined the team this year, has done a great job working with me to finalize the product before publication. My development editor, Danielle Derbenti, as in the many other projects I have worked with her on, brought her enthusiasm and expertise to help me produce a better book. My production manager, Cheryll Linthicum, coordinated the entire project with equal enthusiasm and ensured that the production process ran smoothly from beginning to end. I also wish to thank Martha Emry, Barbara Willette, and Marian Selig for the excellent work they did in the production of this text. Martha spent countless hours working with me to ensure the accuracy and readability of the text and art. Without the help, encouragement, and support of all those mentioned above, I wouldn't have been able to complete this mammoth task.

I wish to express my personal appreciation to each of the following colleagues who reviewed some part of this work. Their many suggestions have helped to make this a much improved book.

Arun Agarwal
Grambling State University

Mazenia Agustin
Southern Illinois University–Edwardsville

Mike Albanese
Central Piedmont Community College

Robert Andersen
University of Wisconsin–Eau Claire

Daniel Anderson
George Mason University

Joan Bell
Northeastern State University

David Bradley
University of Maine–Orono

Bob Bradshaw
Ohlone College

Paul Britt
Louisiana State University

Bob Buchanon
Millersville University

Christine Bush
Palm Beach Community College

Nick Bykov
San Joaquin Delta College

Janette Campbell
Palm Beach Community College

Kevin Charlwood
Washburn University

S.C. Cheng
Creighton University

Vladimir Cherkassky
Diablo Valley College

Charles Cooper
University of Central Oklahoma

Kyle Costello
Salt Lake Community College

Katrina Cunningham
Southern University–Baton Rouge

Eugene Curtin
Texas State University

Wendy Davidson
Georgia Perimeter College–Newton

Steven J. Davidson
San Jacinto College

John Davis
Baylor University

Ann S. DeBoever
Catawba Valley Community College

John Diamantopoulos
Northwestern State University

John Drost
University of Wisconsin–Eau Claire

Joe Fadyn
Southern Polytechnic State University

Tom Fitzkee
Francis Marion University

James Galloway
Collin County Community College

Jason Andrew Geary
Harper College

Don Goral
Northern Virginia Community College

Alan Graves
Collin County Community College

Elton Graves
Rose-Hulman Institute

Ralph Grimaldi
Rose-Hulman Institute

Ron Hammond
Blinn College–Bryan

James Handley
Montana Tech of the University of Montana

Patricia Henry
Drexel University

Irvin Hentzel
Iowa State University

Alfa Heryudono
University of Massachusetts–Dartmouth

Guy Hinman
Brevard Community College–Melbourne

Gloria Hitchcock
Georgia Perimeter College–Newton

Joshua Holden
Rose-Hulman Institute

Martin Isaacs
University of Wisconsin–Madison

Mic Jackson
Earlham College

Hengli Jiao
Ferris State College

Clarence Johnson
Cuyahoga Community College–Metropolitan

Cindy Johnson
Heartland Community College

Phil Johnson
Appalachian State University

Jack Keating
Massasoit Community College

John Khoury
Brevard Community College–Melbourne

Raja Khoury
Collin County Community College

Rethinasamy Kittappa
Millersville University

Carole Krueger
University of Texas–Arlington

Don Krug
Northern Kentucky University

Kouok Law
Georgia Perimeter College–Lawrenceville

Richard Leedy
Polk Community College

Suzanne Lindborg
San Joaquin Delta College

Tristan Londre
Blue River Community College

Ann M. Loving
J. Sargeant Reynolds Community College

Cyrus Malek
Collin County Community College

Robert Maynard
Tidewater Community College

Phillip McCartney
Northern Kentucky University

Robert McCullough
Ferris State University

Shelly McGee
University of Findlay

Rhonda McKee
Central Missouri State University

George McNulty
University of South Carolina–Columbia

Martin Melandro
Sam Houston State University

Mike Montano
Riverside Community College

Humberto Munoz
Southern University–Baton Rouge

Robert Nehs
Texas Southern University

Charlotte Newsom
Tidewater Community College

Jason Pallett
Longview Community College

Joe Perez
Texas A&M University–Kingsville

Paul Plummer
University of Central Missouri

Tom Polaski
Winthrop University

Tammy Potter
Gadsden State Community College

Linda Powers
Virginia Polytechnic Institute

David Price
Tarrant County College–Southeast

Janice Rech
University of Nebraska–Omaha

Ellena Reda
Dutchess Community College

Michael Reed
Tennessee State University

Lynn Foshee Reed
Maggie L. Walker Governor's School

James Reynolds
Clarion University of Pennsylvania

Joe Rody
Arizona State University

Jorge Sarmiento
County College of Morris

Rosa Seyfried
Harrisburg Area Community College

Kyle Siegrist
University of Alabama–Huntsville

Nathan Smale
University of Utah

Teresa Smith
Blinn College

Shing So
University of Central Missouri

Sonya Stephens
Florida A&M University

Andrew Swift
University of Nebraska–Omaha

Arnavaz P. Taraporevala
New York City College of Technology

W.E. Taylor
Texas Southern University

Beimet Teclezghi
New Jersey City University

Tim Teitloff
Clemson University

Jim Thomas
Colorado State University

Fred Thulin
University of Illinois–Chicago

Virginia Toivonen
Montana Tech of the University of Montana

Deborah Upton
Malloy College

Jim Vallade
Monroe Community College

Kathy Vranicar
University of Nebraska–Omaha

Susan Walker
Montana Tech of the University of Montana

Lianwen Wang
University of Central Missouri

Pam Warton
University of Findlay

Rich West
Francis Marion University

Jen Whitfield
Texas A&M University

Board of Advisors

Shawna Haider
Salt Lake Community College

Alice Eiko Pierce
Georgia Perimeter College

James Handley
Montana Tech of the University of Montana

Joe Rody
Arizona State University

Carol Krueger
University of Texas at Arlington

Scott Wilde
Baylor University

■ Note to the Student

The invention of calculus is one of the crowning intellectual achievements of mankind. Its roots can be traced back to the ancient Egyptians, Greeks, and Chinese. The invention of modern calculus is usually credited to both Gottfried Wilhelm Leibniz and Isaac Newton in the seventeenth century. It has widespread applications in many fields, including engineering, the physical and biological sciences, economics, business, and the social sciences. I am constantly amazed not only by the wonderful mathematical content in calculus but also by the enormous reach it has into every practical field of human endeavor. From studying the growth of a population of bacteria, to building a bridge, to exploring the vast expanses of the heavenly bodies, calculus has always played and continues to play an important role in these endeavors.

In writing this book, I have constantly kept you, the student, in mind. I have tried to make the book as interesting and readable as possible. Many mathematical concepts are introduced by using real-life illustrations. On the basis of my many years of teaching the subject, I am convinced that this approach makes it easier for you to understand the definitions and theorems in this book. I have also taken great pains to include as many steps in the examples as are needed for you to read through them smoothly. Finally, I have taken particular care with the graphical illustrations to ensure that they help you to both understand a concept and solve a problem.

The exercises in the book are carefully constructed to help you understand and appreciate the power of calculus. The problems at the beginning of each exercise set are relatively straightforward to solve and are designed to help you become familiar with the material. These problems are followed by others that require a little more effort on your part. Finally, at the end of each exercise set are problems that put the material you have just learned to good use. Here you will find applications of calculus that are drawn from many fields of study. I think you will also enjoy solving real-life problems of general interest that are drawn from many current sources, including magazines and newspapers. The answers often reveal interesting facts.

However interesting and exciting as it may be, reading a calculus book is not an easy task. You might have to go over the definitions and theorems more than once in order to fully understand them. Here you should pay careful attention to the conditions stated in the theorems. Also, it's a good idea to try to understand the definitions, theorems, and procedures as thoroughly as possible before attempting the exercises. Sometimes writing down a formula is a good way to help you remember it. Finally, if you study with a friend, a good test of your mastery of the material is to take turns explaining the topic you are studying to each other.

One more important suggestion: When you write out the solutions to the problems, make sure that you do so neatly, and try to write down each step to explain how you arrive at the solution. Being neat helps you to avoid mistakes that might occur through misreading your own handwriting (a common cause of errors in solving problems), and writing down each step helps you to work through the solution in a logical manner and to find where you went wrong if your answer turns out to be incorrect. Besides, good habits formed here will be of great help when you write reports or present papers in your career later on in life.

Finally, let me say that writing this book has been a labor of love, and I hope that I can convince you to share my love and enthusiasm for the subject.

Soo T. Tan

The shape assumed by the cable in a suspension bridge is a parabola. The parabola, like the ellipse and hyperbola, is a curve that is obtained as a result of the intersection of a plane and a double-napped cone, and is, accordingly, called a conic section. Conic sections appear in various fields of study such as astronomy, physics, engineering, and navigation.

James Osmond/Alamy

10 Conic Sections, Plane Curves, and Polar Coordinates

CONIC SECTIONS ARE curves that can be obtained by intersecting a double-napped right circular cone with a plane. Our immediate goal is to describe conics using algebraic equations. We then turn to applications of conics, which range from the design of suspension bridges to the design of satellite-signal receiving dishes and to the design of whispering galleries, in which a person standing at one spot in a gallery can hear a whisper coming from another spot in the gallery. The orbits of celestial bodies and human-made satellites can also be described by using conics.

Parametric equations afford a way of describing curves in the plane and in space. We will study these representations and use them to describe the motion of projectiles and the motion of other objects.

Polar coordinates provide an alternative way of representing points in the plane. We will see that certain curves have simpler representations with polar equations than with rectangular equations. We will also make use of polar equations to help us find the arc length of a curve, the area of a region bounded by a curve, and the area of a surface obtained by revolving a curve about a given line.

V This symbol indicates that one of the following video types is available for enhanced student learning at **www.academic.cengage.com/login:**
- Chapter lecture videos
- Solutions to selected exercises

10.1   Conic Sections

FIGURE 1
The reflector of a radio telescope

Figure 1 shows the reflector of a radio telescope. The shape of the surface of the reflector is obtained by revolving a plane curve called a *parabola* about its axis of symmetry. (See Figure 2a.) Figure 2b depicts the orbit of a planet P around the sun, S. This curve is called an *ellipse*. Figure 2c depicts the trajectory of an incoming alpha particle heading toward and then repulsed by a massive atomic nucleus located at the point F. The trajectory is one of two branches of a *hyperbola*.

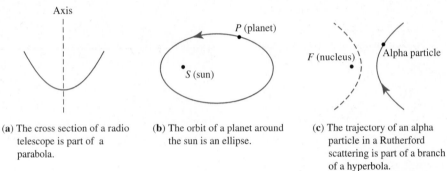

(**a**) The cross section of a radio telescope is part of a parabola.

(**b**) The orbit of a planet around the sun is an ellipse.

(**c**) The trajectory of an alpha particle in a Rutherford scattering is part of a branch of a hyperbola.

FIGURE 2

These curves—parabolas, ellipses, and hyperbolas—are called *conic sections* or, more simply, *conics* because they result from the intersection of a plane and a double-napped cone, as shown in Figure 3.

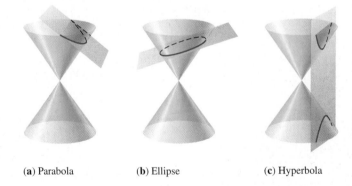

FIGURE 3
The conic sections

(**a**) Parabola   (**b**) Ellipse   (**c**) Hyperbola

In this section we give the geometric definition of each conic section, and we derive an equation for describing each conic section algebraically.

■ Parabola

We first consider a conic section called a *parabola*.

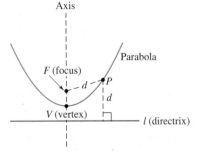

FIGURE 4
The distance between a point P on a parabola and its focus F is the same as the distance between P and the directrix l of the parabola.

> **DEFINITION   Parabola**
>
> A **parabola** is the set of all points in a plane that are equidistant from a fixed point (called the **focus**) and a fixed line (called the **directrix**). (See Figure 4.)

By definition the point halfway between the focus and directrix lies on the parabola. This point V is called the **vertex** of the parabola. The line passing through the focus and perpendicular to the directrix is called the **axis** of the parabola. Observe that the parabola is symmetric with respect to its axis.

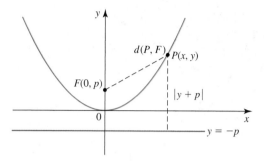

FIGURE 5
The parabola with focus $F(0, p)$ and directrix $y = -p$, where $p > 0$

To find an equation of a parabola, suppose that the parabola is placed so that its vertex is at the origin and its axis is along the y-axis, as shown in Figure 5. Further, suppose that its focus F is at $(0, p)$, and its directrix is the line with equation $y = -p$. If $P(x, y)$ is any point on the parabola, then the distance between P and F is

$$d(P, F) = \sqrt{x^2 + (y - p)^2}$$

whereas the distance between P and the directrix is $|y + p|$. By definition these distances are equal, so

$$\sqrt{x^2 + (y - p)^2} = |y + p|$$

Squaring both sides and simplifying, we obtain

$$x^2 + (y - p)^2 = |y + p|^2 = (y + p)^2$$
$$x^2 + y^2 - 2py + p^2 = y^2 + 2py + p^2$$
$$x^2 = 4py$$

Standard Equation of a Parabola

An equation of the parabola with focus $(0, p)$ and directrix $y = -p$ is

$$x^2 = 4py \tag{1}$$

If we write $a = 1/(4p)$, then Equation (1) becomes $y = ax^2$. Observe that the parabola opens upward if $p > 0$ and opens downward if $p < 0$. (See Figure 6.) Also, the parabola is symmetric with respect to the y-axis (that is, the axis of the parabola coincides with the y-axis), since Equation (1) remains unchanged if we replace x by $-x$.

FIGURE 6
The parabola $x^2 = 4py$ opens upward if $p > 0$ and downward if $p < 0$.

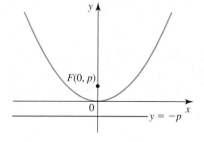

(a) $p > 0$

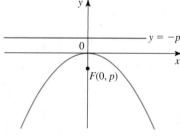

(b) $p < 0$

Interchanging x and y in Equation (1) gives

$$y^2 = 4px \qquad\qquad (2)$$

which is an equation of the parabola with focus $F(p, 0)$ and directrix $x = -p$. The parabola opens to the right if $p > 0$ and opens to the left if $p < 0$. (See Figure 7.) In both cases the axis of the parabola coincides with the x-axis.

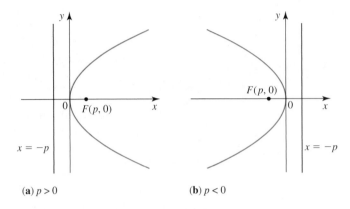

FIGURE 7
The parabola $y^2 = 4px$ opens to the right if $p > 0$ and to the left if $p < 0$.

(a) $p > 0$ (b) $p < 0$

Note   A parabola with vertex at the origin and axis of symmetry lying on the x-axis or y-axis is said to be in **standard position.** (See Figures 6 and 7.) ∎

EXAMPLE 1   Find the focus and directrix of the parabola $y^2 + 6x = 0$, and make a sketch of the parabola.

Solution   Rewriting the given equation in the form $y^2 = -6x$ and comparing it with Equation (2), we see that $4p = -6$ or $p = -\frac{3}{2}$. Therefore, the focus of the parabola is $F\left(-\frac{3}{2}, 0\right)$, and its directrix is $x = \frac{3}{2}$. The parabola is sketched in Figure 8. ∎

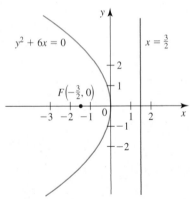

FIGURE 8
The parabola $y^2 + 6x = 0$

EXAMPLE 2   Find an equation of the parabola that has its vertex at the origin with axis of symmetry lying on the y-axis, and passes through the point $P(3, -4)$. What are the focus and directrix of the parabola?

Solution   An equation of the parabola has the form $y = ax^2$. To determine the value of a, we use the condition that the point $P(3, -4)$ lies on the parabola to obtain the equation $-4 = a(3)^2$ giving $a = -\frac{4}{9}$. Therefore, an equation of the parabola is

$$y = -\frac{4}{9}x^2$$

To find the focus of the parabola, observe that it has the form $F(0, p)$. Now

$$p = \frac{1}{4a} = \frac{1}{4\left(-\frac{4}{9}\right)} = -\frac{9}{16}$$

Therefore, the focus is $F\left(0, -\frac{9}{16}\right)$. Its directrix is $y = -p = -\left(-\frac{9}{16}\right)$, or $y = \frac{9}{16}$. The graph of the parabola is sketched in Figure 9. ∎

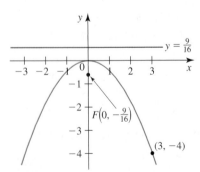

FIGURE 9
The parabola $y = -\frac{4}{9}x^2$

The parabola has many applications. For example, the cables of certain suspension bridges assume shapes that are parabolic.

EXAMPLE 3 **Suspension Bridge Cables**   Figure 10 depicts a bridge, suspended by a flexible cable. If we assume that the weight of the cable is negligible in comparison to the weight of the bridge, then it can be shown that the shape of the cable is described by the equation

$$y = \frac{Wx^2}{2H}$$

where W is the weight of the bridge in pounds per foot and H is the tension at the lowest point of the cable in pounds (the origin). (See Exercise 89.) Suppose that the *span* of the cable is $2a$ ft and the *sag* is h ft.

a. Find an equation describing the shape assumed by the cable in terms of a and h.
b. Find the length of the cable if the span of the cable is 400 ft and the sag is 80 ft.

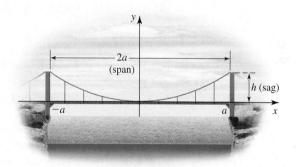

FIGURE 10
A bridge of length $2a$
suspended by a flexible cable

Solution

a. We can write the given equation in the form $y = kx^2$, where $k = W/(2H)$. Since the point (a, h) lies on the parabola $y = kx^2$, we have

$$h = ka^2$$

or $k = h/a^2$, so the required equation is $y = hx^2/a^2$.

b. With $a = 200$ and $h = 80$ an equation that describes the shape of the cable is

$$y = \frac{80x^2}{200^2} = \frac{x^2}{500}$$

Next, the length of the cable is given by

$$s = 2\int_0^{200} \sqrt{1 + (y')^2}\, dx$$

But $y' = x/250$, so

$$s = 2\int_0^{200} \sqrt{1 + \left(\frac{x}{250}\right)^2}\, dx = \frac{1}{125}\int_0^{200} \sqrt{250^2 + x^2}\, dx$$

The easiest way to evaluate this integral is to use Formula 37 from the Table of Integrals found on the reference pages of this book:

$$\int \sqrt{a^2 + u^2}\, du = \frac{u}{2}\sqrt{a^2 + u^2} + \frac{a^2}{2}\ln|u + \sqrt{a^2 + u^2}| + C$$

If we let $a = 250$ and $u = x$, then

$$s = \frac{1}{125} \left[\frac{x}{2} \sqrt{250^2 + x^2} + \frac{250^2}{2} \ln|x + \sqrt{250^2 + x^2}| \right]_0^{200}$$

$$= \frac{1}{125} \left[100 \sqrt{62500 + 40000} + 31250 \ln|200 + \sqrt{62500 + 40000}| - 31250 \ln 250 \right]$$

$$= \frac{4}{5} \sqrt{102500} + 250 \ln\left(\frac{200 + \sqrt{102500}}{250} \right) \approx 439$$

or 439 ft.   ■

Other applications of the parabola include the trajectory of a projectile in the absence of air resistance.

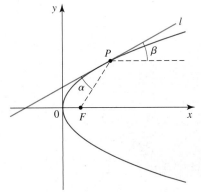

FIGURE 11
The reflective property states that $\alpha = \beta$.

■ Reflective Property of the Parabola

Suppose that P is any point on a parabola with focus F, and let l be the tangent line to the parabola at P. (See Figure 11.) The reflective property states that the angle α that lies between l and the line segment FP is equal to the angle β that lies between l and the line passing through P and parallel to the axis of the parabola. This property is the basis for many applications. (An outline of the proof of this property is given in Exercise 105.)

As was mentioned earlier, the reflector of a radio telescope has a shape that is obtained by revolving a parabola about its axis. Figure 12a shows a cross section of such a reflector. A radio wave coming in from a great distance may be assumed to be parallel to the axis of the parabola. This wave will strike the surface of the reflector and be reflected toward the focus F, where a collector is located. (The angle of incidence is equal to the angle of reflection.)

FIGURE 12
Applications of the reflective property of a parabola

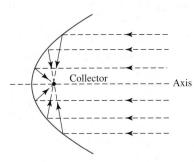

(**a**) A cross section of a radio telescope

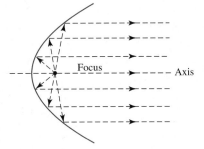

(**b**) A cross section of a headlight

The reflective property of the parabola is also used in the design of headlights of automobiles. Here, a light bulb is placed at the focus of the parabola. A ray of light emanating from the light bulb will strike the surface of the reflector and be reflected outward along a direction parallel to the axis of the parabola (see Figure 12b).

■ Ellipses

Next, we consider a conic section called an ellipse.

> **DEFINITION Ellipse**
>
> An ellipse is the set of all points in a plane the sum of whose distances from two fixed points (called the **foci**) is a constant.

Figure 13 shows an ellipse with foci F_1 and F_2. The line passing through the foci intersects the ellipse at two points, V_1 and V_2, called the **vertices** of the ellipse. The chord joining the vertices is called the **major axis,** and its midpoint is called the **center** of the ellipse. The chord passing through the center of the ellipse and perpendicular to the major axis is called the **minor axis** of the ellipse.

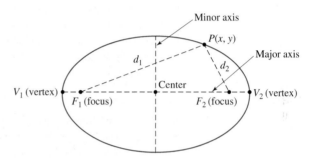

FIGURE 13
An ellipse with foci F_1 and F_2. A point $P(x, y)$ is on the ellipse if and only if $d_1 + d_2 =$ a constant.

Note   We can construct an ellipse on paper in the following way: Place a piece of paper on a flat wooden board. Next, secure the ends of a piece of string to two points (the foci of the ellipse) with thumbtacks. Then trace the required ellipse with a pencil pushed against the string, as shown in Figure 14, making sure that the string is kept taut at all times. ■

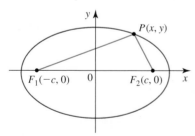

FIGURE 14
Drawing an ellipse on paper using thumbtacks, a string, and a pencil

To find an equation for an ellipse, suppose that the ellipse is placed so that its major axis lies along the x-axis and its center is at the origin, as shown in Figure 15. Then its foci F_1 and F_2 are at the points $(-c, 0)$ and $(c, 0)$, respectively. Let the sum of the distances between any point $P(x, y)$ on the ellipse and its foci be $2a > 2c > 0$. Then, by the definition of an ellipse we have

$$d(P, F_1) + d(P, F_2) = 2a$$

that is,

$$\sqrt{(x + c)^2 + y^2} + \sqrt{(x - c)^2 + y^2} = 2a$$

or

$$\sqrt{(x - c)^2 + y^2} = 2a - \sqrt{(x + c)^2 + y^2}$$

Squaring both sides of this equation, we obtain

$$x^2 - 2cx + c^2 + y^2 = 4a^2 - 4a\sqrt{(x + c)^2 + y^2} + x^2 + 2cx + c^2 + y^2$$

or, upon simplification,

$$a\sqrt{(x + c)^2 + y^2} = a^2 + cx$$

Squaring both sides again, we have

$$a^2(x^2 + 2cx + c^2 + y^2) = a^4 + 2a^2cx + c^2x^2$$

FIGURE 15
The ellipse with foci $F_1(-c, 0)$ and $F_2(c, 0)$

which yields

$$(a^2 - c^2)x^2 + a^2y^2 = a^2(a^2 - c^2)$$

Recall that $a > c$, so $a^2 - c^2 > 0$. Let $b^2 = a^2 - c^2$ with $b > 0$. Then the equation of the ellipse becomes

$$b^2x^2 + a^2y^2 = a^2b^2$$

or, upon dividing both sides by a^2b^2, we obtain

$$\frac{x^2}{a^2} + \frac{y^2}{b^2} = 1$$

By setting $y = 0$, we obtain $x = \pm a$, which gives $(-a, 0)$ and $(a, 0)$ as the vertices of the ellipse. Similarly, by setting $x = 0$, we see that the ellipse intersects the y-axis at the points $(0, -b)$ and $(0, b)$. Since the equation remains unchanged if x is replaced by $-x$ and y is replaced by $-y$, we see that the ellipse is symmetric with respect to both axes.

Observe, too, that $b < a$, since

$$b^2 = a^2 - c^2 < a^2$$

So as the name implies, the length of the major axis, $2a$, is greater than the length of the minor axis, $2b$. Finally, observe that if the foci coincide, then $c = 0$ and $a = b$, so the ellipse is a circle with radius $r = a = b$.

Placing the ellipse so that its major axis lies along the y-axis and its center is at the origin leads to an equation in which the roles of x and y are reversed. To summarize, we have the following.

Standard Equation of an Ellipse

An equation of the ellipse with foci $(\pm c, 0)$ and vertices $(\pm a, 0)$ is

$$\frac{x^2}{a^2} + \frac{y^2}{b^2} = 1 \qquad a \geq b > 0 \qquad \text{(3)}$$

and an equation of the ellipse with foci $(0, \pm c)$ and vertices $(0, \pm a)$ is

$$\frac{x^2}{b^2} + \frac{y^2}{a^2} = 1 \qquad a \geq b > 0 \qquad \text{(4)}$$

where $c^2 = a^2 - b^2$. (See Figure 16.)

FIGURE 16
Two ellipses in standard position with center at the origin

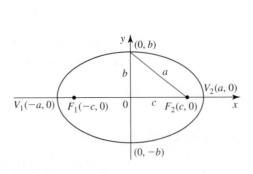

(a) The major axis is along the x-axis.

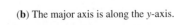

(b) The major axis is along the y-axis.

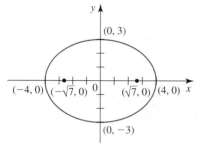

FIGURE 17

The ellipse $\dfrac{x^2}{16} + \dfrac{y^2}{9} = 1$

Note An ellipse with center at the origin and foci lying along the x-axis or the y-axis is said to be in standard position. (See Figure 16.) ▪

EXAMPLE 4 Sketch the ellipse $\dfrac{x^2}{16} + \dfrac{y^2}{9} = 1$. What are the foci and vertices?

Solution Here, $a^2 = 16$ and $b^2 = 9$, so $a = 4$ and $b = 3$. Setting $y = 0$ and $x = 0$ in succession gives the x- and y-intercepts as ± 4 and ± 3, respectively. Also, from

$$c^2 = a^2 - b^2 = 16 - 9 = 7$$

we obtain $c = \sqrt{7}$ and conclude that the foci of the ellipse are $(\pm\sqrt{7}, 0)$. Its vertices are $(\pm 4, 0)$. The ellipse is sketched in Figure 17. ▪

EXAMPLE 5 Find an equation of the ellipse with foci $(0, \pm 2)$ and vertices $(0, \pm 4)$.

Solution Since the foci and therefore the major axis of the ellipse lie along the y-axis, we use Equation (4). Here, $c = 2$ and $a = 4$, so

$$b^2 = a^2 - c^2 = 16 - 4 = 12$$

Therefore, the standard form of the equation for the ellipse is

$$\frac{x^2}{12} + \frac{y^2}{16} = 1$$

or

$$4x^2 + 3y^2 = 48$$ ▪

◾ Reflective Property of the Ellipse

The ellipse, like the parabola, has a reflective property. To describe this property, consider an ellipse with foci F_1 and F_2 as shown in Figure 18. Let P be a point on the ellipse, and let l be the tangent line to the ellipse at P. Then the angle α between the line segment F_1P and l is equal to the angle β between the line segment F_2P and l. You will be asked to establish this property in Exercise 106.

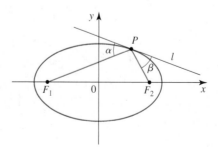

FIGURE 18
The reflective property
states that $\alpha = \beta$.

The reflective property of the ellipse is used to design *whispering galleries*—rooms with elliptical-shaped ceilings, in which a person standing at one focus can hear the whisper of another person standing at the other focus. A whispering gallery can be found in the rotunda of the Capitol Building in Washington, D.C. Also, Paris subway tunnels are almost elliptical, and because of the reflective property of the ellipse, whispering on one platform can be heard on the other. (See Figure 19.)

FIGURE 19
A cross section of a Paris subway
tunnel is almost elliptical.

Yet another application of the reflective property of the ellipse can be found in the field of medicine in a procedure for removing kidney stones called *shock wave lithotripsy*. In this procedue an ellipsoidal reflector is positioned so that a transducer is at one focus and a kidney stone is at the other focus. Shock waves emanating from the transducer are reflected according to the reflective property of the ellipse onto the kidney stone, pulverizing it. This procedure obviates the necessity for surgery.

■ Eccentricity of an Ellipse

To measure the ovalness of an ellipse, we introduce the notion of eccentricity.

DEFINITION   **Eccentricity of an Ellipse**

The **eccentricity** of an ellipse is given by the ratio $e = c/a$.

The eccentricity of an ellipse satisfies $0 < e < 1$, since $0 < c < a$. The closer e is to zero, the more circular is the ellipse.

■ Hyperbolas

The definition of a hyperbola is similar to that of an ellipse. The *sum* of the distances between the foci and a point on an ellipse is fixed, whereas the *difference* of these distances is fixed for a hyperbola.

DEFINITION   **Hyperbola**

A **hyperbola** is the set of all points in a plane the difference of whose distances from two fixed points (called the **foci**) is a constant.

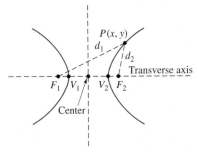

FIGURE 20
A hyperbola with foci F_1 and F_2. A point $P(x, y)$ is on the hyperbola if and only if $|d_1 - d_2|$ is a constant.

Figure 20 shows a hyperbola with foci F_1 and F_2. The line passing through the foci intersects the hyperbola at two points, V_1 and V_2, called the **vertices** of the hyperbola. The line segment joining the vertices is called the **transverse axis** of the hyperbola, and the midpoint of the transverse axis is called the **center** of the hyperbola. Observe that a hyperbola, in contrast to a parabola or an ellipse, has two separate branches.

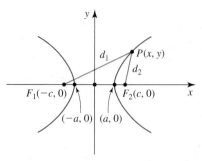

FIGURE 21

An equation of the hyperbola with center $(0, 0)$ and foci $(-c, 0)$ and $(c, 0)$ is $\dfrac{x^2}{a^2} - \dfrac{y^2}{b^2} = 1$

The derivation of an equation of a hyperbola is similar to that of an ellipse. Consider, for example, the hyperbola with center at the origin and foci $F_1(-c, 0)$ and $F_2(c, 0)$ on the x-axis. (See Figure 21.) Using the condition $d(P, F_1) - d(P, F_2) = 2a$, where a is a positive constant, it can be shown that if $P(x, y)$ is any point on the hyperbola, then an equation of the hyperbola is

$$\frac{x^2}{a^2} - \frac{y^2}{b^2} = 1$$

where $b = \sqrt{c^2 - a^2}$ or $c = \sqrt{a^2 + b^2}$.

Observe that the x-intercepts of the hyperbola are $x = \pm a$, giving $(-a, 0)$ and $(a, 0)$ as its vertices. But there are no y-intercepts, since setting $x = 0$ gives $y^2 = -b^2$, which has no real solution. Also, observe that the hyperbola is symmetric with respect to both axes.

If we solve the equation

$$\frac{x^2}{a^2} - \frac{y^2}{b^2} = 1$$

for y, we obtain

$$y = \pm \frac{b}{a} \sqrt{x^2 - a^2}$$

Since $x^2 - a^2 \geq 0$ or, equivalently, $x \leq -a$ or $x \geq a$, we see that the hyperbola actually consists of two separate branches, as was noted earlier. Also, observe that if x is large in magnitude, then $x^2 - a^2 \approx x^2$, so $y = \pm(b/a)x$. This heuristic argument suggests that both branches of the hyperbola approach the slant asymptotes $y = \pm(b/a)x$ as x increases or decreases without bound. (See Figure 22.) You will be asked in Exercise 101 to demonstrate that this is true.

Finally, if the foci of a hyperbola are on the y-axis, then by reversing the roles of x and y, we obtain

$$\frac{y^2}{a^2} - \frac{x^2}{b^2} = 1$$

as an equation of the hyperbola.

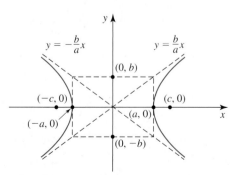

(a) $\dfrac{x^2}{a^2} - \dfrac{y^2}{b^2} = 1$ (The transverse axis is along the x-axis.)

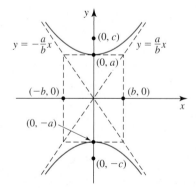

(b) $\dfrac{y^2}{a^2} - \dfrac{x^2}{b^2} = 1$ (The transverse axis is along the y-axis.)

FIGURE 22

Two hyperbolas in standard position with center at the origin

> **Standard Equation of a Hyperbola**
>
> An equation of the hyperbola with foci $(\pm c, 0)$ and vertices $(\pm a, 0)$ is
>
> $$\frac{x^2}{a^2} - \frac{y^2}{b^2} = 1 \qquad (5)$$
>
> where $c = \sqrt{a^2 + b^2}$. The hyperbola has asymptotes $y = \pm(b/a)x$. An equation of the hyperbola with foci $(0, \pm c)$ and vertices $(0, \pm a)$ is
>
> $$\frac{y^2}{a^2} - \frac{x^2}{b^2} = 1 \qquad (6)$$
>
> where $c = \sqrt{a^2 + b^2}$. The hyperbola has asymptotes $y = \pm(a/b)x$.

The line segment of length $2b$ joining the points $(0, -b)$ and $(0, b)$ or $(-b, 0)$ and $(b, 0)$ is called the **conjugate axis** of the hyperbola.

EXAMPLE 6   Find the foci, vertices, and asymptotes of the hyperbola $4x^2 - 9y^2 = 36$.

Solution   Dividing both sides of the given equation by 36 leads to the standard equation

$$\frac{x^2}{9} - \frac{y^2}{4} = 1$$

of a hyperbola. Here, $a^2 = 9$ and $b^2 = 4$, so $a = 3$ and $b = 2$. Setting $y = 0$ gives ± 3 as the x-intercepts, so $(\pm 3, 0)$ are the vertices of the hyperbola. Also, we have $c = \sqrt{a^2 + b^2} = \sqrt{13}$, and conclude that the foci of the hyperbola are $(\pm\sqrt{13}, 0)$. Finally, the asymptotes of the hyperbola are

$$y = \pm\frac{b}{a}x = \pm\frac{2}{3}x$$

When you sketch this hyperbola, draw the asymptotes first so that you can then use them as guides for sketching the hyperbola itself. (See Figure 23.)   ∎

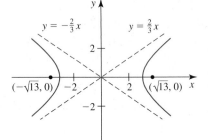

FIGURE 23
The graph of the hyperbola
$4x^2 - 9y^2 = 36$

EXAMPLE 7   A hyperbola has vertices $(0, \pm 3)$ and passes through the point $(2, 5)$. Find an equation of the hyperbola. What are its foci and asymptotes?

Solution   Here, the foci lie along the y-axis, so the standard equation of the hyperbola has the form

$$\frac{y^2}{9} - \frac{x^2}{b^2} = 1 \qquad \text{Note that } a = 3.$$

To determine b, we use the condition that the hyperbola passes through the point $(2, 5)$ to write

$$\frac{25}{9} - \frac{4}{b^2} = 1$$

$$\frac{4}{b^2} = \frac{25}{9} - 1 = \frac{16}{9}$$

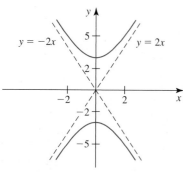

FIGURE 24
The graph of the hyperbola
$y^2 - 4x^2 = 9$

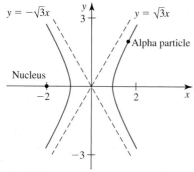

FIGURE 25
The trajectory of an alpha particle in a
Rutherford scattering is a branch of a
hyperbola.

or $b^2 = \frac{9}{4}$. Therefore, a required equation of the hyperbola is

$$\frac{y^2}{9} - \frac{x^2}{\frac{9}{4}} = 1$$

or, equivalently, $y^2 - 4x^2 = 9$. To find the foci of the hyperbola, we compute

$$c^2 = a^2 + b^2 = 9 + \frac{9}{4} = \frac{45}{4}$$

or $c = \pm\sqrt{45/4} = \pm 3\sqrt{5}/2$, from which we see that the foci are $\left(0, \pm\frac{3\sqrt{5}}{2}\right)$. Finally, the asymptotes are obtained by substituting $a = 3$ and $b = \frac{3}{2}$ into the equations $y = \pm(a/b)x$, giving $y = \pm 2x$. The graph of the hyperbola is shown in Figure 24. ∎

EXAMPLE 8 **A Rutherford Scattering** A massive atomic nucleus used as a target for incoming alpha particles is located at the point $(-2, 0)$, as shown in Figure 25. Suppose that an alpha particle approaching the nucleus has a trajectory that is a branch of the hyperbola shown with asymptotes $y = \pm\sqrt{3}x$ and foci $(\pm 2, 0)$. Find an equation of the trajectory.

Solution The asymptotes of a hyperbola with center at the origin and foci lying on the x-axis have equations of the form $y = \pm(b/a)x$. Since the asymptotes of the trajectory are $y = \pm\sqrt{3}x$, we see that

$$\frac{b}{a} = \sqrt{3} \qquad \text{or} \qquad b = \sqrt{3}a$$

Next, since the foci of the hyperbola are $(\pm 2, 0)$, we know that $c = 2$. But $c^2 = a^2 + b^2$, and this gives

$$4 = a^2 + (\sqrt{3}a)^2 = a^2 + 3a^2 = 4a^2$$

or $a = 1$, so $b = \sqrt{3}$. Therefore, an equation of the trajectory is

$$\frac{x^2}{1} - \frac{y^2}{3} = 1$$

or $3x^2 - y^2 = 3$, where $x > 0$. ∎

DEFINITION **Eccentricity of a Hyperbola**
The **eccentricity** of a hyperbola is given by the ratio $e = c/a$.

Since $c > a$, the eccentricity of a hyperbola satisfies $e > 1$. The larger the eccentricity is, the flatter are the branches of the hyperbola.

■ Shifted Conics

By using the techniques in Section 0.4, we can obtain the equations of conics that are translated from their standard positions. In fact, by replacing x by $x - h$ and y by $y - k$ in their standard equations, we obtain the equation of a parabola whose vertex is translated from the origin to the point (h, k) and the equation of an ellipse (or hyperbola) whose center is translated from the origin to the point (h, k).

We summarize these results in Table 1. Figure 26 shows the graphs of these conics.

TABLE 1

Conic	Orientation of axis	Equation of conic		
Parabola	Axis horizontal	$(y - k)^2 = 4p(x - h)$	**(7)**	(See Figure 26a.)
Parabola	Axis vertical	$(x - h)^2 = 4p(y - k)$	**(8)**	(See Figure 26b.)
Ellipse	Major axis horizontal	$\dfrac{(x - h)^2}{a^2} + \dfrac{(y - k)^2}{b^2} = 1$	**(9)**	(See Figure 26c.)
Ellipse	Major axis vertical	$\dfrac{(x - h)^2}{b^2} + \dfrac{(y - k)^2}{a^2} = 1$	**(10)**	(See Figure 26d.)
Hyperbola	Transverse axis horizontal	$\dfrac{(x - h)^2}{a^2} - \dfrac{(y - k)^2}{b^2} = 1$	**(11)**	(See Figure 26e.)
Hyperbola	Transverse axis vertical	$\dfrac{(y - k)^2}{a^2} - \dfrac{(x - h)^2}{b^2} = 1$	**(12)**	(See Figure 26f.)

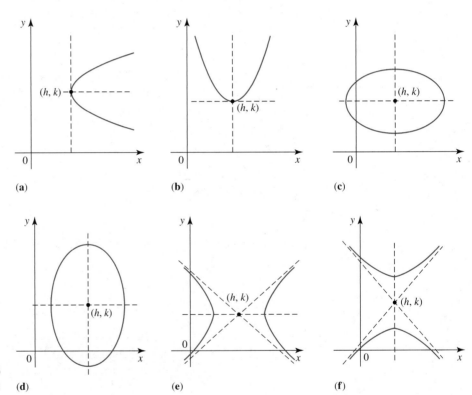

FIGURE 26
Shifted conics with centers at (h, k)

Observe that if $h = k = 0$, then each of the equations listed in Table 1 reduces to the corresponding standard equation of a conic centered at the origin, as expected.

EXAMPLE 9 Find the standard equation of the ellipse with foci at $(1, 2)$ and $(5, 2)$ and major axis of length 6. Sketch the ellipse.

Solution Since the foci $(1, 2)$ and $(5, 2)$ have the same y-coordinate, we see that they lie along the line $y = 2$ parallel to the x-axis. The midpoint of the line segment joining $(1, 2)$ to $(5, 2)$ is $(3, 2)$, and this is the center of the ellipse. From this we can see that the distance from the center of the ellipse to each of the foci is 2, so $c = 2$. Next,

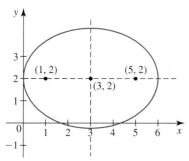

FIGURE 27

The ellipse $\dfrac{(x - 3)^2}{9} + \dfrac{(y - 2)^2}{5} = 1$

since the major axis of the ellipse is known to have length 6, we have $2a = 6$, or $a = 3$. Finally, from the relation $c^2 = a^2 - b^2$, we obtain $4 = 9 - b^2$, or $b^2 = 5$. Therefore, using Equation (9) from Table 1 with $h = 3$, $k = 2$, $a = 3$, and $b = \sqrt{5}$, we obtain the desired equation:

$$\frac{(x - 3)^2}{9} + \frac{(y - 2)^2}{5} = 1$$

The ellipse is sketched in Figure 27. ∎

If you expand and simplify each equation in Table 1, you will see that these equations have the general form

$$Ax^2 + By^2 + Dx + Ey + F = 0$$

where the coefficients are real numbers. Conversely, given such an equation, we can obtain an equivalent equation in the form listed in Table 1 by using the technique of completing the square. The latter can then be analyzed readily to obtain the properties of the conic that it represents.

EXAMPLE 10 Find the standard equation of the hyperbola

$$3x^2 - 4y^2 + 6x + 16y - 25 = 0$$

Find its foci, vertices, and asymptotes, and sketch its graph.

Solution We complete the squares in x and y:

$$3(x^2 + 2x) - 4(y^2 - 4y) = 25$$
$$3[x^2 + 2x + (1)^2] - 4[y^2 - 4y + (-2)^2] = 25 + 3 - 16$$
$$3(x + 1)^2 - 4(y - 2)^2 = 12$$

Then, dividing both sides of this equation by 12 gives the desired equation

$$\frac{(x + 1)^2}{4} - \frac{(y - 2)^2}{3} = 1$$

Comparing this equation with Equation (11) in Table 1, we see that it is an equation of a hyperbola with center $(-1, 2)$ and transverse axis parallel to the x-axis. We also see that $a^2 = 4$ and $b^2 = 3$, from which it follows that $c^2 = a^2 + b^2 = 4 + 3 = 7$. We can think of this hyperbola as one that is obtained by shifting a similar hyperbola, centered at the origin, one unit to the left and two units upward. Then the required foci, vertices, and asymptotes are obtained by shifting the foci, vertices, and asymptotes of this latter hyperbola accordingly. The results are as follows:

Foci	$(-\sqrt{7} - 1, 2)$ and $(\sqrt{7} - 1, 2)$
Vertices	$(-3, 2)$ and $(1, 2)$
Asymptotes	$y - 2 = \pm\frac{\sqrt{3}}{2}(x + 1)$

The hyperbola is sketched in Figure 28.

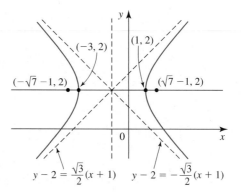

FIGURE 28
The hyperbola
$$3x^2 - 4y^2 + 6x + 16y - 25 = 0$$

The properties of the hyperbola are exploited in the navigational system LORAN (Long Range Navigation). This system utilizes two sets of transmitters: one set located at F_1 and F_2 and another set located at G_1 and G_2. (See Figure 29.) Suppose that synchronized signals sent out by the transmitters located at F_1 and F_2 reach a ship that is located at P. The difference in the times of arrival of the signals are converted by an onboard computer into the difference in the distance $d(P, F_1) - d(P, F_2)$. Using the definition of the hyperbola, we see that this places the ship on a branch of a hyperbola with foci F_1 and F_2 (Figure 29). Similarly, we see that the ship must also lie on a branch of a hyperbola with foci G_1 and G_2. Thus, the position of P is given by the intersection of these two branches of the hyperbolas.

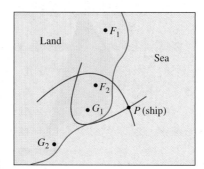

FIGURE 29
In the LORAN navigational system the position of a ship is the point of intersection of two branches of hyperbolas.

10.1   CONCEPT QUESTIONS

1. a. Give the definition of a parabola. What are the focus, directrix, vertex, and axis of a parabola? Illustrate with a sketch.

b. Write the standard equation of (i) a parabola whose axis lies on the y-axis and (ii) a parabola whose axis lies on the x-axis. Illustrate with sketches.

2. a. Give the definition of an ellipse. What are the foci, vertices, center, major axis, and minor axis of an ellipse? Illustrate with a sketch.

b. Write the standard equation of (i) an ellipse with foci $(\pm c, 0)$ and vertices $(\pm a, 0)$ and (ii) an ellipse with foci $(0, \pm c)$ and vertices $(0, \pm a)$. Illustrate with sketches.

3. a. Give the definition of a hyperbola. What are the center, the foci, and the transverse axis of the hyperbola? Illustrate with a sketch.

b. Write the standard equation of (i) a hyperbola with foci $(\pm c, 0)$ and vertices $(\pm a, 0)$ and (ii) a hyperbola with foci $(0, \pm c)$ and vertices $(0, \pm a)$. Illustrate with sketches.

10.1 EXERCISES

In Exercises 1–8, match the equation with one of the conics labeled (a)–(h). If the conic is a parabola, find its vertex, focus and directrix. If it is an ellipse or a hyperbola, find its vertices, foci, and eccentricity.

1. $x^2 = -4y$

2. $y = \dfrac{x^2}{8}$

3. $y^2 = 8x$

4. $x = -\dfrac{1}{4}y^2$

5. $\dfrac{x^2}{9} + \dfrac{y^2}{4} = 1$

6. $x^2 + \dfrac{y^2}{4} = 1$

7. $\dfrac{x^2}{16} - \dfrac{y^2}{9} = 1$

8. $y^2 - \dfrac{x^2}{4} = 1$

(a)
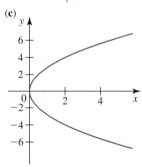

(b)
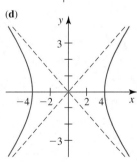

(c)
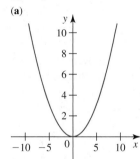

(d)
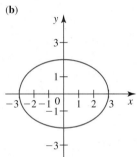

(e)
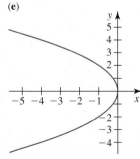

(f)
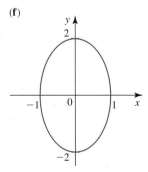

(g)
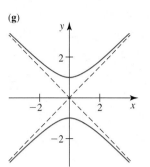

(h)
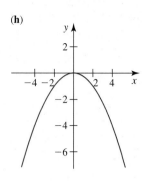

In Exercises 9–14, find the vertex, focus, and directrix of the parabola with the given equation, and sketch the parabola.

9. $y = 2x^2$

10. $x^2 = -12y$

11. $x = 2y^2$

12. $y^2 = -8x$

13. $5y^2 = 12x$

14. $y^2 = -40x$

In Exercises 15–20, find the foci and vertices of the ellipse, and sketch its graph.

15. $\dfrac{x^2}{4} + \dfrac{y^2}{25} = 1$

16. $\dfrac{x^2}{16} + \dfrac{y^2}{9} = 1$

17. $4x^2 + 9y^2 = 36$

18. $25x^2 + 16y^2 = 400$

19. $x^2 + 4y^2 = 4$

20. $2x^2 + y^2 = 4$

In Exercises 21–26, find the vertices, foci, and asymptotes of the hyperbola, and sketch its graph using its asymptotes as an aid.

21. $\dfrac{x^2}{25} - \dfrac{y^2}{144} = 1$

22. $\dfrac{y^2}{16} - \dfrac{x^2}{81} = 1$

23. $x^2 - y^2 = 1$

24. $4y^2 - x^2 = 4$

25. $y^2 - 5x^2 = 25$

26. $x^2 - 2y^2 = 8$

In Exercises 27–30, find an equation of the parabola that satisfies the conditions.

27. Focus $(3, 0)$, directrix $x = -3$

28. Focus $(0, -2)$, directrix $y = 2$

29. Focus $\left(-\tfrac{5}{2}, 0\right)$, directrix $x = \tfrac{5}{2}$

30. Focus $\left(0, \tfrac{3}{2}\right)$, directrix $y = -\tfrac{3}{2}$

In Exercises 31–38, find an equation of the ellipse that satisfies the given conditions.

31. Foci $(\pm 1, 0)$, vertices $(\pm 3, 0)$

32. Foci $(0, \pm 3)$, vertices $(0, \pm 5)$

33. Foci $(0, \pm1)$, length of major axis 6

34. Vertices $(0, \pm5)$, length of minor axis 5

35. Vertices $(\pm3, 0)$, passing through $(1, \sqrt{2})$

36. Passes through $(1, 5)$ and $(2, 4)$ and its center is at $(0, 0)$

37. Passes through $\left(2, \frac{3\sqrt{3}}{2}\right)$ with vertices at $(0, \pm5)$

38. x-intercepts ±3, y-intercepts $\pm\frac{1}{2}$

In Exercises 39–44, find an equation of the hyperbola centered at the origin that satisfies the given conditions.

39. foci $(\pm5, 0)$, vertices $(\pm3, 0)$

40. foci $(0, \pm8)$, vertices $(0, \pm4)$

41. foci $(0, \pm5)$, conjugate axis of length 4

42. vertices $(\pm4, 0)$ passing through $\left(5, \frac{9}{4}\right)$

43. vertices $(\pm2, 0)$, asymptotes $y = \pm\frac{3}{2}x$

44. y-intercepts ±1, asymptotes $y = \pm\frac{1}{2\sqrt{2}}x$

In Exercises 45–48, match the equation with one of the conic sections labeled (a)–(d).

45. $(x + 3)^2 = -2(y - 4)$ **46.** $\dfrac{(x - 2)^2}{16} + \dfrac{(y + 3)^2}{4} = 1$

47. $\dfrac{(y - 3)^2}{16} - \dfrac{(x + 1)^2}{9} = 1$ **48.** $(y - 1)^2 = -4(x - 2)$

(a)

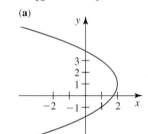

(b)

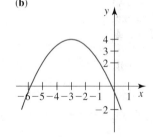

(c)

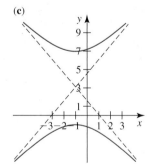

(d)

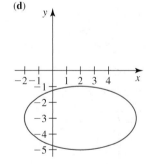

In Exercises 49–66, find an equation of the conic satisfying the given conditions.

49. Parabola, focus $(3, 1)$, directrix $x = 1$

50. Parabola, focus $(-2, 3)$, directrix $y = 5$

51. Parabola, vertex $(2, 2)$, focus $\left(\frac{3}{2}, 2\right)$

52. Parabola, vertex $(1, -2)$, directrix $y = 1$

53. Parabola, axis parallel to the y-axis, passes through $(-3, 2)$, $\left(0, -\frac{5}{2}\right)$, and $(1, -6)$

54. Parabola, axis parallel to the x-axis, passes through $(-6, 6)$, $(0, 0)$, and $(2, 2)$

55. Ellipse, foci $(\pm1, 3)$, vertices $(\pm3, 3)$

56. Ellipse, foci $(0, 2)$ and $(4, 2)$, vertices $(-1, 2)$ and $(5, 2)$

57. Ellipse, foci $(\pm1, 2)$, length of major axis 8

58. Ellipse, foci $(1, \pm3)$, length of minor axis 2

59. Ellipse, center $(2, 1)$, focus $(0, 1)$, vertex $(5, 1)$

60. Ellipse, foci $(2 - \sqrt{3}, -1)$ and $(2 + \sqrt{3}, -1)$, passes through $(2, 0)$

61. Hyperbola, foci $(-2, 2)$ and $(8, 2)$, vertices $(0, 2)$ and $(6, 2)$

62. Hyperbola, foci $(-4, 5)$ and $(-4, -15)$, vertices $(-4, -3)$ and $(-4, -7)$

63. Hyperbola, foci $(6, -3)$ and $(-4, -3)$, asymptotes $y + 3 = \pm\frac{4}{3}(x - 1)$

64. Hyperbola, foci $(2, 2)$ and $(2, 6)$, asymptotes $x = -2 + y$ and $x = 6 - y$

65. Hyperbola, vertices $(4, -2)$ and $(4, 4)$, asymptotes $y - 1 = \pm\frac{3}{2}(x - 4)$

66. Hyperbola, vertices $(0, -2)$ and $(4, -2)$, asymptotes $x = -y$ and $x = y + 4$

In Exercises 67–72, find the vertex, focus, and directrix of the parabola, and sketch its graph.

67. $y^2 - 2y - 4x + 9 = 0$ **68.** $y^2 - 4y - 2x - 4 = 0$

69. $x^2 + 6x - y + 11 = 0$ **70.** $2x^2 - 8x - y + 5 = 0$

71. $4y^2 - 4y - 32x - 31 = 0$

72. $9x^2 + 6x + 9y - 8 = 0$

In Exercises 73–78, find the center, foci, and vertices of the ellipse, and sketch its graph.

73. $(x - 1)^2 + 4(y + 2)^2 = 1$

74. $2x^2 + y^2 - 20x + 2y + 43 = 0$

75. $x^2 + 4y^2 - 2x + 16y + 13 = 0$

76. $2x^2 + y^2 + 12x - 6y + 25 = 0$

77. $4x^2 + 9y^2 - 18x - 27 = 0$

78. $9x^2 + 36y^2 - 36x + 48y + 43 = 0$

In Exercises 79–84, find the center, foci, vertices, and equations of the asymptotes of the hyperbola with the given equation, and sketch its graph using its asymptotes as an aid.

79. $3x^2 - 4y^2 - 8y - 16 = 0$

80. $4x^2 - 9y^2 - 16x - 54y + 79 = 0$

81. $2x^2 - 3y^2 - 4x + 12y + 8 = 0$

82. $4y^2 - 9x^2 + 18x + 16y + 43 = 0$

83. $4x^2 - 2y^2 + 8x + 8y - 12 = 0$

84. $4x^2 - 3y^2 - 12y - 3 = 0$

85. Parabolic Reflectors The following figure shows the cross section of a parabolic reflector. If the reflector is 2 ft wide at the opening and 1 ft deep, how far from the vertex should the light source be placed along the axis of symmetry of the parabola?

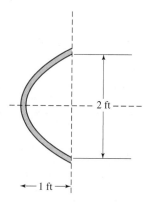

86. Length of Suspension Bridge Cable The figure below depicts a suspension bridge. The shape of the cable is described by the equation

$$y = \frac{hx^2}{a^2}$$

where $2a$ ft is the span of the bridge and h ft is the sag. (See Example 3.) Assuming that the sag is small in comparison to the span (that is, h/a is small), show that the length of the cable is

$$s \approx 2a\left(1 + \frac{2h^2}{3a^2}\right)$$

Use this approximation to estimate the length of the cable in Example 3, where $a = 200$ and $h = 80$. Compare your result with that obtained in the example.

Hint: Retain the first two terms of a binomial series.

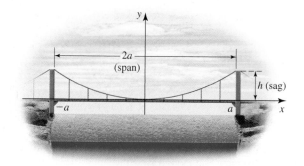

87. Length of Suspension Bridge Cable The cable of the suspension bridge shown in the figure below has the shape of a parabola. If the span of the bridge is 600 ft and the sag is 60 ft, what is the length of the cable?

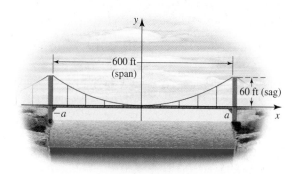

88. Surface Area of a Satellite Dish An 18-in. satellite dish is obtained by revolving the parabola with equation $y = \frac{4}{81}x^2$ about the y-axis. Find the surface area of the dish.

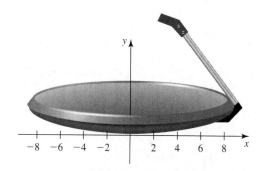

89. Shape of a Suspension Bridge Cable Consider a bridge of weight W lb/ft suspended by a flexible cable. Assume that the weight of the cable is negligible in comparison to the weight of the bridge. The following figure shows a portion of such a structure with the lowest point of the cable located at the origin. Let P be any point on the cable, and suppose that the tension of the cable at P is T lb and lies along the tangent at P (this is the case with flexible cables).

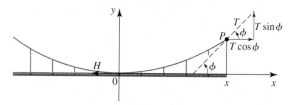

Referring to the figure, we see that

$$\frac{dy}{dx} = \tan \phi = \frac{\sin \phi}{\cos \phi} = \frac{T \sin \phi}{T \cos \phi}$$

But since the bridge is in equilibrium, the horizontal component of T must be equal to H, the tension at the lowest point of the cable (the origin); that is, $T \cos \phi = H$. Similarly, the vertical component of T must be equal to Wx, the load carried over that section of the cable from 0 to P; that is, $T \sin \phi = Wx$. Therefore,

$$\frac{dy}{dx} = \frac{Wx}{H}$$

Finally, since the lowest point of the cable is located at the origin, we have $y(0) = 0$. Solve this initial-value problem to show that the shape of the cable is a parabola.

Note: Observe that in a suspension bridge, the cable supports a load that is *uniformly distributed horizontally*. A cable supporting a load distributed *uniformly along its length* (for example, a cable supporting its own weight) assumes the shape of a catenary, as we saw in Section 6.6.

90. **Shape of a Suspension Bridge Cable** Refer to Exercise 89. Suppose that the span of the cable is $2a$ ft and the sag is h ft. (See Figure 10.) Show that the tension (in pounds) at the highest point of the cable has magnitude

$$T = \frac{Wa\sqrt{a^2 + 4h^2}}{2h}$$

91. **Arch of a Bridge** A bridge spanning the Charles River has three arches that are semielliptical in shape. The base of the center arch is 24 ft across, and the maximum height of the arch is 8 ft. What is the height of the arch 6 ft from the center of the base?

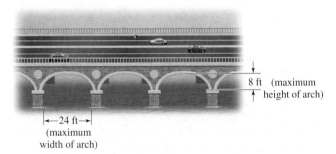

8 ft (maximum height of arch)

←24 ft→
(maximum
width of arch)

92. **a.** Find an equation of the tangent line to the parabola $y = ax^2$ at the point where $x = x_0$.
    b. Use the result of part (a) to show that the x-intercept of this tangent line is $x_0/2$.
    c. Use the result of part (b) to draw the tangent line.

93. Prove that any two distinct tangent lines to a parabola must intersect at one and only one point.

94. **a.** Show that an equation of the tangent line to the parabola $y^2 = 4px$ at the point (x_0, y_0) can be written in the form
$$y_0 y = 2p(x + x_0)$$
    b. Use the result of part (a) to show that the x-intercept of this tangent line is $-x_0$.
    c. Use the result of part (b) to draw the tangent line for $p > 0$.

95. Show that an equation of the tangent line to the ellipse
$$\frac{x^2}{a^2} + \frac{y^2}{b^2} = 1$$
at the point (x_0, y_0) can be written in the form
$$\frac{xx_0}{a^2} + \frac{yy_0}{b^2} = 1$$

96. Use the result of Exercise 95 to find an equation of the tangent line to the ellipse
$$\frac{x^2}{4} + \frac{y^2}{25} = 1$$
at the point $\left(1, \frac{5\sqrt{3}}{2}\right)$.

97. Show that an equation of the tangent line to the hyperbola
$$\frac{x^2}{a^2} - \frac{y^2}{b^2} = 1$$
at the point (x_0, y_0) can be written in the form
$$\frac{xx_0}{a^2} - \frac{yy_0}{b^2} = 1$$

98. Use the result of Exercise 97 to find an equation of the tangent line to the hyperbola
$$\frac{x^2}{4} - \frac{y^2}{9} = 1$$
at the point $(4, 3\sqrt{3})$.

99. Show that the ellipse
$$\frac{x^2}{a^2} + \frac{2y^2}{b^2} = 1$$
and the hyperbola
$$\frac{x^2}{a^2 - b^2} - \frac{2y^2}{b^2} = 1$$
intersect at right angles.

100. Use the definition of a hyperbola to derive Equation (5) for a hyperbola with foci $F_1(-c, 0)$ and $F_2(c, 0)$ and vertices $V_1(a, 0)$ and $V_2(-a, 0)$.

101. Show that the lines $y = (b/a)x$ and $y = -(b/a)x$ are slant asymptotes of the hyperbola
$$\frac{x^2}{a^2} - \frac{y^2}{b^2} = 1$$

102. A transmitter B is located 200 miles due east of a transmitter A on a straight coastline. The two transmitters send out signals simultaneously to a ship that is located at P. Suppose that the ship receives the signal from B, 800 microseconds (μ sec) before it receives the signal from A.
    a. Assuming that radio waves travel at a speed of 980 ft/μ sec, find an equation of the hyperbola on which the ship lies (see page 842).
    Hint: $d(P, A) - d(P, B) = 2a$.

b. If the ship is sailing in a direction parallel to and 20 mi north of the coastline, locate the position of the ship at that instant of time.

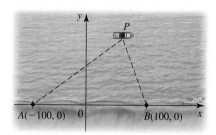

cas 103. Use a computer algebra system (CAS) to find an approximation of the circumference of the ellipse

$$4x^2 + 25y^2 = 100$$

cas 104. The dwarf planet Pluto has an elliptical orbit with the sun at one focus. The length of the major axis of the ellipse is 7.33×10^9 miles, and the length of the minor axis is 7.08×10^9 miles. Use a CAS to approximate the distance traveled by the planet during one complete orbit around the sun.

105. The Reflective Property of the Parabola The figure shows a parabola with equation $y^2 = 4px$. The line l is tangent to the parabola at the point $P(x_0, y_0)$.

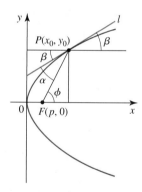

Show that $\alpha = \beta$ by establishing the following:

a. $\tan \beta = \dfrac{2p}{y_0}$ **b.** $\tan \phi = \dfrac{y_0}{x_0 - p}$

c. $\tan \alpha = \dfrac{2p}{y_0}$

Hint: $\tan \alpha = \dfrac{\tan \phi - \tan \beta}{1 + \tan \phi \tan \beta}$

106. The Reflective Property of the Ellipse Establish the reflective property of the ellipse by showing that $\alpha = \beta$ in Figure 18, page 835.
Hint: Use the trigonometric formula

$$\tan(\theta_1 - \theta_2) = \frac{\tan \theta_1 - \tan \theta_2}{1 + \tan \theta_1 \tan \theta_2}$$

107. Reflective Property of the Hyperbola The hyperbola also has the reflective property that the other two conics enjoy. Consider a mirror that has the shape of one branch of a hyperbola as shown in Figure (a). A ray of light aimed at a focus F_2 will be reflected toward the other focus, F_1. To establish the reflective property of the hyperbola, let $P(x_0, y_0)$ be a point on the hyperbola $x^2/a^2 - y^2/b^2 = 1$ with foci F_1 and F_2, and let α and β be the angles between the lines PF_1 and PF_2, as shown in Figure (b). Show that $\alpha = \beta$.

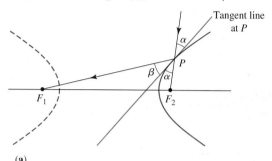

(a)

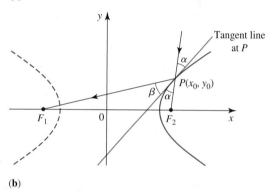

(b)

108. Reflecting Telescopes The reflective properties of the parabola and the hyperbola are exploited in designing a reflecting telescope. A hyperbolic mirror and a parabolic mirror are placed so that one focus of the hyperbola coincides with the focus F of the parabola, as shown in the figure. Use the reflective properties of the two conics to explain why rays of light coming from great distances are finally focused at the eyepiece placed at F', the other focus of the hyperbola.

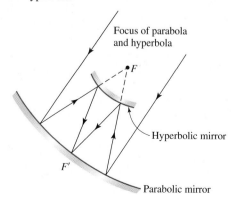

In Exercises 109–114, determine whether the statement is true or false. If it is true, explain why it is true. If it is false, give an example to show why it is false.

109. The graph of $2x^2 - y^2 + F = 0$ is a hyperbola, provided that $F \neq 0$.

110. The graph of $y^4 = 16ax^2$, where $a > 0$, is a parabola.

111. The ellipse $b^2x^2 + a^2y^2 = a^2b^2$, where $a > b > 0$, is contained in the circle $x^2 + y^2 = a^2$ and contains the circle $x^2 + y^2 = b^2$.

112. The asymptotes of the hyperbola $x^2/a^2 - y^2/b^2 = 1$ are perpendicular to each other if and only if $a = b$.

113. If A and C are both positive constants, then

$$Ax^2 + Cy^2 + Dx + Ey + F = 0$$

is an ellipse.

114. If A and C have opposite signs, then

$$Ax^2 + Cy^2 + Dx + Ey + F = 0$$

is a hyperbola.

10.2   Plane Curves and Parametric Equations

■ Why We Use Parametric Equations

Figure 1a gives a bird's-eye view of a proposed training course for a yacht. In Figure 1b we have introduced an xy-coordinate system in the plane to describe the position of the yacht. With respect to this coordinate system the position of the yacht is given by the point $P(x, y)$, and the course itself is the graph of the rectangular equation $4x^4 - 4x^2 + y^2 = 0$, which is called a *lemniscate*. But representing the lemniscate in terms of a rectangular equation in this instance has three major drawbacks.

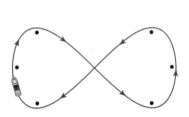

FIGURE 1   **(a)** The dots give the position of markers.   **(b)** An equation of the curve C is $4x^4 - 4x^2 + y^2 = 0$.

First, the equation does not define y explicitly as a function of x or x as a function of y. You can also convince yourself that this is not the graph of a function by applying the vertical and horizontal line tests to the curve in Figure 1b (see Section 0.2). Because of this, we cannot make direct use of many of the results for functions developed earlier. Second, the equation does not tell us when the yacht is at a given point (x, y). Third, the equation gives no inkling as to the direction of motion of the yacht.

To overcome these drawbacks when we consider the motion of an object in the plane or plane curves that are not graphs of functions, we turn to the following representation. If (x, y) is a point on a curve in the xy-plane, we write

$$x = f(t) \qquad y = g(t)$$

where f and g are functions of an auxiliary variable t with (common) domain some interval I. These equations are called **parametric equations,** t is called a **parameter,** and the interval I is called a **parameter interval.**

If we think of t on the closed interval $[a, b]$ as representing time, then we can interpret the parametric equations in terms of the motion of a particle as follows: At $t = a$ the particle is at the **initial point** $(f(a), g(a))$ of the curve or **trajectory** C. As t increases from $t = a$ to $t = b$, the particle traverses the curve in a specific direction called the **orientation** of the curve, eventually ending up at the **terminal point** $(f(b), g(b))$ of the curve. (See Figure 2.)

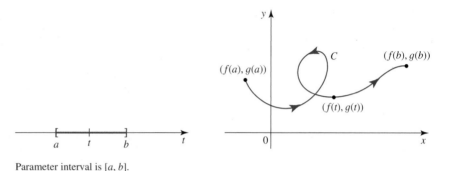

FIGURE 2
As t increases from a to b, the particle traces the curve from $(f(a), g(a))$ to $(f(b), g(b))$ in a specific direction.

Parameter interval is $[a, b]$.

We can also interpret the parametric equations in geometric terms as follows: We take the line segment $[a, b]$ and, by a process of stretching, bending, and twisting, make it conform geometrically to the curve C.

Sketching Curves Defined by Parametric Equations

Before looking at some examples, let's define the following term.

DEFINITION   **Plane Curve**

A plane curve is a set C of ordered pairs (x, y) defined by the parametric equations

$$x = f(t) \qquad \text{and} \qquad y = g(t)$$

where f and g are continuous functions on a parameter interval I.

EXAMPLE 1   Sketch the curve described by the parametric equations

$$x = t^2 - 4 \qquad \text{and} \qquad y = 2t \qquad -1 \leq t \leq 2$$

Solution   By plotting and connecting the points (x, y) for selected values of t (Table 1), we obtain the curve shown in Figure 3.

TABLE 1

t	-1	$-\frac{1}{2}$	0	$\frac{1}{2}$	1	2
(x, y)	$(-3, -2)$	$\left(-\frac{15}{4}, -1\right)$	$(-4, 0)$	$\left(-\frac{15}{4}, 1\right)$	$(-3, 2)$	$(0, 4)$

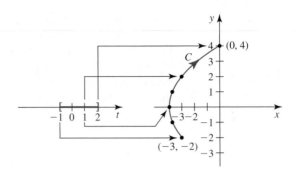

FIGURE 3

As t increases from -1 to 2, the curve
C is traced from the initial point
$(-3, -2)$ to the terminal point $(0, 4)$.

Alternative Solution   We eliminate the parameter t by solving the second of the two given parametric equations for t, obtaining $t = \frac{1}{2}y$. We then substitute this value of t into the first equation to obtain

$$x = \left(\frac{1}{2}y\right)^2 - 4 \qquad \text{or} \qquad x = \frac{1}{4}y^2 - 4$$

This is an equation of a parabola that has the x-axis as its axis of symmetry and its vertex at $(-4, 0)$. Now observe that $t = -1$ gives $(-3, -2)$ as the initial point of the curve and that $t = 2$ gives $(0, 4)$ as the terminal point of the curve. So tracing the graph from the initial point to the terminal point gives the desired curve, as obtained earlier. ■

We will adopt the convention here, just as we did with the domain of a function, that the parameter interval for $x = f(t)$ and $y = g(t)$ will consist of all values of t for which $f(t)$ and $g(t)$ are real numbers, unless otherwise noted.

EXAMPLE 2   Sketch the curves represented by

a. $x = \sqrt{t}$ and $y = t$
b. $x = t$ and $y = t^2$

Solution

a. We eliminate the parameter t by squaring the first equation to obtain $x^2 = t$. Substituting this value of t into the second equation, we obtain $y = x^2$, which is an equation of a parabola. But note that the first parametric equation implies that $t \geq 0$, so $x \geq 0$. Therefore, the desired curve is the right portion of the parabola shown in Figure 4. Finally, note that the parameter interval is $[0, \infty)$, and as t increases from 0, the desired curve starts at the initial point $(0, 0)$ and moves away from it along the parabola.

FIGURE 4

As t increases from 0, the curve
starts out at $(0, 0)$ and follows
the right portion of the parabola
with indicated orientation.

Parameter interval is $[0, \infty)$.

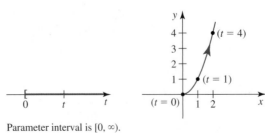

t	(x, y)
0	$(0, 0)$
1	$(1, 1)$
2	$(\sqrt{2}, 2)$
4	$(2, 4)$

b. Substituting the first equation into the second yields $y = x^2$. Although the rectangular equation is the same as that in part (a), the curve described by the parametric equations here is different from that of part (a), as we will now see. In this instance the parameter interval is $(-\infty, \infty)$. Furthermore, as t increases from $-\infty$ to ∞, the curve runs along the parabola $y = x^2$ from left to right, as you can see by plotting the points corresponding to, say, $t = -1$, 0, and 1. You can also see this by examining the parametric equation $x = t$, which tells us that x increases as t increases. (See Figure 5.)

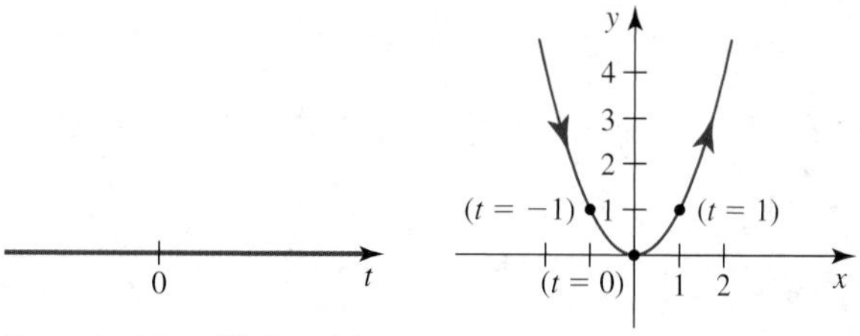

FIGURE 5
As t increases from $-\infty$ to ∞, the entire parabola is traced out, from left to right.

Parameter interval is $(-\infty, \infty)$.

t	(x, y)
-1	$(-1, 1)$
0	$(0, 0)$
1	$(1, 1)$

For problems involving motion, it is natural to use the parameter t to represent time. But other situations call for different representations or interpretations of the parameters, as the next two examples show. Here, we use an *angle* as a parameter.

EXAMPLE 3 Describe the curves represented by the parametric equations

$$x = a \cos \theta \qquad \text{and} \qquad y = a \sin \theta \qquad a > 0$$

with parameter intervals

a. $[0, \pi]$
b. $[0, 2\pi]$
c. $[0, 4\pi]$

Solution We have $\cos \theta = x/a$ and $\sin \theta = y/a$. So

$$1 = \cos^2 \theta + \sin^2 \theta = \left(\frac{x}{a}\right)^2 + \left(\frac{y}{a}\right)^2$$

giving us

$$x^2 + y^2 = a^2$$

This tells us that each of the curves under consideration is contained in a circle of radius a, centered at the origin.

a. If $\theta = 0$, then $x = a$ and $y = 0$, giving $(a, 0)$ as the initial point on the curve. As θ increases from 0 to π, the required curve is traced out in a counterclockwise direction, terminating at the point $(-a, 0)$. (See Figure 6a.)
b. Here, the curve is a complete circle that is traced out in a counterclockwise direction, starting at $(a, 0)$ and terminating at the same point (see Figure 6b).
c. The curve here is a circle that is traced out *twice* in a counterclockwise direction starting at $(a, 0)$ and terminating at the same point (see Figure 6c).

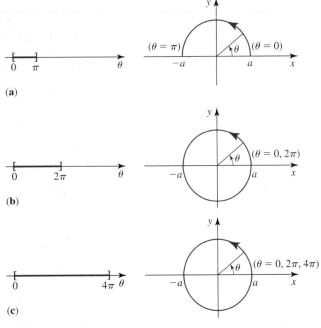

(a)

(b)

(c)

Parameter interval

FIGURE 6

The curve is (a) a semicircle, (b) a complete circle, and (c) a complete circle traced out twice. All curves are traced in a counterclockwise direction.

Historical Biography

MARIA GAËTANA AGNESI
(1718-1799)

Maria Gaetana Agnesi's exceptional academic talents surfaced at an early age, and her wealthy father provided her with the best tutors. By the age of nine, she had learned many languages in addition to her native Italian, including Greek and Hebrew. At that same age, she translated into Latin an article her tutor had written in Italian defending higher education for women. She then delivered it, from memory, to one of the gatherings of intellectuals her father hosted in their home. Agnesi developed a deep interest in Newtonian physics, but her primary interests became religion and mathematics. After Agnesi wrote a book on differential calculus, her talents attracted the attention of Pope Benedict XIV, who appointed her to a position at the University of Bologna. Despite being offered the chair of mathematics at Bologna, Agnesi left the academic world in 1752 so that she could fulfill a desire to serve others. She devoted the rest of her life to religious charitable projects, including running a home for the poor.

Later, in Exercises 10.2, you will learn that the curve in Problem 41 is called the **witch of Agnesi.** Why was this given such a peculiar name? It is actually because of a mistranslation of Maria Agnesi's 1748 work *Instituzione analitiche ad uso della gioventu italiana*. In her discussion of the curve represented by the rectangular equation $y(x^2 + a^2) = a^3$, Agnesi used the Italian word *versiera*, which is derived from the Latin *vertere*, meaning "to turn." However, this word was confused with *avversiera*, meaning "witch" or "devil's wife," and the curve became known as the *witch of Agnesi*.

EXAMPLE 4   Describe the curve represented by

$$x = 4 \cos \theta \qquad \text{and} \qquad y = 3 \sin \theta \qquad 0 \le \theta \le 2\pi$$

Solution   Solving the first equation for $\cos \theta$ and the second equation for $\sin \theta$ gives

$$\cos \theta = \frac{x}{4}$$

and

$$\sin \theta = \frac{y}{3}$$

Squaring each equation and adding the resulting equations, we obtain

$$\cos^2 \theta + \sin^2 \theta = \left(\frac{x}{4}\right)^2 + \left(\frac{y}{3}\right)^2$$

Since $\cos^2 \theta + \sin^2 \theta = 1$, we end up with the rectangular equation

$$\frac{x^2}{16} + \frac{y^2}{9} = 1$$

From this we see that the curve is contained in an ellipse centered at the origin. If $\theta = 0$, then $x = 4$ and $y = 0$, giving $(4, 0)$ as the initial point of the curve. As θ increases from 0 to 2π, the elliptical curve is traced out in a counterclockwise direction, terminating at $(4, 0)$. (See Figure 7.)

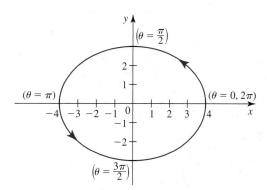

FIGURE 7
As θ increases from 0 to 2π, the curve that is traced out in a counterclockwise direction beginning and ending at $(4, 0)$ is an ellipse.

EXAMPLE 5 A proposed training course for a yacht is represented by the parametric equations

$$x = \sin t \quad \text{and} \quad y = \sin 2t \quad 0 \le t \le 2\pi$$

where x and y are measured in miles.

a. Show that the rectangular equation of the course is $4x^4 - 4x^2 + y^2 = 0$.
b. Describe the course.

Solution
a. Using the trigonometric identity $\sin 2t = 2 \sin t \cos t$, we rewrite the second of the parametric equations in the form

$$y = 2 \sin t \cos t = 2x \cos t \qquad \text{Since } x = \sin t$$

Solving for $\cos t$, we have

$$\cos t = \frac{y}{2x}$$

Then, using the identity $\sin^2 t + \cos^2 t = 1$, we obtain

$$x^2 + \left(\frac{y}{2x}\right)^2 = 1$$

$$x^2 + \frac{y^2}{4x^2} = 1$$

or

$$4x^4 - 4x^2 + y^2 = 0$$

b. From the results of part (a) we see that the required curve is symmetric with respect to the x-axis, the y-axis, and the origin. Therefore, it suffices to concentrate first on drawing the part of the curve that lies in the first quadrant and then make use of symmetry to complete the curve. Since both $\sin t$ and $\sin 2t$ are nonnegative only for $0 \le t \le \frac{\pi}{2}$, we first sketch the curve corresponding to values of t in $\left[0, \frac{\pi}{2}\right]$. With the help of the following table we obtain the curve shown in Figure 8. The direction of the yacht is indicated by the arrows.

t	0	$\frac{\pi}{6}$	$\frac{\pi}{4}$	$\frac{\pi}{3}$	$\frac{\pi}{2}$
(x, y)	$(0, 0)$	$\left(\frac{1}{2}, \frac{\sqrt{3}}{2}\right)$	$\left(\frac{\sqrt{2}}{2}, 1\right)$	$\left(\frac{\sqrt{3}}{2}, \frac{\sqrt{3}}{2}\right)$	$(1, 0)$

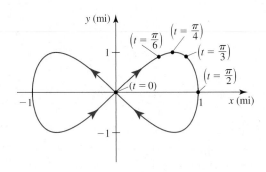

FIGURE 8
The training course for the yacht

EXAMPLE 6   **Cycloids**   Let P be a fixed point on the rim of a wheel. If the wheel is allowed to roll along a straight line without slipping, then the point P traces out a curve called a **cycloid** (see Figure 9). Suppose that the wheel has radius a and rolls along the x-axis. Find parametric equations for the cycloid.

FIGURE 9
The cycloid is the curve traced out by a fixed point P on the rim of a rolling wheel.

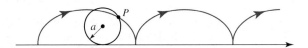

Solution   Suppose that the wheel rolls in a positive direction with the point P initially at the origin of the coordinate system. Figure 10 shows the position of the wheel after it has rotated through θ radians. Because there is no slippage, the distance the wheel has rolled from the origin is

$$d(O, M) = \text{length of arc } PM = a\theta$$

giving its center as $C(a\theta, a)$. Also, from Figure 10 we see that the coordinates of $P(x, y)$ satisfy

$$x = d(O, M) - a \sin \theta = a\theta - a \sin \theta = a(\theta - \sin \theta)$$

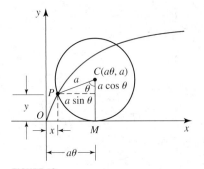

FIGURE 10
The position of the wheel after it has rotated through θ radians

and

$$y = d(C, M) - a \cos \theta = a - a \cos \theta = a(1 - \cos \theta)$$

Although these results are derived under the tacit assumption that $0 < \theta < \frac{\pi}{2}$, it can be demonstrated that they are valid for other values of θ. Therefore, the required parametric equations of the cycloid are

$$x = a(\theta - \sin \theta) \qquad \text{and} \qquad y = a(1 - \cos \theta) \qquad -\infty < \theta < \infty$$

The cycloid provides the solution to two famous problems in mathematics:

1. *The brachistochrone problem:* Find the curve along which a moving particle (under the influence of gravity) will slide from a point A to another point B, not directly beneath A, in the shortest time (see Figure 11a).
2. *The tautochrone problem:* Find the curve having the property that it takes the same time for a particle to slide to the bottom of the curve no matter where the particle is placed on the curve (see Figure 11b).

The brachistochrone problem—the problem of finding the curve of quickest descent—was advanced in 1696 by the Swiss mathematician Johann Bernoulli. Offhand, one might conjecture that such a curve should be a straight line, since it yields

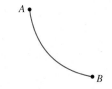

FIGURE 11
The cycloid provides the solution
to both the brachistochrone
and the tautochrone problem.

(a) The brachistochrone problem   (b) The tautochrone problem

the shortest distance between the two points. But the velocity of the particle moving on the straight line will build up comparatively slowly, whereas if we take a curve that is steeper near A, even though the path becomes longer, the particle will cover a large portion of the distance at a greater speed. The problem was solved by Johann Bernoulli, his older brother Jacob Bernoulli, Leibniz, Newton, and l'Hôpital. They found that the curve of quickest descent is an inverted arc of a cycloid (Figure 11a). As it turns out, this same curve is also the solution to the tautochrone problem.

10.2   CONCEPT QUESTIONS

1. What is a plane curve? Give an example of a plane curve that is not the graph of a function.
2. What is the difference between the curve C_1 with parametric representation $x = \cos t$ and $y = \sin t$, where $0 \le t \le 2\pi$, and the curve C_2 with parametric representation $x = \sin t$ and $y = \cos t$, where $0 \le t \le 2\pi$?

3. Describe the relationship between the curve C_1 with parametric equations $x = f(t)$ and $y = g(t)$, where $0 \le t \le 1$, and the curve C_2 with parametric equations $x = f(1 - t)$ and $y = g(1 - t)$, where $0 \le t \le 1$.

10.2   EXERCISES

In Exercises 1–28, (a) find a rectangular equation whose graph contains the curve C with the given parametric equations, and (b) *sketch the curve C and indicate its orientation.*

1. $x = 2t + 1, \quad y = t - 3$
2. $x = t - 2, \quad y = 2t - 1; \quad -1 \le t \le 5$
3. $x = \sqrt{t}, \quad y = 9 - t$
4. $x = t^2, \quad y = t - 1; \quad 0 \le t \le 3$
5. $x = t^2 + 1, \quad y = 2t^2 - 1; \quad -2 \le t \le 2$
6. $x = t^3, \quad y = 2t + 1$
7. $x = t^2, \quad y = t^3; \quad -2 \le t \le 2$
8. $x = 1 + \dfrac{1}{t}, \quad y = t + 1$
9. $x = 2 \sin \theta, \quad y = 2 \cos \theta; \quad 0 \le \theta \le 2\pi$
10. $x = \cos 2\theta, \quad y = 3 \sin \theta; \quad 0 \le \theta \le 2\pi$
11. $x = 2 \sin \theta, \quad y = 3 \cos \theta; \quad 0 \le \theta \le 2\pi$
12. $x = \cos \theta + 1, \quad y = \sin \theta - 2; \quad 0 \le \theta \le 2\pi$
13. $x = 2 \cos \theta + 2, \quad y = 3 \sin \theta - 1; \quad 0 \le \theta \le 2\pi$
14. $x = \sin \theta + 3, \quad y = 3 \cos \theta + 1; \quad 0 \le \theta \le 2\pi$
15. $x = \cos \theta, \quad y = \cos 2\theta$
16. $x = \sec \theta, \quad y = \cos \theta$
17. $x = \sec \theta, \quad y = \tan \theta; \quad -\frac{\pi}{2} < \theta < \frac{\pi}{2}$
18. $x = \cos^3 \theta, \quad y = \sin^3 \theta$
19. $x = \sin^2 \theta, \quad y = \sin^4 \theta; \quad 0 \le \theta \le \frac{\pi}{2}$
20. $x = e^t, \quad y = e^{-t}$
21. $x = -e^t, \quad y = e^{2t}$
22. $x = t^3, \quad y = 3 \ln t$
23. $x = \ln 2t, \quad y = t^2$
24. $x = e^t, \quad y = \ln t$
25. $x = \cosh t, \quad y = \sinh t$
26. $x = 3 \sinh t, \quad y = 2 \cosh t$
27. $x = (t - 1)^2, \quad y = (t - 1)^3; \quad 1 \le t \le 2$
28. $x = \dfrac{2t}{1 + t^2}, \quad y = \dfrac{1 - t^2}{1 + t^2}$

V Videos for selected exercises are available online at **www.academic.cengage.com/login**.

In Exercises 29–34, the position of a particle at time t is (x, y). Describe the motion of the particle as t varies over the time interval [a, b].

29. $x = t + 1$, $y = \sqrt{t}$; $[0, 4]$

30. $x = \sin \pi t$, $y = \cos \pi t$; $[0, 6]$

31. $x = 1 + \cos t$, $y = 2 + \sin t$; $[0, 2\pi]$

32. $x = 1 + 2 \sin 2t$, $y = 2 + 4 \sin 2t$; $[0, 2\pi]$

33. $x = \sin t$, $y = \sin^2 t$; $[0, 3\pi]$

34. $x = e^{-t}$, $y = e^{2t-1}$; $[0, \infty)$

35. Flight Path of an Aircraft The position (x, y) of an aircraft flying in a fixed direction t seconds after takeoff is given by $x = \tan(0.025\pi t)$ and $y = \sec(0.025\pi t) - 1$, where x and y are measured in miles. Sketch the flight path of the aircraft for $0 \le t \le \frac{40}{3}$.

36. Trajectory of a Shell A shell is fired from a howitzer with a muzzle speed of v_0 ft/sec. If the angle of elevation of the howitzer is α, then the position of the shell after t sec is described by the parametric equations

$$x = (v_0 \cos \alpha)t \quad \text{and} \quad y = (v_0 \sin \alpha)t - \frac{1}{2} gt^2$$

where g is the acceleration due to gravity (32 ft/sec^2).
a. Find the range of the shell.
b. Find the maximum height attained by the shell.
c. Show that the trajectory of the shell is a parabola by eliminating the parameter t.

37. Let $P_1(x_1, y_1)$ and $P_2(x_2, y_2)$ be two distinct points in the plane. Show that the parametric equations

$$x = x_1 + (x_2 - x_1)t \quad \text{and} \quad y = y_1 + (y_2 - y_1)t$$

describe (a) the line passing through P_1 and P_2 if $-\infty < t < \infty$ and (b) the line segment joining P_1 and P_2 if $0 \le t \le 1$.

38. Show that

$$x = a \cos t + h \quad \text{and} \quad y = b \sin t + k \quad 0 \le t \le 2\pi$$

are parametric equations of an ellipse with center at (h, k) and axes of lengths $2a$ and $2b$.

39. Show that

$$x = a \sec t + h \quad \text{and} \quad y = b \tan t + k$$
$$t \in \left(-\frac{\pi}{2}, \frac{\pi}{2}\right) \cup \left(\frac{\pi}{2}, \frac{3\pi}{2}\right)$$

are parametric equations of a hyperbola with center at (h, k) and transverse and conjugate axes of lengths $2a$ and $2b$, respectively.

40. Let P be a point located a distance d from the center of a circle of radius r. The curve traced out by P as the circle rolls without slipping along a straight line is called a **trochoid.** (The cycloid is the special case of a trochoid with $d = r$.) Suppose that the circle rolls along the x-axis in the positive direction with $\theta = 0$ when the point P is at one of the lowest points on the trochoid. Show that the parametric equations of the trochoid are

$$x = r\theta - d \sin \theta \quad \text{and} \quad y = r - d \cos \theta$$

where θ is the same parameter as that for the cycloid. Sketch the trochoid for the cases in which $d < r$ and $d > r$.

41. The **witch of Agnesi** is the curve shown in the following figure. Show that the parametric equations of this curve are

$$x = 2a \cot \theta \quad \text{and} \quad y = 2a \sin^2 \theta$$

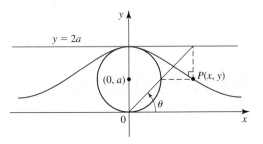

42. If a string is unwound from a circle of radius a in such a way that it is held taut in the plane of the circle, then its end P will trace a curve called the **involute of the circle.** Referring to the following figure, show that the parametric equations of the involute are

$$x = a(\cos \theta + \theta \sin \theta) \quad \text{and} \quad y = a(\sin \theta - \theta \cos \theta)$$

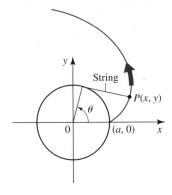

In Exercises 43–46, use a graphing utility to plot the curve with the given parametric equations.

43. $x = 2 \sin 3t$, $y = 3 \sin 1.5t$; $t \ge 0$

44. $x = \cos t + 5 \cos 3t$, $y = 6 \cos t - 5 \sin 3t$; $0 \le t \le 2\pi$

45. $x = 2 \cos t + \cos 2t$, $y = 2 \sin t - \sin 2t$; $0 \le t \le 2\pi$

46. $x = 3 \cos t + \cos 3t$, $y = 3 \sin t - \sin 3t$; $0 \le t \le 2\pi$

47. The **butterfly catastrophe curve,** which is described by the parametric equations

$$x = c(8at^3 + 24t^5) \quad \text{and} \quad y = c(-6at^2 - 15t^4)$$

occurs in the study of catastrophe theory. Plot the curve with $a = -7$ and $c = 0.03$ for t in the parameter interval $[-1.629, 1.629]$.

48. The **swallowtail catastrophe curve,** which is described by the parametric equations

$$x = c(-2at - 4t^3) \quad \text{and} \quad y = c(at^2 + 3t^4)$$

occurs in the study of catastrophe theory. Plot the curve with $a = -2$ and $c = 0.5$ for t in the parameter interval $[-1.25, 1.25]$.

49. The **Lissajous curves,** also known as **Bowditch curves,** have applications in physics, astronomy, and other sciences. They are described by the parametric equations

$$x = \sin(at + b\pi), \quad a \text{ a rational number,} \quad \text{and} \quad y = \sin t$$

Plot the curve with $a = 0.75$ and $b = 0$ for t in the parameter interval $[0, 8\pi]$.

50. The **prolate cycloid** is the path traced out by a fixed point at a distance $b > a$ from the center of a rolling circle, where a

is the radius of the circle. The prolate cycloid is described by the parametric equations

$$x = a(t - b \sin t) \quad \text{and} \quad y = c(1 - d \cos t)$$

Plot the curve with $a = 0.1$, $b = 2$, $c = 0.25$, and $d = 2$ for t in the parameter interval $[-10, 10]$.

In Exercises 51–54, determine whether the statement is true or false. If it is true, explain why it is true. If it is false, give an example to show why it is false.

51. The parametric equations $x = \cos^2 t$ and $y = \sin^2 t$, where $-\infty < t < \infty$, have the same graph as $x + y = 1$.

52. The graph of a function $y = f(x)$ can always be represented by a pair of parametric equations.

53. The curve with parametric equations $x = f(t) + a$ and $y = g(t) + b$ is obtained from the curve C with parametric equations $x = f(t)$ and $y = g(t)$ by shifting the latter horizontally and vertically.

54. The ellipse with center at the origin and major and minor axes a and b, respectively, can be obtained from the circle with equations $x = f(t) = \cos t$ and $y = g(t) = \sin t$ by multiplying $f(t)$ and $g(t)$ by appropriate nonzero constants.

10.3 The Calculus of Parametric Equations

■ Tangent Lines to Curves Defined by Parametric Equations

Suppose that C is a smooth curve that is parametrized by the equations $x = f(t)$ and $y = g(t)$ with parameter interval I and we wish to find the slope of the tangent line to the curve at the point P. (See Figure 1.) Let t_0 be the point in I that corresponds to P, and let (a, b) be the subinterval of I containing t_0 corresponding to the highlighted portion of the curve C in Figure 1. This subset of C is the graph of a function of x, as you can verify using the Vertical Line Test. (The general conditions that f and g must satisfy for this to be true are given in Exercise 66.)

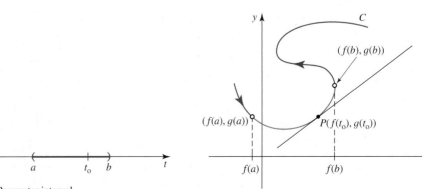

FIGURE 1
We want to find the slope of the tangent line to the curve at the point P.

Parameter interval

Let's denote this function by F so that $y = F(x)$, where $f(a) < x < f(b)$. Since $x = f(t)$ and $y = g(t)$, we may rewrite this equation in the form

$$g(t) = F[f(t)]$$

Using the Chain Rule, we obtain

$$g'(t) = F'[f(t)]f'(t)$$
$$= F'(x)f'(t) \qquad \text{Replace } f(t) \text{ by } x.$$

If $f'(t) \neq 0$, we can solve for $F'(x)$, obtaining

$$F'(x) = \frac{g'(t)}{f'(t)}$$

which can also be written

$$\frac{dy}{dx} = \frac{\dfrac{dy}{dt}}{\dfrac{dx}{dt}} \qquad \text{if} \quad \frac{dx}{dt} \neq 0 \tag{1}$$

The required slope of the tangent line at P is then found by evaluating Equation (1) at t_0. Observe that Equation (1) enables us to solve the problem without eliminating t.

EXAMPLE 1 Find an equation of the tangent line to the curve

$$x = \sec t \qquad y = \tan t \qquad -\frac{\pi}{2} < t < \frac{\pi}{2}$$

at the point where $t = \pi/4$. (See Figure 2.)

Solution The slope of the tangent line at any point (x, y) on the curve is

$$\frac{dy}{dx} = \frac{\dfrac{dy}{dt}}{\dfrac{dx}{dt}}$$

$$= \frac{\sec^2 t}{\sec t \tan t} = \frac{\sec t}{\tan t}$$

In particular, the slope of the tangent line at the point where $t = \pi/4$ is

$$\left.\frac{dy}{dx}\right|_{t=\pi/4} = \frac{\sec \dfrac{\pi}{4}}{\tan \dfrac{\pi}{4}} = \frac{\sqrt{2}}{1} = \sqrt{2}$$

Also, when $t = \pi/4$, we have $x = \sec(\pi/4) = \sqrt{2}$ and $y = \tan(\pi/4) = 1$ giving $(\sqrt{2}, 1)$ as the point of tangency. Finally, using the point-slope form of the equation of a line, we obtain the required equation:

$$y - 1 = \sqrt{2}(x - \sqrt{2}) \qquad \text{or} \qquad y = \sqrt{2}x - 1 \qquad \blacksquare$$

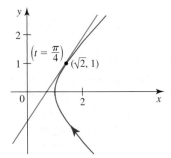

FIGURE 2
The tangent line to the curve at $(\sqrt{2}, 1)$

Horizontal and Vertical Tangents

A curve C represented by the parametric equations $x = f(t)$ and $y = g(t)$ has a **horizontal** tangent at a point (x, y) on C where $dy/dt = 0$ and $dx/dt \neq 0$ and a **vertical** tangent where $dx/dt = 0$ and $dy/dt \neq 0$, so that dy/dx is undefined there. Points where both dy/dt and dx/dt are equal to zero are candidates for horizontal or vertical tangents and may be investigated by using l'Hôpital's rule.

EXAMPLE 2   A curve C is defined by the parametric equations $x = t^2$ and $y = t^3 - 3t$.

a. Find the points on C where the tangent lines are horizontal or vertical.
b. Find the x- and y-intercepts of C.
c. Sketch the graph of C.

Solution

a. Setting $dy/dt = 0$ gives $3t^2 - 3 = 0$, or $t = \pm 1$. Since $dx/dt = 2t \neq 0$ at these values of t, we conclude that C has horizontal tangents at the points on C corresponding to $t = \pm 1$, that is, at $(1, -2)$ and $(1, 2)$. Next, setting $dx/dt = 0$ gives $2t = 0$, or $t = 0$. Since $dy/dt \neq 0$ for this value of t, we conclude that C has a vertical tangent at the point corresponding to $t = 0$, or at $(0, 0)$.

b. To find the x-intercepts, we set $y = 0$, which gives $t^3 - 3t = t(t^2 - 3) = 0$, or $t = -\sqrt{3}, 0,$ and $\sqrt{3}$. Substituting these values of t into the expression for x gives 0 and 3 as the x-intercepts. Next, setting $x = 0$ gives $t = 0$, which, when substituted into the expression for y, gives 0 as the y-intercept.

c. Using the information obtained in parts (a) and (b), we obtain the graph of C shown in Figure 3.   ■

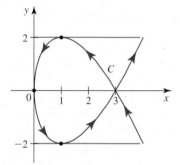

FIGURE 3
The graph of $x = t^2$, $y = t^3 - 3t$, and the tangent lines at $t = \pm 1$.

Finding d^2y/dx^2 from Parametric Equations

Suppose that the parametric equations $x = f(t)$ and $y = g(t)$ define y as a twice-differentiable function of x over some suitable interval. Then d^2y/dx^2 may be found from Equation (1) with another application of the Chain Rule.

$$\frac{d^2y}{dx^2} = \frac{d}{dx}\left(\frac{dy}{dx}\right) = \frac{\dfrac{d}{dt}\left(\dfrac{dy}{dx}\right)}{\dfrac{dx}{dt}} \qquad \text{if} \quad \frac{dx}{dt} \neq 0 \qquad \textbf{(2)}$$

Higher-order derivatives are found in a similar manner.

EXAMPLE 3   Find $\dfrac{d^2y}{dx^2}$ if $x = t^2 - 4$ and $y = t^3 - 3t$.

Solution   First, we use Equation (1) to compute

$$\frac{dy}{dx} = \frac{\dfrac{dy}{dt}}{\dfrac{dx}{dt}} = \frac{3t^2 - 3}{2t}$$

Then, using Equation (2), we obtain

$$\frac{d^2 y}{dx^2} = \frac{\dfrac{d}{dt}\left(\dfrac{dy}{dx}\right)}{\dfrac{dx}{dt}} = \frac{\dfrac{d}{dt}\left(\dfrac{3t^2 - 3}{2t}\right)}{2t}$$

$$= \frac{\dfrac{(2t)(6t) - (3t^2 - 3)(2)}{4t^2}}{2t} \qquad \text{Use the Quotient Rule.}$$

$$= \frac{6t^2 + 6}{8t^3} = \frac{3(t^2 + 1)}{4t^3} \qquad \blacksquare$$

■ The Length of a Smooth Curve

In Section 5.4 we showed that the length L of the graph of a smooth function f on an interval $[a, b]$ can be found by using the formula

$$L = \int_a^b \sqrt{1 + [f'(x)]^2}\, dx \qquad (3)$$

We now generalize this result to include curves defined by parametric equations. We begin by explaining what is meant by a *smooth* curve defined parametrically. Suppose that C is represented by $x = f(t)$ and $y = g(t)$ on a parameter interval I. Then C is **smooth** if f' and g' are continuous on I and are not simultaneously zero, except possibly at the endpoints of I. A smooth curve is devoid of corners or cusps. For example, the cycloid that we discussed in Section 10.2 (see Figure 9 in that section) has sharp corners at the values $x = 2n\pi a$ and, therefore, is not smooth. However, it is smooth between these points.

Now let $P = \{t_0, t_1, \dots, t_n\}$ be a regular partition of the parameter interval $[a, b]$. Then the point $P_k(f(t_k), g(t_k))$ lies on C, and the length of C is approximated by the length of the polygonal curve with vertices $P_0, P_1, \dots, P_n$. (See Figure 4.) Thus,

$$L \approx \sum_{k=1}^{n} d(P_{k-1}, P_k) \qquad (4)$$

where

$$d(P_{k-1}, P_k) = \sqrt{[f(t_k) - f(t_{k-1})]^2 + [g(t_k) - g(t_{k-1})]^2}$$

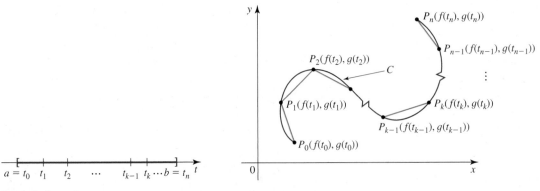

Parameter interval

FIGURE 4

The length of C is approximated by the length of the polygonal curve (the red lines).

Now, since f and g both have continuous derivatives, we can use the Mean Value Theorem to write

$$f(t_k) - f(t_{k-1}) = f'(t_k^*)(t_k - t_{k-1})$$

and

$$g(t_k) - g(t_{k-1}) = g'(t_k^{**})(t_k - t_{k-1})$$

where t_k^* and t_k^{**} are numbers in (t_{k-1}, t_k). Substituting these expressions into Equation (4) gives

$$L \approx \sum_{k=1}^{n} d(P_{k-1}, P_k) = \sum_{k=1}^{n} \sqrt{[f'(t_k^*)]^2 + [g'(t_k^{**})]^2} \, \Delta t \qquad (5)$$

As in Section 5.4, we define

$$L = \lim_{n \to \infty} \sum_{k=1}^{n} d(P_{k-1}, P_k)$$

$$= \lim_{n \to \infty} \sum_{k=1}^{n} \sqrt{[f'(t_k^*)]^2 + [g'(t_k^{**})]^2} \, \Delta t \qquad (6)$$

The sum in Equation (6) looks like a Riemann sum of the function $\sqrt{[f']^2 + [g']^2}$, but it is not, because t_k^* is not necessarily equal to t_k^{**}. But it can be shown that the limit in Equation (6) is the same as that of an expression in which $t_k^* = t_k^{**}$. Therefore,

$$L = \int_a^b \sqrt{[f'(t)]^2 + [g'(t)]^2} \, dt$$

and we have the following result.

THEOREM 1   Length of a Smooth Curve

Let C be a smooth curve represented by the parametric equations $x = f(t)$ and $y = g(t)$ with parameter interval $[a, b]$. If C does not intersect itself, except possibly for $t = a$ and $t = b$, then the length of C is

$$L = \int_a^b \sqrt{[f'(t)]^2 + [g'(t)]^2} \, dt = \int_a^b \sqrt{\left(\frac{dx}{dt}\right)^2 + \left(\frac{dy}{dt}\right)^2} \, dt \qquad (7)$$

Note   Equation (7) is consistent with Equation (4) of Section 5.4. Both have the form $L = \int ds$, where $(ds)^2 = (dx)^2 + (dy)^2$. ∎

EXAMPLE 4   Find the length of one arch of the cycloid

$$x = a(\theta - \sin\theta) \qquad y = a(1 - \cos\theta)$$

(See Example 6 in Section 10.2.)

Solution   One arch of the cycloid is traced out by letting θ run from 0 to 2π. Now

$$\frac{dx}{d\theta} = a(1 - \cos\theta) \qquad \text{and} \qquad \frac{dy}{d\theta} = a\sin\theta$$

Therefore, using Equation (7), we find the required length to be

$$L = \int_0^{2\pi} \sqrt{\left(\frac{dx}{d\theta}\right)^2 + \left(\frac{dy}{d\theta}\right)^2}\, d\theta = \int_0^{2\pi} \sqrt{a^2(1 - \cos\theta)^2 + a^2 \sin^2\theta}\, d\theta$$

$$= \int_0^{2\pi} \sqrt{a^2 - 2a^2 \cos\theta + a^2 \cos^2\theta + a^2 \sin^2\theta}\, d\theta$$

$$= a\int_0^{2\pi} \sqrt{2(1 - \cos\theta)}\, d\theta \qquad \sin^2\theta + \cos^2\theta = 1$$

To evaluate this integral, we use the identity $\sin^2 x = \frac{1}{2}(1 - \cos 2x)$ with $\theta = 2x$. This gives $1 - \cos\theta = 2\sin^2(\theta/2)$, so

$$L = a\int_0^{2\pi} \sqrt{4\sin^2\frac{\theta}{2}}\, d\theta$$

$$= 2a\int_0^{2\pi} \sin\frac{\theta}{2}\, d\theta \qquad \sin\frac{\theta}{2} \geq 0 \text{ on } [0, 2\pi]$$

$$= -4a\left[\cos\frac{\theta}{2}\right]_0^{2\pi}$$

$$= -4a(-1 - 1) = 8a \qquad\blacksquare$$

■ The Area of a Surface of Revolution

Recall that the formulas $S = 2\pi\int y\, ds$ and $S = 2\pi\int x\, ds$ (Formulas 11 and 12 of Section 5.4) give the area of the surface of revolution that is obtained by revolving the graph of a function about the x- and y-axes, respectively. These formulas are valid for finding the area of the surface of revolution that is obtained by revolving a curve described by parametric equations about the x- and the y-axes, provided that we replace the element of arc length ds by the appropriate expression. These results, which may be derived by using the method used to derive Equation (7), are stated in the next theorem.

THEOREM 2   Area of a Surface of Revolution

Let C be a smooth curve represented by the parametric equations $x = f(t)$ and $y = g(t)$ with parameter interval $[a, b]$, and suppose that C does not intersect itself, except possibly for $t = a$ and $t = b$. If $g(t) \geq 0$ for all t in $[a, b]$, then the area S of the surface obtained by revolving C about the x-axis is

$$S = 2\pi\int_a^b y\sqrt{[f'(t)]^2 + [g'(t)]^2}\, dt = 2\pi\int_a^b y\sqrt{\left(\frac{dx}{dt}\right)^2 + \left(\frac{dy}{dt}\right)^2}\, dt \qquad \textbf{(8)}$$

If $f(t) \geq 0$ for all t in $[a, b]$, then the area S of the surface that is obtained by revolving C about the y-axis is

$$S = 2\pi\int_a^b x\sqrt{[f'(t)]^2 + [g'(t)]^2}\, dt = 2\pi\int_a^b x\sqrt{\left(\frac{dx}{dt}\right)^2 + \left(\frac{dy}{dt}\right)^2}\, dt \qquad \textbf{(9)}$$

EXAMPLE 5   Show that the surface area of a sphere of radius r is $4\pi r^2$.

Solution   We obtain this sphere by revolving the semicircle

$$x = r\cos t \qquad y = r\sin t \qquad 0 \leq t \leq \pi$$

about the x-axis. Using Equation (8), the surface area of the sphere is

$$S = 2\pi \int_0^\pi r \sin t \sqrt{(-r \sin t)^2 + (r \cos t)^2} \, dt$$

$$= 2\pi r \int_0^\pi \sin t \sqrt{r^2(\sin^2 t + \cos^2 t)} \, dt$$

$$= 2\pi r \int_0^\pi r \sin t \, dt \qquad \sin^2 t + \cos^2 t = 1$$

$$= 2\pi r^2 \big[-\cos t\big]_0^\pi = 2\pi r^2[-(-1) + 1] = 4\pi r^2$$ ■

10.3 CONCEPT QUESTIONS

1. Suppose that C is a smooth curve with parametric equations $x = f(t)$ and $y = g(t)$ and parameter interval I. Write an expression for the slope of the tangent line to C at the point (x_0, y_0) corresponding to t_0 in I.

2. Suppose that C is a smooth curve with parametric equations $x = f(t)$ and $y = g(t)$ and parameter interval $[a, b]$. Furthermore, suppose that C does not cross itself, except possibly for $t = a$. Write an expression giving the length of C.

3. Suppose that C is a smooth curve with parametric equations $x = f(t)$ and $y = g(t)$ and parameter interval $[a, b]$. Suppose,

further, that C does not intersect itself, except possibly for $t = a$ and $t = b$.

a. Write an integral giving the area of the surface obtained by revolving C about the x-axis assuming that $g(t) \geq 0$ for all t in $[a, b]$.

b. Write an integral giving the area of the surface obtained by revolving C about the y-axis assuming that $f(t) \geq 0$ for all t in $[a, b]$.

10.3 EXERCISES

In Exercises 1–6, find the slope of the tangent line to the curve at the point corresponding to the value of the parameter.

1. $x = t^2 + 1$, $\quad y = t^2 - t$; $\quad t = 1$

2. $x = t^3 - t$, $\quad y = t^2 - 2t + 2$; $\quad t = 2$

3. $x = \sqrt{t}$, $\quad y = \dfrac{1}{t}$; $\quad t = 1$

4. $x = e^{2t}$, $\quad y = \ln t$; $\quad t = 1$

5. $x = 2 \sin \theta$, $\quad y = 3 \cos \theta$; $\quad \theta = \dfrac{\pi}{4}$

6. $x = 2(\theta - \sin \theta)$, $\quad y = 2(1 - \cos \theta)$; $\quad \theta = \dfrac{\pi}{6}$

In Exercises 7 and 8, find an equation of the tangent line to the curve at the point corresponding to the value of the parameter.

7. $x = 2t - 1$, $\quad y = t^3 - t^2$; $\quad t = 1$

8. $x = \theta \cos \theta$, $\quad y = \theta \sin \theta$; $\quad \theta = \dfrac{\pi}{2}$

In Exercises 9 and 10, find an equation of the tangent line to the curve at the given point. Then sketch the curve and the tangent line(s).

9. $x = t^2 + t$, $\quad y = t^2 - t^3$; $\quad (0, 2)$

10. $x = e^t$, $\quad y = e^{-t}$; $\quad (1, 1)$

In Exercises 11 and 12, find the points on the curve at which the slope of the tangent line is m.

11. $x = 2t^2 - 1$, $\quad y = t^3$; $\quad m = 3$

12. $x = t^3$, $\quad y = t^2 + t$; $\quad m = 1$

In Exercises 13–16, find the points on the curve at which the tangent line is either horizontal or vertical. Sketch the curve.

13. $x = t^2 - 4$, $\quad y = t^3 - 3t$

14. $x = t^3 - 3t$, $\quad y = t^2$

15. $x = 1 + 3 \cos t$, $\quad y = 2 - 2 \sin t$

16. $x = \sin t$, $\quad y = \sin 2t$

In Exercises 17–24, find dy/dx and d^2y/dx^2.

17. $x = 3t^2 + 1$, $y = 2t^3$ **18.** $x = t^3 - t$, $y = t^3 + 2t^2$

19. $x = \sqrt{t}$, $y = \dfrac{1}{t}$ **20.** $x = \sin 2t$, $y = \cos 2t$

21. $x = \theta + \cos \theta$, $y = \theta - \sin \theta$

22. $x = e^{-t}$, $y = e^{2t}$

23. $x = \cosh t$, $y = \sinh t$

24. $x = \sqrt{t^2 + 1}$, $y = t \ln t$

25. Let C be the curve defined by the parametric equations $x = t^2$ and $y = t^3 - 3t$ (see Example 2). Find d^2y/dx^2, and use this result to determine the intervals where C is concave upward and where it is concave downward.

26. Show that the curve defined by the parametric equations $x = t^2$ and $y = t^3 - 3t$ crosses itself. Find equations of the tangent lines to the curve at that point (see Example 2).

27. The parametric equations of the astroid $x^{2/3} + y^{2/3} = a^{2/3}$ are $x = a \cos^3 t$ and $y = a \sin^3 t$. (Verify this!) Find an expression for the slope of the tangent line to the astroid in terms of t. At what points on the astroid is the slope of the tangent line equal to -1? Equal to 1?

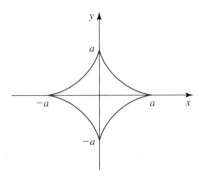

28. Find dy/dx and d^2y/dx^2 if

$$x = \int_1^t \frac{\sin u}{u}\, du \quad \text{and} \quad y = \int_2^{\ln t} e^u \, du$$

29. The function $y = f(x)$ is defined by the parametric equations

$$x = t^5 + 5t^3 + 10t + 2 \quad \text{and} \quad y = 2t^3 - 3t^2 - 12t + 1$$
$$-2 \le t \le 2$$

Find the absolute maximum and the absolute minimum values of f.

30. Find the points on the curve with parametric equations $x = t^3 - t$ and $y = t^2$ at which the tangent line is parallel to the line with parametric equations $x = 2t$ and $y = 2t + 4$.

In Exercises 31–36, find the length of the curve defined by the parametric equations.

31. $x = 2t^2$, $y = 3t^3$; $0 \le t \le 1$

32. $x = 2t^{3/2}$, $y = 3t + 1$; $0 \le t \le 4$

33. $x = \sin^2 t$, $y = \cos 2t$; $0 \le t \le \pi$

34. $x = e^t \cos t$, $y = e^t \sin t$; $0 \le t \le \pi$

35. $x = a(\cos t + t \sin t)$, $y = a(\sin t - t \cos t)$; $0 \le t \le \frac{\pi}{2}$

36. $x = (t^2 - 2)\sin t + 2t \cos t$, $y = (2 - t^2)\cos t + 2t \sin t$; $0 \le t \le \pi$

37. Find the length of the cardioid with parametric equations

$$x = a(2 \cos t - \cos 2t) \quad \text{and} \quad y = a(2 \sin t - \sin 2t)$$

38. Find the length of the astroid with parametric equations

$$x = a \cos^3 t \quad \text{and} \quad y = a \sin^3 t$$

(See the figure for Exercise 27. Compare with Exercise 25 in Section 5.4.)

39. The position of an object at any time t is (x, y), where $x = \cos^2 t$ and $y = \sin^2 t$, $0 \le t \le 2\pi$. Find the distance covered by the object as t runs from $t = 0$ to $t = 2\pi$.

40. The following figure shows the course taken by a yacht during a practice run. The parametric equations of the course are

$$x = 4\sqrt{2} \sin t \quad y = \sin 2t \quad 0 \le t \le 2\pi$$

where x and y are measured in miles. Find the length of the course.

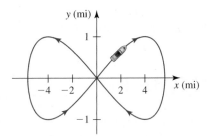

41. Path of a Boat Two towns, A and B, are located directly opposite each other on the banks of a river that is 1600 ft wide and flows east with a constant speed of 4 ft/sec. A boat leaving Town A travels with a constant speed of 18 ft/sec always aimed toward Town B. It can be shown that the path of the boat is given by the parametric equations

$$x = 800(t^{7/9} - t^{11/9}) \quad y = 1600t \quad 0 \le t \le 1$$

Find the distance covered by the boat in traveling from A to B.

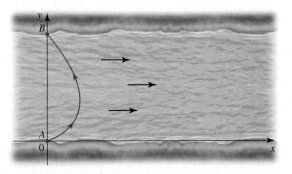

42. Trajectory of an Electron An electron initially located at the origin of a coordinate system is projected horizontally into a uniform electric field with magnitude E and directed upward. If the initial speed of the electron is v_0, then its trajectory is

$$x = v_0 t \qquad y = -\frac{1}{2}\left(\frac{eE}{m}\right)t^2$$

where e is the charge of the electron and m is its mass. Show that the trajectory of the electron is a parabola.

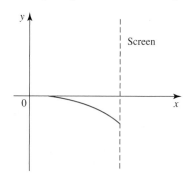

Note: The deflection of electrons by an electric field is used to control the direction of an electron beam in an electron gun.

43. Refer to Exercise 42. If a screen is placed along the vertical line $x = a$, at what point will the electron beam hit the screen?

 44. Find the point that is located one quarter of the way along the arch of the cycloid

$$x = a(t - \sin t) \qquad y = a(1 - \cos t) \qquad 0 \le t \le 2\pi$$

as measured from the origin. What is the slope of the tangent line to the cycloid at that point? Plot the arch of the cycloid and the tangent line on the same set of axes.

 45. The **cornu spiral** is a curve defined by the parametric equations

$$x = C(t) = \int_0^t \cos(\pi u^2/2)\, du \qquad y = S(t) = \int_0^t \sin(\pi u^2/2)\, du$$

where C and S are the Fresnel functions discussed in Section 6.7.
  a. Plot the spiral. Describe the behavior of the curve as $t \to \infty$ and as $t \to -\infty$.
  b. Find the length of the spiral from $t = 0$ to $t = a$.

46. Suppose that the graph of a nonnegative function F on an interval $[a, b]$ is represented by the parametric equations $x = f(t)$ and $y = g(t)$ for t in $[\alpha, \beta]$. Show that the area of the region under the graph of F is given by

$$\int_\alpha^\beta g(t)f'(t)\, dt \qquad \text{or} \qquad \int_\beta^\alpha g(t)f'(t)\, dt$$

47. Use the result of Exercise 46 to find the area of the region under one arch of the cycloid $x = a(\theta - \sin \theta)$, $y = a(1 - \cos \theta)$.

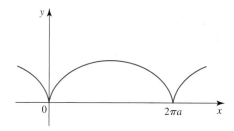

48. Use the result of Exercise 46 to find the area of the region enclosed by the ellipse with parametric equations $x = a \cos \theta$, $y = b \sin \theta$, where $0 \le \theta \le 2\pi$.

49. Use the result of Exercise 46 to find the area of the region enclosed by the astroid $x = a \cos^3 \theta$, $y = a \sin^3 \theta$. (See the figure for Exercise 27.)

50. Use the result of Exercise 46 to find the area of the region enclosed by the curve $x = a \sin t$, $y = b \sin 2t$.

51. Use the result of Exercise 46 to find the area of the region lying inside the course taken by the yacht of Exercise 40.

In Exercises 52–57, find the area of the surface obtained by revolving the curve about the x-axis.

52. $x = t, \quad y = 2 - t; \quad 0 \le t \le 2$

53. $x = t^3, \quad y = t^2; \quad 0 \le t \le 1$

54. $x = \sqrt{3}t^2, \quad y = t - t^3; \quad 0 \le t \le 1$

55. $x = \frac{1}{3}t^3, \quad y = 4 - \frac{1}{2}t^2; \quad 0 \le t \le 2\sqrt{2}$

56. $x = e^t \sin t, \quad y = e^t \cos t; \quad 0 \le t \le \frac{\pi}{2}$

57. $x = t - \sin t, \quad y = 1 - \cos t; \quad 0 \le t \le 2\pi$

In Exercises 58–61, find the area of the surface obtained by rotating the curve about the y-axis.

58. $x = t, \quad y = 2t; \quad 0 \le t \le 4$

59. $x = 3t^2, \quad y = 2t^3; \quad 0 \le t \le 1$

60. $x = a \cos t, \quad y = b \sin t; \quad -\frac{\pi}{2} \le t \le \frac{\pi}{2}$

61. $x = e^t - t, \quad y = 4e^{t/2}; \quad 0 \le t \le 1$

62. Find the area of the surface obtained by revolving the cardioid

$$x = a(2 \cos t - \cos 2t) \qquad y = a(2 \sin t - \sin 2t)$$

about the x-axis.

63. Find the area of the surface obtained by revolving the astroid

$$x = a \cos^3 t \qquad y = a \sin^3 t$$

about the x-axis.

64. Find the areas of the surface obtained by revolving one arch of the cycloid $x = a(\theta - \sin \theta)$, $y = a(1 - \cos \theta)$ about the x- and y-axes.

65. Find the surface area of the torus obtained by revolving the circle $x^2 + (y - b)^2 = r^2$ ($0 < r < b$) about the x-axis.
Hint: Represent the equation of the circle in parametric form: $x = r \cos t$, $y = b + r \sin t$, $0 \le t \le 2\pi$.

66. Show that if f' is continuous and $f'(t) \ne 0$ for $a \le t \le b$, then the parametric curve defined by $x = f(t)$ and $y = g(t)$ for $a \le t \le b$ can be put in the form $y = F(x)$.

cas *In Exercises 67–70, (a) plot the curve defined by the parametric equations and (b) estimate the arc length of the curve accurate to four decimal places.*

67. $x = 2t^2$, $y = t - t^3$; $0 \le t \le 1$

68. $x = \sin(0.5t + 0.4\pi)$, $y = \sin t$; $0 < t \le 4\pi$

69. $x = 0.2(6 \cos t - \cos 6t)$, $y = 0.2(6 \sin t - \sin 6t)$; $0 \le t \le \pi$

70. $x = 2t(1 - t^2)$, $y = -t^2\left(1 - \dfrac{3}{2}t^2\right)$; $-2 < t < 2$
(swallowtail castastrophe)

cas **71.** Use a calculator or computer to approximate the area of the surface obtained by revolving the curve

$$x = 4 \sin 2t \qquad y = 2 \cos 3t \qquad 0 \le t \le \tfrac{\pi}{6}$$

about the x-axis.

72. a. Find an expression for the arc length of the curve defined by the parametric equations

$$x = f''(t)\cos t + f'(t)\sin t \qquad y = -f''(t)\sin t + f'(t)\cos t$$

where $a \le t \le b$ and f has continuous third-order derivatives.

b. Use the result of part (a) to find the arc length of the curve $x = 6t \cos t + 3t^2 \sin t$ and $y = -6t \sin t + 3t^2 \cos t$, where $0 \le t \le 1$.

73. Show that

$$x = \frac{2at}{1 + t^2} \qquad y = \frac{a(1 - t^2)}{1 + t^2}$$

where $a > 0$ and $-\infty < t < \infty$, are parametric equations of a circle. What are its center and radius?

74. Use the parametric representation of a circle in Exercise 73 to show that the circumference of a circle of radius a is $2\pi a$.

75. Find parametric equations for the *Folium of Descartes*, $x^3 + y^3 = 3axy$ with parameter $t = y/x$.

cas **76.** Use the parametric representation of the *Folium of Descartes* to estimate the length of the loop.

cas **77.** Show that the length of the ellipse $x = a \cos t$, $y = b \sin t$, $0 \le t \le 2\pi$, where $a > b > 0$, is given by

$$L = 4a \int_0^{\pi/2} \sqrt{1 - e^2 \sin^2 t} \, dt$$

where

$$e = \frac{c}{a} = \frac{\sqrt{a^2 - b^2}}{a}$$

is the eccentricity of the ellipse.
Note: The integral is called an *elliptical integral of the second kind.*

cas **78.** Use a computer or calculator and the result of Exercise 77 to estimate the circumference of the ellipse

$$\frac{x^2}{100} + \frac{y^2}{36} = 1$$

accurate to three decimal places.

In Exercises 79–80, determine whether the statement is true or false. If it is true, explain why it is true. If it is false, give an example to show why it is false.

79. If $x = f(t)$ and $y = g(t)$, f and g have second-order derivatives, and $f'(t) \ne 0$, then

$$\frac{d^2y}{dx^2} = \frac{f'(t)g''(t) - g'(t)f''(t)}{[f'(t)]^2}$$

80. The curve with parametric equations $x = f(t)$ and $y = g(t)$ is a line if and only if f and g are both linear functions of t.

10.4   Polar Coordinates

The curve shown in Figure 1a is a lemniscate, and the one shown in Figure 1b is called a cardioid. The rectangular equations of these curves are

$$(x^2 + y^2)^2 = 4(x^2 - y^2) \qquad \text{and} \qquad x^4 - 2x^3 + 2x^2y^2 - 2xy^2 - y^2 + y^4 = 0$$

respectively. As you can see, these equations are somewhat complicated. For example, they will not prove very helpful if we want to calculate the area enclosed by the two loops of the lemniscate shown in Figure 1a or the length of the cardioid shown in Figure 1b.

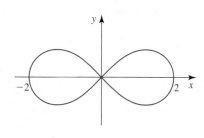
(**a**) A lemniscate

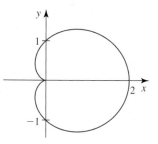
(**b**) A cardioid

FIGURE 1

A rectangular equation of the lemniscate in part (a) is $(x^2 + y^2)^2 = 4(x^2 - y^2)$, and an equation of the cardioid in part (b) is $x^4 - 2x^3 + 2x^2y^2 - 2xy^2 - y^2 + y^4 = 0$.

A question that arises naturally is: Is there a coordinate system other than the rectangular system that we can use to give a simpler representation for curves such as the lemniscate and cardioid? One such system is the *polar coordinate system.*

■ The Polar Coordinate System

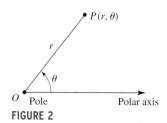

FIGURE 2

To construct the polar coordinate system, we fix a point O called the **pole** (or **origin**) and draw a ray (half-line) emanating from O called the **polar axis.** Suppose that P is any point in the plane, let r denote the distance from O to P, and let θ denote the angle (in degrees or radians) between the polar axis and the line segment OP. (See Figure 2.) Then the point P is represented by the ordered pair (r, θ), also written $P(r, \theta)$, where the numbers r and θ are called the **polar coordinates** of P.

The **angular coordinate** θ is positive if it is measured in the counterclockwise direction from the polar axis and negative if it is measured in the clockwise direction. The **radial coordinate** r may assume positive as well as negative values. If $r > 0$, then $P(r, \theta)$ is on the terminal side of θ and at a distance r from the origin. If $r < 0$, then $P(r, \theta)$ lies on the ray that is opposite the terminal side of θ and at a distance of $|r| = -r$ from the pole. (See Figure 3.) Also, by convention the pole O is represented by the ordered pair $(0, \theta)$ for *any* value of θ. Finally, a plane that is endowed with a polar coordinate system is referred to as an $r\theta$-plane.

FIGURE 3

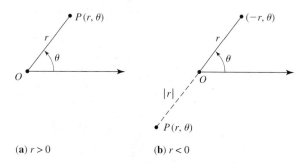

(**a**) $r > 0$ (**b**) $r < 0$

EXAMPLE 1 Plot the following points in the $r\theta$-plane.

a. $\left(1, \frac{2\pi}{3}\right)$ **b.** $\left(2, -\frac{\pi}{4}\right)$ **c.** $\left(-2, \frac{\pi}{3}\right)$ **d.** $(2, -3\pi)$

Solution The points are plotted in Figure 4.

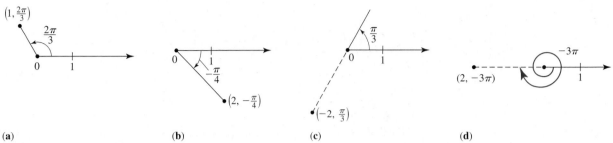

(a)

(b)

(c)

(d)

FIGURE 4
The points in Example 1

Unlike the representation of points in the rectangular system, the representation of points using polar coordinates is *not* unique. For example, the point (r, θ) can also be written as $(r, \theta + 2n\pi)$ or $(-r, \theta + (2n + 1)\pi)$, where n is any integer. Figures 5a and 5b illustrate this for the case $n = 1$ and $n = 0$, respectively.

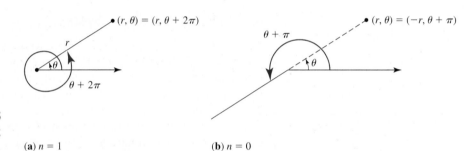

FIGURE 5
Representation of points using
polar coordinates is not unique.

(a) $n = 1$

(b) $n = 0$

■ Relationship Between Polar and Rectangular Coordinates

To establish the relationship between polar and rectangular coordinates, let's superimpose an xy-plane on an $r\theta$-plane in such a way that the origins coincide and the positive x-axis coincides with the polar axis. Let P be any point in the plane other than the origin with rectangular representation (x, y) and polar representation (r, θ). Figure 6a shows a situation in which $r > 0$, and Figure 6b shows a situation in which $r < 0$. If $r > 0$, we see immediately from the figure that

$$\cos \theta = \frac{x}{r} \qquad \sin \theta = \frac{y}{r}$$

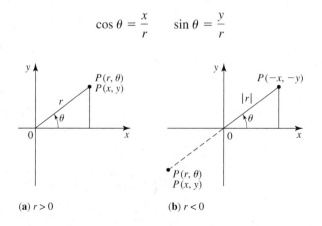

FIGURE 6
The relationship between polar
and rectangular coordinates

(a) $r > 0$

(b) $r < 0$

so $x = r \cos \theta$ and $y = r \sin \theta$. If $r < 0$, we see by referring to Figure 6b that

$$\cos \theta = \frac{-x}{|r|} = \frac{-x}{-r} = \frac{x}{r} \qquad \sin \theta = \frac{-y}{|r|} = \frac{-y}{-r} = \frac{y}{r}$$

so again $x = r \cos \theta$ and $y = r \sin \theta$. Finally, in either case we have

$$x^2 + y^2 = r^2 \qquad \text{and} \qquad \tan \theta = \frac{y}{x} \qquad \text{if } x \neq 0$$

Relationship Between Rectangular and Polar Coordinates

Suppose that a point P (other than the origin) has representation (r, θ) in polar coordinates and (x, y) in rectangular coordinates. Then

$$x = r \cos \theta \qquad \text{and} \qquad y = r \sin \theta \qquad \qquad \textbf{(1)}$$

$$r^2 = x^2 + y^2 \qquad \text{and} \qquad \tan \theta = \frac{y}{x} \qquad \text{if } x \neq 0 \qquad \textbf{(2)}$$

EXAMPLE 2 The point $\left(4, \frac{\pi}{6}\right)$ is given in polar coordinates. Find its representation in rectangular coordinates.

Solution Here, $r = 4$ and $\theta = \pi/6$. Using Equation (1), we obtain

$$x = r \cos \theta = 4 \cos \frac{\pi}{6} = 4 \cdot \frac{\sqrt{3}}{2} = 2\sqrt{3}$$

$$y = r \sin \theta = 4 \sin \frac{\pi}{6} = 4 \cdot \frac{1}{2} = 2$$

Therefore, the given point has rectangular representation $(2\sqrt{3}, 2)$. ∎

EXAMPLE 3 The point $(-1, 1)$ is given in rectangular coordinates. Find its representation in polar coordinates.

Solution Here, $x = -1$ and $y = 1$. Using Equation (2), we have
$$r^2 = x^2 + y^2 = (-1)^2 + 1^2 = 2$$

and

$$\tan \theta = \frac{y}{x} = -1$$

Let's choose r to be positive; that is, $r = \sqrt{2}$. Next, observe that the point $(-1, 1)$ lies in the second quadrant and so we choose $\theta = 3\pi/4$ (other choices are $\theta = (3\pi/4) \pm 2n\pi$, where n is an integer). Therefore, one representation of the given point is $\left(\sqrt{2}, \frac{3\pi}{4}\right)$. ∎

■ Graphs of Polar Equations

The graph of a **polar equation** $r = f(\theta)$ or, more generally, $F(r, \theta) = 0$ is the set of all points (r, θ) whose coordinates satisfy the equation.

EXAMPLE 4 Sketch the graphs of the polar equations, and reconcile your results by finding the corresponding rectangular equations.

a. $r = 2$ **b.** $\theta = \dfrac{2\pi}{3}$

Solution

a. The graph of $r = 2$ consists of all points $P(r, \theta)$ where $r = 2$ and θ can assume *any* value. Since r gives the distance between P and the pole O, we see that the graph consists of all points that are located a distance of 2 units from the pole; in other words, the graph of $r = 2$ is the circle of radius 2 centered at the pole. (See Figure 7a.) To find the corresponding rectangular equation, square both sides of the given equation obtaining $r^2 = 4$. But by Equation (2), $r^2 = x^2 + y^2$, and this gives the desired equation $x^2 + y^2 = 4$. Since this is a rectangular equation of a circle with center at the origin and radius 2, the result obtained earlier has been confirmed.

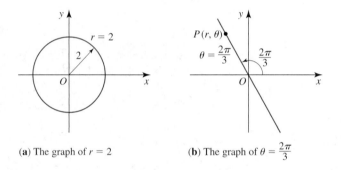

FIGURE 7   (a) The graph of $r = 2$ (b) The graph of $\theta = \dfrac{2\pi}{3}$

b. The graph of $\theta = 2\pi/3$ consists of all points $P(r, \theta)$ where $\theta = 2\pi/3$ and r can assume *any* value. Since θ measures the angle the line segment OP makes with the polar axis, we see that the graph consists of all points that are located on the straight line passing through the pole O and making an angle of $2\pi/3$ radians with the polar axis. (See Figure 7b.) Observe that the half-line in the second quadrant consists of points for which $r > 0$, whereas the half-line in the fourth quadrant consists of points for which $r < 0$. To find the corresponding rectangular equation, we use Equation (2), $\tan \theta = y/x$, to obtain

$$\tan \frac{2\pi}{3} = \frac{y}{x} \qquad \text{or} \qquad \frac{y}{x} = -\sqrt{3}$$

or $y = -\sqrt{3}x$. This equation confirms that the graph of $\theta = 2\pi/3$ is a straight line with slope $-\sqrt{3}$. ∎

As in the case with rectangular equations, we can often obtain a sketch of the graph of a simple polar equation by plotting and connecting some points that lie on the graph.

EXAMPLE 5 Sketch the graph of the polar equation $r = 2 \sin \theta$. Find a corresponding rectangular equation and reconcile your results.

Solution The following table shows the values of r corresponding to some convenient values of θ. It suffices to restrict the values of θ to those lying between 0 and π, since values of θ beyond π will give the same points (r, θ) again.

θ	0	$\frac{\pi}{6}$	$\frac{\pi}{4}$	$\frac{\pi}{3}$	$\frac{\pi}{2}$	$\frac{2\pi}{3}$	$\frac{3\pi}{4}$	$\frac{5\pi}{6}$	π
r	0	1	$\sqrt{2} \approx 1.4$	$\sqrt{3} \approx 1.7$	2	$\sqrt{3} \approx 1.7$	$\sqrt{2} \approx 1.4$	1	0

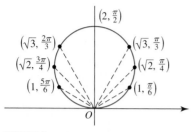

FIGURE 8

The graph of $r = 2 \sin \theta$ is a circle. To plot the points, first draw the ray with the desired angle, then locate the point by measuring off the required distance from the pole.

The graph of $r = 2 \sin \theta$ is sketched in Figure 8. To find a corresponding rectangular equation, we multiply both sides of $r = 2 \sin \theta$ by r to obtain $r^2 = 2r \sin \theta$ and then use the relationships $r^2 = x^2 + y^2$ (Equation (2)) and $y = r \sin \theta$ (Equation (1)), to obtain the desired equation

$$x^2 + y^2 = 2y \qquad \text{or} \qquad x^2 + y^2 - 2y = 0$$

Finally, completing the square in y, we have

$$x^2 + y^2 - 2y + (-1)^2 = 1$$

or

$$x^2 + (y - 1)^2 = 1$$

which is an equation of the circle with center $(0, 1)$ and radius 1, as obtained earlier. ∎

It might have occurred to you that in the last several examples we could have obtained the graphs of the polar equations by first converting them to the corresponding rectangular equations. But as you will see, some curves are easier to graph using polar coordinates.

■ Symmetry

Just as the use of symmetry is helpful in graphing rectangular equations, its use is equally helpful in graphing polar equations. Three types of symmetry are illustrated in Figure 9. The test for each type of symmetry follows.

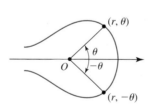

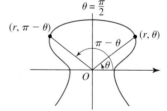

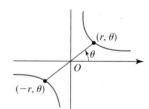

FIGURE 9

Symmetries of graphs of polar equations

(a) Symmetry with respect to the polar axis

(b) Symmetry with respect to the line $\theta = \frac{\pi}{2}$

(c) Symmetry with respect to the pole

Tests for Symmetry

a. The graph of $r = f(\theta)$ is **symmetric with respect to the polar axis** if the equation is unchanged when θ is replaced by $-\theta$.

b. The graph of $r = f(\theta)$ is **symmetric with respect to the vertical line** $\theta = \pi/2$ if the equation is unchanged when θ is replaced by $\pi - \theta$.

c. The graph of $r = f(\theta)$ is **symmetric with respect to the pole** if the equation is unchanged when r is replaced by $-r$ or when θ is replaced by $\theta + \pi$.

To illustrate the use of the tests for symmetry, consider the equation $r = 2 \sin \theta$ of Example 5. Here, $f(\theta) = 2 \sin \theta$, and since

$$f(\pi - \theta) = 2 \sin(\pi - \theta) = 2(\sin \pi \cos \theta - \cos \pi \sin \theta) = 2 \sin \theta = f(\theta)$$

we conclude that the graph of $r = 2 \sin \theta$ is symmetric with respect to the vertical line $\theta = \pi/2$ (Figure 8).

EXAMPLE 6 Sketch the graph of the polar equation $r = 1 + \cos \theta$. This is the polar form of the rectangular equation $x^4 - 2x^3 + 2x^2 y^2 - 2xy^2 - y^2 + y^4 = 0$ of the cardioid that was mentioned at the beginning of this section (Figure 1b).

Solution Writing $f(\theta) = 1 + \cos \theta$ and observing that

$$f(-\theta) = 1 + \cos(-\theta) = 1 + \cos \theta = f(\theta)$$

we conclude that the graph of $r = 1 + \cos \theta$ is symmetric with respect to the polar axis. In view of this, we need only to obtain that part of the graph between $\theta = 0$ and $\theta = \pi$. We can then complete the graph using symmetry.

To sketch the graph of $r = 1 + \cos \theta$ for $0 \le \theta \le \pi$, we can proceed as we did in Example 5 by first plotting some points lying on that part of the graph, or we may proceed as follows: Treat r and θ as *rectangular* coordinates, and make use of our knowledge of graphing rectangular equations to obtain the graph of $r = f(\theta) = 1 + \cos \theta$ on the interval $[0, \pi]$. (See Figure 10a.) Then recalling that θ is the angular coordinate and r is the radial coordinate, we see that as θ increases from 0 to π, the points on the respective rays shrink to 0. (See Figure 10b, where the corresponding points are shown.)

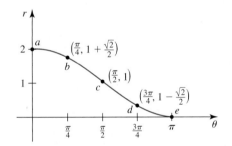

(a) $r = f(\theta)$, treating r and θ as rectangular coordinates
FIGURE 10
Two steps in sketching the graph of the polar equation $r = 1 + \cos \theta$

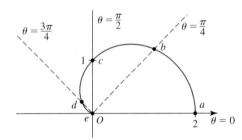

(b) $r = f(\theta)$, treating r and θ as polar coordinates

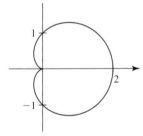

FIGURE 11
The graph of $r = 1 + \cos \theta$
is a cardioid.

Finally, using symmetry, we complete the graph of $r = 1 + \cos \theta$, as shown in Figure 11. It is called a **cardioid** because it is heart-shaped. ∎

EXAMPLE 7 Sketch the graph of the polar equation $r = 2 \cos 2\theta$.

Solution Write $f(\theta) = 2 \cos 2\theta$, and observe that

$$f(-\theta) = 2 \cos 2(-\theta) = 2 \cos 2\theta = f(\theta)$$

and

$$f(\pi - \theta) = 2 \cos 2(\pi - \theta) = 2 \cos(2\pi - 2\theta)$$

$$= 2[\cos 2\pi \cos 2\theta + \sin 2\pi \sin 2\theta] = 2 \cos 2\theta = f(\theta)$$

Therefore, the graph of the given equation is symmetric with respect to both the polar axis and the vertical line $\theta = \pi/2$. It suffices, therefore, to obtain an accurate sketch of that part of the graph for $0 \le \theta \le \frac{\pi}{2}$ and then complete the sketch of the graph using symmetry. Proceeding as in Example 6, we first sketch the graph of $r = 2 \cos 2\theta$ for $0 \le \theta \le \frac{\pi}{2}$ treating r and θ as rectangular coordinates (Figure 12a), and then transcribe the information contained in this graph onto the graph in the $r\theta$-plane for $0 \le \theta \le \frac{\pi}{2}$. (See Figure 12b.)

FIGURE 12
Two steps in sketching the graph of $r = 2 \cos 2\theta$

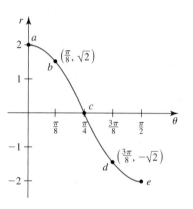

(**a**) $r = f(\theta)$ treating r and θ as rectangular coordinates

(**b**) $r = f(\theta)$, treating r and θ as polar coordinates

Finally, using the symmetry that was established earlier (Figure 13a), we complete the graph of $r = 2 \cos 2\theta$ as shown in Figure 13b. This graph is called a **four-leaved rose.**

FIGURE 13
The graph of $r = 2 \cos 2\theta$ is a four-leaved rose.

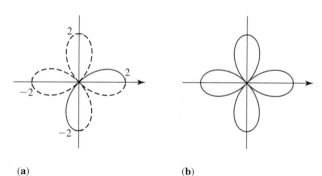

(**a**)

(**b**)

The next example shows how the graph of a rectangular equation can be sketched more easily by first converting it to polar form.

EXAMPLE 8 Sketch the graph of the equation $(x^2 + y^2)^2 = 4(x^2 - y^2)$ by first converting it to polar form. This is an equation of the lemniscate that was mentioned at the beginning of this section.

Solution To convert the given equation to polar form, we use Equations (1) and (2), obtaining

$$(r^2)^2 = 4(r^2 \cos^2 \theta - r^2 \sin^2 \theta)$$

$$= 4r^2(\cos^2 \theta - \sin^2 \theta)$$

$$r^4 = 4r^2 \cos 2\theta$$

or

$$r^2 = 4 \cos 2\theta$$

Observe that $f(\theta) = 2\sqrt{\cos 2\theta}$ is defined for $-\frac{\pi}{4} \le \theta \le \frac{\pi}{4}$ and $\frac{3\pi}{4} \le \theta \le \frac{5\pi}{4}$. Also, observe that $f(-\theta) = f(\theta)$ and $f(\pi - \theta) = f(\theta)$. (These computations are similar to those in Example 7.) So the graph of $r = 2\sqrt{\cos 2\theta}$ is symmetric with respect to the polar axis and the line $\theta = \pi/2$. The graph of $r = f(\theta)$ for $0 \le \theta \le \frac{\pi}{4}$, where r and θ are treated as rectangular coordinates, is shown in Figure 14a. This leads to the part of the required graph for $0 \le \theta \le \frac{\pi}{4}$ shown in Figure 14b. Then, using symmetry, we obtain the graph of $r = 2\sqrt{\cos 2\theta}$ and, therefore, that of $(x^2 + y^2)^2 = 4(x^2 - y^2)$, as shown in Figure 15.

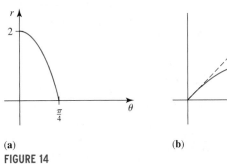

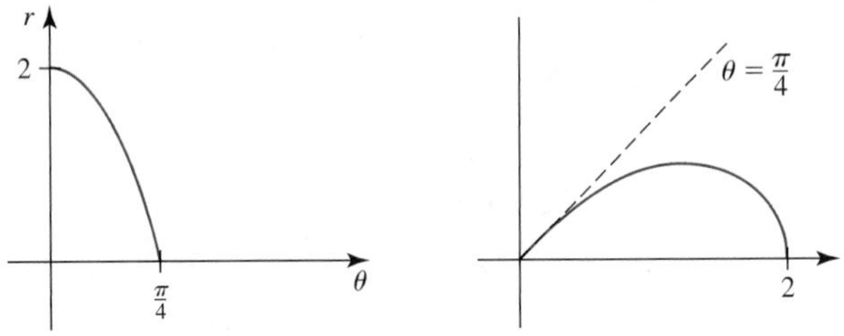

(a)

(b)

FIGURE 14

Two steps in sketching the graph of $r = 2\sqrt{\cos 2\theta}$

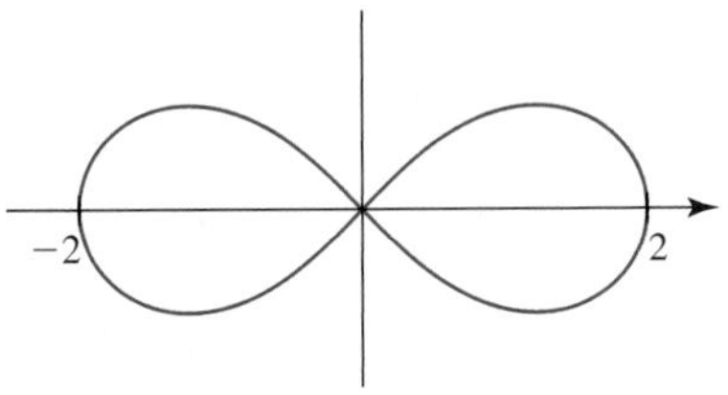

FIGURE 15

The graph of $r = 2\sqrt{\cos 2\theta}$ is a lemniscate.

■ Tangent Lines to Graphs of Polar Equations

To find the slope of the tangent line to the graph of $r = f(\theta)$ at the point $P(r, \theta)$, let $P(x, y)$ be the rectangular representation of P. Then

$$x = r \cos \theta = f(\theta) \cos \theta$$

$$y = r \sin \theta = f(\theta) \sin \theta$$

We can view these equations as parametric equations for the graph of $r = f(\theta)$ with parameter θ. Then, using Equation (1) of Section 10.3, we have

$$\frac{dy}{dx} = \frac{\dfrac{dy}{d\theta}}{\dfrac{dx}{d\theta}} = \frac{\dfrac{dr}{d\theta} \sin \theta + r \cos \theta}{\dfrac{dr}{d\theta} \cos \theta - r \sin \theta} \qquad \text{if} \quad \frac{dx}{d\theta} \ne 0 \qquad (3)$$

and this gives the slope of the tangent line to the graph of $r = f(\theta)$ at any point $P(r, \theta)$.

The horizontal tangent lines to the graph of $r = f(\theta)$ are located at the points where $dy/d\theta = 0$ and $dx/d\theta \ne 0$. The vertical tangent lines are located at the points where $dx/d\theta = 0$ and $dy/d\theta \ne 0$ (so that dy/dx is undefined). Also, points where both $dy/d\theta$ and $dx/d\theta$ are equal to zero are candidates for horizontal or vertical tangent lines, respectively, and may be investigated using l'Hôpital's Rule.

Equation (3) can be used to help us find the tangent lines to the graph of $r = f(\theta)$ at the pole. To see this, suppose that the graph of f passes through the pole when $\theta = \theta_0$. Then $f(\theta_0) = 0$. If $f'(\theta_0) \ne 0$, then Equation (3) reduces to

$$\frac{dy}{dx} = \frac{f'(\theta_0) \sin \theta_0 + f(\theta_0) \cos \theta_0}{f'(\theta_0) \cos \theta_0 - f(\theta_0) \sin \theta_0} = \frac{\sin \theta_0}{\cos \theta_0} = \tan \theta_0$$

This shows that $\theta = \theta_0$ is a tangent line to the graph of $r = f(\theta)$ at the pole $(0, \theta_0)$. The following summarizes this discussion.

> $\theta = \theta_0$ is a tangent line to the graph of $r = f(\theta)$ at the pole if $f(\theta_0) = 0$ and $f'(\theta_0) \neq 0$.

EXAMPLE 9   Consider the cardioid $r = 1 + \cos\theta$ of Example 6.

a. Find the slope of the tangent line to the cardioid at the point where $\theta = \pi/6$.
b. Find the points on the cardioid where the tangent lines are horizontal and where the tangent lines are vertical.

Solution
a. The slope of the tangent line to the cardioid $r = 1 + \cos\theta$ at any point $P(r, \theta)$ is given by

$$\frac{dy}{dx} = \frac{\dfrac{dr}{d\theta}\sin\theta + r\cos\theta}{\dfrac{dr}{d\theta}\cos\theta - r\sin\theta} = \frac{(-\sin\theta)(\sin\theta) + (1 + \cos\theta)\cos\theta}{(-\sin\theta)(\cos\theta) - (1 + \cos\theta)\sin\theta}$$

$$= \frac{(\cos^2\theta - \sin^2\theta) + \cos\theta}{-2\sin\theta\cos\theta - \sin\theta} = -\frac{\cos 2\theta + \cos\theta}{\sin 2\theta + \sin\theta}$$

At the point on the cardioid where $\theta = \pi/6$, the slope of the tangent line is

$$\left.\frac{dy}{dx}\right|_{\theta=\pi/6} = -\frac{\cos\left(\dfrac{\pi}{3}\right) + \cos\left(\dfrac{\pi}{6}\right)}{\sin\left(\dfrac{\pi}{3}\right) + \sin\left(\dfrac{\pi}{6}\right)} = -\frac{\dfrac{1}{2} + \dfrac{\sqrt{3}}{2}}{\dfrac{\sqrt{3}}{2} + \dfrac{1}{2}} = -1$$

b. Observe that $dy/d\theta = 0$ if

$$\cos 2\theta + \cos\theta = 0$$

$$2\cos^2\theta + \cos\theta - 1 = 0$$

$$(2\cos\theta - 1)(\cos\theta + 1) = 0$$

that is, if $\cos\theta = \frac{1}{2}$ or $\cos\theta = -1$. This gives

$$\theta = \frac{\pi}{3}, \quad \pi, \quad \text{or} \quad \frac{5\pi}{3}$$

Next, $dx/d\theta = 0$ if

$$\sin 2\theta + \sin\theta = 0$$

$$2\sin\theta\cos\theta + \sin\theta = 0$$

$$\sin\theta\,(2\cos\theta + 1) = 0$$

that is, if $\sin\theta = 0$ or $\cos\theta = -\frac{1}{2}$. This gives

$$\theta = 0, \quad \pi, \quad \frac{2\pi}{3}, \quad \text{or} \quad \frac{4\pi}{3}$$

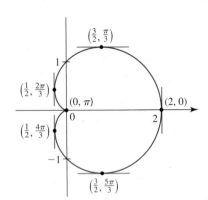

FIGURE 16
The horizontal and vertical tangents to the graph of $r = 1 + \cos \theta$

In view of the remarks following Equation (3), we see that $\theta = \pi/3$ and $\theta = 5\pi/3$ give rise to horizontal tangents. To investigate the candidate $\theta = \pi$, where both $dy/d\theta$ and $dx/d\theta$ are equal to zero, we use l'Hôpital's Rule. Thus,

$$\lim_{\theta \to \pi^-} \frac{dy}{dx} = -\lim_{\theta \to \pi^-} \frac{\cos 2\theta + \cos \theta}{\sin 2\theta + \sin \theta}$$

$$= -\lim_{\theta \to \pi^-} \frac{-2 \sin 2\theta - \sin \theta}{2 \cos 2\theta + \cos \theta} = 0$$

Similarly, we see that

$$\lim_{\theta \to \pi^+} \frac{dy}{dx} = 0$$

Therefore, $\theta = \pi$ also gives rise to a horizontal tangent. Thus, the horizontal tangent lines occur at

$$\left(\tfrac{3}{2}, \tfrac{\pi}{3}\right), \quad (0, \pi), \quad \text{and} \quad \left(\tfrac{3}{2}, \tfrac{5\pi}{3}\right)$$

The vertical tangent lines occur at $\theta = 0$, $2\pi/3$, and $4\pi/3$. The points are $(2, 0)$, $\left(\tfrac{1}{2}, \tfrac{2\pi}{3}\right)$, and $\left(\tfrac{1}{2}, \tfrac{4\pi}{3}\right)$. These tangent lines are shown in Figure 16. ∎

EXAMPLE 10 Find the tangent lines of $r = \cos 2\theta$ at the origin.

Solution Setting $f(\theta) = \cos 2\theta = 0$, we find that

$$2\theta = \frac{\pi}{2}, \quad \frac{3\pi}{2}, \quad \frac{5\pi}{2}, \quad \text{or} \quad \frac{7\pi}{2}$$

or

$$\theta = \frac{\pi}{4}, \quad \frac{3\pi}{4}, \quad \frac{5\pi}{4}, \quad \text{or} \quad \frac{7\pi}{4}$$

Next, we compute $f'(\theta) = -2 \sin 2\theta$. Since $f'(\theta) \neq 0$ for each of these values of θ, we see that $\theta = \pi/4$ and $\theta = 3\pi/4$ (that is, $y = x$ and $y = -x$) are tangent lines to the graph of $r = \cos 2\theta$ at the pole (see Figure 17). ∎

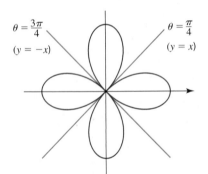

FIGURE 17
The tangent lines to the graph of $r = \cos 2\theta$ at the origin

10.4 CONCEPT QUESTIONS

1. Let $P(r, \theta)$ be a point in the plane with polar coordinates r and θ. Find all possible representations of $P(r, \theta)$.
2. Suppose that P has representation (r, θ) in polar coordinates and (x, y) in rectangular coordinates. Express (a) x and y in terms of r and θ and (b) r and θ in terms of x and y.
3. Explain how you would determine whether the graph of $r = f(\theta)$ is symmetric with respect to (a) the polar axis, (b) the vertical line $\theta = \pi/2$, and (c) the pole.

4. Suppose that $r = f(\theta)$, where f is differentiable.
   a. Write an expression for dy/dx.
   b. How do you find the points on the graph of $r = f(\theta)$ where the tangent lines are horizontal and where the tangent lines are vertical?
   c. How do you find the tangent lines to the graph of $r = f(\theta)$ (if they exist) at the pole?

10.4 EXERCISES

In Exercises 1–8, plot the point with the polar coordinates. Then find the rectangular coordinates of the point.

1. $\left(4, \frac{\pi}{4}\right)$

2. $\left(2, \frac{\pi}{6}\right)$

3. $\left(4, \frac{3\pi}{2}\right)$

4. $(6, 3\pi)$

5. $\left(-\sqrt{2}, \frac{\pi}{4}\right)$

6. $\left(-1, \frac{\pi}{3}\right)$

7. $\left(-4, -\frac{3\pi}{4}\right)$

8. $\left(5, -\frac{5\pi}{6}\right)$

In Exercises 9–16, plot the point with the rectangular coordinates. Then find the polar coordinates of the point taking $r > 0$ and $0 \le \theta < 2\pi$.

9. $(2, 2)$

10. $(1, -1)$

11. $(0, 5)$

12. $(3, -4)$

13. $(-\sqrt{3}, -\sqrt{3})$

14. $(2\sqrt{3}, -2)$

15. $(5, -12)$

16. $(3, -1)$

In Exercises 17–24, sketch the region comprising points whose polar coordinates satisfy the given conditions.

17. $r \ge 1$

18. $r > 1$

19. $0 \le r \le 2$

20. $1 \le r < 2$

21. $0 \le \theta \le \frac{\pi}{4}$

22. $0 \le r \le 3, \quad 0 \le \theta \le \frac{\pi}{3}$

23. $1 \le r \le 3, \quad -\frac{\pi}{6} \le \theta \le \frac{\pi}{6}$

24. $2 < r < 4, \quad -\frac{\pi}{2} < \theta < \frac{\pi}{2}$

In Exercises 25–32, convert the polar equation to a rectangular equation.

25. $r \cos \theta = 2$

26. $r \sin \theta = -3$

27. $2r \cos \theta + 3r \sin \theta = 6$

28. $r \sin \theta = 2r \cos \theta$

29. $r^2 = 4r \cos \theta$

30. $r^2 = \sin 2\theta$

31. $r = \dfrac{1}{1 - \sin \theta}$

32. $r = \dfrac{3}{4 - 5 \cos \theta}$

In Exercises 33–38, convert the rectangular equation to a polar equation.

33. $x = 4$

34. $x + 2y = 3$

35. $x^2 + y^2 = 9$

36. $x^2 - y^2 = 1$

37. $xy = 4$

38. $y^2 - x^2 = 4\sqrt{x^2 + y^2}$

In Exercises 39–64, sketch the curve with the polar equation.

39. $r = 3$

40. $r = -2$

41. $\theta = \dfrac{\pi}{3}$

42. $\theta = -\dfrac{\pi}{6}$

43. $r = 3 \cos \theta$

44. $r = -4 \sin \theta$

45. $r = 3 \cos \theta - 2 \sin \theta$

46. $r = 2 \sin \theta + 4 \cos \theta$

47. $r = 1 + \cos \theta$

48. $r = 1 + \sin \theta$

49. $r = 4(1 - \sin \theta)$

50. $r = 3 - 3 \cos \theta$

51. $r = 2 \csc \theta$

52. $r = -3 \sec \theta$

53. $r = \theta, \quad \theta \ge 0$ (spiral)

54. $r = \dfrac{1}{\theta}$ (spiral)

55. $r = e^{\theta}, \quad \theta \ge 0$ (logarithmic spiral)

56. $r^2 = \dfrac{1}{\theta}$ (lituus)

57. $r^2 = 4 \sin 2\theta$ (lemniscate)

58. $r = 1 - 2 \cos \theta$ (limaçon)

59. $r = 3 + 2 \sin \theta$ (limaçon)

60. $r = \sin 2\theta$ (four-leaved rose)

61. $r = \sin 3\theta$ (three-leaved rose)

62. $r = 2 \cos 4\theta$ (eight-leaved rose)

63. $r = 4 \sin 4\theta$ (eight-leaved rose)

64. $r = 2 \sin 5\theta$ (five-leaved rose)

In Exercises 65–72, find the slope of the tangent line to the curve with the polar equation at the point corresponding to the given value of θ.

65. $r = 4 \cos \theta, \quad \theta = \dfrac{\pi}{3}$

66. $r = 3 \sin \theta, \quad \theta = \dfrac{\pi}{4}$

67. $r = \sin \theta + \cos \theta, \quad \theta = \dfrac{\pi}{4}$

68. $r = 1 + 3 \cos \theta, \quad \theta = \dfrac{\pi}{2}$

69. $r = \theta, \quad \theta = \pi$

70. $r = \sin 3\theta, \quad \theta = \dfrac{\pi}{3}$

71. $r^2 = 4 \cos 2\theta, \quad \theta = \dfrac{\pi}{6}$

72. $r = 2 \sec \theta, \quad \theta = \dfrac{\pi}{4}$

In Exercises 73–78, find the points on the curve with the given polar equation where the tangent line is horizontal or vertical.

73. $r = 4 \cos \theta$

74. $r = \sin \theta + \cos \theta$

75. $r = \sin 2\theta$

76. $r^2 = 4 \cos 2\theta$

77. $r = 1 + 2 \cos \theta$

78. $r = 1 + \sin \theta$

79. Show that the rectangular equation

$$x^4 - 2x^3 + 2x^2y^2 - 2xy^2 - y^2 + y^4 = 0$$

is an equation of the cardioid with polar equation $r = 1 + \cos \theta$.

80. Show that the polar equation $r = a \sin \theta + b \cos \theta$, where a and b are nonzero, represents a circle. What are the center and radius of the circle?

ⓥ Videos for selected exercises are available online at **www.academic.cengage.com/login**.

81. a. Show that the distance between the points with polar coordinates (r_1, θ_1) and (r_2, θ_2) is given by

$$d = \sqrt{r_1^2 + r_2^2 - 2r_1 r_2 \cos(\theta_1 - \theta_2)}$$

b. Find the distance between the points with polar coordinates $\left(4, \frac{2\pi}{3}\right)$ and $\left(2, \frac{\pi}{3}\right)$.

82. Show that the curves with polar equations $r = a \sin \theta$ and $r = a \cos \theta$ intersect at right angles.

 83. a. Plot the graphs of the cardioids $r = a(1 + \cos \theta)$ and $r = a(1 - \cos \theta)$.

b. Show that the cardioids intersect at right angles except at the pole.

84. Let ψ be the angle between the radial line OP and the tangent line to the curve with polar equation $r = f(\theta)$ at P (see the figure). Show that

$$\tan \psi = r \frac{d\theta}{dr}$$

Hint: Observe that $\psi = \phi - \theta$. Then use the trigonometric identity

$$\tan(a - b) = \frac{\tan a - \tan b}{1 + \tan a \tan b}$$

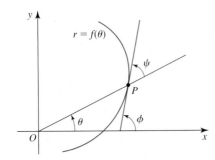

 In Exercises 85–92, use a graphing utility to plot the curve with the polar equation.

85. $r = \cos \theta(4 \sin^2 \theta - 1), \quad 0 \le \theta < 2\pi$

86. $r = 3 \sin \theta \cos^2 \theta, \quad 0 \le \theta < 2\pi$

87. $r = 0.3\left[1 + 2 \sin\left(\frac{\theta}{2}\right)\right], \quad 0 \le \theta < 4\pi$

(nephroid of Freeth)

88. $r = \dfrac{1 - 10 \cos \theta}{1 + 10 \cos \theta}, \quad 0 \le \theta < 2\pi$

89. $r^2 = 0.8(1 - 0.8 \sin^2 \theta), \quad 0 \le \theta < 2\pi$ (hippopede curve)

90. $r^2 = \dfrac{\frac{1}{4} \sin^2 \theta - 3.6 \cos^2 \theta}{\sin^2 \theta - \cos^2 \theta}, \quad 0 \le \theta < 2\pi$ (devil's curve)

91. $r = \dfrac{0.1}{\cos 3\theta}, \quad 0 \le \theta < \pi$ (epi-spiral)

92. $r = \dfrac{\sin \theta}{\theta}, \quad -6\pi \le \theta < 6\pi$ (cochleoid)

In Exercises 93–95, determine whether the statement is true or false. If it is true, explain why it is true. If it is false, give an example to show why it is false.

93. If $P(r_1, \theta_1)$ and $P(r_2, \theta_2)$ represent the same point in polar coordinates, then $r_1 = r_2$.

94. If $P(r_1, \theta_1)$ and $P(r_2, \theta_2)$ represent the same point in polar coordinates, then $\theta_1 = \theta_2$.

95. The graph of $r = f(\theta)$ has a horizontal tangent line at a point on the graph if $dy/d\theta = 0$, where $y = f(\theta)\sin \theta$.

10.5   Areas and Arc Lengths in Polar Coordinates

In this section we see how the use of polar equations to represent curves such as lemniscates and cardioids will simplify the task of finding the areas of the regions enclosed by these curves as well as the lengths of these curves.

◼ Areas in Polar Coordinates

To develop a formula for finding the area of a region bounded by a curve defined by a polar equation, we need the formula for the area of a sector of a circle

$$A = \frac{1}{2}r^2\theta \tag{1}$$

where r is the radius of the circle and θ is the central angle measured in radians. (See Figure 1.) This formula follows by observing that the area of a sector is $\theta/(2\pi)$ times that of the area of a circle; that is,

$$A = \frac{\theta}{2\pi} \cdot \pi r^2 = \frac{1}{2}r^2\theta$$

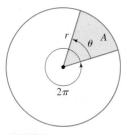

FIGURE 1
The area of a sector of a circle is $A = \frac{1}{2}r^2\theta$.

Now let R be a region bounded by the graph of the polar equation $r = f(\theta)$ and the rays $\theta = \alpha$ and $\theta = \beta$, where f is a nonnegative continuous function and $0 \le \beta - \alpha < 2\pi$, as shown in Figure 2a. Let P be a regular partition of the interval $[\alpha, \beta]$:

$$\alpha = \theta_0 < \theta_1 < \theta_2 < \cdots < \theta_n = \beta$$

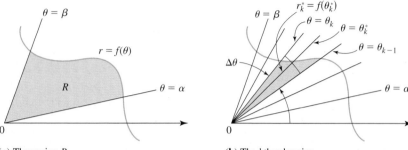

FIGURE 2   (**a**) The region R (**b**) The kth subregion

The rays $\theta = \theta_k$ divide R into n subregions $R_1, R_2, \ldots, R_n$ of area $\Delta A_1, \Delta A_2, \ldots, \Delta A_n$, respectively. If we choose θ_k^* in the interval $[\theta_{k-1}, \theta_k]$, then the area of ΔA_k of the kth subregion bounded by the rays $\theta = \theta_{k-1}$ and $\theta = \theta_k$ is approximated by the sector of a circle with central angle

$$\Delta\theta = \frac{\beta - \alpha}{n}$$

and radius $f(\theta_k^*)$ (highlighted in Figure 2b). Using Equation (1), we have

$$\Delta A_k \approx \frac{1}{2} [f(\theta_k^*)]^2 \Delta\theta$$

Therefore, an approximation of the area A of R is

$$A = \sum_{k=1}^{n} \Delta A_k \approx \sum_{k=1}^{n} \frac{1}{2} [f(\theta_k^*)]^2 \Delta\theta \tag{2}$$

But the sum in Equation (2) is a Riemann sum of the continuous function $\frac{1}{2}f^2$ over the interval $[\alpha, \beta]$. Therefore, it is true, although we will not prove it here, that

$$A = \lim_{n\to\infty} \sum_{k=1}^{n} \frac{1}{2} [f(\theta_k^*)]^2 \Delta\theta = \int_{\alpha}^{\beta} \frac{1}{2} [f(\theta)]^2 \, d\theta$$

THEOREM 1   Area Bounded by a Polar Curve

Let f be a continuous, nonnegative function on $[\alpha, \beta]$ where $0 \le \beta - \alpha < 2\pi$. Then the area A of the region bounded by the graphs of $r = f(\theta)$, $\theta = \alpha$, and $\theta = \beta$ is given by

$$A = \int_{\alpha}^{\beta} \frac{1}{2} [f(\theta)]^2 \, d\theta = \int_{\alpha}^{\beta} \frac{1}{2} r^2 \, d\theta$$

Note   When you determine the limits of integration, keep in mind that the region R is swept out in a counterclockwise direction by the ray emanating from the origin, starting at the angle α and terminating at the angle β. ▪

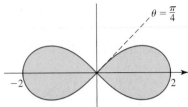

FIGURE 3
The region enclosed by the lemniscate $r^2 = 4 \cos 2\theta$

EXAMPLE 1 Find the area of the region enclosed by the lemniscate $r^2 = 4 \cos 2\theta$. This lemniscate has rectangular equation $x^4 + 2x^2y^2 - 4x^2 + 4y^2 + y^4 = 0$, as you can verify.

Solution The lemniscate is shown in Figure 3. Making use of symmetry, we see that the required area A is four times that of the area swept out by the ray emanating from the origin as θ increases from 0 to $\pi/4$. In other words,

$$A = 4 \int_0^{\pi/4} \frac{1}{2} r^2 \, d\theta = 8 \int_0^{\pi/4} \cos 2\theta \, d\theta$$

$$= \left[4 \sin 2\theta \right]_0^{\pi/4} = 4 \qquad \blacksquare$$

EXAMPLE 2 Find the area of the region enclosed by the cardioid $r = 1 + \cos \theta$.

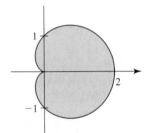

FIGURE 4
The region enclosed by the cardioid $r = 1 + \cos \theta$

Solution The graph of the cardioid $r = 1 + \cos \theta$, sketched previously in Example 6 in Section 10.4, is reproduced in Figure 4. Observe that the ray emanating from the origin sweeps out the required region exactly once as θ increases from 0 to 2π. Therefore, the required area A is

$$A = \int_0^{2\pi} \frac{1}{2} r^2 \, d\theta = \int_0^{2\pi} \frac{1}{2} (1 + \cos \theta)^2 \, d\theta$$

$$= \frac{1}{2} \int_0^{2\pi} (1 + 2 \cos \theta + \cos^2 \theta) \, d\theta$$

$$= \frac{1}{2} \int_0^{2\pi} \left(1 + 2 \cos \theta + \frac{1 + \cos 2\theta}{2} \right) d\theta$$

$$= \frac{1}{2} \int_0^{2\pi} \left(\frac{3}{2} + 2 \cos \theta + \frac{1}{2} \cos 2\theta \right) d\theta$$

$$= \frac{1}{2} \left[\frac{3}{2} \theta + 2 \sin \theta + \frac{1}{4} \sin 2\theta \right]_0^{2\pi} = \frac{3}{2} \pi \qquad \blacksquare$$

EXAMPLE 3 Find the area inside the smaller loop of the limaçon $r = 1 + 2 \cos \theta$.

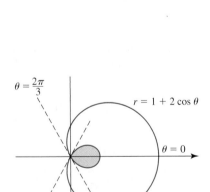

FIGURE 5
The limaçon $r = 1 + 2 \cos \theta$

Solution We first sketch the limaçon $r = 1 + 2 \cos \theta$ (Figure 5). Observe that the region of interest is swept out by the ray emanating from the origin as θ runs from $2\pi/3$ to $4\pi/3$. We can also take advantage of symmetry by observing that the required area is double the area of the smaller loop lying below the polar axis. Since this region is swept out by the ray emanating from the origin as θ runs from $2\pi/3$ to π, we see that the required area is

$$A = 2 \int_{2\pi/3}^{\pi} \frac{1}{2} r^2 \, d\theta = \int_{2\pi/3}^{\pi} r^2 \, d\theta$$

$$= \int_{2\pi/3}^{\pi} (1 + 2 \cos \theta)^2 \, d\theta$$

$$= \int_{2\pi/3}^{\pi} (1 + 4 \cos \theta + 4 \cos^2 \theta) \, d\theta$$

$$= \int_{2\pi/3}^{\pi} \left[1 + 4 \cos \theta + 4 \left(\frac{1 + \cos 2\theta}{2} \right) \right] d\theta$$

$$= \int_{2\pi/3}^{\pi} (3 + 4 \cos \theta + 2 \cos 2\theta) \, d\theta$$

$$= \left[3\theta + 4 \sin \theta + \sin 2\theta \right]_{2\pi/3}^{\pi}$$

$$= 3\pi - \left(2\pi + 4 \cdot \frac{\sqrt{3}}{2} - \frac{\sqrt{3}}{2} \right) = \pi - \frac{3\sqrt{3}}{2} \qquad \blacksquare$$

Area Bounded by Two Graphs

Consider the region R bounded by the graphs of the polar equations $r = f(\theta)$ and $r = g(\theta)$, and the rays $\theta = \alpha$ and $\theta = \beta$, where $f(\theta) \geq g(\theta) \geq 0$ and $0 \leq \beta - \alpha < 2\pi$. (See Figure 6.) From the figure we can see that the area A of R is found by subtracting the area of the region inside $r = g(\theta)$ from the area of the region inside $r = f(\theta)$. Using Theorem 1, we obtain the following theorem.

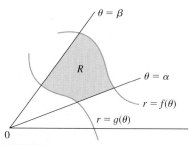

FIGURE 6

R is the region bounded by the graphs of $r = f(\theta)$ and $r = g(\theta)$ for $\alpha \leq \theta \leq \beta$.

> **THEOREM 2   Area Bounded by Two Polar Curves**
>
> Let f and g be continuous on $[\alpha, \beta]$, where $0 \leq g(\theta) \leq f(\theta)$ and $0 \leq \beta - \alpha < 2\pi$. Then the area A of the region bounded by the graphs of $r = g(\theta)$, $r = f(\theta)$, $\theta = \alpha$, and $\theta = \beta$ is given by
>
> $$A = \frac{1}{2} \int_{\alpha}^{\beta} \{ [f(\theta)]^2 - [g(\theta)]^2 \} \, d\theta$$

EXAMPLE 4 Find the area of the region that lies outside the circle $r = 3$ and inside the cardioid $r = 2 + 2 \cos \theta$.

Solution We first sketch the circle $r = 3$ and the cardioid $r = 2 + 2 \cos \theta$. The required region is shown shaded in Figure 7.

To find the points of intersection of the two curves, we solve the two equations simultaneously. We have $2 + 2 \cos \theta = 3$ or $\cos \theta = \frac{1}{2}$, which gives $\theta = \pm \pi/3$. Since the region of interest is swept out by the ray emanating from the origin as θ varies from $-\pi/3$ to $\pi/3$, we see that the required area is, by Theorem 2,

$$A = \frac{1}{2} \int_{\alpha}^{\beta} \{ [f(\theta)]^2 - [g(\theta)]^2 \} \, d\theta$$

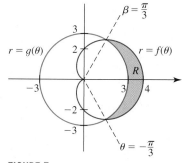

FIGURE 7

R is the region outside the circle $r = 3$ and inside the cardioid $r = 2 + 2 \cos \theta$.

where $f(\theta) = 2 + 2 \cos \theta = 2(1 + \cos \theta)$, $g(\theta) = 3$, $\alpha = -\pi/3$, and $\beta = \pi/3$. If we take advantage of symmetry, we can write

$$A = 2 \left(\frac{1}{2} \right) \int_{0}^{\pi/3} \{ [2(1 + \cos \theta)]^2 - 3^2 \} \, d\theta$$

$$= \int_{0}^{\pi/3} (4 + 8 \cos \theta + 4 \cos^2 \theta - 9) \, d\theta$$

$$= \int_{0}^{\pi/3} \left(-5 + 8 \cos \theta + 4 \cdot \frac{1 + \cos 2\theta}{2} \right) d\theta$$

$$= \int_{0}^{\pi/3} (-3 + 8 \cos \theta + 2 \cos 2\theta) \, d\theta$$

$$= \left[-3\theta + 8 \sin \theta + \sin 2\theta \right]_{0}^{\pi/3}$$

$$= \left(-\pi + 8 \left(\frac{\sqrt{3}}{2} \right) + \frac{\sqrt{3}}{2} \right) = \frac{9\sqrt{3}}{2} - \pi \qquad \blacksquare$$

■ Arc Length in Polar Coordinates

To find the length of a curve C defined by a polar equation $r = f(\theta)$ for $\alpha \le \theta \le \beta$, we use Equation (1) in Section 10.4 to write the parametric equations

$$x = r \cos \theta = f(\theta) \cos \theta \quad\text{and}\quad y = r \sin \theta = f(\theta) \sin \theta \qquad \alpha \le \theta \le \beta$$

for the curve, regarding θ as the parameter. Then

$$\frac{dx}{d\theta} = f'(\theta) \cos \theta - f(\theta) \sin \theta \quad\text{and}\quad \frac{dy}{d\theta} = f'(\theta) \sin \theta + f(\theta) \cos \theta$$

Therefore,

$$
\begin{aligned}
\left(\frac{dx}{d\theta}\right)^2 + \left(\frac{dy}{d\theta}\right)^2 &= [f'(\theta)]^2 \cos^2 \theta - 2f'(\theta)f(\theta) \cos \theta \sin \theta + [f(\theta)]^2 \sin^2 \theta \\
&\quad + [f'(\theta)]^2 \sin^2 \theta + 2f'(\theta)f(\theta) \cos \theta \sin \theta + [f(\theta)]^2 \cos^2 \theta \\
&= [f'(\theta)]^2 + [f(\theta)]^2 \qquad \scriptstyle \sin^2 \theta + \cos^2 \theta = 1
\end{aligned}
$$

Consequently, if f' is continuous, then Theorem 1 in Section 10.3 gives the arc length of C as

$$L = \int_\alpha^\beta \sqrt{\left(\frac{dx}{d\theta}\right)^2 + \left(\frac{dy}{d\theta}\right)^2}\, d\theta = \int_\alpha^\beta \sqrt{[f'(\theta)]^2 + [f(\theta)]^2}\, d\theta$$

THEOREM 3   Arc Length

Let f be a function with a continuous derivative on an interval $[\alpha, \beta]$. If the graph C of $r = f(\theta)$ is traced exactly once as θ increases from α to β, then the length L of C is given by

$$L = \int_\alpha^\beta \sqrt{[f'(\theta)]^2 + [f(\theta)]^2}\, d\theta = \int_\alpha^\beta \sqrt{\left(\frac{dr}{d\theta}\right)^2 + r^2}\, d\theta$$

EXAMPLE 5   Find the length of the cardioid $r = 1 + \cos \theta$.

Solution   The cardioid is shown in Figure 8. Observe that the cardioid is traced exactly once as θ runs from θ to 2π. However, we can also take advantage of symmetry to see that the required length is twice that of the length of the cardioid lying above the polar axis. Thus,

$$L = 2 \int_0^\pi \sqrt{\left(\frac{dr}{d\theta}\right)^2 + r^2}\, d\theta$$

But $r = 1 + \cos \theta$, so

$$\frac{dr}{d\theta} = -\sin \theta$$

Therefore,

$$
\begin{aligned}
L &= 2 \int_0^\pi \sqrt{(-\sin \theta)^2 + (1 + \cos \theta)^2}\, d\theta \\
&= 2 \int_0^\pi \sqrt{\sin^2 \theta + 1 + 2\cos \theta + \cos^2 \theta}\, d\theta
\end{aligned}
$$

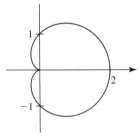

FIGURE 8
The cardioid $r = 1 + \cos \theta$

$$= 2 \int_0^\pi \sqrt{2 + 2 \cos \theta} \, d\theta \qquad \sin^2 \theta + \cos^2 \theta = 1$$

$$= 2\sqrt{2} \int_0^\pi \sqrt{1 + \cos \theta} \, d\theta = 2\sqrt{2} \int_0^\pi \sqrt{2 \cos^2 \frac{\theta}{2}} \, d\theta$$

$$= 4 \int_0^\pi \left| \cos \frac{\theta}{2} \right| d\theta = 4 \int_0^\pi \cos \frac{\theta}{2} \, d\theta \qquad \cos \frac{\theta}{2} \geq 0 \text{ on } [0, \pi]$$

$$= \left[4(2) \sin \frac{\theta}{2} \right]_0^\pi = 8 \qquad \blacksquare$$

Area of a Surface of Revolution

The formulas for finding the area of a surface obtained by revolving a curve defined by a polar equation about the polar axis or about the line $\theta = \pi/2$ can be derived by using Equations (8) and (9) of Section 10.3 and the equations $x = r \cos \theta$ and $y = r \sin \theta$.

THEOREM 4   Area of a Surface of a Revolution

Let f be a function with a continuous derivative on an interval $[\alpha, \beta]$. If the graph C of $r = f(\theta)$ is traced exactly once as θ increases from α to β, then the area of the surface obtained by revolving C about the indicated line is given by

a. $S = 2\pi \displaystyle\int_\alpha^\beta r \sin \theta \sqrt{\left(\frac{dr}{d\theta} \right)^2 + r^2} \, d\theta$ (about the polar axis)

b. $S = 2\pi \displaystyle\int_\alpha^\beta r \cos \theta \sqrt{\left(\frac{dr}{d\theta} \right)^2 + r^2} \, d\theta$ (about the line $\theta = \pi/2$)

Note   In using Theorem 4, we must choose $[\alpha, \beta]$ so that the surface is only traced once when C is revolved about the line. $\blacksquare$

EXAMPLE 6   Find the area S of the surface obtained by revolving the circle $r = \cos \theta$ about the line $\theta = \pi/2$. (See Figure 9.)

Solution   Observe that the circle is traced exactly once as θ increases from 0 to π. Therefore, using Theorem 4 with $r = \cos \theta$, $\alpha = 0$, and $\beta = \pi$, we obtain

$$S = 2\pi \int_\alpha^\beta f(\theta) \cos \theta \sqrt{\left(\frac{dr}{d\theta} \right)^2 + r^2} \, d\theta$$

$$= 2\pi \int_0^\pi \cos \theta (\cos \theta) \sqrt{(-\sin \theta)^2 + (\cos \theta)^2} \, d\theta$$

$$= 2\pi \int_0^\pi \cos^2 \theta \, d\theta = \pi \int_0^\pi (1 + \cos 2\theta) \, d\theta$$

$$= \pi \left[\theta + \frac{\sin 2\theta}{2} \right]_0^\pi = \pi^2 \qquad \blacksquare$$

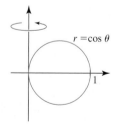

(a)

(b)

FIGURE 9
The solid obtained by revolving the circle $r = \cos \theta$ (a) about the line $\theta = \pi/2$ is a torus (b).

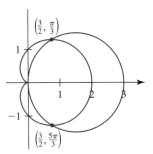

FIGURE 10
The graphs of the cardioid
$r = 1 + \cos\theta$ and the circle
$r = 3\cos\theta$

■ Points of Intersection of Graphs in Polar Coordinates

In Example 4 we were able to find the points of intersection of two curves with representations in polar coordinates by solving a system of two equations simultaneously. This is not always the case. Consider for example, the graphs of the cardioid $r = 1 + \cos\theta$ and the circle $r = 3\cos\theta$ shown in Figure 10. Solving the two equations simultaneously, we obtain

$$3\cos\theta = 1 + \cos\theta$$

$$\cos\theta = \frac{1}{2} \tag{3}$$

or $\theta = \pi/3$ and $5\pi/3$. Therefore, the points of intersection are $\left(\frac{3}{2}, \frac{\pi}{3}\right)$ and $\left(\frac{3}{2}, \frac{5\pi}{3}\right)$. But one glance at Figure 10 shows the pole as a third point of intersection that is not revealed in our calculation. To see how this can happen, think of the cardioid as being traced by the point (r, θ) satisfying

$$r = f(\theta) = 1 + \cos\theta \qquad 0 \le \theta \le 2\pi$$

with θ as a parameter. If we think of θ as representing time, then as θ runs from $\theta = 0$ through $\theta = 2\pi$, the point (r, θ) starts at $(2, 0)$ and traverses the cardioid in a counterclockwise direction, eventually returning to the point $(2, 0)$. (See Figure 11a.) Similarly, the circle is traced *twice* in the counterclockwise direction, by the point (r, θ), where

$$r = g(\theta) = 3\cos\theta \qquad 0 \le \theta \le 2\pi$$

and the parameter θ, once again representing time, runs from $\theta = 0$ through $\theta = 2\pi$ (see Figure 11b).

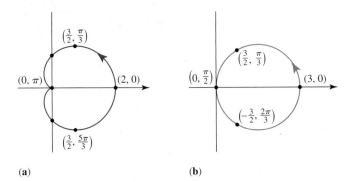

FIGURE 11   **(a)**   **(b)**

Observe that the point tracing the cardioid arrives at the point $\left(\frac{3}{2}, \frac{\pi}{3}\right)$ on the cardioid at precisely the same time that the point tracing the circle arrives at the point $\left(\frac{3}{2}, \frac{\pi}{3}\right)$ on the circle. A similar observation holds at the point $\left(\frac{3}{2}, \frac{5\pi}{3}\right)$ on each of the two curves. These are the points of intersection found earlier.

Next, observe that the point tracing the cardioid arrives at the origin when $\theta = \pi$. But the point tracing the circle first arrives at the origin when $\theta = \pi/2$ and then again when $\theta = 3\pi/2$. In other words, these two points arrive at the origin at *different* times, so there is no (common) value of θ corresponding to the origin that satisfies both Equations (3) simultaneously. Thus, although the origin is a point of intersection of the two curves, this fact will not show up in the solution of the system of equations. For this reason it is recommended that we sketch the graphs of polar equations when finding their points of intersection.

EXAMPLE 7 Find the points of intersection of $r = \cos\theta$ and $r = \cos 2\theta$.

Solution We solve the system of equations

$$r = \cos\theta$$

$$r = \cos 2\theta$$

We set $\cos\theta = \cos 2\theta$ and use the identity $\cos 2\theta = 2\cos^2\theta - 1$. We obtain

$$2\cos^2\theta - \cos\theta - 1 = 0$$

$$(2\cos\theta + 1)(\cos\theta - 1) = 0$$

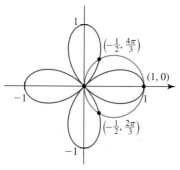

FIGURE 12

So

$$\cos\theta = -\frac{1}{2} \quad\text{or}\quad \cos\theta = 1$$

that is,

$$\theta = \frac{2\pi}{3},\quad \frac{4\pi}{3},\quad \text{or}\quad 0$$

These values of θ give $\left(-\frac{1}{2}, \frac{2\pi}{3}\right)$, $\left(-\frac{1}{2}, \frac{4\pi}{3}\right)$, and $(1, 0)$ as the points of intersection. Since both graphs also pass through the pole, we conclude that the pole is also a point of intersection. (See Figure 12.) ■

10.5 CONCEPT QUESTIONS

1. **a.** Let f be nonnegative and continuous on $[\alpha, \beta]$, where $0 \le \beta - \alpha < 2\pi$. Write an integral giving the area of the region bounded by the graphs of $r = f(\theta)$, $\theta = \alpha$, and $\theta = \beta$. Make a sketch of the region.
   b. If f and g are continuous on $[\alpha, \beta]$ and $0 \le g(\theta) \le f(\theta)$, where $0 \le \alpha \le \beta \le 2\pi$, write an integral giving the area of the region bounded by the graphs of $r = g(\theta)$, $r = f(\theta)$, $\theta = \alpha$, and $\theta = \beta$. Make a sketch of the region.

2. Suppose that f has a continuous derivative on an interval $[\alpha, \beta]$. If the graph C of $r = f(\theta)$ is traced exactly once as θ increases from α to β, write an integral giving the length of C.

3. Suppose that f is a function with a continuous derivative on $[\alpha, \beta]$ and the graph C of $r = f(\theta)$ is traced exactly once as θ increases from α to β. Write an integral giving the area of the surface obtained by revolving C about (a) the polar axis, $y \ge 0$, and (b) the line $\theta = \pi/2$, $x \ge 0$.

10.5 EXERCISES

1. **a.** Find a rectangular equation of the circle $r = 4\cos\theta$, and use it to find its area.
   b. Find the area of the circle of part (a) by integration.

2. **a.** By finding a rectangular equation, show that the polar equation $r = 2\cos\theta - 2\sin\theta$ represents a circle. Then find the area of the circle.
   b. Find the area of the circle of part (a) by integration.

In Exercises 3–8, find the area of the region bounded by the curve and the rays.

3. $r = \theta$, $\theta = 0$, $\theta = \pi$

4. $r = \dfrac{1}{\theta}$, $\theta = \dfrac{\pi}{6}$, $\theta = \dfrac{\pi}{3}$

5. $r = e^{\theta}$, $\theta = -\dfrac{\pi}{2}$, $\theta = 0$

6. $r = e^{-2\theta}$, $\theta = 0$, $\theta = \dfrac{\pi}{4}$

7. $r = \sqrt{\cos\theta}$, $\theta = 0$, $\theta = \dfrac{\pi}{2}$

8. $r = \cos 2\theta$, $\theta = 0$, $\theta = \dfrac{\pi}{16}$

In Exercises 9–12, find the area of the shaded region.

9.

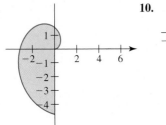

10.

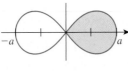

$r = a\sqrt{\cos 2\theta}$

$r = \theta$

11.

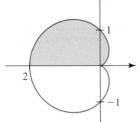

$r = 1 - \cos \theta$

12.

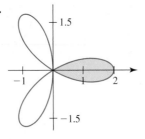

$r = 2 \cos 3\theta$

In Exercises 13–18, sketch the curve, and find the area of the region enclosed by it.

13. $r = 3 \sin \theta$

14. $r = 2(1 - \cos \theta)$

15. $r^2 = \sin \theta$

16. $r^2 = 3 \sin 2\theta$

17. $r = 2 \sin 2\theta$

18. $r = 2 \sin 3\theta$

In Exercises 19–22, find the area of the region enclosed by one loop of the curve.

19. $r = \cos 2\theta$

20. $r = 2 \cos 3\theta$

21. $r = \sin 4\theta$

22. $r = 2 \cos 4\theta$

In Exercises 23–24, find the area of the region described.

23. The inner loop of the limaçon $r = 1 + 2 \cos \theta$

24. Between the loops of the limaçon $r = 1 + 2 \sin \theta$

In Exercises 25–28, find the area of the shaded region.

25.

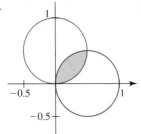

$r = \sin \theta, r = \cos \theta$

26.

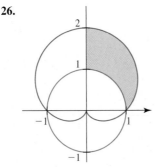

$r = 1, r = 1 + \sin \theta$

27.

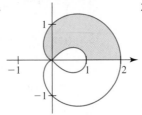

$r = 1 + \cos \theta, r = \sqrt{\cos 2\theta}$

28.

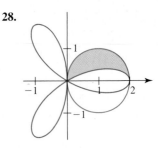

$r = 2 \cos 3\theta, r = 2 \cos \theta$

In Exercises 29–34, find all points of intersection of the given curves.

29. $r = 1$ and $r = 1 + \cos \theta$

30. $r = 3$ and $r = 2 + 2 \cos \theta$

31. $r = 2$ and $r = 4 \cos 2\theta$

32. $r = 1$ and $r^2 = 2 \cos 2\theta$

33. $r = \sin \theta$ and $r = \sin 2\theta$

34. $r = \cos \theta$ and $r = \cos 2\theta$

In Exercises 35–40, find the area of the region that lies outside the first curve and inside the second curve.

35. $r = 1 + \cos \theta$, $r = 3 \cos \theta$

36. $r = 1 - \sin \theta$, $r = 1$

37. $r = 4 \cos \theta$, $r = 2$

38. $r = 3 \sin \theta$, $r = 2 - \sin \theta$

39. $r = 1 - \cos \theta$, $r = \dfrac{3}{2}$

40. $r = 2 \cos 3\theta$, $r = 1$

In Exercises 41–46, find the area of the region that is enclosed by both of the curves.

41. $r = 1$, $r = 2 \sin \theta$

42. $r = \cos \theta$, $r = \sqrt{3} \sin \theta$

43. $r = \sin \theta$, $r = 1 - \sin \theta$

44. $r = \cos \theta$, $r = 1 - \cos \theta$

45. $r^2 = 4 \cos 2\theta$, $r = \sqrt{2}$

46. $r = \sqrt{3} \sin \theta$, $r = 1 + \cos \theta$

In Exercises 47–54, find the length of the given curve.

47. $r = 5 \sin \theta$

48. $r = 2\theta$; $0 \le \theta \le 2\pi$

49. $r = e^{-\theta}$; $0 \le \theta \le 4\pi$

50. $r = 1 + \sin \theta$; $0 \le \theta \le 2\pi$

51. $r = \sin^3 \dfrac{\theta}{3}$; $0 \le \theta \le \pi$

52. $r = \cos^2 \dfrac{\theta}{2}$

53. $r = a \sin^4 \dfrac{\theta}{4}$

54. $r = \sec \theta; \quad 0 \le \theta \le \frac{\pi}{3}$

In Exercises 55–60, find the area of the surface obtained by revolving the given curve about the given line.

55. $r = 4 \cos \theta$ about the polar axis

56. $r = 2 \cos \theta$ about the line $\theta = \dfrac{\pi}{2}$

57. $r = 2 + 2 \cos \theta$ about the polar axis

58. $r^2 = \cos 2\theta$ about the polar axis

59. $r^2 = \cos 2\theta$ about the line $\theta = \dfrac{\pi}{2}$

60. $r = e^{a\theta}, \quad 0 \le \theta \le \frac{\pi}{2}$ about the line $\theta = \dfrac{\pi}{2}$

In Exercises 61 and 62, find the area of the region enclosed by the given curve. (Hint: Convert the rectangular equation to a polar equation.)

61. $(x^2 + y^2)^3 = 16x^2y^2$

62. $x^4 + y^4 = 4(x^2 + y^2)$

63. Let P be a point other than the origin lying on the curve $r = f(\theta)$. If ψ is the angle between the tangent line to the curve at P and the radial line OP, then $\tan \psi = \dfrac{r}{dr/d\theta}$. (See Section 10.4, Exercise 84.)

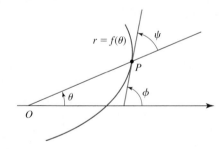

a. Show that the angle between the tangent line to the logarithmic spiral $r = e^{m\theta}$ and the radial line at the point of tangency is a constant.

b. Suppose the curve with polar equation $r = f(\theta)$ has the property that at any point on the curve, the angle ψ between the tangent line to the curve at that point and the radial line from the origin to that point is a constant. Show that $f(\theta) = Ce^{m\theta}$, where C and m are constants.

64. Find the length of the logarithmic spiral $r = ae^{m\theta}$ between the point (r_0, θ_0) and the point (r, θ), and use this result to deduce that the length of a logarithmic spiral is proportional to the difference between the radial coordinates of the points.

65. Show that the length of the parabola $y = (1/2p)x^2$ on the interval $[0, a]$ is the same as the length of the spiral $r = p\theta$ for $0 \le r \le a$.

cas 66. Plot the curve $r = \sin(3 \cos \theta)$, and find an approximation of the area enclosed by the curve accurate to four decimal places.

cas *In Exercises 67–69, (a) plot the curve, and (b) find an approximation of its length accurate to two decimal places.*

67. $r = \sqrt{1 + \theta^2}$, where $0 \le \theta \le 2\pi$ (involute of a circle)

68. $r = 0.2\sqrt{\theta} + 1$, where $0 \le \theta \le 6\pi$ (parabolic spiral)

69. $r = 3 \sin \theta \cos^2 \theta$, where $0 \le \theta \le \pi$ (bifolia)

70. a. Let f be a function with a continuous derivative in an interval $[\alpha, \beta]$. If the graph C of $r = f(\theta)$ is traced exactly once as θ increases from α to β, show that the rectangular coordinates of the centroid of C are

$$\bar{x} = \frac{\displaystyle\int_\alpha^\beta r \cos \theta \sqrt{(r')^2 + r^2}\, d\theta}{\displaystyle\int_\alpha^\beta \sqrt{(r')^2 + r^2}\, d\theta}$$

and

$$\bar{y} = \frac{\displaystyle\int_\alpha^\beta r \sin \theta \sqrt{(r')^2 + r^2}\, d\theta}{\displaystyle\int_\alpha^\beta \sqrt{(r')^2 + r^2}\, d\theta}$$

Hint: See the directions for Exercises 45 and 46 in Exercises 5.7.

b. Use the result of part (a) to find the centroid of the upper semicircle $r = a$, where $a > 0$ and $0 \le \theta \le \pi$.

cas 71. a. Plot the curve with polar equation $r = 2 \cos^3 \theta$ where $-\frac{\pi}{2} \le \theta \le \frac{\pi}{2}$.

b. Find the Cartesian coordinates of the centroid of the region bounded by the curve of part (a).

cas 72. a. Plot the graphs of $r = 1 + \cos \theta$ and $r = 3 \cos \theta$ for $0 \le \theta \le 2\pi$, treating r and θ as rectangular coordinates.

b. Refer to page 884. Reconcile your results with the discussion of finding the points of intersection of graphs in polar coordinates.

cas *In Exercises 73 and 74, (a) find the polar representation of the curve given in rectangular coordinates, (b) plot the curve, and (c) find the area of the region enclosed by a loop (or loops) of the curve.*

73. $x^3 - 3xy + y^3 = 0$ (folium of Descartes)

74. $(x^2 + y^2)^{1/2} - \cos\left[4 \tan^{-1}\left(\dfrac{y}{x}\right)\right] = 0$ (rhodenea)

In Exercises 75 and 76, determine whether the statement is true or false. If it is true, explain why it is true. If it is false, give an example to show why it is false.

75. If there exists a θ_0 such that $f(\theta_0) = g(\theta_0)$, then the graphs of $r = f(\theta)$ and $r = g(\theta)$ have at least one point of intersection.

76. If P is a point of intersection of the graphs of $r = f(\theta)$ and $r = g(\theta)$, then there must exist a θ_0 such that $f(\theta_0) = g(\theta_0)$.

10.6   Conic Sections in Polar Coordinates

In Section 10.1 we obtained representations of the conic sections—the parabola, the ellipse, and the hyperbola—in terms of rectangular equations. In this section we will show that all three types of conic sections can be represented by a single polar equation. As you saw in the preceding sections, some problems can be solved more easily using polar coordinates rather than rectangular coordinates.

We begin by proving the following theorem, which gives an equivalent definition of each conic section in terms of its focus and directrix. As a corollary, we will obtain the desired representation of the conic sections in polar form.

THEOREM 1

Let F be a fixed point, let l be a fixed line in the plane, and let e be a fixed positive number. Then the set of all points P in the plane satisfying

$$\frac{d(P, F)}{d(P, l)} = e$$

is a conic section. The point F is the **focus** of the conic section, and the line l is its **directrix**. The number e, which is the ratio of the distance between P and F and the distance between P and l, is called the **eccentricity** of the conic. The conic is an ellipse if $e < 1$, a parabola if $e = 1$, or a hyperbola if $e > 1$.

The three types of conics are illustrated in Figure 1.

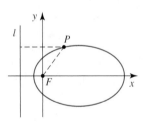

(a) $\dfrac{d(P, F)}{d(P, l)} = e < 1$ (ellipse)

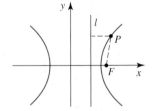

(b) $\dfrac{d(P, F)}{d(P, l)} = e = 1$ (parabola)

(c) $\dfrac{d(P, F)}{d(P, l)} = e > 1$ (hyperbola)

FIGURE 1

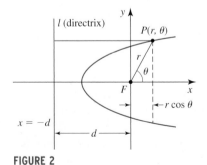

FIGURE 2

PROOF   Observe that if $e = 1$, then $d(P, F) = d(P, l)$. That is, the distance between a point on the curve and the focus is equal to the distance between the point and the directrix. But this is just the definition of a parabola, so the curve is a conic section.

In what follows, we will assume that $e \neq 1$. Refer to Figure 2, where we have placed the focus F at the origin and the directrix l parallel to and d units to the left of the y-axis. Therefore, the directrix has equation $x = -d$, where $d > 0$. If $P(r, \theta)$ is any point lying on the curve, then you can see from Figure 2 that

$$d(P, F) = r \qquad \text{and} \qquad d(P, l) = d + r \cos \theta$$

Therefore, the condition $d(P, F)/d(P, l) = e$ or, equivalently, $d(P, F) = e \cdot d(P, l)$, implies that

$$r = e(d + r \cos \theta) \tag{1}$$

Converting this equation to rectangular coordinates gives

$$\sqrt{x^2 + y^2} = e(d + x)$$

which, upon squaring, yields

$$x^2 + y^2 = e^2(d + x)^2 = e^2(d^2 + 2dx + x^2)$$

$$(1 - e^2)x^2 - 2de^2x + y^2 = e^2d^2$$

or

$$x^2 - \left(\frac{2e^2d}{1 - e^2}\right)x + \frac{y^2}{1 - e^2} = \frac{e^2d^2}{1 - e^2}$$

Completing the square in x, we obtain

$$\left(x - \frac{e^2d}{1 - e^2}\right)^2 + \frac{y^2}{1 - e^2} = \frac{e^2d^2}{1 - e^2} + \frac{e^4d^2}{(1 - e^2)^2} = \frac{e^2d^2}{(1 - e^2)^2} \tag{2}$$

Now, if $e < 1$, then $1 - e^2 > 0$. Dividing both sides by $e^2d^2/(1 - e^2)^2$, we can write Equation (2) in the form

$$\frac{(x - h)^2}{a^2} + \frac{y^2}{b^2} = 1$$

where

$$h = \frac{e^2d}{1 - e^2}, \qquad a^2 = \frac{e^2d^2}{(1 - e^2)^2}, \qquad \text{and} \qquad b^2 = \frac{e^2d^2}{1 - e^2} \tag{3}$$

This is an equation of an ellipse centered at the point $(h, 0)$ on the x-axis.

Next, we compute

$$c^2 = a^2 - b^2 = \frac{e^4d^2}{(1 - e^2)^2} \tag{4}$$

from which we obtain

$$c = \frac{e^2d}{1 - e^2} = h$$

Recalling that the foci of an ellipse are located at a distance c from its center, we have shown that F is indeed the focus of the ellipse. It also follows from Equations (3) and (4) that the eccentricity of the ellipse is given by

$$e = \frac{c}{a} \tag{5}$$

where $c^2 = a^2 - b^2$.

If $e > 1$, then $1 - e^2 < 0$. Proceeding in a similar manner as before, we can write Equation (2) in the form

$$\frac{(x - h)^2}{a^2} - \frac{y^2}{b^2} = 1$$

Historical Biography

SPL/Photo Researchers, Inc.

JOHANNES KEPLER
(1571-1630)

Born in 1571 in Weil der Stadt, Germany, Johannes Kepler was introduced to the wonders of the universe at a young age when his mother took him to observe the comet of 1577. Kepler completed a master's degree in theology at the University of Tübingen and established a reputation as a talented mathematician and astronomer, but an unorthodox Lutheran. Given the volatile religious situation of the time, Kepler was advised not to pursue a career in the ministry and was instead recommended for a position teaching mathematics and astronomy at a school in Graz. He took that position in 1594, and a year later he published his first major work, *Mysterium cosmographicum* ("The Mystery of the Cosmos"). In this work, he explained his discovery of the three laws of planetary motion that now bear his name (see page 893), and he became the first person to correctly explain planetary orbits within our solar system. Kepler's First Law is that every planet's orbit is an ellipse with the sun at one focus.

While Kepler was writing his famous work *The Harmony of the World* (1619), his mother was charged with witchcraft. Kepler enlisted the help of the legal faculty at the University of Tübingen in his effort to prevent his mother from being convicted. Katharina Kepler was eventually released as the result of technical objections on the part of the defense, arising from the authorities' failure to follow the legal procedures of the time regarding the use of torture.

which is an equation of a hyperbola. We also see that the eccentricity of the hyperbola is

$$e = \frac{c}{a} \tag{6}$$

where $c^2 = a^2 + b^2$.

If we solve Equation (1) for r, we obtain the polar equation

$$r = \frac{ed}{1 - e\cos\theta}$$

of the conic shown in Figure 2. If the directrix is chosen so that it lies to the right of the focus, say, $x = d$, where $d > 0$, then the polar equation of the conic is

$$r = \frac{ed}{1 + e\cos\theta}$$

Similarly, we can show that if the directrix $y = \pm d$ is chosen to be parallel to the polar axis, then the polar equation of the conic is

$$r = \frac{ed}{1 \pm e\sin\theta}$$

(See Exercises 28–30.) ■

The conics are illustrated in Figure 3.

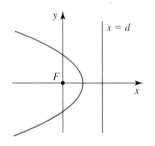

(a) $r = \dfrac{ed}{1 + e\cos\theta}$

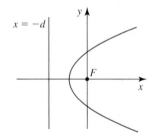

(b) $r = \dfrac{ed}{1 - e\cos\theta}$

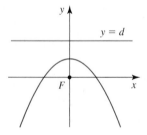

(c) $r = \dfrac{ed}{1 + e\sin\theta}$

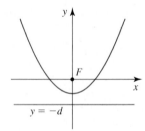

(d) $r = \dfrac{ed}{1 - e\sin\theta}$

FIGURE 3
Polar equations of conics

THEOREM 2

A polar equation of the form

$$r = \frac{ed}{1 \pm e\cos\theta} \qquad \text{or} \qquad r = \frac{ed}{1 \pm e\sin\theta}$$

represents a conic section with eccentricity e. The conic is a parabola if $e = 1$, an ellipse if $e < 1$, and a hyperbola if $e > 1$.

EXAMPLE 1 Find a polar equation of a parabola that has its focus at the pole and the line $y = 2$ as its directrix.

Solution   Since this conic section is a parabola, we see that $e = 1$. Next, observe that its directrix, $y = 2$, is parallel to and lies above the polar axis. So letting $d = 2$ and referring to Figure 3c, we see that a required equation of the parabola is

$$r = \frac{2}{1 + \sin\theta}$$ ∎

EXAMPLE 2   A conic has polar equation

$$r = \frac{15}{3 + 2\cos\theta}$$

Find the eccentricity and the directrix of the conic section, and sketch the conic section.

Solution   We begin by rewriting the given equation in standard form by dividing both its numerator and denominator by 3, obtaining

$$r = \frac{5}{1 + \frac{2}{3}\cos\theta}$$

Then using Theorem 2, we see that $e = \frac{2}{3}$. Since $ed = 5$, we have

$$d = \frac{5}{e} = \frac{5}{\frac{2}{3}} = \frac{15}{2}$$

Since $e < 1$, we conclude that the conic section is an ellipse with focus at the pole and major axis lying along the polar axis. Its directrix has rectangular equation $x = \frac{15}{2}$. Setting $\theta = 0$ and $\theta = \pi$ successively gives $r = 3$ and $r = 15$, giving the vertices of the ellipse in polar coordinates as $(3, 0)$ and $(15, \pi)$. The center of the ellipse is the midpoint $(6, \pi)$ in polar coordinates of the line segment joining the vertices. Since the length of the major axes of the ellipse is 18, we have $2a = 18$, or $a = 9$. Finally, since $e = c/a$, we find that

$$c = ae = 9\left(\frac{2}{3}\right) = 6$$

So

$$b^2 = a^2 - c^2 = 81 - 36 = 45$$

or

$$b = 3\sqrt{5}$$

The graph of the conic is sketched in Figure 4. ∎

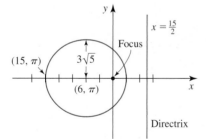

FIGURE 4
The graph of $r = \dfrac{15}{3 + 2\cos\theta}$

EXAMPLE 3   Sketch the graph of the polar equation

$$r = \frac{20}{2 + 3\sin\theta}$$

Solution   By dividing the numerator and the denominator of the given equation by 2, we obtain the equation

$$r = \frac{10}{1 + \frac{3}{2}\sin\theta}$$

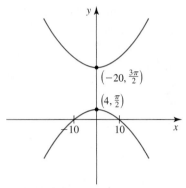

FIGURE 5
The graph of $r = \dfrac{20}{2 + 3 \sin \theta}$

in standard form. We see that $e = \frac{3}{2}$, so the equation represents a hyperbola with one focus at the pole. Comparing this equation with the equation associated with Figure 3c, we see that the transverse axis of the hyperbola lies along the line $\theta = \pi/2$. To find the vertices of the hyperbola, we set $\theta = \pi/2$ and $\theta = 3\pi/2$ successively, giving $\left(4, \frac{\pi}{2}\right)$ and $\left(-20, \frac{3\pi}{2}\right)$ as the required vertices in polar coordinates. The center of the hyperbola in polar coordinates is the midpoint $\left(12, \frac{\pi}{2}\right)$ of the line segment joining the vertices. The x-intercepts (we superimpose the Cartesian system over the polar system) are found by setting $\theta = 0$ and $\theta = \pi$, giving the x-intercepts as 10 and -10. The required graph may be sketched in two steps; first, we sketch the lower branch of the hyperbola, making use of the x-intercepts that we just found. Then, using symmetry, we sketch the upper branch of the hyperbola. (See Figure 5.) ∎

■ Eccentricity of a Conic

As we saw in Theorem 1, the nature of a conic section is determined by its eccentricity e. To see in greater detail the role that is played by the eccentricity of a conic, let's first consider the case in which $e < 1$, so that the conic under consideration is an ellipse. Now by Equation (5) we have

$$e = \frac{c}{a} = \frac{\sqrt{a^2 - b^2}}{a}$$

If e is close to 0, then $\sqrt{a^2 - b^2}$ is close to 0, or a is close to b. This means that the ellipse is almost circular (see Figure 6a). On the other hand, if e is close to 1, then $\sqrt{a^2 - b^2} \approx a$, $a^2 - b^2 \approx a^2$, or b is small. This means that the ellipse is very flat (see Figure 6b).

FIGURE 6
The ellipse is almost circular if e is close to 0 and is very flat if e is close to 1.

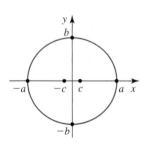

(**a**) e is close to 0.

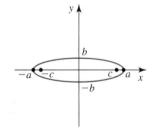

(**b**) e is close to 1.

If $e = 1$, then the conic is a parabola. We leave it to you to perform a similar analysis in the case in which $e > 1$ (so that the conic is a hyperbola).

In Figure 7 we show two hyperbolas: In part (a) the eccentricity e is close to but greater than 1. In part (b) the eccentricity e is much larger than 1.

FIGURE 7
The eccentricity of the hyperbola in part (a) is close to 1, whereas the eccentricity of the hyperbola in part (b) is much larger than 1.

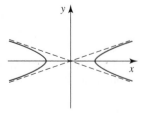

(**a**)

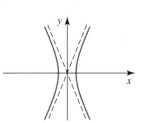

(**b**)

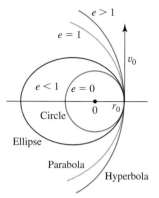

FIGURE 8
The speed v_0 determines
the orbit of the body.

■ Motion of Celestial Bodies

In the last few sections we have seen numerous applications of conics. Yet another important application of the conics arises in the motion of celestial bodies.

Figure 8 shows a body a distance r_0 from the origin 0 moving with a speed v_0 and in a direction perpendicular to the line passing through 0 and v_0. It can be shown, although we will not do so here, that the orbit of the body about the origin depends on the magnitude of v_0. For the planets in the solar system (with the sun at the origin) and for certain comets such as Halley's comet, the speed v_0 is such that they remain captive and will never leave the system; their orbits are ellipses. However, if the speed v_0 of a body is sufficiently large, then its orbit about the sun is a parabola ($e = 1$) or a branch of a hyperbola ($e > 1$). In both these cases the body makes but a single pass about the sun!

The orbits of the planets about the sun, moreover, are described by Kepler's Laws.

Kepler's Laws

1. Planets move in orbits that are ellipses with the sun at one focus.
2. The line from the sun to a planet sweeps out equal areas in equal times. (See Figure 9.)
3. The square of a planet's period is proportional to the cube of the length of the semimajor axis of its orbit.

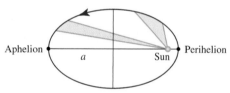

FIGURE 9
Equal areas are swept out in equal times, $T^2 \propto a^3$, where T is the period.

The positions of a planet that are closest to and farthest from the sun are called the *perihelion* and *aphelion,* respectively.

EXAMPLE 4 **The Orbit of Halley's Comet** Halley's comet has an elliptical orbit with an eccentricity of 0.967. Its perihelion distance (shortest distance from the sun) is 8.9×10^7 km.

a. Find a polar equation for the orbit.
b. Find the distance of the comet from the sun when it is at the aphelion.

Solution
a. Suppose that the axis is horizontal as shown in Figure 10. Then the polar equation can be chosen to have the form

$$r = \frac{ed}{1 + e \cos \theta}$$

The distance of the comet from the sun when it is at the perihelion is given by

$$a - c = a - ea = a(1 - e)$$

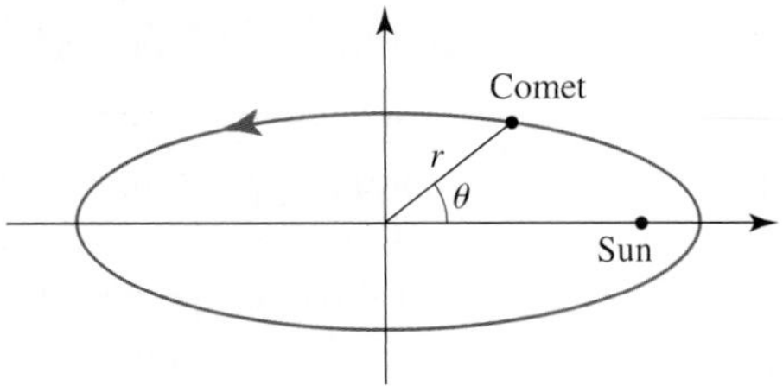

FIGURE 10
In actuality the trajectory is much flatter.

But we are given that at the perihelion the distance from the sun is 8.9×10^7 km and $e = 0.967$. So

$$a(1 - 0.967) = 8.9 \times 10^7$$

or

$$a = \frac{8.9 \times 10^7}{1 - 0.967} \approx 2.697 \times 10^9$$

Next, from Equation (3) we see that

$$ed = a(1 - e^2)$$
$$= (2.697 \times 10^9)(1 - 0.967^2) = 1.75 \times 10^8$$

So the required equation is

$$r = \frac{1.75 \times 10^8}{1 + 0.967 \cos \theta}$$

b. The aphelion distance (farthest distance from the sun) is

$$a + c = a + ea = a(1 + e) \approx (2.697 \times 10^9)(1 + 0.967) \approx 5.305 \times 10^9$$

kilometers.

10.6 CONCEPT QUESTIONS

1. Consider the polar equations

$$r = \frac{ed}{1 \pm e \cos \theta} \quad \text{and} \quad r = \frac{ed}{1 \pm e \sin \theta}$$

Explain the role of the numbers d and e. Illustrate each with a sketch.

2. Give a classification of the conics in terms of their eccentricities.
3. Identify the conic:

  a. $r = \dfrac{3}{1 + 2 \sin \theta}$ **b.** $r = \dfrac{6}{3 + \cos \theta}$

  c. $r = \dfrac{2}{3(1 + \cos \theta)}$ **d.** $r = \dfrac{5}{3 - 2 \sin \theta}$

10.6 EXERCISES

In Exercises 1–8, write a polar equation of the conic that has a focus at the origin and the given properties. Identify the conic.

1. Eccentricity 1, directrix $x = -2$
2. Eccentricity $\frac{1}{3}$, directrix $x = 3$

3. Eccentricity $\frac{1}{2}$, directrix $y = -2$
4. Eccentricity 1, directrix $y = -3$
5. Eccentricity $\frac{3}{2}$, directrix $x = 1$
6. Eccentricity $\frac{5}{4}$, directrix $y = -2$

7. Eccentricity 0.4, directrix $y = 0.4$

8. Eccentricity $\frac{1}{2}$, directrix $r = -2 \sec \theta$

In Exercises 9–20, (a) find the eccentricity and an equation of the directrix of the conic, (b) identify the conic, and (c) sketch the curve.

9. $r = \dfrac{8}{6 + 2 \sin \theta}$

10. $r = \dfrac{8}{6 - 2 \sin \theta}$

11. $r = \dfrac{10}{4 + 6 \cos \theta}$

12. $r = \dfrac{10}{4 - 6 \cos \theta}$

13. $r = \dfrac{5}{2 + 2 \cos \theta}$

14. $r = \dfrac{5}{2 - 2 \sin \theta}$

15. $r = \dfrac{1}{3 - 2 \cos \theta}$

16. $r = \dfrac{12}{3 + \cos \theta}$

17. $r = \dfrac{1}{1 - \sin \theta}$

18. $r = \dfrac{1}{1 + \cos \theta}$

19. $r = -\dfrac{6}{\sin \theta - 2}$

20. $r = -\dfrac{2}{\cos \theta - 3}$

In Exercises 21–26, use Equation (5) or Equation (6) to find the eccentricity of the conic with the given rectangular equation.

21. $\dfrac{x^2}{9} + \dfrac{y^2}{16} = 1$

22. $\dfrac{x^2}{5} - \dfrac{y^2}{3} = 1$

23. $x^2 - y^2 = 1$

24. $9x^2 + 25y^2 = 225$

25. $x^2 - 9y^2 + 2x - 54y = 105$

26. $2x^2 + y^2 + 4x - 6y + 7 = 0$

27. Show that the parabolas with polar equations

$$r = \frac{c}{1 + \sin \theta} \quad \text{and} \quad r = \frac{d}{1 - \sin \theta}$$

intersect at right angles.

28. Show that a conic with focus at the origin, eccentricity e, and directrix $x = d$ has polar equation

$$r = \frac{ed}{1 + e \cos \theta}$$

29. Show that a conic with focus at the origin, eccentricity e, and directrix $y = d$ has polar equation

$$r = \frac{ed}{1 + e \sin \theta}$$

30. Show that a conic with focus at the origin, eccentricity e, and directrix $y = -d$ has polar equation

$$r = \frac{ed}{1 - e \sin \theta}$$

31. a. Show that the polar equation of an ellipse with one focus at the pole and major axis lying along the polar axis is given by

$$r = \frac{a(1 - e^2)}{1 - e \cos \theta}$$

where e is the eccentricity of the ellipse and $2a$ is the length of its major axis.

b. The planets revolve about the sun in elliptical orbits with the sun at one focus. The points on the orbit where a planet is nearest to and farthest from the sun are called the *perihelion* and the *aphelion* of the orbit, respectively. Use the result of part (a) to show that the perihelion distance (minimum distance from the planet to the sun) is $a(1 - e)$.

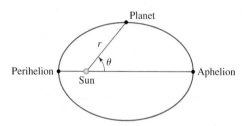

In Exercises 32 and 33, use the results of Exercise 31 to find a polar equation describing the approximate orbit of the given planet and to find the perihelion and aphelion distances.

32. Earth: $e = 0.017$, $a = 92.957 \times 10^6$ mi

33. Saturn: $e = 0.056$, $a = 1.427 \times 10^9$ km

34. The dwarf planet Pluto revolves about the sun in an elliptical orbit. The eccentricity of the orbit is 0.249, and its perihelion distance is 4.43×10^9 km. Use the results of Exercise 31 to find a polar equation for the orbit of Pluto and find its aphelion distance.

35. The planet Mercury revolves about the sun in an elliptical orbit. Its perihelion distance is approximately 4.6×10^7 km, and its aphelion distance is approximately 7.0×10^7 km. Use the results of Exercise 31 to estimate the eccentricity of Mercury's orbit.

CHAPTER 10 REVIEW

CONCEPT REVIEW

In Exercises 1–16, fill in the blanks.

1. a. A parabola is the set of all points in the plane that are _____ from a fixed _____ and a fixed _____. The fixed _____ is called the _____, and the fixed _____ is called the _____.

  b. The point halfway between the focus and the directrix of a parabola is called its _____. The axis of the parabola is the line passing through the _____ and perpendicular to the _____.

2. a. An equation of a parabola with focus $(0, p)$ and directrix $y = -p$ is _____.

  b. An equation of a parabola with focus _____ and directrix _____ is $y^2 = 4px$.

3. a. An ellipse is the set of all points in a plane the _____ of whose distances from two fixed points (called the _____) is a _____.

  b. The vertices of an ellipse are the points of intersection of the line passing through the _____ and the ellipse. The chord joining the vertices is called the _____ _____, and its midpoint is called the _____ of the ellipse. The chord passing through the center of the ellipse and perpendicular to the major axis is called the _____ _____ of the ellipse.

4. a. An equation of the ellipse with foci $(\pm c, 0)$ and vertices $(\pm a, 0)$ is _____, where $c^2 = $ _____.

  b. An equation of the ellipse with foci _____ and vertices _____ is $x^2/b^2 + y^2/a^2 = 1$.

5. a. A hyperbola is the set of all points in a plane the _____ of whose distances from two fixed points (called the _____) is a _____.

  b. The line passing through the foci intersects the hyperbola at two points called the _____ of the hyperbola. The line segment joining the vertices is called the _____ axis of the hyperbola, and the midpoint of the _____ axis is called the _____ of the hyperbola. A hyperbola has _____ _____ branches.

6. a. An equation of a hyperbola with foci $(\pm c, 0)$ and vertices $(\pm a, 0)$ is _____, where $c^2 = $ _____. The hyperbola has asymptotes _____.

  b. An equation of the hyperbola with foci _____ and vertices _____ is $y^2/a^2 - x^2/b^2 = 1$, where $c^2 = $ _____. The hyperbola has asymptotes _____.

7. A plane curve is a set C of ordered pairs (x, y) defined by the parametric equations _____, where f and g are continuous functions on an interval I; I is called the _____ interval.

8. a. If $x = f(t)$ and $y = g(t)$, where f and g are differentiable and $f'(t) \neq 0$, then $dy/dx = $ _____.

  b. If $x = f(t)$ and $y = g(t)$ define y as a twice-differentiable function of x over some suitable interval, then $d^2y/dx^2 = $ _____.

9. a. A curve C represented by $x = f(t)$ and $y = g(t)$ on a parameter interval I is smooth if _____ and _____ are continuous on I and are not _____ _____, except possibly at the _____ of I.

  b. If C is a smooth curve represented by $x = f(t)$ and $y = g(t)$ with parameter interval $[a, b]$, then the length of C is $L = $ _____.

10. If C is a smooth curve as described in Question 9b, C does not intersect itself, except possibly at _____, and $g(t) \geq 0$, then the area of the surface obtained by revolving C about the x-axis is $S = $ _____. If $f(t) \geq 0$ for all t in $[a, b]$, then the area of the surface obtained by revolving C about the y-axis is $S = $ _____.

11. a. The rectangular coordinates (x, y) of a point P are related to the polar coordinates of P by the equations $x = $ _____ and $y = $ _____.

  b. The polar coordinates (r, θ) of a point P are related to the rectangular coordinates of P by the equations $r^2 = $ _____ and $\tan \theta = $ _____.

12. The horizontal tangent lines to the graph of $r = f(\theta)$ are located at the points where $dy/d\theta$ _____ and $dx/d\theta$ _____. The vertical tangent lines are located at the points where $dx/d\theta$ _____ and $dy/d\theta$ _____. Horizontal and vertical tangent lines may also be located at points where $dy/d\theta$ and $dx/d\theta$ are both equal to _____.

13. a. If f is nonnegative and continuous on $[\alpha, \beta]$, where $0 \leq \alpha < \beta \leq 2\pi$, then the area of the region bounded by the graphs of $r = f(\theta)$, $\theta = \alpha$, and $\theta = \beta$ is given by _____.

  b. Let f and g be continuous on $[\alpha, \beta]$, where $0 \leq g(\theta) \leq f(\theta)$ and $0 \leq \alpha < \beta \leq 2\pi$. Then the area of the region bounded by the graphs of $r = g(\theta)$, $r = f(\theta)$, $\theta = \alpha$, and $\theta = \beta$ is given by _____.

14. Suppose that f has a continuous derivative on $[\alpha, \beta]$. If the graph C of $r = f(\theta)$ is traced exactly _____ as θ increases from α to β, then the length of C is given by _____.

15. Let F be a fixed point, let l be a fixed line in the plane, and let e be a fixed positive number. Then a conic section defined by the equation _____ is an ellipse if e satisfies

_____, a parabola if e satisfies _____, and a hyperbola if e satisfies _____.

16. A conic section can be represented by a polar equation of the form _____ or _____. It is an ellipse, a parabola, or a hyperbola depending on whether e satisfies _____, _____, or _____, respectively.

REVIEW EXERCISES

In Exercises 1–6, find the vertices and the foci of the conic and sketch its graph.

1. $\dfrac{x^2}{4} + \dfrac{y^2}{9} = 1$

2. $\dfrac{(x-1)^2}{2} + \dfrac{(y+1)^2}{4} = 1$

3. $x^2 - 9y^2 = 9$

4. $y^2 - 2y - 8x - 15 = 0$

5. $y^2 - 9x^2 + 8y + 7 = 0$

6. $4x^2 + 25y^2 - 16x + 50y - 59 = 0$

In Exercises 7–12, find a rectangular equation of the conic satisfying the given conditions.

7. parabola, focus $(-2, 0)$, directrix $x = 2$

8. parabola, vertex $(-2, 2)$, directrix $y = 4$

9. ellipse, vertices $(\pm 7, 0)$, foci $(\pm 2, 0)$

10. ellipse, foci $(\pm 2, 3)$, major axis has length 8

11. hyperbola, foci $\left(0, \pm\frac{3}{2}\sqrt{5}\right)$, vertices $(0, \pm 3)$

12. hyperbola, vertices $(-2, 0)$ and $(2, 0)$, asymptotes $y = \pm\dfrac{3}{2}x$

13. Show that if m is any real number, then there is exactly one line of slope m that is tangent to the parabola $x^2 = 4py$ and its equation is $y = mx - pm^2$.

14. Show that if m is any real number, then there are exactly two lines of slope m that are tangent to the ellipse $x^2/a^2 + y^2/b^2 = 1$ and their equations are $y = mx \pm \sqrt{a^2m^2 + b^2}$.

In Exercises 15–18, (a) find a rectangular equation whose graph contains the curve C with the given parametric equations, and (b) sketch the curve C and indicate its orientation.

15. $x = 1 + 2t, \quad y = 3 - 2t$

16. $x = e^t, \quad y = e^{-2t}$

17. $x = 1 + 2\sin t, \quad y = 3 + 2\cos t$

18. $x = \cos^3 t, \quad y = 4\sin^3 t$

In Exercises 19–22, find the slope of the tangent line to the curve at the point corresponding to the value of the parameter.

19. $x = t^3 + 1, \quad y = 2t^2 - 1; \quad t = 1$

20. $x = \sqrt{t+1}, \quad y = \sqrt{16-t}; \quad t = 0$

21. $x = te^{-t}, \quad y = \dfrac{1}{t^2+1}; \quad t = 0$

22. $x = 1 - \sin^2 t, \quad y = \cos^3 t; \quad t = \dfrac{\pi}{4}$

In Exercises 23 and 24, find dy/dx and d^2y/dx^2.

23. $x = t^3 + 1, \quad y = t^4 + 2t^2$

24. $x = e^t \sin t, \quad y = e^t \cos t$

In Exercises 25 and 26, find the points on the curve with the given parametric equations at which the tangent lines are vertical or horizontal.

25. $x = t^3 - 4t, \quad y = t^2 + 2$

26. $x = 1 - 2\cos t, \quad y = 1 - 2\sin t$

In Exercises 27 and 28, find the length of the curve defined by the given parametric equations.

27. $x = \dfrac{1}{6}t^6, \quad y = 2 - \dfrac{1}{4}t^4; \quad 0 \le t \le \sqrt[4]{8}$

28. $x = \sqrt{3}t^2, \quad y = t - t^3; \quad -1 \le t \le 1$

29. The position of a body at time t is (x, y), where $x = e^{-t}\cos t$ and $y = e^{-t}\sin t$. Find the distance covered by the body as t runs from 0 to $\pi/2$.

30. The course taken by an oceangoing racing boat during a practice run is described by the parametric equations

$$x = \sqrt{3}(t-1)^2 \quad \text{and} \quad y = (t-1) - (t-1)^3$$
$$0 \le t \le 2$$

where x and y are measured in miles. Sketch the path of the boat, and find the length of the course.

In Exercises 31 and 32, find the area of the surface obtained by revolving the given curve about the x-axis.

31. $x = t^2, \quad y = \dfrac{t}{3}(3 - t^2); \quad 0 \le t \le \sqrt{3}$

32. $x = \ln(\sec t + \tan t) - \sin t, \quad y = \cos t; \quad 0 \le t \le \frac{\pi}{3}$

In Exercises 33–38, sketch the curve with the given polar equation.

33. $r = 2\sin\theta$

34. $r = 3 - 4\cos\theta$

35. $r = 2 \cos 5\theta$

36. $r = e^{-\theta}$

37. $r^2 = \cos 2\theta$

38. $r = 2 \sin \theta \cos^2 \theta$

In Exercises 39 and 40, find the slope of the tangent line to the curve with the given polar equation at the point corresponding to the given value of θ.

39. $r = e^{2\theta}, \quad \theta = \dfrac{\pi}{2}$

40. $r = 2 - \sin \theta, \quad \theta = \dfrac{\pi}{2}$

In Exercises 41 and 42, find the points of intersection of the given curves.

41. $r = \sin \theta, \quad r = 1 - \sin \theta$

42. $r = \cos \theta, \quad r = \cos 2\theta$

43. Find the area of the region enclosed by the curve with polar equation $r = 2 + \cos \theta$.

44. Find the area of the region enclosed by the curve with polar equation $r = 1 + \sin \theta$.

45. Find the area of the region that is enclosed between the petals of the curves with polar equations $r = 2 \sin 2\theta$ and $r = 2 \cos 2\theta$.

46. Find the area of the region that is enclosed between the curves with polar equations $r = 3 + 2 \sin \theta$ and $r = 2$.

In Exercises 47 and 48, find the length of the given curve.

47. $r = \theta^2, \quad 0 \le \theta \le 2\pi$

48. $r = 2(\sin \theta + \cos \theta), \quad 0 \le \theta \le 2\pi$

In Exercises 49 and 50, find the area of the surface obtained by revolving the curve about the given line.

49. $r = 2 \sin \theta$ about the polar axis

50. $r = \sqrt{\cos 2\theta}, \quad 0 \le \theta \le \frac{\pi}{4}$, about the line $\theta = \dfrac{\pi}{2}$

In Exercises 51 and 52, sketch the curve with the given equation.

51. $r = \dfrac{1}{1 + \sin \theta}$

52. $r = \dfrac{16}{3 - 5 \cos \theta}$

In Exercises 53–54, plot the curve with the parametric equations.

53. $x = 0.15(2 \cos t + 3 \cos 2t), \quad y = 0.15(2 \sin t - 3 \sin 2t);$
$0 \le t \le 2\pi$ (hypotrochoid)

54. $x = 0.24(-7t^3 + 3t^5), \quad y = 0.09(14t^2 - 5t^4);$
$-1.629 \le t \le 1.629$ (butterfly catastrophe)

In Exercises 55–57, plot the curve with the polar equation.

55. $r = 0.1e^{0.1\theta}$ where $0 \le \theta \le \frac{15\pi}{2}$

56. $r = \dfrac{0.1}{\cos 4\theta}$ where $0 \le \theta \le 2\pi$ (epi-spiral)

57. $r = \dfrac{1}{2 \sinh \theta}$ where $-2\pi \le \theta \le 2\pi$ (spiral of Poinsot)

58. An ant crawls along the curve $x = \frac{1}{2}t^2$, $y = \frac{1}{3}(2t + 1)^{3/2}$ starting at the point $\left(0, \frac{1}{3}\right)$ and ending at the point $\left(\frac{1}{2}, \sqrt{3}\right)$, where x and y are measured in feet. Find the distance traveled by the ant.

59. An egg has the shape of a solid obtained by revolving the upper half of the ellipse $x^2 + 2y^2 = 2$ about the x-axis. What is the surface area of the egg?

60. A piston is attached to a crankshaft by means of a connecting rod of length L, as shown in the figure. If the disk is of radius r, find the parametric equations giving the position of the point P using the angle θ as a parameter.

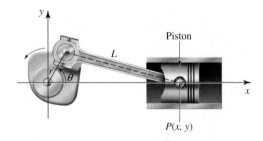

CHALLENGE PROBLEMS

1. a. In the following figure, the axes of an xy-coordinate system have been rotated about the origin through an angle of θ to produce a new $x'y'$-coordinate system.

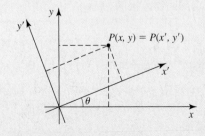

Show that

$$x = x' \cos \theta - y' \sin \theta \qquad y = x' \sin \theta + y' \cos \theta$$

b. Show that the equation $Ax^2 + Bxy + Cy^2 + F = 0$, where $B \ne 0$, will have the form $(A'x')^2 + (C'y')^2 + F = 0$ in the $x'y'$-coordinate system obtained by rotating the xy-system through an angle θ given by

$$\cot 2\theta = \dfrac{A - C}{B}$$

c. Sketch the ellipse $2x^2 + \sqrt{3}xy + y^2 - 20 = 0$.

2. a. Show that the area of the ellipse

$$Ax^2 + Bxy + Cy^2 + F = 0$$

where $B^2 - 4AC < 0$, is given by

$$S = -\frac{2\pi F}{\sqrt{4AC - B^2}}$$

b. Find the area of the ellipse $2x^2 + \sqrt{3}xy + y^2 = 20$.

3. Find the length of the curve with parametric equations

$$x = \int_1^t \frac{\cos u}{u}\, du \qquad y = \int_1^t \frac{\sin u}{u}\, du$$

between the origin and the nearest point from the vertical tangent line.

4. Find the area of the surface obtained by revolving one branch of the lemniscate $r = a\sqrt{\cos 2\theta}$ about the line $\theta = \pi/4$.

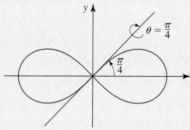

$$r = a\sqrt{\cos 2\theta}$$

5. The curve with equation $x^3 + y^3 = 3axy$, where a is a nonzero constant, is called the **folium of Descartes.**

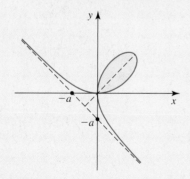

a. Show that the polar equation of the curve is

$$r = \frac{3a \sec \theta \tan \theta}{1 + \tan^3 \theta}$$

b. Find the area of the region enclosed by the loop of the curve.

6. Find the rectangular coordinates of the centroid of the region that is completely enclosed by the curve $r = 3 \cos^3 \theta$.

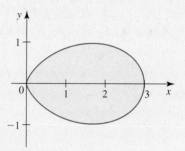

7. An ant is placed at each corner of a square with sides of length a. Starting at the same instant of time, all four ants begin to move counterclockwise at the same speed and in such a way that each ant moves toward the next at all times. The resulting path of each ant is a spiral curve that converges to the center of the square.

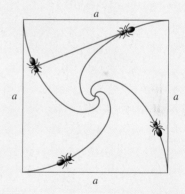

a. Taking the pole to be the center of the square, find the polar equation describing the path taken by one of the ants. Hint: The line passing through the position of two adjacent ants is tangent to the path of one of them.

b. Find the distance traveled by an ant as it moves from a corner of the square to its center.

The photograph shows an Olympian throwing a javelin. Once the javelin has been launched, its direction of motion and speed at any point in its flight through the air can be described by a *vector* at that point. In this chapter we will study vectors—objects that possess both magnitude and direction—and look at some of their applications.

Stu Forster/Getty Images

11 Vectors and the Geometry of Space

IN THIS CHAPTER we will study *vectors*, quantities that have both direction and magnitude. Vectors can be used to describe the position, velocity, and acceleration of a body moving in a plane or in space. Since a force is determined by the direction along which it acts and by its magnitude, we can also represent a force by a vector.

There are two ways in which vectors can be *multiplied* together. These operations on vectors enable us to find the work that is done by a force in moving an object from one point in space to another point, to find the angle between two lines, to compute the volume of a parallelepiped, and to find the torque exerted by a person pushing on a leveraged-impact lug wrench when changing a tire—just to mention a few applications.

Vectors also facilitate the algebraic representation of lines and planes in space. Using such representations, we can easily find the distance between a point in space and a plane and the distance between two skew lines (lines that are neither parallel nor intersect). Finally, we introduce two alternative coordinate systems in space: the *cylindrical coordinate system* and the *spherical coordinate system*. Each of these systems enables us to obtain relatively simple algebraic representation of certain surfaces.

V This symbol indicates that one of the following video types is available for enhanced student learning at **www.academic.cengage.com/login:**
- Chapter lecture videos
- Solutions to selected exercises

V ✐ This symbol indicates that step-by-step video lessons for hand-drawing certain complex figures are available.

11.1   Vectors in the Plane

■ Vectors

Some physical quantities, such as force and velocity, possess both magnitude (size) and direction. These quantities are called **vectors** and can be represented by arrows or directed line segments. The arrow points in the direction of the vector, and the length of the arrow gives the magnitude of the vector.

Figure 1a gives an aerial view of a tugboat trying to free a cruise liner that has run aground in shallow waters. The magnitude and direction of the force exerted by the tugboat are represented by the vector shown in the figure.

FIGURE 1

(a) The vector represents the force exerted by a tugboat on a ship.

(b) The vectors represent the velocity of blood cells flowing through an artery.

In Figure 1b the vectors (arrows) give the magnitude and direction of blood cells flowing through an artery. Observe that the lengths of the vectors vary; this reflects the fact that the blood cells near the central axis have a greater velocity than those near the walls of the artery.

Vectors are customarily denoted by lowercase boldface type such as **v** and **w**. However, if a vector **v** is defined by a directed line segment from the **initial point** A of the vector to the **terminal point** B of the vector, then it is written $\mathbf{v} = \overrightarrow{AB}$. (See Figure 2.)*

Two vectors, **v** and **w**, that have the same magnitude and direction are said to be **equal,** written $\mathbf{v} = \mathbf{w}$. Thus, the vectors $\mathbf{v} = \overrightarrow{AB}$ and $\mathbf{w} = \overrightarrow{CD}$ shown in Figure 3 are equal.

FIGURE 2

v is the directed line segment from A to B.

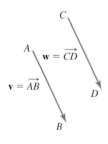

FIGURE 3

v and **w** have the same length and direction.

■ Scalar Multiples

In contrast to a vector, a **scalar** is a quantity that has magnitude but no direction. Real numbers and complex numbers are examples of scalars. In this text, however, the term *scalar* will always refer to a real number. A vector can be multiplied by a scalar. If $c \neq 0$ is a scalar and **v** is a vector, then the **scalar multiple** of c and **v** is a vector $c\mathbf{v}$. The magnitude of $c\mathbf{v}$ is $|c|$ times the magnitude of **v**, and the direction of $c\mathbf{v}$ is the same as that of **v** if $c > 0$ and opposite that of **v** if $c < 0$. (See Figure 4.) Observe that two nonzero vectors are **parallel** if they are scalar multiples of one another.

For convenience we define the **zero vector,** denoted by **0**, to be the vector with length zero and having *no* direction. If $c = 0$, then $c\mathbf{v} = \mathbf{0}$ for any vector **v**.

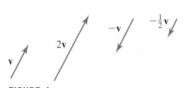

FIGURE 4

Scalar multiples of **v**

———————

*If the vector **v** is handwritten, it is more convenient to write it in the form $\vec{v}$.

■ Vector Addition: The Parallelogram Law

Two vectors may be added together. To see how, consider the two nonzero vectors **v** and **w** shown in Figure 5a. Translate the vector **w** (move **w** without changing its magnitude or direction) so that the initial point of **w** coincides with the terminal point of **v**. (See Figure 5b.) Then the **sum** of **v** and **w**, written **v** + **w**, is the vector represented by the arrow with tail at the initial point of **v** and head at the terminal point of **w** (Figure 5c). If you examine Figure 5d, you will see that the line segment representing the vector **v** + **w** coincides with the diagonal of the parallelogram determined by **v** and **w**. For this reason we say that vector addition obeys the Parallelogram Law. Try to translate the vector **v** instead of **w**, and convince yourself that the result is the same.

  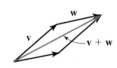

(**a**) The vectors **v** and **w**   (**b**) **w** translated   (**c**) **v** + **w**   (**d**) **v** + **w** lies on the diagonal of the parallelogram determined by **v** and **w**.

FIGURE 5
Geometric construction of **v** + **w**

The difference of two vectors **v** and **w**, written **v** − **w**, is defined by

$$\mathbf{v} - \mathbf{w} = \mathbf{v} + (-\mathbf{w}) \qquad \text{Difference of two vectors}$$

To describe this operation geometrically, consider once again the two vectors **v** and **w** of Figure 5a, which are reproduced in Figure 6a. If we translate **w**, reverse it to obtain −**w**, and then use the parallelogram law to add **v** to −**w**, we obtain **v** − **w**, as shown in Figure 6b.

FIGURE 6
Geometric construction of **v** − **w**

(**a**) The vectors **v** and **w**   (**b**) The vector **v** − **w**

■ Vectors in the Coordinate Plane

Just as the introduction of a rectangular coordinate system in the plane enabled us to describe geometric objects in algebraic terms, we will see that the introduction of a rectangular coordinate system in a "vector space" will enable us to represent vectors algebraically.

EXAMPLE 1 Let **a** be a vector with initial point $A(0, 0)$ and terminal point $B(3, 2)$, and let **b** be a vector with initial point $C(1, 3)$ and terminal point $D(4, 5)$. Show that **a** = **b**.

Solution The vectors $\mathbf{a} = \overrightarrow{AB}$ and $\mathbf{b} = \overrightarrow{CD}$ are shown in Figure 7. To show that **a** = **b**, we need to show that both vectors have the same length and direction. Using the distance formula, we find

$$\text{length of } \overrightarrow{AB} = \sqrt{(3 - 0)^2 + (2 - 0)^2} = \sqrt{9 + 4} = \sqrt{13}$$

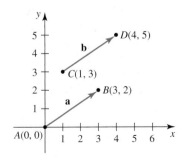

FIGURE 7
a = **b** because both vectors have the same length and direction.

and

$$\text{length of } \overrightarrow{CD} = \sqrt{(4-1)^2 + (5-3)^2} = \sqrt{9+4} = \sqrt{13}$$

so **a** and **b** have the same length. Next, we find

$$\text{slope of } \overrightarrow{AB} = \frac{2-0}{3-0} = \frac{2}{3}$$

and

$$\text{slope of } \overrightarrow{CD} = \frac{5-3}{4-1} = \frac{2}{3}$$

so **a** and **b** have the same direction. This proves that **a** = **b**. ■

In Example 1 we saw that the vector **b** may be represented by the vector **a** that has its initial point at the origin. In general, it is true that any vector in the plane can be represented by such a vector. To see this, suppose that $\mathbf{b} = \overrightarrow{P_1 P_2}$ is any vector with initial point $P_1(x_1, y_1)$ and terminal point $P_2(x_2, y_2)$. (See Figure 8.) Let $a_1 = x_2 - x_1$ and $a_2 = y_2 - y_1$. Then the vector $\mathbf{a} = \overrightarrow{OP}$ with $O(0, 0)$ and $P(a_1, a_2)$ is the required vector, since the length of **b** is

$$\sqrt{a_1^2 + a_2^2} = \sqrt{(x_2 - x_1)^2 + (y_2 - y_1)^2}$$

which is also the length of $\mathbf{b} = \overrightarrow{P_1 P_2}$. Similarly, the slope of **a** is

$$\frac{a_2}{a_1} = \frac{y_2 - y_1}{x_2 - x_1} \qquad x_1 \neq x_2$$

which is also the slope of **b**. (We leave the proof of the case in which $x_1 = x_2$ to you.)

The vector **a** with initial point at the origin and terminal point $P(a_1, a_2)$ is called the **position vector** of the point $P(a_1, a_2)$ and is denoted by $\langle a_1, a_2 \rangle$. Thus, we have shown that any vector in a coordinate plane is equal to a position vector. Since the zero vector has length zero, its terminal point must coincide with its initial point; therefore, it is equal to the position vector of the point $(0, 0)$.

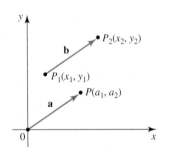

FIGURE 8
Any vector **b** in the plane can be represented by a vector **a** emanating from the origin.

DEFINITIONS   A Vector in the Coordinate Plane

A **vector** in the plane is an ordered pair $\mathbf{a} = \langle a_1, a_2 \rangle$ of real numbers, a_1 and a_2, called the **scalar components** of **a**. The **zero vector** is $\mathbf{0} = \langle 0, 0 \rangle$.

We have also established the following result.

DEFINITION

Given the points $P_1(x_1, y_1)$ and $P_2(x_2, y_2)$, the vector $\overrightarrow{P_1 P_2}$ is represented by the **position vector**

$$\mathbf{a} = \overrightarrow{P_1 P_2} = \langle x_2 - x_1, y_2 - y_1 \rangle \tag{1}$$

Thus, the components of a vector are found by subtracting the respective coordinates of its initial point from the coordinates of its terminal point.

EXAMPLE 2 Find the vector **a** with initial point $A(-1, -2)$ and terminal point $B(3, 2)$.

Solution Using Equation (1), we find the vector **a** to be

$$\mathbf{a} = \langle 3 - (-1), 2 - (-2) \rangle = \langle 4, 4 \rangle$$ ∎

■ Length of a Vector

The **length** or the **magnitude** of a vector $\mathbf{a} = \langle a_1, a_2 \rangle$, denoted by the symbol $|\mathbf{a}|$, is found by using the Pythagorean Theorem. (See Figure 9.)

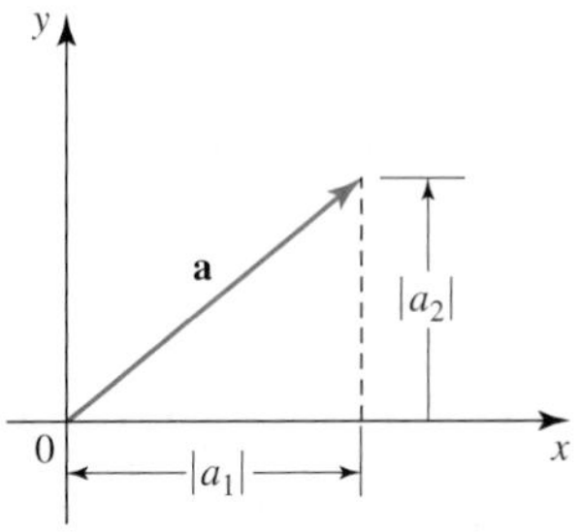

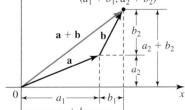

FIGURE 9
The length of **a** is $|\mathbf{a}| = \sqrt{a_1^2 + a_2^2}$.

> **DEFINITION**
> The **length** or **magnitude** of $\mathbf{a} = \langle a_1, a_2 \rangle$ is
> $$|\mathbf{a}| = \sqrt{a_1^2 + a_2^2} \tag{2}$$

■ Vector Addition in the Coordinate Plane

Vector addition is carried out componentwise. To add the two vectors $\mathbf{a} = \langle a_1, a_2 \rangle$ and $\mathbf{b} = \langle b_1, b_2 \rangle$, we add their components. (See Figure 10.)

FIGURE 10
If $\mathbf{a} = \langle a_1, a_2 \rangle$ and $\mathbf{b} = \langle b_1, b_2 \rangle$, then $\mathbf{a} + \mathbf{b} = \langle a_1 + b_1, a_2 + b_2 \rangle$.

> **Parallelogram Law for Vector Addition**
> If $\mathbf{a} = \langle a_1, a_2 \rangle$ and $\mathbf{b} = \langle b_1, b_2 \rangle$, then
> $$\mathbf{a} + \mathbf{b} = \langle a_1 + b_1, a_2 + b_2 \rangle \tag{3}$$

EXAMPLE 3 If $\mathbf{a} = \langle 3, -2 \rangle$, and $\mathbf{b} = \langle -1, 3 \rangle$, then

$$\mathbf{a} + \mathbf{b} = \langle 3, -2 \rangle + \langle -1, 3 \rangle = \langle 3 + (-1), -2 + 3 \rangle = \langle 2, 1 \rangle$$ ∎

> **Scalar Multiplication**
> If $\mathbf{a} = \langle a_1, a_2 \rangle$ and c is a scalar, then
> $$c\mathbf{a} = \langle ca_1, ca_2 \rangle \tag{4}$$
> (See Figure 11.)

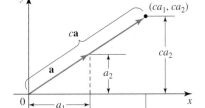

FIGURE 11
If $\mathbf{a} = \langle a_1, a_2 \rangle$, then $c\mathbf{a} = \langle ca_1, ca_2 \rangle$.

Recall that the **difference** of **a** and **b** is defined by $\mathbf{a} - \mathbf{b} = \mathbf{a} + (-\mathbf{b})$, so if $\mathbf{a} = \langle a_1, a_2 \rangle$ and $\mathbf{b} = \langle b_1, b_2 \rangle$, then

$$\mathbf{a} - \mathbf{b} = \mathbf{a} + (-\mathbf{b}) = \mathbf{a} + (-1)\mathbf{b} = \langle a_1, a_2 \rangle + \langle -b_1, -b_2 \rangle$$
$$= \langle a_1 - b_1, a_2 - b_2 \rangle \quad \text{Vector subtraction}$$

EXAMPLE 4 Let $\mathbf{a} = \langle 1, -2 \rangle$ and $\mathbf{b} = \langle -2, 5 \rangle$. Find

a. $\mathbf{a} + \mathbf{b}$ **b.** $\mathbf{a} - \mathbf{b}$ **c.** $5\mathbf{a}$ **d.** $3\mathbf{a} + 2\mathbf{b}$ **e.** $|3\mathbf{a} + 2\mathbf{b}|$

Solution

a. $\mathbf{a} + \mathbf{b} = \langle 1, -2 \rangle + \langle -2, 5 \rangle = \langle 1 - 2, -2 + 5 \rangle = \langle -1, 3 \rangle$

b. $\mathbf{a} - \mathbf{b} = \langle 1, -2 \rangle - \langle -2, 5 \rangle = \langle 1 - (-2), -2 - 5 \rangle = \langle 3, -7 \rangle$

c. $5\mathbf{a} = 5\langle 1, -2 \rangle = \langle 5(1), 5(-2) \rangle = \langle 5, -10 \rangle$

d. $3\mathbf{a} + 2\mathbf{b} = 3\langle 1, -2 \rangle + 2\langle -2, 5 \rangle = \langle 3, -6 \rangle + \langle -4, 10 \rangle = \langle -1, 4 \rangle$

e. $|3\mathbf{a} + 2\mathbf{b}| = |\langle -1, 4 \rangle|$ Use part (d).

$$= \sqrt{(-1)^2 + 4^2}$$
$$= \sqrt{1 + 16} = \sqrt{17}$$ ∎

■ Properties of Vectors

The operations of vector addition and scalar multiplication obey the following rules.

THEOREM 1   Rules for Vector Addition and Scalar Multiplication

Suppose that $\mathbf{a}$, $\mathbf{b}$, and $\mathbf{c}$ are vectors and that c and d are scalars. Then

1. $\mathbf{a} + \mathbf{b} = \mathbf{b} + \mathbf{a}$
2. $(\mathbf{a} + \mathbf{b}) + \mathbf{c} = \mathbf{a} + (\mathbf{b} + \mathbf{c})$
3. $\mathbf{a} + \mathbf{0} = \mathbf{0} + \mathbf{a} = \mathbf{a}$
4. $\mathbf{a} + (-\mathbf{a}) = \mathbf{0}$
5. $c(\mathbf{a} + \mathbf{b}) = c\mathbf{a} + c\mathbf{b}$
6. $c(d\mathbf{a}) = (cd)\mathbf{a}$
7. $(c + d)\mathbf{a} = c\mathbf{a} + d\mathbf{a}$
8. $1\mathbf{a} = \mathbf{a}$

We will prove the first of these rules and leave the proofs of the others as exercises.

PROOF OF 1   Let $\mathbf{a} = \langle a_1, a_2 \rangle$ and $\mathbf{b} = \langle b_1, b_2 \rangle$. Then

$$\mathbf{a} + \mathbf{b} = \langle a_1, a_2 \rangle + \langle b_1, b_2 \rangle = \langle a_1 + b_1, a_2 + b_2 \rangle$$
$$= \langle b_1 + a_1, b_2 + a_2 \rangle = \mathbf{b} + \mathbf{a}$$ ∎

■ Unit Vectors

A **unit vector** is a vector of length 1. Unit vectors are primarily used as indicators of direction. For example, if $\mathbf{a}$ is a nonzero vector, then the vector

$$\mathbf{u} = \frac{\mathbf{a}}{|\mathbf{a}|}$$

is a unit vector having the same direction as $\mathbf{a}$. (See Figure 12.) Furthermore, by writing $\mathbf{a}$ in the form

$$\mathbf{a} = |\mathbf{a}|\left(\frac{\mathbf{a}}{|\mathbf{a}|}\right) = |\mathbf{a}|\,\mathbf{u} \tag{5}$$

Magnitude of $\mathbf{a}$ ⎯⎯⎯⎯⎯⎯ Direction of $\mathbf{a}$

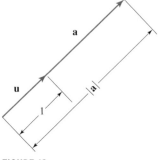

FIGURE 12
The unit vector $\mathbf{u}$ indicates the direction of $\mathbf{a}$.

the two properties of magnitude and direction that define a vector are clearly displayed.

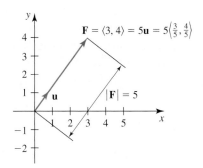

FIGURE 13
The vector $\mathbf{F} = \langle 3, 4 \rangle$ can be written in the alternate form $\mathbf{F} = 5\mathbf{u}$, where $\mathbf{u}$ is the unit vector in the direction of $\mathbf{F}$.

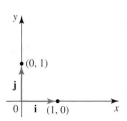

FIGURE 14
The unit vectors $\mathbf{i}$ and $\mathbf{j}$ point in the positive x- and y-direction, respectively.

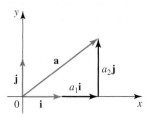

FIGURE 15
Any vector in the plane can be expressed in terms of the standard basis vectors $\mathbf{i}$ and $\mathbf{j}$.

EXAMPLE 5 Let $\mathbf{F} = \langle 3, 4 \rangle$ be a vector describing a force acting on a particle. Express $\mathbf{F}$ in terms of its magnitude (in dynes) and a unit vector having the same direction as $\mathbf{F}$.

Solution The magnitude of $\mathbf{F}$ is

$$|\mathbf{F}| = \sqrt{3^2 + 4^2} = \sqrt{9 + 16} = 5$$

or 5 dynes. Its direction is

$$\mathbf{u} = \frac{\mathbf{F}}{|\mathbf{F}|} = \frac{1}{5}\langle 3, 4 \rangle = \left\langle \frac{3}{5}, \frac{4}{5} \right\rangle$$

So

$$\mathbf{F} = \langle 3, 4 \rangle = 5\left\langle \frac{3}{5}, \frac{4}{5} \right\rangle$$

(See Figure 13.)

Standard Basis Vectors

There are two unit vectors in the coordinate plane that are singled out for a special role. They are the vectors $\mathbf{i}$ and $\mathbf{j}$ defined by

$$\mathbf{i} = \langle 1, 0 \rangle \qquad \text{and} \qquad \mathbf{j} = \langle 0, 1 \rangle$$

The vector $\mathbf{i}$ points in the positive x-direction, whereas the vector $\mathbf{j}$ points in the positive y-direction. (See Figure 14.)

Let $\mathbf{a} = \langle a_1, a_2 \rangle$ be a vector in the coordinate plane. Then

$$
\begin{aligned}
\mathbf{a} = \langle a_1, a_2 \rangle &= \langle a_1, 0 \rangle + \langle 0, a_2 \rangle && \text{By the definition of vector addition}\\
&= a_1\langle 1, 0 \rangle + a_2\langle 0, 1 \rangle && \text{By the definition of scalar multiplication}\\
&= a_1\mathbf{i} + a_2\mathbf{j}
\end{aligned}
$$

This shows that any vector in the plane can be expressed in terms of the vectors $\mathbf{i}$ and $\mathbf{j}$. (See Figure 15.) For this reason the vectors $\mathbf{i}$ and $\mathbf{j}$ are referred to as **standard basis vectors.** The vectors $a_1\mathbf{i}$ and $a_2\mathbf{j}$ are the **horizontal** and **vertical vector components** of $\mathbf{a}$. We also say that $\mathbf{a}$ is *resolved* into a (vector) sum of $a_1\mathbf{i}$ and $a_2\mathbf{j}$.

EXAMPLE 6 Let $\mathbf{F} = \langle 3, 4 \rangle$ be the force vector of Example 5. Express $\mathbf{F}$ in terms of the standard basis vectors $\mathbf{i}$ and $\mathbf{j}$, and identify the horizontal and vertical vector components of $\mathbf{F}$.

Solution Since

$$
\begin{aligned}
\mathbf{F} &= \langle 3, 4 \rangle\\
&= 3\mathbf{i} + 4\mathbf{j}
\end{aligned}
$$

the horizontal vector component of $\mathbf{F}$ is $3\mathbf{i}$, and the vertical vector component of $\mathbf{F}$ is $4\mathbf{j}$.

Note By using standard basis vectors, we are able to express any vector in the coordinate plane in two ways:

$$\mathbf{a} = \langle a_1, a_2 \rangle \qquad \text{and} \qquad \mathbf{a} = a_1\mathbf{i} + a_2\mathbf{j}$$

We will use these representations interchangeably.

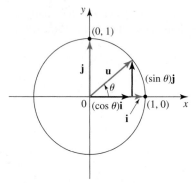

FIGURE 16
Every unit vector **u** can be expressed in the form **u** = $(\cos \theta)$**i** + $(\sin \theta)$**j**.

■ Angular Form of the Unit Vector

Let θ be the angle that the unit vector **u** makes with the positive x-axis. (See Figure 16.) Then resolving **u** into a sum of horizontal and vertical vector components gives

$$\mathbf{u} = (\cos \theta)\mathbf{i} + (\sin \theta)\mathbf{j} \tag{6}$$

EXAMPLE 7 Find an expression for the vector **a** of length 5 that makes an angle of $\pi/6$ radians with the positive axis.

Solution Using Equation (6), we see that the unit vector making an angle of $\pi/6$ with the positive axis is

$$\mathbf{u} = \left(\cos \frac{\pi}{6}\right)\mathbf{i} + \left(\sin \frac{\pi}{6}\right)\mathbf{j} = \frac{\sqrt{3}}{2}\mathbf{i} + \frac{1}{2}\mathbf{j} = \frac{1}{2}(\sqrt{3}\mathbf{i} + \mathbf{j})$$

Therefore, the required expression is

$$\mathbf{a} = 5\mathbf{u} = \frac{5}{2}(\sqrt{3}\mathbf{i} + \mathbf{j})$$ ■

EXAMPLE 8 **Finding the True Course and Ground Speed of an Airplane** An airplane, on level flight, is headed in a direction that makes an angle of 45° with the north (measured in a clockwise direction) and has an airspeed of 500 mph. It is subjected to a tailwind blowing at 80 mph in a direction that makes an angle of 75° with the north. (See Figure 17.) The **true course** and **ground speed** of the airplane are given by the direction and magnitude of the resultant **v** + **w**, where **v** is the velocity of the plane and **w** is the velocity of the wind. (See Figure 18.) Find the true course and ground speed of the airplane.

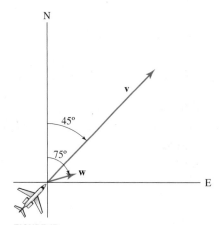

FIGURE 17
v is the velocity of the plane and **w** is the velocity of the wind.

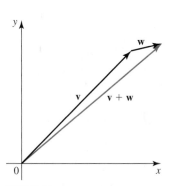

FIGURE 18
v + **w** gives the true course and ground speed of the airplane.

Solution With respect to the coordinate system shown in Figure 17, we can represent **v** and **w** as

$$\mathbf{v} = (500 \cos 45°)\mathbf{i} + (500 \sin 45°)\mathbf{j} \qquad \text{Velocity of the plane}$$

and

$$\mathbf{w} = (80 \cos 15°)\mathbf{i} + (80 \sin 15°)\mathbf{j} \qquad \text{Velocity of the wind}$$

Therefore,

$$\mathbf{v} + \mathbf{w} = [(500 \cos 45°)\mathbf{i} + (500 \sin 45°)\mathbf{j}] + [(80 \cos 15°)\mathbf{i} + (80 \sin 15°)\mathbf{j}]$$

$$= (500 \cos 45° + 80 \cos 15°)\mathbf{i} + (500 \sin 45° + 80 \sin 15°)\mathbf{j}$$

$$\approx 430.8\mathbf{i} + 374.3\mathbf{j}$$

The magnitude of $\mathbf{v} + \mathbf{w}$ is

$$|\mathbf{v} + \mathbf{w}| \approx \sqrt{(430.8)^2 + (374.3)^2} \approx 570.7$$

and this gives the ground speed of the airplane. To find its true course, we compute the unit vector $\mathbf{u}$ having the same direction as $\mathbf{v} + \mathbf{w}$. Thus,

$$\mathbf{u} = \frac{\mathbf{v} + \mathbf{w}}{|\mathbf{v} + \mathbf{w}|} \approx \frac{1}{570.7}(430.8\mathbf{i} + 374.3\mathbf{j}) \approx 0.7549\mathbf{i} + 0.6559\mathbf{j}$$

If we write $\mathbf{u}$ in the form $\mathbf{u} = (\cos\theta)\mathbf{i} + (\sin\theta)\mathbf{j}$ then we see that

$$\cos\theta \approx 0.7549 \qquad \text{or} \qquad \theta \approx \cos^{-1} 0.7549 \approx 41.0°$$

Since θ is measured from the positive x-axis, we conclude that the true course of the airplane is approximately $(90 - 41)°$, or $49°$ with the north. ■

11.1 CONCEPT QUESTIONS

1. **a.** What is a vector? Give examples.
   b. What is the scalar multiple of a vector $\mathbf{v}$ and a scalar c? Give a geometric interpretation.
   c. How are two nonzero vectors added? Illustrate geometrically.
   d. What is the difference of the vectors $\mathbf{v}$ and $\mathbf{w}$? Illustrate geometrically.

2. **a.** What is a vector in the xy-plane? Give an example.
   b. What is the position vector $\overrightarrow{P_1P_2}$ of the vector with initial point $P_1(a_1, a_2)$ and terminal point $P_2(b_1, b_2)$?
   c. What is the length of the vector of part (b)?

3. State the rule for vector addition and scalar multiplication.

4. **a.** What is a unit vector?
   b. If $\mathbf{a} \neq \mathbf{0}$, write $\mathbf{a}$ in terms of its magnitude and direction.
   c. What are the standard basis vectors in the xy-plane?
   d. If the vector $\mathbf{a}$ has magnitude 3 and makes an angle of $2\pi/3$ radians with the positive x-axis, what are its horizontal and vertical vector components? Make a sketch.

11.1 EXERCISES

1. Which quantity is a vector and which is a scalar? Explain.
   a. The amount of water in a swimming pool
   b. The speed and direction of a jet stream (a current of rapidly moving air found in the upper levels of the atmosphere) at a certain point
   c. The population of Los Angeles
   d. The initial speed and direction of a bullet as it leaves a gun

2. Classify each of the following as a scalar or a vector.
   a. temperature   **b.** momentum
   c. specific heat   **d.** weight
   e. work   **f.** density

3. State whether the expression makes sense. Explain your answer.
   a. $\mathbf{a} + c$ **b.** $|\mathbf{a}|\mathbf{b}$
   c. $2\mathbf{a} + \mathbf{b} + 0$ **d.** $\dfrac{\mathbf{a}}{\mathbf{b}}$
   e. $\dfrac{\mathbf{a}}{|\mathbf{b}|}, \quad \mathbf{b} \neq \mathbf{0}$ **f.** $\dfrac{|\mathbf{a}|\mathbf{b} - |\mathbf{b}|\mathbf{a}}{|\mathbf{a}|^2}, \quad \mathbf{a} \neq \mathbf{0}$

In Exercises 4–7, show that the vector **a** *and the position vector* **b** *are equal.*

In Exercises 8–11, find the vector $\overrightarrow{AB}$. *Then sketch the position vector that is equal to* $\overrightarrow{AB}$.

4.

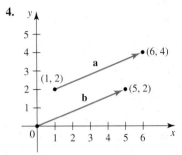

8.

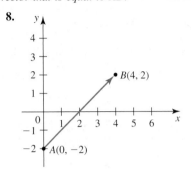

5.

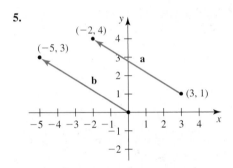

9.

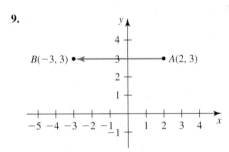

6.

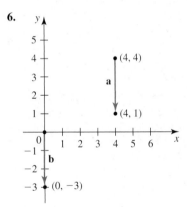

10.

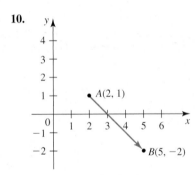

7.

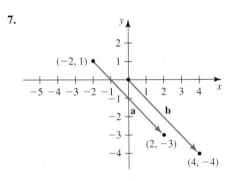

11.

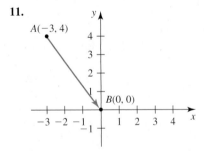

In Exercises 12–15, use the geometric interpretation of scalar multiplication and the parallelogram law of vector addition to sketch the indicated vector.

12. $\mathbf{a} + 2\mathbf{b}$

13. $2\mathbf{a} - 3\mathbf{b}$

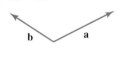

14. $\frac{1}{2}\mathbf{a} + \mathbf{b}$

15. $(\mathbf{a} + 2\mathbf{b}) + \mathbf{c}$

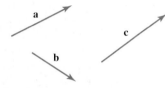

In Exercises 16–19, express the vector **v** *in terms of the vectors shown.*

16.

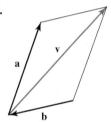

17.

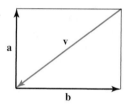

18.

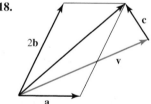

19.

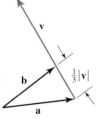

In Exercises 20–23, find the vector $\overrightarrow{AB}$. Sketch $\overrightarrow{AB}$ and the position vector that is equal to $\overrightarrow{AB}$.

20. $A(1, 3)$, $B(3, 4)$

21. $A(3, 4)$, $B(1, 3)$

22. $A\left(-\frac{1}{2}, -\frac{3}{2}\right)$, $B\left(2, \frac{1}{2}\right)$

23. $A(0.1, 0.5)$, $B(-0.2, 0.4)$

24. Suppose that $\overrightarrow{AB} = \langle 2, 3 \rangle$ and $A(-1, 1)$. Find B.

25. Suppose that $\overrightarrow{AB} = \langle -1, 4 \rangle$ and $B(0, 2)$. Find A.

In Exercises 26–31, find $2\mathbf{a}$, $\mathbf{a} + \mathbf{b}$, $\mathbf{a} - \mathbf{b}$, and $|2\mathbf{a} + \mathbf{b}|$.

26. $\mathbf{a} = \langle 1, 3 \rangle$ and $\mathbf{b} = \langle -2, 1 \rangle$

27. $\mathbf{a} = \langle -1, 2 \rangle$ and $\mathbf{b} = \langle 3, 1 \rangle$

28. $\mathbf{a} = 2\mathbf{i} - \mathbf{j}$ and $\mathbf{b} = 3\mathbf{i} + \mathbf{j}$

29. $\mathbf{a} = 3\mathbf{i} - 2\mathbf{j}$ and $\mathbf{b} = 2\mathbf{i}$

30. $\mathbf{a} = \langle 1, 2.4 \rangle$ and $\mathbf{b} = \langle -1, 0.4 \rangle$

31. $\mathbf{a} = \frac{1}{2}\mathbf{i} + \frac{3}{2}\mathbf{j}$ and $\mathbf{b} = \frac{3}{4}\mathbf{i} - \frac{1}{4}\mathbf{j}$

32. If $\mathbf{a} = \langle a_1, a_2 \rangle$ and $\mathbf{b} = \langle b_1, b_2 \rangle$ and if c is a scalar, what are $\mathbf{a} - \mathbf{b}$ and $c(\mathbf{a} + 2\mathbf{b})$?

In Exercises 33 and 34, find $2\mathbf{a} - 3\mathbf{b}$ and $\frac{1}{2}\mathbf{a} + \frac{1}{3}\mathbf{b}$.

33. $\mathbf{a} = 2\mathbf{i}$ and $\mathbf{b} = -6\mathbf{j}$

34. $\mathbf{a} = 2\mathbf{i} - 4\mathbf{j}$ and $\mathbf{b} = 2\mathbf{i} - \mathbf{j}$

35. Let $\mathbf{u} = \langle -1, 3 \rangle$, $\mathbf{v} = \langle 2, 4 \rangle$, and $\mathbf{w} = \langle 6, 4 \rangle$. Find a and b if $a\mathbf{u} + b\mathbf{v} = \mathbf{w}$.

36. Let $\mathbf{u} = \langle 2, 1 \rangle$, $\mathbf{v} = \langle 3, -1 \rangle$ and $\mathbf{w} = \langle 2, 4 \rangle$. Find a and b if $a\mathbf{u} - b\mathbf{v} = 2\mathbf{w}$.

In Exercises 37–42, determine whether **b** *is parallel to* $\mathbf{a} = 3\mathbf{i} - 2\mathbf{j}$.

37. $\mathbf{b} = 9\mathbf{i} - 6\mathbf{j}$

38. $\mathbf{b} = -\frac{3}{2}\mathbf{i} + \mathbf{j}$

39. $\mathbf{b} = \mathbf{i} - \mathbf{j}$

40. $\mathbf{b} = 6\mathbf{i} - 5\mathbf{j}$

41. $\mathbf{b} = \langle -1, 2 \rangle + 4\langle 1, -1 \rangle$

42. $\mathbf{b} = 6\langle -1, 2 \rangle + 4\langle 3, -4 \rangle$

43. Determine the value of c such that $\mathbf{a} = c\mathbf{i} - 2\mathbf{j}$ and $\mathbf{b} = -4\mathbf{i} + 3\mathbf{j}$ are parallel.

44. Prove that $\mathbf{a} = \langle a_1, a_2 \rangle$ and $\mathbf{b} = \langle b_1, b_2 \rangle$ are parallel if and only if $a_1 b_2 - a_2 b_1 = 0$.

In Exercises 45–48, find a unit vector that has (a) the same direction as **a** *and (b) a direction opposite to that of* **a**.

45. $\mathbf{a} = \langle 2, 1 \rangle$

46. $\mathbf{a} = -3\mathbf{i} + 4\mathbf{j}$

47. $\mathbf{a} = \langle -\sqrt{3}, 1 \rangle$

48. $\mathbf{a} = \langle 0, 3 \rangle$

In Exercises 49–52, find a vector **a** *with the given length and having the same direction as* **b**.

49. $|\mathbf{a}| = 5$; $\mathbf{b} = \langle 1, 1 \rangle$

50. $|\mathbf{a}| = 2$; $\mathbf{b} = -3\mathbf{i} + 4\mathbf{j}$

51. $|\mathbf{a}| = \sqrt{3}$; $\mathbf{b} = 3\mathbf{i}$

52. $|\mathbf{a}| = 4$; $\mathbf{b} = \langle \sqrt{3}, -1 \rangle$

53. Let $\mathbf{a} = \langle -3, 4 \rangle$ and $\mathbf{b} = \langle 1, 2 \rangle$. Find a vector with length 3 and having the same direction as $2\mathbf{a} - 3\mathbf{b}$.

54. Let $\mathbf{a} = \langle 1, -2 \rangle$ and $\mathbf{b} = \langle -1, 3 \rangle$. Find a vector with length 4 and having a direction opposite to that of $\mathbf{a} - 3\mathbf{b}$.

In Exercises 55–58, **F** represents a force acting on a particle. Express **F** in terms of its magnitude and direction. What are the horizontal and vertical vector components of **F**?

55. $\mathbf{F} = \langle 3, 1 \rangle$

56. $\mathbf{F} = \langle -3, 4 \rangle$

57. $\mathbf{F} = \sqrt{3}\mathbf{i} + 6\mathbf{j}$

58. $\mathbf{F} = \langle 0, -5 \rangle$

In Exercises 59–62, find a vector **a** that has the given length and makes an angle θ with the positive x-axis.

59. $|\mathbf{a}| = 2; \quad \theta = 0$

60. $|\mathbf{a}| = 5; \quad \theta = \dfrac{\pi}{3}$

61. $|\mathbf{a}| = 3; \quad \theta = \dfrac{5\pi}{3}$

62. $|\mathbf{a}| = 1; \quad \theta = \dfrac{\pi}{2}$

63. Production Planning The Acrosonic Company manufactures two different loudspeaker systems in two locations. Suppose that it produced a_1 Model A systems and b_1 Model B systems in Location I last year. Then we can record this data by writing the production vector $\mathbf{v}_1 = \langle a_1, b_1 \rangle$. Suppose further that the company also produced a_2 Model A systems and b_2 Model B systems in Location II in the same year. Then we can record this using the vector $\mathbf{v}_2 = \langle a_2, b_2 \rangle$.
   a. Find $\mathbf{v}_1 + \mathbf{v}_2$, and interpret your result.
   b. For the next year the company wishes to boost the production of both speaker systems by 10%. Write a vector reflecting the desired level of output.

64. Pulling a Sled A child's sled is pulled with a constant force **F** of magnitude 10 lb that makes an angle of 30° with the horizontal. Find the horizontal and vertical vector components $\mathbf{F}_1$ and $\mathbf{F}_2$ of the force.

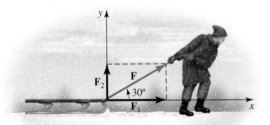

65. Velocity of a Shell A shell is fired from a howitzer at an angle of elevation of 45° and with an initial speed of 800 ft/sec. Find the horizontal and vertical vector components of its velocity.

66. Towing a Cruise Ship The following figure gives an aerial view of two tugboats attempting to free a cruise ship that ran aground during a storm. Tugboat I exerts a force of magnitude 3000 lb, and Tugboat II exerts a force of magnitude 2400 lb.
   a. Find expressions for the forces $\mathbf{F}_1$ and $\mathbf{F}_2$.
   b. Find the resultant force **F** acting on the cruise ship.
   c. Find the angle θ and the magnitude of **F** if **F** acts along the positive x-axis.

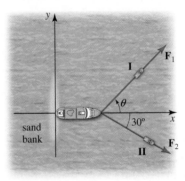

67. An object located at the origin is acted upon by three forces, $\mathbf{F}_1$, $\mathbf{F}_2$, and $\mathbf{F}_3$, as shown in the following figure. Find $\mathbf{F}_3$ if the object is in equilibrium.

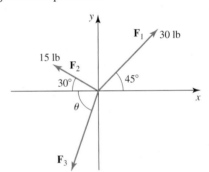

68. A model airplane on display in a hobby store is suspended by two wires attached to the ceiling as shown in the following figure. The model airplane weighs 2 lb. Find the tensions $\mathbf{T}_1$ and $\mathbf{T}_2$ of the wires.

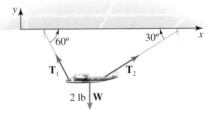

69. A river $\frac{1}{2}$ mi wide flows parallel to the shore at the rate of 5 mph. If a motorboat can move at 10 mph in still water, at what angle with respect to the shore should the boat be pointed to move in a direction perpendicular to the shore? How long will the boat take to cross the river?

70. Finding the True Course and Ground Speed of an Airplane An airplane on level flight has an airspeed of 300 mph and is headed in a direction that makes an angle of 30° with the north (measured in a clockwise direction). The plane is subjected to a headwind blowing at 60 mph in a direction that makes an angle of 270° with the north. Find the true course and ground speed of the plane.

71. Finding the True Course and Ground Speed of an Airplane An airplane pilot wishes to maintain a true course in the direction that makes an angle of 60° with the north (measured in a clockwise direction) and a ground speed of 240 mph when the wind is blowing directly west at 40 mph. Find the required airspeed and compass heading of the airplane.

72. Use vectors to prove that the line segment joining the midpoints of two sides of a triangle is parallel to the third side and half its length.

73. Use vectors to prove that the diagonals of a parallelogram bisect each other.

74. The opposite sides of a quadrilateral are parallel and of equal length. Use vectors to prove that the other two sides must be parallel and of equal length.

In Exercises 75–82, prove the stated property if $\mathbf{a} = \langle a_1, a_2 \rangle$, $\mathbf{b} = \langle b_1, b_2 \rangle$, $\mathbf{c} = \langle c_1, c_2 \rangle$, *and c and d are scalars.*

75. $(\mathbf{a} + \mathbf{b}) + \mathbf{c} = \mathbf{a} + (\mathbf{b} + \mathbf{c})$

76. $\mathbf{a} + \mathbf{0} = \mathbf{a}$

77. $\mathbf{a} + (-\mathbf{a}) = \mathbf{0}$

78. $c(\mathbf{a} + \mathbf{b}) = c\mathbf{a} + c\mathbf{b}$

79. $c(d\mathbf{a}) = (cd)\mathbf{a}$

80. $(c + d)\mathbf{a} = c\mathbf{a} + d\mathbf{a}$

81. $1\mathbf{a} = \mathbf{a}$

82. $c\mathbf{a} = \mathbf{0}$ if and only if $c = 0$ or $\mathbf{a} = \mathbf{0}$

In Exercises 83–88, determine whether the statement is true or false. If it is true, explain why. If it is false, explain why or give an example that shows it is false.

83. $\mathbf{v} - \mathbf{v} = 0$.

84. If $|\mathbf{v}| = |\mathbf{w}|$, then $\mathbf{v} = \mathbf{w}$.

85. If $\mathbf{u}$ is a unit vector having the same direction as $\mathbf{v}$, then $\mathbf{v} = |\mathbf{v}|\mathbf{u}$.

86. If $\mathbf{v} = a\mathbf{i} + b\mathbf{j}$, $\mathbf{w} = b\mathbf{i} - a\mathbf{j}$, and $\mathbf{v} = \mathbf{w}$, then $a = b = 0$.

87. If $\mathbf{v} = a\mathbf{i} + b\mathbf{j}$, where not both a and b are equal to zero, then

$$\mathbf{u} = \pm \left(\frac{a}{\sqrt{a^2 + b^2}}\mathbf{i} + \frac{b}{\sqrt{a^2 + b^2}}\mathbf{j} \right)$$

are two unit vectors having the same direction as $\mathbf{v}$.

88. If $P_1(x_1, y_1)$ and $P_2(x_2, y_2)$ are distinct points and the vector $\overrightarrow{P_1P_2}$ is parallel to $\mathbf{v} = \langle 1, 2 \rangle$, then there exists a nonzero constant c such that $x_2 = x_1 + c$ and $y_2 = y_1 + 2c$.

11.2 Coordinate Systems and Vectors in 3-Space

■ Coordinate Systems in Space

The plane curve C shown in Figure 1a gives the path taken by an aircraft as it taxis to the runway. The position of the plane may be specified by the coordinates of the point $P(x, y)$ lying on the curve C. Figure 1b shows the flight path of the plane shortly *after* takeoff. Because the aircraft is now in the air, we also need to specify its altitude when giving its position. This can be done by introducing an axis that is perpendicular to the x- and y-axes at the origin. The position of the plane may then be specified by giving the *three* **coordinates** x, y, and z of the point P represented by the **ordered triple** (x, y, z). Here, the number z gives the altitude of the plane.

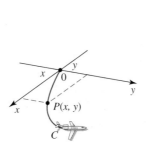

(a) The path of a plane taxiing on the ground

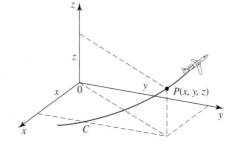

(b) The plane's path after takeoff

FIGURE 1

The three positive axes that we have just drawn in Figure 1b are part of a three-dimensional coordinate system. Figure 2 shows a **three-dimensional rectangular coordinate system** along with the points $A(2, 4, 5)$, $B(3, -4, -2)$, and $C(-2, -3, 3)$.

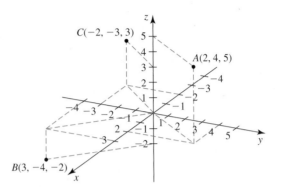

FIGURE 2

The points $A(2, 4, 5)$, $B(3, -4, -2)$, and $C(-2, -3, 3)$

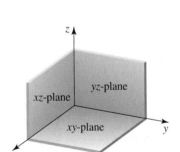

FIGURE 3
The right-handed system

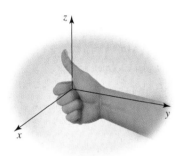

FIGURE 4
The three coordinate planes

The coordinate systems shown in Figure 1 and Figure 2 are **right-handed:** If you start by pointing the fingers of your right hand in the direction of the positive x-axis and then curl them toward the positive y-axis, your thumb will point in the positive direction of the z-axis (see Figure 3).

The three coordinate axes determine the three coordinate planes: The xy-plane is determined by the x- and y-axes, the yz-plane is determined by the y- and z-axes, and the xz-plane is determined by the x- and z-axes. (See Figure 4.) These coordinate planes divide 3-space into eight **octants.** The first octant is the one determined by the positive axes.

Just as an equation in x and y represents a curve in the plane, an equation in x, y, and z represents a *surface* in 3-space. The simplest surfaces in 3-space, other than the coordinate planes, are the planes that are parallel to the coordinate planes.

EXAMPLE 1 Sketch the surface represented by the equation

a. $x = 3$ **b.** $z = 4$

Solution

a. The equation $x = 3$ tells us that the surface consists of the set of points in 3-space whose x-coordinate is held fast at 3 while y and z are allowed to range over all real numbers, written $\{(x, y, z) \mid x = 3\}$. This surface is the plane that is parallel to the yz-plane and located three units in front of it. (See Figure 5.)

b. Similarly, we see that the equation $z = 4$ represents the set $\{(x, y, z) \mid z = 4\}$ and is the plane that is parallel to the xy-plane and located four units above it.

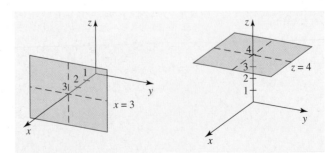

FIGURE 5
The planes $x = 3$ and $z = 4$

⚠ Keep in mind the dimensions you are working in. In the *xy*-plane (2-space) the equation $x = 3$ represents the vertical line parallel to the *y*-axis; in the *xyz*-plane (3-space) the equation $x = 3$ represents a plane parallel to the *yz*-plane, as we have just seen.

In general, if *k* is a constant, then $x = k$ represents a plane that is parallel to the *yz*-plane; $y = k$ represents a plane that is parallel to the *xz*-plane; and $z = k$ represents a plane that is parallel to the *xy*-plane.

■ The Distance Formula

To find a formula for the distance between two points $P_1(x_1, y_1, z_1)$ and $P_2(x_2, y_2, z_2)$ in 3-space, refer to Figure 6. First, apply the distance formula in 2-space to see that the distance between $P_1'(x_1, y_1, 0)$ and $P_2'(x_2, y_2, 0)$, the respective projections of $P_1(x_1, y_1, z_1)$ and $P_2(x_2, y_2, z_2)$ onto the *xy*-plane, is

$$d(P_1', P_2') = \sqrt{(x_2 - x_1)^2 + (y_2 - y_1)^2}$$

But this is also the distance $d(P_1, Q)$ between the points $P_1(x_1, y_1, z_1)$ and $Q(x_2, y_2, z_1)$. Then applying the Pythagorean Theorem to the right triangle P_1QP_2, we have

$$[d(P_1, P_2)]^2 = [d(P_1, Q)]^2 + [d(P_2, Q)]^2 = (x_2 - x_1)^2 + (y_2 - y_1)^2 + (z_2 - z_1)^2$$

which is equivalent to the following.

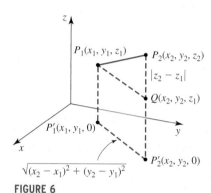

FIGURE 6

The Distance Formula

$$d(P_1, P_2) = \sqrt{(x_2 - x_1)^2 + (y_2 - y_1)^2 + (z_2 - z_1)^2} \qquad \textbf{(1)}$$

EXAMPLE 2 Find the distance between $(3, -2, 1)$ and $(1, 0, 3)$.

Solution Using the distance formula (1) with $P_1(3, -2, 1)$ and $P_2(1, 0, 3)$, we find that the required distance is

$$d = \sqrt{(1 - 3)^2 + [0 - (-2)]^2 + (3 - 1)^2}$$
$$= \sqrt{4 + 4 + 4} = \sqrt{12} = 2\sqrt{3} \qquad ■$$

■ The Midpoint Formula

The formula for finding the coordinates of the midpoint of the line segment joining two points $P_1(x_1, y_1, z_1)$ and $P_2(x_2, y_2, z_2)$ in 3-space is just an extension of that for finding the coordinates of the midpoint of the line segment joining two points in the plane. (See Exercise 77.)

The Midpoint Formula

$$\left(\frac{x_1 + x_2}{2}, \frac{y_1 + y_2}{2}, \frac{z_1 + z_2}{2} \right) \qquad \textbf{(2)}$$

EXAMPLE 3 Find the midpoint of the line segment joining $(3, -2, 1)$ and $(1, 0, 3)$.

Solution   Using the midpoint formula (2) with $P_1(3, -2, 1)$ and $P_2(1, 0, 3)$, we find that the midpoint is

$$\left(\frac{3+1}{2}, \frac{-2+0}{2}, \frac{1+3}{2} \right) \quad \text{or} \quad (2, -1, 2) \qquad ▪$$

As the next example illustrates, we can also use the distance formula to help us find an equation of a sphere.

EXAMPLE 4

a. Find an equation of the sphere with center $C(h, k, l)$ and radius r.
b. Find an equation of the sphere that has a diameter with endpoints $(3, -2, 1)$ and $(1, 0, 3)$.

Solution
a. The sphere is the set of all points $P(x, y, z)$ whose distance from $C(h, k, l)$ is r, or, equivalently, the square of the distance from P to C is r^2. Using the distance formula, we see that an equation of the sphere is

$$(x - h)^2 + (y - k)^2 + (z - l)^2 = r^2$$

b. From Example 2 we see that the distance between $(3, -2, 1)$ and $(1, 0, 3)$ is $2\sqrt{3}$, so the radius of the sphere is $\frac{1}{2}(2\sqrt{3})$, or $\sqrt{3}$. Next, from Example 3 we see that the midpoint of the line segment joining $(3, -2, 1)$ and $(1, 0, 3)$ is $(2, -1, 2)$. This point is the center of the sphere. Finally, using the result of part (a), we obtain the equation of the sphere:

$$(x - 2)^2 + (y + 1)^2 + (z - 2)^2 = 3 \qquad ▪$$

The equation that we obtained in Example 4a is called the *standard equation* of a sphere.

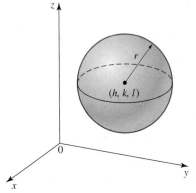

FIGURE 7
The sphere with center $C(h, k, l)$ and radius r

The Standard Equation of a Sphere with Center (h, k, l) and Radius r
$$(x - h)^2 + (y - k)^2 + (z - l)^2 = r^2 \qquad\qquad (3)$$

The graph of this equation appears in Figure 7.

EXAMPLE 5 Show that $x^2 + y^2 + z^2 - 4x + 2y + 6z + 5 = 0$ is an equation of a sphere, and find its center and radius.

Solution   By completing the squares in x, y, and z, we can write the given equation in the form

$$[x^2 - 4x + (-2)^2] + (y^2 + 2y + 1) + (z^2 + 6z + 9) = -5 + 4 + 1 + 9$$

or

$$(x - 2)^2 + (y + 1)^2 + (z + 3)^2 = 3^2$$

Comparing this equation with Equation (3), we conclude that it is an equation of the sphere of radius 3 with center at $(2, -1, -3)$. ▪

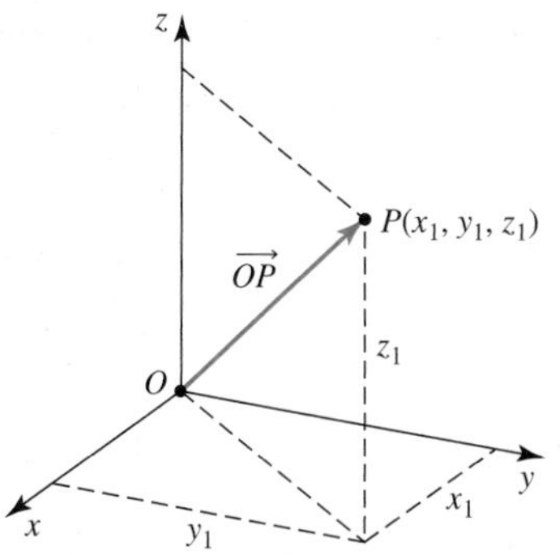

z

$P(x_1, y_1, z_1)$

$\overrightarrow{OP}$

O

z_1

x_1

y

y_1

x

FIGURE 8
The position vector $\overrightarrow{OP}$ has initial point
O and terminal point P.

■ Vectors in 3-Space

A vector in 3-space is an ordered triple of real numbers

$$\mathbf{a} = \langle a_1, a_2, a_3 \rangle$$

where a_1, a_2, and a_3 are the **components** of the vector. In particular, the **position vector** of a point $P(x_1, y_1, z_1)$ is the vector $\overrightarrow{OP} = \langle x_1, y_1, z_1 \rangle$ with initial point at the origin and terminal point $P(x_1, y_1, z_1)$. (See Figure 8.)

The basic definitions and operations of vectors in 3-space are natural generalizations of those of vectors in the plane.

> **DEFINITION** **Vectors in 3-Space**
>
> If $\mathbf{a} = \langle a_1, a_2, a_3 \rangle$ and $\mathbf{b} = \langle b_1, b_2, b_3 \rangle$ are vectors in 3-space and c is a scalar, then
>
> 1. $\mathbf{a} = \mathbf{b}$ if and only if $a_1 = b_1, a_2 = b_2$ and $a_3 = b_3$ Equality
> 2. $\mathbf{a} + \mathbf{b} = \langle a_1 + b_1, a_2 + b_2, a_3 + b_3 \rangle$ Vector addition
> 3. $c\mathbf{a} = \langle ca_1, ca_2, ca_3 \rangle$ Scalar multiplication
> 4. $|\mathbf{a}| = \sqrt{a_1^2 + a_2^2 + a_3^2}$ Length

Also, the rules of vector addition and scalar multiplication stated in Theorem 1 of Section 11.1 are valid for vectors in 3-space. The proofs are similar.

The following representation of a vector $\mathbf{a}$ in 3-space is a natural extension of the representation of vectors in the plane.

> The vector with initial point $P_1(x_1, y_1, z_1)$ and terminal point $P_2(x_2, y_2, z_2)$ is
>
> $$\overrightarrow{P_1P_2} = \langle x_2 - x_1, y_2 - y_1, z_2 - z_1 \rangle \tag{4}$$

Thus, we can find the components of a vector by subtracting the respective coordinates of its initial point from the coordinates of its terminal point, as illustrated in Figure 9. The vectors $\overrightarrow{OP_1}$ and $\overrightarrow{OP_2}$ are the position vectors of the point $P_1(x_1, y_1, z_1)$ and $P_2(x_2, y_2, z_2)$. As a natural extension of the definition of vector subtraction in 2-space into 3-space, we have

$$\overrightarrow{P_1P_2} = \overrightarrow{OP_2} - \overrightarrow{OP_1} = \langle x_2, y_2, z_2 \rangle - \langle x_1, y_1, z_1 \rangle = \langle x_2 - x_1, y_2 - y_1, z_2 - z_1 \rangle$$

FIGURE 9
$\overrightarrow{P_1P_2} = \langle x_2 - x_1, y_2 - y_1, z_2 - z_1 \rangle$
is represented by the position
vector $\overrightarrow{OP}$ of the point
$P(x_2 - x_1, y_2 - y_1, z_2 - z_1)$.

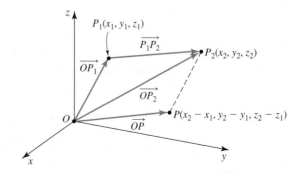

By considering the parallelogram OPP_2P_1 in Figure 9, you can convince yourself that $\overrightarrow{P_1P_2}$ is represented by the position vector $\overrightarrow{OP}$ of the point $(x_2 - x_1, y_2 - y_1, z_2 - z_1)$.

EXAMPLE 6   Let $P(2, -1, 2)$ and $Q(1, 4, 5)$ be two points in 3-space.

a. Find the vector $\overrightarrow{PQ}$.
b. Find $|\overrightarrow{PQ}|$.
c. Find a unit vector having the same direction as $\overrightarrow{PQ}$.

Solution

a. Using Equation (4) with $P_1 = P$ and $P_2 = Q$, we have

$$\overrightarrow{PQ} = \langle 1 - 2, 4 - (-1), 5 - 2 \rangle = \langle -1, 5, 3 \rangle$$

b. Using the result of part (a), we have

$$|\overrightarrow{PQ}| = \sqrt{(-1)^2 + 5^2 + 3^2} = \sqrt{35}$$

c. Using the results of parts (a) and (b), we obtain the unit vector

$$\mathbf{u} = \frac{\overrightarrow{PQ}}{|\overrightarrow{PQ}|} = \frac{1}{\sqrt{35}}\langle -1, 5, 3 \rangle$$

The vector $\overrightarrow{PQ}$, the position vector $\mathbf{a}$ which equals $\overrightarrow{PQ}$, and the unit (position) vector $\mathbf{u}$ are shown in Figure 10.

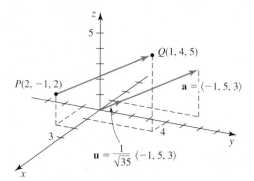

FIGURE 10
The vector $\overrightarrow{PQ}$, its equivalent position vector $\mathbf{a}$, and the (position) unit vector $\mathbf{u}$

■ Standard Basis Vectors in Space

In Section 11.1 we saw that any vector in the plane can be expressed in terms of the standard basis vectors $\mathbf{i} = \langle 1, 0 \rangle$ and $\mathbf{j} = \langle 0, 1 \rangle$. In three-dimensional space, the 3-space vectors

$$\mathbf{i} = \langle 1, 0, 0 \rangle, \qquad \mathbf{j} = \langle 0, 1, 0 \rangle, \qquad \text{and} \qquad \mathbf{k} = \langle 0, 0, 1 \rangle$$

form a basis for the space, in the sense that any vector in the space can be expressed in terms of these vectors. In fact, if $\mathbf{a} = \langle a_1, a_2, a_3 \rangle$ is a vector in 3-space, we can write

$$\mathbf{a} = \langle a_1, a_2, a_3 \rangle = \langle a_1, 0, 0 \rangle + \langle 0, a_2, 0 \rangle + \langle 0, 0, a_3 \rangle$$

$$= a_1\langle 1, 0, 0 \rangle + a_2\langle 0, 1, 0 \rangle + a_3\langle 0, 0, 1 \rangle$$

$$= a_1\mathbf{i} + a_2\mathbf{j} + a_3\mathbf{k}$$

The standard basis vectors **i**, **j**, and **k** are shown in Figure 11a. Figure 11b shows the vector **a** and its three vector components $a_1\mathbf{i}$, $a_2\mathbf{j}$, and $a_3\mathbf{k}$ in the x-, y-, and z-directions, respectively.

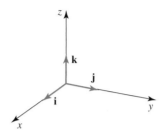

(**a**) The standard basis vectors **i**, **j**, and **k**

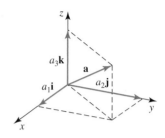

(**b**) The vectors $a_1\mathbf{i}$, $a_2\mathbf{j}$, and $a_3\mathbf{k}$ are the vector components of **a** in the x-, y-, and z-directions.

FIGURE 11

EXAMPLE 7 Write $\mathbf{a} = \langle -1, 2, -3 \rangle$ and $\mathbf{b} = \langle 2, 0, 4 \rangle$ in terms of the standard basis vectors **i**, **j**, and **k**. Then compute $2\mathbf{a} - 3\mathbf{b}$.

Solution We have

$$\mathbf{a} = \langle -1, 2, -3 \rangle = -\mathbf{i} + 2\mathbf{j} - 3\mathbf{k}$$

and

$$\mathbf{b} = \langle 2, 0, 4 \rangle = 2\mathbf{i} + 0\mathbf{j} + 4\mathbf{k} = 2\mathbf{i} + 4\mathbf{k}$$

Next, we find

$$2\mathbf{a} - 3\mathbf{b} = 2(-\mathbf{i} + 2\mathbf{j} - 3\mathbf{k}) - 3(2\mathbf{i} + 4\mathbf{k})$$

$$= -2\mathbf{i} + 4\mathbf{j} - 6\mathbf{k} - 6\mathbf{i} - 12\mathbf{k}$$

$$= -8\mathbf{i} + 4\mathbf{j} - 18\mathbf{k} \qquad \text{Add components.}$$

which can also be written $\langle -8, 4, -18 \rangle$. ∎

11.2 CONCEPT QUESTIONS

1. **a.** What is a right-handed rectangular coordinate system in a three-dimensional space? Plot the point $(-2, 3, 4)$ in this system.
   b. What is the distance between the points $P_1(a_1, b_1, c_1)$ and $P_2(a_2, b_2, c_2)$?
   c. What is the point midway between P_1 and P_2 of part (b)?
2. What is the standard equation of a sphere with center $C(h, k, l)$ and radius r? What happens if $r < 0$? If $r = 0$?

3. **a.** If $\mathbf{a} = \langle a_1, a_2, a_3 \rangle$ and $\mathbf{b} = \langle b_1, b_2, b_3 \rangle$, what is the sum of **a** and **b**, the scalar multiple of **a** by the scalar c, and the length of **a**?
   b. If $P_1(a_1, a_2, a_3)$ and $P_2(b_1, b_2, b_3)$ are points in 3-space, what is $\overrightarrow{P_1P_2}$? $\overrightarrow{P_2P_1}$? How are the two vectors related?
4. **a.** What are the standard basis vectors in a three-dimensional space? Make a sketch.
   b. If $P_1(a_1, a_2, a_3)$ and $P_2(b_1, b_2, b_3)$ are points in 3-space, write $\overrightarrow{P_1P_2}$ in terms of the standard basis vectors.

11.2 EXERCISES

In Exercises 1–6, plot the given points in a three-dimensional coordinate system.

1. (3, 2, 4)   **2.** (2, 3, 2)

3. (3, −1, 4)   **4.** (0, 2, 4)

5. (−3, −2, 4)   **6.** (−2, 0, 4)

In Exercises 7 and 8, find the coordinates of the indicated points.

7.

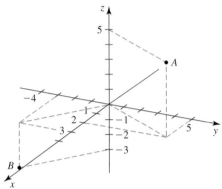

8.

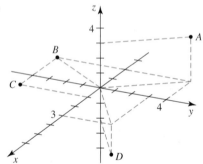

In Exercises 9–12, sketch the plane in three-dimensional space represented by the equation.

9. $y = 5$ **10.** $x = -4$

11. $z = 4$ **12.** $z = -4$

In Exercises 13–16, describe the region in three-dimensional space represented by the inequality.

13. $x \geq 3$ **14.** $y \leq -4$

15. $z > 3$ **16.** $y < 4$

17. Air Traffic Control Suppose that the control tower of a municipal airport is located at the origin of a three-dimensional coordinate system with orientation as shown in the figure. At an instant of time, Plane A is 1000 ft west and 2000 ft south of the tower and flying at an altitude of 3000 ft, and Plane B is 4000 ft east and 1000 ft north of the tower and flying at an altitude of 1000 ft.

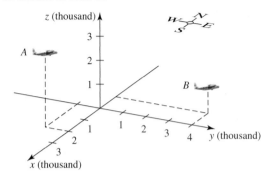

a. Write the coordinates of Plane A and Plane B.

b. How far apart are the planes at that instant of time?

18. After holding a short conversation with each other, Jack and Jill proceeded to return to their downtown offices. Jack walked 1 block east, then 2 blocks north, then took the elevator to the thirtieth floor. At the moment Jack emerged from the elevator, Jill had walked $1\frac{1}{2}$ blocks south and $1\frac{1}{4}$ blocks west. If the length of a city block is 1000 ft and a story in Jack's office building is 10 ft high, how far apart are Jack and Jill at that moment?

In Exercises 19–22, find the length of each side of the triangle ABC and determine whether the triangle is an isosceles triangle, a right triangle, both, or neither.

19. $A(0, 1, 2)$, $B(4, 3, 3)$, $C(3, 4, 5)$

20. $A(3, 4, 1)$, $B(4, 4, 6)$, $C(3, 1, 2)$

21. $A(-1, 0, 1)$, $B(1, 1, -1)$, $C(1, 1, 1)$

22. $A(-1, 5, 2)$, $B(1, -1, 2)$, $C(-3, 1, -2)$

In Exercises 23 and 24, determine whether the given points are collinear.

23. $A(2, 3, 2)$, $B(-4, 0, 5)$, and $C(4, 4, 1)$

24. $A(-1, 3, -2)$, $B(2, 1, -1)$, and $C(8, -3, 1)$

In Exercises 25 and 26, find the midpoint of the line segment joining the given points.

25. (2, 4, −6) and (−4, 2, 4)   **26.** $\left(\frac{1}{2}, -4, 2\right)$ and $\left(\frac{3}{2}, 2, -4\right)$

In Exercises 27–30, find the standard equation of the sphere with center C and radius r.

27. $C(2, 1, 3)$; $r = 3$ **28.** $C(3, 2, 0)$; $r = \sqrt{3}$

29. $C(3, -1, 2)$; $r = 4$ **30.** $C(1, \sqrt{2}, -2)$; $r = 5$

31. Find an equation of the sphere that has the points $(2, -3, 4)$ and $(3, 2, 1)$ at opposite ends of its diameter.

32. Find an equation of the sphere centered at the point $(2, -3, 4)$ and tangent to the xy-plane.

33. Find an equation of the sphere that contains the point $(1, 3, 5)$ and is centered at the point $(-1, 2, 4)$.

34. Find an equation of the sphere centered at the point $(2, 3, 6)$ and tangent to the sphere with equation $x^2 + y^2 + z^2 = 9$.

In Exercises 35–40, find the center and the radius of the sphere that has the given equation.

35. $x^2 + y^2 + z^2 - 2x - 4y - 6z + 10 = 0$

36. $x^2 + y^2 + z^2 + 4x - 5y + 2z + 5 = 0$

37. $x^2 + y^2 + z^2 - 4x + 6y = 0$

38. $x^2 + y^2 + z^2 = y$

39. $2x^2 + 2y^2 + 2z^2 - 6x - 4y + 2z = 1$

40. $3x^2 + 3y^2 + 3z^2 = 6z + 1$

In Exercises 41–44, describe the region in 3-space satisfying the inequality or inequalities.

41. $x^2 + y^2 + z^2 < 4$

42. $x^2 + y^2 + z^2 - 2x - 4y + 2z + 5 \geq 0$

43. $1 \leq x^2 + y^2 + z^2 \leq 9$

44. $x^2 + y^2 + z^2 \leq 4, \quad z \geq 0$

In Exercises 45–48, find the vector $\overrightarrow{AB}$. Then sketch the position vector that is equal to $\overrightarrow{AB}$.

45.

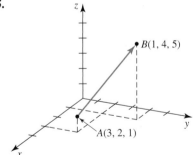

46.

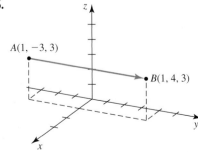

47.

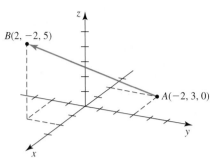

48.

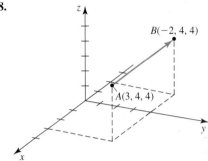

In Exercises 49 and 50, find the vector $\overrightarrow{AB}$. Then sketch $\overrightarrow{AB}$ and the position vector that is equal to $\overrightarrow{AB}$.

49. $A(2, 1, 0), B(1, 4, 5)$

50. $A(-2, -1, 2), B(1, 3, 4)$

51. Suppose that $\overrightarrow{AB} = \langle -1, 3, 4 \rangle$ and $B(2, -3, 1)$. Find A.

52. Suppose that $\overrightarrow{AB} = \langle 3, 0, 4 \rangle$ and $A(-1, -2, -4)$. Find the midpoint of the line segment joining A and B.

In Exercises 53–56, find $\mathbf{a} + \mathbf{b}$, $2\mathbf{a} - 3\mathbf{b}$, $|3\mathbf{a}|$, $|-2\mathbf{b}|$, and $|\mathbf{a} - \mathbf{b}|$.

53. $\mathbf{a} = \langle -1, 2, 0 \rangle$ and $\mathbf{b} = \langle 2, 3, -1 \rangle$

54. $\mathbf{a} = 2\mathbf{i} - \mathbf{j} + \mathbf{k}$ and $\mathbf{b} = 3\mathbf{i} + 2\mathbf{k}$

55. $\mathbf{a} = \langle 0, 2.1, 3.4 \rangle$ and $\mathbf{b} = \langle 1, 4.1, -5.6 \rangle$

56. $\mathbf{a} = -2\mathbf{i} + 4\mathbf{k}$ and $\mathbf{b} = 2\mathbf{j} - \mathbf{k}$

57. Let $\mathbf{u} = \langle -1, 3, -2 \rangle$, $\mathbf{v} = \langle 2, 1, 4 \rangle$, and $\mathbf{w} = \langle 3, -2, 1 \rangle$. Find a, b, and c if $a\mathbf{u} + b\mathbf{v} + c\mathbf{w} = \langle 2, 0, 2 \rangle$.

58. Refer to Exercise 57. Show that $\langle 2, 0, 2 \rangle$ cannot be written in the form $a\mathbf{u} + b\mathbf{v}$ for any choice of a and b.

In Exercises 59–62, determine whether $\mathbf{b}$ is parallel to $\mathbf{a} = \mathbf{i} - 2\mathbf{j} + 5\mathbf{k}$.

59. $\mathbf{b} = \langle 3, -6, 15 \rangle$ **60.** $\mathbf{b} = \dfrac{1}{3}\mathbf{i} - \dfrac{2}{3}\mathbf{j} + \dfrac{5}{3}\mathbf{k}$

61. $\mathbf{b} = 2\mathbf{i} - 3\mathbf{j} + 10\mathbf{k}$ **62.** $\mathbf{b} = \langle -2, 4, -10 \rangle$

In Exercises 63–66, find a unit vector that has (a) *the same direction as* **a** *and* (b) *a direction opposite to that of* **a**.

63. $\mathbf{a} = \langle 1, 2, 2 \rangle$ **64.** $\mathbf{a} = -3\mathbf{i} + 4\mathbf{j} + 5\mathbf{k}$

65. $\mathbf{a} = -\mathbf{i} + 3\mathbf{j} - \mathbf{k}$ **66.** $\mathbf{a} = -2\langle 0, -3, 4 \rangle$

In Exercises 67–70, find a vector **a** *that has the given length and the same direction as* **b**.

67. $|\mathbf{a}| = 10$; $\mathbf{b} = \langle 1, 1, 1 \rangle$ **68.** $|\mathbf{a}| = 2$; $\mathbf{b} = \mathbf{i} - 2\mathbf{j} + 3\mathbf{k}$

69. $|\mathbf{a}| = 3$; $\mathbf{b} = 2\mathbf{i} + 4\mathbf{j}$ **70.** $|\mathbf{a}| = 4$; $\mathbf{b} = \langle -1, 0, 1 \rangle$

71. Let $\mathbf{a} = \langle 3, -1, 2 \rangle$ and $\mathbf{b} = \langle 1, 0, -1 \rangle$. Find a vector that has length 2 and the same direction as $\mathbf{a} - 2\mathbf{b}$.

72. Let $\mathbf{a} = \langle 1, 0, -2 \rangle$ and $\mathbf{b} = \langle 3, 4, 1 \rangle$. Find a vector that has length $|2\mathbf{a} + \mathbf{b}|$ and a direction opposite to that of $\mathbf{a} - \mathbf{b}$.

In Exercises 73 *and* 74, **F** *represents a force acting on a particle. Express* **F** *in terms of its magnitude and direction.*

73. $\mathbf{F} = \langle -3, 4, 5 \rangle$ **74.** $\mathbf{F} = 2\mathbf{i} + 3\mathbf{j} - 4\mathbf{k}$

75. Refer to the figure below. Show that

$$|\mathbf{a} - \mathbf{b}|^2 = |\mathbf{a}|^2 + |\mathbf{b}|^2 - 2|\mathbf{a}||\mathbf{b}| \cos \theta$$

where θ is the angle between **a** and **b**.

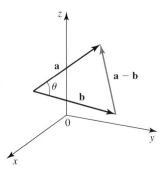

76. Refer to the figure below. Show that

$$|\mathbf{a} + \mathbf{b}|^2 = |\mathbf{a}|^2 + |\mathbf{b}|^2 + 2|\mathbf{a}||\mathbf{b}| \cos \theta$$

where θ is the angle between **a** and **b**.

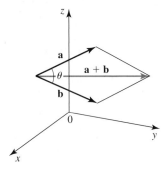

77. Prove that the midpoint of the line segment joining the points $P_1(x_1, y_1, z_1)$ and $P_2(x_2, y_2, z_2)$ is

$$\left(\frac{x_1 + x_2}{2}, \frac{y_1 + y_2}{2}, \frac{z_1 + z_2}{2} \right)$$

78. A person standing on a bridge watches a canoe go by. The canoe moves at a constant speed of 5 ft/sec in a direction parallel to the y-axis. Find a formula for the distance between the spectator and the canoe. How fast is this distance changing when the canoe is 60 ft ($y = 60$) from the bridge?

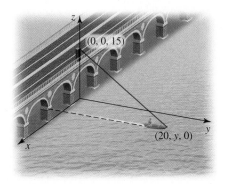

79. Newton's Law of Gravitation Newton's law of gravitation states that every particle of matter in the universe attracts every other particle with a force whose magnitude is proportional to the product of the masses of the particles and inversely proportional to the square of the distance between them. Show that if a particle of mass m_1 is located at a point A and a particle of mass m_2 is located at a point B, then the force of attraction exerted by the particle located at A on the particle located at B is

$$\mathbf{F} = \frac{Gm_1m_2}{|\overrightarrow{BA}|^3} \overrightarrow{BA}$$

where G is a positive constant.

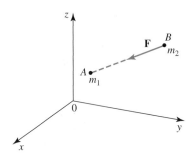

80. Refer to Exercise 79. Suppose that two particles of mass m_1 and m_2 are located at the origin and the point $(d, 0)$, respectively, in a two-dimensional coordinate system. Further, suppose that a third particle of mass m is located at the point $P(x, y)$.

a. Write a vector $\mathbf{F}$ giving the force exerted by the particles of mass m_1 and m_2, respectively, on m.

b. Where should the mass m be located so that the system is in equilibrium?

81. **Coulomb's Law** Coulomb's law states that the force of attraction or repulsion between two point charges (that is, charged bodies whose sizes are small in comparison to the distance between them) is directly proportional to the product of the charges and inversely proportional to the square of the distance between them. Show that if the charges q_1 and q_2 are located at the points A and B (see the figure below), respectively, then the force $\mathbf{F}_1$ exerted by the charge located at A on the charge located at B is given by

$$\mathbf{F}_1 = \frac{kq_1q_2}{|\overrightarrow{AB}|^3}\overrightarrow{AB}$$

and the force $\mathbf{F}_2$ exerted by the charge located at B on the charge located at A is given by

$$\mathbf{F}_2 = \frac{kq_1q_2}{|\overrightarrow{BA}|^3}\overrightarrow{BA}$$

where k is the constant of proportionality.

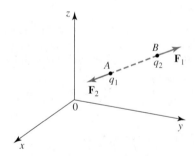

82. Refer to Exercise 81. Suppose that point charges $q_1, q_2, \ldots, q_n$ are placed at the points $P_1, P_2, \ldots, P_n$, respectively.

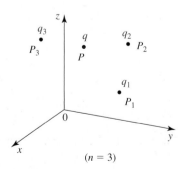

$(n = 3)$

According to the principle of superposition, the total force $\mathbf{F}$ exerted by these charges on the charge q located at the point P is given by

$$\mathbf{F} = kq\left(\frac{q_1}{|\overrightarrow{P_1P}|^3}\overrightarrow{P_1P} + \frac{q_2}{|\overrightarrow{P_2P}|^3}\overrightarrow{P_2P} + \cdots + \frac{q_n}{|\overrightarrow{P_nP}|^3}\overrightarrow{P_nP}\right)$$

$$= kq\sum_{i=1}^{n}\frac{q_i}{|\overrightarrow{P_iP}|^3}\overrightarrow{P_iP}$$

where k is a constant of proportionality.

a. Four equal charges are placed at the points $P_1(d, 0, 0)$, $P_2(0, d, 0)$, $P_3(-d, 0, 0)$, and $P_4(0, -d, 0)$, where $d > 0$ (see the following figure). Find the total force $\mathbf{F}$ exerted by these charges on the charge q_0 placed at the point $(0, 0, z)$ on the z-axis.

b. Does the result of part (a) agree with what you might expect if z is very large in comparison to d?

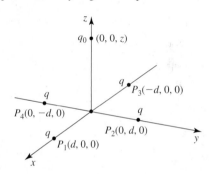

In Exercises 83–86, determine whether the statement is true or false. If it is true, explain why. If it is false, explain why or give an example that shows it is false.

83. The equation $x^2 + 2y^2 + z^2 + Ex + Fy + G = 0$, where E, F, and G are constants, cannot be that of a sphere.

84. The set of points (x, y, z) satisfying $(x - 1)^2 + (y - 2)^2 + (z - 3)^2 \leq 1$ and $(x - 1)^2 + (y - 2)^2 + (z - 3)^2 > 2$ is the empty set.

85. If P_1, P_2, Q_1 and Q_2 are points in 3-space and $\overrightarrow{P_1P_2} = \overrightarrow{Q_1Q_2}$, then $P_1 = Q_1$ and $P_2 = Q_2$.

86. If $|c\mathbf{a}| = 1$ and $\mathbf{a} \neq \mathbf{0}$, then $c = 1/|\mathbf{a}|$.

11.3 The Dot Product

■ Finding the Dot Product

So far, we have looked at two operations involving vectors: vector addition and scalar multiplication. In vector addition, two vectors are combined to yield another vector, and in scalar multiplication a scalar and a vector are combined to yield another vector. In this section we will look at another way of combining two vectors. This operation, called the *dot product,* combines two vectors to yield a *scalar.* As we shall see shortly, the dot product plays a role in the computation of many quantities: the length of a vector, the angle between two vectors, and the work done by a force in moving an object from one point to another, just to mention a few.

DEFINITION　**Dot Product**

Let $\mathbf{a} = \langle a_1, a_2, a_3 \rangle$ and $\mathbf{b} = \langle b_1, b_2, b_3 \rangle$ be any two vectors in space. Then the **dot product** of $\mathbf{a}$ and $\mathbf{b}$ is the number $\mathbf{a} \cdot \mathbf{b}$ defined by

$$\mathbf{a} \cdot \mathbf{b} = a_1 b_1 + a_2 b_2 + a_3 b_3$$

Thus, we can find the dot product of the two vectors $\mathbf{a}$ and $\mathbf{b}$ by adding the products of their corresponding components.

Notes

1. The dot product of two vectors is also called the **inner product,** or **scalar product,** of the two vectors.
2. The definition just given pertains to the dot product of two three-dimensional vectors. For vectors in two-dimensional space, the definition is

$$\mathbf{a} \cdot \mathbf{b} = \langle a_1, a_2 \rangle \cdot \langle b_1, b_2 \rangle = a_1 b_1 + a_2 b_2 \qquad ■$$

EXAMPLE 1　Find the dot product of each pair of vectors:

a. $\mathbf{a} = \langle 1, 3 \rangle$　and　$\mathbf{b} = \langle -1, 2 \rangle$　　　**b.** $\mathbf{a} = \langle 1, -2, 4 \rangle$　and　$\mathbf{b} = \langle -1, -2, 3 \rangle$

Solution

a. $\mathbf{a} \cdot \mathbf{b} = \langle 1, 3 \rangle \cdot \langle -1, 2 \rangle$
$\qquad = (1)(-1) + (3)(2) = 5$

b. $\mathbf{a} \cdot \mathbf{b} = \langle 1, -2, 4 \rangle \cdot \langle -1, -2, 3 \rangle$
$\qquad = (1)(-1) + (-2)(-2) + (4)(3) = 15 \qquad ■$

The dot product obeys the following rules.

Properties of the Dot Product

Let $\mathbf{a}$, $\mathbf{b}$, and $\mathbf{c}$ be vectors in 2- or 3-space and let c be a scalar. Then

1. $\mathbf{a} \cdot \mathbf{b} = \mathbf{b} \cdot \mathbf{a}$
2. $\mathbf{a} \cdot (\mathbf{b} + \mathbf{c}) = \mathbf{a} \cdot \mathbf{b} + \mathbf{a} \cdot \mathbf{c}$
3. $(c\mathbf{a}) \cdot \mathbf{b} = c(\mathbf{a} \cdot \mathbf{b}) = \mathbf{a} \cdot (c\mathbf{b})$
4. $\mathbf{a} \cdot \mathbf{a} = |\mathbf{a}|^2$
5. $\mathbf{0} \cdot \mathbf{a} = 0$

We will prove Properties 1 and 4 here and leave the proofs of the other three as an exercise.

PROOF OF 1 Let $\mathbf{a} = \langle a_1, a_2, a_3 \rangle$ and $\mathbf{b} = \langle b_1, b_2, b_3 \rangle$. Then

$$\begin{aligned}\mathbf{a} \cdot \mathbf{b} &= \langle a_1, a_2, a_3 \rangle \cdot \langle b_1, b_2, b_3 \rangle \\ &= a_1b_1 + a_2b_2 + a_3b_3 = b_1a_1 + b_2a_2 + b_3a_3 \\ &= \langle b_1, b_2, b_3 \rangle \cdot \langle a_1, a_2, a_3 \rangle = \mathbf{b} \cdot \mathbf{a}\end{aligned}$$

■

PROOF OF 4 Let $\mathbf{a} = \langle a_1, a_2, a_3 \rangle$. Then

$$\begin{aligned}\mathbf{a} \cdot \mathbf{a} &= \langle a_1, a_2, a_3 \rangle \cdot \langle a_1, a_2, a_3 \rangle \\ &= a_1^2 + a_2^2 + a_3^2 = |\mathbf{a}|^2\end{aligned}$$

■

Property 4 of dot products gives us the following formula for computing the length of a vector **a** in terms of the dot product of **a** with itself. Thus,

$$|\mathbf{a}| = \sqrt{\mathbf{a} \cdot \mathbf{a}} \tag{1}$$

EXAMPLE 2 Let $\mathbf{a} = \langle 1, 2, 4 \rangle$, $\mathbf{b} = \langle -1, -2, 3 \rangle$, and $\mathbf{c} = \langle 3, 1, 2 \rangle$. Compute

a. $(\mathbf{a} + \mathbf{b}) \cdot \mathbf{c}$ **b.** $(3\mathbf{a}) \cdot \mathbf{c}$ **c.** $(\mathbf{a} \cdot \mathbf{b})\mathbf{c}$ **d.** $|\mathbf{a} - 2\mathbf{b}|$

Solution

a. $\mathbf{a} + \mathbf{b} = \langle 1, 2, 4 \rangle + \langle -1, -2, 3 \rangle = \langle 0, 0, 7 \rangle$. Therefore,

$$(\mathbf{a} + \mathbf{b}) \cdot \mathbf{c} = \langle 0, 0, 7 \rangle \cdot \langle 3, 1, 2 \rangle = 0(3) + 0(1) + 7(2) = 14$$

b. $3\mathbf{a} = 3\langle 1, 2, 4 \rangle = \langle 3, 6, 12 \rangle$. Therefore,

$$(3\mathbf{a}) \cdot \mathbf{c} = \langle 3, 6, 12 \rangle \cdot \langle 3, 1, 2 \rangle = 3(3) + 6(1) + 12(2) = 39$$

c. $\mathbf{a} \cdot \mathbf{b} = \langle 1, 2, 4 \rangle \cdot \langle -1, -2, 3 \rangle = 1(-1) + 2(-2) + 4(3) = 7$. Therefore,

$$(\mathbf{a} \cdot \mathbf{b})\mathbf{c} = 7\langle 3, 1, 2 \rangle = \langle 21, 7, 14 \rangle$$

d. $\mathbf{a} - 2\mathbf{b} = \langle 1, 2, 4 \rangle - 2\langle -1, -2, 3 \rangle = \langle 1, 2, 4 \rangle - \langle -2, -4, 6 \rangle = \langle 3, 6, -2 \rangle$. Therefore,

$$\begin{aligned}|\mathbf{a} - 2\mathbf{b}| &= \sqrt{(\mathbf{a} - 2\mathbf{b}) \cdot (\mathbf{a} - 2\mathbf{b})} = \sqrt{\langle 3, 6, -2 \rangle \cdot \langle 3, 6, -2 \rangle} \\ &= \sqrt{9 + 36 + 4} = \sqrt{49} = 7\end{aligned}$$

■

The Angle Between Two Vectors

The **angle between two nonzero vectors** is the angle θ between their corresponding position vectors, where $0 \le \theta \le \pi$. (See Figure 1.)

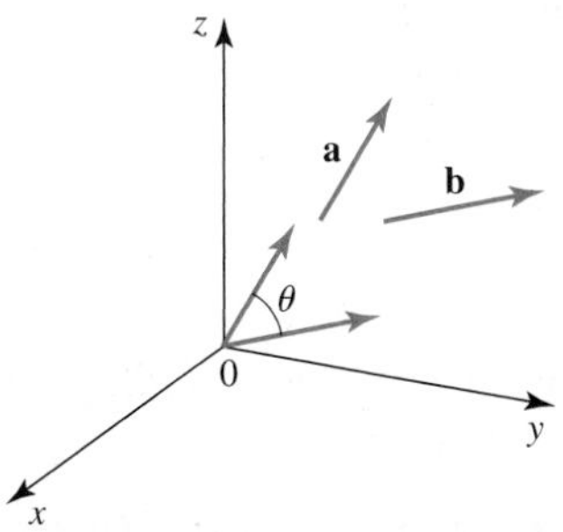

FIGURE 1
The angle between **a** and **b** is the angle between their corresponding position vectors.

Notes

1. If two vectors are parallel, then $\theta = 0$ or $\theta = \pi$.
2. The angle between the zero vector and another vector is not defined. ■

THEOREM 1 The Angle Between Two Vectors

Let θ be the angle between two nonzero vectors **a** and **b**. Then

$$\cos\theta = \frac{\mathbf{a} \cdot \mathbf{b}}{|\mathbf{a}||\mathbf{b}|} \tag{2}$$

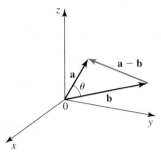

FIGURE 2
The angle between **a** and **b** is θ.

PROOF   Consider the triangle determined by the vectors **a**, **b**, and **a** − **b** as shown in Figure 2. Using the law of cosines, we have

$$|\mathbf{a} - \mathbf{b}|^2 = |\mathbf{a}|^2 + |\mathbf{b}|^2 - 2|\mathbf{a}||\mathbf{b}| \cos \theta \quad c^2 = a^2 + b^2 - 2ab \cos C$$

But

$$|\mathbf{a} - \mathbf{b}|^2 = (\mathbf{a} - \mathbf{b}) \cdot (\mathbf{a} - \mathbf{b}) \qquad \text{Equation (1)}$$

$$= \mathbf{a} \cdot \mathbf{a} - \mathbf{a} \cdot \mathbf{b} - \mathbf{b} \cdot \mathbf{a} + \mathbf{b} \cdot \mathbf{b}$$

$$= |\mathbf{a}|^2 - 2\mathbf{a} \cdot \mathbf{b} + |\mathbf{b}|^2 \qquad \mathbf{a} \cdot \mathbf{b} = \mathbf{b} \cdot \mathbf{a}$$

so we have

$$|\mathbf{a}|^2 - 2\mathbf{a} \cdot \mathbf{b} + |\mathbf{b}|^2 = |\mathbf{a}|^2 + |\mathbf{b}|^2 - 2|\mathbf{a}||\mathbf{b}| \cos \theta$$

$$-2\mathbf{a} \cdot \mathbf{b} = -2|\mathbf{a}||\mathbf{b}| \cos \theta$$

or

$$\cos \theta = \frac{\mathbf{a} \cdot \mathbf{b}}{|\mathbf{a}||\mathbf{b}|}$$ ∎

Note   Because of Equation (2), the dot product of two vectors **a** and **b** can also be defined by the equation $\mathbf{a} \cdot \mathbf{b} = |\mathbf{a}||\mathbf{b}| \cos \theta$, where θ is the angle between **a** and **b**.   ∎

EXAMPLE 3   Find the angle between the vectors $\mathbf{a} = \langle 2, 1, 3 \rangle$ and $\mathbf{b} = \langle 3, -2, 2 \rangle$.

Solution   We have

$$|\mathbf{a}| = \sqrt{2^2 + 1^2 + 3^2} = \sqrt{14} \qquad |\mathbf{b}| = \sqrt{3^2 + (-2)^2 + 2^2} = \sqrt{17}$$

and

$$\mathbf{a} \cdot \mathbf{b} = 2(3) + 1(-2) + 3(2) = 10$$

so upon using Equation (2), we have

$$\cos \theta = \frac{\mathbf{a} \cdot \mathbf{b}}{|\mathbf{a}||\mathbf{b}|} = \frac{10}{\sqrt{14}\sqrt{17}}$$

and

$$\theta = \cos^{-1}\left(\frac{10}{\sqrt{14}\,\sqrt{17}}\right) \approx 49.6°$$ ∎

▪ Orthogonal Vectors

Two nonzero vectors **a** and **b** are said to be **perpendicular,** or **orthogonal,** if the angle between them is a right angle. Now, suppose that **a** and **b** are orthogonal so that the angle between them is $\pi/2$. Then Equation (2) gives

$$\mathbf{a} \cdot \mathbf{b} = |\mathbf{a}||\mathbf{b}| \cos \frac{\pi}{2} = 0$$

Conversely, if $\mathbf{a} \cdot \mathbf{b} = 0$, then $\cos \theta = 0$ (because $|\mathbf{a}|$ and $|\mathbf{b}|$ are both nonzero), so $\theta = \pi/2$. We have proved the following.

THEOREM 2

Two nonzero vectors **a** and **b** are orthogonal if and only if $\mathbf{a} \cdot \mathbf{b} = 0$.

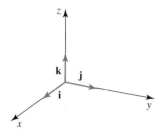

FIGURE 3
The standard basis vectors **i**, **j**, and **k** are mutually orthogonal.

The standard basis vectors $\mathbf{i} = \langle 1, 0, 0 \rangle$, $\mathbf{j} = \langle 0, 1, 0 \rangle$, and $\mathbf{k} = \langle 0, 0, 1 \rangle$ are mutually orthogonal; that is, any two of them are orthogonal. This is evident if you examine Figure 3.

For example,

$$\mathbf{i} \cdot \mathbf{j} = \langle 1, 0, 0 \rangle \cdot \langle 0, 1, 0 \rangle = 0 \qquad \text{and} \qquad \mathbf{j} \cdot \mathbf{k} = \langle 0, 1, 0 \rangle \cdot \langle 0, 0, 1 \rangle = 0$$

EXAMPLE 4 Determine whether the vectors $\mathbf{a} = 2\mathbf{i} + 3\mathbf{j} + 3\mathbf{k}$ and $\mathbf{b} = 3\mathbf{i} - 4\mathbf{j} + 2\mathbf{k}$ are orthogonal.

Solution We compute

$$\mathbf{a} \cdot \mathbf{b} = (2\mathbf{i} + 3\mathbf{j} + 3\mathbf{k})(3\mathbf{i} - 4\mathbf{j} + 2\mathbf{k})$$
$$= 2(3) + 3(-4) + 3(2) = 0$$

and conclude that **a** and **b** are indeed orthogonal. ∎

■ Direction Cosines

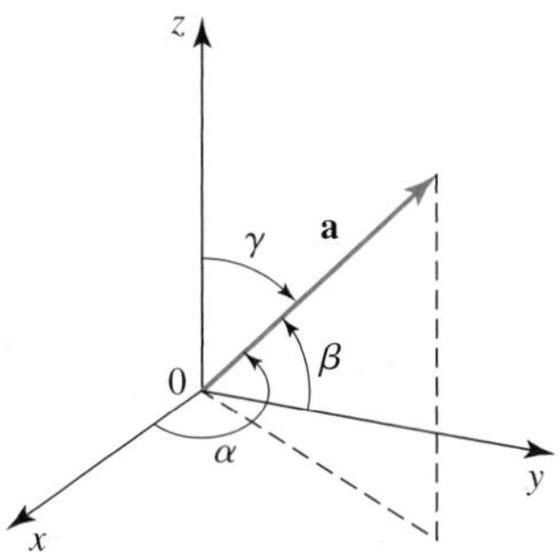

We can describe the direction of a nonzero vector **a** by giving the angles α, β, and γ that **a** makes with the positive x-, y-, and z-axes, respectively. (See Figure 4.) These angles are called the **direction angles** of **a**. The cosines of these angles, $\cos \alpha$, $\cos \beta$, and $\cos \gamma$, are called the **direction cosines** of the vector **a**.

Let $\mathbf{a} = a_1\mathbf{i} + a_2\mathbf{j} + a_3\mathbf{k}$ be a nonzero vector in 3-space. Then

$$\mathbf{a} \cdot \mathbf{i} = (a_1\mathbf{i} + a_2\mathbf{j} + a_3\mathbf{k}) \cdot \mathbf{i} = a_1 \qquad \mathbf{i} \cdot \mathbf{i} = 1, \quad \mathbf{j} \cdot \mathbf{i} = \mathbf{k} \cdot \mathbf{i} = 0$$

So

$$\cos \alpha = \frac{\mathbf{a} \cdot \mathbf{i}}{|\mathbf{a}||\mathbf{i}|} = \frac{a_1}{|\mathbf{a}|}$$

Similarly,

$$\cos \beta = \frac{a_2}{|\mathbf{a}|} \qquad \text{and} \qquad \cos \gamma = \frac{a_3}{|\mathbf{a}|}$$

FIGURE 4
The direction angles of a vector

By squaring and adding the three direction cosines, we obtain

$$\cos^2 \alpha + \cos^2 \beta + \cos^2 \gamma = \frac{a_1^2}{|\mathbf{a}|^2} + \frac{a_2^2}{|\mathbf{a}|^2} + \frac{a_3^2}{|\mathbf{a}|^2} = \frac{|\mathbf{a}|^2}{|\mathbf{a}|^2} = 1$$

THEOREM 3

The three direction cosines of a nonzero vector $\mathbf{a} = a_1\mathbf{i} + a_2\mathbf{j} + a_3\mathbf{k}$ in 3-space are

$$\cos \alpha = \frac{a_1}{|\mathbf{a}|} \qquad \cos \beta = \frac{a_2}{|\mathbf{a}|} \qquad \cos \gamma = \frac{a_3}{|\mathbf{a}|} \qquad (3)$$

The direction cosines satisfy

$$\cos^2 \alpha + \cos^2 \beta + \cos^2 \gamma = 1 \qquad (4)$$

Notes

1. If $\mathbf{a} = a_1\mathbf{i} + a_2\mathbf{j} + a_3\mathbf{k}$ is nonzero, then the unit vector having the same direction as $\mathbf{a}$ is

$$\mathbf{u} = \frac{\mathbf{a}}{|\mathbf{a}|} = \frac{a_1}{|\mathbf{a}|}\mathbf{i} + \frac{a_2}{|\mathbf{a}|}\mathbf{j} + \frac{a_3}{|\mathbf{a}|}\mathbf{k}$$

$$= (\cos\alpha)\mathbf{i} + (\cos\beta)\mathbf{j} + (\cos\gamma)\mathbf{k} \qquad (5)$$

This shows that the direction cosines of $\mathbf{a}$ are the components of the unit vector in the direction of $\mathbf{a}$. This augments the statement made earlier that the direction cosines of a vector define the direction of that vector.

2. From Equation (5) we see that

$$\mathbf{a} = |\mathbf{a}|\bigl[(\cos\alpha)\mathbf{i} + (\cos\beta)\mathbf{j} + (\cos\gamma)\mathbf{k}\bigr]$$

$$\underset{\text{Magnitude}}{\uparrow} \qquad\qquad \underset{\text{Direction}}{\uparrow}$$

EXAMPLE 5 Find the direction angles of the vector $\mathbf{a} = 2\mathbf{i} + 3\mathbf{j} + \mathbf{k}$.

Solution We have

$$|\mathbf{a}| = \sqrt{2^2 + 3^2 + 1^2} = \sqrt{14}$$

so by Equation (3),

$$\cos\alpha = \frac{2}{\sqrt{14}} \qquad \cos\beta = \frac{3}{\sqrt{14}} \qquad \cos\gamma = \frac{1}{\sqrt{14}}$$

Therefore,

$$\alpha = \cos^{-1}\!\left(\frac{2}{\sqrt{14}}\right) \approx 58° \qquad \beta = \cos^{-1}\!\left(\frac{3}{\sqrt{14}}\right) \approx 37° \qquad \gamma = \cos^{-1}\!\left(\frac{1}{\sqrt{14}}\right) \approx 74°$$

■ Vector Projections and Components

Figure 5a depicts a child pulling a sled with a constant force represented by the vector $\mathbf{F}$. The force $\mathbf{F}$ can be expressed as the sum of two forces: a horizontal component $\mathbf{F}_1$ and a vertical component $\mathbf{F}_2$, as shown in Figure 5b.

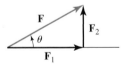

FIGURE 5

(a) $\mathbf{F}$ makes an angle θ with the line of motion.

(b) $\mathbf{F} = \mathbf{F}_1 + \mathbf{F}_2$

Observe that $\mathbf{F}_1$ acts in the direction of motion, whereas $\mathbf{F}_2$ acts in a direction perpendicular to the direction of motion. We will see why it is useful to look at $\mathbf{F}$ in this way when we study the work done by $\mathbf{F}$ in moving the sled.

More generally, we are interested in the component of one vector $\mathbf{b}$ in the direction of another nonzero vector $\mathbf{a}$. The vector that is obtained by projecting $\mathbf{b}$ onto the

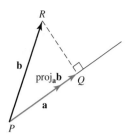

FIGURE 6
$\text{proj}_\mathbf{a}\mathbf{b}\colon\left(\overrightarrow{PQ}\right)$ is the vector
projection of **b** onto **a**.

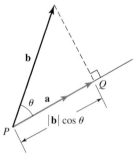

FIGURE 7
$\text{comp}_\mathbf{a}\mathbf{b} = |\mathbf{b}| \cos \theta$

line containing the vector **a** is called the **vector projection of b onto a** (also called the **vector component of b along a**) and denoted by

$$\text{proj}_\mathbf{a}\mathbf{b}$$

(See Figure 6.)

The **scalar projection of b onto a** (also called the **scalar component of b along a**) is the length of $\text{proj}_\mathbf{a}\mathbf{b}$ if the projection has the same direction as **a** and the negative of the length of $\text{proj}_\mathbf{a}\mathbf{b}$ if the projection has the opposite direction. We denote this scalar projection by

$$\text{comp}_\mathbf{a}\mathbf{b}$$

As you can see from Figure 7, it is just the number $|\mathbf{b}| \cos \theta$. Observe that if $\frac{\pi}{2} < \theta \le \pi$, then $|\mathbf{b}| \cos \theta$ is negative. We encourage you to make a sketch of this situation.

Since

$$\cos \theta = \frac{\mathbf{b} \cdot \mathbf{a}}{|\mathbf{b}||\mathbf{a}|}$$

we can write

$$|\mathbf{b}| \cos \theta = \frac{|\mathbf{b}|(\mathbf{b} \cdot \mathbf{a})}{|\mathbf{b}||\mathbf{a}|} = \frac{\mathbf{b} \cdot \mathbf{a}}{|\mathbf{a}|}$$

Therefore, the scalar component of **b** along **a** is

$$\text{comp}_\mathbf{a}\mathbf{b} = \frac{\mathbf{b} \cdot \mathbf{a}}{|\mathbf{a}|} \tag{6}$$

Note   Writing

$$\frac{\mathbf{b} \cdot \mathbf{a}}{|\mathbf{a}|} = \mathbf{b} \cdot \left(\frac{\mathbf{a}}{|\mathbf{a}|}\right)$$

we see that the scalar component of **b** along **a** can also be calculated by taking the dot product of **b** with the unit vector in the direction of **a**.   ■

The vector projection of **b** onto **a** is the scalar component of **b** along **a** times the direction of **a**. (See Figures 6 and 7.) Thus, we have

$$\text{proj}_\mathbf{a}\mathbf{b} = \left(\frac{\mathbf{b} \cdot \mathbf{a}}{|\mathbf{a}|}\right)\frac{\mathbf{a}}{|\mathbf{a}|} = \left(\frac{\mathbf{b} \cdot \mathbf{a}}{|\mathbf{a}|^2}\right)\mathbf{a} = \left(\frac{\mathbf{b} \cdot \mathbf{a}}{\mathbf{a} \cdot \mathbf{a}}\right)\mathbf{a} \tag{7}$$

(See Figure 8.)

FIGURE 8
$\text{proj}_\mathbf{a}\mathbf{b}$ points in the same direction as **a** if θ is acute and points in the opposite direction as **a** if θ is obtuse.

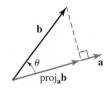 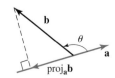

(**a**) θ is acute.   (**b**) θ is obtuse.

EXAMPLE 6 Let $\mathbf{b} = 2\mathbf{i} + 3\mathbf{j} - 4\mathbf{k}$, and let $\mathbf{a} = 3\mathbf{i} - 2\mathbf{j} + \mathbf{k}$. Find the scalar component of $\mathbf{b}$ along $\mathbf{a}$ and the vector projection of $\mathbf{b}$ onto $\mathbf{a}$.

Solution The scalar component of $\mathbf{b}$ along $\mathbf{a}$ is

$$\text{comp}_{\mathbf{a}}\mathbf{b} = \frac{\mathbf{b} \cdot \mathbf{a}}{|\mathbf{a}|} = \frac{2(3) + (3)(-2) + (-4)(1)}{\sqrt{3^2 + (-2)^2 + 1^2}} = -\frac{4}{\sqrt{14}}$$

Next, we compute the unit vector in the direction of $\mathbf{a}$. Thus,

$$\frac{\mathbf{a}}{|\mathbf{a}|} = \frac{3\mathbf{i} - 2\mathbf{j} + \mathbf{k}}{\sqrt{14}} = \frac{1}{\sqrt{14}}(3\mathbf{i} - 2\mathbf{j} + \mathbf{k})$$

Therefore,

$$\begin{aligned}
\text{proj}_{\mathbf{a}}\mathbf{b} &= \left(\frac{\mathbf{b} \cdot \mathbf{a}}{|\mathbf{a}|}\right)\frac{\mathbf{a}}{|\mathbf{a}|} \\
&= -\frac{4}{\sqrt{14}} \cdot \frac{1}{\sqrt{14}}(3\mathbf{i} - 2\mathbf{j} + \mathbf{k}) \\
&= -\frac{6}{7}\mathbf{i} + \frac{4}{7}\mathbf{j} - \frac{2}{7}\mathbf{k}
\end{aligned}$$

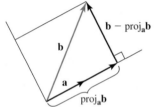

FIGURE 9
The vector $\mathbf{b}$ can be written as the sum of a vector parallel to $\mathbf{a}$ and a vector orthogonal to $\mathbf{a}$.

Using vector projections, we can express any vector $\mathbf{b}$ as the sum of a vector parallel to a vector $\mathbf{a}$ and a vector perpendicular to $\mathbf{a}$. In fact, from Figure 9 we see that

$$\begin{aligned}
\mathbf{b} &= \text{proj}_{\mathbf{a}}\mathbf{b} + (\mathbf{b} - \text{proj}_{\mathbf{a}}\mathbf{b}) \\
&= \underbrace{\left(\frac{\mathbf{b} \cdot \mathbf{a}}{\mathbf{a} \cdot \mathbf{a}}\right)\mathbf{a}}_{\text{Parallel to } \mathbf{a}} + \underbrace{\left[\mathbf{b} - \left(\frac{\mathbf{b} \cdot \mathbf{a}}{\mathbf{a} \cdot \mathbf{a}}\right)\mathbf{a}\right]}_{\text{Orthogonal to } \mathbf{a}}
\end{aligned} \qquad (8)$$

EXAMPLE 7 Write $\mathbf{b} = 3\mathbf{i} - \mathbf{j} + 2\mathbf{k}$ as the sum of a vector parallel to $\mathbf{a} = 2\mathbf{i} - \mathbf{j} + \mathbf{k}$ and a vector perpendicular to $\mathbf{a}$.

Solution Using Equation (8) with

$$\mathbf{b} \cdot \mathbf{a} = (3)(2) + (-1)(-1) + (2)(1) = 9$$

and

$$\mathbf{a} \cdot \mathbf{a} = 2^2 + (-1)^2 + 1^2 = 6$$

gives

$$\begin{aligned}
\mathbf{b} &= \left(\frac{\mathbf{b} \cdot \mathbf{a}}{\mathbf{a} \cdot \mathbf{a}}\right)\mathbf{a} + \left[\mathbf{b} - \left(\frac{\mathbf{b} \cdot \mathbf{a}}{\mathbf{a} \cdot \mathbf{a}}\right)\mathbf{a}\right] \\
&= \frac{9}{6}(2\mathbf{i} - \mathbf{j} + \mathbf{k}) + \left[3\mathbf{i} - \mathbf{j} + 2\mathbf{k} - \frac{9}{6}(2\mathbf{i} - \mathbf{j} + \mathbf{k})\right] \\
&= \underbrace{\left(3\mathbf{i} - \frac{3}{2}\mathbf{j} + \frac{3}{2}\mathbf{k}\right)}_{\text{Parallel to } \mathbf{a}} + \underbrace{\left(\frac{1}{2}\mathbf{j} + \frac{1}{2}\mathbf{k}\right)}_{\text{Perpendicular to } \mathbf{a}}
\end{aligned}$$

◼ Work

One application of vector projections lies in the computation of the work done by a force. Recall that the work W done by a constant force $\mathbf{F}$ acting along the line of motion in moving an object a distance d is given by $W = |\mathbf{F}|d$. But if the constant force $\mathbf{F}$ acts in a direction that is different from the direction of motion, as in the case we mentioned earlier of a child pulling a sled, then work is done only by that component of the force in the direction of motion.

To derive an expression for the work done in this situation, suppose that $\mathbf{F}$ moves an object from P to Q. (See Figure 10.) Then, letting $\mathbf{d}$ denote the **displacement vector** $\overrightarrow{PQ}$, we see that the work done by $\mathbf{F}$ in moving an object from P to Q is given by

$$W = \left(|\mathbf{F}|\cos\theta\right)|\mathbf{d}| \qquad \text{component of } \mathbf{F} \text{ along } \mathbf{d} \cdot \text{distance moved}$$

$$= |\mathbf{F}||\mathbf{d}|\cos\theta$$

$$= \mathbf{F} \cdot \mathbf{d} \qquad\qquad\qquad\qquad (9)$$

Thus, the work done by a constant force $\mathbf{F}$ in moving an object through a displacement $\mathbf{d}$ is the dot product of $\mathbf{F}$ and $\mathbf{d}$.

FIGURE 10
The work done by $\mathbf{F}$ in moving an object from P to Q is $\mathbf{F} \cdot \mathbf{d}$.

EXAMPLE 8   A force $\mathbf{F} = 2\mathbf{i} + 3\mathbf{j} + 4\mathbf{k}$ moves a particle along the line segment from the point $P(1, 2, 1)$ to the point $Q(3, 6, 5)$. (See Figure 11.) Find the work done by the force if $|\mathbf{F}|$ is measured in newtons and $|\mathbf{d}|$ is measured in meters.

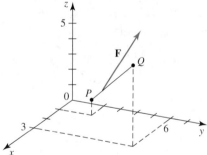

FIGURE 11
The force $\mathbf{F}$ moves the particle from the point P to the point Q.

Solution   The displacement vector is $\mathbf{d} = \overrightarrow{PQ} = \langle 2, 4, 4 \rangle = 2\mathbf{i} + 4\mathbf{j} + 4\mathbf{k}$. Therefore, using Equation (9), we see that the work done by $\mathbf{F}$ is

$$\mathbf{F} \cdot \mathbf{d} = (2\mathbf{i} + 3\mathbf{j} + 4\mathbf{k}) \cdot (2\mathbf{i} + 4\mathbf{j} + 4\mathbf{k})$$

$$= 4 + 12 + 16 = 32$$

or 32 joules.   ◼

◼ 11.3   CONCEPT QUESTIONS

1. a. What is the dot product of the vectors $\mathbf{a} = \langle a_1, a_2, a_3 \rangle$ and $\mathbf{b} = \langle b_1, b_2, b_3 \rangle$?
   b. State the properties of the dot product.
   c. Express $|\mathbf{a}|$ in terms of the dot product.
   d. What is the angle between two nonzero vectors $\mathbf{a}$ and $\mathbf{b}$ in terms of the dot product?
2. a. What does it mean for two nonzero vectors to be orthogonal?
   b. State the condition(s) for two nonzero vectors to be orthogonal.

3. a. What are the direction cosines of a vector $\mathbf{a}$ in 3-space?
   b. Write the nonzero vector $\mathbf{a}$ in terms of its magnitude and its direction cosines.
4. a. What is the scalar component of $\mathbf{b}$ along the nonzero vector $\mathbf{a}$? Illustrate with a diagram.
   b. What is the vector projection of $\mathbf{b}$ onto $\mathbf{a}$? Illustrate.
5. What is the work done by a constant force $\mathbf{F}$ in moving an object along a straight line from the point P to the point Q?

■ 11.3 EXERCISES

1. State whether the expression makes sense. Explain.
   a. $(\mathbf{a} \cdot \mathbf{b})\mathbf{c}$
   b. $\mathbf{a} \cdot (\mathbf{b} \cdot \mathbf{c})$
   c. $\mathbf{a} + \mathbf{b} \cdot \mathbf{c}$
   d. $\mathbf{a} \cdot \mathbf{b} - |\mathbf{a}\|\mathbf{b}|$
   e. $|\mathbf{a}|\mathbf{b} + (\mathbf{a} \cdot \mathbf{b})\mathbf{a}$
   f. $\mathbf{a} \cdot \left(\dfrac{\mathbf{b}}{|\mathbf{b}|}\right)$

2. a. If $\mathbf{a} \cdot \mathbf{a} = 0$ and $\mathbf{a} \cdot \mathbf{b} = 0$, what can you say about $\mathbf{b}$?
   b. If $\mathbf{a} \cdot \mathbf{a} \neq 0$ and $\mathbf{a} \cdot \mathbf{b} = 0$, what can you say about $\mathbf{b}$?

In Exercises 3–8, find $\mathbf{a} \cdot \mathbf{b}$.

3. $\mathbf{a} = \langle 1, 3 \rangle$, $\mathbf{b} = \langle 2, -1 \rangle$

4. $\mathbf{a} = \langle 1, 3, -2 \rangle$, $\mathbf{b} = \langle -1, 1, -1 \rangle$

5. $\mathbf{a} = 2\mathbf{i} + 3\mathbf{j}$, $\mathbf{b} = \mathbf{i} - 2\mathbf{j}$

6. $\mathbf{a} = 2\mathbf{i} - 3\mathbf{j} + \mathbf{k}$, $\mathbf{b} = -\mathbf{i} + 2\mathbf{j} - 2\mathbf{k}$

7. $\mathbf{a} = \langle 0, 1, -3 \rangle$, $\mathbf{b} = \langle 10, \pi, -\pi \rangle$

8. $\mathbf{a} = 2\mathbf{i} + 3\mathbf{k}$, $\mathbf{b} = -\mathbf{i} + 0.2\mathbf{j} - \sqrt{3}\mathbf{k}$

In Exercises 9–16, $\mathbf{a} = \langle 1, -3, 2 \rangle$, $\mathbf{b} = \langle -2, 4, 1 \rangle$, *and* $\mathbf{c} = \langle 2, -4, 1 \rangle$. *Find the indicated quantity.*

9. $\mathbf{a} \cdot (\mathbf{b} + \mathbf{c})$

10. $\mathbf{b} \cdot (\mathbf{a} - \mathbf{c})$

11. $(2\mathbf{a} + 3\mathbf{b}) \cdot (3\mathbf{c})$

12. $(\mathbf{a} - \mathbf{b}) \cdot (\mathbf{a} + \mathbf{b})$

13. $(\mathbf{a} \cdot \mathbf{b})\mathbf{c}$

14. $(\mathbf{a} \cdot \mathbf{b})\mathbf{c} - (\mathbf{b} \cdot \mathbf{c})\mathbf{a}$

15. $|\mathbf{a} - \mathbf{b}|^2 + |\mathbf{a} + \mathbf{b}|^2$

16. $\left(\dfrac{\mathbf{a} \cdot \mathbf{b}}{\mathbf{b} \cdot \mathbf{b}}\right)\mathbf{b}$

In Exercises 17–22, find the angle between the vectors.

17. $\mathbf{a} = \langle 2, 1 \rangle$, $\mathbf{b} = \langle 3, 4 \rangle$

18. $\mathbf{a} = \mathbf{i} + 3\mathbf{j}$, $\mathbf{b} = -\mathbf{i} + 2\mathbf{j}$

19. $\mathbf{a} = \langle 1, 1, 1 \rangle$, $\mathbf{b} = \langle 2, 3, -6 \rangle$

20. $\mathbf{a} = \mathbf{i} + 2\mathbf{j} + \mathbf{k}$, $\mathbf{b} = 8\mathbf{i} - 4\mathbf{j} - 3\mathbf{k}$

21. $\mathbf{a} = -2\mathbf{j} + 3\mathbf{k}$, $\mathbf{b} = \mathbf{i} + \mathbf{j} + 2\mathbf{k}$

22. $\mathbf{a} = \langle -2, 1, 1 \rangle$, $\mathbf{b} = \langle -3, 2, 1 \rangle$

23. Find the value of c such that the angle between $\mathbf{a} = \langle 1, c \rangle$ and $\mathbf{b} = \langle 1, 2 \rangle$ is $45°$.

24. Find the value(s) of c such that the angle between $\mathbf{a} = \mathbf{i} + c\mathbf{j} + 2\mathbf{k}$ and $\mathbf{b} = -\mathbf{i} + 2\mathbf{j} - \mathbf{k}$ is $60°$.

In Exercises 25–30, determine whether the vectors are orthogonal, parallel, or neither.

25. $\mathbf{a} = \langle 1, 2 \rangle$, $\mathbf{b} = \langle 3, 0 \rangle$

26. $\mathbf{a} = \langle 3, -4 \rangle$, $\mathbf{b} = \langle 4, 3 \rangle$

27. $\mathbf{a} = \mathbf{i} - 2\mathbf{j} + \mathbf{k}$, $\mathbf{b} = 3\mathbf{i} + 2\mathbf{j} - 2\mathbf{k}$

28. $\mathbf{a} = 2\mathbf{i} + 4\mathbf{j} - \mathbf{k}$, $\mathbf{b} = 6\mathbf{i} + 12\mathbf{j} - 3\mathbf{k}$

29. $\mathbf{a} = \langle 2, 3, -1 \rangle$, $\mathbf{b} = \langle 2, -1, 1 \rangle$

30. $\mathbf{a} = \langle 2, 3, -3 \rangle$, $\mathbf{b} = \langle \frac{4}{3}, 2, -2 \rangle$

31. Find a value of c such that $\langle c, 2, -1 \rangle$ and $\langle 2, 3, c \rangle$ are orthogonal.

32. Find a unit vector that is orthogonal to both $\mathbf{a} = \mathbf{i} + \mathbf{j} + \mathbf{k}$ and $\mathbf{b} = -2\mathbf{i} + \mathbf{k}$.

In Exercises 33–36, find the direction cosines and direction angles of the vector.

33. $\mathbf{a} = \langle 1, 2, 3 \rangle$

34. $\mathbf{a} = 2\mathbf{i} + 2\mathbf{j} - \mathbf{k}$

35. $\mathbf{a} = -\mathbf{i} + 3\mathbf{j} + 5\mathbf{k}$

36. $\mathbf{a} = \langle 3, -4, 5 \rangle$

37. A vector has direction angles $\alpha = \pi/3$ and $\gamma = \pi/4$. Find the direction angle β.

38. Find a unit vector whose direction angles are all equal.

In Exercises 39–44, find (a) $\text{proj}_{\mathbf{a}}\mathbf{b}$ *and* (b) $\text{proj}_{\mathbf{b}}\mathbf{a}$.

39. $\mathbf{a} = \langle 2, 3 \rangle$, $\mathbf{b} = \langle 1, 4 \rangle$

40. $\mathbf{a} = -\mathbf{i} + 2\mathbf{j}$, $\mathbf{b} = -3\mathbf{i} + 4\mathbf{j}$

41. $\mathbf{a} = 2\mathbf{i} + \mathbf{j} + 4\mathbf{k}$, $\mathbf{b} = 3\mathbf{i} + \mathbf{k}$

42. $\mathbf{a} = \langle 1, 2, 0 \rangle$, $\mathbf{b} = \langle -3, 0, -4 \rangle$

43. $\mathbf{a} = \langle -3, 4, -2 \rangle$, $\mathbf{b} = \langle 0, 1, 0 \rangle$

44. $\mathbf{a} = \langle -1, 3, -2 \rangle$, $\mathbf{b} = \langle 0, 3, 1 \rangle$

In Exercises 45–48, write $\mathbf{b}$ *as the sum of a vector parallel to* $\mathbf{a}$ *and a vector perpendicular to* $\mathbf{a}$.

45. $\mathbf{a} = \langle 1, 3 \rangle$, $\mathbf{b} = \langle 2, 4 \rangle$

46. $\mathbf{a} = -\mathbf{i} + 2\mathbf{j}$, $\mathbf{b} = 2\mathbf{i} + 3\mathbf{j}$

47. $\mathbf{a} = \mathbf{i} + 2\mathbf{j} + 3\mathbf{k}$, $\mathbf{b} = 2\mathbf{i} - \mathbf{j} + \mathbf{k}$

48. $\mathbf{a} = \mathbf{i} + 2\mathbf{k}$, $\mathbf{b} = 2\mathbf{i} - \mathbf{j}$

In Exercises 49 and 50, find the work done by the force $\mathbf{F}$ *in moving a particle from the point* P *to the point* Q.

49. $\mathbf{F} = 2\mathbf{i} + 3\mathbf{j} - \mathbf{k}$; $P(-1, -2, 2)$; $Q(2, 1, 5)$

50. $\mathbf{F} = \langle 1, 4, 5 \rangle$; $P(2, 3, 1)$; $Q(-1, 2, -4)$

51. Find the angle between a diagonal of a cube and one of its edges.

52. Find the angle between a diagonal of a cube and a diagonal of one of its sides.

53. Refer to the following figure. Find the angles θ and ψ.

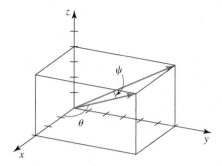

54. Bond Angle of a Molecule of Methane The following figure gives the configuration of a molecule of methane, CH_4. The four hydrogen atoms are located at the vertices of a regular tetrahedron and the carbon atom is located at the centroid. The *bond angle* for the molecule is the angle between the line segments joining the carbon atom to two of the hydrogen atoms. Show that this angle is approximately $109.5°$
Hint: The centroid of the tetrahedron is $\left(\frac{k}{2}, \frac{k}{2}, \frac{k}{2}\right)$.

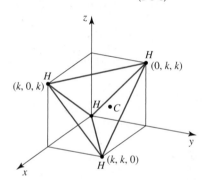

55. A passenger in an airport terminal pulls his luggage with a constant force of magnitude 24 lb. If the handle makes an angle of $30°$ with the horizontal surface, find the work done by the passenger in pulling the luggage a distance of 50 ft.

56. A child pulls a toy wagon up a straight incline that makes an angle of $15°$ with the horizontal. If the handle makes an angle of $30°$ with the incline and she exerts a constant force of 15 lb on the handle, find the work done in pulling the wagon a distance of 30 ft along the incline.

57. The following figure gives an aerial view of two tugboats attempting to free a cruise ship that ran aground during a storm. Tugboat I exerts a force of magnitude 3000 lb, and Tugboat II exerts a force of magnitude 2400 lb. If the resultant force acts along the positive x-axis and the cruise ship is towed a distance of 100 ft in that direction, find the work done by each tugboat. (See Exercise 66 in Section 11.1.)

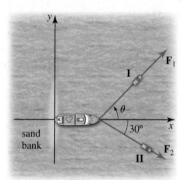

58. Prove Property 2 of the dot product:
$\mathbf{a} \cdot (\mathbf{b} + \mathbf{c}) = \mathbf{a} \cdot \mathbf{b} + \mathbf{a} \cdot \mathbf{c}$.

59. Prove Property 3 of the dot product:
$(c\mathbf{a}) \cdot \mathbf{b} = c(\mathbf{a} \cdot \mathbf{b}) = \mathbf{a} \cdot (c\mathbf{b})$.

60. Prove that if $\mathbf{a}$ is orthogonal to both $\mathbf{b}$ and $\mathbf{c}$, then $\mathbf{a}$ is orthogonal to $p\mathbf{b} + q\mathbf{c}$ for any scalars p and q. Give a geometric interpretation of this result.

61. Let $\mathbf{a}$ and $\mathbf{b}$ be nonzero vectors. Then the vector $\mathbf{b} - \text{proj}_{\mathbf{a}}\mathbf{b}$ is called the *vector component of* $\mathbf{b}$ *orthogonal to* $\mathbf{a}$.
   a. Make a sketch of the vectors $\mathbf{a}$, $\mathbf{b}$, $\text{proj}_{\mathbf{a}}\mathbf{b}$ and
      $\mathbf{b} - \text{proj}_{\mathbf{a}}\mathbf{b}$.
   b. Show that $\mathbf{b} - \text{proj}_{\mathbf{a}}\mathbf{b}$ is orthogonal to $\mathbf{a}$.

62. Refer to Exercise 61. Let $\mathbf{a} = \langle 2, 1 \rangle$ and $\mathbf{b} = \langle 4, 5 \rangle$.
   a. Find $\mathbf{b} - \text{proj}_{\mathbf{a}}\mathbf{b}$.
   b. Sketch the vectors $\mathbf{a}$, $\mathbf{b}$, $\text{proj}_{\mathbf{a}}\mathbf{b}$ and $\mathbf{b} - \text{proj}_{\mathbf{a}}\mathbf{b}$.
   c. Show that $\mathbf{b} - \text{proj}_{\mathbf{a}}\mathbf{b}$ is orthogonal to $\mathbf{a}$.

63. a. Show that the vector $\mathbf{n} = \langle a, b \rangle$ is orthogonal to the line $ax + by + c = 0$.

b. Use the result of part (a) to show that the distance from a point $P_1(x_1, y_1)$ to the line $ax + by + c = 0$ is

$$\frac{|ax_1 + by_1 + c|}{\sqrt{a^2 + b^2}}$$

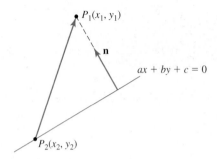

Hint: Let $P_2(x_2, y_2)$ be a point on the line, and consider the scalar projection of $\overrightarrow{P_1P_2}$ onto $\mathbf{n}$.

c. Use the formula found in part (b) to find the distance from the point $(1, -3)$ to the line $2x + 3y - 6 = 0$.

64. Prove that $(\mathbf{a} \cdot \mathbf{b})\mathbf{c} - (\mathbf{b} \cdot \mathbf{c})\mathbf{a}$ is orthogonal to $\mathbf{b}$, where $\mathbf{a}$, $\mathbf{b}$, and $\mathbf{c}$ are any three vectors.

In Exercises 65–68, determine whether the statement is true or false. If it is true, explain why. If it is false, explain why or give an example that shows it is false.

65. $(\mathbf{a} \cdot \mathbf{b})^2 = |\mathbf{a}|^2 |\mathbf{b}|^2$

66. If $\mathbf{a} \neq \mathbf{0}$, and $\mathbf{a} \cdot \mathbf{b} = \mathbf{a} \cdot \mathbf{c}$, then $\mathbf{b} = \mathbf{c}$.

67. If $\mathbf{u}$, $\mathbf{v}$, and $\mathbf{w}$ are nonzero vectors, and both $\mathbf{u}$ and $\mathbf{v}$ are orthogonal to $\mathbf{w}$, then $2\mathbf{u} + 3\mathbf{v}$ is orthogonal to $\mathbf{w}$.

68. If $\mathbf{a}$ and $\mathbf{b}$ are nonzero vectors, then $\text{proj}_{\mathbf{a}}\mathbf{b} = \mathbf{0}$ if and only if $\mathbf{a}$ is orthogonal to $\mathbf{b}$.

11.4 The Cross Product

The Cross Product of Two Vectors

In the preceding section we saw how an operation called the dot product combines two vectors to yield a *scalar*. In this section we will look at yet another operation on vectors. This operation, called the *cross product*, combines two vectors to yield a vector.

DEFINITION   The Cross Product of Two Vectors in Space

Let $\mathbf{a} = a_1\mathbf{i} + a_2\mathbf{j} + a_3\mathbf{k}$ and $\mathbf{b} = b_1\mathbf{i} + b_2\mathbf{j} + b_3\mathbf{k}$ be any two vectors in space. Then the **cross product** of $\mathbf{a}$ and $\mathbf{b}$ is the vector

$$\mathbf{a} \times \mathbf{b} = (a_2b_3 - a_3b_2)\mathbf{i} + (a_3b_1 - a_1b_3)\mathbf{j} + (a_1b_2 - a_2b_1)\mathbf{k} \qquad (1)$$

The cross product is used in computing quantities as diverse as the volume of a parallelepiped and the rate of rotation of an incompressible fluid.

Before giving a geometric interpretation of the cross product of two vectors, let's find a simpler way to remember the cross product. Recall that a **determinant of order 2** is defined by

$$\begin{vmatrix} a & b \\ c & d \end{vmatrix} = ad - bc$$

For example,

$$\begin{vmatrix} 2 & 1 \\ 3 & -4 \end{vmatrix} = 2(-4) - 1(3) = -11$$

A **determinant of order 3** is defined in terms of second-order determinants as follows:

$$\begin{vmatrix} a_1 & a_2 & a_3 \\ b_1 & b_2 & b_3 \\ c_1 & c_2 & c_3 \end{vmatrix} = a_1 \begin{vmatrix} b_2 & b_3 \\ c_2 & c_3 \end{vmatrix} - a_2 \begin{vmatrix} b_1 & b_3 \\ c_1 & c_3 \end{vmatrix} + a_3 \begin{vmatrix} b_1 & b_2 \\ c_1 & c_2 \end{vmatrix}$$

In this definition the determinant is said to be *expanded about the first row.* Observe that each term on the right involves the product of a term from the first row and a second-order determinant that is obtained by deleting the row and column containing that term. Also note how the signs of the terms alternate.

As an example,

$$\begin{vmatrix} 1 & 2 & 4 \\ 3 & -1 & 2 \\ 4 & 0 & 3 \end{vmatrix} = 1 \begin{vmatrix} -1 & 2 \\ 0 & 3 \end{vmatrix} - 2 \begin{vmatrix} 3 & 2 \\ 4 & 3 \end{vmatrix} + 4 \begin{vmatrix} 3 & -1 \\ 4 & 0 \end{vmatrix}$$

$$= 1(-3 - 0) - 2(9 - 8) + 4(0 + 4) = 11$$

As a mnemonic device for remembering the expression for the cross product of **a** and **b**, where $\mathbf{a} = a_1\mathbf{i} + a_2\mathbf{j} + a_3\mathbf{k}$ and $\mathbf{b} = b_1\mathbf{i} + b_2\mathbf{j} + b_3\mathbf{k}$, let's expand the following expression as if it were a determinant. (Technically it is not a determinant, because **i**, **j**, and **k** are not real numbers.) Thus,

$$\begin{vmatrix} \mathbf{i} & \mathbf{j} & \mathbf{k} \\ a_1 & a_2 & a_3 \\ b_1 & b_2 & b_3 \end{vmatrix} = \mathbf{i} \begin{vmatrix} a_2 & a_3 \\ b_2 & b_3 \end{vmatrix} - \mathbf{j} \begin{vmatrix} a_1 & a_3 \\ b_1 & b_3 \end{vmatrix} + \mathbf{k} \begin{vmatrix} a_1 & a_2 \\ b_1 & b_2 \end{vmatrix}$$

$$= (a_2 b_3 - a_3 b_2)\mathbf{i} + (a_3 b_1 - a_1 b_3)\mathbf{j} + (a_1 b_2 - a_2 b_1)\mathbf{k}$$

Comparing the last expression with Equation (1), we are led to the following result.

Let $\mathbf{a} = a_1\mathbf{i} + a_2\mathbf{j} + a_3\mathbf{k}$ and $\mathbf{b} = b_1\mathbf{i} + b_2\mathbf{j} + b_3\mathbf{k}$. Then

$$\mathbf{a} \times \mathbf{b} = \begin{vmatrix} \mathbf{i} & \mathbf{j} & \mathbf{k} \\ a_1 & a_2 & a_3 \\ b_1 & b_2 & b_3 \end{vmatrix} \tag{2}$$

⚠ Note the order in which the scalar components of the vectors are written.

EXAMPLE 1   Let $\mathbf{a} = 2\mathbf{i} + \mathbf{j} - \mathbf{k}$ and $\mathbf{b} = -3\mathbf{i} - 2\mathbf{j} + \mathbf{k}$. Find $\mathbf{a} \times \mathbf{b}$ and $\mathbf{b} \times \mathbf{a}$.

Solution

$$\mathbf{a} \times \mathbf{b} = \begin{vmatrix} \mathbf{i} & \mathbf{j} & \mathbf{k} \\ 2 & 1 & -1 \\ -3 & -2 & 1 \end{vmatrix} = \mathbf{i} \begin{vmatrix} 1 & -1 \\ -2 & 1 \end{vmatrix} - \mathbf{j} \begin{vmatrix} 2 & -1 \\ -3 & 1 \end{vmatrix} + \mathbf{k} \begin{vmatrix} 2 & 1 \\ -3 & -2 \end{vmatrix}$$

$$= -\mathbf{i} + \mathbf{j} - \mathbf{k}$$

and

$$\mathbf{b} \times \mathbf{a} = \begin{vmatrix} \mathbf{i} & \mathbf{j} & \mathbf{k} \\ -3 & -2 & 1 \\ 2 & 1 & -1 \end{vmatrix} = \mathbf{i} \begin{vmatrix} -2 & 1 \\ 1 & -1 \end{vmatrix} - \mathbf{j} \begin{vmatrix} -3 & 1 \\ 2 & -1 \end{vmatrix} + \mathbf{k} \begin{vmatrix} -3 & -2 \\ 2 & 1 \end{vmatrix}$$

$$= \mathbf{i} - \mathbf{j} + \mathbf{k} \qquad \blacksquare$$

Note that $\mathbf{b} \times \mathbf{a} = -\mathbf{a} \times \mathbf{b}$ in Example 1. This is true in general if we recall the property of determinants that states that if two rows of a determinant are interchanged, then the sign of the determinant is changed.

■ Geometric Properties of the Cross Product

The cross product $\mathbf{a} \times \mathbf{b}$ being a vector, has both magnitude and direction. The following theorem tells us the direction of the vector $\mathbf{a} \times \mathbf{b}$.

THEOREM 1

Let $\mathbf{a}$ and $\mathbf{b}$ be nonzero vectors in 3-space. Then $\mathbf{a} \times \mathbf{b}$ is orthogonal to both $\mathbf{a}$ and $\mathbf{b}$.

PROOF Let $\mathbf{a} = a_1\mathbf{i} + a_2\mathbf{j} + a_3\mathbf{k}$ and $\mathbf{b} = b_1\mathbf{i} + b_2\mathbf{j} + b_3\mathbf{k}$. Then by Equation (1) or by expanding Equation (2), we have

$$\mathbf{a} \times \mathbf{b} = (a_2b_3 - a_3b_2)\mathbf{i} + (a_3b_1 - a_1b_3)\mathbf{j} + (a_1b_2 - a_2b_1)\mathbf{k}$$

Therefore,

$$(\mathbf{a} \times \mathbf{b}) \cdot \mathbf{a} = [(a_2b_3 - a_3b_2)\mathbf{i} + (a_3b_1 - a_1b_3)\mathbf{j} + (a_1b_2 - a_2b_1)\mathbf{k}]$$
$$\cdot (a_1\mathbf{i} + a_2\mathbf{j} + a_3\mathbf{k})$$
$$= (a_2b_3 - a_3b_2)a_1 + (a_3b_1 - a_1b_3)a_2 + (a_1b_2 - a_2b_1)a_3$$
$$= a_1a_2b_3 - a_1b_2a_3 + b_1a_2a_3 - a_1a_2b_3 + a_1b_2a_3 - b_1a_2a_3 = 0$$

which shows that $\mathbf{a} \times \mathbf{b}$ is orthogonal to $\mathbf{a}$. Similarly, by showing that $(\mathbf{a} \times \mathbf{b}) \cdot \mathbf{b} = 0$, we prove that $\mathbf{a} \times \mathbf{b}$ is orthogonal to $\mathbf{b}$. Therefore, $\mathbf{a} \times \mathbf{b}$ is orthogonal to both $\mathbf{a}$ and $\mathbf{b}$. $\qquad \blacksquare$

Let $\mathbf{a}$ and $\mathbf{b}$ be vectors in 3-space, and suppose that $\mathbf{a}$ and $\mathbf{b}$ have the same initial point. Then Theorem 1 tells us that $\mathbf{a} \times \mathbf{b}$ has a direction that is perpendicular to the plane determined by $\mathbf{a}$ and $\mathbf{b}$. (See Figure 1.) The direction of $\mathbf{a} \times \mathbf{b}$ is determined by the right-hand rule: Point the fingers of your open right hand in the direction of $\mathbf{a}$, then curl them towards the vector $\mathbf{b}$. Your thumb will then point in the direction of $\mathbf{a} \times \mathbf{b}$.

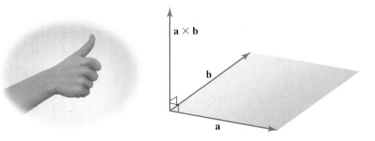

FIGURE 1
The vector $\mathbf{a} \times \mathbf{b}$ is orthogonal to both $\mathbf{a}$ and $\mathbf{b}$ with direction determined by the right-hand rule.

The next theorem gives the magnitude of $\mathbf{a} \times \mathbf{b}$.

THEOREM 2

Let $\mathbf{a}$ and $\mathbf{b}$ be vectors in 3-space. Then

$$|\mathbf{a} \times \mathbf{b}| = |\mathbf{a}||\mathbf{b}| \sin \theta$$

where θ is the angle between $\mathbf{a}$ and $\mathbf{b}$ and $0 \le \theta \le \pi$.

PROOF Let $\mathbf{a} = a_1\mathbf{i} + a_2\mathbf{j} + a_3\mathbf{k}$ and $\mathbf{b} = b_1\mathbf{i} + b_2\mathbf{j} + b_3\mathbf{k}$. Then, from Equation (1) we have

$$\mathbf{a} \times \mathbf{b} = (a_2b_3 - a_3b_2)\mathbf{i} + (a_3b_1 - a_1b_3)\mathbf{j} + (a_1b_2 - a_2b_1)\mathbf{k}$$

Next, using Property (4) of dot products we have

$$\begin{aligned}
|\mathbf{a} \times \mathbf{b}|^2 &= (\mathbf{a} \times \mathbf{b}) \cdot (\mathbf{a} \times \mathbf{b}) \\
&= a_2^2b_3^2 - 2a_2a_3b_2b_3 + a_3^2b_2^2 + a_3^2b_1^2 - 2a_1a_3b_1b_3 + a_1^2b_3^2 \\
&\quad + a_1^2b_2^2 - 2a_1a_2b_1b_2 + a_2^2b_1^2 \\
&= (a_1^2 + a_2^2 + a_3^2)(b_1^2 + b_2^2 + b_3^2) - (a_1b_1 + a_2b_2 + a_3b_3)^2 \\
&= |\mathbf{a}|^2|\mathbf{b}|^2 - (\mathbf{a} \cdot \mathbf{b})^2 \\
&= |\mathbf{a}|^2|\mathbf{b}|^2 - |\mathbf{a}|^2|\mathbf{b}|^2 \cos^2 \theta \qquad \text{Use Equation (2) of Section 11.3.} \\
&= |\mathbf{a}|^2|\mathbf{b}|^2 (1 - \cos^2 \theta) = |\mathbf{a}|^2|\mathbf{b}|^2 \sin^2 \theta
\end{aligned}$$

Finally, taking the square root on both sides and observing that $\sin \theta \ge 0$ for $0 \le \theta \le \pi$, we obtain

$$|\mathbf{a} \times \mathbf{b}| = |\mathbf{a}||\mathbf{b}| \sin \theta \qquad \blacksquare$$

We can combine the results of Theorems 1 and 2 to express the vector $\mathbf{a} \times \mathbf{b}$ in the following form.

An Alternative Definition of $\mathbf{a} \times \mathbf{b}$

$$\mathbf{a} \times \mathbf{b} = \underbrace{(|\mathbf{a}||\mathbf{b}| \sin \theta)}_{\text{Length of } \mathbf{a} \times \mathbf{b}} \overset{\uparrow}{\underset{\text{Direction of } \mathbf{a} \times \mathbf{b}}{\mathbf{n}}}$$

where θ is the angle between $\mathbf{a}$ and $\mathbf{b}$, and $\mathbf{n}$ is a unit vector orthogonal to both $\mathbf{a}$ and $\mathbf{b}$. (See Figure 2.)

FIGURE 2
The vector $\mathbf{a} \times \mathbf{b}$ has length $|\mathbf{a}||\mathbf{b}| \sin \theta$ and direction given by $\mathbf{n}$, the unit vector perpendicular to the plane determined by $\mathbf{a}$ and $\mathbf{b}$.

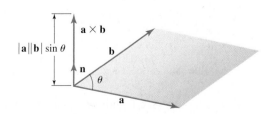

Note Since $\mathbf{b} \times \mathbf{a} = -\mathbf{a} \times \mathbf{b}$ we see that the vector $\mathbf{b} \times \mathbf{a}$ has the same length as $\mathbf{a} \times \mathbf{b}$ but points in the direction opposite to that of $\mathbf{a} \times \mathbf{b}$. ∎

EXAMPLE 2 Let $\mathbf{a} = 2\mathbf{i} + 3\mathbf{j}$ and $\mathbf{b} = 2\mathbf{j} + \mathbf{k}$.

a. Find a unit vector $\mathbf{n}$ that is orthogonal to both $\mathbf{a}$ and $\mathbf{b}$.
b. Express $\mathbf{a} \times \mathbf{b}$ in terms of $|\mathbf{a} \times \mathbf{b}|$ and $\mathbf{n}$.

Solution

a. A vector that is orthogonal to both $\mathbf{a}$ and $\mathbf{b}$ is

$$\mathbf{a} \times \mathbf{b} = \begin{vmatrix} \mathbf{i} & \mathbf{j} & \mathbf{k} \\ 2 & 3 & 0 \\ 0 & 2 & 1 \end{vmatrix} = 3\mathbf{i} - 2\mathbf{j} + 4\mathbf{k}$$

The length of $\mathbf{a} \times \mathbf{b}$ is

$$|\mathbf{a} \times \mathbf{b}| = \sqrt{3^2 + (-2)^2 + 4^2} = \sqrt{29}$$

Therefore, a unit vector that is orthogonal to both $\mathbf{a}$ and $\mathbf{b}$ is

$$\mathbf{n} = \frac{\mathbf{a} \times \mathbf{b}}{|\mathbf{a} \times \mathbf{b}|} = \frac{3}{\sqrt{29}}\mathbf{i} - \frac{2}{\sqrt{29}}\mathbf{j} + \frac{4}{\sqrt{29}}\mathbf{k}$$

b. We can write $\mathbf{a} \times \mathbf{b}$ as

$$\mathbf{a} \times \mathbf{b} = |\mathbf{a} \times \mathbf{b}|\mathbf{n}$$

$$= \sqrt{29}\left(\frac{3}{\sqrt{29}}\mathbf{i} - \frac{2}{\sqrt{29}}\mathbf{j} + \frac{4}{\sqrt{29}}\mathbf{k}\right)$$

The vectors $\mathbf{a}$, $\mathbf{b}$, $\mathbf{a} \times \mathbf{b}$ and $\mathbf{n}$ are shown in Figure 3. ∎

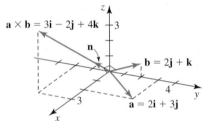

FIGURE 3
The vector $\mathbf{a} \times \mathbf{b}$ and the unit vector $\mathbf{n}$ are orthogonal to both $\mathbf{a}$ and $\mathbf{b}$.

EXAMPLE 3 **An Application in Mechanics** Figure 4a depicts a force $\mathbf{F}$ applied at a point (the terminal point of the position vector $\mathbf{r}$) on a wrench. This force produces a torque that acts along the axis of the bolt and has the effect of driving the bolt forward. To derive an expression for the torque, we recall that the magnitude of the torque $\boldsymbol{\tau}$ is given by

$|\boldsymbol{\tau}| =$ the length of the moment arm $\times$ the magnitude of the vertical component of $\mathbf{F}$

$\qquad = |\mathbf{r}||\mathbf{F}| \sin\theta$

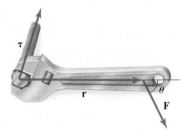

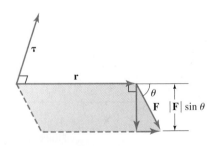

(a) The force $\mathbf{F}$ applied on the wrench produces a torque that acts along the axis of the bolt.

(b) The torque $\boldsymbol{\tau} = \mathbf{r} \times \mathbf{F}$

FIGURE 4

(See Figure 4b.) Since the direction of the torque is along the axis of the bolt (which is orthogonal to the plane determined by $\mathbf{r}$ and $\mathbf{F}$), we conclude that $\boldsymbol{\tau} = \mathbf{r} \times \mathbf{F}$. For example, if a force of magnitude 3 lb is applied to the wrench at a point $1\frac{1}{2}$ ft from the bolt at an angle of 60°, then the magnitude of the torque exerted on the bolt will be

$$\left(\frac{3}{2}\right)(3)\sin 60° \qquad \text{or} \qquad \frac{9\sqrt{3}}{4} \text{ ft-lb} \qquad \blacksquare$$

The following test for determining whether two vectors are parallel is an immediate consequence of Theorem 2.

Test for Parallel Vectors

Two nonzero vectors **a** and **b** are parallel if and only if $\mathbf{a} \times \mathbf{b} = \mathbf{0}$.

PROOF   Two nonzero vectors **a** and **b** are parallel if and only if $\theta = 0$ or π. In either case, $|\mathbf{a} \times \mathbf{b}| = |\mathbf{a}||\mathbf{b}|\sin\theta = 0$, so $\mathbf{a} \times \mathbf{b} = \mathbf{0}$. $\blacksquare$

■ Finding the Area of a Triangle

Consider the parallelogram determined by the vectors **a** and **b** shown in Figure 5a. The altitude of the parallelogram is $|\mathbf{b}|\sin\theta$ and the length of its base is $|\mathbf{a}|$, so its area is

$$A = |\mathbf{a}||\mathbf{b}|\sin\theta = |\mathbf{a} \times \mathbf{b}| \qquad \text{By Theorem 2}$$

Thus, the *length of the cross product* $\mathbf{a} \times \mathbf{b}$ *and the area of the parallelogram determined by* **a** *and* **b** have the same numerical value. (See Figure 5b.)

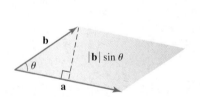
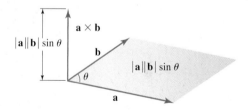

FIGURE 5

(a) The parallelogram determined by **a** and **b**

(b) The length of $\mathbf{a} \times \mathbf{b}$ is numerically equal to the area of the parallelogram determined by **a** and **b**.

Therefore, the area of the triangle determined by **a** and **b** is $\frac{1}{2}|\mathbf{a} \times \mathbf{b}|$.

EXAMPLE 4   Find the area of the triangle with vertices $P(3, -3, 0)$, $Q(1, 2, 2)$, and $R(1, -2, 5)$.

Solution   The area of $\triangle PQR$ is half the area of the parallelogram determined by the vectors $\overrightarrow{PQ}$ and $\overrightarrow{PR}$. Now $\overrightarrow{PQ} = \langle -2, 5, 2 \rangle$ and $\overrightarrow{PR} = \langle -2, 1, 5 \rangle$, so

$$\overrightarrow{PQ} \times \overrightarrow{PR} = \begin{vmatrix} \mathbf{i} & \mathbf{j} & \mathbf{k} \\ -2 & 5 & 2 \\ -2 & 1 & 5 \end{vmatrix}$$

$$= \mathbf{i}\begin{vmatrix} 5 & 2 \\ 1 & 5 \end{vmatrix} - \mathbf{j}\begin{vmatrix} -2 & 2 \\ -2 & 5 \end{vmatrix} + \mathbf{k}\begin{vmatrix} -2 & 5 \\ -2 & 1 \end{vmatrix}$$

$$= 23\mathbf{i} + 6\mathbf{j} + 8\mathbf{k}$$

Therefore, the area of the parallelogram is

$$|\overrightarrow{PQ} \times \overrightarrow{PR}| = \sqrt{23^2 + 6^2 + 8^2} = \sqrt{629} \approx 25.1$$

so the area of the required triangle is $\frac{1}{2}\sqrt{629}$ or approximately 12.5. (See Figure 6.)

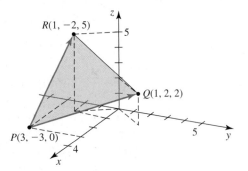

FIGURE 6
The triangle with vertices
at $P(3, -3, 0)$, $Q(1, 2, 2)$,
and $R(1, -2, 5)$

■ Properties of the Cross Product

The cross product obeys the following rules.

THEOREM 3   Properties of the Cross Product

If **a**, **b**, and **c** are vectors and c is a scalar, then

1. $\mathbf{a} \times \mathbf{b} = -\mathbf{b} \times \mathbf{a}$
2. $\mathbf{a} \times (\mathbf{b} + \mathbf{c}) = \mathbf{a} \times \mathbf{b} + \mathbf{a} \times \mathbf{c}$
3. $(\mathbf{a} + \mathbf{b}) \times \mathbf{c} = \mathbf{a} \times \mathbf{c} + \mathbf{b} \times \mathbf{c}$
4. $c(\mathbf{a} \times \mathbf{b}) = (c\mathbf{a}) \times \mathbf{b} = \mathbf{a} \times (c\mathbf{b})$
5. $\mathbf{a} \times \mathbf{0} = \mathbf{0} \times \mathbf{a} = \mathbf{0}$
6. $\mathbf{a} \times \mathbf{a} = \mathbf{0}$
7. $\mathbf{a} \cdot (\mathbf{b} \times \mathbf{c}) = (\mathbf{a} \times \mathbf{b}) \cdot \mathbf{c}$
8. $\mathbf{a} \times (\mathbf{b} \times \mathbf{c}) = (\mathbf{a} \cdot \mathbf{c})\mathbf{b} - (\mathbf{a} \cdot \mathbf{b})\mathbf{c}$

PROOF   Each of these properties may be proved by applying the definition of the cross product. For example, to prove Property 1, let $\mathbf{a} = a_1\mathbf{i} + a_2\mathbf{j} + a_3\mathbf{k}$ and $b = b_1\mathbf{i} + b_2\mathbf{j} + b_3\mathbf{k}$. Then

$$\mathbf{a} \times \mathbf{b} = \begin{vmatrix} \mathbf{i} & \mathbf{j} & \mathbf{k} \\ a_1 & a_2 & a_3 \\ b_1 & b_2 & b_3 \end{vmatrix} = (a_2 b_3 - a_3 b_2)\mathbf{i} - (a_1 b_3 - a_3 b_1)\mathbf{j} + (a_1 b_2 - a_2 b_1)\mathbf{k}$$

and

$$\mathbf{b} \times \mathbf{a} = \begin{vmatrix} \mathbf{i} & \mathbf{j} & \mathbf{k} \\ b_1 & b_2 & b_3 \\ a_1 & a_2 & a_3 \end{vmatrix} = (b_2 a_3 - b_3 a_2)\mathbf{i} - (b_1 a_3 - b_3 a_1)\mathbf{j} + (b_1 a_2 - b_2 a_1)\mathbf{k}$$

So $\mathbf{a} \times \mathbf{b} = -\mathbf{b} \times \mathbf{a}$. See Example 1 for an illustration of this property.   ■

The proofs of the other properties are left as exercises.

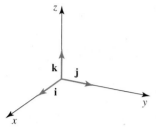

FIGURE 7
Using the right-hand rule, we can see the validity of the relationships in Example 5.

FIGURE 8
The cross product of two consecutive vectors in the counterclockwise direction is the next vector and has a positive direction; in the clockwise direction it is the next vector and has a negative direction. For example, $\mathbf{i} \times \mathbf{j} = \mathbf{k}$ and $\mathbf{j} \times \mathbf{i} = -\mathbf{k}$.

EXAMPLE 5 By direct computation or using Property 6 of Theorem 3, we can show that

$$\mathbf{i} \times \mathbf{i} = \mathbf{0}, \qquad \mathbf{j} \times \mathbf{j} = \mathbf{0}, \qquad \text{and} \qquad \mathbf{k} \times \mathbf{k} = \mathbf{0} \qquad (3)$$

Next, by direct computation, we can verify that

$$\mathbf{i} \times \mathbf{j} = \mathbf{k}, \qquad \mathbf{j} \times \mathbf{k} = \mathbf{i}, \qquad \text{and} \qquad \mathbf{k} \times \mathbf{i} = \mathbf{j} \qquad (4)$$

so by Property 1 of Theorem 3 we also have

$$\mathbf{j} \times \mathbf{i} = -\mathbf{k}, \qquad \mathbf{k} \times \mathbf{j} = -\mathbf{i}, \qquad \text{and} \qquad \mathbf{i} \times \mathbf{k} = -\mathbf{j} \qquad (5)$$

These results are also evident if you interpret each of the cross products (3)–(5) geometrically while looking at Figure 7. ∎

You can use a simple mnemonic device to help remember the cross products (3)–(5). Consider the circle shown in Figure 8. The cross product of two consecutive vectors in the counterclockwise direction is the next vector, and its direction is positive. Likewise, the cross product of two consecutive vectors in the clockwise direction is the next vector but with a negative direction.

■ The Scalar Triple Product

Suppose that $\mathbf{a}$, $\mathbf{b}$, and $\mathbf{c}$ are three vectors in three-dimensional space. The dot product of $\mathbf{a}$ and $\mathbf{b} \times \mathbf{c}$, $\mathbf{a} \cdot (\mathbf{b} \times \mathbf{c})$, is called the **scalar triple product.** If we write $\mathbf{a} = a_1\mathbf{i} + a_2\mathbf{j} + a_3\mathbf{k}$, $\mathbf{b} = b_1\mathbf{i} + b_2\mathbf{j} + b_3\mathbf{k}$, and $\mathbf{c} = c_1\mathbf{i} + c_2\mathbf{j} + c_3\mathbf{k}$, then by direct computation,

$$\mathbf{a} \cdot (\mathbf{b} \times \mathbf{c}) = \begin{vmatrix} a_1 & a_2 & a_3 \\ b_1 & b_2 & b_3 \\ c_1 & c_2 & c_3 \end{vmatrix} \qquad (6)$$

(We leave it as an exercise for you to verify this computation.)

The geometric significance of the scalar triple product can be seen by examining the parallelepiped determined by the vectors $\mathbf{a}$, $\mathbf{b}$, and $\mathbf{c}$. (See Figure 9.) The base of the parallelepiped is a parallelogram with adjacent sides determined by $\mathbf{b}$ and $\mathbf{c}$ with area $|\mathbf{b} \times \mathbf{c}|$. If θ is the angle between $\mathbf{a}$ and $\mathbf{b} \times \mathbf{c}$, then the height of the parallelepiped is given by $h = |\mathbf{a}||\cos \theta|$. Therefore, the volume of the parallelepiped is

$$V = |\mathbf{b} \times \mathbf{c}||\mathbf{a}||\cos \theta| \qquad \text{area of base · height}$$

$$= |\mathbf{a} \cdot (\mathbf{b} \times \mathbf{c})| \qquad \text{By Equation (2) of Section 11.3}$$

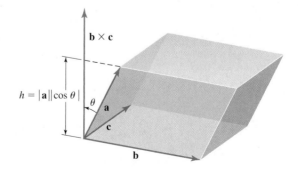

FIGURE 9
The volume V of the parallelepiped is equal to $|\mathbf{a} \cdot (\mathbf{b} \times \mathbf{c})|$.

We have established the following result.

THEOREM 4   Geometric Interpretation of the Scalar Triple Product

The volume V of the parallelepiped determined by the vectors $\mathbf{a}$, $\mathbf{b}$, and $\mathbf{c}$ is given by

$$V = |\mathbf{a} \cdot (\mathbf{b} \times \mathbf{c})|$$

EXAMPLE 6 Find the volume of the parallelepiped determined by the vectors $\mathbf{a} = \mathbf{i} + 2\mathbf{j} + 3\mathbf{k}$, $\mathbf{b} = \mathbf{i} - \mathbf{j} + \mathbf{k}$, and $\mathbf{c} = 3\mathbf{i} + \mathbf{j} - 2\mathbf{k}$.

Solution By Theorem 4 the volume of the parallelepiped is $V = |\mathbf{a} \cdot (\mathbf{b} \times \mathbf{c})|$. But by Equation (6),

$$\mathbf{a} \cdot (\mathbf{b} \times \mathbf{c}) = \begin{vmatrix} 1 & 2 & 3 \\ 1 & -1 & 1 \\ 3 & 1 & -2 \end{vmatrix} = 1 \begin{vmatrix} -1 & 1 \\ 1 & -2 \end{vmatrix} - 2 \begin{vmatrix} 1 & 1 \\ 3 & -2 \end{vmatrix} + 3 \begin{vmatrix} 1 & -1 \\ 3 & 1 \end{vmatrix}$$

$$= 1(1) - 2(-5) + 3(4) = 23$$

Therefore, the required volume is $|23|$, or 23. ∎

Because the volume of a parallelepiped is zero if and only if the vectors $\mathbf{a}$, $\mathbf{b}$, and $\mathbf{c}$ forming the adjacent sides of the parallelepiped are coplanar, we have the following result.

Test for Coplanar Vectors

The vectors $\mathbf{a} = a_1\mathbf{i} + a_2\mathbf{j} + a_3\mathbf{k}$, $\mathbf{b} = b_1\mathbf{i} + b_2\mathbf{j} + b_3\mathbf{k}$, and $\mathbf{c} = c_1\mathbf{i} + c_2\mathbf{j} + c_3\mathbf{k}$ are coplanar if and only if

$$\mathbf{a} \cdot (\mathbf{b} \times \mathbf{c}) = \begin{vmatrix} a_1 & a_2 & a_3 \\ b_1 & b_2 & b_3 \\ c_1 & c_2 & c_3 \end{vmatrix} = 0$$

11.4   CONCEPT QUESTIONS

1. **a.** Give the definition of $\mathbf{a} \times \mathbf{b}$.
   b. What is the length or magnitude of $\mathbf{a} \times \mathbf{b}$ in terms of the angle between $\mathbf{a}$ and $\mathbf{b}$? Give a geometric interpretation of $|\mathbf{a} \times \mathbf{b}|$.
   c. What is the direction of $\mathbf{a} \times \mathbf{b}$?
   d. Write $\mathbf{a} \times \mathbf{b}$ in terms of $\mathbf{a}$, $\mathbf{b}$, and θ (the angle between $\mathbf{a}$ and $\mathbf{b}$).

2. **a.** How would you use the cross product of two nonzero vectors $\mathbf{a}$ and $\mathbf{b}$ to determine whether $\mathbf{a}$ and $\mathbf{b}$ are parallel?
   b. What can you say about the three nonzero vectors $\mathbf{a}$, $\mathbf{b}$, and $\mathbf{c}$ if $(\mathbf{c} \times \mathbf{b}) \cdot \mathbf{a} = 0$?

11.4 EXERCISES

1. State whether the expression makes sense. Explain.
   a. $(\mathbf{a} \times \mathbf{b}) \cdot \mathbf{c}$ **b.** $\mathbf{a} \times (\mathbf{b} \cdot \mathbf{c})$
   c. $(\mathbf{a} + \mathbf{b}) \cdot (\mathbf{c} \times \mathbf{d})$ **d.** $(\mathbf{a} \times \mathbf{b}) \cdot (\mathbf{c} \times \mathbf{d})$
   e. $\mathbf{a} \times [(\mathbf{b} \cdot \mathbf{c})\mathbf{d}]$ **f.** $(\mathbf{a} \times \mathbf{b}) \times (\mathbf{c} \times \mathbf{d})$

2. If $\mathbf{a} \cdot \mathbf{b} = 0$ and $\mathbf{a} \times \mathbf{b} = \mathbf{0}$ what can you say about $\mathbf{a}$ and $\mathbf{b}$?

In Exercises 3–10, find $\mathbf{a} \times \mathbf{b}$.

3. $\mathbf{a} = \mathbf{i} + \mathbf{j}, \quad \mathbf{b} = 2\mathbf{j} + 3\mathbf{k}$

4. $\mathbf{a} = \langle 0, 1, 2 \rangle, \quad \mathbf{b} = \langle 2, 1, 3 \rangle$

5. $\mathbf{a} = \langle 1, -2, 1 \rangle, \quad \mathbf{b} = \langle 3, 1, -2 \rangle$

6. $\mathbf{a} = 2\mathbf{i} - 3\mathbf{j} + 4\mathbf{k}, \quad \mathbf{b} = -\mathbf{i} - 2\mathbf{j} + 3\mathbf{k}$

7. $\mathbf{a} = 2\mathbf{i} + 3\mathbf{k}, \quad \mathbf{b} = -3\mathbf{i} + 2\mathbf{j} - \mathbf{k}$

8. $\mathbf{a} = \langle 0, 1, 0 \rangle, \quad \mathbf{b} = \langle -3, 0, 2 \rangle$

9. $\mathbf{a} = 2\mathbf{i} + \mathbf{j} - 3\mathbf{k}, \quad \mathbf{b} = \frac{2}{3}\mathbf{i} + \frac{1}{3}\mathbf{j} - \mathbf{k}$

10. $\mathbf{a} = \langle 1, 1, 2 \rangle, \quad \mathbf{b} = \langle \frac{1}{2}, 2, -\frac{1}{2} \rangle$

In Exercises 11 and 12, find $\mathbf{a} \times \mathbf{b}$ *and* $\mathbf{b} \times \mathbf{a}$.

11. $\mathbf{a} = \mathbf{i} + 2\mathbf{j} + 3\mathbf{k}, \quad \mathbf{b} = 2\mathbf{i} - \mathbf{j} - \mathbf{k}$

12. $\mathbf{a} = \langle 1, 1, 2 \rangle, \quad \mathbf{b} = \langle -1, 3, -1 \rangle$

In Exercises 13 and 14, find two vectors that are orthogonal to both $\mathbf{a}$ *and* $\mathbf{b}$.

13. $\mathbf{a} = 2\mathbf{i} - 3\mathbf{j} + 4\mathbf{k}, \quad \mathbf{b} = -\mathbf{i} + \mathbf{j} - 2\mathbf{k}$

14. $\mathbf{a} = \langle 1, -2, 1 \rangle, \quad \mathbf{b} = \langle 2, 3, -4 \rangle$

In Exercises 15 and 16, find two unit vectors that are orthogonal to both $\mathbf{a}$ *and* $\mathbf{b}$.

15. $\mathbf{a} = -3\mathbf{i} + \mathbf{j} - 2\mathbf{k}, \quad \mathbf{b} = \mathbf{i} + \mathbf{j} + \mathbf{k}$

16. $\mathbf{a} = \langle -1, 1, -1 \rangle, \quad \mathbf{b} = \langle 0, 3, 4 \rangle$

In Exercises 17–20, find the area of the triangle with the given vertices.

17. $P(1, 0, 0), Q(0, 1, 0), R(0, 0, 1)$

18. $P(1, 1, 1), Q(1, 2, 1), R(2, 2, 3)$

19. $P(1, -1, 2), Q(2, 3, 1), R(-2, 3, 4)$

20. $P(0, 0, 0), Q(1, 3, 2), R(-1, -2, 3)$

In Exercises 21–26, let $\mathbf{a} = \mathbf{i} - \mathbf{j} + \mathbf{k}$, $\mathbf{b} = 2\mathbf{i} + 3\mathbf{j} - \mathbf{k}$, *and* $\mathbf{c} = -\mathbf{i} + \mathbf{j} + 2\mathbf{k}$. *Find the indicated quantity.*

21. $(2\mathbf{a}) \times \mathbf{b}$ **22.** $(\mathbf{a} + \mathbf{b}) \times \mathbf{c}$

23. $(\mathbf{a} \times \mathbf{b}) \times \mathbf{c}$ **24.** $\mathbf{a} \cdot (\mathbf{b} \times \mathbf{c})$

25. $(\mathbf{a} \times \mathbf{b}) \cdot \mathbf{c}$ **26.** $(3\mathbf{a}) \times (\mathbf{a} - 2\mathbf{b} + 3\mathbf{c})$

In Exercises 27 and 28, find the volume of the parallelepiped determined by the vectors, $\mathbf{a}$, $\mathbf{b}$, *and* $\mathbf{c}$.

27. $\mathbf{a} = \mathbf{i} + \mathbf{j}, \mathbf{b} = \mathbf{j} - 2\mathbf{k}, \mathbf{c} = \mathbf{i} + 2\mathbf{j} + 3\mathbf{k}$

28. $\mathbf{a} = \langle 1, 3, 2 \rangle, \mathbf{b} = \langle 2, -1, 3 \rangle, \mathbf{c} = \langle 1, -1, -2 \rangle$

In Exercises 29 and 30, find the volume of the parallelepiped with adjacent edges PQ, PR, and PS.

29. $P(0, 0, 0), Q(3, -2, 1), R(1, 2, 2), S(1, 1, 4)$

30. $P(1, 1, 1), Q(2, 1, 3), R(-1, 0, 3), S(4, -1, 2)$

31. Find the height of a parallelepiped determined by $\mathbf{a} = \mathbf{i} + 2\mathbf{j} + \mathbf{k}, \mathbf{b} = 2\mathbf{i} + \mathbf{j} - \mathbf{k}$, and $\mathbf{c} = \mathbf{i} + \mathbf{j} + 3\mathbf{k}$ if its base is determined by $\mathbf{a}$ and $\mathbf{b}$.

32. Find c such that $\mathbf{a} = 2\mathbf{i} + 3\mathbf{j} + \mathbf{k}, \mathbf{b} = \mathbf{i} + 2\mathbf{j} + 3\mathbf{k}$, and $\mathbf{c} = \mathbf{i} - 3\mathbf{j} + c\mathbf{k}$ are coplanar.

33. Determine whether the vectors $\mathbf{a} = \mathbf{i} + 2\mathbf{j} + 4\mathbf{k}$, $\mathbf{b} = -2\mathbf{i} + 3\mathbf{j} - \mathbf{k}$, and $\mathbf{c} = \mathbf{j} + \mathbf{k}$ are coplanar.

34. Find the value of c such that the vectors $\mathbf{u} = \mathbf{i} - 2\mathbf{j} + 3\mathbf{k}$, $\mathbf{v} = \mathbf{i} + 2\mathbf{j} - \mathbf{k}$, and $\mathbf{w} = 2\mathbf{i} + 3\mathbf{j} + c\mathbf{k}$ are coplanar.

35. Determine whether the points $P(1, 0, 1), Q(2, 3, 1)$, $R(-1, 2, -3)$, and $S\left(\frac{2}{3}, -1, 1\right)$ are coplanar.

36. Find the area of the parallelogram shown in the figure in terms of a, b, c, and d.

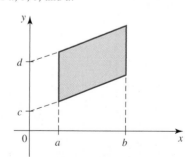

37. A force of magnitude 50 lb is applied at the end of an 18-in.-long leveraged-impact lug wrench in the direction as shown in the figure. Find the magnitude of the torque about P.

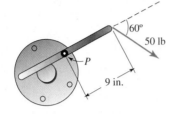

38. A 15-lb force is applied to a stapler at the point shown. Find the magnitude of the torque about P.

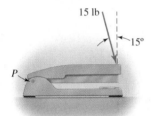
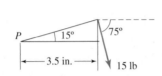

39. **Force on a Proton Moving Through a Magnetic Field** The force $\mathbf{F}$ acting on a charge q moving with velocity $\mathbf{v}$ in a magnetic field $\mathbf{B}$ is given by $\mathbf{F} = q\mathbf{v} \times \mathbf{B}$, where q is measured in coulombs, $|\mathbf{B}|$ in tesla, and $|\mathbf{v}|$ in meters per second. Suppose that a proton beam moves through a region of space where there is a uniform magnetic field $\mathbf{B} = 2\mathbf{k}$. The protons have velocity

$$\mathbf{v} = \left(\frac{3}{2} \times 10^5\right)\mathbf{i} + \left(\frac{3\sqrt{3}}{2} \times 10^5\right)\mathbf{k}$$

Find the force on a proton if its charge is $q = 1.6 \times 10^{-19}$ C.

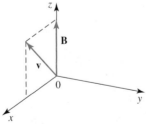

Directions of $\mathbf{v}$ and $\mathbf{B}$

40. Find $(\mathbf{a} \times \mathbf{b}) \times \mathbf{c}$ and $\mathbf{a} \times (\mathbf{b} \times \mathbf{c})$ given that $\mathbf{a} = 2\mathbf{i} + \mathbf{j}$, $\mathbf{b} = 3\mathbf{i} + \mathbf{j} - \mathbf{k}$, and $\mathbf{c} = \mathbf{i} + \mathbf{k}$. Does the associative law hold for vector products?

41. Let $\mathbf{a} = a_1\mathbf{i} + a_2\mathbf{j} + a_3\mathbf{k}$, $\mathbf{b} = b_1\mathbf{i} + b_2\mathbf{j} + b_3\mathbf{k}$ and $\mathbf{c} = c_1\mathbf{i} + c_2\mathbf{j} + c_3\mathbf{k}$. Prove that

$$\mathbf{a} \cdot (\mathbf{b} \times \mathbf{c}) = (\mathbf{a} \times \mathbf{b}) \cdot \mathbf{c} = \begin{vmatrix} a_1 & a_2 & a_3 \\ b_1 & b_2 & b_3 \\ c_1 & c_2 & c_3 \end{vmatrix}$$

42. Prove that $(\mathbf{a} + \mathbf{b}) \times (\mathbf{a} - \mathbf{b}) = 2(\mathbf{b} \times \mathbf{a})$.

43. Prove that $\mathbf{a} \times (\mathbf{b} \times \mathbf{c}) + \mathbf{b} \times (\mathbf{c} \times \mathbf{a}) + \mathbf{c} \times (\mathbf{a} \times \mathbf{b}) = \mathbf{0}$.

44. Prove that

$$(\mathbf{a} \times \mathbf{b}) \cdot (\mathbf{c} \times \mathbf{d}) = \begin{vmatrix} \mathbf{a} \cdot \mathbf{c} & \mathbf{b} \cdot \mathbf{c} \\ \mathbf{a} \cdot \mathbf{d} & \mathbf{b} \cdot \mathbf{d} \end{vmatrix}$$

45. Prove Lagrange's identity:

$$|\mathbf{a} \times \mathbf{b}|^2 = |\mathbf{a}|^2|\mathbf{b}|^2 - (\mathbf{a} \cdot \mathbf{b})^2$$

46. Refer to the following figure.

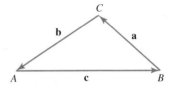

a. Show that $\mathbf{a} + \mathbf{b} + \mathbf{c} = \mathbf{0}$.

b. Show that $\mathbf{a} \times \mathbf{b} = \mathbf{b} \times \mathbf{c} = \mathbf{c} \times \mathbf{a}$, and hence deduce the law of sines for plane triangles:

$$\frac{\sin A}{a} = \frac{\sin B}{b} = \frac{\sin C}{c} \qquad a = |\mathbf{a}|, \quad b = |\mathbf{b}|, \quad c = |\mathbf{c}|$$

In Exercises 47–52, prove the given property of the cross-product.

47. $\mathbf{a} \times (\mathbf{b} + \mathbf{c}) = \mathbf{a} \times \mathbf{b} + \mathbf{a} \times \mathbf{c}$

48. $c(\mathbf{a} \times \mathbf{b}) = (c\mathbf{a}) \times \mathbf{b}$

49. $\mathbf{a} \times \mathbf{0} = \mathbf{0} \times \mathbf{a} = \mathbf{0}$

50. $\mathbf{a} \times \mathbf{a} = \mathbf{0}$

51. $(\mathbf{a} + \mathbf{b}) \times \mathbf{c} = \mathbf{a} \times \mathbf{c} + \mathbf{b} \times \mathbf{c}$

52. $\mathbf{a} \times (\mathbf{b} \times \mathbf{c}) = (\mathbf{a} \cdot \mathbf{c})\mathbf{b} - (\mathbf{a} \cdot \mathbf{b})\mathbf{c}$

53. **Angular Velocity** Consider a rigid body rotating about a fixed axis with a constant angular speed ω. The *angular velocity* is represented by a vector $\boldsymbol{\omega}$ of magnitude ω lying along the axis of rotation as shown in the figure. If we place the origin 0 on the axis of rotation and let $\mathbf{R}$ denote the position vector of a particle in the body, then the velocity $\mathbf{v}$ of the particle is given by

$$\mathbf{v} = \boldsymbol{\omega} \times \mathbf{R}$$

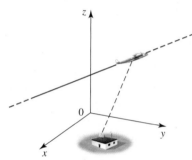

Suppose that the axis of rotation is parallel to the vector $2\mathbf{i} + 2\mathbf{j} + \mathbf{k}$. What is the speed of a particle at the instant it passes through the point $(3, 5, 2)$?

54. Find a unit vector in the plane that contains the vectors $\mathbf{a} = \mathbf{i} + 3\mathbf{j} + 2\mathbf{k}$ and $\mathbf{b} = -\mathbf{i} + 2\mathbf{j} + 4\mathbf{k}$, and is perpendicular to the vector $\mathbf{c} = 3\mathbf{i} - \mathbf{j} + 2\mathbf{k}$.

55. Find a and b if $\mathbf{a} = \langle -1, a, 3 \rangle$ and $\mathbf{b} = \langle 2, 3, b \rangle$ are parallel.

56. Find s and t such that $(\mathbf{a} \times \mathbf{b}) \times \mathbf{a} = s\mathbf{a} + t\mathbf{b}$, where $\mathbf{a} = \langle 2, 1, 3 \rangle$ and $\mathbf{b} = \langle 1, 3, 4 \rangle$.

In Exercises 57–62, determine whether the statement is true or false. If it is true, explain why. If it is false, explain why or give an example that shows it is false.

57. $\mathbf{a} \times \mathbf{b} + \mathbf{b} \times \mathbf{a} = \mathbf{0}$

58. $[(\mathbf{a} \times \mathbf{b}) \times \mathbf{c}] \times [(\mathbf{a} \times \mathbf{b}) \times \mathbf{c}] = \mathbf{0}$

59. $\mathbf{a} \cdot (\mathbf{a} \times \mathbf{b}) = 0$

60. $\mathbf{a} \times \mathbf{b} = \mathbf{a} \times \mathbf{c}$ if and only if $\mathbf{b} = \mathbf{c}$

61. $(\mathbf{a} - \mathbf{b}) \times (\mathbf{a} + \mathbf{b}) = 2\mathbf{a} \times \mathbf{b}$

62. If $\mathbf{a} \neq \mathbf{0}$, $\mathbf{a} \cdot \mathbf{b} = \mathbf{a} \cdot \mathbf{c}$, and $\mathbf{a} \times \mathbf{b} = \mathbf{a} \times \mathbf{c}$ then $\mathbf{b} = \mathbf{c}$.

11.5 Lines and Planes in Space

Equations of Lines in Space

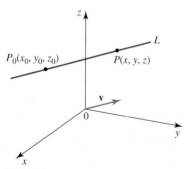

FIGURE 1
The path of the airplane is a straight line.

Figure 1 depicts an airplane flying in a straight line above a ground radar station. How fast is the distance between the airplane and the radar station changing at any time? How close to the radar station does the airplane get? To answer questions such as these, we need to be able to describe the path of the airplane. More specifically, we want to be able to represent a line in space algebraically.

In this section we will see how lines as well as planes in space can be described in algebraic terms. We begin by considering a line in space. Such a line is uniquely determined by specifying its direction and a point through which it passes. The direction may be specified by a vector that has the same direction as the line. So suppose that the line L passes through the point $P_0(x_0, y_0, z_0)$ and has the same direction as the vector $\mathbf{v} = \langle a, b, c \rangle$. (See Figure 2.)

Let $P(x, y, z)$ be *any* point on L. Then the vector $\overrightarrow{P_0P}$ is parallel to $\mathbf{v}$. But two vectors are parallel if and only if one is a scalar multiple of the other. Therefore, there exists some number t, called a *parameter,* such that

$$\overrightarrow{P_0P} = t\mathbf{v}$$

or, since $\overrightarrow{P_0P} = \langle x - x_0, y - y_0, z - z_0 \rangle$, we have

$$\langle x - x_0, y - y_0, z - z_0 \rangle = t\langle a, b, c \rangle = \langle ta, tb, tc \rangle$$

Equating the corresponding components of the two vectors then yields

$$x - x_0 = ta, \qquad y - y_0 = tb, \qquad \text{and} \qquad z - z_0 = tc$$

FIGURE 2
The line L passes through P_0 and is parallel to the vector $\mathbf{v}$.

Solving these equations for x, y, and z, respectively, gives the following standard *parametric equations* of the line L.

DEFINITION   Parametric Equations of a Line

The parametric equations of the line passing through the point $P_0(x_0, y_0, z_0)$ and parallel to the vector $\mathbf{v} = \langle a, b, c \rangle$ are

$$x = x_0 + at, \qquad y = y_0 + bt, \qquad \text{and} \qquad z = z_0 + ct \qquad \textbf{(1)}$$

Each value of the parameter t corresponds to a point $P(x, y, z)$ on L. As t takes on all values in the parameter interval $(-\infty, \infty)$, the line L is traced out. (See Figure 3.)

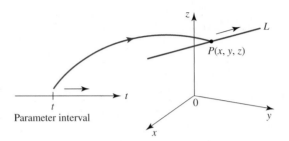

FIGURE 3
As t runs through all values in the parameter interval $(-\infty, \infty)$, L is traced out.

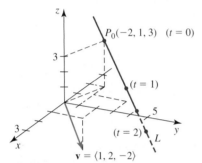

FIGURE 4
The line L and some points on L corresponding to selected values of t. Note the orientation of the line.

EXAMPLE 1 Find parametric equations for the line passing through the point $P_0(-2, 1, 3)$ and parallel to the vector $\mathbf{v} = \langle 1, 2, -2 \rangle$.

Solution We use Equation (1) with $x_0 = -2, y_0 = 1, z_0 = 3, a = 1, b = 2$, and $c = -2$, obtaining

$$x = -2 + t, \qquad y = 1 + 2t, \qquad \text{and} \qquad z = 3 - 2t$$

The line L in question is sketched in Figure 4.   ∎

Suppose that the vector $\mathbf{v} = \langle a, b, c \rangle$ defines the direction of a line L. Then the numbers a, b, and c are called the **direction numbers** of L. Observe that if a line L is described by a set of parametric equations (1), then the direction numbers of L are precisely the coefficients of t in each of the parametric equations.

There is another way of describing a line in space. We start with the parametric equations of the line L,

$$x = x_0 + at, \qquad y = y_0 + bt, \qquad \text{and} \qquad z = z_0 + ct$$

If the direction numbers a, b, and c are all nonzero, then we can solve each of these equations for t. Thus,

$$t = \frac{x - x_0}{a}, \qquad t = \frac{y - y_0}{b}, \qquad \text{and} \qquad t = \frac{z - z_0}{c}$$

which gives the following *symmetric equations* of L.

DEFINITION   Symmetric Equations of a Line

The **symmetric equations** of the line L passing through the point $P_0(x_0, y_0, z_0)$ and parallel to the vector $\mathbf{v} = \langle a, b, c \rangle$ are

$$\frac{x - x_0}{a} = \frac{y - y_0}{b} = \frac{z - z_0}{c} \tag{2}$$

Note   Suppose $a = 0$ and both b and c are not equal to zero, then the parametric equations of the line take the form

$$x = x_0, \qquad y = y_0 + bt, \qquad \text{and} \qquad z = z_0 + ct$$

and the line lies in the plane $x = x_0$ (parallel to the yz-plane). Solving the second and third equations for t leads to

$$x = x_0, \qquad \frac{y - y_0}{b} = \frac{z - z_0}{c}$$

which are the symmetric equations of the line. We leave it to you to consider and interpret the other cases. ■

EXAMPLE 2

a. Find parametric equations and symmetric equations for the line L passing through the points $P(-3, 3, -2)$ and $Q(2, -1, 4)$.
b. At what point does L intersect the xy-plane?

Solution
a. The direction of L is the same as that of the vector $\overrightarrow{PQ} = \langle 5, -4, 6 \rangle$. Since L passes through $P(-3, 3, -2)$, we can use Equation (1) with $a = 5$, $b = -4$, $c = 6$, $x_0 = -3$, $y_0 = 3$, and $z_0 = -2$, to obtain the parametric equations

$$x = -3 + 5t, \qquad y = 3 - 4t, \qquad \text{and} \qquad z = -2 + 6t$$

Next, using Equation (2), we obtain the following symmetric equations for L:

$$\frac{x + 3}{5} = \frac{y - 3}{-4} = \frac{z + 2}{6}$$

b. At the point where the line intersects the xy-plane, we have $z = 0$. So setting $z = 0$ in the third parametric equation, we obtain $t = \frac{1}{3}$. Substituting this value of t into the other parametric equations gives the required point as $\left(-\frac{4}{3}, \frac{5}{3}, 0\right)$. (See Figure 5.)

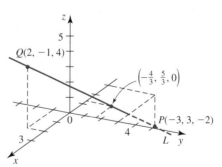

FIGURE 5
The line L intersects the xy-plane at the point $\left(-\frac{4}{3}, \frac{5}{3}, 0\right)$.

■

Suppose that L_1 and L_2 are lines having the same directions as the vectors $\mathbf{v}_1$ and $\mathbf{v}_2$, respectively. Then L_1 is **parallel** to L_2 if $\mathbf{v}_1$ is parallel to $\mathbf{v}_2$.

EXAMPLE 3 Let L_1 be the line with parametric equations

$$x = 1 + 2t, \qquad y = 2 - 3t, \qquad \text{and} \qquad z = 2 + t$$

and let L_2 be the line with parametric equations

$$x = 3 - 4t, \qquad y = 1 + 4t, \qquad \text{and} \qquad z = -3 + 4t$$

a. Show that the lines L_1 and L_2 are not parallel to each other.
b. Do the lines L_1 and L_2 intersect? If so, find their point of intersection.

Solution

a. By inspection the direction numbers of L_1 are $2, -3$, and 1. Therefore, L_1 has the same direction as the vector $\mathbf{v}_1 = \langle 2, -3, 1 \rangle$. Similarly, we see that L_2 has direction given by the vector $\mathbf{v}_2 = \langle -4, 4, 4 \rangle = -4\langle 1, -1, -1 \rangle$. Since $\mathbf{v}_1$ is not a scalar multiple of $\mathbf{v}_2$, the vectors are not parallel, so L_1 and L_2 are not parallel as well.

b. Suppose that L_1 and L_2 intersect at the point $P_0(x_0, y_0, z_0)$. Then there must exist parameter values t_1 and t_2 such that

$$x_0 = 1 + 2t_1, \qquad y_0 = 2 - 3t_1, \qquad \text{and} \qquad z_0 = 2 + t_1$$

t_1 corresponds to P_0 on L_1.

and

$$x_0 = 3 - 4t_2, \qquad y_0 = 1 + 4t_2, \qquad \text{and} \qquad z_0 = -3 + 4t_2$$

t_2 corresponds to P_0 on L_2.

This leads to the system of three linear equations

$$1 + 2t_1 = 3 - 4t_2$$
$$2 - 3t_1 = 1 + 4t_2$$
$$2 + t_1 = -3 + 4t_2$$

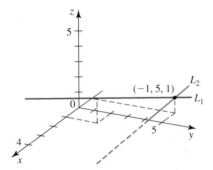

that must be satisfied by t_1 and t_2. Adding the first two equations gives $3 - t_1 = 4$, or $t_1 = -1$. Substituting this value of t_1 into either the first or the second equation then gives $t_2 = 1$. Finally, substituting these values of t_1 and t_2 into the third equation gives $2 - 1 = -3 + 4(1) = 1$, which shows that the third equation is also satisfied by these values. We conclude that L_1 and L_2 do indeed intersect at a point.

To find the point of intersection, substitute $t_1 = -1$ into the parametric equations defining L_1, or, equivalently substitute $t_2 = 1$ into the parametric equations defining L_2. In both cases we find that $x_0 = -1$, $y_0 = 5$, and $z_0 = 1$, so the point of intersection is $(-1, 5, 1)$. (See Figure 6.) ■

FIGURE 6
The lines L_1 and L_2 intersect at the point $(-1, 5, 1)$.

FIGURE 7
The lines L_1 and L_2 are skew lines.

Two lines in space are said to be **skew** if they do not intersect and are not parallel. (See Figure 7.)

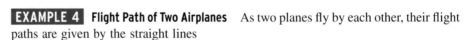

EXAMPLE 4 **Flight Path of Two Airplanes** As two planes fly by each other, their flight paths are given by the straight lines

$$L_1: \quad x = 1 - t \qquad y = -2 - 3t \qquad z = 4 + t$$

and

$$L_2: \quad x = 2 - 2t \qquad y = -4 + 3t \qquad z = 1 + 4t$$

Show that the lines are skew and, therefore, that there is no danger of the planes colliding.

Solution The directions of L_1 and L_2 are given by the directions of the vectors $\mathbf{v}_1 = \langle -1, -3, 1 \rangle$ and $\mathbf{v}_2 = \langle -2, 3, 4 \rangle$, respectively. Since one vector is not a scalar multiple of the other, the lines L_1 and L_2 are not parallel. Next, suppose that the two lines do intersect at some point $P_0(x_0, y_0, z_0)$. Then

$$x_0 = 1 - t_1 \qquad y_0 = -2 - 3t_1 \qquad z_0 = 4 + t_1$$

and

$$x_0 = 2 - 2t_2 \qquad y_0 = -4 + 3t_2 \qquad z_0 = 1 + 4t_2$$

for some t_1 and t_2. Equating the values of x_0, y_0, and z_0 then gives

$$1 - t_1 = 2 - 2t_2$$
$$-2 - 3t_1 = -4 + 3t_2$$
$$4 + t_1 = 1 + 4t_2$$

Solving the first two equations for t_1 and t_2 yields $t_1 = \frac{1}{9}$ and $t_2 = \frac{5}{9}$. Substituting these values of t_1 and t_2 into the third equation gives $\frac{37}{9} = \frac{29}{9}$, a contradiction. This shows that there are no values of t_1 and t_2 that satisfy the three equations simultaneously. Thus, L_1 and L_2 do not intersect. We have shown that L_1 and L_2 are skew lines, so there is no possibility of the planes colliding. ∎

◼ Equations of Planes in Space

A plane in space is uniquely determined by specifying a point $P_0(x_0, y_0, z_0)$ lying in the plane and a vector $\mathbf{n} = \langle a, b, c \rangle$ that is **normal** (perpendicular) to it. (See Figure 8.) To find an equation of the plane, let $P(x, y, z)$ be *any* point in the plane. Then the vector $\overrightarrow{P_0P}$ must be orthogonal to $\mathbf{n}$. But two vectors are orthogonal if and only if their dot product is equal to zero. Therefore, we must have

$$\mathbf{n} \cdot \overrightarrow{P_0P} = 0 \tag{3}$$

Since $\overrightarrow{P_0P} = \langle x - x_0, y - y_0, z - z_0 \rangle$, we can also write Equation (3) as

$$\langle a, b, c \rangle \cdot \langle x - x_0, y - y_0, z - z_0 \rangle = 0$$

or

$$a(x - x_0) + b(y - y_0) + c(z - z_0) = 0$$

FIGURE 8
The vector $\overrightarrow{P_0P}$ lying in the plane must be orthogonal to the normal $\mathbf{n}$ so that $\mathbf{n} \cdot \overrightarrow{P_0P} = 0$.

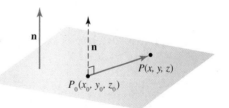

DEFINITION The Standard Form of the Equation of a Plane

The **standard form of the equation of a plane** containing the point $P_0(x_0, y_0, z_0)$ and having the normal vector $\mathbf{n} = \langle a, b, c \rangle$ is

$$a(x - x_0) + b(y - y_0) + c(z - z_0) = 0 \tag{4}$$

EXAMPLE 5 Find an equation of the plane containing the point $P_0(3, -3, 2)$ and having a normal vector $\mathbf{n} = \langle 4, 2, 3 \rangle$. Find the x-, y-, and z-intercepts, and make a sketch of the plane.

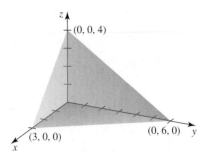

FIGURE 9
The portion of the plane
$4x + 2y + 3z = 12$ in the
first octant

Solution   We use Equation (4) with $a = 4$, $b = 2$, $c = 3$, $x_0 = 3$, $y_0 = -3$, and $z_0 = 2$, obtaining

$$4(x - 3) + 2(y + 3) + 3(z - 2) = 0$$

or

$$4x + 2y + 3z = 12$$

To find the x-intercept, we note that any point on the x-axis must have both its y- and z-coordinates equal to zero. Setting $y = z = 0$ in the equation of the plane, we find that $x = 3$. Therefore, 3 is the x-intercept. Similarly, we find that the y- and z-intercepts are 6 and 4, respectively. By connecting the points $(3, 0, 0)$, $(0, 6, 0)$, and $(0, 0, 4)$ with straight line segments, we obtain a sketch of that portion of the plane lying in the first octant. (See Figure 9.)   ■

EXAMPLE 6   Find an equation of the plane containing the points $P(3, -1, 1)$, $Q(1, 4, 2)$, and $R(0, 1, 4)$.

Solution   To use Equation (4), we need to find a vector normal to the plane in question. Observe that both of the vectors $\overrightarrow{PQ} = \langle -2, 5, 1 \rangle$ and $\overrightarrow{PR} = \langle -3, 2, 3 \rangle$ lie in the plane, so the vector $\overrightarrow{PQ} \times \overrightarrow{PR}$ is normal to the plane. Denoting this vector by $\mathbf{n}$, we have

$$\mathbf{n} = \overrightarrow{PQ} \times \overrightarrow{PR} = \begin{vmatrix} \mathbf{i} & \mathbf{j} & \mathbf{k} \\ -2 & 5 & 1 \\ -3 & 2 & 3 \end{vmatrix} = 13\mathbf{i} + 3\mathbf{j} + 11\mathbf{k}$$

Finally, using the point $P(3, -1, 1)$ in the plane (any of the other two points will also do) and the normal vector $\mathbf{n}$ just found, with $a = 13$, $b = 3$, $c = 11$, $x_0 = 3$, $y_0 = -1$, and $z_0 = 1$, Equation (4) gives

$$13(x - 3) + 3(y + 1) + 11(z - 1) = 0$$

or, upon simplification,

$$13x + 3y + 11z = 47$$

The plane is sketched in Figure 10.

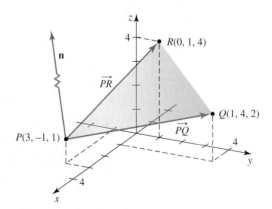

FIGURE 10
The normal to the plane
is $\mathbf{n} = \overrightarrow{PQ} \times \overrightarrow{PR}$.

By expanding Equation (4) and regrouping the terms, as we did in Examples 5 and 6, we obtain the **general form** of the equation of a plane in space,

$$ax + by + cz = d \qquad \qquad \textbf{(5)}$$

where $d = ax_0 + by_0 + cz_0$. Conversely, given $ax + by + cz = d$ with a, b, and c not all equal to zero, we can choose numbers x_0, y_0, and z_0 such that $ax_0 + by_0 + cz_0 = d$. For example, if $c \neq 0$, we can pick x_0 and y_0 arbitrarily and solve the equation $ax_0 + by_0 + cz_0 = d$ for z_0, obtaining $z_0 = (d - ax_0 - by_0)/c$. Therefore, with these choices of x_0, y_0, and z_0, Equation (5) takes the form

$$ax + by + cz = ax_0 + by_0 + cz_0$$

or

$$a(x - x_0) + b(y - y_0) + c(z - z_0) = 0$$

which we recognize to be an equation of the plane containing the point (x_0, y_0, z_0) and having a normal vector $\mathbf{n} = \langle a, b, c \rangle$. (See Equation (4).) An equation of the form $ax + by + cz = d$, with a, b, and c not all zero, is called a **linear equation in the three variables** x, y, and z.

THEOREM 1

Every plane in space can be represented by a linear equation $ax + by + cz = d$, where a, b, and c are not all equal to zero. Conversely, every linear equation $ax + by + cz = d$ represents a plane in space having a normal vector $\langle a, b, c \rangle$.

Note Notice that the coefficients of x, y, and z are precisely the components of the normal vector $\mathbf{n} = \langle a, b, c \rangle$. Thus, we can write a normal vector to a plane by simply inspecting its equation. ∎

■ Parallel and Orthogonal Planes

Two planes with normal vectors $\mathbf{m}$ and $\mathbf{n}$ are **parallel** to each other if $\mathbf{m}$ and $\mathbf{n}$ are parallel; the planes are orthogonal if $\mathbf{m}$ and $\mathbf{n}$ are orthogonal. (See Figure 11.)

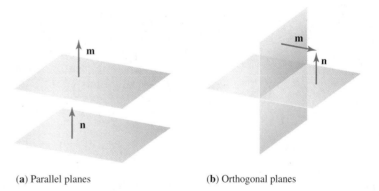

FIGURE 11
Two planes are parallel if **m** and **n** are parallel and orthogonal if **m** and **n** are orthogonal.

(**a**) Parallel planes   (**b**) Orthogonal planes

EXAMPLE 7 Find an equation of the plane containing $P(2, -1, 3)$ and parallel to the plane defined by $2x - 3y + 4z = 6$.

Solution By Theorem 1 the normal vector of the given plane is $\mathbf{n} = \langle 2, -3, 4 \rangle$. Since the required plane is parallel to the given plane, it also has $\mathbf{n}$ as a normal vector. Therefore, using Equation (4), we obtain

$$2(x - 2) - 3(y + 1) + 4(z - 3) = 0$$

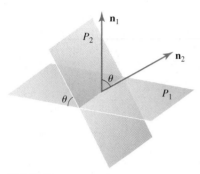

FIGURE 12
The angle between two planes is the angle between their normal vectors.

or

$$2x - 3y + 4z = 19$$

as an equation of the plane.

■ The Angle Between Two Planes

Two distinct planes in space are either parallel to each other or intersect in a straight line. If they do intersect, then the **angle between the two planes** is defined to be the acute angle between their normal vectors (see Figure 12).

EXAMPLE 8   Find the angle between the two planes defined by $3x - y + 2z = 1$ and $2x + 3y - z = 4$.

Solution   The normal vectors of these planes are

$$\mathbf{n}_1 = \langle 3, -1, 2 \rangle \qquad \text{and} \qquad \mathbf{n}_2 = \langle 2, 3, -1 \rangle$$

Therefore, the angle θ between the planes is given by

$$\cos \theta = \frac{\mathbf{n}_1 \cdot \mathbf{n}_2}{|\mathbf{n}_1||\mathbf{n}_2|} \qquad \text{Use Equation (2) of Section 11.3.}$$

$$= \frac{\langle 3, -1, 2 \rangle \cdot \langle 2, 3, -1 \rangle}{\sqrt{9 + 1 + 4} \sqrt{4 + 9 + 1}} = \frac{3(2) + (-1)(3) + 2(-1)}{\sqrt{14} \sqrt{14}} = \frac{1}{14}$$

or

$$\theta = \cos^{-1}\left(\frac{1}{14}\right) \approx 86°$$

EXAMPLE 9   Find parametric equations for the line of intersection of the planes defined by $3x - y + 2z = 1$ and $2x + 3y - z = 4$.

Solution   We need the direction of the line of intersection L as well as a point on L. To find the direction of L, we observe that a vector $\mathbf{v}$ is parallel to L if and only if it is orthogonal to the normal vectors of both planes. (See Figure 13 for the general case.) In other words, $\mathbf{v} = \mathbf{n}_1 \times \mathbf{n}_2$, where $\mathbf{n}_1$ and $\mathbf{n}_2$ are the normal vectors of the two planes. Here, the normal vectors are $\mathbf{n}_1 = \langle 3, -1, 2 \rangle$ and $\mathbf{n}_2 = \langle 2, 3, -1 \rangle$, so the vector $\mathbf{v}$ is given by

$$\mathbf{v} = \mathbf{n}_1 \times \mathbf{n}_2 = \begin{vmatrix} \mathbf{i} & \mathbf{j} & \mathbf{k} \\ 3 & -1 & 2 \\ 2 & 3 & -1 \end{vmatrix} = -5\mathbf{i} + 7\mathbf{j} + 11\mathbf{k}$$

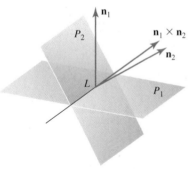

FIGURE 13
The vector $\mathbf{n}_1 \times \mathbf{n}_2$ has the same direction as L, the line of intersection of the two planes.

To find a point on L, let's set $z = 0$ in both of the equations defining the planes. (This will give us the point where L intersects the xy-plane.) We obtain

$$3x - y = 1 \qquad \text{and} \qquad 2x + 3y = 4$$

Solving these equations simultaneously gives $x = \frac{7}{11}$ and $y = \frac{10}{11}$. Finally, by using Equation (1), the required parametric equations are

$$x = \frac{7}{11} - 5t, \qquad y = \frac{10}{11} + 7t, \qquad \text{and} \qquad z = 11t$$

■ The Distance Between a Point and a Plane

To find a formula for the distance between a point and a plane, suppose that $P_1(x_1, y_1, z_1)$ is a point *not* lying in the plane $ax + by + cz = d$. Let $P_0(x_0, y_0, z_0)$ be any point lying in the plane. Then, as you can see in Figure 14, the distance D between P_1 and the plane is given by the length of the vector projection of $\overrightarrow{P_0P_1}$ onto the normal vector $\mathbf{n} = \langle a, b, c \rangle$ of the plane. Equivalently, D is the absolute value of the scalar component of $\overrightarrow{P_0P_1}$ along $\mathbf{n}$. Using Equation (6) of Section 11.3 (and taking the absolute value), we obtain

$$D = \frac{|\overrightarrow{P_0P_1} \cdot \mathbf{n}|}{|\mathbf{n}|}$$

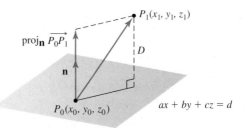

FIGURE 14
The distance from P_1 to the plane is the length of $\text{proj}_\mathbf{n} \overrightarrow{P_0P_1}$.

But $\overrightarrow{P_0P_1} = \langle x_1 - x_0, y_1 - y_0, z_1 - z_0 \rangle$, so we have

$$D = \frac{|\langle x_1 - x_0, y_1 - y_0, z_1 - z_0 \rangle \cdot \langle a, b, c \rangle|}{\sqrt{a^2 + b^2 + c^2}}$$

$$= \frac{|a(x_1 - x_0) + b(y_1 - y_0) + c(z_1 - z_0)|}{\sqrt{a^2 + b^2 + c^2}}$$

$$= \frac{|ax_1 + by_1 + cz_1 - (ax_0 + by_0 + cz_0)|}{\sqrt{a^2 + b^2 + c^2}}$$

Since $P_0(x_0, y_0, z_0)$ lies in the plane, its coordinates must satisfy the equation of the plane, that is, $ax_0 + by_0 + cz_0 = d$; so we can write D in the following form:

$$D = \frac{|ax_1 + by_1 + cz_1 - d|}{\sqrt{a^2 + b^2 + c^2}} \tag{6}$$

EXAMPLE 10 Find the distance between the point $(-2, 1, 3)$ and the plane $2x - 3y + z = 1$.

Solution Using Equation (6) with $x_1 = -2$, $y_1 = 1$, $z_1 = 3$, $a = 2$, $b = -3$, $c = 1$ and $d = 1$, we obtain

$$D = \frac{|2(-2) - 3(1) + 1(3) - 1|}{\sqrt{2^2 + (-3)^2 + 1^2}} = \frac{5}{\sqrt{14}} = \frac{5\sqrt{14}}{14} \qquad ■$$

11.5 CONCEPT QUESTIONS

1. **a.** Write the parametric equations of the line passing through the point $P_0(x_0, y_0, z_0)$ and having the same direction as the vector $\mathbf{v} = \langle a, b, c \rangle$.
   b. What are the symmetric equations of the line of part (a)?
   c. Write the parametric and symmetric equations of the line that passes through the point $P(x_0, y_0, z_0)$ and has the same direction as the vector $\mathbf{v} = \langle a, b, 0 \rangle$.

2. If you are given two lines L_1 and L_2 in space, how do you determine whether they are (a) parallel to each other, (b) perpendicular to each other, or (c) skew?

3. **a.** Write the standard form of an equation of the plane containing the point $P_0(x_0, y_0, z_0)$ and having the normal vector $\mathbf{n} = \langle a, b, c \rangle$.
   b. What is the general form of the equation of a plane in space?

4. **a.** What is the angle between two planes in space? How do you find it?
   b. Write the formula giving the distance between a point and a plane in space.

11.5 EXERCISES

In Exercises 1 and 2 describe in your own words the strategy you might adopt to solve the problem. For example, to find the parametric equations of the line passing through two distinct points, you might use this strategy: Let P and Q denote the two points, and write the vector $\overrightarrow{PQ}$ that gives the direction numbers of the line. Then using this information and either P or Q, write the desired equations using Equation (1).

1. Find parametric equations of a line, given that the line
   a. Passes through a given point and is parallel to a given line.
   b. Passes through a given point and is perpendicular to two distinct lines passing through that point.
   c. Passes through a given point lying in a given plane and is perpendicular to the plane.
   d. Is the intersection of two given nonparallel planes.

2. Find an equation of a plane, given that the plane
   a. Contains three distinct points.
   b. Contains a given line and a point not lying on the line.
   c. Contains a given point and is parallel to a given plane.
   d. Contains two intersecting nonparallel lines.
   e. Contains two parallel and distinct lines.

In Exercises 3–6, find parametric and symmetric equations for the line passing through the point P that is parallel to the vector $\mathbf{v}$.

3. $P(1, 3, 2);\quad \mathbf{v} = \langle 2, 4, 5 \rangle$
4. $P(1, -4, 2);\quad \mathbf{v} = 2\mathbf{i} - 3\mathbf{j} + \mathbf{k}$
5. $P(3, 0, -2);\quad \mathbf{v} = 2\mathbf{i} - \mathbf{j} + 3\mathbf{k}$
6. $P(0, 1, 3);\quad \mathbf{v} = \langle 2, -3, 4 \rangle$

In Exercises 7–10, find parametric and symmetric equations for the line passing through the given points.

7. $(2, 1, 4)$ and $(1, 3, 7)$
8. $(3, -2, 1)$ and $(3, 4, 4)$
9. $\left(-1, -2, -\frac{1}{2}\right)$ and $\left(1, \frac{3}{2}, -3\right)$
10. $\left(\frac{1}{2}, -\frac{1}{3}, \frac{1}{4}\right)$ and $\left(\frac{1}{2}, -\frac{1}{3}, \frac{3}{4}\right)$

11. Find parametric and symmetric equations of the line passing through the point $(1, 2, -1)$ and parallel to the line with parametric equations $x = -1 + t$, $y = 2 + 2t$, and $z = -2 - 3t$. At what points does the line intersect the coordinate planes?

12. Find parametric equations of the line passing through the point $(-1, 3, -2)$ and parallel to the line with symmetric equation

$$\frac{x - 2}{3} = \frac{y + 1}{-3} = z + 2$$

   At what point does the line intersect the yz-plane?

13. Determine whether the point $(-3, 6, 1)$ lies on the line L passing through the point $(-1, 4, 3)$ and parallel to the vector $\mathbf{v} = -\mathbf{i} + \mathbf{j} - \mathbf{k}$.

14. Find parametric equations of the line that is parallel to the line with equation

$$\frac{x - 1}{4} = \frac{y + 4}{5} = \frac{z + 1}{2}$$

   and contains the point of intersection of the lines

$$L_1:\quad x = 4 + t \qquad y = 5 + t \qquad z = -1 + 2t$$
$$L_2:\quad x = 6 + 2t \qquad y = 11 + 4t \qquad z = -3 + t$$

In Exercises 15–18, determine whether the lines L_1 and L_2 are parallel, are skew, or intersect each other. If they intersect, find the point of intersection.

15. $L_1: x = -1 + 3t, \ y = -2 + 3t, \ z = 3 + t$
    $L_2: x = 1 + 4t, \ y = -2 + 6t, \ z = 4 + t$

16. $L_1: x = 1 - 2t, \ y = -1 - 3t, \ z = -2 + t$
    $L_2: x = -3 + t, \ y = -2 + 2t, \ z = 3 - t$

17. L_1: $\dfrac{x-2}{4} = \dfrac{z-1}{-1}$, $y = 3$

$\quad$ L_2: $\dfrac{x-2}{2} = \dfrac{y-3}{2} = z - 1$

18. L_1: $\dfrac{x-4}{-1} = \dfrac{y+1}{6} = z - 4$

$\quad$ L_2: $\dfrac{x-1}{2} = \dfrac{y-1}{-4} = \dfrac{z-5}{-1}$

In Exercises 19–22, determine whether the lines L_1 and L_2 intersect. If they do intersect, find the angle between them.

19. L_1: $x = 1 - t$, $y = 3 - 2t$, $z = t$
$\quad$ L_2: $x = 2 + 3t$, $y = 3 + 2t$, $z = 1 + t$

20. L_1: $x = 2 + 3t$, $y = -2 - 2t$, $z = 3 + t$
$\quad$ L_2: $x = -3 - t$, $y = 1 + t$, $z = -4 + 5t$

21. L_1: $\dfrac{x-1}{-3} = \dfrac{y+2}{2} = \dfrac{z+1}{4}$

$\quad$ L_2: $\dfrac{x+2}{2} = \dfrac{y-4}{4} = z - 3$

22. L_1: $\dfrac{x+4}{3} = \dfrac{y+1}{-2} = \dfrac{z-3}{3}$

$\quad$ L_2: $\dfrac{x+32}{6} = \dfrac{y-8}{-2}$, $z = 4$

In Exercises 23–26, find an equation of the plane that has the normal vector $\mathbf{n}$ and passes through the given point.

23. $(2, 1, 5)$; $\mathbf{n} = \langle 1, 2, 4 \rangle$

24. $(-1, 3, -2)$; $\mathbf{n} = \mathbf{i} - 2\mathbf{j} + \mathbf{k}$

25. $(1, 3, 0)$; $\mathbf{n} = 2\mathbf{i} - 4\mathbf{k}$

26. $(3, 0, 3)$; $\mathbf{n} = \langle 0, 0, 1 \rangle$

In Exercises 27–30, find an equation of the plane that passes through the given point and is parallel to the given plane.

27. $(3, 6, -2)$; $2x + 3y - z = 4$

28. $(2, -1, 0)$; $x - 2y - 3z = 1$

29. $(-1, -2, -3)$; $x - 3z = 1$

30. $(0, 2, -1)$; $\dfrac{1}{2}x - \dfrac{1}{3}y + \dfrac{1}{4}z = 2$

In Exercises 31 and 32, find an equation of the plane that passes through the three given points.

31. $(1, 0, -2)$, $(1, 3, 2)$, $(2, 3, 0)$

32. $(2, 3, -1)$, $(1, -2, 3)$, $(-1, 2, 4)$

In Exercises 33–36, find an equation of the plane that passes through the given point and contains the given line.

33. $(1, 3, 2)$; $x = 1 + t$, $y = -1 - 2t$, $z = 3 + 2t$

34. $(-1, 2, 3)$; $x = -1 + 2t$, $y = -2 + 3t$, $z = 3 - t$

35. $(3, -4, 5)$; $\dfrac{x-2}{2} = \dfrac{y+1}{-3} = \dfrac{z+3}{5}$

36. $(1, 3, 0)$; $\dfrac{x+4}{-3} = \dfrac{y-3}{5}$, $z = 2$

In Exercises 37 and 38, find an equation of the plane passing through the given points and perpendicular to the given plane.

37. $(2, 1, 1)$ and $(-1, 3, 2)$; $2x + 3y - 4z = 3$

38. $(-1, 3, 0)$ and $(2, -1, 4)$; $3x - 4y + 5z = 1$

In Exercises 39–42, determine whether the planes are parallel, orthogonal, or neither. If they are neither parallel nor orthogonal, find the angle between them.

39. $x + 2y + z = 1$, $2x - 3y + 4z = 3$

40. $2x - y + 4z = 7$, $6x - 3y + 12z = -1$

41. $3x - y + 2z = 2$, $2x + 3y + z = 4$

42. $4x - 4y + 2z = 7$, $3x + 2y - 2z = 5$

In Exercises 43 and 44, find the angle between the plane and the line.

43. $x + y + 2z = 6$; $x = 1 + t$, $y = 2 + t$, $z = -1 + t$

44. $2x - 3y + 4z = 12$; $\dfrac{x-1}{2} = \dfrac{y+1}{3} = \dfrac{z}{2}$

In Exercises 45 and 46, find parametric equations for the line of intersection of the planes.

45. $2x - 3y + 4z = 3$, $x + 4y - 2z = 7$

46. $3x + y - 2z = 4$, $2x - y - 3z = 6$

47. Find parametric equations of the line that passes through the point $(2, 3, -1)$ and is perpendicular to the plane $2x + 4y - 3z = 4$.

48. Find an equation of the plane that passes through the point $(3, -2, 4)$ and is perpendicular to the line

$$\dfrac{x+1}{-2} = \dfrac{y-2}{3} = \dfrac{z+4}{4}$$

49. Find an equation of the plane that contains the lines given by

$$x = -1 + 2t \qquad y = 2 - 3t \qquad z = 1 + t$$

$$x = 2 - t \qquad y = 1 - 2t \qquad z = 5 - 3t$$

50. Find an equation of the plane that passes through the point $(3, 4, 1)$ and contains the line of intersection of the planes $x - y + 2z = 1$ and $2x + 3y - z = 2$.

51. Find an equation of the plane that is orthogonal to the plane $3x + 2y - 4z = 7$ and contains the line of intersection of the planes $2x - 3y + z = 3$ and $x + 2y - 3z = 5$.

52. Find an equation of the plane that is parallel to the line of intersection of the planes $x - y + 2z = 3$ and $2x + 3y - z = 4$ and contains the points $(2, 3, 5)$ and $(3, 4, 1)$.

In Exercises 53 and 54, find the point of intersection, if any, of the plane and the line.

53. $2x + 3y - z = 9$; $\quad x = 2 + 3t$, $\quad y = -1 + t$, $z = 3 - 2t$

54. $x - y + 2z = 13$; $\quad \dfrac{x + 1}{3} = \dfrac{y - 2}{4} = z + 1$

In Exercises 55 and 56, find the distance between the point and the plane.

55. $(3, 1, 2)$; $\quad 2x - 3y + 4z = 7$

56. $(-1, 3, -2)$; $\quad 3x + y + z = 2$

In Exercises 57 and 58, show that the two planes are parallel and find the distance between them.

57. $x + 2y - 4z = 1$, $\quad x + 2y - 4z = 7$

58. $2x - 3y + z = 2$, $\quad 4x - 6y + 2z = 8$

59. Let P be a point that is not on the line L. Show that the distance D between the point P and the line L is

$$D = \frac{|\overrightarrow{QP} \times \mathbf{u}|}{|\mathbf{u}|}$$

where $\mathbf{u}$ is a vector having the same direction as L, and Q is any point on L.

In Exercises 60 and 61, use the result of Exercise 59 to find the distance between the point and the line.

60. $(3, 4, 6)$; $\quad x = 2 + t, y = 1 - 2t, z = 3 - t$

61. $(1, -2, 3)$; $\quad \dfrac{x + 2}{3} = \dfrac{y - 1}{1} = \dfrac{z + 3}{2}$

62. Find the distance between the point $(1, 4, 2)$ and the line passing through the points $(-1, 3, -1)$ and $(1, 2, 3)$.

63. Show that the distance D between the parallel planes $ax + by + cz = d_1$ and $ax + by + cz = d_2$ is

$$D = \frac{|d_1 - d_2|}{\sqrt{a^2 + b^2 + c^2}}$$

In Exercises 64 and 65, find the distance between the skew lines. (Hint: Use the result of Exercise 63.)

64. $x = 1 + 5t, y = -1 + 2t, z = 2 - 3t$ $\quad$ and $x = 1 + t, y = -1 - 2t, z = 1 + t$

65. $\dfrac{x - 1}{-2} = \dfrac{y - 4}{-6} = \dfrac{z - 3}{-2}$ $\quad$ and $\quad x - 2 = \dfrac{y + 2}{-5} = \dfrac{z - 1}{-3}$

66. Find the distance between the line given by

$$x = 1 + 3t, \qquad y = 2 - 6t, \qquad \text{and} \qquad z = -1 + 2t$$

and the plane that passes through the point $(2, 3, 1)$ and is perpendicular to the line containing the points $(2, 1, 4)$ and $(4, 4, 10)$.

In Exercises 67–72, determine whether the statement is true or false. If it is true, explain why. If it is false, explain why or give an example that shows it is false.

67. If the lines L_1 and L_2 are both perpendicular to the line L_3, then L_1 must be parallel to L_2.

68. If the lines L_1 and L_2 do not intersect, then they must be parallel to each other.

69. If the planes P_1 and P_2 are both perpendicular to the plane P_3, then P_1 must be perpendicular to P_2.

70. If the planes P_1 and P_2 are both parallel to a line L, then P_1 and P_2 must be parallel to each other.

71. There always exists a unique plane passing through a given point and a given line.

72. Given any two lines that are not coincident, there is a plane containing the two lines.

11.6 Surfaces in Space

In Section 11.5 we saw that the graph of a linear equation in three variables is a plane in space. In general, the graph of an equation in three variables, $F(x, y, z) = 0$, is a surface in 3-space. In this section we will study surfaces called *cylinders* and *quadric surfaces*.

The paraboloidal surface shown in Figure 1a is an example of a quadric surface. A uniformly rotating liquid acquires this shape as a result of the interaction between the force of gravity and centrifugal force. As was explained in Section 10.1, this surface is ideal for radio and optical telescope mirrors. (See Figure 1b.) Mathematically, a paraboloid is obtained by revolving a parabola about its axis of symmetry. (See Figure 1c.)

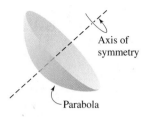

Axis of symmetry

Parabola

(a) Surface of rotating liquid **(b)** Surface of a radio telescope **(c)** Surface obtained by revolving a parabola about its axis

FIGURE 1

▣ Traces

Just as we can use the x- and y-intercepts of a plane curve to help us sketch the graph of a plane curve, so can we use the traces of a surface in the coordinate planes to help us sketch the surface itself. The **trace** of a surface S in a plane is the intersection of the surface and the plane. In particular, the traces of S in the xy-plane, the yz-plane, and the xz-plane are called the **xy-trace,** the **yz-trace** and the **xz-trace,** respectively.

To find the xy-traces, we set $z = 0$ and sketch the graph of the resulting equation in the xy-plane. The other traces are obtained in a similar manner. Of course, if the surface does not intersect the plane, there is no trace in that plane.

EXAMPLE 1 Consider the plane with equation $4x + 2y + 3z = 12$. (See Example 5 in Section 11.5.) Find the traces of the plane in the coordinate planes, and sketch the plane.

Solution To find the xy-trace, we first set $z = 0$ in the given equation to obtain the equation $4x + 2y = 12$. Then we sketch the graph of this equation in the xy-plane. (See Figure 2a.) To find the yz-trace, we set $x = 0$ to obtain the equation $2y + 3z = 12$, whose graph in the yz-plane gives the required trace. (See Figure 2b.) The xz-trace is obtained in a similar manner. (See Figure 2c.) The graph of the plane in the first octant is sketched in Figure 2d.

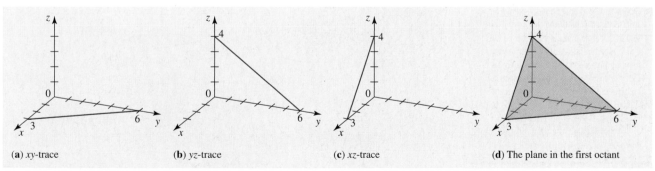

(a) xy-trace **(b)** yz-trace **(c)** xz-trace **(d)** The plane in the first octant

Ⓥ⵼ FIGURE 2
The traces of the plane $4x + 2y + 3z = 12$ in the coordinate planes are shown in parts (a)–(c). ▬

Sometimes it is useful to obtain the traces of a surface in planes that are parallel to the coordinate planes, as illustrated in the next example.

EXAMPLE 2 Let S be the surface defined by $z = x^2 + y^2$.

a. Find the traces of S in the coordinate planes.
b. Find the traces of S in the plane $z = k$, where k is a constant.
c. Sketch the surface S.

Solution

a. Setting $z = 0$ gives $x^2 + y^2 = 0$, from which we see that the xy-trace is the origin $(0, 0)$. (See Figure 3a.) Next, setting $x = 0$ gives $z = y^2$, from which we see that the yz-trace is a parabola. (See Figure 3b.) Finally, setting $y = 0$ gives $z = x^2$, so the xz-trace is also a parabola. (See Figure 3c.)

b. Setting $z = k$, we obtain $x^2 + y^2 = k$, from which we see that the trace of S in the plane $z = k$ is a circle of radius $\sqrt{k}$ centered at the point of intersection of the plane and the z-axis, provided that $k > 0$. (See Figure 3d.) Observe that if $k = 0$, the trace is the point $(0, 0)$ (degenerate circle) obtained in part (a).

c. The graph of $z = x^2 + y^2$ sketched in Figure 3e is called a *circular paraboloid* because its traces in planes parallel to the coordinate planes are either circles or parabolas.

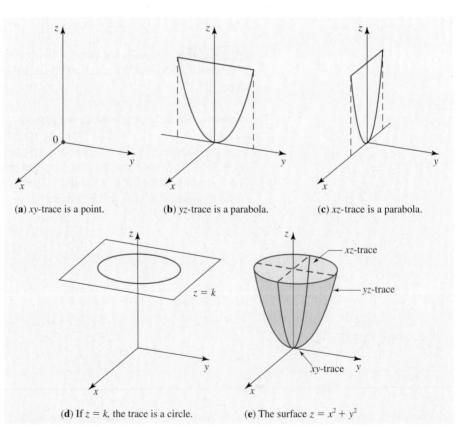

(**a**) xy-trace is a point.   (**b**) yz-trace is a parabola.   (**c**) xz-trace is a parabola.

(**d**) If $z = k$, the trace is a circle.   (**e**) The surface $z = x^2 + y^2$

FIGURE 3
The traces of the surface S

Cylinders

We now turn our attention to a class of surfaces called *cylinders*.

> **DEFINITION Cylinder**
>
> Let C be a curve in a plane, and let l be a line that is not parallel to that plane. Then the set of all points generated by letting a line traverse C while parallel to l at all times is called a cylinder. The curve C is called the **directrix** of the cylinder, and each line through C parallel to l is called a **ruling** of the cylinder. (See Figure 4.)

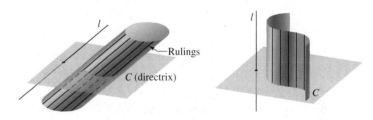

FIGURE 4
Two cylinders. The curve C is the directrix. The rulings are parallel to l.

Cylinders in which the directrix lies in a coordinate plane and the rulings are perpendicular to that plane have relatively simple algebraic representations. Consider, for example, the surface S with equation $f(x, y) = 0$. The xy-trace of S is the graph C of the equation $f(x, y) = 0$ in the xy-plane. (See Figure 5a.) Next, observe that if $(x, y, 0)$ is any point on C, then the point (x, y, z) must satisfy the equation $f(x, y) = 0$ for *any* value of z (since z is not present in the equation). But all such points lie on the line perpendicular to the xy-plane and pass through the point $(x, y, 0)$. This shows that the surface S is a cylinder with directrix $f(x, y) = 0$ and rulings that are parallel to the direction of the axis of the missing variable z. (See Figure 5b.)

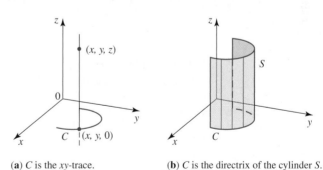

FIGURE 5
The surface $f(x, y) = 0$ is a cylinder with directrix C defined by $f(x, y) = 0$ in the xy-plane and with rulings parallel to the direction of the axis of the missing variable z.

(a) C is the xy-trace.

(b) C is the directrix of the cylinder S. The rulings are parallel to the z-axis.

EXAMPLE 3 Sketch the graph of $y = x^2 - 4$.

Solution The given equation has the form $f(x, y) = 0$, where $f(x, y) = x^2 - y - 4$. Therefore, its graph is a cylinder with directrix given by the graph of $y = x^2 - 4$ in the xy-plane and rulings parallel to the z-axis (corresponding to the variable missing in the equation). The graph of $y = x^2 - 4$ in the xy-plane is the parabola shown in Figure 6a. The required cylinder is shown in Figure 6b. It is called a **parabolic cylinder.**

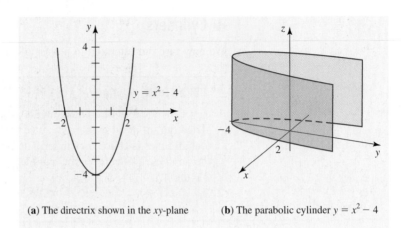

(a) The directrix shown in the xy-plane   (b) The parabolic cylinder $y = x^2 - 4$

FIGURE 6
The graph of $y = x^2 - 4$ is sketched in two steps.

EXAMPLE 4 Sketch the graph of $\dfrac{y^2}{4} + \dfrac{z^2}{9} = 1$.

Solution The given equation has the form $f(y, z) = 0$, where

$$f(y, z) = \frac{y^2}{4} + \frac{z^2}{9} - 1$$

Its graph is a cylinder with directrix given by

$$\frac{y^2}{4} + \frac{z^2}{9} = 1$$

and rulings parallel to the x-axis. The graph of

$$\frac{y^2}{4} + \frac{z^2}{9} = 1$$

in the yz-plane is the ellipse shown in Figure 7a. The required cylinder is shown in Figure 7b. It is called an **elliptic cylinder.**

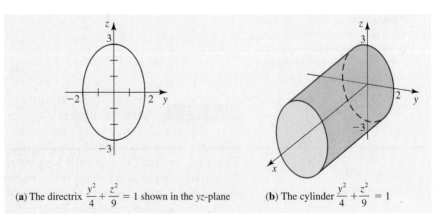

(a) The directrix $\dfrac{y^2}{4} + \dfrac{z^2}{9} = 1$ shown in the yz-plane   (b) The cylinder $\dfrac{y^2}{4} + \dfrac{z^2}{9} = 1$

FIGURE 7

EXAMPLE 5 Sketch the graph of $z = \cos x$.

Solution   The given equation has the form $f(x, z) = 0$, where $f(x, z) = z - \cos x$. Therefore, its graph is a cylinder with directrix given by the graph of $z = \cos x$ in the xz-plane and rulings parallel to the y-axis. The graph of the directrix in the xz-plane is shown in Figure 8a, and the graph of the cylinder is sketched in Figure 8b.

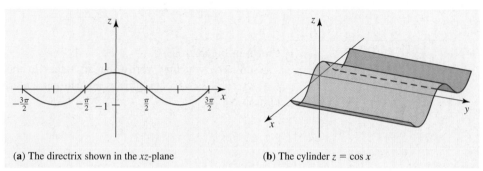

(a) The directrix shown in the xz-plane   **(b)** The cylinder $z = \cos x$

FIGURE 8

> Note that, while an equation in two variables represents a curve in 2-space, the same equation represents a cylinder when we are working in 3-space. For example, the equation $x^2 + y^2 = 1$ represents a circle in the plane, but the same equation represents a right circular cylinder in 3-space.

■ Quadric Surfaces

The equation of a sphere given in Section 11.2 and the equations in Examples 1, 2, and 3 in this section are special cases of the second-degree equation in x, y, and z

$$Ax^2 + By^2 + Cz^2 + Dxy + Exz + Fyz + Gx + Hy + Iz + J = 0$$

where $A, B, C, \ldots, J$ are constants. The graph of this equation is a **quadric surface.** By making a suitable translation and/or rotation of the coordinate system, a quadric surface can always be put in standard position with respect to a new coordinate system. (See Figure 9.) With respect to the new system the equation will assume one of the two standard forms

$$\overline{A}X^2 + \overline{B}Y^2 + \overline{C}Z^2 + \overline{J} = 0 \qquad \text{or} \qquad \overline{A}X^2 + \overline{B}Y^2 + \overline{I}Z = 0$$

For this reason we will restrict our study of quadric surfaces to those represented by the equations

$$AX^2 + BY^2 + CZ^2 + J = 0 \qquad \text{or} \qquad AX^2 + BY^2 + IZ = 0$$

When we sketch the following quadric surfaces, we will find it useful to look at their traces in the coordinate planes as well as planes that are parallel to the coordinate planes.

In the remainder of this section, unless otherwise noted, a, b, and c denote positive real numbers.

Ellipsoids   The graph of the equation

$$\frac{x^2}{a^2} + \frac{y^2}{b^2} + \frac{z^2}{c^2} = 1$$

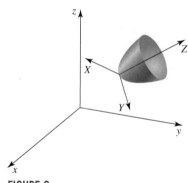

FIGURE 9
By translating and rotating the xyz-system, we have the XYZ-system in which the paraboloid is in standard position with respect to the latter.

is an ellipsoid because its traces in the planes parallel to the coordinate planes are ellipses. In fact, its trace in the plane $z = k$, where $-c < k < c$, is the ellipse

$$\frac{x^2}{a^2} + \frac{y^2}{b^2} = 1 - \frac{k^2}{c^2}$$

and, in particular, its trace in the xy-plane is the ellipse

$$\frac{x^2}{a^2} + \frac{y^2}{b^2} = 1$$

shown in Figure 10a.

Similarly, you may verify that its traces in the planes $x = k\,(-a < k < a)$ and $y = k\,(-b < k < b)$ are ellipses and, in particular, that its yz- and xz-traces are the ellipses

$$\frac{y^2}{b^2} + \frac{z^2}{c^2} = 1 \qquad \text{and} \qquad \frac{x^2}{a^2} + \frac{z^2}{c^2} = 1$$

respectively. (See Figures 10b–c.) The ellipsoid is sketched in Figure 10d.

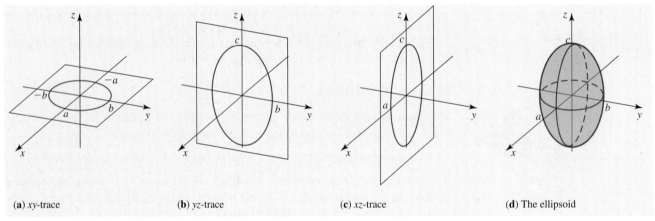

(a) xy-trace   (b) yz-trace   (c) xz-trace   (d) The ellipsoid

FIGURE 10
The traces in the coordinate planes and the ellipsoid $\dfrac{x^2}{a^2} + \dfrac{y^2}{b^2} + \dfrac{z^2}{c^2} = 1$

Note that if $a = b = c$, then the ellipsoid is in fact a sphere of radius a with center at the origin.

Hyperboloids of One Sheet   The graph of the equation

$$\frac{x^2}{a^2} + \frac{y^2}{b^2} - \frac{z^2}{c^2} = 1$$

is a **hyperboloid of one sheet.** The xy-trace of this surface is the ellipse

$$\frac{x^2}{a^2} + \frac{y^2}{b^2} = 1$$

(Figure 11a) whereas both the yz- and xz-traces are hyperbolas (Figures 11b–c), as you may verify. The trace of the surface in the plane $z = k$ is an ellipse

$$\frac{x^2}{a^2} + \frac{y^2}{b^2} = 1 + \frac{k^2}{c^2}$$

As $|k|$ increases, the ellipses grow larger and larger. The hyperboloid is sketched in Figure 11d.

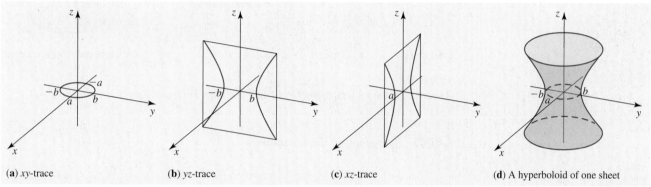

(a) xy-trace   (b) yz-trace   (c) xz-trace   (d) A hyperboloid of one sheet

FIGURE 11
The traces in the coordinate plane and the hyperboloid of one sheet $\dfrac{x^2}{a^2} + \dfrac{y^2}{b^2} - \dfrac{z^2}{c^2} = 1$

The z-axis is called the **axis of the hyperboloid.** Note that the orientation of the axis of the hyperboloid is associated with the term that has a minus sign in front of it. Thus, if the minus sign had been in front of the term involving x, then the surface would have been a hyperboloid of one sheet with the x-axis as its axis.

Hyperboloids of Two Sheets

The graph of the equation

$$-\frac{x^2}{a^2} - \frac{y^2}{b^2} + \frac{z^2}{c^2} = 1$$

is a **hyperboloid of two sheets.** The xz- and yz-traces are the hyperbolas

$$-\frac{x^2}{a^2} + \frac{z^2}{c^2} = 1 \qquad \text{and} \qquad -\frac{y^2}{b^2} + \frac{z^2}{c^2} = 1$$

sketched in Figures 12a–b. The trace of the surface in the plane $z = k$ is an ellipse

$$\frac{x^2}{a^2} + \frac{y^2}{b^2} = \frac{k^2}{c^2} - 1$$

provided that $|k| > c$. There are no values of x and y that satisfy the equation if $|k| < c$, so the surface is made up of two parts, as shown in Figure 12c: one part lying on or above the plane $z = c$ and the other part lying on or below the plane $z = -c$.

The axis of the hyperboloid is the z-axis. Observe that the sign associated with the variable z is positive. Had the positive sign been in front of one of the other variables, then the surface would have been a hyperboloid of two sheets with its axis along the axis associated with that variable.

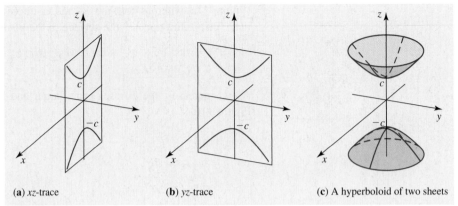

(a) xz-trace   **(b)** yz-trace   **(c)** A hyperboloid of two sheets

FIGURE 12
The traces in the xz- and yz-planes and the hyperboloid of two sheets $-\dfrac{x^2}{a^2} - \dfrac{y^2}{b^2} + \dfrac{z^2}{c^2} = 1$

Cones   The graph of the equation

$$\frac{x^2}{a^2} + \frac{y^2}{b^2} - \frac{z^2}{c^2} = 0$$

is a double-napped **cone.** The xz- and yz-traces are the lines $z = \pm(c/a)x$ and $z = \pm(c/b)y$, respectively. (See Figures 13a–b.) The trace in the plane $z = k$ is an ellipse,

$$\frac{x^2}{a^2} + \frac{y^2}{b^2} = \frac{k^2}{c^2}$$

As $|k|$ increases, so do the lengths of the axes of the resulting ellipses. The traces in planes parallel to the other two coordinate planes are hyperbolas. The cone is sketched in Figure 13c. The **axis of the cone** is the z-axis.

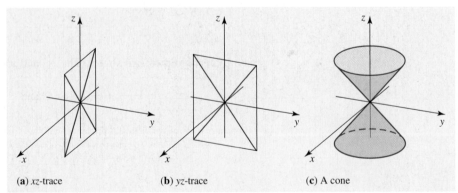

(a) xz-trace   **(b)** yz-trace   **(c)** A cone

FIGURE 13
The traces in the xz- and yz-planes and the cone $\dfrac{x^2}{a^2} + \dfrac{y^2}{b^2} - \dfrac{z^2}{c^2} = 0$

Paraboloids   The graph of the equation

$$\frac{x^2}{a^2} + \frac{y^2}{b^2} = cz$$

where c is a real number, is called an **elliptic paraboloid** because its traces in planes parallel to the xy-coordinate plane are ellipses and its traces in planes parallel to the other two coordinate planes are parabolas. If $a = b$, the surface is called a **circular paraboloid.** We will let you verify these statements. The graph of an elliptic paraboloid with $c > 0$ is sketched in Figure 14a. The **axis** of the paraboloid is the z-axis, and its **vertex** is the origin.

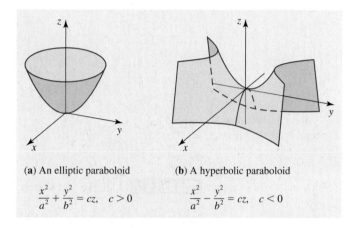

(a) An elliptic paraboloid

$$\frac{x^2}{a^2} + \frac{y^2}{b^2} = cz, \quad c > 0$$

(b) A hyperbolic paraboloid

$$\frac{x^2}{a^2} - \frac{y^2}{b^2} = cz, \quad c < 0$$

FIGURE 14

Hyperbolic Paraboloids   The graph of the equation

$$\frac{x^2}{a^2} - \frac{y^2}{b^2} = cz$$

where c is a real number, is called a **hyperbolic paraboloid** because the xz- and yz-traces are parabolas and the traces in planes parallel to the xy-plane are hyperbolas. The graph of a hyperbolic paraboloid with $c < 0$ is shown in Figure 14b.

EXAMPLE 6 Identify and sketch the surface $12x^2 - 3y^2 + 4z^2 + 12 = 0$.

Solution Rewriting the equation in the standard form

$$-\frac{x^2}{1} + \frac{y^2}{4} - \frac{z^2}{3} = 1$$

we see that it represents a hyperboloid of two sheets with the y-axis as its axis.

To sketch the surface, observe that the surface intersects the y-axis at the points $(0, -2, 0)$ and $(0, 2, 0)$, as you can verify by setting $x = z = 0$ in the given equation. Next, let's find the trace in the plane $y = k$. We obtain

$$\frac{x^2}{1} + \frac{z^2}{3} = \frac{k^2}{4} - 1$$

In particular, the trace in the plane $y = 6$ is the ellipse

$$\frac{x^2}{1} + \frac{z^2}{3} = 8 \qquad \text{or} \qquad \frac{x^2}{8} + \frac{z^2}{24} = 1$$

A sketch of this trace is shown in Figure 15a. The completed sketch of the hyperboloid of two sheets is shown in Figure 15b.

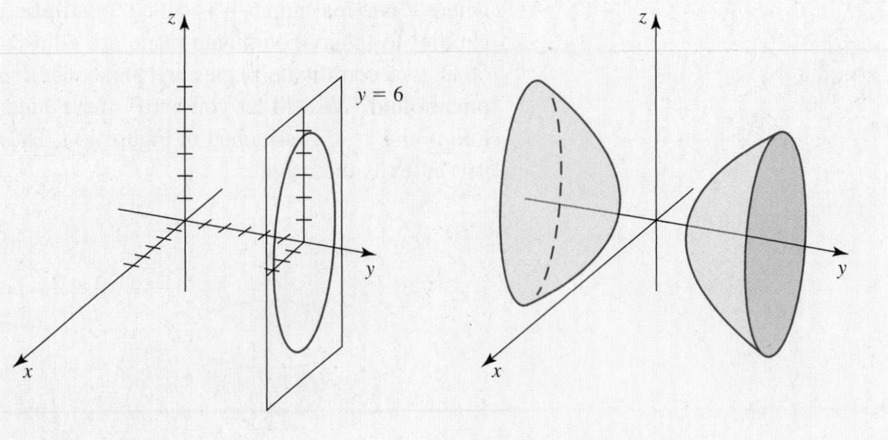

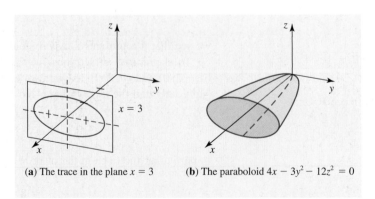

FIGURE 15

Steps in sketching a
hyperboloid of two sheets

(a) The trace in the plane $y = 6$

(b) The hyperboloid $12x^2 - 3y^2 + 4z^2 + 12 = 0$

EXAMPLE 7 Identify and sketch the surface $4x - 3y^2 - 12z^2 = 0$.

Solution Rewriting the equation in the standard form,

$$4x = 3y^2 + 12z^2$$

we see that it represents a paraboloid with the x-axis as its axis. To sketch the surface, let's find the trace in the plane $x = k$. We obtain

$$4k = 3y^2 + 12z^2$$

Letting $k = 3$, we see that the trace in the plane $x = 3$ is the ellipse with equation $12 = 3y^2 + 12z^2$ or, in standard form,

$$\frac{y^2}{4} + \frac{z^2}{1} = 1$$

A sketch of this trace is shown in Figure 16a. The completed sketch of the paraboloid is shown in Figure 16b.

(a) The trace in the plane $x = 3$

(b) The paraboloid $4x - 3y^2 - 12z^2 = 0$

FIGURE 16

We now give a summary of the quadric surfaces and their general shapes, and we also suggest an aid for sketching these surfaces. Note that in many instances, finding the intercepts and using a judiciously chosen trace will be sufficient to help you obtain a good sketch of the surface.

Equation	Surface (computer generated)	Aid for Sketching the Figure
Ellipsoid $$\frac{x^2}{a^2} + \frac{y^2}{b^2} + \frac{z^2}{c^2} = 1$$ **Note:** All signs are positive.		Find x-, y-, and z-intercepts, and then sketch.
Hyperboloid of One Sheet $$\frac{x^2}{a^2} + \frac{y^2}{b^2} - \frac{z^2}{c^2} = 1$$ **Notes:** 1. One sign is negative. 2. The axis lies along the coordinate axis associated with the variable with the negative coefficient.		Sketch the trace on the plane $z = k$ (in this case) for an appropriate value of k and for $z = 0$. Then use symmetry.
Hyperboloid of Two Sheets $$-\frac{x^2}{a^2} - \frac{y^2}{b^2} + \frac{z^2}{c^2} = 1$$ **Notes:** 1. Two signs are negative. 2. The axis lies along the coordinate axis associated with the variable with the positive coefficient.		Sketch the trace on the plane $z = k$ (in this case) for an appropriate value of k. Find the z-intercept (in this case) and use symmetry.

(continued)

Equation	Surface (computer generated)	Aid for Sketching the Figure
Cone $$\frac{x^2}{a^2} + \frac{y^2}{b^2} - \frac{z^2}{c^2} = 0$$ **Notes:** 1. One sign is negative. 2. The constant term is zero. 3. The axis lies along the coordinate axis associated with the variable with the negative coefficient.		Sketch the trace on the plane $z = k$ (in this case) for an appropriate value of k. Then use symmetry.
Paraboloids $$\frac{x^2}{a^2} + \frac{y^2}{b^2} = cz$$ **Notes:** 1. There are two positive signs. 2. The axis lies along the coordinate axis associated with the variable of degree 1. 3. It opens upward if $c > 0$ and opens downward if $c < 0$.		Sketch the trace on the plane $z = k$ (in this case) for an appropriate value of k.

Equation	Surface (computer generated)	Aid for Sketching the Figure
Hyperbolic Paraboloid $$\frac{x^2}{a^2} - \frac{y^2}{b^2} = cz$$ **Note:** There is one positive and one negative sign.	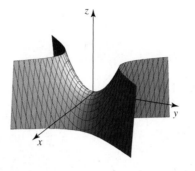	**a.** For the case $c < 0$, sketch the parabolas $$z = \frac{x^2}{ca^2} \quad \text{and} \quad z = -\frac{y^2}{cb^2}$$ 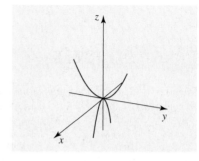 **b.** Sketch the hyperbola $$\frac{x^2}{a^2} - \frac{y^2}{b^2} = ck$$ for an appropriate value of k. 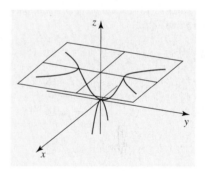 **c.** Complete your sketch. 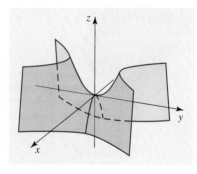

11.6 CONCEPT QUESTIONS

1. What is the trace of a surface in a plane? Illustrate by showing the trace of the surface $z = x^2 + y^2$ in the plane $z = 4$.
2. What is a cylinder? Illustrate by sketching the cylinder $x = y^2 - 4$.

3. **a.** What is a quadric surface?
   b. Write a standard equation for (1) an ellipsoid, (2) a hyperboloid of one sheet, (3) a hyperboloid of two sheets, (4) a cone, (5) an elliptic paraboloid, and (6) a hyperbolic paraboloid.

11.6 EXERCISES

In Exercises 1–12, sketch the graph of the cylinder with the given equation.

1. $x^2 + y^2 = 4$
2. $y^2 + z^2 = 9$
3. $x^2 + z^2 = 16$
4. $y = 4x^2$
5. $z = 4 - x^2$
6. $y = z^2 - 9$
7. $9x^2 + 4y^2 = 36$
8. $x^2 + 4z^2 = 16$
9. $yz = 1$
10. $y^2 - x^2 = 1$
11. $z = \cos y$
12. $y = \sec x, \quad -\frac{\pi}{2} < x < \frac{\pi}{2}$

In Exercises 13–20, match each equation with one of the graphs labeled (a)–(h).

13. $x^2 + \dfrac{y^2}{16} + \dfrac{z^2}{4} = 1$
14. $x^2 - \dfrac{y^2}{9} + z^2 = 9$
15. $-2x^2 - 2y^2 + z^2 = 1$
16. $x^2 - y^2 + z^2 = 0$
17. $x^2 + \dfrac{z^2}{4} = y$
18. $x^2 - \dfrac{y^2}{4} = -z$
19. $\dfrac{x^2}{4} + \dfrac{y^2}{9} + \dfrac{z^2}{25} = 1$
20. $x^2 - z^2 = -y$

(a)

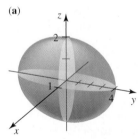

(b)

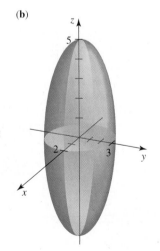

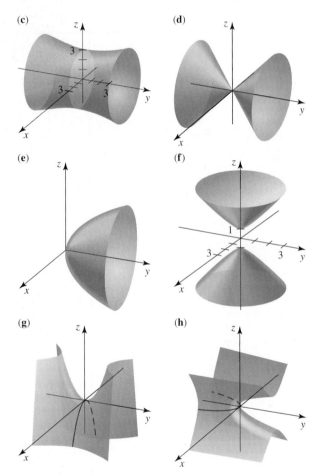

In Exercises 21–44, write the given equation in standard form and sketch the surface represented by the equation.

21. $4x^2 + y^2 + z^2 = 4$
22. $4x^2 + 4y^2 + z^2 = 16$
23. $9x^2 + 4y^2 + z^2 = 36$
24. $36x^2 + 100y^2 + 225z^2 = 900$
25. $4x^2 + 4y^2 - z^2 = 4$

26. $9x^2 + 9z^2 - 4y^2 = 36$ **27.** $x^2 + 4y^2 - z^2 = 4$

28. $9x^2 + 9y^2 - 4z^2 = 36$ **29.** $z^2 - x^2 - y^2 = 1$

30. $y^2 - x^2 - 9z^2 = 9$ **31.** $4x^2 - y^2 + 2z^2 + 4 = 0$

32. $4x^2 - 3y^2 + 12z^2 + 12 = 0$

33. $x^2 + y^2 - z^2 = 0$ **34.** $y^2 + z^2 = x^2$

35. $9x^2 + 4y^2 - z^2 = 0$ **36.** $x^2 - 4y^2 - 16z^2 = 0$

37. $x^2 + y^2 = z$ **38.** $y^2 + z^2 = x$

39. $x^2 + 9y^2 = z$ **40.** $x^2 + z^2 + y = 1$

41. $z = x^2 + y^2 + 4$ **42.** $z = x^2 + 4y^2 - 4$

43. $y^2 - x^2 = z$ **44.** $x^2 - y^2 = z$

In Exercises 45–50, sketch the region bounded by the surfaces with the given equations.

45. $x + 3y + 2z = 6$, $x = 0$, $y = 0$, and $z = 0$

46. $z = \sqrt{x^2 + y^2}$ and $z = 2$

47. $x^2 + y^2 = 4$, $x + z = 2$, $x = 0$, $y = 0$, and $z = 0$

48. $y^2 + z^2 = 1$, $x^2 + z^2 = 1$, $x = 0$, $y = 0$, and $z = 0$

49. $z = \sqrt{x^2 + y^2}$ and $z = 9 - x^2 - y^2$

50. $z = x^2 + y^2$ and $z = 2 - x^2 - y^2$

In Exercises 51 and 52, find an equation of the surface satisfying the conditions. Identify the surface.

51. The set of all points equidistant from the point $(-3, 0, 0)$ and the plane $x = 3$.

52. The set of all points whose distance from the y-axis is twice its distance from the xz-plane.

53. Show that the curve of intersection of the surfaces $2x^2 + y^2 - 3z^2 + 2y = 6$ and $4x^2 + 2y^2 - 6z^2 - 4x = 4$ lies in a plane.

54. Show that the straight lines
$L_1: x = a + t, y = b + t, z = b^2 - a^2 + 2(b - a)t$ and
$L_2: x = a + t, y = b - t, z = b^2 - a^2 - 2(b + a)t$,
passing through each point $(a, b, b^2 - a^2)$ on the hyperbolic paraboloid $z = y^2 - x^2$, both lie entirely on the surface.
Note: This shows that the hyperbolic paraboloid is a **ruled surface,** that is, a surface that can be swept out by a line moving in space. The only other quadric surfaces that are ruled surfaces are cylinders, cones, and hyperboloids of one sheet.

cas *In Exercises 55–60, use a computer algebra system (CAS) to plot the surface with the given equation.*

55. $2x^2 + 3y^2 + 6z^2 = 36$ **56.** $-x^2 + 4y^2 + z^2 = 2$

57. $-2x^2 - 9y^2 + z^2 = 1$ **58.** $-x^2 - 3y^2 + z^2 = 0$

59. $-x^2 + y^2 - z = 0$

60. $x^2 - 2x + 4y^2 - 16y - z = 0$

In Exercises 61–64, determine whether the statement is true or false. If it is true, explain why. If it is false, explain why or give an example that shows it is false.

61. The graph of $y = x + 3$ in 3-space is a line lying in the xy-plane.

62. The graph of $9x^2 + 4z^2 = 1$ and $y = 0$ is the graph of an ellipse lying in the xz-plane.

63. The surface $\dfrac{x^2}{a^2} + \dfrac{y^2}{b^2} + \dfrac{z^2}{c^2} = 4$ is an ellipsoid obtained by stretching the ellipsoid $\dfrac{x^2}{a^2} + \dfrac{y^2}{b^2} + \dfrac{z^2}{c^2} = 1$ by a factor of 2 in each of the x-, y-, and z-directions.

64. The surface $\dfrac{x^2}{a^2} + \dfrac{y^2}{b^2} = c(z - z_0)$ is obtained by translating the paraboloid $\dfrac{x^2}{a^2} + \dfrac{y^2}{b^2} = cz$ vertically.

11.7 Cylindrical and Spherical Coordinates

Just as certain curves in the plane are described more easily by using polar coordinates than by using rectangular coordinates, there are some surfaces in space that can be described more conveniently by using coordinates other than rectangular coordinates. In this section we will look at two such coordinate systems.

■ The Cylindrical Coordinate System

The **cylindrical coordinate system** is just an extension of the polar coordinate system in the plane to a three-dimensional system in space obtained by adding the (perpendicular) z-axis to the system (see Figure 1).

A point P in this system is represented by the ordered triple (r, θ, z), where r and θ are the polar coordinates of the projection of P onto the xy-plane and z is the directed distance from $(r, \theta, 0)$ to P.

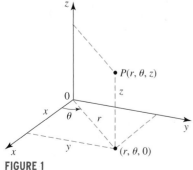

FIGURE 1
The cylindrical coordinate system

The relationship between rectangular coordinates and cylindrical coordinates can be seen by examining Figure 1. If P has representation (x, y, z) in terms of rectangular coordinates, then we have the following equations for converting cylindrical coordinates to rectangular coordinates and vice versa.

Converting Cylindrical to Rectangular Coordinates

$$x = r \cos \theta \qquad y = r \sin \theta \qquad z = z \qquad \textbf{(1)}$$

Converting Rectangular to Cylindrical Coordinates

$$r^2 = x^2 + y^2 \qquad \tan \theta = \frac{y}{x} \qquad z = z \qquad \textbf{(2)}$$

EXAMPLE 1   The point $\left(3, \frac{\pi}{4}, 3\right)$ is expressed in cylindrical coordinates. Find its rectangular coordinates.

Solution   We are given that $r = 3$, $\theta = \pi/4$, and $z = 3$. Using the equations in (1), we have

$$x = r \cos \theta = 3 \cos \frac{\pi}{4} = \frac{3\sqrt{2}}{2}$$

$$y = r \sin \theta = 3 \sin \frac{\pi}{4} = \frac{3\sqrt{2}}{2}$$

and

$$z = 3$$

Therefore, the rectangular coordinates of the given point are $\left(\frac{3\sqrt{2}}{2}, \frac{3\sqrt{2}}{2}, 3\right)$. (See Figure 2.)

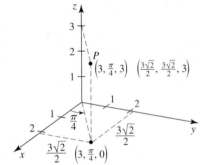

FIGURE 2
The point P can be written as $\left(\frac{3\sqrt{2}}{2}, \frac{3\sqrt{2}}{2}, 3\right)$ in rectangular coordinates.

EXAMPLE 2   The point $(-\sqrt{2}, \sqrt{2}, 2)$ is expressed in rectangular coordinates. Find its cylindrical coordinates.

Solution   We are given that $x = -\sqrt{2}$, $y = \sqrt{2}$, and $z = 2$. Using the equations in (2), we have

$$r^2 = x^2 + y^2 = (-\sqrt{2})^2 + (\sqrt{2})^2 = 2 + 2 = 4$$

and

$$\tan \theta = \frac{y}{x} = \frac{\sqrt{2}}{-\sqrt{2}} = -1$$

So $r = \pm 2$, and

$$\theta = \tan^{-1}(-1) + n\pi = \frac{3}{4}\pi + n\pi$$

where n is an integer; and $z = 2$.

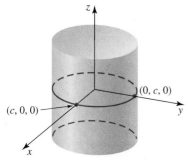

FIGURE 3
The point P can be written as
$\left(2, \frac{3\pi}{4}, 2\right)$ or $\left(-2, \frac{7\pi}{4}, 2\right)$, among
others, in cylindrical coordinates.

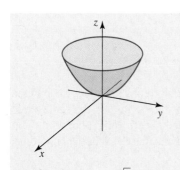

FIGURE 4
The circular cylinder has the simple
representation $r = c$ in cylindrical
coordinates.

We have two choices for r and infinitely many choices for θ. For example, we have the representations

$$\left(2, \frac{3\pi}{4}, 2\right) \qquad \text{If we pick } r > 0$$

and

$$\left(-2, \frac{7\pi}{4}, 2\right) \qquad \text{If we pick } r < 0$$

The point is shown in Figure 3. Note that neither the combination $r = 2$ and $\theta = 7\pi/4$ nor $r = -2$ and $\theta = 3\pi/4$ in Example 2 will do. (Why?) ∎

Cylindrical coordinates are especially useful in describing surfaces that are symmetric about the z-axis. For example, the circular cylinder with rectangular equation $x^2 + y^2 = c^2$ has the simple representation $r = c$ in the cylindrical coordinate system. (See Figure 4.)

EXAMPLE 3 Find an equation in cylindrical coordinates of the surface with the given rectangular equation.

a. $x^2 + y^2 = 9z$ **b.** $x^2 + y^2 = z^2$ **c.** $9x^2 + 9y^2 + 4z^2 = 36$

Solution In each case we use the relationship $r^2 = x^2 + y^2$.
a. We obtain $r^2 = 9z$ as the required equation.
b. Here, $r^2 = z^2$.
c. Here we have $9(x^2 + y^2) + 4z^2 = 36$,

$$9r^2 + 4z^2 = 36$$

The surfaces are shown in Figure 5.

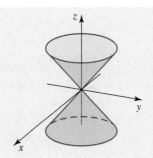

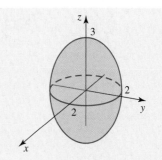

(a) $x^2 + y^2 = 9z$; $r = 3\sqrt{z}$ (paraboloid)   **(b)** $x^2 + y^2 = z^2$; $r = z$ (cone)   **(c)** $9x^2 + 9y^2 + 4z^2 = 36$; $z^2 = 9 - \frac{9}{4}r^2$ (ellipsoid)

FIGURE 5 ∎

EXAMPLE 4 Find an equation in rectangular coordinates for the surface with the given cylindrical equation.

a. $\theta = \dfrac{\pi}{4}$ **b.** $r^2 \cos 2\theta - z^2 = 4$

Solution

a. Using the equations in (2), we have

$$\frac{y}{x} = \tan \theta = \tan \frac{\pi}{4} = 1$$

or

$$y = x$$

b. First, we use the trigonometric identity $\cos 2\theta = \cos^2 \theta - \sin^2 \theta$ to rewrite the given equation in the form

$$r^2(\cos^2 \theta - \sin^2 \theta) - z^2 = 4$$

$$r^2 \cos^2 \theta - r^2 \sin^2 \theta - z^2 = 4$$

Using the equations in (1), we then obtain

$$x^2 - y^2 - z^2 = 4$$

The surfaces are shown in Figure 6. Note that Figure 6a shows only the part of the plane in the first octant.

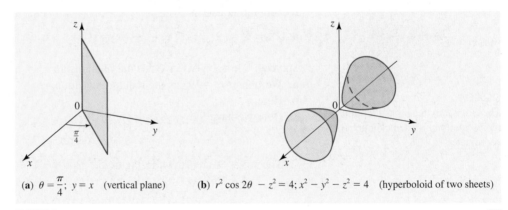

(a) $\theta = \frac{\pi}{4}$; $y = x$ (vertical plane)   **(b)** $r^2 \cos 2\theta - z^2 = 4$; $x^2 - y^2 - z^2 = 4$ (hyperboloid of two sheets)

FIGURE 6

■ The Spherical Coordinate System

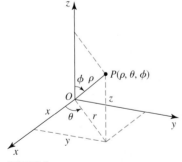

FIGURE 7
The spherical coordinate system

In the **spherical coordinate system** a point P is represented by an ordered triple (ρ, θ, ϕ), where ρ is the distance between P and the origin, θ is the same angle as the one used in the cylindrical coordinate system, and ϕ is the angle between the positive z-axis and the line segment OP. (See Figure 7.) Note that the spherical coordinates satisfy $\rho \geq 0$, $0 \leq \theta < 2\pi$, and $0 \leq \phi \leq \pi$.

The relationship between rectangular coordinates and spherical coordinates can be seen by examining Figure 7. If P has representation (x, y, z) in terms of rectangular coordinates, then

$$x = r \cos \theta \quad \text{and} \quad y = r \sin \theta$$

Since

$$r = \rho \sin \phi \quad \text{and} \quad z = \rho \cos \phi$$

we have the following equations for converting spherical coordinates to rectangular coordinates and vice versa.

Converting Spherical to Rectangular Coordinates

$$x = \rho \sin \phi \cos \theta \qquad y = \rho \sin \phi \sin \theta \qquad z = \rho \cos \phi \qquad \textbf{(3)}$$

Converting Rectangular to Spherical Coordinates

$$\rho^2 = x^2 + y^2 + z^2 \qquad \tan \theta = \frac{y}{x} \qquad \cos \phi = \frac{z}{\rho} \qquad \textbf{(4)}$$

EXAMPLE 5 The point $\left(3, \frac{\pi}{3}, \frac{\pi}{4}\right)$ is expressed in spherical coordinates. Find its rectangular coordinates.

Solution Using the equations in (3) with $\rho = 3$, $\theta = \pi/3$, and $\phi = \pi/4$, we have

$$x = \rho \sin \phi \cos \theta = 3 \sin \frac{\pi}{4} \cos \frac{\pi}{3} = 3\left(\frac{\sqrt{2}}{2}\right)\left(\frac{1}{2}\right) = \frac{3\sqrt{2}}{4}$$

$$y = \rho \sin \phi \sin \theta = 3 \sin \frac{\pi}{4} \sin \frac{\pi}{3} = 3\left(\frac{\sqrt{2}}{2}\right)\left(\frac{\sqrt{3}}{2}\right) = \frac{3\sqrt{6}}{4}$$

and

$$z = \rho \cos \phi = 3 \cos \frac{\pi}{4} = 3\left(\frac{\sqrt{2}}{2}\right) = \frac{3\sqrt{2}}{2}$$

Thus, in terms of rectangular coordinates the given point is $\left(\frac{3\sqrt{2}}{4}, \frac{3\sqrt{6}}{4}, \frac{3\sqrt{2}}{2}\right)$. ∎

EXAMPLE 6 The point $(\sqrt{3}, 3, 2)$ is given in rectangular coordinates. Find its spherical coordinates.

Solution We use the equations in (4). First, we have

$$\rho^2 = x^2 + y^2 + z^2 = (\sqrt{3})^2 + 3^2 + 2^2 = 3 + 9 + 4 = 16$$

so $\rho = 4$. (Remember that $\rho \geq 0$.) Next, from

$$\tan \theta = \frac{y}{x} = \frac{3}{\sqrt{3}} = \sqrt{3}$$

we see that $\theta = \pi/3$. Finally, from

$$\cos \phi = \frac{z}{\rho} = \frac{2}{4} = \frac{1}{2}$$

we see that $\phi = \pi/3$. Therefore, in terms of spherical coordinates, the given point is $\left(4, \frac{\pi}{3}, \frac{\pi}{3}\right)$. ∎

Spherical coordinates are particularly useful in describing surfaces that are symmetric about the origin. For example, the sphere with rectangular equation $x^2 + y^2 + z^2 = c^2$ has the simple representation $\rho = c$ in the spherical coordinate system. (See Figure 8a.) Also, shown in Figures 8b–8c are surfaces described by the equations $\theta = c$ and $\phi = c$, where $0 < c < \frac{\pi}{2}$.

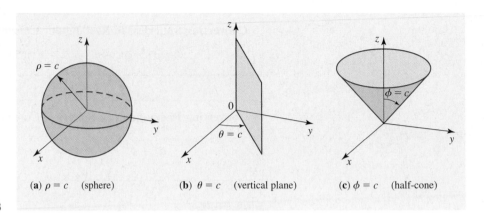

(a) $\rho = c$ (sphere)   **(b)** $\theta = c$ (vertical plane)   **(c)** $\phi = c$ (half-cone)

FIGURE 8

EXAMPLE 7 Find an equation in spherical coordinates for the paraboloid with rectangular equation $4z = x^2 + y^2$.

Solution Using the equations in (3), we obtain

$$4\rho \cos \phi = \rho^2 \sin^2 \phi \cos^2 \theta + \rho^2 \sin^2 \phi \sin^2 \theta$$
$$= \rho^2 \sin^2 \phi (\cos^2 \theta + \sin^2 \theta)$$
$$= \rho^2 \sin^2 \phi$$

or

$$\rho \sin^2 \phi = 4 \cos \phi$$

EXAMPLE 8 Find an equation in rectangular coordinates for the surface with spherical equation $\rho = 4 \cos \phi$.

Solution Multiplying both sides of the given equation by ρ gives

$$\rho^2 = 4\rho \cos \phi$$

Then, using the equations in (4), we have

$$x^2 + y^2 + z^2 = 4z$$

or, upon completing the square in z, we obtain

$$x^2 + y^2 + (z - 2)^2 = 4$$

which is an equation of the sphere with center $(0, 0, 2)$ and radius 2.

11.7 CONCEPT QUESTIONS

1. Sketch the cylindrical coordinate system, and use it as an aid to help you give the equations (a) for converting cylindrical coordinates to rectangular coordinates and (b) for converting rectangular coordinates to cylindrical coordinates.

2. Sketch the spherical coordinate system, and use it as an aid to help you give the equations (a) for converting spherical coordinates to rectangular coordinates and (b) for converting rectangular coordinates to spherical coordinates.

11.7   EXERCISES

In Exercises 1–6, the point is expressed in cylindrical coordinates. Write it in terms of rectangular coordinates.

1. $\left(3, \frac{\pi}{2}, 2\right)$ **2.** $(4, 0, -3)$

3. $\left(\sqrt{2}, \frac{\pi}{4}, \sqrt{3}\right)$ **4.** $\left(2, \frac{\pi}{3}, 5\right)$

5. $\left(3, -\frac{\pi}{6}, 2\right)$ **6.** $(1, \pi, \pi)$

In Exercises 7–12, the point is expressed in rectangular coordinates. Write it in terms of cylindrical coordinates.

7. $(2, 0, 3)$ **8.** $(3, 3, 3)$

9. $(1, \sqrt{3}, 5)$ **10.** $(\sqrt{2}, -\sqrt{2}, 4)$

11. $(\sqrt{3}, 1, -2)$ **12.** $(\sqrt{3}, -1, 4)$

In Exercises 13–18, the point is expressed in spherical coordinates. Write it in terms of rectangular coordinates.

13. $(5, 0, 0)$ **14.** $\left(2, \frac{\pi}{2}, \frac{\pi}{6}\right)$

15. $\left(2, 0, \frac{\pi}{4}\right)$ **16.** $\left(3, \frac{\pi}{4}, \frac{3\pi}{4}\right)$

17. $\left(5, \frac{\pi}{6}, \frac{\pi}{4}\right)$ **18.** $\left(1, \pi, \frac{\pi}{2}\right)$

In Exercises 19–24, the point is expressed in rectangular coordinates. Write it in terms of spherical coordinates.

19. $(-2, 0, 0)$ **20.** $(1, 1, 1)$

21. $(\sqrt{3}, 0, 1)$ **22.** $(-2, 2\sqrt{3}, 4)$

23. $(0, 2\sqrt{3}, 2)$ **24.** $(\sqrt{3}, 1, 2\sqrt{3})$

In Exercises 25–30, the point is expressed in cylindrical coordinates. Write it in terms of spherical coordinates.

25. $\left(2, \frac{\pi}{4}, 0\right)$ **26.** $\left(2, \frac{\pi}{2}, -2\right)$

27. $\left(4, \frac{\pi}{3}, -4\right)$ **28.** $(12, \pi, 5)$

29. $\left(4, \frac{\pi}{6}, 6\right)$ **30.** $\left(12, \frac{\pi}{2}, 5\right)$

In Exercises 31–36, the point is expressed in spherical coordinates. Write it in terms of cylindrical coordinates.

31. $(3, 0, 0)$ **32.** $\left(5, \frac{\pi}{6}, \frac{\pi}{2}\right)$

33. $\left(2, \frac{3\pi}{2}, \frac{\pi}{2}\right)$ **34.** $\left(4, -\frac{\pi}{6}, \frac{\pi}{6}\right)$

35. $\left(1, \frac{\pi}{4}, \frac{\pi}{3}\right)$ **36.** $\left(5, \frac{\pi}{4}, \frac{3\pi}{4}\right)$

37. Find the distance between $\left(2, \frac{\pi}{3}, 0\right)$ and $(1, \pi, 2)$, where the points are given in cylindrical coordinates.

38. Find the distance between $\left(4, \frac{\pi}{2}, \frac{2\pi}{3}\right)$ and $\left(3, \pi, \frac{\pi}{2}\right)$, where the points are given in spherical coordinates.

In Exercises 39–58, identify the surface whose equation is given.

39. $r = 2$ **40.** $z = 2$

41. $\rho = 2$ **42.** $\theta = \dfrac{\pi}{6}$ (spherical coordinates)

43. $\phi = \dfrac{\pi}{4}$ **44.** $z = 4r^2$

45. $z = 4 - r^2$ **46.** $r = 6 \sin \theta$

47. $\rho \cos \phi = 3$ **48.** $\rho \sin \phi = 3$

49. $r \sec \theta = 4$ **50.** $r = -\csc \theta$

51. $z = r^2 \sin^2 \theta$ **52.** $\rho = 4 \cos \phi$

53. $r^2 + z^2 = 16$ **54.** $3r^2 - 4z^2 = 12$

55. $\rho = 2 \csc \phi \sec \theta$ **56.** $\rho^2(\sin^2 \phi - 2 \cos^2 \phi) = 1$

57. $r^2 - 3r + 2 = 0$ **58.** $\rho^2 - 4\rho + 3 = 0$

In Exercises 59–66, write the given equation (a) in cylindrical coordinates and (b) in spherical coordinates.

59. $x^2 + y^2 + z^2 = 4$ **60.** $x^2 - y^2 + z^2 = 4$

61. $x^2 + y^2 = 2z$ **62.** $x^2 + y^2 = 9$

63. $2x + 3y - 4z = 12$ **64.** $x^2 + y^2 = 4y$

65. $x^2 + z^2 = 4$ **66.** $x^2 - y^2 - z^2 = 1$

In Exercises 67–70, sketch the region described by the inequalities.

67. $r \le z \le 2$ **68.** $r^2 \le z \le 4 - r^2$

69. $0 \le \theta \le 2\pi$, $0 \le \phi \le \frac{\pi}{6}$, $0 \le \rho \le a \sec \phi$

70. $0 \le \phi \le \frac{\pi}{4}$, $\rho \le 2$

71. Spherical Coordinate System for the Earth A spherical coordinate system for the earth can be set up as follows. Let the origin of the system be at the center of the earth, and choose the positive *x*-axis to pass through the point of intersection of the equator and the prime meridian and the positive *z*-axis to pass through the North Pole. Recall that the parallels of latitude are measured from 0° to 90° degrees north and south of the equator and the meridians of longitude are measured from 0° to 180° east and west of the prime meridian.

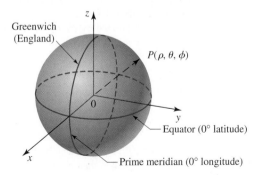

a. Express the locations of Los Angeles (latitude 34.06° North, longitude 118.25° West) and Paris (latitude 48.52° North, longitude 2.20° East) in terms of spherical coordinates. Take the radius of the earth to be 3960 miles.

b. Express the points found in part (a) in terms of rectangular coordinates.

c. Find the great-circle distance between Los Angeles and Paris. (A *great circle* is the circle obtained by intersecting a sphere with a plane passing through the center of the sphere.)

72. A **geodesic** on a surface is the curve that minimizes the distance between any two points on the surface. Suppose a cylinder of radius r is oriented so that its axis coincides with the z-axis of a cylindrical coordinate system. If $P_1(r, \theta_1, z_1)$ and $P_2(r, \theta_2, z_2)$ are two points on the cylinder, show that the length of the geodesic joining P_1 to P_2 is

$$\sqrt{r^2(\theta_2 - \theta_1)^2 + (z_2 - z_1)^2}$$

In Exercises 73–76, determine whether the statement is true or false. If it is true, explain why. If it is false, explain why or give an example that shows it is false.

73. The representation of a point in the cylindrical coordinate system is not unique.

74. The equation $\theta = \pi/4$ in cylindrical coordinates represents the plane with rectangular equation $y = x$.

75. The equation $\theta = c$, where c is a constant, in cylindrical coordinates and the equation $\theta = c$ in spherical coordinates represent different surfaces.

76. The surface defined by the spherical equation $\phi = \pi/4$ is the same as the surface defined by the rectangular equation $x^2 + y^2 = z^2$.

CHAPTER 11 REVIEW

CONCEPT REVIEW

In Exercises 1–14, fill in the blanks.

1. a. A vector is a quantity that possesses both _____ and _____.

b. A vector can be represented by an _____; the _____ points in the _____ of the vector, and the _____ of the arrow gives its magnitude.

c. The vector $\mathbf{v} = \overrightarrow{AB}$ has _____ point _____ and _____ point _____.

d. Two vectors $\mathbf{v}$ and $\mathbf{w}$ are equal if they have the same _____ and _____.

2. a. The scalar multiple of a scalar c and a vector $\mathbf{v}$ is the vector _____ whose magnitude is _____ times that of $\mathbf{v}$ and the direction is the same as that of $\mathbf{v}$ if _____ and opposite that of $\mathbf{v}$ if _____.

b. The vector $\mathbf{v} + \mathbf{w}$ is represented by the arrow with tail at the _____ point of $\mathbf{v}$ and head at the _____ point of $\mathbf{w}$.

c. The vector $\mathbf{v} - \mathbf{w}$ is defined to be the vector _____.

3. a. A vector in the plane is an ordered pair $\mathbf{a} =$ _____ of real numbers _____ and _____, called the _____ components of $\mathbf{a}$. The zero vector is $\mathbf{0} =$ _____.

b. If $\mathbf{a} = \langle a_1, a_2 \rangle$ and $\mathbf{b} = \langle b_1, b_2 \rangle$, then $\mathbf{a} + \mathbf{b} =$ _____. If $\mathbf{a} = \langle a_1, a_2 \rangle$ and c is a scalar, then $c\mathbf{a} =$ _____.

4. a. If $\mathbf{a} \neq \mathbf{0}$, then a unit vector having the same direction as $\mathbf{a}$ is $\mathbf{u} =$ _____.

b. The vectors $\mathbf{i} =$ _____ and $\mathbf{j} =$ _____ are called the _____ _____ vectors. If $\mathbf{a}$ is any vector in the plane, then $\mathbf{a}$ can be expressed in terms of $\mathbf{i}$ and $\mathbf{j}$ as $\mathbf{a} =$ _____.

c. If θ is the angle that the unit vector $\mathbf{u}$ makes with the positive x-axis, then $\mathbf{u}$ can be written in terms of θ as $\mathbf{u} =$ _____.

5. a. The standard equation of a sphere with center _____ and radius _____ is $(x - h)^2 + (y - k)^2 + (z - l)^2 = r^2$.

b. The midpoint of the line segment joining the points $P_1(x_1, y_1, z_1)$ and $P_2(x_2, y_2, z_2)$ is _____.

6. a. The vector with initial point $P_1(x_1, y_1, z_1)$ and terminal point $P_2(x_2, y_2, z_2)$ is $\overrightarrow{P_1 P_2} =$ _____.

b. A vector $\mathbf{a} = \langle a_1, a_2, a_3 \rangle$ in 3-space can be written in terms of the basis vectors $\mathbf{i} =$ _____, $\mathbf{j} =$ _____, and $\mathbf{k} =$ _____ as $\mathbf{a} =$ _____.

7. a. The dot product of $\mathbf{a} = \langle a_1, a_2, a_3 \rangle$ and $\mathbf{b} = \langle b_1, b_2, b_3 \rangle$ is $\mathbf{a} \cdot \mathbf{b} =$ _____ and is a _____.

b. The magnitude of a vector $\mathbf{a}$ can be written in terms of the dot product as $|\mathbf{a}| =$ _____.

c. The angle between two nonzero vectors $\mathbf{a}$ and $\mathbf{b}$ is given by $\cos \theta =$ _____.

8. a. Two nonzero vectors $\mathbf{a}$ and $\mathbf{b}$ are orthogonal if and only if _____.

b. The angles α, β, and γ that a nonzero vector $\mathbf{a}$ makes with the positive x-, y-, and z-axes, respectively, are called the _____ angles of $\mathbf{a}$.

c. The direction cosines of a nonzero vector $\mathbf{a} = a\mathbf{i} + b\mathbf{j} + c\mathbf{k}$ satisfy _____.

9. a. The vector obtained by projecting $\mathbf{b}$ onto the line containing the vector $\mathbf{a}$ is called the _____ _____ of $\mathbf{b}$ onto $\mathbf{a}$; it is also called the _____ _____ of $\mathbf{b}$ along $\mathbf{a}$.

b. The length of $\text{proj}_\mathbf{a}\mathbf{b}$ is called the _____ _____ of $\mathbf{b}$ along $\mathbf{a}$.

c. $\text{proj}_\mathbf{a}\mathbf{b} =$ _____.

d. The work done by a constant force $\mathbf{F}$ in moving an object along a straight line from P to Q is $W =$ _____.

10. a. If $\mathbf{a}$ and $\mathbf{b}$ are nonzero vectors in 3-space, then $\mathbf{a} \times \mathbf{b}$ is _____ to both $\mathbf{a}$ and $\mathbf{b}$.

b. If $0 \leq \theta \leq \pi$ is the angle between $\mathbf{a}$ and $\mathbf{b}$, then $|\mathbf{a} \times \mathbf{b}| =$ _____.

c. Two nonzero vectors $\mathbf{a}$ and $\mathbf{b}$ are parallel if and only if $\mathbf{a} \times \mathbf{b} =$ _____.

11. a. The scalar triple product of $\mathbf{a}$, $\mathbf{b}$, and $\mathbf{c}$ is _____.

b. The volume V of the parallelepiped determined by $\mathbf{a}$, $\mathbf{b}$, and $\mathbf{c}$ is $V =$ _____.

12. a. The parametric equations of the line passing through $P_0(x_0, y_0, z_0)$ and parallel to $\mathbf{v} = \langle a, b, c \rangle$ are _____.

b. The symmetric equations of the line passing through $P_0(x_0, y_0, z_0)$ and parallel to $\mathbf{v} = \langle a, b, c \rangle$ are _____.

13. a. The standard form of the equation of the plane containing the points $P_0(x_0, y_0, z_0)$ and having the normal vector $\mathbf{n} = \langle a, b, c \rangle$ is _____.

b. The linear equation $ax + by + cz = d$ represents a _____ in space having the normal vector _____. The acute angle between two intersecting planes is the angle between their _____ _____.

14. a. If a point has rectangular coordinates (x, y, z) and cylindrical coordinates (r, θ, z), then $x =$ _____, $y =$ _____, and $z =$ _____; and $r^2 =$ _____, $\tan \theta =$ _____, and $z =$ _____.

b. If a point has rectangular coordinates (x, y, z) and spherical coordinates (ρ, θ, ϕ), then $x =$ _____, $y =$ _____, and $z =$ _____; and $\rho^2 =$ _____, $\tan \theta =$ _____, $\cos \phi =$ _____.

REVIEW EXERCISES

In Exercises 1–17, let $\mathbf{a} = 2\mathbf{i} - \mathbf{j} + 3\mathbf{k}$, $\mathbf{b} = \mathbf{i} + 2\mathbf{j} - \mathbf{k}$, *and* $\mathbf{c} = 3\mathbf{i} - 2\mathbf{j} + \mathbf{k}$. *Find the given quantities.*

1. $2\mathbf{a} - 3\mathbf{b}$

2. $\mathbf{a} \cdot (\mathbf{b} + \mathbf{c})$

3. $|3\mathbf{a} + 2\mathbf{b}|$

4. $|\mathbf{a}| + |\mathbf{c}|$

5. $\mathbf{a} \times \mathbf{c}$

6. $|\mathbf{b} \times (\mathbf{c} \times \mathbf{a})|$

7. $\mathbf{a} \cdot (\mathbf{b} \times \mathbf{c})$

8. $|\mathbf{a} \times \mathbf{a}|$

9. $\mathbf{a} \times (\mathbf{b} + \mathbf{c})$

10. The angle between $\mathbf{a}$ and $\mathbf{b}$

11. Two unit vectors having the same direction as $\mathbf{c}$

12. A vector having twice the magnitude of $\mathbf{a}$ and direction opposite to that of $\mathbf{a}$

13. The direction cosines of $\mathbf{b}$

14. The scalar projection of $\mathbf{b}$ onto $\mathbf{a}$

15. The vector projection of $\mathbf{b}$ onto $\mathbf{a}$

16. The scalar projection of $\mathbf{b} \times \mathbf{c}$ onto $\mathbf{a}$

17. The volume of the parallelepiped determined by $\mathbf{a}$, $\mathbf{b}$, and $\mathbf{c}$

18. Which of the following are legitimate operations?
  a. $\mathbf{a} \cdot (\mathbf{b} - \mathbf{c})$
  b. $\mathbf{a} \times (\mathbf{b} \cdot \mathbf{c})$
  c. $(|\mathbf{a}|\mathbf{b} - |\mathbf{b}|\mathbf{a} \times \mathbf{c})$

19. Show that $\mathbf{a} = 2\mathbf{i} - 3\mathbf{j} + 4\mathbf{k}$ and $\mathbf{b} = 3\mathbf{i} + 6\mathbf{j} + 3\mathbf{k}$ are orthogonal.

20. Find the value of x such that $3\mathbf{i} + x\mathbf{j} - 2\mathbf{k}$ and $2x\mathbf{i} - 3\mathbf{j} + 6\mathbf{k}$ are orthogonal.

21. Find c such that $\mathbf{a} = 2\mathbf{i} + 3\mathbf{j} + \mathbf{k}$, $\mathbf{b} = \mathbf{i} + 2\mathbf{j} + 3\mathbf{k}$, and $\mathbf{c} = \mathbf{i} - 3\mathbf{j} + c\mathbf{k}$ are coplanar.

22. Find two unit vectors that are orthogonal to $\langle 3, 1, -2 \rangle$ and $\langle 2, 3, -2 \rangle$.

23. Find the acute angle between two diagonals of a cube.

24. Find the volume of the parallelepiped with adjacent sides AB, AC, and AD, where $A(2, -1, 1)$, $B(2, 0, 1)$, $C(4, -1, 1)$, and $D(5, -2, 0)$.

25. a. Find a vector perpendicular to the plane passing through the points $P(-1, 2, -2)$, $Q(2, 3, 1)$, and $R(3, 2, 1)$.
  b. What is the area of the triangle with vertices P, Q, and R?

26. A force $\mathbf{F} = \mathbf{i} + 2\mathbf{j} + 4\mathbf{k}$ moves an object along the line segment from $(-3, -1, 1)$ to $(2, 1, 1)$. Find the work done by the force if the distance is measured in feet and the force is measured in pounds.

27. A constant force has a magnitude of 20 newtons and acts in the direction of the vector $\mathbf{a} = 2\mathbf{i} - \mathbf{j} + 3\mathbf{k}$. If this force moves a particle along the line segment from $(1, 2, 1)$ to $(2, 1, 4)$ and the distance is measured in meters, find the work done by the force.

28. Two men wish to push a crate in the x-direction as shown in the figure. If one man pushes with a force $\mathbf{F}_1$ of magnitude 80 N in the direction indicated in the figure and the second man pushes in the indicated direction, find the force $\mathbf{F}_2$ with which he must push.

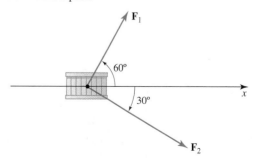

29. Two forces $\mathbf{F}_1$ and $\mathbf{F}_2$ with magnitude 10 N and 8 N, respectively, are applied to a bar as shown in the figure. Find the resultant torque about the point 0.

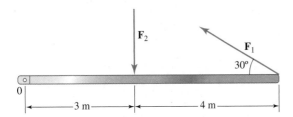

In Exercises 30–33, find (a) parametric equations and (b) symmetric equations for the line satisfying the given conditions.

30. Passes through $(2, 3, 1)$ and has the direction of
$\mathbf{v} = \mathbf{i} - 2\mathbf{j} + 3\mathbf{k}$

31. Passes through $(-1, 2, -4)$ and $(2, -1, 3)$

32. Passes through $(2, -1, 3)$ and is parallel to the line with parametric equations $x = 1 - 2t$, $y = 2 + 3t$, $z = -1 - t$

33. Passes through $(1, 2, 4)$ and is perpendicular to
$\mathbf{u} = \langle 1, -2, 1 \rangle$ and $\mathbf{v} = \langle 3, 2, 5 \rangle$

In Exercises 34–37, find an equation of the plane satisfying the given conditions.

34. Passes through $(-1, 2, 3)$ and has a normal vector
$2\mathbf{i} - 3\mathbf{j} + 5\mathbf{k}$

35. Passes through $(-2, 4, 3)$ and is parallel to the plane with equation $2x + 4y - 3z = 12$

36. Passes through $(-2, 1, 1)$, $(2, -2, 4)$, and $(3, 1, 5)$

37. Passes through $(3, 2, 2)$ and is parallel to the xz-plane

38. Find the point of intersection (if any) of the line with parametric equations $x = 1 + 2t$, $y = -1 + t$, and $z = 2 + 3t$ and the plane $2x + 3y - 4z = 6$

39. Find the distance between the point $(2, 1, 4)$ and the plane $2x - 3y + 4z = 12$

40. Determine whether the lines with parametric equations $x = 3 - 3t$, $y = -1 + 2t$, $z = 3 - 2t$, and $x = 1 + 2t$, $y = 2 - 3t$, $z = 2 + t$ are parallel, are skew, or intersect. If they intersect, find the point of intersection.

41. Show that the lines with symmetric equations

$$\frac{x - 1}{-2} = \frac{y - 3}{-1} = \frac{z}{2}$$

and

$$\frac{x - 2}{-3} = \frac{y - 1}{1} = \frac{z + 1}{3}$$

intersect, and find the angle between the two lines.

42. Determine whether the planes $2x + 3y - 4z = 12$ and $2x - 3y - 5z = 8$ are parallel, perpendicular, or neither. If they are neither parallel nor perpendicular, find the angle between them.

In Exercises 43 and 44, find the distance between the parallel planes.

43. $x + 2y - 3z = 2$; $2x + 4y - 6z = 6$

44. $2x - 3y - z = 2$; $6x - 9y - 3z = 10$

45. Find the distance between the point $(3, 4, 5)$ and the plane $2x + 4y - 3z = 12$.

46. Find the curve of intersection of the plane $x + z = 5$ and the cylinder $x^2 + y^2 = 9$.

In Exercises 47–50, describe and sketch the region in 3-space defined by the inequality or inequalities.

47. $x^2 + y^2 \leq 4$

48. $1 \leq x^2 + z^2 \leq 4$

49. $y \leq x$, $0 \leq x \leq 1$, $0 \leq y \leq 1$, $0 \leq z \leq 1$

50. $9x^2 + 4y^2 = 36$, $0 \leq z \leq 2$

In Exercises 51–58, identify and sketch the surface represented by the given equation.

51. $2x - y = 6$
52. $x = 9 - y^2$
53. $x = y^2 + z^2$
54. $9x^2 + 4y^2 - z^2 = 36$
55. $4x^2 + 9z^2 = y^2$
56. $225x^2 + 100y^2 + 36z^2 = 900$
57. $x^2 - z^2 = y$
58. $z = \sin y$

59. The point $(1, 1, \sqrt{2})$ is expressed in rectangular coordinates. Write it in terms of cylindrical and spherical coordinates.

60. The point $\left(2, \frac{\pi}{6}, 4\right)$ is expressed in cylindrical coordinates. Write it in terms of rectangular and spherical coordinates.

61. The point $\left(2, \frac{\pi}{4}, \frac{\pi}{3}\right)$ is expressed in spherical coordinates. Write it in terms of rectangular and cylindrical coordinates.

In Exercises 62–66, identify the surface whose equation is given.

62. $z = -2$
63. $\theta = \dfrac{\pi}{3}$ (spherical coordinates)
64. $\phi = \dfrac{\pi}{3}$
65. $r = 2 \sin \theta$
66. $\rho = 2 \sec \phi$

In Exercises 67–70, write the given equation (a) in cylindrical coordinates and (b) in spherical coordinates.

67. $x^2 + y^2 = 2$
68. $x^2 + y^2 + z^2 = 9$
69. $x^2 + y^2 + 2z^2 = 1$
70. $x^2 + y^2 + z^2 = 2y$

In Exercises 71 and 72, sketch the region described by the given inequalities.

71. $0 \leq r \leq z$, $0 \leq \theta \leq \frac{\pi}{2}$

72. $0 \leq \phi \leq \frac{\pi}{3}$, $\rho \leq 2$

CHALLENGE PROBLEMS

1. Prove the Cauchy-Schwarz inequality $|\mathbf{a} \cdot \mathbf{b}| \leq |\mathbf{a}||\mathbf{b}|$, without using trigonometry, by demonstrating the following:
   a. $|t\mathbf{a} + \mathbf{b}|^2 = |\mathbf{a}|^2 t^2 + 2(\mathbf{a} \cdot \mathbf{b})t + |\mathbf{b}|^2 \geq 0$ for all values of t
   b. $4(\mathbf{a} \cdot \mathbf{b})^2 - 4|\mathbf{a}|^2|\mathbf{b}|^2 \leq 0$
   Hint: Recall the relationship between the discriminant $b^2 - 4ac$ of the quadratic equation $at^2 + bt + c = 0$ and the nature of the roots of the equation.

2. Use vectors to prove that if A and B are endpoints of a diameter of a circle and C is any other point on the circle, then the triangle $\triangle ABC$ is a right triangle.

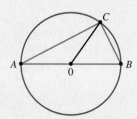

3. Let $\mathbf{a} = a_1\mathbf{i} + a_2\mathbf{j} + a_3\mathbf{k}$ and $\mathbf{b} = b_1\mathbf{i} + b_2\mathbf{j} + b_3\mathbf{k}$ be nonparallel vectors in three-dimensional space, and let $\mathbf{c} = x\mathbf{i} + y\mathbf{j} + z\mathbf{k}$ be a vector that is perpendicular to both $\mathbf{a}$ and $\mathbf{b}$.
   a. Show that x, y, and z satisfy the system of two equations
   $$a_1 x + a_2 y + a_3 z = 0$$
   $$b_1 x + b_2 y + b_3 z = 0$$
   b. Solving the system in part (a) for x and y in terms of z, show that
   $$x = \frac{a_2 b_3 - a_3 b_2}{a_1 b_2 - a_2 b_1} z \quad \text{and} \quad y = \frac{a_3 b_1 - a_1 b_3}{a_1 b_2 - a_2 b_1} z$$
   c. Show that $\mathbf{c}$ can be written in the form
   $$\mathbf{c} = x\mathbf{i} + y\mathbf{j} + z\mathbf{k}$$
   $$= (a_2 b_3 - a_3 b_2)\mathbf{i} + (a_3 b_1 - a_1 b_3)\mathbf{j} + (a_1 b_2 - a_2 b_1)\mathbf{k}$$
   $$= \mathbf{a} \times \mathbf{b}$$
   (This is the definition of the cross-product of $\mathbf{a}$ and $\mathbf{b}$ given in Section 11.4.)

4. The area of a triangle with sides of lengths a, b, and c is given by
   $$A = \sqrt{s(s-a)(s-b)(s-c)}$$

where $s = \frac{1}{2}(a + b + c)$ is the semiperimeter of the triangle. Derive this formula, known as Heron's formula.
Hint: Let a and b denote two sides of the triangle. Then $A = \frac{1}{2}|\mathbf{a} \times \mathbf{b}|$. Use the result of Exercise 44 in Section 11.4.

5. **a.** Let $\mathbf{a}$, $\mathbf{b}$, and $\mathbf{c}$ be noncoplanar vectors and let $\mathbf{v}$ be an arbitrary vector. Show that there exist constants α, β, and γ such that $\mathbf{v} = \alpha\mathbf{a} + \beta\mathbf{b} + \gamma\mathbf{c}$.
   Hint: To find α, take the dot product of $\mathbf{v}$ with $\mathbf{b} \times \mathbf{c}$.
   b. Let $\mathbf{a} = \langle 1, 3, 1 \rangle$, $\mathbf{b} = \langle 2, -1, 1 \rangle$, and $\mathbf{c} = \langle 3, 1, 2 \rangle$. Express $\mathbf{v} = \langle 3, 2, 4 \rangle$ in terms of $\mathbf{a}$, $\mathbf{b}$, and $\mathbf{c}$ as suggested in part (a).

6. **a.** Consider the portion of the plane $ax + by + cz = d$ lying in the first octant, where a, b, and c, are positive real constants. Show that its area is given by
   $$\frac{d^2 \sqrt{a^2 + b^2 + c^2}}{2abc}$$
   b. Use the result of part (a) to find the area of the plane $x + 2y + 3z = 6$ that lies in the first octant.

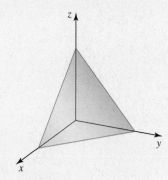

7. Find the points of intersection of the line
   $$x - 2 = \frac{y}{2} = \frac{z - 24}{16}$$
   and the elliptic paraboloid $z = 4x^2 + y^2$.

8. **a.** Let $\mathbf{a}$, $\mathbf{b}$, and $\mathbf{c}$ be vectors in 3-space. Show that there exist scalars s and t such that
   $$\mathbf{a} \times (\mathbf{b} \times \mathbf{c}) = s\mathbf{b} + t\mathbf{c}$$
   b. Let $\mathbf{a} = \langle 1, -2, 4 \rangle$, $\mathbf{b} = \langle 2, 3, 2 \rangle$, and $\mathbf{c} = \langle -2, 4, 5 \rangle$. Use the result of part (a) to write $\mathbf{a} \times (\mathbf{b} \times \mathbf{c})$ in the form $s\mathbf{b} + t\mathbf{c}$, where s and t are scalars.

The "human cannonball" is a popular attraction at circuses. The trajectory of the person shot out of the cannon can be described by a *vector function*—a function whose domain is a set of real numbers and whose range is a set of vectors. We will look at an exercise involving a human cannonball in Section 12.4.

Anthony Reynolds/Corbis

12 Vector-Valued Functions

IN THIS CHAPTER we will study functions whose values are vectors in the plane or in space. These vector-valued functions can be used to describe plane curves and space curves, and they also allow us to study the motion of objects along such curves.

We will also develop formulas for computing the arc length of plane and space curves and for finding the *curvature* of a curve. (The curvature measures the rate at which a curve bends.)

We end the chapter by demonstrating how vector calculus can be used to prove Kepler's laws of planetary motion.

Ⓥ This symbol indicates that one of the following video types is available for enhanced student learning at **www.academic.cengage.com/login:**
- Chapter lecture videos
- Solutions to selected exercises

12.1   Vector-Valued Functions and Space Curves

In Section 10.2 we saw that the position of an object such as a boat or a car moving in the xy-plane can be described by a pair of parametric equations

$$x = f(t) \qquad y = g(t)$$

where f and g are continuous functions on a parameter interval I.

Using vector notation, we can denote the position of the object in an equivalent and somewhat abbreviated form via its *position vector* **r** as follows: For each t in I, the position vector **r** of the object is the vector with initial point at the origin and terminal point $(f(t), g(t))$. In other words,

$$\mathbf{r}(t) = \langle f(t), g(t) \rangle = f(t)\mathbf{i} + g(t)\mathbf{j} \qquad t \in I$$

As t takes on increasing values, the terminal point of $\mathbf{r}(t)$ traces the path of the object which is a plane curve C. This is illustrated in Figure 1 for the parameter interval $I = [a, b]$.

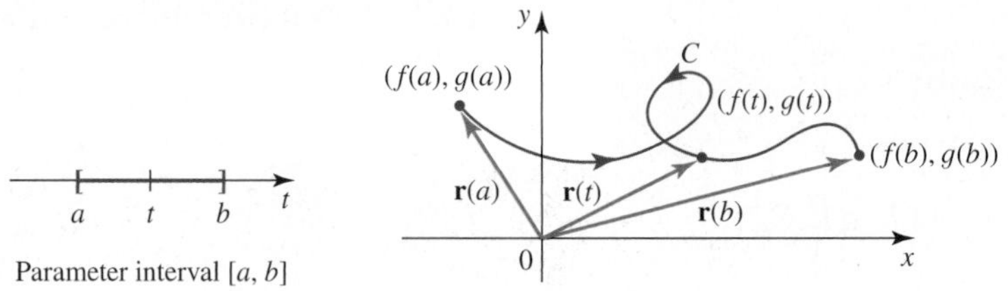

FIGURE 1

As t increases from a to b, the terminal point of **r** traces the curve C.

Similarly, in 3-space we can describe the position of an object such as a plane or a satellite using the parametric equations

$$x = f(t) \qquad y = g(t) \qquad z = h(t)$$

where f, g, and h are continuous functions on a parameter interval I. Equivalently, we can describe its position using the *position vector* **r** defined by

$$\mathbf{r}(t) = \langle f(t), g(t), h(t) \rangle = f(t)\mathbf{i} + g(t)\mathbf{j} + h(t)\mathbf{k} \qquad t \in I$$

As t takes on increasing values, the terminal point of $\mathbf{r}(t)$ traces the path of the object, which is a **space curve** C. (See Figure 2.)

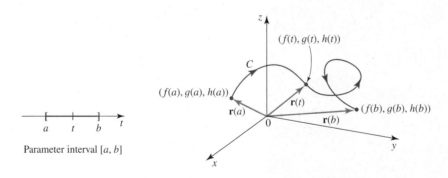

FIGURE 2

As t increases from a to b, the terminal point of **r** traces the curve C.

The function **r** is called a **vector-valued function,** or **vector function,** of a real variable t because its value $\mathbf{r}(t)$ is a vector and its **domain** (parameter interval) is a subset of the real numbers.

> **DEFINITION**   **Vector Function**
>
> A **vector-valued function**, or **vector function**, is a function **r** defined by
>
> $$\mathbf{r}(t) = f(t)\mathbf{i} + g(t)\mathbf{j} + h(t)\mathbf{k}$$
>
> where the component functions f, g, and h of **r** are real-valued functions of the parameter t lying in a **parameter interval I.**

Unless otherwise specified, the parameter interval will be taken to be the intersection of the domains of the real-valued functions f, g, and h.

EXAMPLE 1   Find the domain (parameter interval) of the vector function

$$\mathbf{r}(t) = \left\langle \frac{1}{t}, \sqrt{t-1}, \ln t \right\rangle$$

Solution   The component functions of **r** are $f(t) = 1/t$, $g(t) = \sqrt{t-1}$, and $h(t) = \ln t$. Observe that f is defined for all values of t except $t = 0$, g is defined for all $t \geq 1$, and h is defined for all $t > 0$. Therefore, f, g, and h are all defined if $t \geq 1$, and we conclude that the domain of **r** is $[1, \infty)$. ∎

■ Curves Defined by Vector Functions

As was mentioned earlier, a plane or space curve is the curve traced out by the terminal point of $\mathbf{r}(t)$ of a vector function **r** as t takes on all values in a parameter interval.

EXAMPLE 2   Sketch the curve defined by the vector function

$$\mathbf{r}(t) = \langle 3 \cos t, -2 \sin t \rangle \qquad 0 \leq t \leq 2\pi$$

Solution   The parametric equations for the curve are

$$x = 3 \cos t \qquad \text{and} \qquad y = -2 \sin t$$

Solving the first equation for $\cos t$ and the second equation for $\sin t$ and using the identity $\cos^2 t + \sin^2 t = 1$, we obtain the rectangular equation

$$\frac{x^2}{9} + \frac{y^2}{4} = 1$$

The curve described by this equation is the ellipse shown in Figure 3. As t increases from 0 to 2π, the terminal point of **r** traces the ellipse in a clockwise direction.

FIGURE 3
As t increases from 0 to 2π, the terminal point of the vector $\mathbf{r}(t)$ traces the ellipse $\dfrac{x^2}{9} + \dfrac{y^2}{4} = 1$ in a clockwise direction, starting and ending at $(3, 0)$.

Parameter interval $[0, 2\pi]$

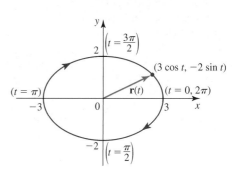

∎

EXAMPLE 3 Sketch the curve defined by the vector function

$$\mathbf{r}(t) = (2 - 4t)\mathbf{i} + (-1 + 3t)\mathbf{j} + (3 + 2t)\mathbf{k} \qquad 0 \le t \le 1$$

Solution The parametric equations for the curve are

$$x = 2 - 4t \qquad y = -1 + 3t \qquad z = 3 + 2t$$

which are parametric equations of the line passing through the point $(2, -1, 3)$ with direction numbers -4, 3, and 2. Because the parameter interval is the closed interval $[0, 1]$, we see that the curve is a straight line segment: Its initial point $(2, -1, 3)$ is the terminal point of the vector $\mathbf{r}(0) = 2\mathbf{i} - \mathbf{j} + 3\mathbf{k}$, and its terminal point $(-2, 2, 5)$ is the terminal point of the vector $\mathbf{r}(1) = -2\mathbf{i} + 2\mathbf{j} + 5\mathbf{k}$. (See Figure 4.)

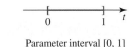
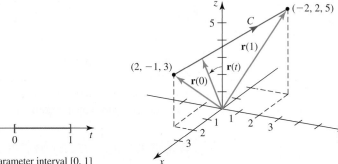

FIGURE 4
As t increases from 0 to 1, the tip of $\mathbf{r}(t)$ traces the straight line segment from $(2, -1, 3)$ to $(-2, 2, 5)$.

Parameter interval $[0, 1]$

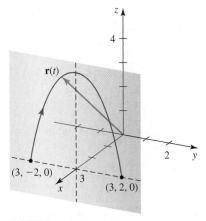

FIGURE 5
As t increases from -2 to 2, the terminal point of $\mathbf{r}(t)$ traces the part of the parabola lying in the plane $x = 3$ from the point $(3, -2, 0)$ to the point $(3, 2, 0)$.

EXAMPLE 4 Sketch the curve defined by the vector function

$$\mathbf{r}(t) = 3\mathbf{i} + t\mathbf{j} + (4 - t^2)\mathbf{k} \qquad -2 \le t \le 2$$

Solution The parametric equations for the curve are

$$x = 3 \qquad y = t \qquad z = 4 - t^2$$

Eliminating t from the second and third equations, we obtain

$$z = 4 - y^2$$

Since the x-coordinate of any point on the curve must always be 3, as implied by the equation $x = 3$, we conclude that the desired curve is contained in the parabola $z = 4 - y^2$, which lies in the plane $x = 3$. In fact, as t runs from -2 to 2, the terminal point of $\mathbf{r}$ traces the part of the parabola starting at the point $(3, -2, 0)$ [since $\mathbf{r}(-2) = 3\mathbf{i} - 2\mathbf{j}$] and ending at the point $(3, 2, 0)$ [since $\mathbf{r}(2) = 3\mathbf{i} + 2\mathbf{j}$], as shown in Figure 5.

EXAMPLE 5 Sketch the curve defined by the vector function

$$\mathbf{r}(t) = 2 \cos t\,\mathbf{i} + 2 \sin t\,\mathbf{j} + t\mathbf{k} \qquad 0 \le t \le 2\pi$$

Solution The parametric equations for the curve are

$$x = 2 \cos t \qquad y = 2 \sin t \qquad z = t$$

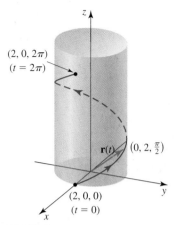

FIGURE 6

As t increases from 0 to 2π, the terminal point of $\mathbf{r}(t)$ traces the helix beginning at $(2, 0, 0)$ and terminating at $(2, 0, 2\pi)$.

From the first two equations we obtain

$$\left(\frac{x}{2}\right)^2 + \left(\frac{y}{2}\right)^2 = \cos^2 t + \sin^2 t = 1 \qquad \text{or} \qquad x^2 + y^2 = 4$$

This says that the curve lies on the right circular cylinder of radius 2, whose axis is the z-axis. At $t = 0$, $\mathbf{r}(0) = 2\mathbf{i}$, and this gives $(2, 0, 0)$ as the starting point of the curve. Since $z = t$, the z-coordinate of the point on the curve increases (linearly) as t increases, and the curve spirals upward around the cylinder in a counterclockwise direction, terminating at the point $(2, 0, 2\pi)$ [$\mathbf{r}(2\pi) = 2\mathbf{i} + 2\pi\mathbf{k}$]. The curve, called a **helix,** is shown in Figure 6. ■

EXAMPLE 6 Find a vector function that describes the curve of intersection of the cylinder $x^2 + y^2 = 4$ and the plane $x + y + 2z = 4$. (See Figure 7.)

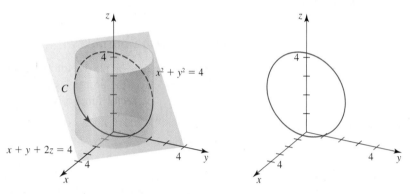

FIGURE 7   (**a**) Intersection of the plane and the cylinder   (**b**) Curve of intersection

Solution If $P(x, y, z)$ is any point on the curve of intersection C, then the x- and y-coordinates lie on the right circular cylinder of radius 2 and axis lying along the z-axis. Therefore,

$$x = 2 \cos t \qquad \text{and} \qquad y = 2 \sin t$$

To find the z-coordinate of the point, we substitute these values of x and y into the equation of the plane, obtaining

$$2 \cos t + 2 \sin t + 2z = 4 \qquad \text{or} \qquad z = 2 - \cos t - \sin t$$

So a required vector function is

$$\mathbf{r}(t) = 2 \cos t\mathbf{i} + 2 \sin t\mathbf{j} + (2 - \cos t - \sin t)\mathbf{k} \qquad 0 \le t \le 2\pi \qquad ■$$

You might have noticed that the space curves in Examples 4, 5, and 6 are relatively easy to sketch by hand. This is partly because they are relatively simple and partly because they lie in a plane. For more complicated curves we turn to computers.

EXAMPLE 7 Use a computer to plot the curve represented by

$$\mathbf{r}(t) = (0.2 \sin 20t + 0.8) \cos t\mathbf{i} + (0.2 \sin 20t + 0.8) \sin t\mathbf{j} + 0.2 \cos 20t\mathbf{k}$$

$$0 \le t \le 2\pi$$

Solution   The curve is shown in Figure 8.

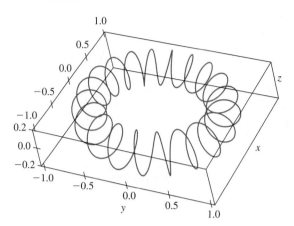

FIGURE 8
The curve in Example 7 is
called a **toroidal spiral**
because it lies on a torus.

Limits and Continuity

Because the range of the vector function $\mathbf{r}$ is a subset of vectors in two- or three-dimensional space, the properties of vectors given in Chapter 11 can be used to study the properties of vector functions. For example, we add two vector functions componentwise. Thus, if

$$\mathbf{r}_1(t) = f_1(t)\mathbf{i} + g_1(t)\mathbf{j} + h_1(t)\mathbf{k} \qquad \text{and} \qquad \mathbf{r}_2(t) = f_2(t)\mathbf{i} + g_2(t)\mathbf{j} + h_2(t)\mathbf{k}$$

then

$$(\mathbf{r}_1 + \mathbf{r}_2)(t) = \mathbf{r}_1(t) + \mathbf{r}_2(t) = [f_1(t) + f_2(t)]\mathbf{i} + [g_1(t) + g_2(t)]\mathbf{j} + [h_1(t) + h_2(t)]\mathbf{k}$$

Similarly, if c is a scalar, then the scalar multiple of $\mathbf{r}$ by c is

$$(c\mathbf{r})(t) = c\mathbf{r}(t) = cf(t)\mathbf{i} + cg(t)\mathbf{j} + ch(t)\mathbf{k}$$

Next, because the components f, g, and h of the vector function $\mathbf{r}$ are real-valued functions, we can investigate the notions of limits and continuity involving $\mathbf{r}$ using the properties of such functions. As you might expect, the limit of $\mathbf{r}(t)$ is defined in terms of the limits of its component functions.

> **DEFINITION   The Limit of a Vector Function**
> Let $\mathbf{r}$ be a function defined by $\mathbf{r}(t) = f(t)\mathbf{i} + g(t)\mathbf{j} + h(t)\mathbf{k}$. Then
>
> $$\lim_{t \to a} \mathbf{r}(t) = \left[\lim_{t \to a} f(t) \right]\mathbf{i} + \left[\lim_{t \to a} g(t) \right]\mathbf{j} + \left[\lim_{t \to a} h(t) \right]\mathbf{k}$$
>
> provided that the limits of the component functions exist.

To obtain a geometric interpretation of $\lim_{t \to a} \mathbf{r}(t)$, suppose that the limit exists. Let $\lim_{t \to a} f(t) = L_1$, $\lim_{t \to a} g(t) = L_2$, and $\lim_{t \to a} h(t) = L_3$, and let $\mathbf{L} = L_1\mathbf{i} + L_2\mathbf{j} + L_3\mathbf{k}$. Then, by definition, $\lim_{t \to a} \mathbf{r}(t) = \mathbf{L}$. This says that as t approaches a, the vector $\mathbf{r}(t)$ approaches the constant vector $\mathbf{L}$. (See Figure 9.)

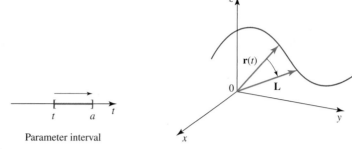

Parameter interval

FIGURE 9

$\lim_{t \to a} \mathbf{r}(t) = \mathbf{L}$ means that as t approaches a, $\mathbf{r}(t)$ approaches $\mathbf{L}$.

EXAMPLE 8 Find $\lim_{t \to 0} \mathbf{r}(t)$, where $\mathbf{r}(t) = \sqrt{t + 2}\,\mathbf{i} + t \cos 2t\,\mathbf{j} + e^{-t}\mathbf{k}$.

Solution

$$\lim_{t \to 0} \mathbf{r}(t) = \left[\lim_{t \to 0} \sqrt{t + 2}\right]\mathbf{i} + \left[\lim_{t \to 0} t \cos 2t\right]\mathbf{j} + \left[\lim_{t \to 0} e^{-t}\right]\mathbf{k}$$

$$= \sqrt{2}\,\mathbf{i} + \mathbf{k}$$

The notion of continuity is extended to vector functions via the following definition.

DEFINITION   Continuity of a Vector Function

A vector function $\mathbf{r}$ is continuous at a if

$$\lim_{t \to a} \mathbf{r}(t) = \mathbf{r}(a)$$

A vector function $\mathbf{r}$ **is continuous on an interval** I if it is continuous at every number in I.

It follows from this definition that a vector function is continuous at a if and only if each of its component functions is continuous at a.

EXAMPLE 9 Find the interval(s) on which the vector function $\mathbf{r}$ defined by

$$\mathbf{r}(t) = \sqrt{t}\,\mathbf{i} + \left(\frac{1}{t^2 - 1}\right)\mathbf{j} + \ln t\,\mathbf{k}$$

is continuous.

Solution   The component functions of $\mathbf{r}$ are $f(t) = \sqrt{t}$, $g(t) = 1/(t^2 - 1)$, and $h(t) = \ln t$. Observe that f is continuous for $t \geq 0$, g is continuous for all values of t except $t = \pm 1$, and h is continuous for $t > 0$. Therefore, $\mathbf{r}$ is continuous on the intervals $(0, 1)$ and $(1, \infty)$.

12.1 CONCEPT QUESTIONS

1. **a.** What is a vector-valued function?
   b. Give an example of a vector function. What is the parameter interval of the function that you picked?
2. Let $\mathbf{r}(t)$ be a vector function defined by
   $\mathbf{r}(t) = \langle f(t), g(t), h(t) \rangle$.
   a. Define $\lim_{t \to a} \mathbf{r}(t)$.
   b. Give an example of a vector function $\mathbf{r}(t)$ such that $\lim_{t \to 1} \mathbf{r}(t)$ does not exist.

3. **a.** What does it mean for a vector function $\mathbf{r}(t)$ to be continuous at a? Continuous on an interval I?
   b. Give an example of a function $\mathbf{r}(t)$ that is defined on the interval $(-1, 1)$ but fails to be continuous at 0.

12.1 EXERCISES

In Exercises 1–6, find the domain of the vector function.

1. $\mathbf{r}(t) = t\mathbf{i} + \dfrac{1}{t}\mathbf{j}$

2. $\mathbf{r}(t) = \cos t\mathbf{i} + 2 \sin t\mathbf{j} + \sqrt{t+1}\,\mathbf{k}$

3. $\mathbf{r}(t) = \left\langle \sqrt{t}, \dfrac{1}{t-1}, \ln t \right\rangle$ 4. $\mathbf{r}(t) = \left\langle \dfrac{1}{\sqrt{t-1}}, e^{-t} \right\rangle$

5. $\mathbf{r}(t) = \ln t\mathbf{i} + \cosh t\mathbf{j} + \tanh t\mathbf{k}$

6. $\mathbf{r}(t) = \sqrt[3]{t}\,\mathbf{i} + e^{1/t}\mathbf{j} + \dfrac{1}{t+2}\mathbf{k}$

In Exercises 7–12, match the vector functions with the curves labeled (a)–(f). Explain your choices.

7. $\mathbf{r}(t) = t^2\mathbf{i} + t^2\mathbf{j} + t^2\mathbf{k}$

8. $\mathbf{r}(t) = 2 \cos 2t\mathbf{i} + t\mathbf{j} + 2 \sin 2t\mathbf{k}$

9. $\mathbf{r}(t) = t\mathbf{i} + t\mathbf{j} + \left(\dfrac{1}{t^2+1}\right)\mathbf{k}$

10. $\mathbf{r}(t) = t \sin t\mathbf{i} + t \cos t\mathbf{j} + t\mathbf{k}, \quad 0 \le t \le 10\pi$

11. $\mathbf{r}(t) = 2 \cos t\mathbf{i} + 3 \sin t\mathbf{j} + e^{0.1t}\mathbf{k}, \quad t \ge 0$

12. $\mathbf{r}(t) = \cos t\mathbf{i} + \sin t\mathbf{j} + \sin 3t\mathbf{k}$

(a)   **(b)**

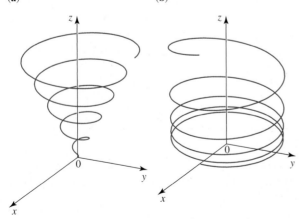

(c)   **(d)**

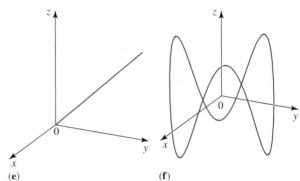

(e)   **(f)**

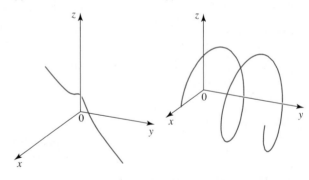

In Exercises 13–26, sketch the curve with the given vector function, and indicate the orientation of the curve.

13. $\mathbf{r}(t) = 2t\mathbf{i} + (3t+1)\mathbf{j}, \quad -1 \le t \le 2$

14. $\mathbf{r}(t) = \sqrt{t}\,\mathbf{i} + (4-t)\mathbf{j}, \quad t \ge 0$

15. $\mathbf{r}(t) = \langle t^2, t^3 \rangle, \quad -1 \le t \le 2$

16. $\mathbf{r}(t) = 2 \sin t\mathbf{i} + 3 \cos t\mathbf{j}, \quad 0 \le t \le 2\pi$

17. $\mathbf{r}(t) = e^t\mathbf{i} + e^{2t}\mathbf{j}, \quad -\infty < t < \infty$

18. $\mathbf{r}(t) = \langle 1 + 2 \cos t, 3 + 2 \sin t \rangle, \quad 0 \le t \le 2\pi$

19. $\mathbf{r}(t) = (1+t)\mathbf{i} + (2-t)\mathbf{j} + (3-2t)\mathbf{k}, \quad -\infty < t < \infty$

20. $\mathbf{r}(t) = (2+t)\mathbf{i} + (3-2t)\mathbf{j} + (2+4t)\mathbf{k}, \quad 0 \le t \le 1$

21. $\mathbf{r}(t) = \langle t, t^2, t^3 \rangle, \quad t \ge 0$

22. $\mathbf{r}(t) = 2 \cos t\mathbf{i} + 4 \sin t\mathbf{j} + 3\mathbf{k}, \quad 0 \leq t \leq 2\pi$

23. $\mathbf{r}(t) = 2 \cos t\mathbf{i} + 4 \sin t\mathbf{j} + t\mathbf{k}, \quad 0 \leq t \leq 2\pi$

24. $\mathbf{r}(t) = t\mathbf{i} + 2t\mathbf{j} + \sin 2t\mathbf{k}, \quad -\infty < t < \infty$

25. $\mathbf{r}(t) = \langle t \cos t, t \sin t, t \rangle, \quad -\infty < t < \infty$
 Hint: Show that it lies on a cone.

26. $\mathbf{r}(t) = e^t \cos t\mathbf{i} + e^t \sin t\mathbf{j} + e^t\mathbf{k}, \quad -\infty < t < \infty$

cas *In Exercises 27–30 use a computer to graph the curve described by the function.*

27. $\mathbf{r}(t) = 2 \sin \pi t\mathbf{i} + 3 \cos \pi t\mathbf{j} + 0.1t\mathbf{k}, \quad 0 \leq t \leq 10$

28. $\mathbf{r}(t) = (t^2 - t + 1)\mathbf{i} + (t^2 + 1)\mathbf{j} + t^3\mathbf{k}, \quad -3 \leq t \leq 3$

29. $\mathbf{r}(t) = \sin 3t \cos t\mathbf{i} + \sin 3t \sin t\mathbf{j} + \dfrac{t}{2\pi}\mathbf{k}, \quad -2\pi \leq t \leq 2\pi$

 (rotating sine wave)

30. $\mathbf{r}(t) = \dfrac{1}{2} \sin t\mathbf{i} + \dfrac{1}{2} \cos t\mathbf{j} + \dfrac{t^2}{100\pi^2}\mathbf{k}, \quad 0 \leq t \leq 10\pi$

 (Fresnel integral spiral)

31. a. Show that the curve

$$\mathbf{r}(t) = \sqrt{1 - 0.09 \cos^2 10t} \, \cos t\mathbf{i}$$
$$+ \sqrt{1 - 0.09 \cos^2 10t} \, \sin t\mathbf{j} + 0.3 \cos 10t\mathbf{k}$$

 lies on a sphere.

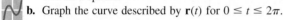 **b.** Graph the curve described by $\mathbf{r}(t)$ for $0 \leq t \leq 2\pi$.

32. a. Show that the curve

$$\mathbf{r}(t) = (1 + \cos 12t) \cos t\mathbf{i} + (1 + \cos 12t) \sin t\mathbf{j}$$
$$+ (1 + \cos 12t)\mathbf{k}$$

 lies on a cone.

 b. Graph the curve described by $\mathbf{r}(t)$ for $0 \leq t \leq 2\pi$.

In Exercises 33–36, find a vector function describing the curve of intersection of the two surfaces.

33. The cylinder $x^2 + y^2 = 1$ and the plane $x + y + 2z = 1$

34. The cylinder $x^2 + y^2 = 4$ and the surface $z = xy$

35. The cone $z = \sqrt{x^2 + y^2}$ and the plane $x + y + z = 1$

36. The paraboloid $z = x^2 + y^2$ and the sphere $x^2 + y^2 + z^2 = 1$

In Exercises 37–42, find the given limit.

37. $\lim\limits_{t \to 0} [(t^2 + 1)\mathbf{i} + \cos t\mathbf{j} - 3\mathbf{k}]$

38. $\lim\limits_{t \to 0} \left\langle e^{-t}, \dfrac{\sin t}{t}, \cos t \right\rangle$

39. $\lim\limits_{t \to 2} \left[\sqrt{t}\mathbf{i} + \left(\dfrac{t^2 - 4}{t - 2} \right)\mathbf{j} + \left(\dfrac{t}{t^2 + 1} \right)\mathbf{k} \right]$

40. $\lim\limits_{t \to 0^+} \left[\cos t\mathbf{i} + \dfrac{\tan t}{t}\mathbf{j} + t \ln t\mathbf{k} \right]$

41. $\lim\limits_{t \to \infty} \left\langle e^{-t}, \dfrac{1}{t}, \dfrac{2t^2}{t^2 + 1} \right\rangle$

42. $\lim\limits_{t \to -\infty} \left[\left(\dfrac{t - 1}{2t + 1} \right)\mathbf{i} + e^{2t}\mathbf{j} + \tan^{-1} t\mathbf{k} \right]$

In Exercises 43–48 find the interval(s) on which the vector function is continuous.

43. $\mathbf{r}(t) = \sqrt{t + 1}\mathbf{i} + \dfrac{1}{t}\mathbf{j}$

44. $\mathbf{r}(t) = \sin t\mathbf{i} + \cos t\mathbf{j} - \tan^{-1} t\mathbf{k}$

45. $\mathbf{r}(t) = \left\langle \dfrac{\cos t - 1}{t}, \dfrac{\sqrt{t}}{1 + 2t}, te^{-1/t} \right\rangle$

46. $\mathbf{r}(t) = \left(\dfrac{2t}{t^2 - 4} \right)\mathbf{i} + \sin^{-1} t\mathbf{j} + \sqrt[3]{t}\mathbf{k}$

47. $\mathbf{r}(t) = e^{-t}\mathbf{i} + \cos\sqrt{4 - t}\mathbf{j} + \dfrac{1}{t^2 - 1}\mathbf{k}$

48. $\mathbf{r}(t) = \dfrac{1}{\sqrt{t}}\mathbf{i} + \tan t\mathbf{j} + e^{-t} \cos t\mathbf{k}$

49. Trajectory of a Plane An airplane is circling an airport in a holding pattern. Suppose that the airport is located at the origin of a three-dimensional coordinate system and that the trajectory of the plane traveling at a constant speed is described by

$$\mathbf{r}(t) = 44{,}000 \cos 60t\mathbf{i} + 44{,}000 \sin 60t\mathbf{j} + 10{,}000\mathbf{k}$$

 where the distance is measured in feet and the time is measured in hours. What is the distance covered by the plane over a 2-min period?

50. Temperature at a Point Suppose that the temperature at a point (x, y, z) in 3-space is $T(x, y, z) = x^2 + 2y^2 + 3z^2$ and that the position of a particle at time t is described by $\mathbf{r}(t) = \langle t, t^2, e^t \rangle$. What is the temperature at the point occupied by the particle when $t = 1$?

In Exercises 51–54, suppose that $\mathbf{u}$ and $\mathbf{v}$ are vector functions such that $\lim_{t \to a} \mathbf{u}(t)$ and $\lim_{t \to a} \mathbf{v}(t)$ exist and c is a constant. Prove the given property.

51. $\lim\limits_{t \to a} [\mathbf{u}(t) + \mathbf{v}(t)] = \lim\limits_{t \to a} \mathbf{u}(t) + \lim\limits_{t \to a} \mathbf{v}(t)$

52. $\lim\limits_{t \to a} c\mathbf{u}(t) = c \lim\limits_{t \to a} \mathbf{u}(t)$

53. $\lim\limits_{t \to a} [\mathbf{u}(t) \cdot \mathbf{v}(t)] = \lim\limits_{t \to a} \mathbf{u}(t) \cdot \lim\limits_{t \to a} \mathbf{v}(t)$

54. $\lim\limits_{t \to a} [\mathbf{u}(t) \times \mathbf{v}(t)] = \lim\limits_{t \to a} \mathbf{u}(t) \times \lim\limits_{t \to a} \mathbf{v}(t)$

55. a. Prove that if $\mathbf{r}$ is a vector function that is continuous at a, then $|\mathbf{r}|$ is also continuous at a.

 b. Show that the converse is false by exhibiting a vector function $\mathbf{r}$ such that $|\mathbf{r}|$ is continuous at a but $\mathbf{r}$ is not continuous at a.

56. Evaluate

$$\lim_{h \to 0} \left\langle \frac{(t + h)^2 - t^2}{h}, \frac{\cos(t + h) - \cos t}{h}, \frac{e^{t+h} - e^t}{h} \right\rangle$$

57. Evaluate

$$\lim_{t \to 0} \left[\frac{\sin t}{t} \mathbf{i} + \frac{1 - \cos t}{t^2} \mathbf{j} + \frac{\ln(1 + t^2)}{\cos t - e^{-t}} \mathbf{k} \right]$$

58. a. Find a vector function describing the curve of intersection of the plane $x + y + 2z = 2$ and the paraboloid $z = x^2 + y^2$.
 b. Find the point(s) on the curve of part (a) that are closest to and farthest from the origin.

In Exercises 59–62, determine whether the statement is true or false. If it is true, explain why. If it is false, explain why or give an example to show that it is false.

59. The curve defined by $\mathbf{r}_1(t) = t^2\mathbf{i} + t^2\mathbf{j} + t^2\mathbf{k}$ is the same as the curve defined by $\mathbf{r}_2(\theta) = \theta\mathbf{i} + \theta\mathbf{j} + \theta\mathbf{k}$.

60. If f, g, and h are linear functions of t for t in $(-\infty, \infty)$, then $\mathbf{r}(t) = f(t)\mathbf{i} + g(t)\mathbf{j} + h(t)\mathbf{k}$ defines a line in 3-space.

61. The curve defined by $\mathbf{r}(t) = f(t)\mathbf{i} + g(t)\mathbf{j} + c\mathbf{k}$, where c is a constant, is a curve lying in the plane $z = c$.

62. If $\mathbf{r}$ is continuous on an interval I and if a is any number in I, then $\lim_{t \to a} \mathbf{r}(t) = \mathbf{r}(a)$.

12.2 Differentiation and Integration of Vector-Valued Functions

■ The Derivative of a Vector Function

The derivative of a vector function is defined in much the same way as the derivative of a real-valued function of a real variable.

DEFINITION **Derivative of a Vector Function**

The derivative of a **vector function** $\mathbf{r}$ is the vector function $\mathbf{r}'$ defined by

$$\mathbf{r}'(t) = \frac{d\mathbf{r}}{dt} = \lim_{h \to 0} \frac{\mathbf{r}(t + h) - \mathbf{r}(t)}{h}$$

provided that the limit exists.

To obtain a geometric interpretation of this derivative, let $\mathbf{r}$ be a vector function, and let C be the curve traced by the tip of $\mathbf{r}$. Let t be a fixed but otherwise arbitrary number in the parameter interval I. If $h > 0$, then the vector $\mathbf{r}(t + h) - \mathbf{r}(t)$ lies on the secant line passing through the points P and Q, the terminal points of the vectors $\mathbf{r}(t)$ and $\mathbf{r}(t + h)$, respectively. (See Figure 1.)

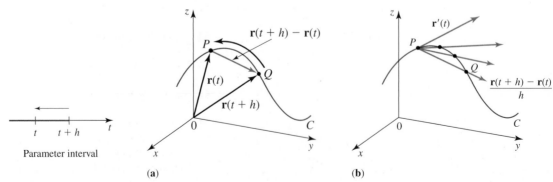

FIGURE 1
As h approaches 0, Q approaches P along C, and the vector $\dfrac{\mathbf{r}(t + h) - \mathbf{r}(t)}{h}$ approaches the tangent vector $\mathbf{r}'(t)$.

The vector $[\mathbf{r}(t + h) - \mathbf{r}(t)]/h$, which is a scalar multiple of $\mathbf{r}(t + h) - \mathbf{r}(t)$, also lies on the secant line. (See Figure 1b.) As h approaches 0, the number $t + h$ approaches t along the parameter interval, and the point Q, in turn, approaches the point P along the curve C. As a consequence, the vector $[\mathbf{r}(t + h) - \mathbf{r}(t)]/h$ approaches the fixed vector $\mathbf{r}'(t)$, which lies on the tangent line to the curve at P. In other words, the derivative $\mathbf{r}'$ of the vector $\mathbf{r}$ may be interpreted as the **tangent vector** to the curve defined by $\mathbf{r}$ at the point P, provided that $\mathbf{r}'(t) \neq \mathbf{0}$. If we divide $\mathbf{r}'(t)$ by its length, we obtain the **unit tangent vector**

$$\mathbf{T}(t) = \frac{\mathbf{r}'(t)}{|\mathbf{r}'(t)|}$$

which has unit length and the direction of $\mathbf{r}'$.

The following theorem tells us that the derivative $\mathbf{r}'$ of a vector function can be found by differentiating the components of $\mathbf{r}$.

THEOREM 1 **Differentiation of Vector Functions**

Let $\mathbf{r}(t) = f(t)\mathbf{i} + g(t)\mathbf{j} + h(t)\mathbf{k}$, where f, g, and h are differentiable functions of t. Then

$$\mathbf{r}'(t) = f'(t)\mathbf{i} + g'(t)\mathbf{j} + h'(t)\mathbf{k}$$

PROOF We compute

$$\mathbf{r}'(t) = \lim_{\Delta t \to 0} \frac{\mathbf{r}(t + \Delta t) - \mathbf{r}(t)}{\Delta t} \qquad \text{We use } \Delta t \text{ instead of } h \text{ so as not to confuse the increment of } t \text{ with the component function } h.$$

$$= \lim_{\Delta t \to 0} \left[\frac{f(t + \Delta t)\mathbf{i} + g(t + \Delta t)\mathbf{j} + h(t + \Delta t)\mathbf{k} - [f(t)\mathbf{i} + g(t)\mathbf{j} + h(t)\mathbf{k}]}{\Delta t} \right]$$

$$= \lim_{\Delta t \to 0} \left[\frac{f(t + \Delta t) - f(t)}{\Delta t}\mathbf{i} + \frac{g(t + \Delta t) - g(t)}{\Delta t}\mathbf{j} + \frac{h(t + \Delta t) - h(t)}{\Delta t}\mathbf{k} \right]$$

$$= \left[\lim_{\Delta t \to 0} \frac{f(t + \Delta t) - f(t)}{\Delta t} \right]\mathbf{i} + \left[\lim_{\Delta t \to 0} \frac{g(t + \Delta t) - g(t)}{\Delta t} \right]\mathbf{j} + \left[\lim_{\Delta t \to 0} \frac{h(t + \Delta t) - h(t)}{\Delta t} \right]\mathbf{k}$$

$$= f'(t)\mathbf{i} + g'(t)\mathbf{j} + h'(t)\mathbf{k} \qquad \blacksquare$$

EXAMPLE 1

a. Find the derivative of $\mathbf{r}(t) = (t^2 + 1)\mathbf{i} + e^{-t}\mathbf{j} - \sin 2t\mathbf{k}$.
b. Find the point of tangency and the unit tangent vector at the point on the curve corresponding to $t = 0$.

Solution
a. Using Theorem 1, we obtain

$$\mathbf{r}'(t) = 2t\mathbf{i} - e^{-t}\mathbf{j} - 2\cos 2t\mathbf{k}$$

b. Since $\mathbf{r}(0) = \mathbf{i} + \mathbf{j}$, we see that the point of tangency is $(1, 1, 0)$. Next, since $\mathbf{r}'(0) = -\mathbf{j} - 2\mathbf{k}$, we find the unit tangent vector at $(1, 1, 0)$ to be

$$\mathbf{T}(0) = \frac{\mathbf{r}'(0)}{|\mathbf{r}'(0)|} = \frac{-\mathbf{j} - 2\mathbf{k}}{\sqrt{1 + 4}} = -\frac{1}{\sqrt{5}}\mathbf{j} - \frac{2}{\sqrt{5}}\mathbf{k} \qquad \blacksquare$$

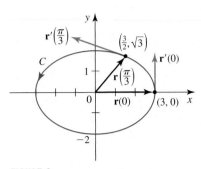

FIGURE 2
The vectors $\mathbf{r}'(0)$ and $\mathbf{r}'(\pi/3)$ are tangent to the curve at the points $(3, 0)$ and $\left(\frac{3}{2}, \sqrt{3}\right)$, respectively.

EXAMPLE 2 Find the tangent vectors to the plane curve C defined by the vector function $\mathbf{r}(t) = 3 \cos t\mathbf{i} + 2 \sin t\mathbf{j}$ at the points where $t = 0$ and $t = \pi/3$. Make a sketch of C, and display the position vectors $\mathbf{r}(0)$ and $\mathbf{r}(\pi/3)$ and the tangent vectors $\mathbf{r}'(0)$ and $\mathbf{r}'(\pi/3)$.

Solution The tangent vector to the curve C at any point is given by

$$\mathbf{r}'(t) = -3 \sin t\mathbf{i} + 2 \cos t\mathbf{j}$$

In particular, the tangent vectors at the points where $t = 0$ and $t = \pi/3$ are

$$\mathbf{r}'(0) = 2\mathbf{j} \quad \text{and} \quad \mathbf{r}'\left(\frac{\pi}{3}\right) = -\frac{3\sqrt{3}}{2}\mathbf{i} + \mathbf{j}$$

These vectors are shown emanating from their points of tangency at $(3, 0)$ and $\left(\frac{3}{2}, \sqrt{3}\right)$ in Figure 2. ■

EXAMPLE 3 Find parametric equations for the tangent line to the helix with parametric equations

$$x = 3 \cos t \quad y = 2 \sin t \quad z = t$$

at the point where $t = \pi/6$.

Solution The vector function that describes the helix is

$$\mathbf{r}(t) = 3 \cos t\mathbf{i} + 2 \sin t\mathbf{j} + t\mathbf{k}$$

The tangent vector at any point on the helix is

$$\mathbf{r}'(t) = -3 \sin t\mathbf{i} + 2 \cos t\mathbf{j} + \mathbf{k}$$

In particular, the tangent vector at the point $\left(\frac{3\sqrt{3}}{2}, 1, \frac{\pi}{6}\right)$, where $t = \pi/6$, is

$$\mathbf{r}'\left(\frac{\pi}{6}\right) = -\frac{3}{2}\mathbf{i} + \sqrt{3}\mathbf{j} + \mathbf{k}$$

Finally, we observe that the required tangent line passes through the point $\left(\frac{3\sqrt{3}}{2}, 1, \frac{\pi}{6}\right)$ and has the same direction as the tangent vector $\mathbf{r}'(\pi/6)$. Using Equation (1) of Section 11.5, we see that the parametric equations of this line are

$$x = \frac{3\sqrt{3}}{2} - \frac{3}{2}t, \quad y = 1 + \sqrt{3}t, \quad \text{and} \quad z = \frac{\pi}{6} + t \quad ■$$

■ Higher-Order Derivatives

Higher-order derivatives of vector functions are obtained by successive differentiation of the lower-order derivatives of the function. For example, the **second derivative** of $\mathbf{r}(t)$ is

$$\mathbf{r}''(t) = \frac{d}{dt}\mathbf{r}'(t) = f''(t)\mathbf{i} + g''(t)\mathbf{j} + h''(t)\mathbf{k}$$

EXAMPLE 4 Find $\mathbf{r}''(t)$ if $\mathbf{r}(t) = 2e^{3t}\mathbf{i} + \ln t\mathbf{j} + \sin t\mathbf{k}$.

Solution We have

$$\mathbf{r}'(t) = 6e^{3t}\mathbf{i} + \frac{1}{t}\mathbf{j} + \cos t\mathbf{k}$$

and

$$\mathbf{r}''(t) = 18e^{3t}\mathbf{i} - \frac{1}{t^2}\mathbf{j} - \sin t\mathbf{k} \qquad \blacksquare$$

■ Rules of Differentiation

The following theorem gives the rules of differentiation for vector functions. As you might expect, some of the rules are similar to the differentiation rules of Chapter 2.

THEOREM 2   Rules of Differentiation

Suppose that **u** and **v** are differentiable vector functions, f is a differentiable real-valued function, and c is a scalar. Then

1. $\dfrac{d}{dt}[\mathbf{u}(t) \pm \mathbf{v}(t)] = \mathbf{u}'(t) \pm \mathbf{v}'(t)$

2. $\dfrac{d}{dt}[c\mathbf{u}(t)] = c\mathbf{u}'(t)$

3. $\dfrac{d}{dt}[f(t)\mathbf{u}(t)] = f'(t)\mathbf{u}(t) + f(t)\mathbf{u}'(t)$

4. $\dfrac{d}{dt}[\mathbf{u}(t) \cdot \mathbf{v}(t)] = \mathbf{u}'(t) \cdot \mathbf{v}(t) + \mathbf{u}(t) \cdot \mathbf{v}'(t)$

5. $\dfrac{d}{dt}[\mathbf{u}(t) \times \mathbf{v}(t)] = \mathbf{u}'(t) \times \mathbf{v}(t) + \mathbf{u}(t) \times \mathbf{v}'(t)$

6. $\dfrac{d}{dt}[\mathbf{u}(f(t))] = \mathbf{u}'(f(t))f'(t)$ \qquad Chain Rule

---

We will prove Rule 4 and leave the proofs of the other rules as exercises.

PROOF Let

$$\mathbf{u}(t) = f_1(t)\mathbf{i} + g_1(t)\mathbf{j} + h_1(t)\mathbf{k} \qquad \text{and} \qquad \mathbf{v}(t) = f_2(t)\mathbf{i} + g_2(t)\mathbf{j} + h_2(t)\mathbf{k}$$

Then

$$\mathbf{u}(t) \cdot \mathbf{v}(t) = f_1(t)f_2(t) + g_1(t)g_2(t) + h_1(t)h_2(t)$$

Therefore

$$\begin{aligned}
\frac{d}{dt}[\mathbf{u}(t) \cdot \mathbf{v}(t)] &= [f_1'(t)f_2(t) + g_1'(t)g_2(t) + h_1'(t)h_2(t)] \\
&\quad + [f_1(t)f_2'(t) + g_1(t)g_2'(t) + h_1(t)h_2'(t)] \\
&= \mathbf{u}'(t) \cdot \mathbf{v}(t) + \mathbf{u}(t) \cdot \mathbf{v}'(t) \qquad \blacksquare
\end{aligned}$$

EXAMPLE 5 Suppose that **v** is a differentiable vector function of constant length c. Show that $\mathbf{v} \cdot \mathbf{v}' = 0$. In other words, the vector **v** and its tangent vector $\mathbf{v}'$ must be orthogonal.

Solution The condition on **v** implies that

$$\mathbf{v} \cdot \mathbf{v} = |\mathbf{v}|^2 = c^2$$

Differentiating both sides of this equation with respect to t and using Rule 4 of differentiation, we obtain

$$\frac{d}{dt} (\mathbf{v} \cdot \mathbf{v}) = \mathbf{v} \cdot \mathbf{v}' + \mathbf{v}' \cdot \mathbf{v} = \frac{d}{dt} (c^2) = 0$$

But $\mathbf{v}' \cdot \mathbf{v} = \mathbf{v} \cdot \mathbf{v}'$, so we have

$$2\mathbf{v} \cdot \mathbf{v}' = 0 \qquad \text{or} \qquad \mathbf{v} \cdot \mathbf{v}' = 0 \qquad \blacksquare$$

The result of Example 5 has the following geometric interpretation: If a curve lies on a sphere with center at the origin, then the tangent vector $\mathbf{r}'(t)$ is always perpendicular to the position vector $\mathbf{r}(t)$.

EXAMPLE 6 Let $\mathbf{r}(s) = 2 \cos 2s\mathbf{i} + 3 \sin 2s\mathbf{j} + 4s\mathbf{k}$, where $s = f(t) = t^2$. Find $\dfrac{d\mathbf{r}}{dt}$.

Solution Using the Chain Rule, we obtain

$$\frac{d}{dt} [\mathbf{r}(s)] = \frac{d}{ds} (2 \cos 2s\mathbf{i} + 3 \sin 2s\mathbf{j} + 4s\mathbf{k}) \left(\frac{ds}{dt} \right)$$

$$= (-4 \sin 2s\mathbf{i} + 6 \cos 2s\mathbf{j} + 4\mathbf{k})(2t)$$

$$= -8t \sin 2t^2\mathbf{i} + 12t \cos 2t^2\mathbf{j} + 8t\mathbf{k} \qquad \text{Replace } s \text{ by } t^2. \qquad \blacksquare$$

■ Integration of Vector Functions

As with the differentiation of vector functions, integration of vector functions is done component-wise, so we have the following definitions.

DEFINITIONS **Integration of Vector Functions**

Let $\mathbf{r}(t) = f(t)\mathbf{i} + g(t)\mathbf{j} + h(t)\mathbf{k}$, where f, g, and h are integrable. Then

1. The **indefinite integral of r** with respect to t is

$$\int \mathbf{r}(t) \, dt = \left[\int f(t) \, dt \right] \mathbf{i} + \left[\int g(t) \, dt \right] \mathbf{j} + \left[\int h(t) \, dt \right] \mathbf{k}$$

2. The **definite integral of r** over the interval $[a, b]$ is

$$\int_a^b \mathbf{r}(t) \, dt = \left[\int_a^b f(t) \, dt \right] \mathbf{i} + \left[\int_a^b g(t) \, dt \right] \mathbf{j} + \left[\int_a^b h(t) \, dt \right] \mathbf{k}$$

EXAMPLE 7 Find $\int \mathbf{r}(t) \, dt$ if $\mathbf{r}(t) = (t + 1)\mathbf{i} + \cos 2t\mathbf{j} + e^{3t}\mathbf{k}$.

Solution

$$\int \mathbf{r}(t) \, dt = \int [(t + 1)\mathbf{i} + \cos 2t\mathbf{j} + e^{3t}\mathbf{k}] \, dt$$

$$= \left[\int (t + 1) \, dt \right] \mathbf{i} + \left[\int \cos 2t \, dt \right] \mathbf{j} + \left[\int e^{3t} \, dt \right] \mathbf{k}$$

$$= \left(\frac{1}{2} t^2 + t + C_1 \right) \mathbf{i} + \left(\frac{1}{2} \sin 2t + C_2 \right) \mathbf{j} + \left(\frac{1}{3} e^{3t} + C_3 \right) \mathbf{k}$$

where C_1, C_2, and C_3 are constants of integration. We can rewrite the last expression as

$$\left(\frac{1}{2}t^2 + t\right)\mathbf{i} + \frac{1}{2}\sin 2t\mathbf{j} + \frac{1}{3}e^{3t}\mathbf{k} + C_1\mathbf{i} + C_2\mathbf{j} + C_3\mathbf{k}$$

or, upon letting $\mathbf{C} = C_1\mathbf{i} + C_2\mathbf{j} + C_3\mathbf{k}$,

$$\int \mathbf{r}(t)\, dt = \left(\frac{1}{2}t^2 + t\right)\mathbf{i} + \frac{1}{2}\sin 2t\mathbf{j} + \frac{1}{3}e^{3t}\mathbf{k} + \mathbf{C}$$

where $\mathbf{C}$ is a constant (vector) of integration. ∎

Note In general, the indefinite integral of $\mathbf{r}$ can be written as

$$\int \mathbf{r}(t)\, dt = \mathbf{R}(t) + \mathbf{C}$$

where $\mathbf{C}$ is an arbitrary constant vector and $\mathbf{R}'(t) = \mathbf{r}(t)$. ∎

EXAMPLE 8 Find the antiderivative of $\mathbf{r}'(t) = \cos t\mathbf{i} + e^{-t}\mathbf{j} + \sqrt{t}\mathbf{k}$ satisfying the initial condition $\mathbf{r}(0) = \mathbf{i} + 2\mathbf{j} + 3\mathbf{k}$.

Solution We have

$$\mathbf{r}(t) = \int \mathbf{r}'(t)\, dt = \int (\cos t\mathbf{i} + e^{-t}\mathbf{j} + t^{1/2}\mathbf{k})\, dt$$

$$= \sin t\mathbf{i} - e^{-t}\mathbf{j} + \frac{2}{3}t^{3/2}\mathbf{k} + \mathbf{C}$$

where $\mathbf{C}$ is a constant (vector) of integration. To determine $\mathbf{C}$, we use the condition $\mathbf{r}(0) = \mathbf{i} + 2\mathbf{j} + 3\mathbf{k}$ to obtain

$$\mathbf{r}(0) = 0\mathbf{i} - \mathbf{j} + 0\mathbf{k} + \mathbf{C} = \mathbf{i} + 2\mathbf{j} + 3\mathbf{k}$$

from which we find $\mathbf{C} = \mathbf{i} + 3\mathbf{j} + 3\mathbf{k}$. Therefore,

$$\mathbf{r}(t) = \sin t\mathbf{i} - e^{-t}\mathbf{j} + \frac{2}{3}t^{3/2}\mathbf{k} + \mathbf{i} + 3\mathbf{j} + 3\mathbf{k}$$

$$= (1 + \sin t)\mathbf{i} + (3 - e^{-t})\mathbf{j} + \left(3 + \frac{2}{3}t^{3/2}\right)\mathbf{k} \qquad ∎$$

EXAMPLE 9 Evaluate $\int_0^1 \mathbf{r}(t)\, dt$ if $\mathbf{r}(t) = t^2\mathbf{i} + \dfrac{1}{t + 1}\mathbf{j} + e^{-t}\mathbf{k}$.

Solution

$$\int_0^1 \mathbf{r}(t)\, dt = \int_0^1 \left(t^2\mathbf{i} + \frac{1}{t + 1}\mathbf{j} + e^{-t}\mathbf{k}\right)dt$$

$$= \left[\int_0^1 t^2\, dt\right]\mathbf{i} + \left[\int_0^1 \frac{1}{t + 1}\, dt\right]\mathbf{j} + \left[\int_0^1 e^{-t}\, dt\right]\mathbf{k}$$

$$= \left[\frac{1}{3}t^3\right]_0^1 \mathbf{i} + \left[\ln(t + 1)\right]_0^1 \mathbf{j} + \left[-e^{-t}\right]_0^1 \mathbf{k}$$

$$= \frac{1}{3}\mathbf{i} + \ln 2\mathbf{j} + \left(1 - \frac{1}{e}\right)\mathbf{k} \qquad ∎$$

12.2 CONCEPT QUESTIONS

1. **a.** What is the derivative of a vector function?
   b. If $\mathbf{r}(t) = f(t)\mathbf{i} + g(t)\mathbf{j} + h(t)\mathbf{k}$, what is $\mathbf{r}'(t)$?
   c. Give an example of a function $\mathbf{r}(t)$ such that $\mathbf{r}'(0)$ does not exist.
2. If $\mathbf{w}(t) = \mathbf{u}(f(t)) \times \mathbf{v}(f(t))$, what is $\mathbf{w}'(t)$? Assume that $\mathbf{u}$, $\mathbf{v}$, and f are all differentiable.

3. Let $\mathbf{r}(t) = f(t)\mathbf{i} + g(t)\mathbf{j} + h(t)\mathbf{k}$.
   a. What is the indefinite integral of $\mathbf{r}$ with respect to t?
   b. What is the definite integral of $\mathbf{r}$ over the interval $[a, b]$?

12.2 EXERCISES

In Exercises 1–8, find $\mathbf{r}'(t)$ and $\mathbf{r}''(t)$.

1. $\mathbf{r}(t) = t\mathbf{i} + t^2\mathbf{j} + t^3\mathbf{k}$

2. $\mathbf{r}(t) = \sqrt{t}\mathbf{i} + \dfrac{1}{t}\mathbf{j} + \ln t\mathbf{k}$

3. $\mathbf{r}(t) = \langle t^2 - 1, \sqrt{t^2 + 1}\rangle$

4. $\mathbf{r}(t) = \langle t\cos t, t\sin t, \tan t\rangle$

5. $\mathbf{r}(t) = \langle t\cos t - \sin t, t\sin t + \cos t\rangle$

6. $\mathbf{r}(t) = e^{-t}\mathbf{i} + te^t\mathbf{j} + e^{-2t}\mathbf{k}$

7. $\mathbf{r}(t) = e^{-t}\sin t\mathbf{i} + e^{-t}\cos t\mathbf{j} + \tan^{-1}t\mathbf{k}$

8. $\mathbf{r}(t) = \langle \sin^{-1}t, \sec t, \ln|t|\rangle$

In Exercises 9–16, (a) find $\mathbf{r}(a)$ and $\mathbf{r}'(a)$ at the given value of a. (b) Sketch the curve defined by $\mathbf{r}$ and the vectors $\mathbf{r}(a)$ and $\mathbf{r}'(a)$ on the same set of axes.

9. $\mathbf{r}(t) = \sqrt{t}\mathbf{i} + (t - 4)\mathbf{j}; \quad a = 2$

10. $\mathbf{r}(t) = \sin t\mathbf{i} + \cos t\mathbf{j}; \quad a = \dfrac{\pi}{4}$

11. $\mathbf{r}(t) = \langle 4\cos t, 2\sin t\rangle; \quad a = \dfrac{\pi}{3}$

12. $\mathbf{r}(t) = t^2\mathbf{i} + t^3\mathbf{j}; \quad a = 1$

13. $\mathbf{r}(t) = (2 + 3t)\mathbf{i} + (1 - 2t)\mathbf{j}; \quad a = 1$

14. $\mathbf{r}(t) = \langle e^t, e^{-2t}\rangle; \quad a = 0$

15. $\mathbf{r}(t) = \sec t\mathbf{i} + 2\tan t\mathbf{j}; \quad a = \dfrac{\pi}{4}$

16. $\mathbf{r}(t) = b\cos^3 t\mathbf{i} + b\sin^3 t\mathbf{j}; \quad a = \dfrac{\pi}{4}$

In Exercises 17–20, find the unit tangent vector $\mathbf{T}(t)$ at the point corresponding to the given value of the parameter t.

17. $\mathbf{r}(t) = t\mathbf{i} + 2t\mathbf{j} + 3t\mathbf{k}; \quad t = 1$

18. $\mathbf{r}(t) = \langle e^t, te^{-t}, (t + 1)e^{2t}\rangle; \quad t = 0$

19. $\mathbf{r}(t) = 2\sin 2t\mathbf{i} + 3\cos 2t\mathbf{j} + 3\mathbf{k}; \quad t = \dfrac{\pi}{6}$

20. $\mathbf{r}(t) = t\sin t\mathbf{i} + t\cos t\mathbf{j} + t\mathbf{k}; \quad t = \dfrac{\pi}{2}$

In Exercises 21–26, find parametric equations for the tangent line to the curve with the given parametric equations at the point with the indicated value of t.

21. $x = t, \quad y = t^2, \quad z = t^3; \quad t = 1$

22. $x = 1 + t, \quad y = t^2 - 4, \quad z = \sqrt{t}; \quad t = 4$

23. $x = \sqrt{t + 2}, \quad y = \dfrac{1}{t + 1}, \quad z = \dfrac{2}{t^2 + 4}; \quad t = 2$

24. $x = 2\cos t, \quad y = t^2, \quad z = 2\sin t; \quad t = \dfrac{\pi}{4}$

25. $x = t\cos t, \quad y = t\sin t, \quad z = te^t; \quad t = \dfrac{\pi}{6}$

26. $x = e^{-t}\cos t, \quad y = e^{-t}\sin t, \quad z = \sin^{-1}t; \quad t = 0$

In Exercises 27–34, find or evaluate the integral.

27. $\displaystyle\int (t\mathbf{i} + 2t^2\mathbf{j} + 3\mathbf{k})\,dt$

28. $\displaystyle\int_0^1 (t\mathbf{i} + t^2\mathbf{j} + t^3\mathbf{k})\,dt$

29. $\displaystyle\int \left(\sqrt{t}\mathbf{i} + \dfrac{1}{t}\mathbf{j} - t^{3/2}\mathbf{k}\right)dt$

30. $\displaystyle\int_1^2 \left[\sqrt{t - 1}\mathbf{i} + \dfrac{1}{\sqrt{t}}\mathbf{j} + (2t - 1)^5\mathbf{k}\right]dt$

31. $\displaystyle\int (\sin 2t\mathbf{i} + \cos 2t\mathbf{j} + e^{-t}\mathbf{k})\,dt$

32. $\displaystyle\int (te^t\mathbf{i} + 2\mathbf{j} - \sec^2 t\mathbf{k})\,dt$

33. $\displaystyle\int (t\cos t\mathbf{i} + t\sin t^2\mathbf{j} - te^{t^2}\mathbf{k})\,dt$

34. $\displaystyle\int \left[\dfrac{1}{1 + t^2}\mathbf{i} + \dfrac{t}{1 + 2t^2}\mathbf{j} - \dfrac{1}{\sqrt{1 - t^2}}\mathbf{k}\right]dt$

In Exercises 35–40, find $\mathbf{r}(t)$ *satisfying the given conditions.*

35. $\mathbf{r}'(t) = 2\mathbf{i} + 4t\mathbf{j} - 6t^2\mathbf{k}; \quad \mathbf{r}(0) = \mathbf{i} + \mathbf{k}$

36. $\mathbf{r}'(t) = 2\sin 2t\mathbf{i} + 3\cos 2t\mathbf{j} + t\mathbf{k}; \quad \mathbf{r}(0) = \mathbf{i} + 2\mathbf{j} + \dfrac{1}{2}\mathbf{k}$

37. $\mathbf{r}'(t) = 2e^{2t}\mathbf{i} + 3e^{-t}\mathbf{j} + e^t\mathbf{k}; \quad \mathbf{r}(0) = \mathbf{i} - \mathbf{j} + \mathbf{k}$

38. $\mathbf{r}'(t) = \sqrt{t+1}\,\mathbf{i} + \dfrac{t}{t^2+1}\mathbf{j} + \dfrac{1}{t}\mathbf{k}; \quad \mathbf{r}(3) = \mathbf{i} + \mathbf{j} + 2\mathbf{k}$

39. $\mathbf{r}''(t) = \sqrt{t}\,\mathbf{i} + \sec^2 t\mathbf{j} + e^t\mathbf{k}; \quad \mathbf{r}'(0) = \mathbf{i} + \mathbf{k},$
$\mathbf{r}(0) = 2\mathbf{i} + \mathbf{j} - \mathbf{k}$

40. $\mathbf{r}''(t) = 3\cos 2t\mathbf{i} + 4\sin 2t\mathbf{j} + \mathbf{k}; \quad \mathbf{r}'(0) = \mathbf{i} + 2\mathbf{j},$
$\mathbf{r}(0) = 2\mathbf{i} + \mathbf{j} - \mathbf{k}$

In Exercises 41–46, let $\mathbf{u}(t) = t^2\mathbf{i} - 2t\mathbf{j} + 2\mathbf{k},$
$\mathbf{v}(t) = \cos t\mathbf{i} + \sin t\mathbf{j} + t^2\mathbf{k},$ *and* $f(t) = e^{2t}.$

41. Show that $\dfrac{d}{dt}[\mathbf{u}(t) + \mathbf{v}(t)] = \mathbf{u}'(t) + \mathbf{v}'(t).$

42. Show that $\dfrac{d}{dt}[3\mathbf{u}(t)] = 3\mathbf{u}'(t).$

43. Show that $\dfrac{d}{dt}[f(t)\mathbf{u}(t)] = f'(t)\mathbf{u}(t) + f(t)\mathbf{u}'(t).$

44. Show that $\dfrac{d}{dt}[\mathbf{u}(t) \cdot \mathbf{v}(t)] = \mathbf{u}'(t) \cdot \mathbf{v}(t) + \mathbf{u}(t) \cdot \mathbf{v}'(t).$

45. Show that $\dfrac{d}{dt}[\mathbf{u}(t) \times \mathbf{v}(t)] = \mathbf{u}'(t) \times \mathbf{v}(t) + \mathbf{u}(t) \times \mathbf{v}'(t).$

46. Show that $\dfrac{d}{dt}[\mathbf{u}(f(t))] = \mathbf{u}'[f(t)]f'(t).$

In Exercises 47–52, suppose $\mathbf{u}$ *and* $\mathbf{v}$ *are differentiable vector functions, f is a differentiable real-valued function, and c is a scalar. Prove each rule.*

47. $\dfrac{d}{dt}[\mathbf{u}(t) + \mathbf{v}(t)] = \mathbf{u}'(t) + \mathbf{v}'(t)$

48. $\dfrac{d}{dt}[\mathbf{u}(t) - \mathbf{v}(t)] = \mathbf{u}'(t) - \mathbf{v}'(t)$

49. $\dfrac{d}{dt}[c\mathbf{u}(t)] = c\mathbf{u}'(t)$

50. $\dfrac{d}{dt}[f(t)\mathbf{u}(t)] = f'(t)\mathbf{u}(t) + f(t)\mathbf{u}'(t)$

51. $\dfrac{d}{dt}[\mathbf{u}(t) \times \mathbf{v}(t)] = \mathbf{u}'(t) \times \mathbf{v}(t) + \mathbf{u}(t) \times \mathbf{v}'(t)$

52. $\dfrac{d}{dt}[\mathbf{u}(f(t))] = \mathbf{u}'(f(t))f'(t)$

53. Prove that $\dfrac{d}{dt}[\mathbf{r}(t) \times \mathbf{r}'(t)] = \mathbf{r}(t) \times \mathbf{r}''(t).$

54. Prove that
$$\dfrac{d}{dt}[\mathbf{r}(t) \cdot (\mathbf{u}(t) \times \mathbf{v}(t))]$$
$$= \mathbf{r}'(t) \cdot [\mathbf{u}(t) \times \mathbf{v}(t)] + \mathbf{r}(t) \cdot [\mathbf{u}'(t) \times \mathbf{v}(t)]$$
$$+ \mathbf{r}(t) \cdot [\mathbf{u}(t) \times \mathbf{v}'(t)]$$

In Exercises 55–58, find the indicated derivative.

55. $\dfrac{d}{dt}\left[\mathbf{r}(-t) + \mathbf{r}\left(\dfrac{1}{t}\right)\right]$

56. $\dfrac{d}{dt}[\mathbf{r}(2t) \cdot \mathbf{r}(t^2)]$

57. $\dfrac{d}{dt}[\mathbf{r}(t) \cdot (\mathbf{r}'(t) \times \mathbf{r}''(t))]$

58. $\dfrac{d}{dt}\{\mathbf{u}(t) \times [\mathbf{v}(t) \times \mathbf{w}(t)]\}$

In Exercises 59 and 60, suppose that $\mathbf{u}$ *and* $\mathbf{v}$ *are integrable on* $[a, b]$ *and that c is a scalar. Prove each property.*

59. $\displaystyle\int_a^b [\mathbf{u}(t) + \mathbf{v}(t)]\, dt = \int_a^b \mathbf{u}(t)\, dt + \int_a^b \mathbf{v}(t)\, dt$

60. $\displaystyle\int_a^b c\mathbf{u}(t)\, dt = c\int_a^b \mathbf{u}(t)\, dt$

61. a. Suppose that $\mathbf{r}$ is integrable on $[a, b]$ and that $\mathbf{c}$ is a constant vector. Prove that
$$\int_a^b \mathbf{c} \cdot \mathbf{r}(t)\, dt = \mathbf{c} \cdot \int_a^b \mathbf{r}(t)\, dt$$

b. Verify this property directly for the vector function
$$\mathbf{r}(t) = \sin t\mathbf{i} + \cos t\mathbf{j} + t\mathbf{k},$$
$$\mathbf{c} = 2\mathbf{i} + 3\mathbf{j} - \mathbf{k}, \quad \text{and} \quad a = 0, \quad b = \pi$$

In Exercises 62–65, determine whether the statement is true or false. If it is true, explain why. If it is false, explain why or give an example to show that it is false.

62. If $\mathbf{c}$ is a constant vector, then $\dfrac{d}{dt}(\mathbf{c}) = 0.$

63. $\dfrac{d}{dt}(|\mathbf{u}|^2) = 2\mathbf{u} \cdot \mathbf{u}'$

64. If $\mathbf{r}'(t) = \mathbf{0},$ then $\mathbf{r}(t) = \mathbf{c},$ where $\mathbf{c}$ is an arbitrary constant vector.

65. If $\mathbf{r}$ is differentiable and $\mathbf{r}(t) \cdot \mathbf{r}'(t) = 0$ for all t, then $\mathbf{r}$ must have constant length.

12.3   Arc Length and Curvature

■ Arc Length

In Section 10.3 we saw that the length of the plane curve given by the parametric equations $x = f(t)$ and $y = g(t)$, where $a \leq t \leq b$, is

$$L = \int_a^b \sqrt{\left(\frac{dx}{dt}\right)^2 + \left(\frac{dy}{dt}\right)^2}\, dt = \int_a^b \sqrt{[f'(t)]^2 + [g'(t)]^2}\, dt$$

Now, suppose that C is described by the vector function $\mathbf{r}(t) = f(t)\mathbf{i} + g(t)\mathbf{j}$ instead. Then

$$\mathbf{r}'(t) = f'(t)\mathbf{i} + g'(t)\mathbf{j}$$

and

$$|\mathbf{r}'(t)| = \sqrt{\mathbf{r}'(t) \cdot \mathbf{r}'(t)} = \sqrt{[f'(t)]^2 + [g'(t)]^2}$$

from which we see that L can also be written in the form

$$L = \int_a^b |\mathbf{r}'(t)|\, dt$$

A similar formula for calculating the length of a space curve is contained in the following theorem.

THEOREM 1   Arc Length of a Space Curve

Let C be a curve given by the vector function

$$\mathbf{r}(t) = f(t)\mathbf{i} + g(t)\mathbf{j} + h(t)\mathbf{k} \qquad a \leq t \leq b$$

where f', g', and h' are continuous. If C is traversed exactly once as t increases from a to b, then its length is given by

$$L = \int_a^b \sqrt{[f'(t)]^2 + [g'(t)]^2 + [h'(t)]^2}\, dt = \int_a^b |\mathbf{r}'(t)|\, dt$$

FIGURE 1
The length of the arc of the helix
for $0 \leq t \leq 2\pi$ is $2\sqrt{5}\pi$.

EXAMPLE 1   Find the length of the arc of the helix C given by the vector function $\mathbf{r}(t) = 2\cos t\,\mathbf{i} + 2\sin t\,\mathbf{j} + t\mathbf{k}$, where $0 \leq t \leq 2\pi$, as shown in Figure 1.

Solution   We first compute

$$\mathbf{r}'(t) = -2\sin t\,\mathbf{i} + 2\cos t\,\mathbf{j} + \mathbf{k}$$

Then, using Theorem 1, we see that the length of the arc in question is

$$L = \int_0^{2\pi} |\mathbf{r}'(t)|\, dt = \int_0^{2\pi} \sqrt{4\sin^2 t + 4\cos^2 t + 1}\, dt$$

$$= \int_0^{2\pi} \sqrt{5}\, dt = 2\sqrt{5}\pi$$

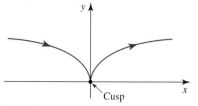

FIGURE 2
The curve defined by $\mathbf{r}(t) = t^3\mathbf{i} + t^2\mathbf{j}$
is smooth everywhere except at $(0, 0)$.

Smooth Curves

A curve that is defined by a vector function $\mathbf{r}$ on a parameter interval I is said to be **smooth** if $\mathbf{r}'(t)$ is continuous and $\mathbf{r}'(t) \neq \mathbf{0}$ for all t in I with the possible exception of the endpoints. For example, the plane curve defined by $\mathbf{r}(t) = t^3\mathbf{i} + t^2\mathbf{j}$ is smooth everywhere except at the point $(0, 0)$ corresponding to $t = 0$. To see this, we compute $\mathbf{r}'(t) = 3t^2\mathbf{i} + 2t\mathbf{j}$ and note that $\mathbf{r}'(0) = \mathbf{0}$. The curve is shown in Figure 2. The point $(0, 0)$ where the curve has a sharp corner is called a **cusp.**

Arc Length Parameter

The curve C described by the vector function $\mathbf{r}(t)$ with parameter t in some parameter interval I is said to be **parametrized** by t. A curve C can have more than one parametrization. For example, the helix represented by the vector function

$$\mathbf{r}_1(t) = 2\cos t\mathbf{i} + 3\sin t\mathbf{j} + t\mathbf{k} \qquad 2\pi \leq t \leq 4\pi$$

with parameter t is also represented by the function

$$\mathbf{r}_2(u) = 2\cos e^u\mathbf{i} + 3\sin e^u\mathbf{j} + e^u\mathbf{k} \qquad \ln 2\pi \leq u \leq \ln 4\pi$$

with parameter u, where t and u are related by $t = e^u$.

A useful parametrization of a curve C is obtained by using the arc length of C as its parameter. To see how this is done, we need the following definition.

DEFINITION Arc Length Function
Suppose that C is a smooth curve described by $\mathbf{r}(t) = f(t)\mathbf{i} + g(t)\mathbf{j} + h(t)\mathbf{k}$, where $a \leq t \leq b$. Then the **arc length function** s is defined by

$$s(t) = \int_a^t |\mathbf{r}'(u)| \, du \qquad (1)$$

Thus, $s(t)$ is the length of that part of C (shown in red) between $\mathbf{r}(a)$ and $\mathbf{r}(t)$. (See Figure 3.) Because $s(a) = 0$, we see that the length L of C from $t = a$ to $t = b$ is

$$s(b) = \int_a^b |\mathbf{r}'(t)| \, dt$$

FIGURE 3
The arc length function $s(t)$ gives the length of that part of C corresponding to the parameter interval $[a, t]$.

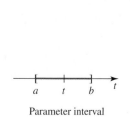

Parameter interval

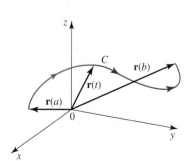

If we differentiate both sides of Equation (1) with respect to t and use the Fundamental Theorem of Calculus, Part 2, we obtain

$$\frac{ds}{dt} = |\mathbf{r}'(t)| \tag{2}$$

or, in differential form,

$$ds = |\mathbf{r}'(t)|\, dt \tag{3}$$

The following example shows how to parametrize a curve in terms of its arc length.

EXAMPLE 2   Find the arc length function $s(t)$ for the circle C in the plane described by

$$\mathbf{r}(t) = 2\cos t\mathbf{i} + 2\sin t\mathbf{j} \qquad 0 \le t \le 2\pi$$

Then use your result to find a parametrization of C in terms of s.

Solution   We first compute $\mathbf{r}'(t) = -2\sin t\mathbf{i} + 2\cos t\mathbf{j}$, and then compute

$$|\mathbf{r}'(t)| = \sqrt{4\sin^2 t + 4\cos^2 t} = 2$$

Using Equation (1), we obtain

$$s(t) = \int_0^t |\mathbf{r}'(u)|\, du = \int_0^t 2\, du = 2t \qquad 0 \le t \le 2\pi$$

Writing s for $s(t)$, we have $s = 2t$, where $0 \le t \le 2\pi$, which when solved for t, yields $t = t(s) = s/2$. Substituting this value of t into the equation for $\mathbf{r}(t)$ gives

$$\mathbf{r}(t(s)) = 2\cos\left(\frac{s}{2}\right)\mathbf{i} + 2\sin\left(\frac{s}{2}\right)\mathbf{j}$$

Finally, since $s(0) = 0$ and $s(2\pi) = 4\pi$, we see that the parameter interval for this parametrization by the arc length s is $[0, 4\pi]$. (See Figure 4.)

FIGURE 4
The curve C is described by
$\mathbf{r}(t) = 2\cos t\mathbf{i} + 2\sin t\mathbf{j}$,
where $0 \le t \le 2\pi$, and
$\mathbf{r}(t(s)) = 2\cos(s/2)\mathbf{i} + 2\sin(s/2)\mathbf{j}$,
where $0 \le s \le 4\pi$.

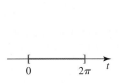

Parameter interval for $\mathbf{r}(t)$

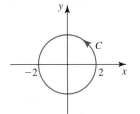

Parameter interval for $\mathbf{r}(s)$

One reason for using the arc length of a curve C as the parameter stems from the fact that its tangent vector $\mathbf{r}'(s)$ has unit length; that is, $\mathbf{r}'(s)$ is a unit tangent vector. Consider the circle of Example 2. Here,

$$\mathbf{r}'(s) = -\sin\left(\frac{s}{2}\right)\mathbf{i} + \cos\left(\frac{s}{2}\right)\mathbf{j}$$

so

$$|\mathbf{r}'(s)| = \sqrt{\sin^2\left(\frac{s}{2}\right) + \cos^2\left(\frac{s}{2}\right)} = 1$$

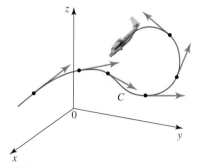

FIGURE 5
The unit tangent vector $\mathbf{T}(s)$ turns faster along the stretch of the path where the turn is sharper.

Curvature

Figure 5 depicts the flight path C of an aerobatic plane as it executes a maneuver. Suppose that the smooth curve C is defined by the vector function $\mathbf{r}(s)$, where s is the arc length parameter. Then the unit tangent vector function $\mathbf{T}(s) = \mathbf{r}'(s)$ gives the direction of the plane at the point on C corresponding to the parameter value s.

In Figure 5 we have drawn the unit tangent vector $\mathbf{T}(s)$ to C corresponding to several values of s. Observe that $\mathbf{T}(s)$ turns rather slowly along a stretch of the flight path that is relatively straight but turns more quickly along a stretch of the curve where the plane executes a sharp turn.

To measure how quickly a curve bends, we introduce the notion of the *curvature* of a curve. Specifically, we define the curvature at a point on a curve C to be the magnitude of the rate of change of the unit tangent vector with respect to arc length at that point.

> **DEFINITION** Curvature
>
> Let C be a smooth curve defined by $\mathbf{r}(s)$, where s is the arc length of the parameter. Then the **curvature** of C at s is
>
> $$\kappa(s) = \left| \frac{d\mathbf{T}}{ds} \right| = |\mathbf{T}'(s)|$$
>
> where $\mathbf{T}$ is the unit tangent vector.

Note   The Greek letter κ is read "kappa." ■

Although the use of the arc length parameter s provides us with a natural way for defining the curvature of a curve, it is generally easier to find the curvature in terms of the parameter t. To see how this is done, let's apply the Chain Rule (Rule 6 in Section 12.2) to write

$$\frac{d\mathbf{T}}{dt} = \frac{d\mathbf{T}}{ds}\frac{ds}{dt}$$

Then

$$\kappa(s) = \left| \frac{d\mathbf{T}}{ds} \right| = \frac{\left| \dfrac{d\mathbf{T}}{dt} \right|}{\left| \dfrac{ds}{dt} \right|}$$

Since $ds/dt = |\mathbf{r}'(t)|$ by Equation (2), we are led to the following formula:

$$\kappa(t) = \frac{|\mathbf{T}'(t)|}{|\mathbf{r}'(t)|} \tag{4}$$

EXAMPLE 3   Find the curvature of a circle of radius a.

Solution   Without loss of generality we may take the circle C with center at the origin. This circle is represented by the vector function

$$\mathbf{r}(t) = a \cos t\,\mathbf{i} + a \sin t\,\mathbf{j} \qquad 0 \le t \le 2\pi$$

Now

$$\mathbf{r}'(t) = -a \sin t \mathbf{i} + a \cos t \mathbf{j}$$

so

$$|\mathbf{r}'(t)| = \sqrt{a^2 \sin^2 t + a^2 \cos^2 t} = a$$

Therefore,

$$\mathbf{T}(t) = \frac{\mathbf{r}'(t)}{|\mathbf{r}'(t)|} = -\sin t \mathbf{i} + \cos t \mathbf{j}$$

Next, we compute

$$\mathbf{T}'(t) = -\cos t \mathbf{i} - \sin t \mathbf{j}$$

and

$$|\mathbf{T}'(t)| = \sqrt{\cos^2 t + \sin^2 t} = 1$$

Finally, using Equation (4), we obtain

$$\kappa(t) = \frac{|\mathbf{T}'(t)|}{|\mathbf{r}'(t)|} = \frac{1}{a}$$

Therefore, the curvature at every point on the circle of radius a is $1/a$. This result agrees with our intuition: A big circle has a small curvature and vice versa. ∎

The following formula expresses the curvature in terms of the vector function $\mathbf{r}$ and its derivatives.

THEOREM 2 **Formula for Finding Curvature**

Let C be a smooth curve given by the vector function $\mathbf{r}$. Then the curvature of C at any point on C corresponding to t is given by

$$\kappa(t) = \frac{|\mathbf{r}'(t) \times \mathbf{r}''(t)|}{|\mathbf{r}'(t)|^3}$$

PROOF We begin by recalling that

$$\mathbf{T}(t) = \frac{\mathbf{r}'(t)}{|\mathbf{r}'(t)|}$$

Since $|\mathbf{r}'(t)| = ds/dt$, we have

$$\mathbf{r}'(t) = \frac{ds}{dt} \mathbf{T}(t)$$

Differentiating both sides of this equation with respect to t and using Rule 3 in Section 12.2, we obtain

$$\mathbf{r}''(t) = \frac{d^2 s}{dt^2} \mathbf{T}(t) + \frac{ds}{dt} \mathbf{T}'(t)$$

Next, we use the fact that $\mathbf{T} \times \mathbf{T} = \mathbf{0}$ (Property 6 of Theorem 3 in Section 11.4) to obtain

$$\mathbf{r}'(t) \times \mathbf{r}''(t) = \left(\frac{ds}{dt}\right)^2 (\mathbf{T}(t) \times \mathbf{T}'(t))$$

Also, $|\mathbf{T}(t)| = 1$ for all t implies that $\mathbf{T}(t)$ and $\mathbf{T}'(t)$ are orthogonal. (See Example 5 in Section 12.2.) Therefore, using Theorem 2 in Section 11.4, we have

$$|\mathbf{r}'(t) \times \mathbf{r}''(t)| = \left(\frac{ds}{dt}\right)^2 |\mathbf{T}(t) \times \mathbf{T}'(t)| = \left(\frac{ds}{dt}\right)^2 |\mathbf{T}(t)| |\mathbf{T}'(t)| = \left(\frac{ds}{dt}\right)^2 |\mathbf{T}'(t)|$$

Upon solving for $|\mathbf{T}'(t)|$, we obtain

$$|\mathbf{T}'(t)| = \frac{|\mathbf{r}'(t) \times \mathbf{r}''(t)|}{\left(\dfrac{ds}{dt}\right)^2} = \frac{|\mathbf{r}'(t) \times \mathbf{r}''(t)|}{|\mathbf{r}'(t)|^2}$$

from which we deduce that

$$\kappa(t) = \frac{|\mathbf{T}'(t)|}{|\mathbf{r}'(t)|} = \frac{|\mathbf{r}'(t) \times \mathbf{r}''(t)|}{|\mathbf{r}'(t)|^3} \qquad ■$$

EXAMPLE 4 Find the curvature of the "twisted cubic" described by the vector function

$$\mathbf{r}(t) = t\mathbf{i} + \frac{1}{2}t^2\mathbf{j} + \frac{1}{3}t^3\mathbf{k}$$

Solution Since

$$\mathbf{r}'(t) = \mathbf{i} + t\mathbf{j} + t^2\mathbf{k}$$

and

$$\mathbf{r}''(t) = \mathbf{j} + 2t\mathbf{k}$$

we have

$$\mathbf{r}'(t) \times \mathbf{r}''(t) = \begin{vmatrix} \mathbf{i} & \mathbf{j} & \mathbf{k} \\ 1 & t & t^2 \\ 0 & 1 & 2t \end{vmatrix} = t^2\mathbf{i} - 2t\mathbf{j} + \mathbf{k}$$

so

$$|\mathbf{r}'(t) \times \mathbf{r}''(t)| = \sqrt{t^4 + 4t^2 + 1}$$

Also,

$$|\mathbf{r}'(t)| = \sqrt{1 + t^2 + t^4} = \sqrt{t^4 + t^2 + 1}$$

Therefore,

$$\kappa(t) = \frac{|\mathbf{r}'(t) \times \mathbf{r}''(t)|}{|\mathbf{r}'(t)|^3} = \frac{\sqrt{t^4 + 4t^2 + 1}}{(t^4 + t^2 + 1)^{3/2}} \qquad ■$$

If a plane curve C happens to be contained in the graph of a function defined by $y = f(x)$, then we can use the following formula to compute its curvature.

THEOREM 3 Formula for the Curvature of the Graph of a Function

If C is the graph of a twice differentiable function f, then the curvature at the point (x, y) where $y = f(x)$ is given by

$$\kappa(x) = \frac{|f''(x)|}{[1 + [f'(x)]^2]^{3/2}} = \frac{|y''|}{[1 + (y')^2]^{3/2}} \tag{5}$$

PROOF Using x as the parameter, we can represent C by the vector function $\mathbf{r}(x) = x\mathbf{i} + f(x)\mathbf{j} + 0\mathbf{k}$. Differentiating $\mathbf{r}(x)$ with respect to x successively, we obtain

$$\mathbf{r}'(x) = \mathbf{i} + f'(x)\mathbf{j} + 0\mathbf{k} \qquad \text{and} \qquad \mathbf{r}''(x) = 0\mathbf{i} + f''(x)\mathbf{j} + 0\mathbf{k}$$

from which we obtain

$$\mathbf{r}'(x) \times \mathbf{r}''(x) = \begin{vmatrix} \mathbf{i} & \mathbf{j} & \mathbf{k} \\ 1 & f'(x) & 0 \\ 0 & f''(x) & 0 \end{vmatrix} = f''(x)\mathbf{k}$$

and

$$|\mathbf{r}'(x) \times \mathbf{r}''(x)| = |f''(x)|$$

Also,

$$|\mathbf{r}'(x)| = \sqrt{1 + [f'(x)]^2}$$

Therefore,

$$\kappa(x) = \frac{|\mathbf{r}'(x) \times \mathbf{r}''(x)|}{|\mathbf{r}'(x)|^3} = \frac{|f''(x)|}{[1 + [f'(x)]^2]^{3/2}} \qquad \blacksquare$$

EXAMPLE 5

a. Find the curvature of the parabola $y = \frac{1}{4}x^2$ at the points where $x = 0$ and $x = 1$.
b. Find the point(s) where the curvature is largest.

Solution
a. We first compute $y' = \frac{1}{2}x$ and $y'' = \frac{1}{2}$. Then using Theorem 3, we find the curvature at any point (x, y) on the parabola $y = \frac{1}{2}x^2$ to be

$$\kappa(x) = \frac{|y''|}{[1 + (y')^2]^{3/2}} = \frac{\frac{1}{2}}{\left(1 + \frac{1}{4}x^2\right)^{3/2}} = \frac{4}{(4 + x^2)^{3/2}}$$

In particular, the curvature at the point $(0, 0)$, where $x = 0$, is

$$\kappa(0) = \frac{4}{(4 + x^2)^{3/2}}\bigg|_{x=0} = \frac{1}{2}$$

and the curvature at the point $\left(1, \frac{1}{4}\right)$, where $x = 1$, is

$$\kappa(1) = \frac{4}{(4 + x^2)^{3/2}}\bigg|_{x=1} = \frac{4}{5^{3/2}} \approx 0.358$$

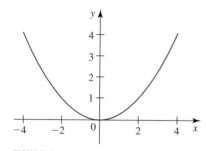

FIGURE 6
The graph of $y = \frac{1}{4}x^2$

b. To find the value of x at which κ is largest, we compute

$$\kappa'(x) = \frac{d}{dx}[4(4 + x^2)^{-3/2}] = -6(4 + x^2)^{-5/2}(2x) = -\frac{12x}{(4 + x^2)^{5/2}}$$

Setting $\kappa'(x) = 0$ yields the sole critical point $x = 0$. We leave it to you to show that $x = 0$ does give the absolute maximum value of $\kappa(x)$.

The graph of $y = \frac{1}{4}x^2$ is shown in Figure 6. ■

■ Radius of Curvature

Suppose that C is a plane curve with curvature κ at the point P. Then the reciprocal of the curvature, $\rho = 1/\kappa$, is called the **radius of curvature** of C at P. The radius of curvature at any point P on a curve C is the radius of the circle that best "fits" the curve at that point. This circle, which lies on the concave side of the curve and shares a common tangent line with the curve at P, is called the **circle of curvature** or **osculating circle.** (See Figure 7.)

The center of the circle is called the **center of curvature.** As an example, the curvature of the parabola $y = \frac{1}{4}x^2$ of Example 5 at the point $(0, 0)$ was found to be $\frac{1}{2}$. Therefore, the radius of curvature of the parabola at $(0, 0)$ is $\rho = 1/(1/2) = 2$. The circle of curvature is shown in Figure 8. Its equation is $x^2 + (y - 2)^2 = 4$.

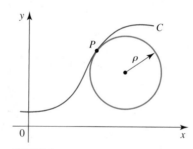

FIGURE 7
The radius of curvature at P is the radius of the circle that best fits the curve C at P.

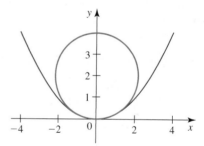

FIGURE 8
The circle of curvature is tangent to the parabola.

12.3 CONCEPT QUESTIONS

1. Give the formula for finding the arc length of the curve C defined by $\mathbf{r}(t) = \langle f(t), g(t), h(t) \rangle$ for $a \le t \le b$. What condition, if any, must be imposed on C?

2. **a.** What is a smooth curve?
   b. Give an example of a curve in 3-space that is not smooth.

3. **a.** What is the arc length function associated with $\mathbf{r}(t) = \langle f(t), g(t), h(t) \rangle$, where $a \le t \le b$?
   b. If a curve is parametrized in terms of its arc length, what is the unit tangent vector $\mathbf{T}(s)$? What is $\mathbf{T}(t)$, where t is not the arc length parameter?

4. **a.** What is the curvature of a smooth curve C at s, where s is the arc length parameter?
   b. If t is not the arc length parameter, what is the curvature of C at t?
   c. What is the radius of curvature of a curve C at a point P on C?

12.3   EXERCISES

In Exercises 1–8, find the length of the curve.

1. $\mathbf{r}(t) = t\mathbf{i} + 2t\mathbf{j} + 3t\mathbf{k}, \quad 0 \le t \le 4$

2. $\mathbf{r}(t) = \langle 5t, 3t^2, 4t^2 \rangle, \quad 0 \le t \le 2$

3. $\mathbf{r}(t) = 4\sin t\mathbf{i} + 3t\mathbf{j} + 4\cos t\mathbf{k}, \quad 0 \le t \le 2\pi$

4. $\mathbf{r}(t) = a\cos t\mathbf{i} + a\sin t\mathbf{j} + bt\mathbf{k}, \quad 0 \le t \le 2\pi$

5. $\mathbf{r}(t) = \langle e^t \cos t, e^t \sin t, e^t \rangle, \quad 0 \le t \le 2\pi$

6. $\mathbf{r}(t) = t^2\mathbf{i} + t\cos t\mathbf{j} + t\sin t\mathbf{k}, \quad 0 \le t \le 1$

7. $\mathbf{r}(t) = 2t\mathbf{i} + t^2\mathbf{j} + \ln t\mathbf{k}, \quad 1 \le t \le e$

8. $\mathbf{r}(t) = (\cos t + t\sin t)\mathbf{i} + (\sin t - t\cos t)\mathbf{j} + t^2\mathbf{k},$
   $0 \le t \le \frac{\pi}{2}$

 In Exercises 9 and 10, use a calculator or computer to graph the curve represented by $\mathbf{r}(t)$, and find the length of the curve for t defined on the indicated interval.

9. $\mathbf{r}(t) = t\sin t\mathbf{i} + t\cos t\mathbf{j} + t\mathbf{k}; \quad [0, 2\pi]$

10. $\mathbf{r}(t) = 2\sin t\mathbf{i} + 2\cos t\mathbf{j} + \frac{1}{2}t^2\mathbf{k}; \quad [0, 2\pi]$

In Exercises 11–14, find the arc length function s(t) for the curve defined by $\mathbf{r}(t)$. Then use this result to find a parametrization of C in terms of s.

11. $\mathbf{r}(t) = (1 + t)\mathbf{i} + (1 + 2t)\mathbf{j} + 3t\mathbf{k}, \quad t \ge 0$

12. $\mathbf{r}(t) = 4\sin t\mathbf{i} + 4\cos t\mathbf{j} + 3t\mathbf{k}, \quad t \ge 0$

13. $\mathbf{r}(t) = e^t \cos t\mathbf{i} + e^t \sin t\mathbf{j} + e^t\mathbf{k}, \quad t \ge 0$

14. $\mathbf{r}(t) = a\cos^3 t\mathbf{i} + a\sin^3 t\mathbf{j} + \mathbf{k}, \quad 0 \le t \le \frac{\pi}{2}$

In Exercises 15–20, use Theorem 2 to find the curvature of the curve.

15. $\mathbf{r}(t) = 2t\mathbf{i} + 2t\mathbf{j} + \mathbf{k}$

16. $\mathbf{r}(t) = t\mathbf{i} + \mathbf{j} + t^2\mathbf{k}$

17. $\mathbf{r}(t) = t\mathbf{i} + \frac{1}{2}t^2\mathbf{j} + t^2\mathbf{k}$

18. $\mathbf{r}(t) = (1 - t)\mathbf{i} + (1 + t)\mathbf{j} + 3t^2\mathbf{k}$

19. $\mathbf{r}(t) = 2\sin t\mathbf{i} + 2\cos t\mathbf{j} + 2t\mathbf{k}$

20. $\mathbf{r}(t) = \langle e^t \cos t, e^t \sin t, e^t \rangle$

In Exercises 21–26, use Theorem 3 to find the curvature of the curve.

21. $y = x^3 + 1$

22. $y = x^4$

23. $y = \sin 2x$

24. $y = \ln x$

25. $y = e^{-x^2}$

26. $y = \sec x$

27. Find the point(s) on the graph of $y = e^{-x^2}$ at which the curvature is zero.

28. Find an equation of the circle of curvature for the graph of $f(x) = x + (1/x)$ at the point $(1, 2)$. Sketch the graph of f and the circle of curvature.

In Exercises 29–32, find the point(s) on the curve at which the curvature is largest.

29. $y = e^x$

30. $y = \ln x$

31. $xy = 1$

32. $4x^2 + 9y^2 = 36$

In Exercises 33–36, match the curve with the graph of its curvature $y = \kappa(x)$ in (a)–(d).

33.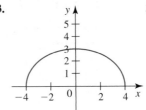

34.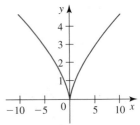

35.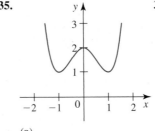

36.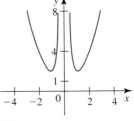

(a)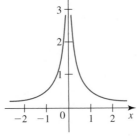

(b)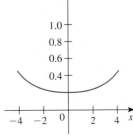

(c)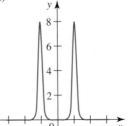

(d)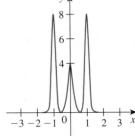

 In Exercises 37 and 38, find the curvature function $\kappa(x)$ of the curve. Then use a calculator or computer to graph both the curve and its curvature function $\kappa(x)$ on the same set of axes.

37. $y = e^{-x^2}$

38. $y = \ln(1 + x^2)$

39. Suppose that C is a smooth curve with parametric equations $x = f(t)$, $y = g(t)$. Using Theorem 2, show that the curvature at the point (x, y) corresponding to any value of t is

$$\kappa(t) = \frac{|f'(t)g''(t) - g'(t)f''(t)|}{\{[f'(t)]^2 + [g'(t)]^2\}^{3/2}}$$

In Exercises 40 and 41, use the formula in Exercise 39 to find the curvature of the curve.

40. $x = \cos t,\quad y = t \sin t$

41. $x = t - \sin t,\quad y = 1 - \cos t$

42. a. The curvature of the curve C at P shown in the figure is 2. Sketch the osculating circle at P. (Use the tangent line shown at P as an aid.)
  b. What is the curvature of C at Q?

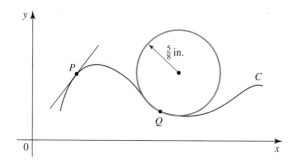

43. a. Find the curvature at the point (x, y) on the ellipse

$$\frac{x^2}{9} + \frac{y^2}{4} = 1$$

  b. Find the curvature and the equation of the osculating circle at the points $(3, 0)$ and $(0, 2)$.
  c. Sketch the graph of the ellipse and the osculating circles of part (b).

44. Find the curvature $\kappa(t)$ for the curve with parametric equations

$$x = t^2 \quad \text{and} \quad y = t^3$$

What happens to $\kappa(t)$ as t approaches 0?
Note: The curve is not smooth at $t = 0$.

45. The spiral of cornu is defined by the parametric equations

$$x = \int_0^t \cos\left(\frac{\pi u^2}{2}\right) du \qquad y = \int_0^t \sin\left(\frac{\pi u^2}{2}\right) du$$

and was encountered in Exercise 45 of Section 10.3. Its graph follows.
  a. Find $\dfrac{dy}{dx}$ and $\dfrac{d^2y}{dx^2}$.
  b. Find the curvature of the spiral.
    Note: The curvature $\kappa(t)$ increases from 0 at a constant rate with respect to t as t increases from $t = 0$. This property of the spiral

of cornu makes the curve useful in highway design: It provides a gradual transition from a straight road (zero curvature) to a curved road (with positive curvature), such as an exit ramp.

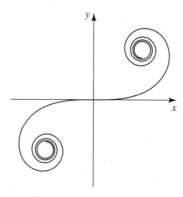

46. Suppose that the curve C is described by a polar equation $r = f(\theta)$. Show that the curvature at the point (r, θ) is given by

$$\kappa(\theta) = \frac{|2(r')^2 - rr'' + r^2|}{[(r')^2 + r^2]^{3/2}}$$

Hint: Represent C by $\mathbf{r}(\theta) = r \cos \theta \mathbf{i} + r \sin \theta \mathbf{j}$.

In Exercises 47 and 48, use the formula in Exercise 46 to find the curvature of the curve.

47. $r = 1 + \sin \theta$ **48.** $r = e^{\theta}$

49. Show that the curvature at every point on the helix

$$x = a \cos t \qquad y = a \sin t \qquad z = bt$$

where $a > 0$, is given by $\kappa(t) = a/(a^2 + b^2)$.

50. Find the curvature at the point (x, y, z) on an elliptic helix with parametric equations

$$x = a \cos t \qquad y = b \sin t \qquad z = ct$$

where a, b, and c are positive and $a \neq b$.

51. Find the arc length of $\mathbf{r}(t) = t \cos t \mathbf{i} + t \sin t \mathbf{j} + t \mathbf{k}$, where $0 \leq t \leq 2\pi$.

52. Find the curvature of the graph of $x^3 + y^3 = 9xy$ (folium of Descartes) at the point $(2, 4)$ accurate to four decimal places.

In Exercises 53–58, determine whether the statement is true or false. If it is true, explain why. If it is false, explain why or give an example to show that it is false.

53. If C is a smooth curve in the xy-plane defined by $\mathbf{r}(t) = x(t)\mathbf{i} + y(t)\mathbf{j}$ on a parameter interval I, then dy/dx is defined at every point on the curve.

54. The curve defined by $\mathbf{r}(t) = \langle t, |t| \rangle$ is smooth.

55. If the graph of a twice differentiable function f has an inflection point at a, then the curvature at the point $(a, f(a))$ is zero.

56. If C is the curve defined by the parametric equations

$$x = 1 - 2t \qquad y = 2 + 3t \qquad z = 4t$$

then $d\mathbf{T}/ds = 0$, where $\mathbf{T}$ is the unit tangent vector to C.

57. The radius of curvature of the plane curve $y = \sqrt{a^2 - x^2}$ is constant at each point on the curve.

58. If $\mathbf{r}'(t)$ is continuous for all t in an interval I, then $\mathbf{r}$ defines a smooth curve.

12.4 Velocity and Acceleration

◼ Velocity, Acceleration, and Speed

The curve C in Figure 1 is the flight path of a fighter plane. We can represent C by the vector function

$$\mathbf{r}(t) = f(t)\mathbf{i} + g(t)\mathbf{j} + h(t)\mathbf{k} \qquad t \in I$$

where we think of the parameter interval I as a time interval and use $\mathbf{r}(t)$ to indicate the position of the plane at time t.

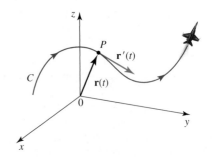

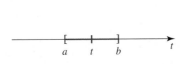

Parameter interval (time interval)

FIGURE 1
The position vector $\mathbf{r}(t)$ gives the position of a fighter plane at time t, and its derivative $\mathbf{r}'(t)$ gives the plane's velocity at time t.

From Sections 12.2 and 12.3 we know that the vector $\mathbf{r}'(t)$ has the following properties:

1. $\mathbf{r}'(t)$ is tangent to C at the point P corresponding to time t.

2. $|\mathbf{r}'(t)| = \dfrac{ds}{dt}$.

Since ds/dt is the rate of change of the distance (measured along the arc) with respect to time, it measures the *speed* of the plane. Thus, the vector $\mathbf{r}'(t)$ gives both the speed and the direction of the plane. In other words, it makes sense to define the *velocity vector* of the plane at time t to be $\mathbf{r}'(t)$, the rate of change of its position vector with respect to time. Similarly, we define the *acceleration vector* of the plane at time t to be $\mathbf{r}''(t)$, the rate of change of its velocity vector with respect to time.

To gain insight into the nature of the acceleration vector, let's refer to Figure 2. Here, t is fixed, and h is a small number. The vector $\mathbf{r}'(t)$ is tangent to the flight path at the tip of the position vector $\mathbf{r}(t)$, and $\mathbf{r}'(t + h)$ is tangent to the flight path at the tip of $\mathbf{r}(t + h)$. The vector

$$\frac{\mathbf{r}'(t + h) - \mathbf{r}'(t)}{h}$$

points in the general direction in which the plane is turning. Therefore, the acceleration vector

$$\mathbf{r}''(t) = \frac{d}{dt}\mathbf{r}'(t) = \lim_{h \to 0} \frac{\mathbf{r}'(t + h) - \mathbf{r}'(t)}{h}$$

points toward the concave side of the flight path as long as the direction of $\mathbf{r}'(t)$ is changing, in agreement with our intuition.

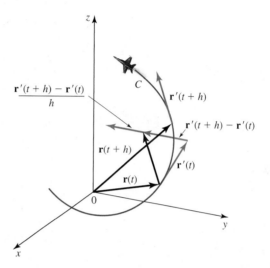

FIGURE 2
To find $\mathbf{r}'(t + h) - \mathbf{r}'(t)$, translate $\mathbf{r}'(t + h)$ so that its tail is at the tip of $\mathbf{r}(t)$.

Let's summarize these definitions.

DEFINITIONS **Velocity, Acceleration, and Speed**

Let $\mathbf{r}(t) = f(t)\mathbf{i} + g(t)\mathbf{j} + h(t)\mathbf{k}$ be the position vector of an object. If f, g, and h are twice differentiable functions of t, then the **velocity vector** $\mathbf{v}(t)$, **acceleration vector** $\mathbf{a}(t)$, and **speed** $|\mathbf{v}(t)|$ of the object at time t are defined by

$$\mathbf{v}(t) = \mathbf{r}'(t) = f'(t)\mathbf{i} + g'(t)\mathbf{j} + h'(t)\mathbf{k}$$

$$\mathbf{a}(t) = \mathbf{r}''(t) = f''(t)\mathbf{i} + g''(t)\mathbf{j} + h''(t)\mathbf{k}$$

$$|\mathbf{v}(t)| = |\mathbf{r}'(t)| = \sqrt{[f'(t)]^2 + [g'(t)]^2 + [h'(t)]^2}$$

EXAMPLE 1 The position of an object moving in a plane is given by

$$\mathbf{r}(t) = t^2\mathbf{i} + t\mathbf{j} \qquad t \geq 0$$

Find its velocity, acceleration, and speed when $t = 2$. Sketch the path of the object and the vectors $\mathbf{v}(2)$ and $\mathbf{a}(2)$.

Solution The velocity and acceleration vectors of the object are

$$\mathbf{v}(t) = \mathbf{r}'(t) = 2t\mathbf{i} + \mathbf{j}$$

and

$$\mathbf{a}(t) = \mathbf{r}''(t) = 2\mathbf{i}$$

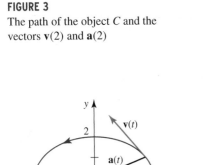

FIGURE 3
The path of the object C and the
vectors $\mathbf{v}(2)$ and $\mathbf{a}(2)$

Therefore, its velocity, acceleration, and speed when $t = 2$ are

$$\mathbf{v}(2) = 4\mathbf{i} + \mathbf{j}$$

$$\mathbf{a}(2) = 2\mathbf{i}$$

and

$$|\mathbf{v}(2)| = \sqrt{16 + 1} = \sqrt{17}$$

respectively.

To sketch the path of the object, observe that the parametric equations of the curve described by $\mathbf{r}(t)$ are $x = t^2$ and $y = t$. By eliminating t from these equations, we obtain the rectangular equation $x = y^2$, where $y \geq 0$, which tells us that the path of the object is contained in the graph of the parabola $x = y^2$. This path together with the vectors $\mathbf{v}(2)$ and $\mathbf{a}(2)$ is shown in Figure 3.   ■

EXAMPLE 2 Find the velocity vector, speed, and acceleration vector of an object that moves along the plane curve C described by the position vector

$$\mathbf{r}(t) = 3 \cos t\mathbf{i} + 2 \sin t\mathbf{j}$$

Solution The velocity vector is

$$\mathbf{v}(t) = -3 \sin t\mathbf{i} + 2 \cos t\mathbf{j}$$

The speed of the object at time t is

$$|\mathbf{v}(t)| = \sqrt{9 \sin^2 t + 4 \cos^2 t}$$

Finally, the acceleration vector is

$$\mathbf{a}(t) = -3 \cos t\mathbf{i} - 2 \sin t\mathbf{j} = -\mathbf{r}(t)$$

which shows the acceleration is directed toward the origin (see Figure 4).   ■

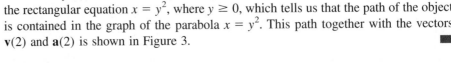

FIGURE 4
The acceleration vector $\mathbf{a}$ points toward
the origin.

EXAMPLE 3 Find the velocity vector, acceleration vector, and speed of a particle with position vector

$$\mathbf{r}(t) = \sqrt{t}\mathbf{i} + t^2\mathbf{j} + e^{2t}\mathbf{k} \qquad t \geq 0$$

Solution The required quantities are

$$\mathbf{v}(t) = \mathbf{r}'(t) = \frac{1}{2}t^{-1/2}\mathbf{i} + 2t\mathbf{j} + 2e^{2t}\mathbf{k} = \frac{1}{2\sqrt{t}}\mathbf{i} + 2t\mathbf{j} + 2e^{2t}\mathbf{k}$$

$$\mathbf{a}(t) = \mathbf{r}''(t) = -\frac{1}{4}t^{-3/2}\mathbf{i} + 2\mathbf{j} + 4e^{2t}\mathbf{k} = -\frac{1}{4\sqrt{t^3}}\mathbf{i} + 2\mathbf{j} + 4e^{2t}\mathbf{k}$$

and

$$|\mathbf{v}(t)| = \sqrt{\frac{1}{4t} + 4t^2 + 4e^{4t}} = \frac{\sqrt{1 + 16t^3 + 16te^{4t}}}{2\sqrt{t}}$$ ■

Suppose that we are given the velocity or acceleration vector of a moving object. Then it is possible to find the position vector of the object by integration, as is shown in the next example.

EXAMPLE 4 A moving object has an initial position and an initial velocity given by the vectors $\mathbf{r}(0) = \mathbf{i} + 2\mathbf{j} + \mathbf{k}$ and $\mathbf{v}(0) = \mathbf{i} + 2\mathbf{k}$. Its acceleration at time t is $\mathbf{a}(t) = 6t\mathbf{i} + \mathbf{j} + 2\mathbf{k}$. Find its velocity and position at time t.

Solution Since $\mathbf{v}'(t) = \mathbf{a}(t)$, we can obtain $\mathbf{v}(t)$ by integrating both sides of this equation with respect to t. Thus,

$$\mathbf{v}(t) = \int \mathbf{a}(t)\,dt = \int (6t\mathbf{i} + \mathbf{j} + 2\mathbf{k})\,dt = 3t^2\mathbf{i} + t\mathbf{j} + 2t\mathbf{k} + \mathbf{C}$$

Letting $t = 0$ in this expression and using the initial condition $\mathbf{v}(0) = \mathbf{i} + 2\mathbf{k}$, we obtain

$$\mathbf{v}(0) = \mathbf{C} = \mathbf{i} + 2\mathbf{k}$$

Therefore, the velocity of the object at any time t is

$$\mathbf{v}(t) = (3t^2\mathbf{i} + t\mathbf{j} + 2t\mathbf{k}) + \mathbf{i} + 2\mathbf{k}$$
$$= (3t^2 + 1)\mathbf{i} + t\mathbf{j} + 2(t + 1)\mathbf{k}$$

Next, integrating the equation $\mathbf{r}'(t) = \mathbf{v}(t)$ with respect to t gives

$$\mathbf{r}(t) = \int \mathbf{v}(t)\,dt = \int [(3t^2 + 1)\mathbf{i} + t\mathbf{j} + 2(t + 1)\mathbf{k}]\,dt$$
$$= (t^3 + t)\mathbf{i} + \frac{1}{2}t^2\mathbf{j} + (t^2 + 2t)\mathbf{k} + \mathbf{D}$$

Letting $t = 0$ in $\mathbf{r}(t)$ and using the initial condition $\mathbf{r}(0) = \mathbf{i} + 2\mathbf{j} + \mathbf{k}$, we have

$$\mathbf{r}(0) = \mathbf{D} = \mathbf{i} + 2\mathbf{j} + \mathbf{k}$$

Therefore, the position of the object at any time t is

$$\mathbf{r}(t) = (t^3 + t)\mathbf{i} + \frac{1}{2}t^2\mathbf{j} + (t^2 + 2t)\mathbf{k} + (\mathbf{i} + 2\mathbf{j} + \mathbf{k})$$
$$= (t^3 + t + 1)\mathbf{i} + \left(\frac{1}{2}t^2 + 2\right)\mathbf{j} + (t^2 + 2t + 1)\mathbf{k}$$
$$= (t^3 + t + 1)\mathbf{i} + \left(\frac{1}{2}t^2 + 2\right)\mathbf{j} + (t + 1)^2\mathbf{k} \qquad ■$$

■ Motion of a Projectile

A projectile of mass m is fired from a height h with an initial velocity $\mathbf{v}_0$ and an angle of elevation α. If we describe the position of the projectile at any time t by the position vector $\mathbf{r}(t)$, then its initial position may be described by the vector

$$\mathbf{r}(0) = h\mathbf{j}$$

and its initial velocity by the vector

$$\mathbf{v}(0) = \mathbf{v}_0 = (v_0 \cos \alpha)\mathbf{i} + (v_0 \sin \alpha)\mathbf{j} \qquad v_0 = |\mathbf{v}_0| \qquad (1)$$

(See Figure 5.)

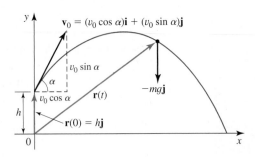

FIGURE 5
The initial position of the projectile
is $\mathbf{r}(0) = h\mathbf{j}$, and its initial velocity
is $\mathbf{v}_0 = (v_0 \cos \alpha)\mathbf{i} + (v_0 \sin \alpha)\mathbf{j}$.

If we assume that air resistance is negligible and that the only external force acting on the projectile is due to gravity, then the force acting on the projectile during its
flight is

$$\mathbf{F} = -mg\mathbf{j}$$

where g is the acceleration due to gravity (32 ft/sec² or 9.8 m/sec²). By Newton's
Second Law of Motion this force is equal to $m\mathbf{a}$, where $\mathbf{a}$ is the acceleration of the projectile. Therefore,

$$m\mathbf{a} = -mg\mathbf{j}$$

giving the acceleration of the projectile as

$$\mathbf{a}(t) = -g\mathbf{j}$$

To find the velocity of the projectile at any time t, we integrate the last equation with
respect to t to obtain

$$\mathbf{v}(t) = \int -g\mathbf{j}\, dt = -gt\mathbf{j} + \mathbf{C}$$

Setting $t = 0$ and using the initial condition $\mathbf{v}(0) = \mathbf{v}_0$, we obtain

$$\mathbf{v}(0) = \mathbf{C} = \mathbf{v}_0$$

Therefore, the velocity of the projectile at any time t is

$$\mathbf{v}(t) = -gt\mathbf{j} + \mathbf{v}_0$$

Integrating this equation then gives

$$\mathbf{r}(t) = \int (-gt\mathbf{j} + \mathbf{v}_0)\, dt = -\frac{1}{2}gt^2\mathbf{j} + \mathbf{v}_0 t + \mathbf{D}$$

Setting $t = 0$ and using the initial condition $\mathbf{r}(0) = h\mathbf{j}$, we obtain

$$\mathbf{r}(0) = \mathbf{D} = h\mathbf{j}$$

Therefore, the position of the projectile at any time t is

$$\mathbf{r}(t) = -\frac{1}{2}gt^2\mathbf{j} + \mathbf{v}_0 t + h\mathbf{j}$$

or, upon using Equation (1),

$$\mathbf{r}(t) = -\frac{1}{2}gt^2\mathbf{j} + [(v_0 \cos \alpha)\mathbf{i} + (v_0 \sin \alpha)\mathbf{j}]t + h\mathbf{j}$$

$$= (v_0 \cos \alpha)t\mathbf{i} + \left[h + (v_0 \sin \alpha)t - \frac{1}{2}gt^2 \right]\mathbf{j}$$

DEFINITION   **Position Function for a Projectile**

The trajectory of a projectile fired from a height h with an initial speed v_0 and an angle of elevation α is given by the position vector function

$$\mathbf{r}(t) = (v_0 \cos \alpha)t\mathbf{i} + \left[h + (v_0 \sin \alpha)t - \frac{1}{2}gt^2 \right]\mathbf{j} \qquad (2)$$

where g is the constant of acceleration due to gravity.

EXAMPLE 5   **Motion of a Projectile**   A shell is fired from a gun located on a hill 100 m above a level terrain. The muzzle speed of the gun is 500 m/sec, and its angle of elevation is 30°.

a. Find the range of the shell.
b. What is the maximum height attained by the shell?
c. What is the speed of the shell at impact?

Solution   Using Equation (2) with $h = 100$, $v_0 = 500$, $\alpha = 30°$, and $g = 9.8$, we see that the position of the shell at any time t is given by

$$\mathbf{r}(t) = (500 \cos 30°)t\mathbf{i} + [100 + (500 \sin 30°)t - 4.9t^2]\mathbf{j}$$
$$= 250\sqrt{3}t\mathbf{i} + (100 + 250t - 4.9t^2)\mathbf{j}$$

The corresponding parametric equations are

$$x = 250\sqrt{3}t \qquad \text{and} \qquad y = 100 + 250t - 4.9t^2$$

a. We first find the time when the shell strikes the ground by solving the equation

$$4.9t^2 - 250t - 100 = 0$$

obtained by setting $y = 0$. Using the quadratic formula, we have

$$t = \frac{250 \pm \sqrt{62{,}500 + 1960}}{9.8} \approx 51.4 \qquad \text{We reject the negative root.}$$

or 51.4 sec. Substituting this value of t into the expression for x we find that the range of the shell is approximately

$$250\sqrt{3}(51.4) \approx 22{,}257$$

or 22,257 m.

b. The height of the shell at any time t is given by

$$y = 100 + 250t - 4.9t^2$$

To find the maximum value of y, we solve

$$y' = 250 - 9.8t = 0$$

to obtain $t \approx 25.5$. Since $y'' = -9.8 < 0$, the Second Derivative Test implies that at approximately 25.5 sec into flight, the shell attains its maximum height

$$y\Big|_{t \approx 25.5} \approx 100 + 250(25.5) - 4.9(25.5)^2 \approx 3289$$

or 3289 m.

c. By differentiating the position function

$$\mathbf{r}(t) = 250\sqrt{3}\,t\mathbf{i} + (100 + 250t - 4.9t^2)\mathbf{j}$$

we obtain the velocity of the shell at any time t. Thus,

$$\mathbf{v}(t) = \mathbf{r}'(t) = 250\sqrt{3}\,\mathbf{i} + (250 - 9.8t)\mathbf{j}$$

From part (a) we know that the time of impact is $t \approx 51.4$. So at the time of impact the velocity of the shell is

$$\mathbf{v}(51.4) \approx 250\sqrt{3}\,\mathbf{i} + [250 - 9.8(51.4)]\mathbf{j}$$
$$= 250\sqrt{3}\,\mathbf{i} - 253.7\mathbf{j}$$

Therefore, its speed at impact is

$$|\mathbf{v}(51.4)| = \sqrt{(250\sqrt{3})^2 + (253.7)^2} \approx 502$$

or 502 m/sec. The trajectory of the shell is shown in Figure 6.

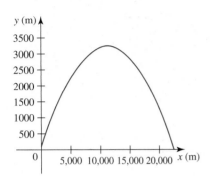

FIGURE 6
The trajectory of the shell

12.4 CONCEPT QUESTIONS

1. a. What are the velocity, acceleration, and speed of an object with position vector $\mathbf{r}(t)$?
  b. Give the expressions for the quantities in part (a) if the position vector has the form $\mathbf{r}(t) = f(t)\mathbf{i} + g(t)\mathbf{j} + h(t)\mathbf{k}$.

2. A projectile of mass m is fired from a height h with an initial velocity $\mathbf{v}_0$ and an angle of elevation α. Write a vector representing
  a. Its initial position.
  b. Its initial velocity in terms of $v_0 = |\mathbf{v}_0|$ and α.
  c. Its velocity at time t.
  d. Its position at time t.

12.4 EXERCISES

In Exercises 1–6, find the velocity, acceleration, and speed of an object with the position function for the given value of t. Sketch the path of the object and its velocity and acceleration vectors.

1. $\mathbf{r}(t) = t\mathbf{i} + (4 - t^2)\mathbf{j}; \quad t = 1$

2. $\mathbf{r}(t) = \langle t^2 - 4, 2t \rangle; \quad t = 1$

3. $\mathbf{r}(t) = \cos t\mathbf{i} + 3\sin t\mathbf{j}; \quad t = \dfrac{\pi}{4}$

4. $\mathbf{r}(t) = e^t\mathbf{i} + e^{-t}\mathbf{j}; \quad t = 0$

5. $\mathbf{r}(t) = \cos t\mathbf{i} + \sin t\mathbf{j} + t\mathbf{k}; \quad t = \dfrac{\pi}{2}$

6. $\mathbf{r}(t) = \langle t, t^2, t^3 \rangle; \quad t = 1$

In Exercises 7–12, find the velocity, acceleration, and speed of an object with the given position vector.

7. $\mathbf{r}(t) = t\mathbf{i} + t^2\mathbf{j} + (t^2 - 4)\mathbf{k}$

8. $\mathbf{r}(t) = \langle \sqrt{t}, 1 + \sqrt{t}, t \rangle$

9. $\mathbf{r}(t) = t\mathbf{i} + t^2\mathbf{j} + \dfrac{1}{t}\mathbf{k}$

10. $\mathbf{r}(t) = e^t\mathbf{i} + e^{-t}\mathbf{j} + t^2\mathbf{k}$

11. $\mathbf{r}(t) = e^t\langle \cos t, \sin t, 1 \rangle$

12. $\mathbf{r}(t) = t\cos t\mathbf{i} + t\sin t\mathbf{j} + t^2\mathbf{k}$

In Exercises 13–18, find the velocity and position vectors of an object with the given acceleration, initial velocity, and position.

13. $\mathbf{a}(t) = -32\mathbf{k}, \quad \mathbf{v}(0) = \mathbf{i} + 2\mathbf{j}, \quad \mathbf{r}(0) = 128\mathbf{k}$

14. $\mathbf{a}(t) = 2\mathbf{i} + t\mathbf{k}, \quad \mathbf{v}(0) = \mathbf{k}, \quad \mathbf{r}(0) = \mathbf{0}$

15. $\mathbf{a}(t) = \mathbf{i} - t\mathbf{j} + (1 + t)\mathbf{k}, \quad \mathbf{v}(0) = \mathbf{i} + \mathbf{k}, \quad \mathbf{r}(0) = \mathbf{j} + \mathbf{k}$

16. $\mathbf{a}(t) = \langle e^t, 0, e^{-t} \rangle, \quad \mathbf{v}(0) = \langle 1, 2, 0 \rangle, \quad \mathbf{r}(0) = \langle 3, 1, 2 \rangle$

17. $\mathbf{a}(t) = -\cos t\mathbf{i} - \sin t\mathbf{j} + \mathbf{k}, \quad \mathbf{v}(0) = 2\mathbf{k}, \mathbf{r}(0) = \mathbf{i}$

18. $\mathbf{a}(t) = \langle \cosh t, \sinh t, 0 \rangle, \quad \mathbf{v}(0) = \langle 0, 1, 1 \rangle, \quad \mathbf{r}(0) = \langle 1, 0, 0 \rangle$

19. An object moves with a constant speed. Show that the velocity and acceleration vectors associated with this motion are orthogonal.
  Hint: Study Example 5 in Section 12.2.

20. Suppose that the acceleration of a moving object is always **0**. Show that its motion is rectilinear (that is, along a straight line).

21. A particle moves in three-dimensional space in such a way that its velocity is always orthogonal to its position vector. Show that its trajectory lies on a sphere centered at the origin.

22. A particle moves in three-dimensional space in such a way that its velocity is always parallel to its position vector. Show that its trajectory lies on a straight line passing through the origin.

23. Motion of a Projectile A projectile is fired from ground level with an initial speed of 1500 ft/sec and an angle of elevation of 30°.
  a. Find the range of the projectile.
  b. What is the maximum height attained by the projectile?
  c. What is the speed of the projectile at impact?

24. Motion of a Projectile Rework Exercise 23 if the projectile is fired with an angle of elevation of 60°.

25. Motion of a Projectile Rework Exercise 23 if the projectile is fired with an angle of elevation of 30° from a height of 200 ft above a level terrain.

26. Motion of a Projectile A shell is fired from a gun situated on a hill 500 ft above level ground. If the angle of elevation of the gun is 0° and the muzzle speed of the shell is 2000 ft/sec, when and where will the shell strike the ground?

27. Motion of a Projectile A mortar shell is fired with a muzzle speed of 500 ft/sec. Find the angle of elevation of the mortar if the shell strikes a target located 1200 ft away.

28. Path of a Baseball A baseball player throws a ball at an angle of 45° with the horizontal. If the ball lands 250 ft away, what is the initial speed of the ball? (Ignore the height of the player.)

29. An object moves in a circular path described by the position vector

$$\mathbf{r}(t) = a \cos \omega t \mathbf{i} + a \sin \omega t \mathbf{j}$$

where $\omega = d\theta/dt$ is the constant angular velocity of the object.
  a. Find the velocity vector, and show that it is orthogonal to $\mathbf{r}(t)$.
  b. Find the acceleration vector, and show that it always points toward the center of the circle.
  c. Find the speed and the magnitude of the acceleration vector of the object.

30. An object located at the origin is to be projected at an initial speed of v_0 m/sec and an angle of elevation of α so that it will strike a target located at the point $(r, 0)$. Neglecting air resistance, find the required angle α.

31. Human Cannonball The following figure shows the trajectory of a "human cannonball" who will be shot out of a cannon located at ground level onto a net. If the angle of elevation of the cannon is 60° and the initial speed of the man is

v_0 ft/sec, determine the range of values of v_0 that will allow the man to land on the net. Neglect air resistance.

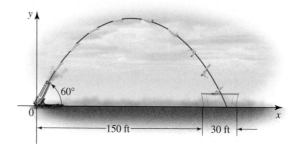

32. Cycloid Motion A particle of charge Q is released at rest from the origin in a region of uniform electric and magnetic fields described by $\mathbf{E} = E\mathbf{k}$ and $\mathbf{B} = B\mathbf{i}$.

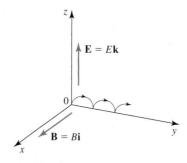

  a. Use the *Lorentz Force Law*, $\mathbf{F} = Q[\mathbf{E} + (\mathbf{v} \times \mathbf{B})]$ where $\mathbf{v}$ is the velocity of the particle, to show that

$$\mathbf{F} = QB\frac{dz}{dt}\mathbf{j} + Q\left(E - B\frac{dy}{dt}\right)\mathbf{k}$$

  b. Use the result of part (a) and Newton's Second Law of Motion to show that the equations of motion of the particle take the form

$$\frac{d^2x}{dt^2} = 0 \qquad \frac{d^2y}{dt^2} = \omega\frac{dz}{dt} \qquad \frac{d^2z}{dt^2} = \omega\left(\frac{E}{B} - \frac{dy}{dt}\right)$$

  where $\omega = \dfrac{QB}{m}$ and m is the mass of the particle.

  c. Show that the general solution of the system in part (b) is $x(t) = C_1 t + C_2$, $y(t) = C_3 \cos \omega t + C_4 \sin \omega t + (E/B)t + C_5$, and $z(t) = C_4 \cos \omega t - C_3 \sin \omega t + C_6$.

  d. Use the initial conditions

$$x(0) = \frac{dx}{dt}(0) = 0, \qquad y(0) = \frac{dy}{dt}(0) = 0,$$

$$\text{and} \qquad z(0) = \frac{dz}{dt}(0) = 0$$

  to determine $C_1, C_2, \ldots, C_6$ and hence show that the trajectory of the particle is the cycloid $x(t) = 0$, $y(t) = (E/(\omega B))(\omega t - \sin \omega t)$, and $z(t) = (E/(\omega B))(1 - \cos \omega t)$. (See Section 10.2.)

33. Newton's Law of Inertia As a model train moves along a straight track at a constant speed of v_0 ft/sec, a ball bearing is ejected vertically from the train at an initial speed of v_1 ft/sec. Show that at some later time the ball bearing will return to the location on the train from which it was released.

Note: This experiment demonstrates Newton's Law of Inertia.

34. Let $\mathbf{r}(t)$ be the position vector of a moving particle and let $r(t) = |\mathbf{r}(t)|$.

   a. Show that $\mathbf{r} \cdot \mathbf{r}' = rr'$.

   b. What can you say about the orbit of the particle if $\mathbf{v} = \mathbf{r}'$ is perpendicular to $\mathbf{r}$?

   c. What can you say about the relationship between the velocity vector and the position vector of the particle if the orbit is circular?

35. A particle has position given by $\mathbf{r}(t) = \langle t, t^2, t^3 \rangle$ at time t for $0 \le t \le 1$. At time $t = 1$ the particle departs the curve and flies off along the line tangent to the curve at $\mathbf{r}(1)$. If the particle maintains a constant speed given by $|\mathbf{v}(1)|$, what is the trajectory of the particle for $t \ge 1$? What is its position at $t = 2$?

36. Motion of a Projectile Suppose that a projectile is fired from the origin of a two-dimensional coordinate system at an angle of elevation of α and an initial speed of v_0.

   a. Show that the position function $\mathbf{r}(t)$ of the projectile is equivalent to the parametric equations $x(t) = (v_0 \cos \alpha)t$ and $y(t) = (v_0 \sin \alpha)t - (1/2)gt^2$.

   b. Eliminate t in the equations in part (a) to find an equation in x and y describing the trajectory of the projectile. What is the shape of the trajectory?

In Exercises 37–38, determine whether the statement is true or false. If it is true, explain why. If it is false, explain why or give an example to show that it is false.

37. If $\mathbf{r}(t)$ gives the position of a particle at time t, then $\mathbf{r}'(t) = |\mathbf{r}'(t)|\mathbf{T}(t)$, where $\mathbf{T}(t)$ is the unit tangent vector to the curve described by $\mathbf{r}$ at t and $\mathbf{r}'(t) \ne \mathbf{0}$.

38. If a particle moves in such a way that its speed is always constant, then its acceleration is zero.

12.5 Tangential and Normal Components of Acceleration

■ The Unit Normal

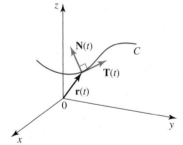

FIGURE 1
At each point on the curve C, the unit normal vector $\mathbf{N}(t)$ is orthogonal to $\mathbf{T}(t)$ and points in the direction the curve is turning.

Suppose that C is a smooth space curve described by the vector function $\mathbf{r}(t)$. Then, as we saw earlier,

$$\mathbf{T}(t) = \frac{\mathbf{r}'(t)}{|\mathbf{r}'(t)|} \qquad \mathbf{r}'(t) \ne \mathbf{0}$$

is the *unit tangent vector* to the curve C at the point corresponding to t. Since $|\mathbf{T}(t)| = 1$ for every t, the result of Example 5 in Section 12.2 tells us that the vector $\mathbf{T}'(t)$ is orthogonal to $\mathbf{T}(t)$. Therefore, if $\mathbf{r}'$ is also smooth, we can *normalize* $\mathbf{T}'(t)$ to obtain a *unit* vector that is orthogonal to $\mathbf{T}(t)$. This vector

$$\mathbf{N}(t) = \frac{\mathbf{T}'(t)}{|\mathbf{T}'(t)|}$$

is called the **principal unit normal vector** (or simply the **unit normal**) to the curve C at the point corresponding to t. (See Figure 1.)

EXAMPLE 1 Let C be the helix defined by

$$\mathbf{r}(t) = 2 \cos t\, \mathbf{i} + 2 \sin t\, \mathbf{j} + t\mathbf{k} \qquad t \ge 0$$

Find $\mathbf{T}(t)$ and $\mathbf{N}(t)$. Sketch C and the vectors $\mathbf{T}(\pi/2)$ and $\mathbf{N}(\pi/2)$.

Solution Since

$$\mathbf{r}'(t) = -2 \sin t\, \mathbf{i} + 2 \cos t\, \mathbf{j} + \mathbf{k}$$

and

$$|\mathbf{r}'(t)| = \sqrt{4 \sin^2 t + 4 \cos^2 t + 1} = \sqrt{5}$$

we have

$$\mathbf{T}(t) = \frac{\mathbf{r}'(t)}{|\mathbf{r}'(t)|} = \frac{1}{\sqrt{5}}(-2\sin t\mathbf{i} + 2\cos t\mathbf{j} + \mathbf{k})$$

Next, differentiating $\mathbf{T}$, we obtain

$$\mathbf{T}'(t) = \frac{1}{\sqrt{5}}(-2\cos t\mathbf{i} - 2\sin t\mathbf{j}) = -\frac{2}{\sqrt{5}}(\cos t\mathbf{i} + \sin t\mathbf{j})$$

Since

$$|\mathbf{T}'(t)| = \frac{2}{\sqrt{5}}\sqrt{\cos^2 t + \sin^2 t} = \frac{2}{\sqrt{5}}$$

it follows that

$$\mathbf{N}(t) = \frac{\mathbf{T}'(t)}{|\mathbf{T}'(t)|} = -(\cos t\mathbf{i} + \sin t\mathbf{j})$$

In particular, at $t = \pi/2$ we have

$$\mathbf{T}\left(\frac{\pi}{2}\right) = \frac{1}{\sqrt{5}}(-2\sin t\mathbf{i} + 2\cos t\mathbf{j} + \mathbf{k})\bigg|_{t=\pi/2} = -\frac{2}{\sqrt{5}}\mathbf{i} + \frac{1}{\sqrt{5}}\mathbf{k}$$

and

$$\mathbf{N}\left(\frac{\pi}{2}\right) = -(\cos t\mathbf{i} + \sin t\mathbf{j})\bigg|_{t=\pi/2} = -\mathbf{j}$$

The curve C and the unit vectors $\mathbf{T}(\pi/2)$ and $\mathbf{N}(\pi/2)$ are shown in Figure 2. Note that, in general, the principal normal vector $\mathbf{N}(t)$ is parallel to the xy-plane and points toward the z-axis. ∎

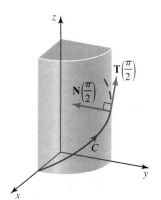

FIGURE 2
The unit vectors $\mathbf{T}(\pi/2)$ and $\mathbf{N}(\pi/2)$ at the point $\left(0, 2, \frac{\pi}{2}\right)$ on the helix

Tangential and Normal Components of Acceleration

Let's return to the study of the motion of an object moving along the curve C described by the vector function $\mathbf{r}$ defined on the parameter interval I. Recall that the speed v of the object at any time t is $v = |\mathbf{v}(t)| = |\mathbf{r}'(t)|$. But

$$\mathbf{T} = \frac{\mathbf{r}'(t)}{|\mathbf{r}'(t)|}$$

so we can write

$$\mathbf{v}(t) = \mathbf{r}'(t) = |\mathbf{r}'(t)|\mathbf{T} = v\mathbf{T} \tag{1}$$

which expresses the velocity of the object in terms of its speed and direction. (See Figure 3.)

FIGURE 3
The velocity of the object at time t is $\mathbf{v}(t) = v\mathbf{T}$.

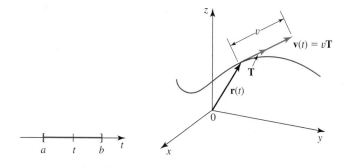

The acceleration of the object at time t is

$$\mathbf{a} = \mathbf{v}' = \frac{d}{dt}(v\mathbf{T}) = v'\mathbf{T} + v\mathbf{T}'$$

To obtain an expression for $\mathbf{T}'$, recall that

$$\mathbf{N} = \frac{\mathbf{T}'}{|\mathbf{T}'|}$$

so $\mathbf{T}' = |\mathbf{T}'|\mathbf{N}$. Now we need an expression for $|\mathbf{T}'|$. But from Equation (4) in Section 12.3 we have

$$\kappa = \frac{|\mathbf{T}'|}{|\mathbf{r}'|}$$

where κ is the curvature of C. This gives

$$|\mathbf{T}'| = \kappa|\mathbf{r}'| = \kappa v$$

so $\mathbf{T}' = |\mathbf{T}'|\mathbf{N} = \kappa v \mathbf{N}$.

Therefore,

$$\mathbf{a} = v'\mathbf{T} + v(\kappa v \mathbf{N})$$
$$= v'\mathbf{T} + \kappa v^2 \mathbf{N} \qquad (2)$$

This result shows that the acceleration vector $\mathbf{a}$ can be resolved into the sum of two vectors—one along the tangential direction and the other along the normal direction. The magnitude of the acceleration along the tangential direction is called the **tangential scalar component of acceleration** and is denoted by $a_\mathbf{T}$, whereas the magnitude of the acceleration along the normal direction is called the **normal scalar component of acceleration** and is denoted by $a_\mathbf{N}$. Thus,

$$\mathbf{a} = a_\mathbf{T}\mathbf{T} + a_\mathbf{N}\mathbf{N} \qquad (3)$$

where

$$a_\mathbf{T} = v' \qquad \text{and} \qquad a_\mathbf{N} = \kappa v^2 \qquad (4)$$

(See Figure 4.)

The following theorem gives formulas for calculating $a_\mathbf{T}$ and $a_\mathbf{N}$ directly from $\mathbf{r}$ and its derivatives.

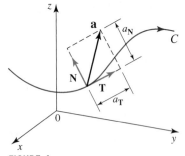

FIGURE 4
The acceleration $\mathbf{a}$ has a component $a_\mathbf{T}\mathbf{T}$ in the tangential direction and a component $a_\mathbf{N}\mathbf{N}$ in the normal direction.

THEOREM 1 **Tangential and Normal Components of Acceleration**

Let $\mathbf{r}(t)$ be the position vector of an object moving along a smooth curve C. Then

$$\mathbf{a} = a_\mathbf{T}\mathbf{T} + a_\mathbf{N}\mathbf{N}$$

where

$$a_\mathbf{T} = \frac{\mathbf{r}'(t) \cdot \mathbf{r}''(t)}{|\mathbf{r}'(t)|} \qquad \text{and} \qquad a_\mathbf{N} = \frac{|\mathbf{r}'(t) \times \mathbf{r}''(t)|}{|\mathbf{r}'(t)|}$$

PROOF If we take the dot product of $\mathbf{v}$ and $\mathbf{a}$ as given by Equations (1) and (2), we obtain

$$\mathbf{v} \cdot \mathbf{a} = (v\mathbf{T}) \cdot (v'\mathbf{T} + \kappa v^2\mathbf{N})$$
$$= vv'\mathbf{T} \cdot \mathbf{T} + \kappa v^3\mathbf{T} \cdot \mathbf{N}$$

But $\mathbf{T} \cdot \mathbf{T} = |\mathbf{T}|^2 = 1$, since $\mathbf{T}$ is a unit vector, and $\mathbf{T} \cdot \mathbf{N} = 0$, since $\mathbf{T}$ and $\mathbf{N}$ are orthogonal. Therefore,

$$\mathbf{v} \cdot \mathbf{a} = vv'$$

or, in view of Equation (4),

$$a_{\mathbf{T}} = v' = \frac{\mathbf{v} \cdot \mathbf{a}}{v} = \frac{\mathbf{r}'(t) \cdot \mathbf{r}''(t)}{|\mathbf{r}'(t)|}$$

Next, using Equation (4) and the formula for curvature (Theorem 2 in Section 12.3), we have

$$a_{\mathbf{N}} = \kappa v^2 = \frac{|\mathbf{r}'(t) \times \mathbf{r}''(t)|}{|\mathbf{r}'(t)|^3} |\mathbf{r}'(t)|^2 = \frac{|\mathbf{r}'(t) \times \mathbf{r}''(t)|}{|\mathbf{r}'(t)|} \qquad\blacksquare$$

EXAMPLE 2 A particle moves along a curve described by the vector function $\mathbf{r}(t) = t\mathbf{i} + t^2\mathbf{j} + t^3\mathbf{k}$. Find the tangential scalar and normal scalar components of acceleration of the particle at any time t.

Solution   We begin by computing

$$\mathbf{r}'(t) = \mathbf{i} + 2t\mathbf{j} + 3t^2\mathbf{k}$$

$$\mathbf{r}''(t) = 2\mathbf{j} + 6t\mathbf{k}$$

Then, using Theorem 1, we obtain

$$a_{\mathbf{T}} = \frac{\mathbf{r}'(t) \cdot \mathbf{r}''(t)}{|\mathbf{r}'(t)|} = \frac{4t + 18t^3}{\sqrt{1 + 4t^2 + 9t^4}}$$

Next, we compute

$$\mathbf{r}'(t) \times \mathbf{r}''(t) = \begin{vmatrix} \mathbf{i} & \mathbf{j} & \mathbf{k} \\ 1 & 2t & 3t^2 \\ 0 & 2 & 6t \end{vmatrix} = 6t^2\mathbf{i} - 6t\mathbf{j} + 2\mathbf{k}$$

Then, using Theorem 1, we have

$$a_{\mathbf{N}} = \frac{|\mathbf{r}'(t) \times \mathbf{r}''(t)|}{|\mathbf{r}'(t)|} = \frac{\sqrt{36t^4 + 36t^2 + 4}}{\sqrt{1 + 4t^2 + 9t^4}} = 2\sqrt{\frac{9t^4 + 9t^2 + 1}{9t^4 + 4t^2 + 1}} \qquad\blacksquare$$

EXAMPLE 3 **Motion of a Projectile**   Refer to Example 5 in Section 12.4. The position function of a shell is given by

$$\mathbf{r}(t) = 250\sqrt{3}\,t\mathbf{i} + (100 + 250t - 4.9t^2)\mathbf{j}$$

a. Find the tangential and normal scalar components of acceleration of the shell at any time t.
b. Find $a_{\mathbf{T}}(t)$ and $a_{\mathbf{N}}(t)$ for $t = 0$, 12.75, 25.5, and 38.25.
c. Is the shell accelerating or decelerating in the tangential direction at the values of t specified in part (b)?

Solution
a. $\mathbf{r}'(t) = 250\sqrt{3}\mathbf{i} + (250 - 9.8t)\mathbf{j}$

$\mathbf{r}''(t) = -9.8\mathbf{j}$

The tangential scalar component acceleration of the shell is

$$a_{\mathbf{T}}(t) = \frac{\mathbf{r}'(t) \cdot \mathbf{r}''(t)}{|\mathbf{r}'(t)|} = \frac{-9.8(250 - 9.8t)}{\sqrt{96.04t^2 - 4900t + 250{,}000}}$$

Next

$$\mathbf{r}'(t) \times \mathbf{r}''(t) = \begin{vmatrix} \mathbf{i} & \mathbf{j} & \mathbf{k} \\ 250\sqrt{3} & 250 - 9.8t & 0 \\ 0 & -9.8 & 0 \end{vmatrix} = -2450\sqrt{3}\,\mathbf{k}$$

So the normal scalar component of acceleration of the shell is

$$a_{\mathbf{N}}(t) = \frac{|\mathbf{r}'(t) \times \mathbf{r}''(t)|}{|\mathbf{r}'(t)|} = \frac{2450\sqrt{3}}{\sqrt{96.04t^2 - 4900t + 250{,}000}}$$

b. The values of $a_{\mathbf{T}}(t)$ and $a_{\mathbf{N}}(t)$ for the specified values of t are shown in Table 1.

TABLE 1

t	0	12.75	25.5	38.25
$a_{\mathbf{T}}(t)$	−4.9	−2.7	0	2.7
$a_{\mathbf{N}}(t)$	8.5	9.4	9.8	9.4

c. Since $a_{\mathbf{T}}(0) = -4.9 < 0$, the shell is decelerating at $t = 0$. Since $a_{\mathbf{T}}(12.75) \approx -2.7 < 0$, the shell is decelerating at $t = 12.75$ but not by as much as it was at $t = 0$. Since $a_{\mathbf{T}}(25.5) \approx 0$, the shell is neither accelerating nor decelerating at $t = 25.5$ (when the shell is at its maximum height). Finally, since $a_{\mathbf{T}}(38.25) \approx 2.7 > 0$, the shell is accelerating at $t = 38.25$ as it continues to plunge toward the earth. ▪

■ Kepler's Laws of Planetary Motion

We close this chapter by demonstrating how calculus can be used to derive Kepler's Laws of Planetary Motion. After laboring for more than 20 years analyzing the empirical data obtained by the Danish astronomer Tycho Brahe, the German astronomer Johannes Kepler (1571–1630) formulated the following three laws describing the motion of the planets around the sun.

Kepler's Laws

1. The orbit of each planet is an ellipse with the sun at one focus. (See Figure 5a.)
2. The line joining the sun to a planet sweeps out equal areas in equal intervals of time. (See Figure 5b.)
3. The square of the period of revolution T of a planet is proportional to the cube of the length of the major axis a of its orbit; that is, $T^2 = ka^3$, where k is a constant.

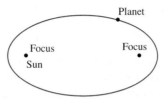

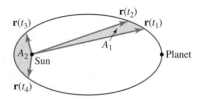

FIGURE 5

(**a**) The orbit of a planet is an ellipse.   (**b**) If the time intervals $[t_1, t_2]$ and $[t_3, t_4]$ are of equal length, then the area of A_1 is equal to the area of A_2.

Sir Isaac Newton proved these laws approximately 50 years after they were formulated. He showed that they were consequences of his own Law of Universal Gravitation and the Second Law of Motion. We will prove Kepler's First Law and leave the derivation of the other two laws as exercises. (See Exercises 37 and 38.)

◼ Derivation of Kepler's First Law

We begin by showing that the orbit of a planet lies in a plane. Let's place the sun at the origin of a coordinate system. By Newton's Law of Gravitation the force **F** of gravitational attraction exerted by the sun on the planet is given by

$$\mathbf{F} = -\frac{GMm}{r^2}\,\mathbf{u}$$

where M and m are the masses of the sun and the planet, respectively; **r** is the position vector of the planet; G is the gravitational constant; $r = |\mathbf{r}|$; and $\mathbf{u} = \mathbf{r}/|\mathbf{r}|$ is the unit vector having the same direction as **r**. This force, which is always directed toward a fixed point O, is an example of a **central force.** But by Newton's Second Law of Motion the acceleration, **a**, of the planet is related to the force, **F**, to which it is subjected by

$$\mathbf{F} = m\mathbf{a}$$

Equating these two expressions for **F** gives

$$m\mathbf{a} = -\frac{GMm}{r^2}\,\mathbf{u}$$

or, upon dividing through by m,

$$\mathbf{a} = -\frac{GM}{r^2}\,\mathbf{u}$$

Next, we will show that for any central force, **r** and **a** satisfy $\mathbf{r} \times \mathbf{a} = \mathbf{0}$. To see this, we compute

$$\mathbf{r} \times \mathbf{a} = \mathbf{r} \times \left(-\frac{GM}{r^2}\,\mathbf{u} \right) = -\frac{GM}{r^2}\,(\mathbf{r} \times \mathbf{u})$$

$$= -\frac{GM}{r^2}\left(\mathbf{r} \times \frac{\mathbf{r}}{|\mathbf{r}|} \right) = -\frac{GM}{r^3}\,(\mathbf{r} \times \mathbf{r}) = \mathbf{0}$$

Using this result, we have

$$\frac{d}{dt}(\mathbf{r} \times \mathbf{v}) = \mathbf{r}' \times \mathbf{v} + \mathbf{r} \times \mathbf{v}' = \mathbf{v} \times \mathbf{v} + \mathbf{r} \times \mathbf{a} = \mathbf{0} + \mathbf{0} = \mathbf{0}$$

Integrating both sides of this equation with respect to t yields

$$\mathbf{r} \times \mathbf{v} = \mathbf{c}$$

where $\mathbf{c}$ is a constant vector. By the definition of the cross-product, $\mathbf{c}$ is orthogonal to both $\mathbf{r}$ and $\mathbf{v}$, and we conclude that both $\mathbf{r}(t)$ and $\mathbf{v}(t)$ lie in a fixed plane containing the point O. This shows that the orbit of the planet is a plane curve, as was claimed earlier. (See Figure 6.)

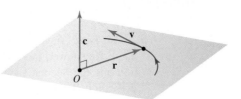

FIGURE 6
The orbit of the planet lies in the plane passing through the origin and orthogonal to $\mathbf{c}$.

To show that this curve is an ellipse with the sun at one focus, we observe that

$$\mathbf{c} = \mathbf{r} \times \mathbf{v} = \mathbf{r} \times \mathbf{r}' = (r\mathbf{u}) \times (r\mathbf{u})'$$

$$= (r\mathbf{u}) \times (r\mathbf{u}' + r'\mathbf{u}) = r^2(\mathbf{u} \times \mathbf{u}') + rr'(\mathbf{u} \times \mathbf{u})$$

$$= r^2(\mathbf{u} \times \mathbf{u}') \qquad \mathbf{u} \times \mathbf{u} = \mathbf{0}.$$

Therefore,

$$\mathbf{a} \times \mathbf{c} = \left(-\frac{GM}{r^2} \mathbf{u} \right) \times [r^2(\mathbf{u} \times \mathbf{u}')] = -GM[\mathbf{u} \times (\mathbf{u} \times \mathbf{u}')]$$

$$= -GM[(\mathbf{u} \cdot \mathbf{u}')\mathbf{u} - (\mathbf{u} \cdot \mathbf{u})\mathbf{u}'] \qquad \text{See Theorem 3 in Section 11.4.}$$

Since $\mathbf{u} \cdot \mathbf{u} = |\mathbf{u}|^2 = 1$, we see that $\mathbf{u} \cdot \mathbf{u}' = 0$. (See Example 5 in Section 12.2.) So the last equation reduces to

$$\mathbf{a} \times \mathbf{c} = GM\mathbf{u}'$$

But $\mathbf{a} \times \mathbf{c}$ can also be written as

$$\mathbf{a} \times \mathbf{c} = \mathbf{v}' \times \mathbf{c} = \frac{d}{dt}(\mathbf{v} \times \mathbf{c}) \qquad \text{Remember that } \mathbf{c} \text{ is a constant vector.}$$

Therefore,

$$\frac{d}{dt}(\mathbf{v} \times \mathbf{c}) = GM\mathbf{u}' = \frac{d}{dt}(GM\mathbf{u})$$

Integrating both sides of this equation with respect to t gives

$$\mathbf{v} \times \mathbf{c} = GM\mathbf{u} + \mathbf{b} \qquad\qquad (5)$$

where $\mathbf{b}$ is a constant vector that depends on the initial conditions. If we take the dot product of both sides of the last equation with $\mathbf{r}$, we have

$$\mathbf{r} \cdot (\mathbf{v} \times \mathbf{c}) = \mathbf{r} \cdot (GM\mathbf{u} + \mathbf{b}) = GM\mathbf{r} \cdot \mathbf{u} + \mathbf{r} \cdot \mathbf{b}$$

$$= GM(r\mathbf{u} \cdot \mathbf{u}) + \mathbf{r} \cdot \mathbf{b} = GMr + \mathbf{r} \cdot \mathbf{b}$$

But

$$\mathbf{r} \cdot (\mathbf{v} \times \mathbf{c}) = \mathbf{c} \cdot (\mathbf{r} \times \mathbf{v}) \qquad \text{See Theorem 3 in Section 11.4.}$$

$$= \mathbf{c} \cdot \mathbf{c} = |\mathbf{c}|^2$$

So

$$GMr + \mathbf{r} \cdot \mathbf{b} = |\mathbf{c}|^2 = c^2 \tag{6}$$

where $c = |\mathbf{c}|$. If $\mathbf{b} = \mathbf{0}$, then Equation (6) reduces to

$$GMr = c^2$$

or

$$r = \frac{c^2}{GM}$$

and the orbit of the planet is a circle. Note that none of the orbits of the planets in our solar system has such an orbit. So we may assume that $\mathbf{b} \neq \mathbf{0}$ for planets in our solar system. In this case, letting θ be the angle between $\mathbf{r}$ and $\mathbf{b}$ and writing $|\mathbf{c}| = c$, we can write Equation (6) in the form

$$GMr + |\mathbf{r}||\mathbf{b}| \cos \theta = c^2$$

$$GMr + rb \cos \theta = c^2$$

or

$$r = \frac{c^2}{GM + b \cos \theta} \tag{7}$$

Dividing both the numerator and denominator of Equation (7) by GM, we obtain

$$r = \frac{\dfrac{c^2}{GM}}{1 + \left(\dfrac{b}{GM}\right) \cos \theta} = \frac{\dfrac{ec^2}{b}}{1 + e \cos \theta}$$

where $e = b/GM$. Finally, if we write $d = c^2/b$, we obtain the equation

$$r = \frac{ed}{1 + e \cos \theta} \tag{8}$$

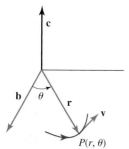

FIGURE 7
In the plane determined by $\mathbf{r}$ and $\mathbf{c}$, r and θ are polar coordinates.

Since $\mathbf{v} \times \mathbf{c}$ and $\mathbf{r}$ are both orthogonal to $\mathbf{c}$, we see from Equation (5) that $\mathbf{b}$ is orthogonal to $\mathbf{c}$. Therefore, in the plane determined by $\mathbf{b}$ and $\mathbf{r}$ (see Figure 7), we can regard $\mathbf{r}$ and θ as polar coordinates of a point P on the orbit of the planet. Comparing Equation (8) with that of Theorem 2 in Section 10.6, we see that it is the polar equation of a conic section with focus at the origin and eccentricity e. Since the orbit of the planet is a closed curve, we know that the conic must be an ellipse. This completes the proof of Kepler's First Law.

12.5 CONCEPT QUESTIONS

1. a. What are the unit tangent and the unit normal vectors at a point on a curve? Illustrate with a sketch.
   b. Suppose that a curve is described by the vector function $\mathbf{r}(t)$. Give formulas for computing the quantities in part (a).

2. a. What are the tangential and normal components of acceleration of an acceleration vector? Illustrate with a sketch.
   b. Write expressions for the quantities in part (a) in terms of $\mathbf{r}(t)$ and its derivatives.

12.5 EXERCISES

In Exercises 1–4, find the unit tangent and unit normal vector $\mathbf{T}(t)$ *and* $\mathbf{N}(t)$ *for the curve C defined by* $\mathbf{r}(t)$. *Sketch the graph of C, and show* $\mathbf{T}(t)$ *and* $\mathbf{N}(t)$ *for the given value of t.*

1. $\mathbf{r}(t) = t\mathbf{i} + 2t^2\mathbf{j}; \quad t = 1$

2. $\mathbf{r}(t) = t^2\mathbf{i} - 2t\mathbf{j}; \quad t = 1$

3. $\mathbf{r}(t) = t^2\mathbf{i} + t^3\mathbf{j}; \quad t = 1$

4. $\mathbf{r}(t) = (2 + \cos t)\mathbf{i} + (3 - \sin t)\mathbf{j}; \quad t = \dfrac{\pi}{4}$

In Exercises 5–10, find the unit tangent and unit normal vectors $\mathbf{T}(t)$ *and* $\mathbf{N}(t)$ *for the curve C defined by* $\mathbf{r}(t)$.

5. $\mathbf{r}(t) = \mathbf{i} + t\mathbf{j} + t^2\mathbf{k}$ 6. $\mathbf{r}(t) = t\mathbf{i} + t^2\mathbf{j} + \dfrac{2}{3}t^3\mathbf{k}$

7. $\mathbf{r}(t) = \langle \sin 2t, \cos 2t, 3t \rangle$

8. $\mathbf{r}(t) = 2\cos t\mathbf{i} + \mathbf{j} + 2\sin t\mathbf{k}$

9. $\mathbf{r}(t) = e^t\langle \cos t, \sin t, 1 \rangle$ 10. $\mathbf{r}(t) = 2t\mathbf{i} + t^2\mathbf{j} + \ln t\mathbf{k}$

In Exercises 11–18, find the scalar tangential and normal components of acceleration of a particle with the given position vector.

11. $\mathbf{r}(t) = t\mathbf{i} + (t^2 + 4)\mathbf{j}$ 12. $\mathbf{r}(t) = (2t^2 - 1)\mathbf{i} + 2t\mathbf{j}$

13. $\mathbf{r}(t) = t\mathbf{i} + t^2\mathbf{j} + t^3\mathbf{k}$

14. $\mathbf{r}(t) = t^2\mathbf{i} + t^3\mathbf{j} + t^2\mathbf{k}; \quad t > 0$

15. $\mathbf{r}(t) = 2\sin t\mathbf{i} + 2\cos t\mathbf{j} + t\mathbf{k}$

16. $\mathbf{r}(t) = \cos^2 t\mathbf{i} + \sin^2 t\mathbf{j} + t\mathbf{k}$

17. $\mathbf{r}(t) = e^t\langle \cos t, \sin t, 1 \rangle$ 18. $\mathbf{r}(t) = \langle t\cos t, t\sin t, 4 \rangle$

19. The accompanying figure shows the path of an object moving in the plane and its acceleration vector $\mathbf{a}$, its unit tangent vector $\mathbf{T}$, and its unit normal vector $\mathbf{N}$ at the points A and B.
   a. Sketch the vectors $a_{\mathbf{T}}\mathbf{T}$ and $a_{\mathbf{N}}\mathbf{N}$ at A and B.
   b. Is the particle accelerating or decelerating at A? At B?

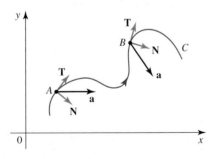

20. At a certain instant of time, the position, velocity, and acceleration of a particle moving in the plane are $\mathbf{r} = 4\mathbf{i} + 2\mathbf{j}$, $\mathbf{v} = 3\mathbf{i} + 4\mathbf{j}$, and $\mathbf{a} = 5\mathbf{i} - 5\mathbf{j}$, respectively.
   a. Sketch $\mathbf{r}$, $\mathbf{v}$, and $\mathbf{a}$.
   b. Is the particle accelerating or decelerating at that instant of time? Explain.
   c. Verify your assertion by computing $a_{\mathbf{T}}$.

21. At a certain instant of time, the velocity and acceleration of a particle are $\mathbf{v} = 2\mathbf{i} + 3\mathbf{j} - 6\mathbf{k}$ and $\mathbf{a} = -6\mathbf{i} - 4\mathbf{j} + 3\mathbf{k}$, respectively.
   a. Find $a_{\mathbf{T}}$ and $a_{\mathbf{N}}$.
   b. Is the particle accelerating or decelerating?

22. At a certain instant of time, the velocity and acceleration of a particle at that time are $\mathbf{v} = \langle 2, 3, 6 \rangle$ and $\mathbf{a} = \langle -6, -4, 3 \rangle$, respectively.
   a. Find $a_{\mathbf{T}}$ and $a_{\mathbf{N}}$.
   b. Is the particle accelerating or decelerating?

23. The position of a particle at time t is $\mathbf{r}(t) = \langle \cos t^2, \sin t^2 \rangle$.
   a. Show that the path of the particle is a circular orbit with center at the origin.
   b. Show that $\mathbf{r} \cdot \mathbf{a} \leq 0$, where $\mathbf{a}$ is the acceleration vector of the particle.
   Hint: Show that $\mathbf{a} \cdot \mathbf{r} + \mathbf{v} \cdot \mathbf{v} = 0$.

24. **Trajectory of a Shell** A shell is fired from a howitzer with a muzzle speed of v_0 m/sec at angle of elevation of α. What are the scalar tangential and normal components of acceleration of the shell?

25. A particle moves along a curve C with a constant speed. Show that the acceleration of the particle is always normal to C.

26. An object moves along a curve C with a constant speed. Show that the magnitude of the acceleration of the object is directly proportional to the curvature of C.

27. Suppose that a particle moves along a plane curve that is the graph of a function f whose second derivative exists. Show that its normal component of acceleration is zero when the particle is at an inflection point of the graph of f.

Let C be a smooth curve defined by $\mathbf{r}(t)$, *and let* $\mathbf{T}(t)$ *and* $\mathbf{N}(t)$ *be the unit tangent vector and unit normal vector to C corresponding to t. The plane determined by* $\mathbf{T}$ *and* $\mathbf{N}$ *is called the* **osculating plane**. *In Exercises 28 and 29, find an equation of the osculating plane of the curve described by* $\mathbf{r}(t)$ *at the point corresponding to the given value of t.*

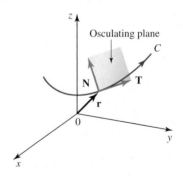

28. $\mathbf{r}(t) = t\mathbf{i} + 2t^2\mathbf{j} + t^3\mathbf{k}; \quad t = 1$

29. $\mathbf{r}(t) = \langle e^t, e^{-t}, \sqrt{2}t \rangle; \quad t = 0$

Let C be a smooth curve defined by $\mathbf{r}(t)$, and let $\mathbf{T}(t)$ and $\mathbf{N}(t)$ be the unit tangent vector and unit normal vector to C corresponding to t. The vector $\mathbf{B}$ defined by $\mathbf{B} = \mathbf{T} \times \mathbf{N}$ is orthogonal to $\mathbf{T}$ and $\mathbf{N}$ and is called the **unit binormal vector**. The vectors, $\mathbf{T}$, $\mathbf{N}$, and $\mathbf{B}$ form a right-handed set of orthogonal unit vectors. In Exercises 30 and 31, find $\mathbf{B}$ for the curve described by $\mathbf{r}(t)$.

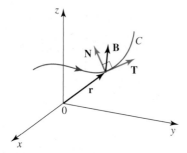

30. $\mathbf{r}(t) = t\mathbf{i} + 2t^2\mathbf{j} + t^3\mathbf{k}$

31. $\mathbf{r}(t) = 2\cosh t\mathbf{i} + 2\sinh t\mathbf{j} + 2t\mathbf{k}$

32. Refer to Exercises 30 and 31. Show that $\mathbf{B}$ can be expressed in terms of r and its derivatives by the formula

$$\mathbf{B} = \frac{\mathbf{r}' \times \mathbf{r}''}{|\mathbf{r}' \times \mathbf{r}''|}$$

33. Rework Exercise 30 using the formula for $\mathbf{B}$ in Exercise 32.

34. Let $\mathbf{T}$, $\mathbf{N}$, and $\mathbf{B}$ be the unit tangent, unit normal, and unit binormal, respectively, associated with a smooth curve C described by $\mathbf{r}(t)$.
   a. Show that $d\mathbf{B}/ds$ is orthogonal to $\mathbf{T}$ and to $\mathbf{B}$.
   b. Use the result of part (a) to show that $d\mathbf{B}/ds = \tau\mathbf{N}$ for some scalar $\tau(t)$. (The number $\tau(t)$ is called the **torsion** of the curve. It measures the rate at which the curve twists out of its osculating plane (the definition of an osculating plane is given on page 1026). We define τ to be equal to 0 for a straight line.)
   c. Use the result of part (b) to show that the torsion of a plane curve is zero.

The torsion of a curve defined by $\mathbf{r}(t)$ is given by

$$\tau = \frac{(\mathbf{r}' \times \mathbf{r}'') \cdot \mathbf{r}'''}{|\mathbf{r}' \times \mathbf{r}''|^2}$$

In Exercises 35 and 36, find the torsion of the curve defined by $\mathbf{r}(t)$.

35. $\mathbf{r}(t) = \cos t\mathbf{i} + \sin t\mathbf{j} + t\mathbf{k}$

36. $\mathbf{r}(t) = (t - \sin t)\mathbf{i} + (1 - \cos t)\mathbf{j} + t\mathbf{k}$

37. Kepler's Second Law Prove Kepler's Second Law using the following steps. (All notation is the same as that used in the text).
   a. Show that if $A(t)$ is the area swept out by the radius vector $\mathbf{r}(t)$ in the time interval $[t_0, t]$, then

$$\frac{dA}{dt} = \frac{1}{2} r^2 \frac{d\theta}{dt}$$

(See the figure below.)
   b. Show that $\mathbf{c} = r^2 \dfrac{d\theta}{dt}\mathbf{k}$, so $r^2 \dfrac{d\theta}{dt} = c$.
   c. Conclude that $\dfrac{dA}{dt} = \dfrac{1}{2}c$, so the rate at which the area is swept out is constant. This is precisely Kepler's Second Law.

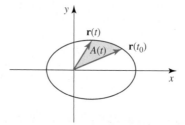

38. Kepler's Third Law Prove Kepler's Third Law by using the following steps. In addition, suppose that the lengths of the major and minor axes of the elliptical orbit are $2a$ and $2b$, respectively. (All notation is the same as that used in the text.)
   a. Use the result of part (c) Exercise 37 and the fact that the area of the ellipse is πab square units to show that $T = 2\pi ab/c$.
   b. Show that $c^2/(GM) = ed = b^2/a$.
   c. Using the result of parts (a) and (b), show that $T^2 = ka^3$, where $k = 4\pi^2/(GM)$.

39. Period of the Earth's Orbit The period of the earth's orbit about the sun is approximately 365.26 days. Also, the mass of the sun is approximately 1.99×10^{30} kg, and the gravitational constant is $G = 6.67 \times 10^{-11}$ Nm²/kg². Find the length of the major axis of the earth's orbit.

40. Artificial Satellites A communications relay satellite is to be placed in *geosynchronous* orbit; that is, its circular orbit about the earth is to have a period of revolution of 24 hr so that the satellite appears to be stationary in the sky. Use the fact that the moon has a period of 27.32 days in a circular orbit of radius 238,850 mi from the center of the earth to determine the radius of the satellite's orbit.

41. Motion of a Projectile A projectile is fired from a height h with an initial speed v_0 and an angle of elevation α.
   a. What are the scalar tangential and normal components of acceleration of the projectile?
   b. What are the scalar tangential and normal components of acceleration of the projectile when the projectile is at its maximum height?

42. Trajectory of a Shell A shell is fired from a gun located on a hill 50 m above a level terrain. The muzzle speed of the gun is 500 m/sec, and its angle of elevation is 45°.
   a. Find the scalar tangential and normal components of acceleration of the shell.
   Hint: Use the result of Exercise 41.
   b. When is the shell accelerating, and when is it decelerating?

43. Derive the following formula for calculating the radius of curvature ρ of a curve C represented by $\mathbf{r}(t) = x(t)\mathbf{i} + y(t)\mathbf{j} + z(t)\mathbf{k}$:

$$\rho = \frac{(x'(t))^2 + (y'(t))^2 + (z'(t))^2}{\sqrt{(x''(t))^2 + (y''(t))^2 + (z''(t))^2 - (v'(t))^2}}$$

44. Use the result of Exercise 43 to find the radius of curvature of the space curve with position vector $\mathbf{r}(t) = t\mathbf{i} + \sin t\mathbf{j} + \cos t\mathbf{k}$.

In Exercises 45–48, determine whether the statement is true or false. If it is true, explain why it is true. If it is false, explain why or give an example to show why it is false.

45. If $|\mathbf{r}'(t)| = c$, where c is a nonzero constant, then the unit normal to the curve C defined by $\mathbf{r}(t)$ is given by $\mathbf{N} = \mathbf{r}''/|\mathbf{r}''|$.

46. If motion takes place along the x-axis, then $a_T = d^2x/dt^2$.

47. If $\mathbf{r}(t)$ is the position vector of a particle with respect to time t and $\mathbf{r}(s)$ is the position vector of the particle with respect to arc length, then $\mathbf{r}''(t)$ is a scalar multiple of $\mathbf{r}''(s)$.

48. If $\mathbf{r}(t)$ is the position vector of a particle moving along a smooth curve C, then $\mathbf{a} = v'\mathbf{T} + (v^2/\rho)\mathbf{N}$, where ρ is the radius of curvature.

CHAPTER 12 REVIEW

CONCEPT REVIEW

In Exercises 1–11, fill in the blanks.

1. a. A vector function is a function of the form $\mathbf{r}(t) =$ _____, where f, g, and h are _____ functions of a variable _____, called a _____.
   b. The domain of $\mathbf{r}(t)$, called the _____ _____ is a subset of the _____ _____.

2. a. A space curve is traced out by the _____ point of $\mathbf{r}(t)$ as t takes on all values in the _____ interval $[a, b]$.
   b. The terminal point of $\mathbf{r}(a)$ corresponds to the _____ point of the curve, and the terminal point of $\mathbf{r}(b)$ gives the _____ point of the curve.

3. a. If $\mathbf{r}(t) = \langle f(t), g(t), h(t) \rangle$, then $\lim_{t \to a} \mathbf{r}(t)$ exists if and only if _____, _____, and _____ exist.
   b. A vector function $\mathbf{r}$ is continuous at a if $\lim_{t \to a} \mathbf{r}(t) =$ _____. The function is _____ on I if it is continuous at each point in I.

4. a. The derivative of $\mathbf{r}$ is $\mathbf{r}'(t) =$ _____, provided that the limit exists.
   b. If $\mathbf{r}(t) = \langle f(t), g(t), h(t) \rangle$, then $\mathbf{r}'(t) =$ _____.

5. a. If $\mathbf{u}$ and $\mathbf{v}$ are differentiable, then $\dfrac{d}{dt}[\mathbf{u}(t) \cdot \mathbf{v}(t)] =$

_____, and $\dfrac{d}{dt}[\mathbf{u}(t) \times \mathbf{v}(t)] =$ _____.

   b. If $\mathbf{u}$ is differentiable and f is differentiable, then

$\dfrac{d}{dt}[\mathbf{u}(f(t))] =$ _____.

6. If f, g, and h are integrable and $\mathbf{r}(t) = \langle f(t), g(t), h(t) \rangle$, then
   a. $\int \mathbf{r}(t)\, dt =$ _____.
   b. $\int_a^b \mathbf{r}(t)\, dt =$ _____.

7. The length of the curve C, $\mathbf{r}(t) = \langle f(t), g(t), h(t) \rangle$, where $a \leq t \leq b$, is given by $L =$ _____.

8. a. A curve C described by $\mathbf{r}(t)$ with parameter t is said to be _____ by t.
   b. The arc length function s associated with a smooth curve C described by $\mathbf{r}(t)$ is $s(t) =$ _____.
   c. The curve C has arc length parametrization if it is parametrized by the _____ _____ function $s(t)$.

9. a. If a smooth curve C is described by $\mathbf{r}(s)$, where s is the arc length parameter, then the curvature of C is $\kappa(s) =$

_____.
   b. If C is parametrized by t, then $\kappa(t) =$ _____.
   c. The curvature of C is also given by $\kappa(t) =$ _____.
   d. If C is a plane curve, then $\kappa(x) =$ _____.
   e. For a plane curve C, the reciprocal of the curvature at the point P is called the _____ _____ _____ of C at P, and the circle with this _____ that shares a common _____ _____ with the curve at P is called the _____ _____ _____.

10. If the position vector of a particle is $\mathbf{r}(t)$, then its velocity is _____, its speed is _____, and its acceleration is _____. The acceleration vector points toward the _____ side of the trajectory of the particle.

11. a. If C is a smooth curve described by the vector function $\mathbf{r}(t)$, then the unit tangent vector is $\mathbf{T}(t) =$ _____, and the principal unit normal vector is $\mathbf{N}(t) =$ _____.
   b. The acceleration of a particle can be resolved into the sum of two vectors—one along the direction of _____ and the other along the direction of _____. In fact, $\mathbf{a} = a_T\mathbf{T} + a_N\mathbf{N}$, where $a_T =$ _____ and $a_N =$ _____; the former is called the scalar _____ component of acceleration, and the latter is called the scalar _____ component of acceleration.
   c. In terms of $\mathbf{r}$ and its derivatives, $a_T =$ _____, and $a_N =$ _____.

REVIEW EXERCISES

In Exercises 1–4, sketch the curve with the given vector equation, and indicate the orientation of the curve.

1. $\mathbf{r}(t) = (2 + 3t)\mathbf{i} + (2t - 1)\mathbf{j}$

2. $\mathbf{r}(t) = t^3\mathbf{i} + t^2\mathbf{j}; \quad 0 \le t \le 2$

3. $\mathbf{r}(t) = (\cos t - 1)\mathbf{i} + (\sin t + 2)\mathbf{j} + 2\mathbf{k}$

4. $\mathbf{r}(t) = 2\cos t\mathbf{i} + 3\sin t\mathbf{j} + t^2\mathbf{k}; \quad 0 \le t \le 2\pi$

5. Find the domain of $\mathbf{r}(t) = \dfrac{1}{\sqrt{5 - t}}\mathbf{i} + \dfrac{\sin t}{t}\mathbf{j} + \ln(1 + t)\mathbf{k}$.

6. Find $\lim\limits_{t \to 0^+} \mathbf{r}(t)$, where $\mathbf{r}(t) = \dfrac{\sqrt{t}}{1 + t^2}\mathbf{i} + \dfrac{t^2}{\sin t}\mathbf{j} + \dfrac{e^t - 1}{t}\mathbf{k}$.

7. Find the interval in which

$$\mathbf{r}(t) = \sqrt{t + 1}\mathbf{i} + \frac{e^t}{\sqrt{2 - t}}\mathbf{j} + \frac{t^2}{(t - 1)^2}\mathbf{k}$$

is continuous.

8. Find $\mathbf{r}'(t)$ if $\mathbf{r}(t) = \left[\displaystyle\int_0^t \cos^2 u\, du\right]\mathbf{i} + \left[\displaystyle\int_0^{t^2} \sin u\, du\right]\mathbf{j}$.

In Exercises 9–12, find $\mathbf{r}'(t)$ and $\mathbf{r}''(t)$.

9. $\mathbf{r}(t) = \sqrt{t}\mathbf{i} + t^2\mathbf{j} + \dfrac{1}{t + 1}\mathbf{k}$

10. $\mathbf{r}(t) = e^{-t}\mathbf{i} + t\cos t\mathbf{j} + t\sin t\mathbf{k}$

11. $\mathbf{r}(t) = (t^2 + 1)\mathbf{i} + 2t\mathbf{j} + \ln t\mathbf{k}$

12. $\mathbf{r}(t) = \langle t\sin t, t\cos t, e^{2t}\rangle$

In Exercises 13 and 14, find parametric equations for the tangent line to the curve with the given parametric equations at the point with the given value of t.

13. $x = t^2 + 1, \quad y = 2t - 3, \quad z = t^3 + 1; \quad t = 0$

14. $x = t\cos t - \sin t, \quad y = t\sin t + \cos t, \quad z = t^2; \quad t = \dfrac{\pi}{2}$

In Exercises 15 and 16, evaluate the integral.

15. $\displaystyle\int \left(\sqrt{t}\mathbf{i} + e^{-2t}\mathbf{j} + \frac{1}{t + 1}\mathbf{k}\right) dt$

16. $\displaystyle\int_0^1 (2t\mathbf{i} + t^2\mathbf{j} + t^{3/2}\mathbf{k})\, dt$

In Exercises 17 and 18, find $\mathbf{r}(t)$ for the vector function $\mathbf{r}'(t)$ or $\mathbf{r}''(t)$ and the given initial condition(s).

17. $\mathbf{r}'(t) = 2\sqrt{t}\mathbf{i} + 3\cos 2\pi t\mathbf{j} - e^{-t}\mathbf{k}; \quad \mathbf{r}(0) = \mathbf{i} + 2\mathbf{j}$

18. $\mathbf{r}''(t) = 2\mathbf{i} + t\mathbf{j} + e^{-t}\mathbf{k}; \quad \mathbf{r}'(0) = \mathbf{i} + \mathbf{k}$,
$\mathbf{r}(0) = 2\mathbf{i} + \mathbf{j} + 3\mathbf{k}$

In Exercises 19 and 20, find the unit tangent and the unit normal vectors for the curve C defined by $\mathbf{r}(t)$ for the given value of t.

19. $\mathbf{r}(t) = t\mathbf{i} + t^2\mathbf{j} + t^3\mathbf{k}; \quad t = 1$

20. $\mathbf{r}(t) = 2\cos t\mathbf{i} + 2\sin t\mathbf{j} + e^t\mathbf{k}; \quad t = 0$

In Exercises 21 and 22, find the length of the curve.

21. $\mathbf{r}(t) = 2\sin 2t\mathbf{i} + 2\cos 2t\mathbf{j} + 3t\mathbf{k}; \quad 0 \le t \le 2$

22. $\mathbf{r}(t) = \sqrt{2}t\mathbf{i} + \dfrac{1}{2}t^2\mathbf{j} + \ln t\mathbf{k}; \quad 1 \le t \le 2$

In Exercises 23 and 24, find the curvature of the curve.

23. $\mathbf{r}(t) = t\mathbf{i} + t^2\mathbf{j} + t^3\mathbf{k}$

24. $\mathbf{r}(t) = t\sin t\mathbf{i} + t\cos t\mathbf{j} + t\mathbf{k}$

In Exercises 25 and 26, find the curvature of the plane curve, and determine the point on the curve at which the curvature is largest.

25. $y = x - \dfrac{1}{4}x^2$ **26.** $y = e^{-x}$

In Exercises 27 and 28, find the velocity, acceleration, and speed of the object with the given position vector.

27. $\mathbf{r}(t) = 2t\mathbf{i} + e^{-2t}\mathbf{j} + \cos t\mathbf{k}$

28. $\mathbf{r}(t) = te^{-t}\mathbf{i} + \cos 2t\mathbf{j} + \sin 2t\mathbf{k}$

In Exercises 29 and 30, find the velocity and position vectors of an object with the given acceleration and the given initial velocity and position.

29. $\mathbf{a}(t) = t\mathbf{i} + \dfrac{1}{3}t^2\mathbf{j} + 3\mathbf{k}; \quad \mathbf{v}(0) = 2\mathbf{i} + 3\mathbf{j} + \mathbf{k}, \quad \mathbf{r}(0) = \mathbf{0}$

30. $\mathbf{a}(t) = e^t\mathbf{i} + e^{-t}\mathbf{j} + t\mathbf{k}; \quad \mathbf{v}(0) = 2\mathbf{i}, \quad \mathbf{r}(0) = \mathbf{i} + \mathbf{k}$

In Exercises 31–34, find the scalar tangential and normal components of acceleration of a particle with the given position vector.

31. $\mathbf{r}(t) = \mathbf{i} + t\mathbf{j} + t^2\mathbf{k}$

32. $\mathbf{r}(t) = 2\cos t\mathbf{i} + 3\sin t\mathbf{j} + t\mathbf{k}$

33. $\mathbf{r}(t) = \cos t\mathbf{i} + \sin 2t\mathbf{j}$

34. $\mathbf{r}(t) = \sqrt{2}t\mathbf{i} + e^t\mathbf{j} + e^{-t}\mathbf{k}$

35. A Shot Put In a track and field meet, a shot putter heaves a shot at an angle of $45°$ with the horizontal. As the shot leaves her hand, it is at a height of 7 ft and moving at a speed of 40 ft/sec. Set up a coordinate system so that the shot putter is at the origin.
   a. What is the position of the shot at time t?
   b. How far is her put?

CHALLENGE PROBLEMS

1. a. Show that the curve C defined by the vector function

$$\mathbf{r}(t) = (a_1t^2 + b_1t + c)\mathbf{i} + (a_2t^2 + b_2t + c)\mathbf{j} + (a_3t^2 + b_3t + c)\mathbf{k}$$

lies in a plane.

b. Show that the plane of part (a) can be written in the form

$$\begin{vmatrix} x - c & y - c & z - c \\ a_1 & a_2 & a_3 \\ b_1 & b_2 & b_3 \end{vmatrix} = 0$$

2. Tracking Planes in a Holding Pattern at an Airport Suppose that an airport is located at the origin of a three-dimensional coordinate system and two airplanes are circling the airport in a holding pattern at an altitude of 2 mi. The planes fly at a constant speed of 300 mph along circular paths of radius 10 mi and are separated by 90°, as shown in the figure.

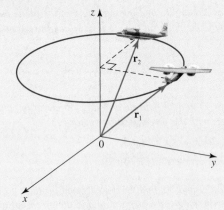

a. Show that the position vectors of the planes are

$$\mathbf{r}_1(t) = 10 \cos 30t\,\mathbf{i} + 10 \sin 30t\,\mathbf{j} + 2\mathbf{k}$$

and

$$\mathbf{r}_2(t) = -10 \sin 30t\,\mathbf{i} + 10 \cos 30t\,\mathbf{j} + 2\mathbf{k}$$

respectively.

b. Let $\mathbf{r} = \mathbf{r}_2 - \mathbf{r}_1$. Interpret your results.

c. Find $\mathbf{r}'$, and interpret your result.

d. Find $\mathbf{r}''$, and interpret your result.

3. Hitting a Moving Target A target is located at a height h over level ground, and a gun, located at ground level and at a distance d from the point directly below the target, is aimed directly at the target. Suppose that the gun is fired at the instant the target is released.

a. Show that the bullet will hit the target if its initial speed v_0 satisfies

$$v_0 \geq \sqrt{\frac{g(d^2 + h^2)}{2h}}$$

b. Assuming that the condition in part (a) is satisfied, find the distance the target has fallen before it was hit.

4. Motion of a Projectile A projectile of mass m is fired from the origin of a coordinate system at an angle of elevation α. Assume that air resistance acting on the projectile is proportional to its velocity. Then by Newton's Second Law of Motion the motion of the projectile is described by the equation

$$m\mathbf{r}'' = -mg\mathbf{j} - k\mathbf{r}' \qquad (1)$$

where $\mathbf{r}(t)$ is the position vector of the projectile and $k > 0$ is the constant of proportionality.

a. By integrating Equation (1), obtain the equation

$$\mathbf{r}' + \frac{k}{m}\mathbf{r} = -gt\mathbf{j} + \mathbf{v}_0 \qquad (2)$$

where $\mathbf{v}_0 = \mathbf{v}(0) = \mathbf{r}'(0)$.

b. Multiply both sides of Equation (2) by $e^{(k/m)t}$, and show that the left-hand side of the resulting equation can be written as $\dfrac{d}{dt}[e^{(k/m)t}\mathbf{r}(t)]$. Make use of this observation to find an expression for $\mathbf{r}(t)$.

5. Motion of a Projectile Refer to Exercise 4.

a. If the initial speed of the projectile is v_0, show that the position function $\mathbf{r}(t)$ is equivalent to the parametric equations

$$x(t) = \frac{mv_0 \cos \alpha}{k}(1 - e^{-(k/m)t})$$

$$y(t) = \left(\frac{m^2 g}{k^2} + \frac{mv_0 \sin \alpha}{k}\right)(1 - e^{-(k/m)t}) - \frac{mg}{k}t$$

b. Solve the first equation in part (a) for t to obtain

$$t = \frac{m}{k}\ln\frac{mv_0 \cos \alpha}{mv_0 \cos \alpha - kx}$$

Then substitute this value into the second equation in part (a) to obtain

$$y = \left(\frac{mg}{kv_0 \cos \alpha} + \tan \alpha\right)x + \frac{m^2 g}{k^2}\ln\left(1 - \frac{kx}{mv_0 \cos \alpha}\right) \quad (3)$$

c. Suppose that a projectile of weight 1600 lb is fired from the origin with an initial speed of 1200 mph and at an angle of elevation of 30°. Draw the trajectories of the projectile for values of k equal to 0.01, 0.1, 0.5, and 1, using the viewing rectangle $[0, 100{,}000] \times [0, 15{,}000]$. Comment on the shape of the trajectories.

d. Expand the expression for y in Equation (3) as a power series to show that

$$y = (\tan \alpha)x - \frac{1}{2}\frac{g}{(v_0 \cos \alpha)^2}x^2 - \frac{1}{3}\frac{kg}{m(v_0 \cos \alpha)^3}x^3 - \cdots$$

and hence deduce that if k is very small, then the trajectory of the projectile is almost parabolic.

6. A particle moves in a circular orbit in the plane given by $\mathbf{r}(t) = R \cos t \mathbf{i} + R \sin t \mathbf{j}$, where R is a constant. At a certain instant of time, the particle is to be released so that it will strike a target located at the point (a, b), where $a^2 + b^2 > R^2$. Find the time at which the particle is to be released.

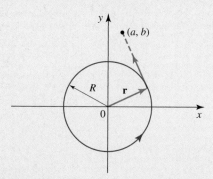

7. **Coriolis Acceleration** Consider the motion of a particle in the xy-plane in which the position of the particle is given in polar coordinates r and θ.

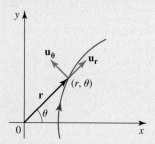

a. If $\mathbf{u_r}$ and $\mathbf{u_\theta}$ are unit vectors that point in the direction of the position vector and at right angles to it (in the direction of increasing θ), respectively, show that

$$\mathbf{u_r} = \cos \theta \mathbf{i} + \sin \theta \mathbf{j}$$

$$\mathbf{u_\theta} = -\sin \theta \mathbf{i} + \cos \theta \mathbf{j}$$

b. If $\mathbf{r} = r\mathbf{u_r}$ is the position vector of a particle located at (r, θ), show that its velocity vector is given by

$$\mathbf{v}(t) = \mathbf{r}'(t) = \frac{dr}{dt} \mathbf{u_r} + r \frac{d\theta}{dt} \mathbf{u_\theta}$$

and its acceleration vector is given by

$$\mathbf{a}(t) = \frac{d\mathbf{v}}{dt} = \left[\frac{d^2 r}{dt^2} - r \left(\frac{d\theta}{dt} \right)^2 \right] \mathbf{u_r} + \left[r \frac{d^2 \theta}{dt^2} + 2 \frac{dr}{dt} \frac{d\theta}{dt} \right] \mathbf{u_\theta}$$

Note: The fourth term in the expression for $\mathbf{a}(t)$

$$2 \frac{dr}{dt} \frac{d\theta}{dt} \mathbf{u_\theta}$$

is called the Coriolis acceleration. It is due partly to the change in the direction of the radial component of velocity and partly to the change in the transverse component of velocity.

8. **Kepler's Second Law of Planetary Motion** Use the result of Exercise 7(b) to prove Kepler's Second Law of Planetary Motion: The radius vector in a central force field (that is, one in which the force is always directed radially toward or away from the origin) sweeps over area at a constant rate. **Hint:** Use Newton's Second Law of Motion, $\mathbf{F} = m\mathbf{a}$, to show that $mr^2 \frac{d\theta}{dt} = C$, where C is constant, and use the fact that the area swept out by $\mathbf{r}$ is

$$\frac{dA}{dt} = \frac{1}{2} r^2 \frac{d\theta}{dt}$$

9. **Coriolis Acceleration** A turntable rotates at a constant angular velocity of 30 rev/min. An ant walks from the center of the turntable outward toward the edge at a speed of 2 cm/sec (relative to the turntable).
   a. What are the speed and the magnitude of the acceleration of the ant 3 sec later?
   b. What is the magnitude of the Coriolis acceleration at that time?
   Hint: Use the results of Challenge Problem 7.

10. **Path of a Boat** The path of a boat is given by

$$\mathbf{r}(t) = \frac{1}{5}(t^2 - 4t + 8)\mathbf{i} + \frac{1}{5}(3t^2 - 6t + 4)\mathbf{j} \qquad 0 \le t \le 3$$

The shoreline lies along the positive x-axis. All distances are measured in miles and time is measured in minutes.
   a. Plot the path of the boat.
   b. At what time is the boat closest to the shoreline? What is the distance of the boat from the shoreline at that time?
   c. What is the velocity, speed, and acceleration of the boat at the time it is closest to the shoreline?

The rules for the new International America's cup class include a formula that governs the basic yacht dimensions. This formula balances the rated length, the sail area, and the displacement of the yacht. It is an example of an expression involving three variables. We will use this formula in this chapter.

Shaun Botterill/Getty Images

13

Functions of Several Variables

UP TO NOW we have dealt primarily with functions involving one independent variable. In this chapter we consider functions involving two or more independent variables. The related notions of limits, continuity, differentiability, and optimization of a function of one variable have their counterparts in the case of a function of several variables, and we will develop these concepts in this chapter. As we will see, many real-life applications of mathematics involve more than one independent variable.

V This symbol indicates that one of the following video types is available for enhanced student learning at **www.academic.cengage.com/login:**

• Chapter lecture videos   • Solutions to selected exercises

V This symbol indicates that step-by-step video lessons for hand-drawing certain complex figures are available.

13.1   Functions of Two or More Variables

■ Functions of Two Variables

Up to now, we have dealt only with functions of one variable. In practice, however, we often encounter situations in which one quantity depends on two or more quantities. For example, consider the following:

- The volume V of a right circular cylinder depends on its radius r and its height h ($V = \pi r^2 h$).
- The volume V of a rectangular box depends on its length l, width w, and height h ($V = lwh$).
- The revenue R from the sale of commodities A, B, C, and D at the unit prices of 10, 14, 20, and 30 dollars, respectively, depends on the number of units x, y, z, and w of commodities A, B, C, and D sold ($R = 10x + 14y + 20z + 30w$).

Just as we used a function of one variable to describe the dependency of one variable on another, we can use the notion of a function of several variables to describe the dependency of one variable on several variables. We begin with the definition of a function of two variables.

DEFINITION   Function of Two Variables

Let $D = \{(x, y) \,|\, x, y \in R\}$ be a subset of the xy-plane. A **function f of two variables** is a rule that assigns to each ordered pair of real numbers (x, y) in D a unique real number z. The set D is called the **domain** of f, and the set of corresponding values of z is called the **range** of f.

The number z is usually written $z = f(x, y)$. The variables x and y are **independent variables,** and z is the **dependent variable.**

As in the case of a function of a single variable, a function of two or more variables can be described verbally, numerically, graphically, or algebraically.

EXAMPLE 1   **Home Mortgage Payments**   In a typical housing loan, the borrower makes periodic payments toward reducing indebtedness to the lender, who charges interest at a fixed rate on the unpaid portion of the debt. In practice, the borrower is required to repay the lender in periodic installments, usually of the same size over a fixed term, so that the loan (principal plus interest charges) is amortized at the end of the term. Table 1 gives the monthly loan repayment on a loan of $1000, $f(t, r)$, where t is the term of the loan in years and r is the interest rate per annum (%/year) compounded monthly. Referring to the table, we see that the monthly installment for a 30-year loan of $1000 when the current interest rate is 7%/year is given by $f(30, 7) = 6.6530$ (dollars). Therefore, if the amount borrowed is $350,000, the monthly repayment is 350(6.6530), or $2328.55.

TABLE 1

					Interest rate %/year				
t \ r	**6**	$6\frac{1}{4}$	$6\frac{1}{2}$	$6\frac{3}{4}$	**7**	$7\frac{1}{4}$	$7\frac{1}{2}$	$7\frac{3}{4}$	**8**
5	19.3328	19.4493	19.5661	19.6835	19.8012	19.9194	20.0379	20.1570	20.2764
10	11.1021	11.2280	11.3548	11.4824	11.6108	11.7401	11.8702	12.0011	12.1328
15	8.4386	8.5742	8.7111	8.8491	8.9883	9.1286	9.2701	9.4128	9.5565
20	7.1643	7.3093	7.4557	7.6036	7.7530	7.9038	8.0559	8.2095	8.3644
25	6.4430	6.5967	6.7521	6.9091	7.0678	7.2281	7.3899	7.5533	7.7182
30	5.9955	6.1572	6.3207	6.4860	6.6530	6.8218	6.9921	7.1641	7.3376
35	5.7019	5.8708	6.0415	6.2142	6.3886	6.5647	6.7424	6.9218	7.1026
40	5.5021	5.6774	5.8546	6.0336	6.2143	6.3967	6.5807	6.7662	6.9531

Term of the loan (years)

Although the monthly installments based on a $1000 loan are displayed in the form of a table for selected values of t and r in Example 1, an algebraic expression for computing $f(t, r)$ also exists:

$$f(t, r) = \frac{10r}{12\left[1 - \left(1 + \dfrac{0.01r}{12}\right)^{-12t}\right]}$$

But as in the case of a single variable, we are primarily interested in functions that can be described by an equation relating the **dependent variable** z to the **independent variables** x and y. Also, as in the case of a single variable, unless otherwise specified, the domain of a function of two variables is the set of all points (x, y) for which $z = f(x, y)$ is a real number.

EXAMPLE 2 Let $f(x, y) = x^2 - xy + 2y$. Find the domain of f, and evaluate $f(1, 2)$, $f(2, 1)$, $f(t, 2t)$, $f(x^2, y)$, and $f(x + y, x - y)$.

Solution Since $x^2 - xy + 2y$ is a real number whenever (x, y) is an ordered pair of real numbers, we see that the domain of f is the entire xy-plane. Next, we have

$$f(1, 2) = 1^2 - (1)(2) + 2(2) = 3$$
$$f(2, 1) = 2^2 - (2)(1) + 2(1) = 4$$
$$f(t, 2t) = t^2 - (t)(2t) + 2(2t) = -t^2 + 4t$$
$$f(x^2, y) = (x^2)^2 - (x^2)(y) + 2y = x^4 - x^2y + 2y$$

and

$$f(x + y, x - y) = (x + y)^2 - (x + y)(x - y) + 2(x - y)$$
$$= x^2 + 2xy + y^2 - x^2 + y^2 + 2x - 2y$$
$$= 2(y^2 + xy + x - y)$$

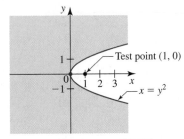

(a) The domain of $f(x, y) = \sqrt{y^2 - x}$

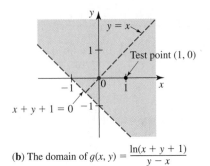

(b) The domain of $g(x, y) = \dfrac{\ln(x + y + 1)}{y - x}$

FIGURE 1

EXAMPLE 3 Find and sketch the domain of the function:

a. $f(x, y) = \sqrt{y^2 - x}$ **b.** $g(x, y) = \dfrac{\ln(x + y + 1)}{y - x}$

Solution

a. $f(x, y)$ is a real number provided that $y^2 - x \geq 0$. Therefore, the domain of f is

$$D = \{(x, y) \mid y^2 - x \geq 0\}$$

To sketch the region D, we first draw the curve $y^2 - x = 0$, or $y^2 = x$, which is a parabola (Figure 1a). Observe that this curve divides the xy-plane into two regions: the points satisfying $y^2 - x > 0$ and the points satisfying $y^2 - x < 0$. To determine the region of interest, we pick a point in one of the regions, say, the point $(1, 0)$. Substituting the coordinates $x = 1$ and $y = 0$ into the inequality $y^2 - x > 0$, we obtain $0 - 1 > 0$, which is false. This shows that the test point is *not* contained in the required region. Therefore, the region that does not contain the test point together with the curve $x = y^2$ is the required domain (Figure 1a).

b. Because the logarithmic function is defined only for positive numbers, we must have $x + y + 1 > 0$. Furthermore, the denominator of the expression cannot be zero, so $y - x \neq 0$, or $y \neq x$. Therefore, the domain of g is

$$D = \{(x, y) \mid x + y + 1 > 0 \quad \text{and} \quad y \neq x\}$$

To sketch the domain of D, we first draw the graph of the equation

$$x + y + 1 = 0$$

which is a straight line. The dashed line is used to indicate that points on the line are not included in D. This line divides the xy-plane into two half-planes. If we pick the test point $(1, 0)$ and substitute the coordinates $x = 1$ and $y = 0$ into the inequality $x + y + 1 > 0$, we obtain $2 > 0$, which is true. This computation tells us that the upper half-plane containing the test point satisfies the inequality $x + y + 1 > 0$. Next, because $y \neq x$, all the points lying on the line $y = x$ in this half-plane must be excluded from D. Again, we indicate this with a dashed line (Figure 1b). ∎

Graphs of Functions of Two Variables

Just as the graph of a function of one variable enables us to visualize the function, so too does the graph of a function of two variables.

DEFINITION **Graph of a Function of Two Variables**

Let f be a function of two variables with domain D. The graph of f is the set

$$S = \{(x, y, z) \mid z = f(x, y), (x, y) \in D\}$$

FIGURE 2
The graph of f is the surface S consisting of all points (x, y, z), where $z = f(x, y)$ and $(x, y) \in D$.

Since each ordered triple (x, y, z) may be represented as a point in three-dimensional space, R^3, the set S is a surface in space (see Figure 2).

EXAMPLE 4 Sketch the graph of $f(x, y) = \sqrt{9 - x^2 - y^2}$. What is the range of f?

Solution The domain of f is $D = \{(x, y) \mid x^2 + y^2 \leq 9\}$, the disk with radius 3, centered at the origin. Writing $z = f(x, y)$, we have

$$z = \sqrt{9 - x^2 - y^2}$$
$$z^2 = 9 - x^2 - y^2$$

or

$$x^2 + y^2 + z^2 = 9$$

The last equation represents a sphere of radius 3 centered at the origin. Since $z \geq 0$, we see that the graph of f is just an upper hemisphere (Figure 3). Furthermore, z must be less than or equal to 3, so the range of f is $[0, 3]$.

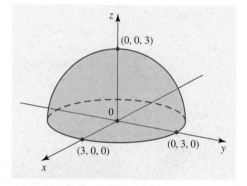

Ⓥ ✐ **FIGURE 3**
The graph of $f(x, y) = \sqrt{9 - x^2 - y^2}$ is the upper hemisphere of radius 3, centered at the origin.

■ Computer Graphics

The graph of a function of two variables can be sketched with the aid of a graphing utility. In most cases the techniques that are used involve plotting the traces of a surface in the vertical planes $x = k$ and $y = k$ for equally spaced values of k. The program uses a "hidden line" routine that determines what parts of certain traces should be eliminated to give the illusion of the surface in three dimensions. In the next example we sketch the graph of a function of two variables and then show a computer-generated version of it.

EXAMPLE 5 Let $f(x, y) = x^2 + 4y^2$.

a. Sketch the graph of f.   **b.** Use a CAS to plot the graph of f.

Solution

a. We recognize that the graph of the function is the surface $z = x^2 + 4y^2$, which is the elliptic paraboloid

$$\frac{x^2}{1} + \frac{y^2}{\left(\dfrac{1}{2}\right)^2} = z$$

Using the drawing skills developed in Section 11.6, we obtain the sketch shown in Figure 4a.

b. The computer-generated graph of f is shown in Figure 4(b).

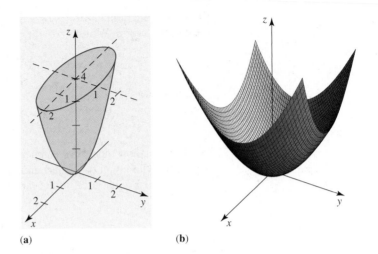

V ✏ **FIGURE 4**
The graph of $f(x, y) = x^2 + 4y^2$

(a) (b)

Figure 5 shows the computer-generated graphs of several functions.

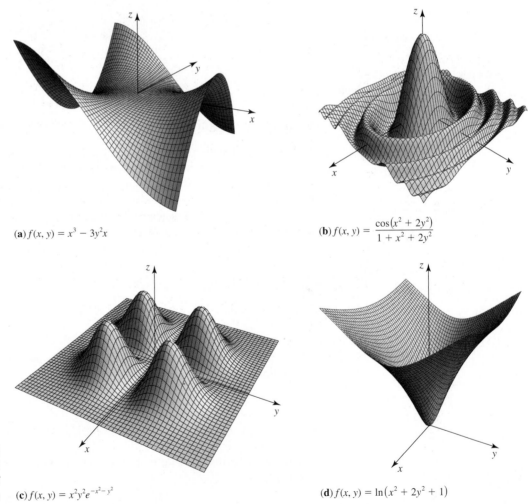

(a) $f(x, y) = x^3 - 3y^2 x$

(b) $f(x, y) = \dfrac{\cos(x^2 + 2y^2)}{1 + x^2 + 2y^2}$

FIGURE 5
Some computer-generated
graphs of functions of
two variables

(c) $f(x, y) = x^2 y^2 e^{-x^2 - y^2}$

(d) $f(x, y) = \ln(x^2 + 2y^2 + 1)$

■ Level Curves

We can visualize the graph of a function of two variables by using *level curves.* To define the level curve of a function f of two variables, let $z = f(x, y)$ and consider the trace of f in the plane $z = k$ (k, a constant), as shown in Figure 6a. If we project this trace onto the xy-plane, we obtain a curve C with equation $f(x, y) = k$, called a *level curve* of f (Figure 6b).

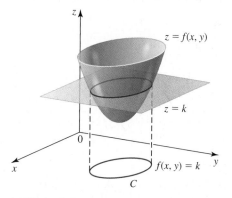

(a) The level curve C with equation $f(x, y) = k$ is the projection of the trace of f in the plane $z = k$ onto the xy-plane.

(b) The level curve C

FIGURE 6

DEFINITION **Level Curves**

The **level curves** of a function f of two variables are the curves in the xy-plane with equations $f(x, y) = k$, where k is a constant in the range of f.

Notice that the level curve with equation $f(x, y) = k$ is the set of all points in the domain of f corresponding to the points on the surface $z = f(x, y)$ having the same height or depth k. By drawing the level curves corresponding to several admissible values of k, we obtain a *contour map.* The map enables us to visualize the surface represented by the graph of $z = f(x, y)$: We simply lift or depress the level curve to see the "cross sections" of the surface. Figure 7a shows a hill, and Figure 7b shows a contour map associated with that hill.

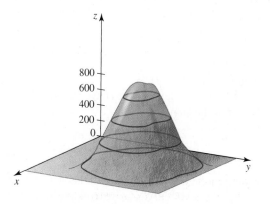

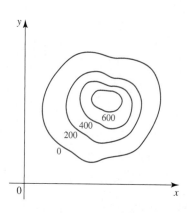

FIGURE 7   **(a)** A hill

(b) A contour map of the hill

EXAMPLE 6 Sketch a contour map for the surface described by $f(x, y) = x^2 + y^2$, using the level curves corresponding to $k = 0, 1, 4, 9$, and 16.

Solution The level curve of f corresponding to each value of k is a circle $x^2 + y^2 = k$ of radius $\sqrt{k}$, centered at the origin. For example, if $k = 4$, the level curve is the circle with equation $x^2 + y^2 = 4$, centered at the origin and having radius 2. The required contour map of f comprises the origin and the four concentric circles shown in Figure 8a. The graph of f is the paraboloid shown in Figure 8b.

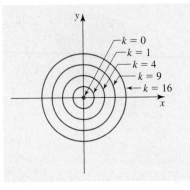

V ✏ **FIGURE 8**   (**a**) Contour map for $f(x, y) = x^2 + y^2$ (**b**) The graph of $z = x^2 + y^2$

EXAMPLE 7 Sketch a contour map for the hyperbolic paraboloid defined by $f(x, y) = y^2 - x^2$.

Solution The level curve corresponding to each value of k is the graph of the equation $y^2 - x^2 = k$. For $k > 0$ the level curves have equations

$$\frac{y^2}{k} - \frac{x^2}{k} = 1$$

or

$$\frac{y^2}{(\sqrt{k})^2} - \frac{x^2}{(\sqrt{k})^2} = 1$$

These curves are a family of hyperbolas with asymptotes $y = \pm x$ and vertices $(0, \pm\sqrt{k})$. For example, if $k = 4$, then the level curve is the hyperbola

$$\frac{y^2}{4} - \frac{x^2}{4} = 1$$

with vertices $(0, \pm 2)$.

If $k < 0$, the level curves have equations $y^2 - x^2 = k$ or $x^2 - y^2 = -k$, which can be put in the standard form

$$\frac{x^2}{(\sqrt{-k})^2} - \frac{y^2}{(\sqrt{-k})^2} = 1$$

and represent a family of hyperbolas with asymptotes $y = \pm x$. The contour map comprising the level curves corresponding to $k = 0, \pm 2, \pm 4, \pm 6$, and ± 8 is sketched in Figure 9a. The graph of $z = y^2 - x^2$ is shown in Figure 9b.

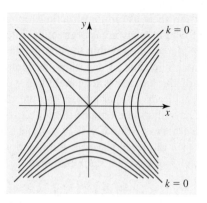

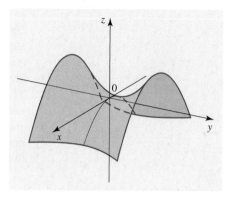

FIGURE 9   (**a**) Contour map for $f(x, y) = y^2 - x^2$ (**b**) The graph of $z = y^2 - x^2$

Figure 10 shows some computer-generated graphs of functions of two variables and their corresponding level curves.

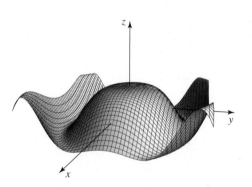

(**a**) Graph of $f(x, y) = \cos\left(\dfrac{x^2 + 2y^2}{4}\right)$

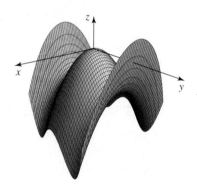

(**b**) Graph of $f(x, y) = y^4 - 8y^2 + 4x^2$

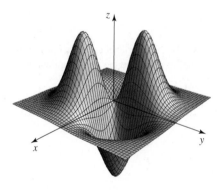

(**c**) Graph of $f(x, y) = -xye^{-x^2 - y^2}$

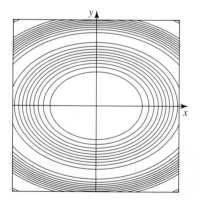

(**d**) Level curves of $f(x, y) = \cos\left(\dfrac{x^2 + 2y^2}{4}\right)$

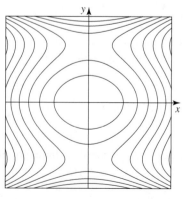

(**e**) Level curves of $f(x, y) = y^4 - 8y^2 + 4x^2$

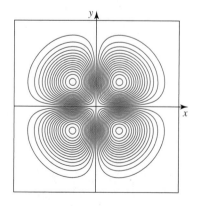

(**f**) Level curves of $f(x, y) = -xye^{-x^2 - y^2}$

FIGURE 10
The graphs of some functions and their level curves

Aside from their use in constructing topographic maps of mountain ranges, level curves are found in many areas of practical interest. For example, if $T(x, y)$ denotes the temperature at a location within the continental United States with longitude x and latitude y at a certain time of day, then the temperature at the point (x, y) is the height (or depth) of the surface with equation $z = T(x, y)$. In this context the level curve $T(x, y) = k$ is a curve superimposed on the map of the United States connecting all points that have the same temperature at a given time (Figure 11). These level curves are called **isotherms.** Similarly, if $P(x, y)$ measures the barometric pressure at the location (x, y), then the level curves of the function P are called **isobars.** All points on an isobar $P(x, y) = k$ have the same barometric pressure at a given time.

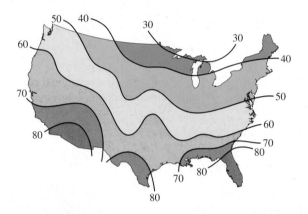

FIGURE 11
Isotherms: level curves connecting points that have the same temperature

Functions of Three Variables and Level Surfaces

A function f of three variables is a rule that assigns to each ordered triple (x, y, z) in a domain $D = \{(x, y, z) \mid x, y, z \in R\}$ a unique real number w denoted by $f(x, y, z)$. For example, the volume V of a rectangular box of length x, width y, and height z can be described by the function f defined by $f(x, y, z) = xyz$.

EXAMPLE 8 Find the domain of the function f defined by

$$f(x, y, z) = \sqrt{x + y - z} + xe^{yz}$$

Solution $f(x, y, z)$ is a real number provided that $x + y - z \geq 0$ or, equivalently, $z \leq x + y$. Therefore, the domain of f is

$$D = \{(x, y, z) \mid z \leq x + y\}$$

This is the half-space consisting of all points lying on or below the plane $z = x + y$. ∎

Since the graph of a function of three variables is composed of the points (x, y, z, w), where $w = f(x, y, z)$, lying in four-dimensional space, we cannot draw the graphs of such functions. But by examining the **level surfaces,** which are the surfaces with equations

$$f(x, y, z) = k \qquad k, \text{a constant}$$

we are often able to gain some insight into the nature of f.

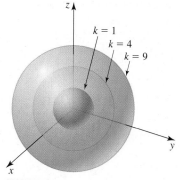

FIGURE 12
The level surfaces of
$f(x, y, z) = x^2 + y^2 + z^2$
corresponding to $k = 1, 4, 9$

EXAMPLE 9 Find the level surfaces of the function f defined by

$$f(x, y, z) = x^2 + y^2 + z^2$$

Solution The required level surfaces of f are the graphs of the equations $x^2 + y^2 + z^2 = k$, where $k \geq 0$. These surfaces are concentric spheres of radius $\sqrt{k}$ centered at the origin (see Figure 12). Observe that f has the same value for all points (x, y, z) lying on any such sphere. ■

13.1 CONCEPT QUESTIONS

1. What is a function of two variables? Give an example of one by stating its rule, domain, and range.
2. What is the graph of a function of two variables? Illustrate with a sketch.
3. What is a level curve of a function of two variables? Illustrate with a sketch.

4. What is a level surface of a function of three variables? If $w = T(x, y, z)$ gives the temperature of a point (x, y, z) in three-dimensional space, what does the level surface $w = k$ describe?

13.1 EXERCISES

1. Let $f(x, y) = x^2 + 3xy - 2x + 3$. Find
   a. $f(1, 2)$ b. $f(2, 1)$
   c. $f(2h, 3k)$ d. $f(x + h, y)$
   e. $f(x, y + k)$

2. Let $g(x, y) = \dfrac{2xy}{2x^2 + 3y^2}$. Find

   a. $g(-1, 2)$ b. $g(2, -1)$
   c. $g(u, -v)$ d. $g(2, a)$
   e. $g(u + v, v)$

3. Let $f(x, y, z) = \sqrt{x^2 + 2y^2 + 3z^2}$. Find
   a. $f(1, 2, 3)$ b. $f(0, 2, -1)$
   c. $f(t, -t, t)$ d. $f(u, u - 1, u + 1)$
   e. $f(-x, x, -2x)$

4. Let $g(r, s, t) = re^{s/t}$. Find
   a. $g(2, 0, 3)$ b. $g(1, \ln 3, 1)$
   c. $g(-1, -1, -1)$ d. $g(t, t, t)$
   e. $g(r + h, s + k, t + l)$

In Exercises 5–14, find the domain and the range of the function.

5. $f(x, y) = x + 3y - 1$ 6. $g(x, y) = x^2 + 2y^2 + 3$

7. $f(u, v) = \dfrac{uv}{u - v}$ 8. $h(x, y) = \sqrt{x - 2y}$

9. $g(x, y) = \sqrt{4 - x^2 - y^2}$ 10. $h(x, y) = \ln(xy - 1)$

11. $f(x, y, z) = \sqrt{9 - x^2 - y^2 - z^2}$

12. $g(x, y, z) = \dfrac{x}{y + z}$

13. $h(u, v, w) = \tan u + v \cos w$

14. $f(x, y, z) = \dfrac{1}{\sqrt{4 - x^2 - y^2 - z^2}}$

In Exercises 15–22, find and sketch the domain of the function.

15. $f(x, y) = \sqrt{y} - \sqrt{x}$ 16. $g(x, y) = \dfrac{xy}{2x - y}$

17. $f(u, v) = \dfrac{uv}{u^2 - v^2}$ 18. $h(x, y) = \sqrt{xy - 1}$

19. $f(x, y) = x \ln y + y \ln x$

20. $h(x, y) = \dfrac{\ln(y - x)}{\sqrt{x - y + 1}}$

21. $f(x, y, z) = \sqrt{9 - x^2 - y^2 - z^2}$

22. $g(x, y, z) = \dfrac{\sqrt{4 - x^2 - y^2}}{z - 3}$

V Videos for selected exercises are available online at **www.academic.cengage.com/login**.

In Exercises 23–30, sketch the graph of the function.

23. $f(x, y) = 4$

24. $f(x, y) = 6 - 2x + 3y$

25. $f(x, y) = x^2 + y^2$

26. $g(x, y) = y^2$

27. $h(x, y) = 9 - x^2 - y^2$

28. $f(x, y) = \sqrt{x^2 + y^2}$

29. $f(x, y) = \dfrac{1}{2}\sqrt{36 - 9x^2 - 36y^2}$

30. $f(x, y) = \cos x$

31. The figure shows the contour map of a hill. The numbers in the figure are measured in feet. Use the figure to answer the questions below.

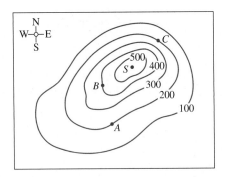

a. What is the altitude of the point on the hill corresponding to the point A? The point B?

b. If you start out from the point on the hill corresponding to point A and move north, will you be ascending or descending? What if you move east from the point on the hill corresponding to point B?

c. Is the hill steeper at the point corresponding to A or at the point corresponding to C? Explain.

32. A contour map of a function f is shown in the figure. Use it to estimate the value of f at P and Q.

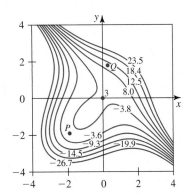

In Exercises 33–38, match the function with one of the graphs labeled a through f.

(a)

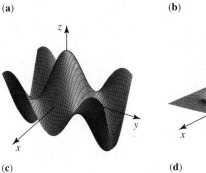

(b)

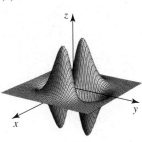

(c)

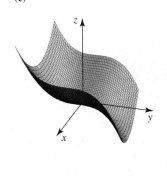

(d)

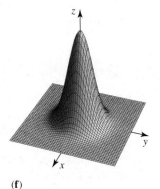

(e)

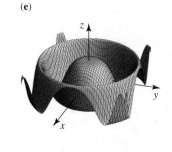

(f)

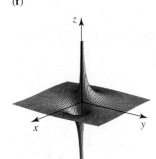

33. $f(x, y) = 2x^2 - y^3$

34. $f(x, y) = \cos(x^2 + y^2)$

35. $f(x, y) = \cos\dfrac{x}{2}\cos y$

36. $f(x, y) = (x^2 - y^2)e^{-x^2 - y^2}$

37. $f(x) = e^{-x^2 - y^2}$

38. $f(x, y) = -\dfrac{x}{2(x^2 + y^2)}$

cas *In Exercises 39–42, use a computer or calculator to plot the graph of the function.*

39. $f(x, y) = 3x^2 - 3y^2 + 2$

40. $f(x, y) = (4x^2 + 9y^2)e^{-x^2 - y^2}$

41. $f(x, y) = \cos x + \cos y$

42. $f(x, y) = \dfrac{1 - 2\sin(x^2 + y^2)}{x^2 + y^2}$

In Exercises 43–52, sketch the level curves $f(x, y) = k$ of the function for the indicated values of k.

43. $f(x, y) = 2x + 3y$; $\quad k = -2, -1, 0, 1, 2$

44. $f(x, y) = x^2 + 4y^2$; $\quad k = 0, 1, 2, 3, 4$

45. $f(x, y) = xy$; $\quad k = -2, -1, 0, 1, 2$

46. $f(x, y) = \sqrt{16 - x^2 - y^2}$; $\quad k = 0, 1, 2, 3, 4$

47. $f(x, y) = \dfrac{x + y}{x - y}$; $\quad k = -2, 0, 1, 2$

48. $f(x, y) = y^2 - x^2$; $\quad k = -2, -1, 0, 1, 2$

49. $f(x, y) = \ln(x + y)$; $\quad k = -2, -1, 0, 1, 2$

50. $f(x, y) = \dfrac{x}{y}$; $\quad k = -2, -1, 0, 1, 2$

51. $f(x, y) = y - x^2$; $\quad k = -2, -1, 0, 1, 2$

52. $f(x, y) = x - \sin y$; $\quad k = -2, -1, 0, 1, 2$

In Exercises 53–56, describe the level surfaces of the function.

53. $f(x, y, z) = 2x + 4y - 3z + 1$

54. $f(x, y, z) = 2x^2 + 3y^2 + 6z^2$

55. $f(x, y, z) = x^2 + y^2 - z^2$

56. $f(x, y, z) = -x^2 - y^2 + z + 2$

In Exercises 57–62, match the graph of the surface with one of the contour maps labeled a through f.

(a)

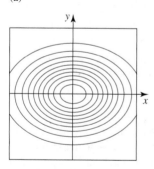

(b)

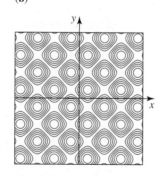

(c)

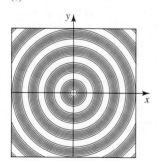

(d)

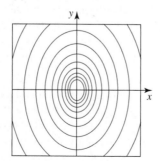

(e)

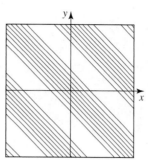

(f)

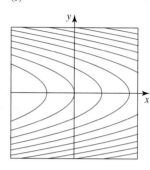

57. $f(x, y) = e^{1 - 2x^2 - 4y^2}$

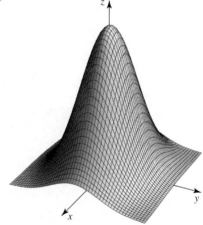

58. $f(x, y) = x + y^2$

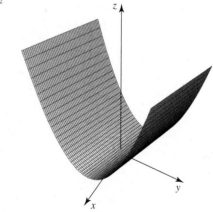

59. $f(x, y) = \cos\sqrt{x^2 + y^2}$

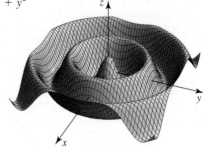

60. $f(x, y) = \sin x + \sin y$

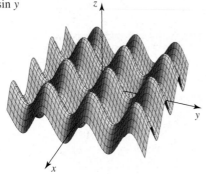

61. $f(x, y) = \sin(x + y)$

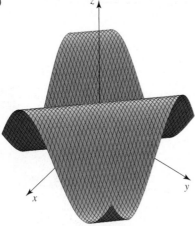

62. $f(x, y) = \ln(2x^2 + y^2)$

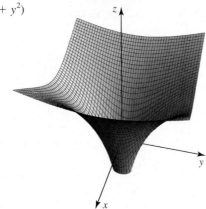

 In Exercises 63–66, (a) *use a computer or calculator to plot the graph of the function f, and* (b) *plot some level curves of f and compare them with the graph obtained in part* (a).

63. $f(x, y) = |x| + |y|$

64. $f(x, y) = \dfrac{xy}{\sqrt{x^2 + y^2}}$

65. $f(x, y) = \dfrac{xy(x^2 - y^2)}{x^2 + y^2}$

66. $f(x, y) = ye^{1 - x^2 - y^2}$

67. Find an equation of the level curve of $f(x, y) = \sqrt{x^2 + y^2}$ that contains the point (3, 4).

68. Find an equation of the level surface of $f(x, y, z) = 2x^2 + 3y^2 - z$ that contains the point $(-1, 2, -3)$.

69. Can two level curves of a function f of two variables x and y intersect? Explain.

70. A *level set* of f is the set $S = \{(x, y) \mid f(x, y) = k$, where k is in the range of $f\}$. Let

$$f(x, y) = \begin{cases} 0 & \text{if } x^2 + y^2 < 1 \\ x^2 + y^2 - 1 & \text{if } x^2 + y^2 \geq 1 \end{cases}$$

Sketch the level set of f for $k = 0$ and 3.

71. Refer to Exercise 70. Let

$$f(x, y) = \begin{cases} 1 - \sqrt{x^2 + y^2} & \text{if } x^2 + y^2 < 1 \\ x^2 + y^2 - 1 & \text{if } x^2 + y^2 \geq 1 \end{cases}$$

(a) Sketch the graph of f and (b) describe the level set of f for $k = 0, \frac{1}{2}, 1$, and 3.

72. Body Mass The body mass index (BMI) is used to identify, evaluate, and treat overweight and obese adults. The BMI value for an adult of weight w (in kilograms) and height h (in meters) is defined to be

$$M = f(w, h) = \frac{w}{h^2}$$

According to federal guidelines, an adult is overweight if he or she has a BMI value between 25 and 29.9 and is "obese" if the value is greater than or equal to 30.
  a. What is the BMI of an adult who weighs in at 80 kg and stands 1.8 m tall?
  b. What is the maximum weight of an adult of height 1.8 m who is not classified as overweight or obese?

73. Poiseuille's Law Poiseuille's Law states that the resistance R, measured in dynes, of blood flowing in a blood vessel of length l and radius r (both in centimeters) is given by

$$R = f(l, r) = \frac{kl}{r^4}$$

where k is the viscosity of blood (in dyne-sec/cm^2). What is the resistance, in terms of k, of blood flowing through an arteriole with radius 0.1 cm and length 4 cm?

74. Surface Area of a Human Body An empirical formula by E.F. Dubois relates the surface area S of a human body (in square meters) to its weight W (in kilograms) and its height h (in centimeters). The formula, given by

$$S = 0.007184W^{0.425}H^{0.725}$$

is used by physiologists in metabolism studies.
  a. Find the domain of the function S.
  b. What is the surface area of a human body that weighs 70 kg and has a height of 178 cm?

75. Cobb-Douglas Production Function Economists have found that the output of a finished product, $f(x, y)$, is sometimes described by the function

$$f(x, y) = ax^b y^{1-b}$$

where x stands for the amount of money expended for labor, y stands for the amount expended on capital, and a and b are positive constants with $0 < b < 1$.

a. If p is a positive number, show that $f(px, py) = pf(x, y)$.

b. Use the result of part (a) to show that if the amount of money expended for labor and capital are both increased by r percent, then the output is also increased by r percent.

76. Continuous Compound Interest If a principal of P dollars is deposited in an account earning interest at the rate of r/year compounded continuously, then the accumulated amount at the end of t years is given by

$$A = f(P, r, t) = Pe^{rt}$$

dollars. Find the accumulated amount at the end of 3 years if \$10,000 is deposited in an account earning interest at the rate of 10%/year.

77. Home Mortgages Suppose a home buyer secures a bank loan of A dollars to purchase a house. If the interest rate charged is r/year and the loan is to be amortized in t years, then the principal repayment at the end of i months is given by

$$B = f(A, r, t, i) = A\left[\frac{\left(1 + \frac{r}{12}\right)^i - 1}{\left(1 + \frac{r}{12}\right)^{12t} - 1}\right] \quad 0 \le i \le 12t$$

Suppose the Blakelys borrow \$280,000 from a bank to help finance the purchase of a house and the bank charges interest at a rate of 6%/year, compounded monthly. If the Blakelys agree to repay the loan in equal installments over 30 years, how much will they owe the bank after the sixtieth payment (5 years)? The 240th payment (20 years)?

78. Wilson Lot-Size Formula The Wilson lot-size formula in economics states that the optimal quantity Q of goods for a store to order is given by

$$Q = f(C, N, h) = \sqrt{\frac{2CN}{h}}$$

where C is the cost of placing an order, N is the number of items the store sells per week, and h is the weekly holding cost for each item. Find the most economical quantity of ten-speed bicycles to order if it costs the store \$20 to place an order and \$5 to hold a bicycle for a week and the store expects to sell 40 bicycles a week.

79. Force Generated by a Centrifuge A centrifuge is a machine designed for the specific purpose of subjecting materials to a sustained centrifugal force. The magnitude of a centrifugal force F in dynes is given by

$$F = f(M, S, R) = \frac{\pi^2 S^2 MR}{900}$$

where S is in revolutions per minute (rpm), M is the mass in grams, and R is the radius in centimeters. Find the centrifugal force generated by an object revolving at the rate of 600 rpm in a circle of radius 10 cm. Express your answer as a multiple of the force of gravity. (Recall that 1 gram of force is equal to 980 dynes.)

80. Temperature of a Thin Metal Plate A thin metal plate located in the xy-plane has a temperature of

$$T(x, y) = \frac{120}{1 + 2x^2 + y^2}$$

degrees Celsius at the point (x, y). Describe the isotherms of T, and sketch those corresponding to $T = 120, 60, 40,$ and 20.

81. International America's Cup Class Drafted by an international committee in 1989, the rules for the new International America's cup class includes a formula that governs the basic yacht dimensions. The formula $f(L, S, D) \le 42$, where

$$f(L, S, D) = \frac{L + 1.25S^{1/2} - 9.80D^{1/3}}{0.388}$$

balances the rated length L (in meters), the rated sail area S (in square meters) and the displacement D (in cubic meters). All changes in the basic dimensions are tradeoffs. For example, if you want to pick up speed by increasing the sail area, you must pay for it by decreasing the length or increasing the displacement, both of which slow the boat down. Show that yacht A of rated length 20.95 m, rated sail area 277.3 m², and displacement 17.56 m³, and the longer and heavier yacht B with $L = 21.87$, $S = 311.78$, and $D = 22.48$ both satisfy the formula.

82. Ideal Gas Law According to the *ideal gas law*, the volume V of an ideal gas is related to its pressure P and temperature T by the formula

$$V = \frac{kT}{P}$$

where k is a positive constant. Describe the level curves of V, and give a physical interpretation of your result.

83. Newton's Law of Gravitation According to Newton's Law of Gravitation a body of mass m_1 located at the origin of an xyz-coordinate system attracts another body of mass m_2 located at the point (x, y, z) with a force of magnitude given by

$$F = \frac{Gm_1m_2}{x^2 + y^2 + z^2}$$

where G is the universal constant of gravitation. Describe the level surfaces of F, and give a physical interpretation of your result.

84. Equipotential Curves Consider the crescent-shaped region R (in the following figure) that lies inside the disk

$$D_1 = \{(x, y) \mid (x - 2)^2 + y^2 \le 4\}$$

and outside the disk

$$D_2 = \{(x, y) \mid (x - 1)^2 + y^2 \le 1\}$$

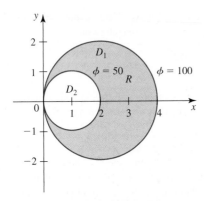

If the electrostatic potential along the inner circle is kept at 50 volts and the electrostatic potential along the outer circle is kept at 100 volts, then the electrostatic potential at any point (x, y) in the region R is given by

$$\phi(x, y) = 150 - \frac{200x}{x^2 + y^2}$$

Show that the equipotential curves of ϕ are arcs of circles that have their centers on the positive x-axis and pass through the origin. Sketch the equipotential curve corresponding to a potential of 75 volts.

85. **The Doppler Effect** Suppose that a sound with frequency f is emitted by an object moving along a straight line with speed u and that a listener is traveling along the same line in the opposite direction with speed v. Then the frequency F heard by the listener is given by

$$F = \left(\frac{c - v}{c + u}\right)f$$

where c is the speed of sound in still air, about 1100 ft/sec. (This phenomenon is called the **Doppler effect.**) Suppose a railroad train is traveling at 100 ft/sec (approximately 68 mph) in still air and the frequency of a note emitted by the locomotive whistle is 500 Hz. What is the frequency of the note heard by a passenger in a train moving at 50 ft/sec in the opposite direction to the first train?

86. A function $f(x, y)$ is *homogeneous of degree n* if it satisfies the equation $f(tx, ty) = t^n f(x, y)$ for all t. Show that

$$f(x, y) = \frac{xy - y^2}{2x + y}$$

is homogeneous of degree 1.

In Exercises 87–90, determine whether the statement is true or false. If it is true, explain why it is true. If it is false, give an example to show why it is false.

87. f is a function of x and y if and only if for any two points $P_1(x_1, y_1)$ and $P_2(x_2, y_2)$ in the domain of f, $f(x_1, y_1) = f(x_2, y_2)$ implies that $P_1(x_1, y_1) = P_2(x_2, y_2)$.

88. The equation $x^2 + y^2 + z^2 = 4$ defines at least two functions of x and y.

89. The level curves of a function f of two variables, $f(x, y) = k$, exist for all values of k.

90. The level surfaces of the function $f(x, y, z) = ax + by + cz + d$ consist of a family of parallel planes that are orthogonal to the vector $\mathbf{n} = a\mathbf{i} + b\mathbf{j} + c\mathbf{k}$.

13.2 Limits and Continuity

■ An Intuitive Definition of a Limit

Figure 1 shows the graph of a function f of two variables. This figure suggests that $f(x, y)$ is close to the number L if the point (x, y) is close to the point (a, b).

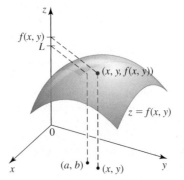

FIGURE 1
The functional value $f(x, y)$ is close to L if (x, y) is close to (a, b).

> **DEFINITION** **Limit of a Function of Two Variables at a Point**
>
> Let f be a function that is defined for all points (x, y) close to the point (a, b) with the possible exception of (a, b) itself. Then the **limit of $f(x, y)$ as (x, y) approaches (a, b)** is L, written
>
> $$\lim_{(x, y) \to (a, b)} f(x, y) = L$$
>
> if $f(x, y)$ can be made as close to L as we please by restricting (x, y) to be sufficiently close to (a, b).

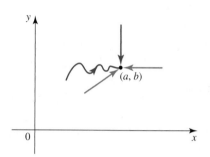

FIGURE 2
There are infinitely many paths the point (x, y) could take in approaching the point (a, b).

At first glance, there appears to be little difference between this definition and the definition of the limit of a function of one variable, with the exception that the points (x, y) and (a, b) lie in the plane. But there are subtle differences. In the case of a function of one variable, the point x can approach the point $x = a$ from only two directions: from the left and from the right. As a consequence, the function f has a limit L as x approaches a if and only if $f(x)$ approaches L from the left ($\lim_{x \to a^-} f(x) = L$) and from the right ($\lim_{x \to a^+} f(x) = L$), a fact that we observed in Section 1.1.

The situation is a little more complicated in the case of a function of two variables because there are infinitely many ways in which we can approach a point (a, b) in the plane (Figure 2). Thus, if f has a limit L as (x, y) approaches (a, b), then $f(x, y)$ must approach L along *every* possible path leading to (a, b).

To see why this is true, suppose that

$$f(x, y) \to L_1 \quad \text{as} \quad (x, y) \to (a, b)$$

along a path C_1 and that

$$f(x, y) \to L_2 \quad \text{as} \quad (x, y) \to (a, b)$$

along another path C_2, where $L_1 \neq L_2$. Then no matter how close (x, y) is to (a, b), $f(x, y)$ will assume values that are close to L_1 and also values that are close to L_2 depending on whether (x, y) is on C_1 or on C_2. Therefore, $f(x, y)$ cannot be made as close as we please to a unique number L by restricting (x, y) to be sufficiently close to (a, b); that is, $\lim_{(x, y) \to (a, b)} f(x, y)$ cannot exist.

An immediate consequence of this observation is the following criterion for demonstrating that a limit does *not* exist.

> **Technique for Showing That $\lim_{(x, y) \to (a, b)} f(x, y)$ Does Not Exist**
>
> If $f(x, y)$ approaches two different numbers as (x, y) approaches (a, b) along two different paths, then $\lim_{(x, y) \to (a, b)} f(x, y) = L$ does not exist.

EXAMPLE 1 Show that $\displaystyle\lim_{(x, y) \to (0, 0)} \frac{x^2 - y^2}{x^2 + y^2}$ does not exist.

Solution The function $f(x, y) = (x^2 - y^2)/(x^2 + y^2)$ is defined everywhere except at $(0, 0)$. Let's approach $(0, 0)$ along the x-axis (see Figure 3). On the path C_1, $y = 0$, so

$$\lim_{\substack{(x, y) \to (0, 0) \\ \text{along } C_1}} f(x, y) = \lim_{x \to 0} f(x, 0) = \lim_{x \to 0} \frac{x^2}{x^2} = \lim_{x \to 0} 1 = 1$$

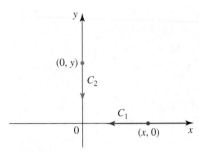

FIGURE 3
A point on C_1 has the form $(x, 0)$, and a point on C_2 has the form $(0, y)$.

Next, let's approach $(0, 0)$ along the y-axis. On the path C_2, $x = 0$ (Figure 3), so

$$\lim_{\substack{(x,\,y)\to(0,\,0) \\ \text{along } C_2}} f(x, y) = \lim_{y\to0} f(0, y) = \lim_{y\to0} \frac{-y^2}{y^2} = \lim_{y\to0}(-1) = -1$$

Since $f(x, y)$ approaches two different numbers as (x, y) approaches $(0, 0)$ along two different paths, we conclude that the given limit does not exist. ∎

EXAMPLE 2 Show that $\displaystyle\lim_{(x,\,y)\to(0,\,0)} \frac{xy}{x^2 + y^2}$ does not exist.

Solution The function $f(x, y) = xy/(x^2 + y^2)$ is defined everywhere except at $(0, 0)$. Let's approach $(0, 0)$ along the x-axis (Figure 4). On the path C_1, $y = 0$, so

$$\lim_{\substack{(x,\,y)\to(0,\,0) \\ \text{along } C_1}} f(x, y) = \lim_{x\to0} f(x, 0) = \lim_{x\to0} \frac{0}{x^2} = \lim_{x\to0} 0 = 0$$

Similarly, you can show that $f(x, y)$ also approaches 0 as (x, y) approaches $(0, 0)$ along the y-axis, path C_2 (Figure 4).

Now consider yet another approach to $(0, 0)$, this time along the line $y = x$ (Figure 4). On the path C_3, $y = x$, so

$$\lim_{\substack{(x,\,y)\to(0,\,0) \\ \text{along } C_3}} f(x, y) = \lim_{x\to0} f(x, x) = \lim_{x\to0} \frac{x^2}{x^2 + x^2} = \lim_{x\to0} \frac{1}{2} = \frac{1}{2}$$

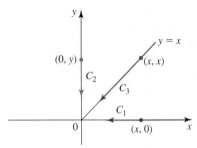

FIGURE 4
$f(x, y) \to 0$ as $(x, y) \to (0, 0)$ along C_1 and C_2, but $f(x, y) \to \frac{1}{2}$ as $(x, y) \to (0, 0)$ along C_3, so $\lim_{(x,\,y)\to(0,\,0)} f(x, y)$ does not exist.

Since $f(x, y)$ approaches two different numbers as (x, y) approaches $(0, 0)$ along two different paths, we conclude that the given limit does not exist.

The graph of f shown in Figure 5 confirms this result visually. Notice the ridge that occurs above the line $y = x$ because $f(x, y) = \frac{1}{2}$ for all points (x, y) on that line except at the origin.

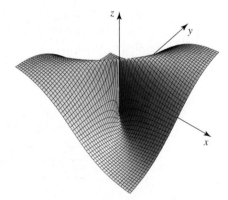

FIGURE 5
The graph of $f(x, y) = \dfrac{xy}{x^2 + y^2}$

∎

Although the method of Examples 1 and 2 is effective in demonstrating when a limit does not exist, it cannot be used to prove the existence of the limit of a function at a point. Using this method, we would have to show that $f(x, y)$ approaches a unique number L as (x, y) approaches the point along *every* path, which is clearly an impossible task. Fortunately, the Limit Laws for a function of a single variable can be extended to functions of two or more variables. For example, the Sum Law, the Product Law, the Quotient Law, and so forth, all hold. So does the Squeeze Theorem.

EXAMPLE 3 Evaluate

a. $\lim_{(x, y)\to(1, 2)}(x^3y^2 - x^2y + x^2 - 2x + 3y)$

b. $\lim_{(x, y)\to(2, 4)} \sqrt[3]{\dfrac{8xy}{2x + y}}$

Solution

a. We have

$$\lim_{(x, y)\to(1, 2)} (x^3y^2 - x^2y + x^2 - 2x + 3y) = (1)^3(2)^2 - (1)^2(2) + (1)^2 - 2(1) + 3(2)$$
$$= 7$$

b. We have

$$\lim_{(x, y)\to(2, 4)} \sqrt[3]{\frac{8xy}{2x + y}} = \sqrt[3]{\lim_{(x, y)\to(2, 4)} \frac{8xy}{2x + y}}$$
$$= \sqrt[3]{\frac{8(2)(4)}{2(2) + 4}} = \sqrt[3]{8} = 2 \qquad \blacksquare$$

The next example utilizes the Squeeze Theorem to show the existence of a limit.

EXAMPLE 4 Find $\lim_{(x, y)\to(0, 0)} \dfrac{2x^2y}{x^2 + y^2}$ if it exists.

Solution   Observe that the numerator of the rational function has degree 3, whereas the denominator has degree 2. This suggests that when x and y are both close to zero, the numerator is much smaller than the denominator, and we suspect that the limit might exist and that it is equal to zero.

To prove our assertion, we observe that $y^2 \geq 0$, so $x^2/(x^2 + y^2) \leq 1$. Therefore,

$$0 \leq \left| \frac{2x^2y}{x^2 + y^2} \right| = \frac{2x^2|y|}{x^2 + y^2} \leq 2|y|$$

Let $f(x, y) = 0$, $g(x, y) = \left| \dfrac{2x^2y}{x^2 + y^2} \right|$, and $h(x, y) = 2|y|$. Then

$$\lim_{(x, y)\to(0, 0)} f(x, y) = \lim_{(x, y)\to(0, 0)} 0 = 0 \quad \text{and} \quad \lim_{(x, y)\to(0, 0)} h(x, y) = \lim_{(x, y)\to(0, 0)} 2|y| = 0$$

By the Squeeze Theorem,

$$\lim_{(x, y)\to(0, 0)} g(x, y) = \lim_{(x, y)\to(0, 0)} \left| \frac{2x^2y}{x^2 + y^2} \right| = 0$$

and this, in turn, implies that

$$\lim_{(x, y)\to(0, 0)} \frac{2x^2y}{x^2 + y^2} = 0 \qquad \blacksquare$$

■ Continuity of a Function of Two Variables

The definition of continuity for a function of two variables is similar to that for a function of one variable.

> **DEFINITION   Continuity at a Point**
>
> Let f be a function that is defined for all points (x, y) close to the point (a, b). Then f is **continuous at the point** (a, b) if
>
> $$\lim_{(x, y) \to (a, b)} f(x, y) = f(a, b)$$

Thus, f is continuous at (a, b) if $f(x, y)$ approaches $f(a, b)$ as (x, y) approaches (a, b) along any path. Loosely speaking, a function f is continuous at a point (a, b) if the graph of f does not have a hole, gap, or jump at (a, b). If f is not continuous at (a, b), then f is said to be **discontinuous** there. For example, the functions f, g, and h whose graphs are shown in Figure 6 are discontinuous at the indicated points.

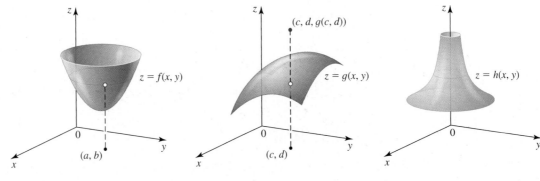

FIGURE 6   **(a)** f is not defined at (a, b).   **(b)** $\lim\limits_{(x, y) \to (c, d)} g(x, y) \neq g(c, d)$ **(c)** $\lim\limits_{(x, y) \to (0, 0)} h(x, y)$ does not exist.

Continuity on a Set

Let's digress a little to introduce some terminology. We define the **δ-neighborhood** about (a, b) to be the set

$$N_\delta = \{(x, y) \mid \sqrt{(x - a)^2 + (y - b)^2} < \delta\}$$

Thus, N_δ is just the set of all points lying inside the circle of radius δ centered at (a, b) (see Figure 7).

Let R be a plane region. A point (a, b) is said to be an **interior point** of R if there exists a δ-neighborhood about (a, b) that lies entirely in R (Figure 8). A point (a, b) is called a **boundary point** of R if every δ-neighborhood of R contains points in R and also points not in R.

A region R is said to be an **open region** if every point of R is an interior point of R. A region is **closed** if it contains all of its boundary points. Finally, a region that contains some but not all of its boundary points is neither open nor closed. For example, the regions

$$A = \left\{(x, y) \,\middle|\, \frac{x^2}{9} + \frac{y^2}{4} < 1\right\}, \qquad B = \left\{(x, y) \,\middle|\, \frac{x^2}{9} + \frac{y^2}{4} \leq 1\right\}$$

and

$$C = \left\{(x, y) \,\middle|\, \frac{x^2}{9} + \frac{y^2}{4} \leq 1; y \geq 0\right\} \bigcup \left\{(x, y) \,\middle|\, \frac{x^2}{9} + \frac{y^2}{4} < 1; y < 0\right\}$$

FIGURE 7
The δ-neighborhood about (a, b)

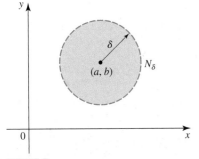

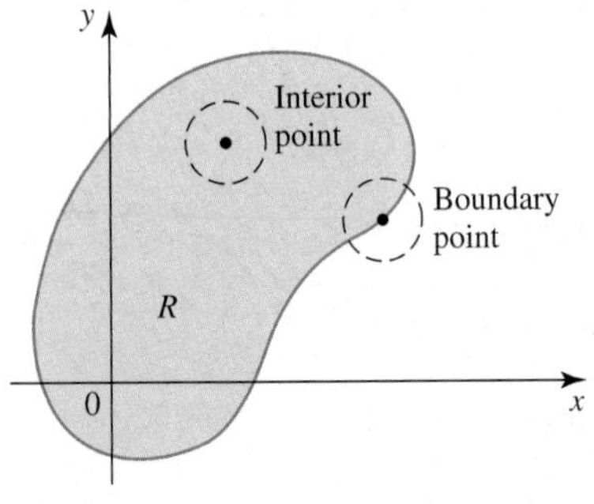

FIGURE 8
An interior point and a boundary point of R

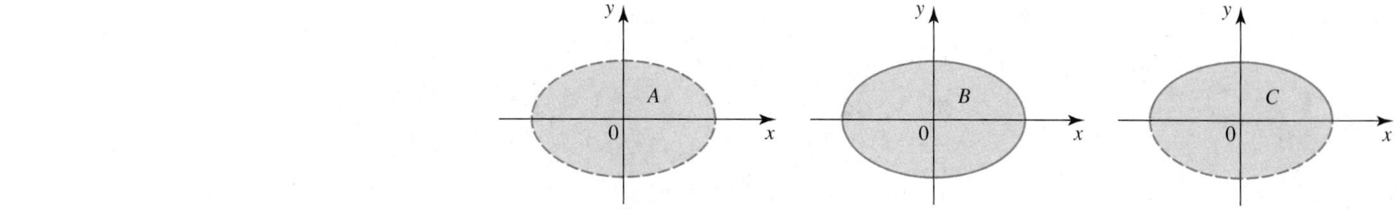

(a) *A* is open. **(b)** *B* is closed. **(c)** *C* is neither open nor closed.

FIGURE 9

Every point in *A* is an interior point; *B* contains all of its boundary points;
C contains some but not all of its boundary points.

shown in Figure 9a–c are open, closed, and neither open nor closed, respectively.

As we mentioned in Section 1.3, continuity is a "local" concept. The following definition explains what we mean by continuity on a region.

DEFINITION   **Continuity on a Region**

Let *R* be a region in the plane. Then *f* is **continuous on *R*** if *f* is continuous at every point (x, y) in *R*. If (a, b) is a boundary point, the condition for continuity is modified to read

$$\lim_{(x, y) \to (a, b)} f(x, y) = f(a, b)$$

where $(x, y) \in R$, that is, (x, y) is restricted to approach (a, b) along paths lying inside *R*.

EXAMPLE 5 Show that the function *f* defined by $f(x, y) = \sqrt{9 - x^2 - y^2}$ is continuous on the closed region $R = \{(x, y) \mid x^2 + y^2 \leq 9\}$, which is the set of all points lying on and inside the circle of radius 3 centered at $(0, 0)$ in the *xy*-plane.

Solution   Observe that the set *R* is precisely the domain of *f*. Now, if (a, b) is any interior point of *R*, then

$$\lim_{(x, y) \to (a, b)} f(x, y) = \lim_{(x, y) \to (a, b)} \sqrt{9 - x^2 - y^2}$$

$$= \sqrt{\lim_{(x, y) \to (a, b)} (9 - x^2 - y^2)}$$

$$= \sqrt{9 - a^2 - b^2}$$

$$= f(a, b)$$

This shows that *f* is continuous at (a, b).

Next, if (c, d) is a boundary point of *R* and (x, y) is restricted to lie inside *R*, we obtain

$$\lim_{(x, y) \to (c, d)} f(x, y) = f(c, d)$$

as before, thus showing that *f* is continuous at (c, d) as well.

The graph of *f* is the upper hemisphere of radius 3 centered at the origin together with the circle in the *xy*-plane having equation $x^2 + y^2 = 9$. (See Figure 10.)

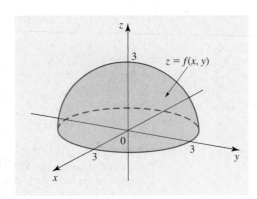

FIGURE 10
The graph of $f(x, y) = \sqrt{9 - x^2 - y^2}$ has no holes, gaps, or jumps.

The following theorem summarizes the properties of continuous functions of two variables. The proofs of these properties follow from the limit laws and will be omitted.

THEOREM 1   Properties of Continuous Functions of Two Variables

If f and g are continuous at (a, b), then the following functions are also continuous at (a, b).

a. $f \pm g$ **b.** fg **c.** cf c, a constant   **d.** f/g $g(a, b) \neq 0$

A consequence of Theorem 1 is that *polynomial* and *rational* functions are continuous.

A **polynomial function** of two variables is a function whose rule can be expressed as a finite sum of terms of the form $cx^m y^n$, where c is a constant and m and n are nonnegative integers. For example, the function f defined by

$$f(x, y) = 2x^2 y^5 - 3xy^3 + 8xy^2 - 3y + 4$$

is a polynomial function in the two variables x and y. A **rational function** is the quotient of two polynomial functions. For example, the function g defined by

$$g(x, y) = \frac{x^3 + xy + y^2}{x^2 - y^2}$$

is a rational function.

THEOREM 2   Continuity of Polynomial and Rational Functions

A polynomial function is continuous everywhere (that is, in the whole plane). A rational function is continuous at all points in its domain (that is, at all points where its denominator is defined and not equal to zero).

EXAMPLE 6 Determine where the function is continuous:

a. $f(x, y) = \dfrac{xy(x^2 - y^2)}{x^2 + y^2}$ **b.** $g(x, y) = \dfrac{1}{y - x^2}$

Solution

a. The function f is a rational function and is therefore continuous everywhere except at $(0, 0)$, where its denominator is equal to zero (Figure 11).

b. The function g is a rational function and is continuous everywhere except along the curve $y = x^2$, where its denominator is equal to zero (Figure 12).

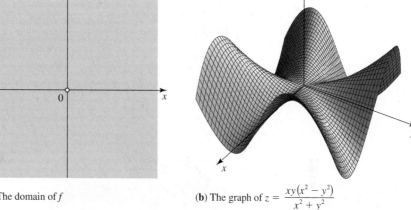

FIGURE 11
The graph of f has a hole at the origin.

(a) The domain of f

(b) The graph of $z = \dfrac{xy(x^2 - y^2)}{x^2 + y^2}$

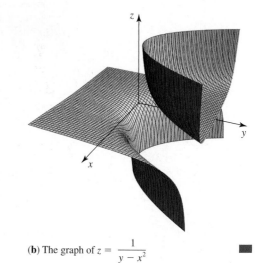

FIGURE 12
As (x, y) approaches the curve $y = x^2$ from the region $y > x^2$, $z = f(x, y)$ approaches infinity; as (x, y) approaches the curve $y = x^2$ from the region $y < x^2$, z approaches minus infinity.

(a) The domain of g

(b) The graph of $z = \dfrac{1}{y - x^2}$

The next theorem tells us that the composite function of two continuous functions is also a continuous function.

THEOREM 3 Continuity of a Composite Function

If f is continuous at (a, b) and g is continuous at $f(a, b)$, then the composite function $h = g \circ f$ defined by $h(x, y) = g(f(x, y))$ is continuous at (a, b).

EXAMPLE 7 Determine where the function is continuous:

a. $F(x, y) = \sin xy$

b. $G(x, y) = \dfrac{\frac{1}{2} \cos(2x^2 + y^2)}{1 + 2x^2 + y^2}$

Solution

a. We can view the function F as the composition $g \circ f$ of the functions f and g defined by $f(x, y) = xy$ and $g(t) = \sin t$. Thus,

$$F(x, y) = g(f(x, y)) = \sin(f(x, y)) = \sin xy$$

Since f is continuous on the whole plane and g is continuous on $(-\infty, \infty)$, we conclude that F is continuous everywhere. The graph of F is shown in Figure 13a.

b. The function G is the quotient of $p(x, y) = \frac{1}{2}\cos(2x^2 + y^2)$ and $q(x, y) = 1 + 2x^2 + y^2$. The function p in turn involves the composition of $g(t) = \frac{1}{2}\cos t$ and $f(x, y) = 2x^2 + y^2$. Since both f and g are continuous everywhere, we see that p is continuous everywhere. The function q is continuous everywhere as well and is never zero. Therefore, by Theorem 3, G is continuous everywhere. The graph of G is shown in Figure 13b.

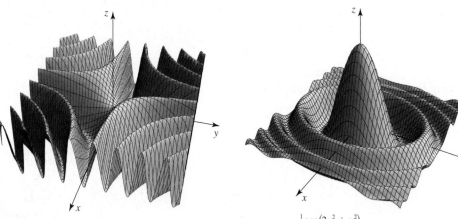

FIGURE 13   **(a)** $F(x, y) = \sin xy$ is continuous everywhere.   **(b)** $G(x, y) = \dfrac{\frac{1}{2}\cos(2x^2 + y^2)}{1 + 2x^2 + y^2}$ is continuous everywhere.

■ Functions of Three or More Variables

The notions of the limit of a function of three or more variables and that of the continuity of a function of three or more variables parallel those of a function of two variables. For example, if f is a function of three variables, then we write

$$\lim_{(x, y, z) \to (a, b, c)} f(x, y, z) = L$$

to mean that there exists a number L such that $f(x, y, z)$ can be made as close to L as we please by restricting (x, y, z) to be sufficiently close to (a, b, c).

EXAMPLE 8 Evaluate $\displaystyle\lim_{(x, y, z) \to (\frac{\pi}{2}, 0, 1)} \frac{e^{2y}(\sin x + \cos y)}{1 + y^2 + z^2}$.

Solution

$$\lim_{(x, y, z) \to (\frac{\pi}{2}, 0, 1)} \frac{e^{2y}(\sin x + \cos y)}{1 + y^2 + z^2} = \frac{e^0[\sin(\pi/2) + \cos 0]}{1 + 0 + 1} = \frac{2}{2} = 1$$

A function f of three variables is continuous at (a, b, c) if

$$\lim_{(x, y, z) \to (a, b, c)} f(x, y, z) = f(a, b, c)$$

EXAMPLE 9 Determine where $f(x, y, z) = \dfrac{\ln z}{\sqrt{1 - x^2 - y^2 - z^2}}$ is continuous.

Solution We require that $z > 0$ and $1 - x^2 - y^2 - z^2 > 0$; that is, $z > 0$ and $x^2 + y^2 + z^2 < 1$. So f is continuous on the set $\{(x, y, z) \mid x^2 + y^2 + z^2 < 1$ and $z > 0\}$, which is the set of points above the xy-plane and inside the upper hemisphere with center at the origin and radius 1. ∎

■ The ε-δ Definition of a Limit (Optional)

The notion of the limit of a function of two variables given earlier can be made more precise as follows.

> **DEFINITION** **Limit of $f(x, y)$**
>
> Let f be a function of two variables that is defined for all points (x, y) on a disk with center at (a, b) with the possible exception of (a, b) itself. Then
>
> $$\lim_{(x, y) \to (a, b)} f(x, y) = L$$
>
> if for every $\varepsilon > 0$, there exists a $\delta > 0$ such that
>
> $$|f(x, y) - L| < \varepsilon \qquad \text{whenever} \qquad 0 < \sqrt{(x - a)^2 + (y - b)^2} < \delta$$

Geometrically speaking, f has the limit L at (a, b) if given *any* $\varepsilon > 0$, we can find a circle of radius δ centered at (a, b) such that $L - \varepsilon < f(x, y) < L + \varepsilon$ for all interior points $(x, y) \neq (a, b)$ of the circle (Figure 14).

EXAMPLE 10 Prove that $\lim_{(x, y) \to (a, b)} x = a$.

Solution Let $\varepsilon > 0$ be given. We need to show that there exists a $\delta > 0$ such that

$$|f(x, y) - a| < \varepsilon$$

whenever $(x, y) \neq (a, b)$ is in the δ-neighborhood about (a, b). To find such a δ, consider

$$|f(x, y) - a| = |x - a| = \sqrt{(x - a)^2} \leq \sqrt{(x - a)^2 + (y - b)^2}$$

Thus, if we pick $\delta = \varepsilon$, we see that $\delta > 0$ and that $\sqrt{(x - a)^2 + (y - b)^2} < \delta$ implies that $|f(x, y) - a| < \varepsilon$ as was to be shown. Since ε is arbitrary, the proof is complete. ∎

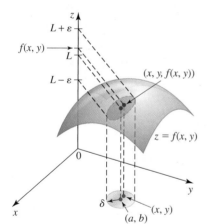

FIGURE 14
$f(x, y)$ lies in the interval $(L - \varepsilon, L + \varepsilon)$ whenever $(x, y) \neq (a, b)$ is in the δ-neighborhood of (a, b).

EXAMPLE 11 Prove that $\displaystyle\lim_{(x, y) \to (0, 0)} \frac{2x^2 y}{x^2 + y^2} = 0$. (See Example 4.)

Solution Let $\varepsilon > 0$ be given. Consider

$$|f(x, y) - 0| = \left| \frac{2x^2 y}{x^2 + y^2} \right| = 2|y| \left(\frac{x^2}{x^2 + y^2} \right) \qquad (x, y) \neq (0, 0)$$

$$\leq 2|y| = 2\sqrt{y^2} \leq 2\sqrt{x^2 + y^2}$$

If we pick $\delta = \varepsilon/2$, then $\delta > 0$, and $\sqrt{x^2 + y^2} < \delta$ implies that $|f(x, y) - 0| < \varepsilon$. Since ε is arbitrary, the proof is complete. ∎

13.2   CONCEPT QUESTIONS

1. **a.** Explain what it means for a function f of two variables to have a limit at (a, b).
   b. Describe a technique that you could use to show that the limit of $f(x, y)$ as (x, y) approaches (a, b) does not exist.
2. Explain what it means for a function of two variables to be continuous (a) at a point (a, b) and (b) on a region in the plane.
3. Determine whether each function f is continuous or discontinuous. Explain your answer.
   a. $f(P, T)$ measures the volume of a balloon ascending into the sky as a function of the atmospheric pressure P and the air temperature T.

b. $f(H, W)$ measures the surface area of a human body as a function of its height H and weight W.
c. $f(d, t)$ measures the fare as a function of distance d and time t for taking a cab from O'Hare Airport to downtown Chicago.
d. $f(T, P)$ measures the volume of a certain mass of gas as a function of the temperature T and the pressure P.

4. Suppose f has the property that it is not defined at the point $(1, 2)$ but $\lim_{(x, y)\to(1, 2)} f(x, y) = 3$. Can you define $f(1, 2)$ so that f is continuous at $(1, 2)$? If so, what should the value of $f(1, 2)$ be?

13.2   EXERCISES

In Exercises 1–12, show that the limit does not exist.

1. $\lim\limits_{(x, y)\to(0, 0)} \dfrac{x^2 - y^2}{2x^2 + y^2}$

2. $\lim\limits_{(x, y)\to(0, 0)} \dfrac{2x^2 - 3xy + 4y^2}{2x^2 + 3y^2}$

3. $\lim\limits_{(x, y)\to(0, 0)} \dfrac{3xy}{3x^2 + y^2}$

4. $\lim\limits_{(x, y)\to(0, 0)} \dfrac{xy^2}{x^2 + y^4}$

5. $\lim\limits_{(x, y)\to(0, 0)} \dfrac{2xy}{\sqrt{x^4 + y^4}}$

6. $\lim\limits_{(x, y)\to(0, 0)} \dfrac{\sin xy}{x^2 + y^2}$

7. $\lim\limits_{(x, y)\to(1, 0)} \dfrac{2xy - 2y}{x^2 + y^2 - 2x + 1}$

8. $\lim\limits_{(x, y)\to(0, 0)} \dfrac{xy^3 \cos x}{2x^2 + y^6}$

9. $\lim\limits_{(x, y, z)\to(0, 0, 0)} \dfrac{xy + yz + xz}{x^2 + y^2 + z^2}$

10. $\lim\limits_{(x, y, z)\to(0, 0, 0)} \dfrac{2xyz}{x^3 + y^3 + z^3}$

11. $\lim\limits_{(x, y, z)\to(0, 0, 0)} \dfrac{xz^2 + 2y^2}{x^2 + 2y^2 + z^4}$

   Hint: Approach $(0, 0, 0)$ along the curve with parametric equations $x = t^2$, $y = t^2$, $z = t$.

12. $\lim\limits_{(x, y, z)\to(0, 0, 0)} \dfrac{xy}{x^2 + y^2 + z^2}$

In Exercises 13–26, find the given limit.

13. $\lim\limits_{(x, y)\to(1, 2)} (x^2 + 2y^2)$

14. $\lim\limits_{(x, y)\to(1, -1)} (2x^2 + xy + 3y + 1)$

15. $\lim\limits_{(x, y)\to(1, 2)} \dfrac{2x^2 - 3y^3 + 4}{3 - xy}$

16. $\lim\limits_{(x, y)\to(-1, 3)} \dfrac{x + 2y^2}{(x - 1)(y + 1)}$

17. $\lim\limits_{(x, y)\to(1, -2)} \dfrac{3xy}{2x^2 - y^2}$

18. $\lim\limits_{(x, y)\to(1, \frac{1}{2})} x^2 \sin \pi(2x + y)$

19. $\lim\limits_{(x, y)\to(0^+, 0^+)} \dfrac{e^{\sqrt{x+y}}}{x + y - 1}$

20. $\lim\limits_{(x, y)\to(0, 1)} \dfrac{\sin^{-1}\left(\dfrac{x}{y}\right)}{1 + \dfrac{x}{y}}$

21. $\lim\limits_{(x, y)\to(1, 1)} \dfrac{\tan^{-1}\left(\dfrac{x}{y}\right)}{\cos^{-1}(x - 2y)}$

22. $\lim\limits_{(x, y)\to(0, 1)} e^{-x} \sin^{-1}(y - x)$

23. $\lim\limits_{(x, y)\to(2, 1)} \ln(x^2 - 3y)$

24. $\lim\limits_{(x, y)\to(3, 4)} e^{\sqrt{x^2 + y^2}}$

25. $\lim\limits_{(x, y, z)\to(1, 2, 3)} \dfrac{xy + yz + xz}{xyz - 3}$

26. $\lim\limits_{(x, y, z)\to(0, 3, 1)} [e^{\sin \pi x} + \ln(\cos \pi(y - z))]$

In Exercises 27–30, use polar coordinates to find the limit. Hint: If $x = r \cos \theta$ and $y = r \sin \theta$, then $(x, y) \to (0, 0)$ if and only if $r \to 0^+$.

27. $\lim\limits_{(x, y)\to(0, 0)} \dfrac{x^3 + y^3}{x^2 + y^2}$

28. $\lim\limits_{(x, y)\to(0, 0)} \dfrac{\sin(2x^2 + 2y^2)}{x^2 + y^2}$

29. $\lim\limits_{(x, y)\to(0, 0)} (x^2 + y^2) \ln(x^2 + y^2)$

30. $\lim\limits_{(x, y)\to(0, 0)} \dfrac{\tan(2x^2 + 2y^2)}{\tanh(3x^2 + 3y^2)}$

In Exercises 31–40, determine where the function is continuous.

31. $f(x, y) = \dfrac{2xy}{2x + 3y - 1}$

32. $f(x, y) = \dfrac{x^3 + xy + y^3}{x^2 + y^2}$

33. $g(x, y) = \sqrt{x + y} - \sqrt{x - y}$

34. $h(x, y) = \sin(2x + 3y)$

35. $F(x, y) = \sqrt{x} e^{x/y}$

36. $G(x, y) = \ln(2x - y)$

37. $f(x, y, z) = \dfrac{xyz}{x^2 + y^2 + z^2 - 4}$

38. $g(x, y, z) = \sqrt{x} + \cos\sqrt{y + z}$

39. $h(x, y, z) = x \ln(yz - 1)$

40. $F(x, y, z) = x \tan \dfrac{y}{z}$

41. Let

$$f(x, y) = \begin{cases} \dfrac{\sin xy}{xy} & \text{if } xy \neq 0 \\ 1 & \text{if } xy = 0 \end{cases}$$

    a. Determine all the points where f is continuous.

cas **b.** Plot the graph of f. Does the graph give a visual confirmation of your conclusion in part (a)?

42. Let

$$f(x, y) = \begin{cases} \dfrac{x}{\sin x} + y & \text{if } x \neq 0 \\ 1 + y & \text{if } x = 0 \end{cases}$$

    a. Determine all the points where f is continuous.

cas **b.** Plot the graph of f. Does the graph give a visual confirmation of your conclusion in part (a)?

In Exercises 43–48, find $h(x, y) = g(f(x, y))$, and determine where h is continuous.

43. $f(x, y) = x^2 - xy + y^2$, $g(t) = t \cos t + \sin t$

44. $f(x, y) = x^3 + xy - xy^2 + y^3$, $g(t) = te^{-t}$

45. $f(x, y) = 2x - y$, $g(t) = \dfrac{t + 2}{t - 1}$

46. $f(x, y) = x - 2y + 3$, $g(t) = \sqrt{t} + \dfrac{1}{t}$

47. $f(x, y) = x \tan y$, $g(t) = \cos t$

48. $f(x, y) = y \ln x$, $g(t) = e^{t^2}$

49. Use the precise definition of a limit to prove that $\lim_{(x, y) \to (a, b)} c = c$ where c is a constant.

50. Use the precise definition of a limit to prove that $\lim_{(x, y) \to (a, b)} y = b$.

51. Use the precise definition of a limit to prove that

$$\lim_{(x, y) \to (0, 0)} \frac{3xy^3}{x^2 + y^2} = 0$$

52. Use the precise definition of a limit to prove that if $\lim_{(x, y) \to (a, b)} f(x, y) = L$ and c is a constant, then $\lim_{(x, y) \to (a, b)} cf(x, y) = cL$.

In Exercises 53–58, determine whether the statement is true or false. If it is true, explain why it is true. If it is false, give an example to show why it is false.

53. If $\lim_{(x, y) \to (a, b)} f(x, y) = L$, then $\lim_{(x, y) \to (a, b) \text{ along } C} f(x, y) = L$, where C is any path leading to (a, b).

54. If $\lim_{(x, y) \to (a, b)} f(x, y) = L$ and f is defined at (a, b), then $f(a, b) = L$.

55. If $f(x, y) = g(x)h(y)$, where g and h are continuous at a and b, respectively, then f is continuous at (a, b).

56. If $f(1, 3) = 4$, then $\lim_{(x, y) \to (1, 3)} f(x, y) = 4$.

57. If f is continuous at $(3, -1)$ and $f(3, -1) = 2$, then $\lim_{(x, y) \to (3, -1)} f(x, y) = 2$.

58. If f is continuous at (a, b) and g is continuous at $f(a, b)$, then $\lim_{(x, y) \to (a, b)} g(f(x, y)) = g(f(a, b))$.

13.3 Partial Derivatives

■ Partial Derivatives of Functions of Two Variables

For a function of one variable x, there is no ambiguity when we speak of the rate of change of $f(x)$ with respect to x. The situation becomes more complicated, however, when we study the rate of change of a function of two or more variables. For example, for the function of two variables defined by the equation $z = f(x, y)$, *both* the independent variables x and y may be allowed to vary in some arbitrary fashion, thus making it unclear what we mean by the phrase "the rate of change of z with respect to x and y."

One way of getting around this difficulty is to hold one variable constant and consider the rate of change of f with respect to the other variable. This approach might be familiar to anyone who has used the expression "everything else being equal" while debating the merits of a complicated issue.

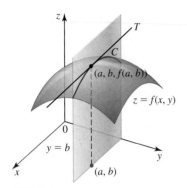

FIGURE 1

$$\lim_{h\to 0} \frac{f(a + h, b) - f(a, b)}{h}$$ measures

the slope of T and the rate of change of $f(x, y)$ in the x-direction when $x = a$ and $y = b$.

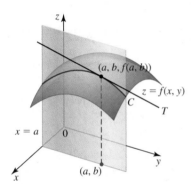

FIGURE 2

$$\lim_{h\to 0} \frac{f(a, b + h) - f(a, b)}{h}$$ measures

the slope of T and the rate of change of $f(x, y)$ in the y-direction when $x = a$ and $y = b$.

Specifically, suppose that (a, b) is a point in the domain of f. Fix $y = b$. Then the function that is defined by $z = f(x, b)$ is a function of the single variable x. Its graph is the curve C formed by the intersection of the vertical plane $y = b$ and the surface $z = f(x, y)$ (Figure 1).

Therefore, the quantity

$$\lim_{h\to 0} \frac{f(a + h, b) - f(a, b)}{h} \tag{1}$$

if it exists, measures both the slope of the tangent line T to the curve C at the point $(a, b, f(a, b))$ as well as the rate of change of $f(x, y)$ with respect to x (in the x-direction) with y held constant when $x = a$ and $y = b$.

Similarly, the quantity

$$\lim_{h\to 0} \frac{f(a, b + h) - f(a, b)}{h} \tag{2}$$

if it exists, measures the slope of the tangent line T to the curve C (formed by the intersection of the vertical plane $x = a$ and the surface $z = f(x, y)$ at $(a, b, f(a, b))$, and the rate of change of $f(x, y)$ with respect to y (in the y-direction) with x held constant when $x = a$ and $y = b$ (Figure 2).

In expressions (1) and (2) the point (a, b) is fixed but otherwise arbitrary. Therefore, we may replace (a, b) by (x, y), leading to the following definitions.

DEFINITIONS   Partial Derivatives of a Function of Two Variables

Let $z = f(x, y)$. Then the **partial derivative of f with respect to x** is

$$\frac{\partial f}{\partial x} = \lim_{h\to 0} \frac{f(x + h, y) - f(x, y)}{h}$$

and the **partial derivative of f with respect to y** is

$$\frac{\partial f}{\partial y} = \lim_{h\to 0} \frac{f(x, y + h) - f(x, y)}{h}$$

provided that each limit exists.

■ Computing Partial Derivatives

The partial derivatives of f can be calculated by using the following rules.

Computing Partial Derivatives

To compute $\partial f/\partial x$, treat y as a constant and differentiate in the usual manner with respect to x (an operation that we denote by $\partial/\partial x$).

To compute $\partial f/\partial y$, treat x as a constant and differentiate in the usual manner with respect to y (an operation that we denote by $\partial/\partial y$).

EXAMPLE 1 Find $\dfrac{\partial f}{\partial x}$ and $\dfrac{\partial f}{\partial y}$ if $f(x, y) = 2x^2y^3 - 3xy^2 + 2x^2 + 3y^2 + 1$.

Solution To compute $\partial f/\partial x$, we think of the variable y as a constant and differentiate with respect to x. Let's write

$$f(x, y) = 2x^2y^3 - 3xy^2 + 2x^2 + 3y^2 + 1$$

where the variable y to be treated as a constant is shown in color. Then

$$\frac{\partial f}{\partial x} = 4xy^3 - 3y^2 + 4x$$

To compute $\partial f/\partial y$, we think of the variable x as a constant and differentiate with respect to y. In this case,

$$f(x, y) = 2x^2y^3 - 3xy^2 + 2x^2 + 3y^2 + 1$$

and

$$\frac{\partial f}{\partial y} = 6x^2y^2 - 6xy + 6y$$ ∎

Before looking at more examples, let's introduce some alternative notations for the partial derivatives of a function. If $z = f(x, y)$, then

$$\frac{\partial}{\partial x} f(x, y) = \frac{\partial f}{\partial x} = f_x = z_x \qquad \text{and} \qquad \frac{\partial}{\partial y} f(x, y) = \frac{\partial f}{\partial y} = f_y = z_y$$

EXAMPLE 2 Find f_x and f_y if $f(x, y) = x \cos xy^2$.

Solution To compute f_x, we think of the variable y as a constant and differentiate with respect to x. Thus,

$$f(x, y) = x \cos xy^2$$

and

$$f_x = \frac{\partial}{\partial x}(x \cos xy^2) = x \frac{\partial}{\partial x}(\cos xy^2) + (\cos xy^2)\frac{\partial}{\partial x}(x) \qquad \text{Use the Product Rule.}$$

$$= x(-\sin xy^2)\frac{\partial}{\partial x}(xy^2) + \cos xy^2 \qquad \text{Use the Chain Rule on the first term.}$$

$$= -xy^2 \sin xy^2 + \cos xy^2$$

Next, to compute f_y, we treat x as a constant and differentiate with respect to y. Thus,

$$f(x, y) = x \cos xy^2$$

and

$$f_y = \frac{\partial}{\partial y}(x \cos xy^2) = x\frac{\partial}{\partial y}(\cos xy^2) + (\cos xy^2)\frac{\partial}{\partial y}(x)$$

$$= x(-\sin xy^2)\frac{\partial}{\partial y}(xy^2) + 0 = -2x^2y \sin xy^2$$ ∎

EXAMPLE 3 Let $f(x, y) = 4 - 2x^2 - y^2$. Find the slope of the tangent line at the point $(1, 1, 1)$ on the curve formed by the intersection of the surface $z = f(x, y)$ and

a. the plane $y = 1$ **b.** the plane $x = 1$

Solution

a. The slope of the tangent line at any point on the curve formed by the intersection of the plane $y = 1$ and the surface $z = 4 - 2x^2 - y^2$ is given by

$$\frac{\partial f}{\partial x} = \frac{\partial}{\partial x}(4 - 2x^2 - y^2) = -4x$$

In particular, the slope of the required tangent line is

$$\left.\frac{\partial f}{\partial x}\right|_{(1, 1)} = -4(1) = -4$$

b. The slope of the tangent line at any point on the curve formed by the intersection of the plane $x = 1$ and the surface $z = 4 - 2x^2 - y^2$ is given by

$$\frac{\partial f}{\partial y} = \frac{\partial}{\partial y}(4 - 2x^2 - y^2) = -2y$$

In particular, the slope of the required tangent line is

$$\left.\frac{\partial f}{\partial y}\right|_{(1, 1)} = -2(1) = -2$$

(See Figure 3.)

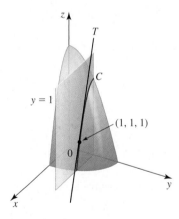

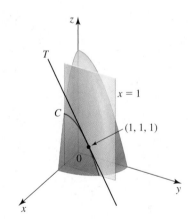

FIGURE 3   **(a)** The slope of the tangent line is -4.   **(b)** The slope of the tangent line is -2.   ■

EXAMPLE 4 Electrostatic Potential Figure 4 shows a crescent-shaped region R that lies inside the disk $D_1 = \{(x, y) \mid (x - 2)^2 + y^2 \le 4\}$ and outside the disk $D_2 = \{(x, y) \mid (x - 1)^2 + y^2 \le 1\}$. Suppose that the electrostatic potential along the inner circle is kept at 50 volts and the electrostatic potential along the outer circle is kept at 100 volts. Then the electrostatic potential at any point (x, y) in R is given by

$$U(x, y) = 150 - \frac{200x}{x^2 + y^2}$$

volts.

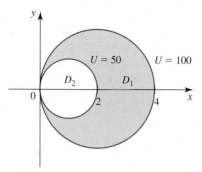

FIGURE 4
The electrostatic potential inside the crescent-shaped region is $U(x, y)$.

a. Compute $U_x(x, y)$ and $U_y(x, y)$.
b. Compute $U_x(3, 1)$ and $U_y(3, 1)$ and interpret your results.

Solution

a. $U_x(x, y) = \dfrac{\partial}{\partial x}\left[150 - \dfrac{200x}{x^2 + y^2}\right] = -\dfrac{\partial}{\partial x}\left(\dfrac{200x}{x^2 + y^2}\right)$

$$= -\frac{(x^2 + y^2)\dfrac{\partial}{\partial x}(200x) - 200x\dfrac{\partial}{\partial x}(x^2 + y^2)}{(x^2 + y^2)^2}$$

$$= -\frac{200(x^2 + y^2) - 200x(2x)}{(x^2 + y^2)^2} = \frac{200(x^2 - y^2)}{(x^2 + y^2)^2}$$

$U_y(x, y) = \dfrac{\partial}{\partial y}\left[150 - \dfrac{200x}{x^2 + y^2}\right] = -\dfrac{\partial}{\partial y}\left(\dfrac{200x}{x^2 + y^2}\right)$

$$= -200x\frac{\partial}{\partial y}(x^2 + y^2)^{-1}$$

$$= -200x(-1)(x^2 + y^2)^{-2}\frac{\partial}{\partial y}(x^2 + y^2)$$

$$= 200x(x^2 + y^2)^{-2}(2y) = \frac{400xy}{(x^2 + y^2)^2}$$

b. $U_x(3, 1) = \dfrac{200(9 - 1)}{(9 + 1)^2} = 16$ and $U_y(3, 1) = \dfrac{400(3)(1)}{(9 + 1)^2} = 12$

This tells us that the rate of change of the electrostatic potential at the point $(3, 1)$ in the x-direction is 16 volts per unit change in x with y held fixed at 1, and the rate of change of the electrostatic potential at the point $(3, 1)$ in the y-direction is 12 volts per unit change in y with x held fixed at 3. ∎

EXAMPLE 5 **A Production Function** The production function of a certain country is given by

$$f(x, y) = 20x^{2/3}y^{1/3}$$

billion dollars, when x billion dollars of labor and y billion dollars of capital are spent.

a. Compute $f_x(x, y)$ and $f_y(x, y)$.
b. Compute $f_x(125, 27)$ and $f_y(125, 27)$, and interpret your results.
c. Should the government encourage capital investment rather than investment in labor to increase the country's productivity?

Solution

a. $f_x(x, y) = \dfrac{\partial}{\partial x}(20x^{2/3}y^{1/3}) = (20)\left(\dfrac{2}{3}x^{-1/3}\right)(y^{1/3}) = \dfrac{40}{3}\left(\dfrac{y}{x}\right)^{1/3}$

$f_y(x, y) = \dfrac{\partial}{\partial y}(20x^{2/3}y^{1/3}) = (20x^{2/3})\left(\dfrac{1}{3}y^{-2/3}\right) = \dfrac{20}{3}\left(\dfrac{x}{y}\right)^{2/3}$

b. $f_x(125, 27) = \dfrac{40}{3}\left(\dfrac{27}{125}\right)^{1/3} = \dfrac{40}{3}\left(\dfrac{3}{5}\right) = 8$

This says that the production is increasing at the rate of $8 billion per billion dollar increase in labor expenditure when the labor expenditure stands at $125 billion (capital expenditure held constant at $27 billion).

Next,

$$f_y(125, 27) = \dfrac{20}{3}\left(\dfrac{125}{27}\right)^{2/3} = \dfrac{20}{3}\left(\dfrac{25}{9}\right) = 18\dfrac{14}{27}$$

This tells us that production is increasing at the rate of approximately $18.5 billion per billion dollar increase in capital outlay when the capital expenditure stands at $27 billion (with labor expenditure held constant at $125 billion).

c. Yes. Since a unit increase in capital expenditure results in a greater increase in production than a unit increase in labor, the government should encourage spending on capital rather than on labor. ∎

Sometimes we have available only the contour map of a function f. In such instances we can use the contour map to help us estimate the partial derivatives of f at a specified point, as the following example shows.

EXAMPLE 6 Figure 5 shows the contour map of a function f. Use it to estimate $f_x(3, 1)$ and $f_y(3, 1)$.

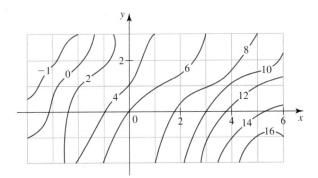

FIGURE 5
A contour map of f

Solution To estimate $f_x(3, 1)$, we start at the point $(3, 1)$, where the value of f at $(3, 1)$ can be read off from the contour map: $f(3, 1) = 8$. Then we proceed along the positive x-axis until we arrive at the point on the next level curve whose location is approximately $(3.8, 1)$. Using the definition of the partial derivative, we find

$$f_x(3, 1) \approx \dfrac{f(3.8, 1) - f(3, 1)}{3.8 - 3} = \dfrac{10 - 8}{0.8} = 2.5$$

Similarly, starting at the point $(3, 1)$ and moving along the positive y-axis, we find

$$f_y(3, 1) \approx \dfrac{f(3, 3) - f(3, 1)}{3 - 1} = \dfrac{6 - 8}{2} = -1 \qquad \blacksquare$$

■ Implicit Differentiation

EXAMPLE 7 Suppose z is a differentiable function of x and y that is defined implicitly by $x^2 + y^3 - z + 2yz^2 = 5$. Find $\partial z/\partial x$ and $\partial z/\partial y$.

Solution Differentiating the given equation implicitly with respect to x, we find

$$\frac{\partial}{\partial x}(x^2 + y^3 - z + 2yz^2) = \frac{\partial}{\partial x}(5)$$

$$2x - \frac{\partial z}{\partial x} + 2y\left(2z\frac{\partial z}{\partial x}\right) = 0 \qquad \text{Remember that } y \text{ is treated as a constant.}$$

$$\frac{\partial z}{\partial x}(4yz - 1) + 2x = 0$$

and

$$\frac{\partial z}{\partial x} = \frac{2x}{1 - 4yz}$$

Next, differentiating the given equation with respect to y, we obtain

$$\frac{\partial}{\partial y}(x^2 + y^3 - z + 2yz^2) = \frac{\partial}{\partial y}(5)$$

$$3y^2 - \frac{\partial z}{\partial y} + 2y\left(2z\frac{\partial z}{\partial y}\right) + 2z^2 = 0$$

$$3y^2 - \frac{\partial z}{\partial y}(1 - 4yz) + 2z^2 = 0$$

and

$$\frac{\partial z}{\partial y} = \frac{3y^2 + 2z^2}{1 - 4yz}$$

■ Partial Derivatives of Functions of More Than Two Variables

The partial derivatives of a function of more than two variables are defined in much the same way as the partial derivatives of a function of two variables. For example, suppose that f is a function of three variables defined by $w = f(x, y, z)$. Then the partial derivative of f with respect to x is defined as

$$\frac{\partial w}{\partial x} = \frac{\partial f}{\partial x} = \lim_{h \to 0} \frac{f(x + h, y, z) - f(x, y, z)}{h}$$

where y and z are held fixed, provided that the limit exists. The other two partial derivatives, $\partial f/\partial y$ and $\partial f/\partial z$, are defined in a similar manner.

Finding the Partial Derivative of a Function of More Than Two Variables

To find the partial derivative of a function of more than two variables with respect to a certain variable, say x, we treat all the other variables as if they are constants and differentiate with respect to x in the usual manner.

EXAMPLE 8 Find

a. f_x if $f(x, y, z) = x^2y + y^2z + xz$ **b.** h_w if $h(x, y, z, w) = \dfrac{xw^2}{y + \sin zw}$

Solution

a. To find f_x, we treat y and z as constants and differentiate f with respect to x to obtain

$$f_x = \frac{\partial}{\partial x}(x^2y + y^2z + xz) = 2xy + z$$

b. To find h_w, we treat x, y, and z as constants and differentiate h with respect to w, obtaining

$$h_w = \frac{\partial}{\partial w}\left(\frac{xw^2}{y + \sin zw}\right)$$

$$= \frac{(y + \sin zw)\dfrac{\partial}{\partial w}(xw^2) - xw^2\dfrac{\partial}{\partial w}(y + \sin zw)}{(y + \sin zw)^2} \qquad \text{Use the Quotient Rule.}$$

$$= \frac{(y + \sin zw)(2xw) - xw^2\left[0 + \cos zw \cdot \dfrac{\partial}{\partial w}(zw)\right]}{(y + \sin zw)^2} \qquad \text{Use the Chain Rule.}$$

$$= \frac{2xw(y + \sin zw) - xw^2 z \cos zw}{(y + \sin zw)^2} = \frac{xw(2y + 2\sin zw - wz \cos zw)}{(y + \sin zw)^2} \quad\blacksquare$$

■ Higher-Order Derivatives

Consider the function $z = f(x, y)$ of two variables. Each of the partial derivatives $\partial f/\partial x$ and $\partial f/\partial y$ are functions of x and y. Therefore, we may take the partial derivatives of these functions to obtain the four **second-order partial derivatives**

$$\frac{\partial^2 f}{\partial x^2} = \frac{\partial}{\partial x}\left(\frac{\partial f}{\partial x}\right), \quad \frac{\partial^2 f}{\partial y\,\partial x} = \frac{\partial}{\partial y}\left(\frac{\partial f}{\partial x}\right), \quad \frac{\partial^2 f}{\partial x\,\partial y} = \frac{\partial}{\partial x}\left(\frac{\partial f}{\partial y}\right), \quad \text{and} \quad \frac{\partial^2 f}{\partial y^2} = \frac{\partial}{\partial y}\left(\frac{\partial f}{\partial y}\right)$$

(See Figure 6.)

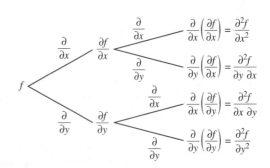

FIGURE 6
The differential operators are shown
on the limbs of the tree diagram.

Before we turn to an example, let's introduce some additional notation for the second-order partial derivatives of f:

$$\frac{\partial^2 f}{\partial x^2} = f_{xx} \qquad \frac{\partial^2 f}{\partial y\,\partial x} = f_{xy} \qquad \frac{\partial^2 f}{\partial x\,\partial y} = f_{yx} \qquad \frac{\partial^2 f}{\partial y^2} = f_{yy}$$

Note the order in which the derivatives are taken: Using the notation $\partial^2 f/(\partial y\,\partial x)$, we differentiate first with respect to x—the independent variable that appears first when read from *right to left*. In the notation f_{xy} we also differentiate first with respect to x—the independent variable that appears first when read from *left to right*. The derivatives f_{xy} and f_{yx} are called **mixed partial derivatives.**

Note   If f is defined by the equation $z = f(x, y)$, then the four partial derivatives of f are also written

$$z_{xx} \quad z_{xy} \quad z_{yx} \quad \text{and} \quad z_{yy} \qquad \blacksquare$$

EXAMPLE 9   Find the second-order partial derivatives of $f(x, y) = 2xy^2 - 3x^2 + xy^3$.

Solution   We first compute the first-order partial derivatives

$$f_x = \frac{\partial}{\partial x}(2xy^2 - 3x^2 + xy^3) = 2y^2 - 6x + y^3$$

and

$$f_y = \frac{\partial}{\partial y}(2xy^2 - 3x^2 + xy^3) = 4xy + 3xy^2$$

Then differentiating each of these functions, we obtain

$$f_{xx} = \frac{\partial}{\partial x}f_x = \frac{\partial}{\partial x}(2y^2 - 6x + y^3) = -6$$

$$f_{xy} = \frac{\partial}{\partial y}f_x = \frac{\partial}{\partial y}(2y^2 - 6x + y^3) = 4y + 3y^2$$

$$f_{yx} = \frac{\partial}{\partial x}f_y = \frac{\partial}{\partial x}(4xy + 3xy^2) = 4y + 3y^2$$

$$f_{yy} = \frac{\partial}{\partial y}f_y = \frac{\partial}{\partial y}(4xy + 3xy^2) = 4x + 6xy \qquad \blacksquare$$

Notice that the mixed derivatives f_{xy} and f_{yx} in Example 9 are equal. The following theorem, which we state without proof, gives the conditions under which this is true.

THEOREM 1   Clairaut's Theorem

If $f(x, y)$ and its partial derivatives f_x, f_y, f_{xy}, and f_{yx} are continuous on an open region R, then

$$f_{xy}(x, y) = f_{yx}(x, y)$$

for all (x, y) in R.

A function u of two variables x and y is called a **harmonic function** if $u_{xx} + u_{yy} = 0$ for all (x, y) in the domain of u. Harmonic functions are used in the study of heat conduction, fluid flow, and potential theory. The *partial differential equation* $u_{xx} + u_{yy} = 0$ is called **Laplace's equation,** named for Pierre Laplace (1749–1827).

Historical Biography

ALEXIS CLAUDE CLAIRAUT
(1713-1765)

Alexis Claude Clairaut was one of twenty children born to his mother but the only to survive to adulthood. His father, a mathematics teacher in Paris, educated his son at home with very high standards: Alexis was taught to read using Euclid's *Elements*. As a result of both nature and nurture, Clairaut turned out to be a very precocious mathematician. He studied calculus by the age of 10 and wrote an original mathematical paper at 13. At 18 he published his first paper; he also became the youngest member ever admitted to the prestigious Academie des Sciences. Clairaut excelled in many areas of mathematics, including geometry, calculus, and celestial mechanics. He was the first to prove the prediction by Isaac Newton and the astronomer Christiaan Huygens that the earth is an oblate ellipsoid. Clairaut also accurately predicted the return of Halley's comet in 1759, a prediction that made him famous. Clairaut developed the notation for partial derivatives that we still use today, and he was the first to prove that mixed second-order partial derivatives of a function at a point are equal if the derivatives are continuous at that point.

EXAMPLE 10 Show that the function $u(x, y) = e^x \cos y$ is harmonic in the xy-plane.

Solution We find

$$u_x = \frac{\partial}{\partial x}(e^x \cos y) = e^x \cos y, \qquad u_y = \frac{\partial}{\partial y}(e^x \cos y) = -e^x \sin y$$

$$u_{xx} = \frac{\partial}{\partial x}(e^x \cos y) = e^x \cos y, \qquad u_{yy} = \frac{\partial}{\partial y}(-e^x \sin y) = -e^x \cos y$$

Therefore,

$$u_{xx} + u_{yy} = e^x \cos y - e^x \cos y = 0$$

This holds for all (x, y) in the plane, so u is harmonic there. ∎

Partial derivatives of order three and higher are defined in a similar manner. For example,

$$f_{xxx} = \frac{\partial}{\partial x}f_{xx} \qquad f_{xxy} = \frac{\partial}{\partial y}f_{xx} \qquad \text{and} \qquad f_{xyx} = \frac{\partial}{\partial x}f_{xy}$$

Also, Theorem 1 is valid for mixed derivatives of higher order. For example, if the third partial derivatives of f are continuous, then the order in which the differentiation is taken does not matter.

EXAMPLE 11 Let $f(x, y, z) = xe^{yz}$. Compute f_{xzy} and f_{yxz}.

Solution We have

$$f_x = \frac{\partial}{\partial x}(xe^{yz}) = e^{yz}$$

$$f_{xz} = \frac{\partial}{\partial z}f_x = \frac{\partial}{\partial z}(e^{yz}) = ye^{yz}$$

so

$$f_{xzy} = \frac{\partial}{\partial y}f_{xz} = \frac{\partial}{\partial y}(ye^{yz}) = e^{yz} + yze^{yz} = (1 + yz)e^{yz}$$

Next, we have

$$f_y = \frac{\partial}{\partial y}(xe^{yz}) = xze^{yz}$$

$$f_{yx} = \frac{\partial}{\partial x}f_y = \frac{\partial}{\partial x}(xze^{yz}) = ze^{yz}$$

so

$$f_{yxz} = \frac{\partial}{\partial z}f_{yx} = \frac{\partial}{\partial z}(ze^{yz}) = e^{yz} + yze^{yz} = (1 + yz)e^{yz}$$

Observe that both f_{xzy} and f_{yxz} are continuous everywhere and are equal. ∎

13.3 CONCEPT QUESTIONS

1. a. Define the partial derivatives of a function of two variables, x and y, with respect to x and with respect to y.
 b. Give a geometric and a physical interpretation of $f_x(x, y)$.
2. Let f be a function of x and y. Describe a procedure for finding f_x and f_y.

3. Suppose $F(x, y, z) = 0$ defines x implicitly as a function of y and z; that is, $x = f(y, z)$. Describe a procedure for finding $\partial x/\partial z$. Illustrate with an example of your choice.
4. If f is a function of x and y, give a condition that will guarantee that $f_{xy}(x, y) = f_{yx}(x, y)$ for all (x, y) in some open region.

13.3 EXERCISES

1. Let $f(x, y) = x^2 + 2y^2$.
 a. Find $f_x(2, 1)$ and $f_y(2, 1)$.
 b. Interpret the numbers in part (a) as slopes.
 c. Interpret the numbers in part (a) as rates of change.

2. Let $f(x, y) = 9 - x^2 + xy - 2y^2$.
 a. Find $f_x(1, 2)$ and $f_y(1, 2)$.
 b. Interpret the numbers in part (a) as slopes.
 c. Interpret the numbers in part (a) as rates of change.

3. Determine the sign of $\partial f/\partial x$ and $\partial f/\partial y$ at the points P, Q, and R on the graph of the function f shown in the figure.

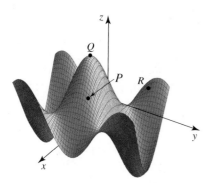

4. The graphs of a function f and its partial derivatives f_x and f_y are labeled (a), (b), and (c). Identify the graphs of f, f_x, and f_y, and give a reason for your answer.

(a)

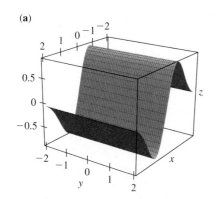

(b)

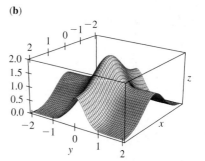

(c)

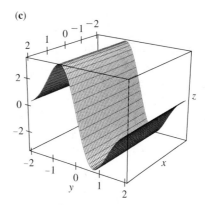

5. The figure below shows the contour map of the function T (measured in degrees Fahrenheit) giving the temperature at each point (x, y) on an 8 in. $\times$ 5 in. rectangular metal plate. Use it to estimate the rate of change of the temperature at the point $(3, 2)$ in the positive x-direction and in the positive y-direction.

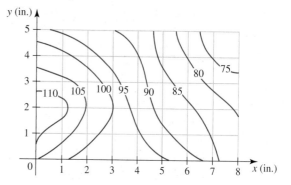

In Exercises 6–29, find the first partial derivatives of the function.

6. $f(x, y) = 3x - 4y + 2$

7. $f(x, y) = 2x^2 - 3xy + y^2$

8. $z = 2x^3 - 3x^2y^3 + xy^2 - 2x$

9. $z = x\sqrt{y}$

10. $f(x, y) = (2x^2 - y^3)^4$

11. $g(r, s) = \sqrt{r + s^2}$

12. $h(u, v) = \ln(u^2 + v^2)$

13. $f(x, y) = xe^{y/x}$

14. $f(x, y) = e^x \cos y + e^y \sin x$

15. $z = \tan^{-1}(x^2 + y^2)$

16. $f(x, y) = \sqrt{3 - 2x^2 - y^2}$

17. $g(u, v) = \dfrac{uv}{u^2 + v^3}$

18. $f(x, y) = \sinh xy$

19. $g(x, y) = x^2 \cosh \dfrac{x}{y}$

20. $z = \ln(e^x + y^2)$

21. $f(x, y) = y^x$

22. $f(x, y) = \displaystyle\int_x^y \cos t \, dt$

23. $f(x, y) = \displaystyle\int_x^y te^{-t} dt$

24. $f(x, y, z) = 2x^3 + 3xy + 2yz - z^2$

25. $g(x, y, z) = \sqrt{xyz}$

26. $f(u, v, w) = ue^v - ve^u + we^u$

27. $u = xe^{y/z} - z^2$

28. $u = x \sin \dfrac{y}{x + z}$

29. $f(r, s, t) = rs \ln st$

In Exercises 30–33, use implicit differentiation to find $\partial z/\partial x$ and $\partial z/\partial y$.

30. $x^2y + xz + yz^2 = 8$

31. $xe^y + ye^{-x} + e^z = 10$

32. $2 \cos(x + 2y) + \sin yz - 1 = 0$

33. $\ln(x^2 + z^2) + yz^3 + 2x^2 = 10$

In Exercises 34–39, find the second partial derivatives of the function.

34. $f(x, y) = x^4 - 2x^2y^3 + y^4 - 3x$

35. $g(x, y) = x^3y^2 + xy^3 - 2x + 3y + 1$

36. $z = xe^{2y} + ye^{2x}$

37. $w = \cos(2u - v) + \sin(2u + v)$

38. $z = \sqrt{x^2 + y^2}$

39. $h(x, y) = \tan^{-1} \dfrac{y}{x}$

In Exercises 40–45, find the indicated partial derivative.

40. $f(x, y) = x^3 + y^3 - 3x^2y^2 + 2x + 3y + 4; \quad f_{xxx}$

41. $f(x, y) = x^4 - 2x^2y^2 + xy^3 + 2y^4; \quad f_{xyx}$

42. $f(x, y, z) = \ln(x^2 + y^2 + z^2); \quad f_{yxz}$

43. $z = x \cos y + y \sin x; \quad \dfrac{\partial^3 z}{\partial x \, \partial y \, \partial x}$

44. $p = e^{uvw}; \quad \dfrac{\partial^3 p}{\partial u \, \partial w \, \partial v}$

45. $h(x, y, z) = e^x \cos(y + 2z); \quad h_{zzy}$

In Exercises 46–49, show that the mixed partial derivatives f_{xy} and f_{yx} are equal.

46. $f(x, y) = x^2 + 2x^2y + y^3$

47. $f(x, y) = x \sin^2 y + y^2 \cos x$

48. $f(x, y) = e^{-2x} \cos 3y$

49. $f(x, y) = \tan^{-1}(x^2 + y^3)$

In Exercises 50–53, show that the mixed partial derivatives f_{xyz}, f_{yxz}, and f_{zyx} are equal.

50. $f(x, y, z) = x^2y^3 - y^2z^3$

51. $f(x, y, z) = \sqrt{9 - x^2 - 2y^2 - z^2}$

52. $f(x, y, z) = \ln(x + 2y + 3z)$

53. $f(x, y, z) = e^{-x} \cos yz$

54. The figure shows the contour map of a function f. Use it to determine the sign of (a) f_x, (b) f_y, (c) f_{xx}, (d) f_{xy}, and (e) f_{yy} at the point P.

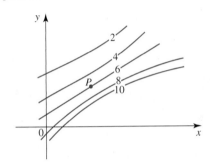

In Exercises 55 and 56, show that the function satisfies the one-dimensional heat equation $u_t = c^2 u_{xx}$.

55. $u = e^{-t} \sin \dfrac{x}{c}$

56. $u = e^{-c^2k^2t} \cos kx$

In Exercises 57 and 58, show that the function satisfies the one-dimensional wave equation $u_{tt} = c^2 u_{xx}$.

57. $u = \cos(x - ct) + 2 \sin(x + ct)$

58. $u = \sin(kct) \sin(kx)$

In Exercises 59–64, show that the function satisfies the two-dimensional Laplace's equation $u_{xx} + u_{yy} = 0$.

59. $u = 3x^2y - y^3$

60. $u = \dfrac{x}{x^2 + y^2}$

61. $u = \ln\sqrt{x^2 + y^2}$

62. $u = e^{-x} \cos y + e^{-y} \cos x$

63. $u = \tan^{-1} \dfrac{y}{x}$

64. $u = \cosh y \sin x + \sinh y \cos x$

In Exercises 65 and 66, show that the function satisfies the three-dimensional Laplace's equation $u_{xx} + u_{yy} + u_{zz} = 0$.

65. $u = x^2 + 3xy + 2y^2 - 3z^2 + 4xyz$

66. $u = (x^2 + y^2 + z^2)^{-1/2}$

67. Show that the function $z = \sqrt{x^2 + y^2}\,\tan^{-1}\dfrac{y}{x}$ satisfies the equation $x\dfrac{\partial z}{\partial x} + y\dfrac{\partial z}{\partial y} = z$.

68. Show that the function $u = 20x^2\cos\dfrac{y}{x}$ satisfies the equation $x\dfrac{\partial u}{\partial x} + y\dfrac{\partial u}{\partial y} = 2u$.

69. According to the ideal gas law, the volume V (in liters) of an ideal gas is related to its pressure P (in pascals) and temperature T (in kelvins) by the formula

$$V = \frac{kT}{P}$$

where k is a constant. Compute $\partial V/\partial T$ and $\partial V/\partial P$ if $k = 8.314$, $T = 300$, and $P = 125$, and interpret your results.

70. Refer to Exercise 69. Show that

$$\frac{\partial V}{\partial T}\cdot\frac{\partial T}{\partial P}\cdot\frac{\partial P}{\partial V} = -1$$

71. The total resistance R (in ohms) of three resistors with resistances of R_1, R_2, and R_3 ohms connected in parallel is given by the formula

$$\frac{1}{R} = \frac{1}{R_1} + \frac{1}{R_2} + \frac{1}{R_3}$$

Find $\partial R/\partial R_1$ and interpret your result.

72. The height of a hill (in feet) is given by

$$h(x, y) = 20(16 - 4x^2 - 3y^2 + 2xy + 28x - 18y)$$

where x is the distance (in miles) east and y the distance (in miles) north of Bolton. If you are at a point on the hill 1 mile north and 1 mile east of Bolton, what is the rate of change of the height of the hill (a) in a northerly direction and (b) in an easterly direction?

73. Profit Versus Inventory and Floor Space The monthly profit (in dollars) of the Barker Department Store depends on its level of inventory x (in thousands of dollars) and the floor space y (in thousands of square feet) available for display of its merchandise, as given by the equation

$$P(x, y) = -0.02x^2 - 15y^2 + xy + 39x + 25y - 15{,}000$$

Find $\partial P/\partial x$ and $\partial P/\partial y$ when $x = 5000$ and $y = 200$, and interpret your result.

74. Steady-State Temperature Consider the upper half-disk $H = \{(x, y)\,|\,x^2 + y^2 \le 1, y \ge 0\}$ (see the figure). If the temperature at points on the upper boundary is kept at $100°C$ and the temperature at points on the lower boundary is kept at $50°C$, then the steady-state temperature at any point (x, y) inside the half-disk is given by

$$T(x, y) = 100 - \frac{100}{\pi}\tan^{-1}\frac{1 - x^2 - y^2}{2y}$$

a. Compute $T_x\!\left(\frac{1}{2}, \frac{1}{2}\right)$, and interpret your result.

b. Find the rate of change of the temperature at the point $P\!\left(\frac{1}{2}, \frac{1}{2}\right)$ in the y-direction.

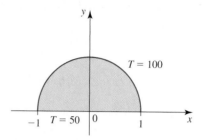

75. Electric Potential A charge Q (in coulombs) located at the origin of a three-dimensional coordinate system produces an electric potential V (in volts) given by

$$V(x, y, z) = \frac{kQ}{\sqrt{x^2 + y^2 + z^2}}$$

where k is a positive constant and x, y, and z are measured in meters. Find the rate of change of the potential at the point $P(1, 2, 3)$ in the x-direction.

76. Surface Area of a Human The formula

$$S = 0.007184W^{0.425}H^{0.725}$$

gives the surface area S of a human body (in square meters) in terms of its weight W (in kilograms) and its height H (in centimeters). Compute $\partial S/\partial W$ and $\partial S/\partial H$ when $W = 70$ and $H = 180$, and interpret your results.

77. Arson Study A study of arson for profit conducted for a certain city found that the number of suspicious fires is approximated by the formula

$$N(x, y) = \frac{120\sqrt{1000 + 0.03x^2 y}}{(5 + 0.2y)^2}$$
$$0 \le x \le 150, \quad 5 \le y \le 35$$

where x denotes the number of persons per census tract and y denotes the level of reinvestment in conventional mortgages by the city's ten largest banks measured in cents per dollars deposited. Compute $\partial N/\partial x$ and $\partial N/\partial y$ when $x = 100$ and $y = 20$, and interpret your results.

78. Production Functions The productivity of a Central American country is given by the function

$$f(x, y) = 20x^{3/4}y^{1/4}$$

when x units of labor and y units of capital are used.

a. What are the marginal productivity of labor and the marginal productivity of capital when the amounts expended on labor and capital are 256 units and 16 units, respectively?

b. Should the government encourage capital investment rather than increased expenditure on labor at this time to increase the country's productivity?

79. Wind Chill Factor A formula that meteorologists use to calculate the wind chill temperature (the temperature that you would feel in still air that is the same as the actual temperature when the presence of wind is taken into consideration) is

$$T = f(t, s) = 35.74 + 0.6125t - 35.75s^{0.16} + 0.4275ts^{0.16}$$
$$s \geq 1$$

where t is the air temperature in degrees Fahrenheit and s is the wind speed in mph.

a. What is the wind chill temperature when the actual air temperature is 32°F and the wind speed is 20 mph?

b. What is the rate of change of the wind chill temperature with respect to the wind speed if the temperature is 32°F and the wind speed is 20 mph?

80. Wind Chill Factor The wind chill temperature is the temperature that you would feel in still air that is the same as the actual temperature when the presence of wind is taken into consideration. The following table gives the wind chill temperature $T = f(t, s)$ in degrees Fahrenheit in terms of the actual air temperature t in degrees Fahrenheit and the wind speed s in mph.

		Wind speed (mph)						
	s t	**10**	**15**	**20**	**25**	**30**	**35**	**40**
Actual air temperature (°F)	**30**	21.2	19.0	17.4	16.0	14.9	13.9	13.0
	32	23.7	21.6	20.0	18.7	17.6	16.6	15.8
	34	26.2	24.2	22.6	21.4	20.3	19.4	18.6
	36	28.7	26.7	25.2	24.0	23.0	22.2	21.4
	38	31.2	29.3	27.9	26.7	25.7	24.9	24.2
	40	33.6	31.8	20.5	29.4	28.5	27.7	26.9

a. Estimate the rate of change of the wind chill temperature T with respect to the actual air temperature when the wind speed is constant at 25 mph and the actual air temperature is 34°F.

Hint: Show that it is given by

$$\frac{\partial T}{\partial t}(34, 25) \approx \frac{f(36, 25) - f(34, 25)}{2}$$

b. Estimate the rate of change of the wind chill temperature T with respect to the wind speed when the actual air temperature is constant at 34°F and the wind speed is 25 mph.

Source: National Weather Service.

81. Let f be a function of two variables.

a. Put $g(x) = f(x, b)$, and use the definition of the derivative of a function of one variable to show that $f_x(a, b) = g'(a)$.

b. Put $h(y) = f(a, y)$, and show that $f_y(a, b) = h'(b)$.

82. a. Use the result of Exercise 81 to find $f_x\left(1, \frac{\pi}{2}\right)$ if $f(x, y) = x^2 \cos xy$.

b. Verify the result of part (a) by evaluating $f_x(x, y)$ at $\left(1, \frac{\pi}{2}\right)$.

cas *In Exercises 83 and 84, use the result of Exercise 81 and a calculator or computer to find the partial derivative.*

83. $f_x(2, 1)$ if $f(x, y) = \ln\left(e^{xy} + \cos\sqrt{x^2 + y^2}\right)$

84. $f_y(2, 1)$ if $f(x, y) = \dfrac{\sin \pi xy}{\left(1 + \sqrt{x^2 + y^3}\right)^{3/2}}$

85. Cobb-Douglas Production Function Show that the Cobb-Douglas production function $P = kx^\alpha y^{1-\alpha}$, where $0 < \alpha < 1$, satisfies the equation

$$x\frac{\partial P}{\partial x} + y\frac{\partial P}{\partial y} = P$$

Note: This equation is called **Euler's equation.**

86. Let S be the surface with equation $z = f(x, y)$, where f has continuous first-order partial derivatives and $P(x_0, y_0, z_0)$ is a point on S (see the figure). Let C_1 and C_2 be the curves obtained by the intersection of the surface S with the planes $x = x_0$ and $y = y_0$, respectively. Let T_1 and T_2 be the tangent lines to the curves C_1 and C_2 at P. Then the *tangent plane* to the surface S at the point P is the plane that contains both tangent lines T_1 and T_2.

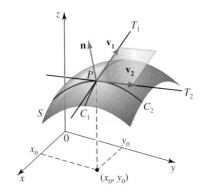

a. Show that the vectors $\mathbf{v}_1 = \mathbf{i} + f_x(x_0, y_0)\mathbf{k}$ and $\mathbf{v}_2 = \mathbf{j} + f_y(x_0, y_0)\mathbf{k}$ are parallel to T_1 and T_2, respectively.

b. Using the result of part (a), find a vector $\mathbf{n}$ that is normal to both $\mathbf{v}_1$ and $\mathbf{v}_2$.

c. Use the result of part (b) to show that an equation of the tangent plane to S at P is

$$z - z_0 = f_x(x_0, y_0)(x - x_0) + f_y(x_0, y_0)(y - y_0)$$

87. Use the result of Exercise 86 to find an equation of the tangent plane to the paraboloid $z = x^2 + \frac{1}{4}y^2$ at the point $(1, 2, 2)$.

88. Engine Efficiency The efficiency of an internal combustion engine is given by

$$E = \left(1 - \frac{v}{V}\right)^{0.4}$$

where V and v are the respective maximum and minimum volumes of air in each cylinder.

a. Show that $\partial E/\partial V > 0$, and interpret your result.

b. Show that $\partial E/\partial v < 0$, and interpret your result.

89. A semi-infinite strip has faces that are insulated. If the edges $x = 0$ and $x = \pi$ of the strip are kept at temperature zero and the base of the strip is kept at temperature 1, then the steady-state temperature (that is, the temperature after a long time) is given by

$$T(x, y) = \frac{2}{\pi} \tan^{-1} \frac{\sin x}{\sinh y}$$

Find $\dfrac{\partial T}{\partial x}\left(\frac{\pi}{2}, 1\right)$ and $\dfrac{\partial T}{\partial y}\left(\frac{\pi}{2}, 1\right)$, and interpret your results.

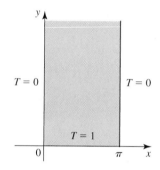

90. Let

$$f(x, y) = \begin{cases} \dfrac{xy(x^2 - y^2)}{x^2 + y^2} & \text{if } (x, y) \neq (0, 0) \\ 0 & \text{if } (x, y) = (0, 0) \end{cases}$$

a. Find $f_x(x, y)$ and $f_y(x, y)$ for $(x, y) \neq (0, 0)$.

b. Use the definition of partial derivatives to find $f_x(0, 0)$ and $f_y(0, 0)$.

c. Show that $f_{xy}(0, 0) = -1$ and $f_{yx}(0, 0) = 1$.

d. Does the result of part (c) contradict Theorem 1? Explain.

91. Does there exist a function f of two variables x and y with continuous second-order partial derivatives such that $f_x(x, y) = e^{2x}(2 \cos xy - y \sin xy)$ and $f_y(x, y) = -ye^{2x} \sin xy$? Explain.

92. Show that if a function f of two variables x and y has continuous third-order partial derivatives, then $f_{xyx} = f_{yxx} = f_{xxy}$.

In Exercises 93–96, determine whether the statement is true or false. If it is true, explain why it is true. If it is false, give an example to show why it is false.

93. If $z = f(x, y)$ has a partial derivative with respect to x at the point (a, b), then

$$\frac{\partial f}{\partial x}(a, b) = \lim_{x \to a} \frac{f(x, b) - f(a, b)}{x - a}$$

94. If $\partial f/\partial y (a, b) = 0$, then the tangent line to the curve formed by the intersection of the plane $x = a$ and the surface $z = f(x, y)$ at the point $(a, b, f(a, b))$ is horizontal; that is, it is parallel to the xy-plane.

95. If $f_{xx}(x, y)$ is defined for all x and y and $f_{xx}(a, b) < 0$ for all x in the interval (a, b), then the curve C formed by the intersection of the plane $y = b$ and the surface $z = f(x, y)$ is concave downward on (a, b).

96. If $f(x, y) = \ln xy$, then $f_{xy}(x, y) = f_{yx}(x, y)$ for all (x, y) in $D = \{(x, y) \,|\, xy > 0\}$.

13.4 Differentials

Increments

Recall that if f is a function of one variable defined by $y = f(x)$, then the *increment* in y is defined to be

$$\Delta y = f(x + \Delta x) - f(x)$$

where Δx is an increment in x (Figure 1a). The increment of a function of two or more variables is defined in an analogous manner. For example, if z is a function of two variables defined by $z = f(x, y)$, then the **increment** in z produced by increments of Δx and Δy in the independent variables x and y, respectively, is defined to be

$$\Delta z = f(x + \Delta x, y + \Delta y) - f(x, y) \tag{1}$$

(See Figure 1b.)

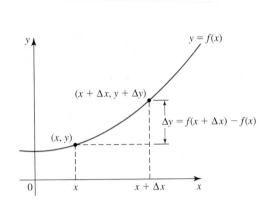

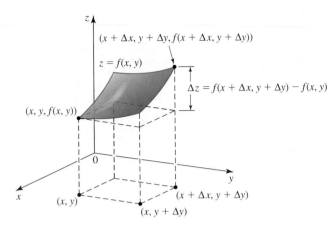

(a) The increment Δy is the change in y as x changes from x to $x + \Delta x$.

(b) The increment Δz is the change in z as x changes from x to $x + \Delta x$ and y changes from y to $y + \Delta y$.

FIGURE 1

EXAMPLE 1 Let $z = f(x, y) = 2x^2 - xy$. Find Δz. Then use your result to find the change in z if (x, y) changes from $(1, 1)$ to $(0.98, 1.03)$.

Solution Using Equation (1), we obtain

$$\Delta z = f(x + \Delta x, y + \Delta y) - f(x, y)$$
$$= [2(x + \Delta x)^2 - (x + \Delta x)(y + \Delta y)] - (2x^2 - xy)$$
$$= 2x^2 + 4x\,\Delta x + 2(\Delta x)^2 - xy - x\,\Delta y - y\,\Delta x - \Delta x\,\Delta y - 2x^2 + xy$$
$$= (4x - y)\,\Delta x - x\,\Delta y + 2(\Delta x)^2 - \Delta x\,\Delta y$$

Next, to find the increment in z if (x, y) changes from $(1, 1)$ to $(0.98, 1.03)$, we note that $\Delta x = 0.98 - 1 = -0.02$ and $\Delta y = 1.03 - 1 = 0.03$. Therefore, using the result obtained earlier with $x = 1$, $y = 1$, $\Delta x = -0.02$, and $\Delta y = 0.03$, we obtain

$$\Delta z = [4(1) - 1](-0.02) - (1)(0.03) + 2(-0.02)^2 - (-0.02)(0.03)$$
$$= -0.0886$$

You can verify the correctness of this result by computing the quantity $f(0.98, 1.03) - f(1, 1)$. ■

■ The Total Differential

Recall from Section 2.9 that if f is a function of one variable defined by $y = f(x)$, then the differential of f at x is defined by

$$dy = f'(x)\,dx$$

where $dx = \Delta x$ is the differential in x. Furthermore,

$$\Delta y \approx dy \tag{2}$$

if Δx is small (see Figure 2).

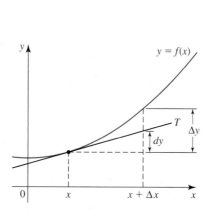

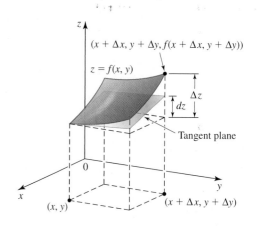

(a) Relationship between dy and Δy

(b) Relationship between dz and Δz. The tangent plane is the analog of the tangent line T in the one-variable case.

FIGURE 2

For an analog of this result for a function of two variables, we begin with the following definition.

DEFINITION **Differentials**

Let $z = f(x, y)$, and let Δx and Δy be increments of x and y, respectively. The **differentials** dx and dy of the independent variables x and y are

$$dx = \Delta x \qquad \text{and} \qquad dy = \Delta y$$

The **differential** dz, or **total differential,** of the dependent variable z is

$$dz = \frac{\partial f}{\partial x}\, dx + \frac{\partial f}{\partial y}\, dy = f_x(x, y)\, dx + f_y(x, y)\, dy$$

Later in this section, we will show that

$$\Delta z = dz + \varepsilon_1 \, \Delta x + \varepsilon_2 \, \Delta y$$

where ε_1 and ε_2 are functions of Δx and Δy that approach 0 as Δx and Δy approach 0. This implies that

$$\Delta z \approx dz \qquad\qquad\qquad \textbf{(3)}$$

if both Δx and Δy are small.

Figure 2b shows the geometric relationship between Δz and dz. Observe that as x changes from x to $x + \Delta x$ and y changes from y to $y + \Delta y$, Δz measures the change in the height of the graph of f, whereas dz measures the change in the height of the tangent plane.*

---

*For now, we will rely on our intuitive definition of the tangent plane. We will define the tangent plane in Section 13.7.

EXAMPLE 2 Let $z = f(x, y) = 2x^2 - xy$.

a. Find the differential dz.
b. Compute the value of dz if (x, y) changes from $(1, 1)$ to $(0.98, 1.03)$, and compare your result with the value of Δz obtained in Example 1.

Solution

a. $dz = \dfrac{\partial f}{\partial x} dx + \dfrac{\partial f}{\partial y} dy = (4x - y)\, dx - x\, dy$

b. Here $x = 1$, $y = 1$, $dx = \Delta x = -0.02$, and $dy = \Delta y = 0.03$. Therefore,

$$dz = [4(1) - 1](-0.02) - 1(0.03) = -0.09$$

The value of Δz obtained in Example 1 was -0.0886, so dz is a good approximation of Δz in this case. Observe that it is easier to compute dz than to compute Δz.

EXAMPLE 3 A storage tank has the shape of a right circular cylinder. Suppose that the radius and height of the tank are measured at 1.5 ft and 5 ft, respectively, with a possible error of 0.05 ft and 0.1 ft, respectively. Use differentials to estimate the maximum error in calculating the capacity of the tank.

Solution The capacity (volume) of the tank is $V = \pi r^2 h$. The error in calculating the capacity of the tank is given by

$$\Delta V \approx dV = \frac{\partial V}{\partial r} dr + \frac{\partial V}{\partial h} dh = 2\pi rh\, dr + \pi r^2\, dh$$

Since the errors in the measurement of r and h are at most 0.05 ft and 0.1 ft, respectively, we have $dr = 0.05$ and $dh = 0.1$. Therefore, taking $r = 1.5$, $h = 5$, $dr = 0.05$, and $dh = 0.1$, we obtain

$$dV = 2\pi rh\, dr + \pi r^2\, dh$$
$$\approx 2\pi(1.5)(5)(0.05) + \pi(1.5)^2(0.1) = 0.975\pi$$

Thus, the maximum error in calculating the volume of the storage tank is approximately 0.975π, or 3.1, ft^3.

EXAMPLE 4 **The Error in Computing the Range of a Projectile** If a projectile is fired with an angle of elevation θ and initial speed of v ft/sec, then its range (in feet) is

$$R = \frac{v^2 \sin 2\theta}{g}$$

where g is the constant of acceleration due to gravity. (See Figure 3.) Suppose that a projectile is launched with an initial speed of 2000 ft/sec at an angle of elevation of $\pi/12$ radians and that the maximum percentage errors in the measurement of v and θ are 0.5% and 1%, respectively.

a. Estimate the maximum error in the computation of the range of the projectile.
b. Find the maximum percentage error in computing the range of this projectile.

FIGURE 3
We want to find the range R of a projectile fired with an angle of elevation θ and initial speed of v ft/sec.

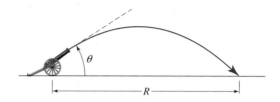

Solution

a. The error in the computation of R is

$$\Delta R \approx dR = \frac{\partial R}{\partial v}\,dv + \frac{\partial R}{\partial \theta}\,d\theta = \frac{2v\sin 2\theta}{g}\,dv + \frac{2v^2\cos 2\theta}{g}\,d\theta$$

The maximum error in the computation of v is $(0.005)(2000)$ or 10 ft/sec; that is, $|dv| \le 10$. Also, the maximum error in the computation of θ is $(0.01)(\pi/12)$ radians. In other words, $|d\theta| \le 0.01(\pi/12)$. Therefore, the maximum error in computing the range of the projectile is approximately

$$|\Delta R| \approx |dR| \le \frac{2v\sin 2\theta}{g}\,|dv| + \frac{2v^2\cos 2\theta}{g}\,|d\theta|$$

$$= \frac{2(2000)\sin\left(\dfrac{\pi}{6}\right)}{32}\,(10) + \frac{2(2000)^2\cos\left(\dfrac{\pi}{6}\right)}{32}\left(\frac{0.01\pi}{12}\right)$$

$$\approx 1192$$

or approximately 1192 ft.

b. Using $v = 2000$ and $\theta = \pi/12$, we find the range of the projectile to be

$$R = \frac{v^2\sin 2\theta}{g} = \frac{(2000)^2\sin\left(\dfrac{\pi}{6}\right)}{32} = 62{,}500$$

Therefore, the maximum percentage error in computing the range of the projectile is

$$100\left|\frac{\Delta R}{R}\right| \approx 100\left(\frac{1192}{62{,}500}\right)$$

or approximately 1.91%. ■

■ Error in Approximating Δz by dz

The following theorem tells us that dz gives a good approximation of Δz if Δx and Δy are small, provided that both f_x and f_y are continuous.

THEOREM 1

Let f be a function defined on an open region R. Suppose that the points (x, y) and $(x + \Delta x, y + \Delta y)$ are in R and that f_x and f_y are continuous at (x, y). Then

$$\Delta z = f_x(x, y)\,\Delta x + f_y(x, y)\,\Delta y + \varepsilon_1\,\Delta x + \varepsilon_2\,\Delta y$$

where ε_1 and ε_2 are functions of Δx and Δy such that

$$\lim_{(\Delta x,\,\Delta y)\to(0,\,0)} \varepsilon_1 = 0 \qquad \text{and} \qquad \lim_{(\Delta x,\,\Delta y)\to(0,\,0)} \varepsilon_2 = 0$$

PROOF   Fix x and y. By adding and subtracting $f(x + \Delta x, y)$ to Δz, we have

$$\Delta z = f(x + \Delta x, y + \Delta y) - f(x, y)$$

$$= [f(x + \Delta x, y) - f(x, y)] + [f(x + \Delta x, y + \Delta y) - f(x + \Delta x, y)]$$

$$= \Delta z_1 + \Delta z_2$$

where Δz_1 is the change in z as (x, y) changes from (x, y) to $(x + \Delta x, y)$ and Δz_2 is the change in z as (x, y) changes from $(x + \Delta x, y)$ to $(x + \Delta x, y + \Delta y)$. (See Figure 4a.)

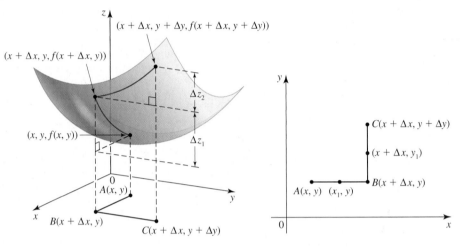

FIGURE 4

(a) $\Delta z_1 = f(x + \Delta x, y) - f(x, y)$ and
$\Delta z_2 = f(x + \Delta x, y + \Delta y) - f(x + \Delta x, y)$

(b) The points A, B, and C shown in the xy-plane.

On the interval between A and B, y is constant, so the function g defined by $g(t) = f(t, y)$ for $x \le t \le x + \Delta x$ is a function of one variable. (See Figure 4b.) Therefore, by the Mean Value Theorem, there exists a point (x_1, y) with $x < x_1 < x + \Delta x$ such that

$$g(x + \Delta x) - g(x) = g'(x_1) \Delta x$$

Since $g'(x_1) = f_x(x_1, y)$, we have

$$\Delta z_1 = f(x + \Delta x, y) - f(x, y) = g(x + \Delta x) - g(x)$$

$$= g'(x_1) \Delta x = f_x(x_1, y) \Delta x \qquad x < x_1 < x + \Delta x$$

Next, on the interval between B and C, both x and Δx are constant, so the function h defined by $h(t) = f(x + \Delta x, t)$ for $y \le t \le y + \Delta y$ is a function of one variable. (See Figure 4b.) Therefore, by the Mean Value Theorem there exists a point $(x + \Delta x, y_1)$ with $y < y_1 < y + \Delta y$ such that

$$h(y + \Delta y) - h(y) = h'(y_1) \Delta y$$

Since $h'(y_1) = f_y(x + \Delta x, y_1)$, we have

$$\Delta z_2 = f(x + \Delta x, y + \Delta y) - f(x + \Delta x, y)$$

$$= h(y + \Delta y) - h(y) = h'(y_1) \Delta y = f_y(x + \Delta x, y_1) \Delta y$$

Therefore,

$$\Delta z = \Delta z_1 + \Delta z_2$$

$$= f_x(x_1, y) \Delta x + f_y(x + \Delta x, y_1) \Delta y$$

Adding and subtracting $f_x(x, y) \Delta x + f_y(x, y) \Delta y$ to the right-hand side of the previous equation and rearranging terms, we obtain

$$\Delta z = f_x(x, y) \Delta x + f_y(x, y) \Delta y + [f_x(x_1, y) - f_x(x, y)] \Delta x + [f_y(x + \Delta x, y_1) - f_y(x, y)] \Delta y$$

$$= f_x(x, y) \Delta x + f_y(x, y) \Delta y + \varepsilon_1 \Delta x + \varepsilon_2 \Delta y$$

where

$$\varepsilon_1 = f_x(x_1, y) - f_x(x, y)$$

and

$$\varepsilon_2 = f_y(x + \Delta x, y_1) - f_y(x, y)$$

Observe that as $(\Delta x, \Delta y) \to (0, 0)$, $x_1 \to x$ and $y_1 \to y$. Therefore, the continuity of f_x and f_y implies that

$$\lim_{(\Delta x, \Delta y) \to (0, 0)} \varepsilon_1 = 0 \qquad \text{and} \qquad \lim_{(\Delta x, \Delta y) \to (0, 0)} \varepsilon_2 = 0$$

and this proves the result. ■

Note Observe that the conclusion of Theorem 1 can be written as

$$\Delta z - dz = \varepsilon_1 \Delta x + \varepsilon_2 \Delta y$$

Therefore, if Δx and Δy are both small, then

$$\Delta z - dz = (\text{small number})(\text{small number}) + (\text{small number})(\text{small number})$$

and this quantity is a *very* small number, which accounts for the closeness of the approximation. Compare this with the case of a function of one variable discussed in Section 2.9. ■

■ Differentiability of a Function of Two Variables

The conclusion of Theorem 1 can be written as

$$\Delta z = dz + \varepsilon_1 \Delta x + \varepsilon_2 \Delta y \tag{4}$$

where $\varepsilon_1 \to 0$ and $\varepsilon_2 \to 0$ as $(\Delta x, \Delta y) \to (0, 0)$. We *define* a function of two variables to be *differentiable* if $z = f(x, y)$ satisfies Equation (4).

DEFINITION **Differentiability of a Function of Two Variables**

Let $z = f(x, y)$. The function f is differentiable at (a, b) if Δz can be expressed in the form

$$\Delta z = f_x(a, b) \Delta x + f_y(a, b) \Delta y + \varepsilon_1 \Delta x + \varepsilon_2 \Delta y$$

where $\varepsilon_1 \to 0$ and $\varepsilon_2 \to 0$ as $(\Delta x, \Delta y) \to (0, 0)$. The function f is differentiable in a region R if it is differentiable at each point of R.

EXAMPLE 5 Show that the function f defined by $f(x, y) = 2x^2 - xy$ is differentiable in the plane.

Solution Write $z = f(x, y) = 2x^2 - xy$, and let (x, y) be any point in the plane. Then using the result of Example 1, we have

$$\Delta z = (4x - y) \Delta x - x \Delta y + 2(\Delta x)^2 - \Delta x \Delta y$$

Since $f_x = 4x - y$ and $f_y = -x$, we can write

$$\Delta z = f_x \Delta x + f_y \Delta y + \varepsilon_1 \Delta x + \varepsilon_2 \Delta y$$

where $\varepsilon_1 = 2 \Delta x$ and $\varepsilon_2 = -\Delta x$. Since $\varepsilon_1 \to 0$ and $\varepsilon_2 \to 0$ as $(\Delta x, \Delta y) \to (0, 0)$, it follows that f is differentiable at (x, y). But (x, y) is any point in the plane, so f is differentiable in the plane.   ∎

The next theorem, which is an immediate consequence of Theorem 1, guarantees when a function of two variables is differentiable.

THEOREM 2   Criterion for Differentiability

Let f be a function of the variables x and y. If f_x and f_y exist and are continuous on an open region R, then f is differentiable in R.

For the function $f(x, y) = 2x^2 - xy$ of Example 5, we have $f_x(x, y) = 4x - y$ and $f_y(x, y) = -x$, both of which are continuous everywhere. Therefore, by Theorem 2 we conclude that f is differentiable in the plane, as demonstrated earlier.

⚠ Remember that the mere existence of the partial derivatives f_x and f_y of a function f at a point (x, y) is not enough to guarantee the differentiability of f at (x, y). (See Exercise 43.)

■ Differentiability and Continuity

Just as a differentiable function of one variable is continuous, the following theorem shows that a differentiable function of two variables is also continuous.

THEOREM 3   Differentiable Functions Are Continuous

Let f be a function of two variables. If f is differentiable at (a, b), then f is continuous at (a, b).

PROOF   Using the result of Theorem 1, we have

$$\Delta z = f(a + \Delta x, b + \Delta y) - f(a, b)$$

$$= f_x(a, b) \Delta x + f_y(a, b) \Delta y + \varepsilon_1 \Delta x + \varepsilon_2 \Delta y$$

Writing $x = a + \Delta x$ and $y = b + \Delta y$, we have

$$f(x, y) - f(a, b) = [f_x(a, b) + \varepsilon_1](x - a) + [f_y(a, b) + \varepsilon_2](y - b)$$

Noting that $\varepsilon_1 \to 0$ and $\varepsilon_2 \to 0$ as $(\Delta x, \Delta y) \to (0, 0)$, we see that

$$f(x, y) - f(a, b) \to 0 \quad \text{as} \quad (\Delta x, \Delta y) \to (0, 0)$$

Equivalently,

$$\lim_{(x, y) \to (a, b)} f(x, y) = f(a, b)$$

Therefore, f is continuous at (a, b).   ∎

Functions of Three or More Variables

The notions of differentiability and the differential of functions of more than two variables are similar to those of functions of two variables. For example, suppose that f is a function of three variables that is defined by $w = f(x, y, z)$. Then the increment Δw of w corresponding to increments of Δx, Δy, and Δz of x, y, and z, respectively, is

$$\Delta w = f(x + \Delta x, y + \Delta y, z + \Delta z) - f(x, y, z)$$

The function f is **differentiable** at (x, y, z) if Δw can be written in the form

$$\Delta w = f_x(x, y, z) \, \Delta x + f_y(x, y, z) \, \Delta y + f_z(x, y, z) \, \Delta z + \varepsilon_1 \, \Delta x + \varepsilon_2 \, \Delta y + \varepsilon_3 \, \Delta z$$

where $\varepsilon_1, \varepsilon_2$, and ε_3 are functions of $\Delta x, \Delta y$, and Δz that approach zero as $(\Delta x, \Delta y, \Delta z) \to (0, 0, 0)$.

The **differential** dw of the dependent variable w is defined to be

$$dw = \frac{\partial w}{\partial x} \, dx + \frac{\partial w}{\partial y} \, dy + \frac{\partial w}{\partial z} \, dz$$

where $dx = \Delta x$, $dy = \Delta y$, and $dz = \Delta z$ are the differentials of the independent variables, x, y, and z. If f has continuous partial derivatives and dx, dy, and dz are all small, then $\Delta w \approx dw$.

EXAMPLE 6 **Maximum Error in Calculating Centrifugal Force** A centrifuge is a machine designed for the specific purpose of subjecting materials to a sustained centrifugal force. The magnitude of a centrifugal force F in dynes is given by

$$F = f(M, S, R) = \frac{\pi^2 S^2 M R}{900}$$

where S is in revolutions per minute (rpm), M is the mass in grams, and R is the radius in centimeters. If the maximum percentage errors in the measurement of M, S, and R are 0.1%, 0.4%, and 0.2%, respectively, use differentials to estimate the maximum percentage error in calculating F.

Solution The error in calculating F is ΔF, and

$$\Delta F \approx dF = \frac{\partial F}{\partial M} \, dM + \frac{\partial F}{\partial S} \, dS + \frac{\partial F}{\partial R} \, dR$$

$$= \frac{\pi^2 S^2 R}{900} \, dM + \frac{2\pi^2 SMR}{900} \, dS + \frac{\pi^2 S^2 M}{900} \, dR$$

Therefore,

$$\frac{\Delta F}{F} \approx \frac{dF}{F} = \frac{dM}{M} + 2\frac{dS}{S} + \frac{dR}{R}$$

and

$$\left| \frac{\Delta F}{F} \right| \approx \left| \frac{dF}{F} \right| \le \left| \frac{dM}{M} \right| + 2\left| \frac{dS}{S} \right| + \left| \frac{dR}{R} \right|$$

Since

$$\left| \frac{dM}{M} \right| \le 0.001, \qquad \left| \frac{dS}{S} \right| \le 0.004, \qquad \text{and} \qquad \left| \frac{dR}{R} \right| \le 0.002$$

we have

$$\left| \frac{dF}{F} \right| \leq 0.001 + 2(0.004) + 0.002 = 0.011$$

Thus, the maximum percentage error in calculating the centrifugal force is approximately 1.1%. ■

13.4 CONCEPT QUESTIONS

1. If $z = f(x, y)$, what is the differential of x? The differential of y? What is the total differential of z?
2. Let $z = f(x, y)$. What is the relationship between the actual change Δz, when x changes from x to $x + \Delta x$ and y changes from y to $y + \Delta y$, and the total differential dz of f at (x, y)?
3. a. What does it mean for a function f of two variables x and y to be differentiable at (a, b)? To be differentiable in a region R?

b. Give a condition that guarantees that a function f of two variables x and y is differentiable in an open region R.
c. If a function f of two variables x and y is differentiable at (a, b), what can you say about the continuity of f at (a, b)?

13.4 EXERCISES

1. Let $z = 2x^2 + 3y^2$, and suppose that (x, y) changes from $(2, -1)$ to $(2.01, -0.98)$.
   a. Compute Δz.   b. Compute dz.
   c. Compare the values of Δz and dz.

2. Let $z = x^2 - 2xy + 3y^2$, and suppose that (x, y) changes from $(2, 1)$ to $(1.97, 1.02)$.
   a. Compute Δz.   b. Compute dz.
   c. Compare the values of Δz and dz.

In Exercises 3–20, find the differential of the function.

3. $z = 3x^2y^3$

4. $z = x^4 - 2x^2y^2 + 3xy^2 + y^3$

5. $z = \dfrac{x + y}{x - y}$

6. $w = \dfrac{xy}{1 + x^2}$

7. $z = (2x^2y + 3y^3)^3$

8. $z = \sqrt{2x^2 + 3y^2}$

9. $w = ye^{x^2 - y^2}$

10. $z = \ln(2x + 3y)$

11. $w = x^2 \ln(x^2 + y^2)$

12. $z = x^2 \sin 2y$

13. $z = e^{2x} \cos 3y$

14. $w = \tan^{-1}\left(\dfrac{y}{x}\right)$

15. $w = x^2 + xy + z^2$

16. $w = \sqrt{x^2 + xy + z^2}$

17. $w = x^2 e^{-yz}$

18. $w = e^{-x^2} \sin(2y + 3z)$

19. $w = x^2 e^y + y \ln z$

20. $w = x \cosh yz$

In Exercises 21–24, use differentials to approximate the change in f due to the indicated change in the independent variables.

21. $f(x, y) = x^4 - 3x^2y^2 + y^3 - 2y + 4$; (x, y) changes from $(2, 2)$ to $(1.98, 2.01)$.

22. $f(x, y) = \sqrt{2x + 3y} - \dfrac{x}{y}$; (x, y) changes from $(3, 1)$ to $(2.96, 1.02)$.

23. $f(x, y, z) = \ln(2x - y) + e^{2xz}$; (x, y, z) changes from $(2, 3, 0)$ to $(2.01, 2.97, 0.04)$.

24. $f(x, y, z) = x^2y \cos \pi z$; (x, y, z) changes from $(1, 3, 2)$ to $(0.98, 2.97, 2.01)$.

25. The dimensions of a closed rectangular box are measured as 30 in., 40 in., and 60 in., with a maximum error of 0.2 in. in each measurement. Use differentials to estimate the maximum error in calculating the volume of the box.

26. Use differentials to estimate the maximum error in calculating the surface area of the box of Exercise 25.

27. A piece of land is triangular in shape. Two of its sides are measured as 80 and 100 ft, and the included angle is measured as $\pi/3$ rad. If the sides are measured with a maximum error of 0.3 ft and the angle is measured with a maximum error of $\pi/180$ rad, what is the approximate maximum error in the calculated area of the land?

28. **Production Functions** The productivity of a certain country is given by the function

$$f(x, y) = 30x^{4/5}y^{1/5}$$

when x units of labor and y units of capital are utilized. What is the approximate change in the number of units produced if the amount expended on labor is decreased from 243 to 240 units and the amount expended on capital is increased from 32 units to 35 units?

29. The pressure P (in pascals), the volume V (in liters), and the temperature T (in kelvins) of an ideal gas are related by the

equation $PV = 8.314T$. Use differentials to find the approximate change in the pressure of the gas if its volume increases from 20 L to 20.2 L and its temperature decreases from 300 K to 295 K.

30. Consider the ideal gas law equation $PV = 8.314T$ of Exercise 29. If T and P are measured with maximum errors of 0.6% and 0.4%, respectively, determine the maximum percentage error in calculating the value of V.

31. **Surface Area of Humans** The surface area S of humans is related to their weight W and height H by the formula $S = 0.1091W^{0.425}H^{0.725}$. If W and H are measured with maximum errors of 3% and 2%, respectively, find the approximate maximum percentage error in the measurement of S.

32. **Specific Gravity** The specific gravity of an object with density greater than that of water can be determined by using the formula

$$S = \frac{A}{A - W}$$

where A and W are the weights of the object in air and in water, respectively. If the measurements of an object are $A = 2.2$ lb and $W = 1.8$ lb with maximum errors of 0.02 lb and 0.04 lb, respectively, find the approximate maximum error in calculating S.

33. **Flow of Blood** The flow of blood through an arteriole measured in cm^3/sec is given by

$$F = \frac{\pi P R^4}{8kL}$$

where L is the length of the arteriole in centimeters, R is the radius in centimeters, P is the difference in pressure between the two ends of the arteriole in dyne-sec/cm^2, and k is the viscosity of blood in dyne-sec/cm^2. Find the approximate maximum percentage error in measuring the flow of blood if an error of at most 1% is made in measuring the length of the arteriole and an error of at most 2% is made in measuring its radius. Assume that P and k are constant.

34. The figure below shows two long, parallel wires that are at a distance of d m apart, carrying currents of I_1 and I_2 amps. It can be shown that the force of attraction per unit length between the two wires as a result of magnetic fields generated by the currents is given by

$$f = \frac{\mu_0}{2\pi}\left(\frac{I_1 I_2}{D}\right)$$

teslas per meter, where μ_0 ($4\pi \times 10^{-7}$ N/amp^2) is a constant called the *permeability of free space*. Use differentials to find the approximate percentage change in f if I_1 increases by 2%, I_2 decreases by 2%, and D decreases by 5%.

35. **Error in Measuring the Period of a Pendulum** The period T of a simple pendulum executing small oscillations is given by $T = 2\pi\sqrt{L/g}$, where L is the length of the pendulum and g is the constant of acceleration due to gravity. If T is computed by using $L = 4$ ft and $g = 32$ ft/sec^2, find the approximate percentage error in T if the true values for L and g are 4.05 ft and 32.2 ft/sec^2.

36. **Error in Calculating the Power of a Battery** Suppose the source of current in an electric circuit is a battery. Then the power output P (in watts) obtained if the circuit has a resistance of R ohms is given by

$$P = \frac{E^2 R}{(R + r)^2}$$

where E is the electromotive force (EMF) in volts and r is the internal resistance of the battery. Estimate the maximum percentage error in calculating the power if an EMF of 12 volts is applied in a circuit with a resistance of 100 ohms, the internal resistance of the battery is 5 ohms, and the possible maximum percentage errors in measuring E, R, and r are 2%, 3%, and 1%, respectively.

37. **Error in Measuring the Resistance of a Circuit** The total resistance R (in ohms) of three resistors with resistances of R_1, R_2, and R_3 ohms connected in parallel is given by

$$\frac{1}{R} = \frac{1}{R_1} + \frac{1}{R_2} + \frac{1}{R_3}$$

If R_1, R_2, and R_3 are measured as 20, 30, and 50 ohms, respectively, with a maximum error of 0.5 in each measurement, estimate the maximum error in the calculated value of R.

38. A container with a constant cross section of A ft^2 is filled with water to a height of h ft. The water is then allowed to flow out through an orifice of cross section a $in.^2$ located at the base of the container. It can be shown that the time (in seconds) that it takes to empty the tank is given by

$$T = f(A, a, h) = \frac{A}{a}\sqrt{\frac{2h}{g}}$$

where g is the constant of acceleration. Suppose that the measurements of A, a, and h are 5 ft^2, 2 $in.^2$, and 16 ft with errors of 0.05 ft^2, -0.04 $in.^2$, and 0.2 ft, respectively. Find the error in computing T. (Take g to be 32 ft/sec^2.)

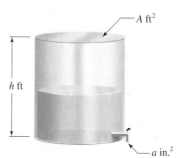

39. Suspension Bridge Cables The supports of a cable of a suspension bridge are at the same level and at a distance of L ft apart. The supports are a feet higher than the lowest point of the cable (see the figure). If the weight of the cable is negligible and the bridge has a uniform weight of W lb/ft, then the tension (in lb) in the cable at its lowest point is given by

$$H = \frac{WL^2}{8a}$$

If W, L, and a are measured with possible maximum errors of 1%, 2%, and 2%, respectively, determine the maximum percentage error in calculating H.

40. Flight of a Projectile A projectile is fired with a muzzle velocity of v ft/sec at an angle α radians above the horizontal. If the launch site is located at a height of h ft above the target (see the figure below), then the time of the flight of the projectile in seconds is given by

$$T = \frac{v \sin \alpha + \sqrt{(v \sin \alpha)^2 + 2gh}}{g}$$

Suppose that the projectile is fired with an initial speed of 800 ft/sec at an angle of elevation of $\pi/4$ radians from a site

that is located 400 ft above the target. If the initial speed of the projectile, the angle of elevation of the cannon, and the height of the site above the target are measured with maximum possible percentage errors of 0.05%, 0.02%, and 0.5%, respectively, find the maximum error in computing the time of flight of the projectile. (Take g to be 32 ft/sec².)

In Exercises 41 and 42, show that the function is differentiable in the plane. (See Example 5.)

41. $f(x, y) = x^2 - y^2$

42. $f(x, y) = 2xy - y^2$

43. Let f be defined by

$$f(x, y) = \begin{cases} \dfrac{xy}{x^2 + y^2} & \text{if } (x, y) \neq 0 \\ 0 & \text{if } (x, y) = (0, 0) \end{cases}$$

Show that $f_x(0, 0)$ and $f_y(0, 0)$ both exist but that f is not differentiable at $(0, 0)$.

Hint: Use the result of Theorem 3.

In Exercises 44–47, determine whether the statement is true or false. If it is true, explain why it is true. If it is false, give an example to show why it is false.

44. If $z = f(x, y)$ and $dz = 0$ for all x and y and for all differentials dx and dy, then $f_x(x, y) = 0$ and $f_y(x, y) = 0$ for all x and y.

45. If $f(x, y)$ is differentiable at (a, b), then $f(a, b) = \lim_{(x, y) \to (a, b)} f(x, y)$.

46. If $F(x, y) = f(x) + g(y)$, where f and g are differentiable in the interval (a, b), then F is differentiable on $R = \{(x, y) \mid a < x < b, a < y < b\}$.

47. The function

$$f(x, y) = \begin{cases} x^2 + y^2 & \text{if } (x, y) \neq (0, 0) \\ 1 & \text{if } (x, y) = (0, 0) \end{cases}$$

is differentiable everywhere.

13.5 The Chain Rule

The Chain Rule for Functions Involving One Independent Variable

In this section we extend the Chain Rule to functions of two or more variables. First, let's recall the Chain Rule for functions of one variable: If y is a differentiable function of x and x is a differentiable function of t (so that y is a function of t), then

$$\frac{dy}{dt} = \frac{dy}{dx} \frac{dx}{dt}$$

This rule is easily recalled by using the diagram shown in Figure 1.

We begin by looking at the Chain Rule for the case in which a variable w depends on two *intermediate* variables x and y, which in turn depend on a third variable t (so w is a function of one independent variable t).

FIGURE 1
To find dy/dt, compute dy/dx (y depends on x), compute dx/dt (x depends on t), and then multiply the two quantities together.

THEOREM 1 **The Chain Rule for Functions Involving One Independent Variable**

Let $w = f(x, y)$, where f is a differentiable function of x and y. If $x = g(t)$ and $y = h(t)$, where g and h are differentiable functions of t, then w is a differentiable function of t, and

$$\frac{dw}{dt} = \frac{\partial w}{\partial x}\frac{dx}{dt} + \frac{\partial w}{\partial y}\frac{dy}{dt}$$

Note Observe that the derivative of w with respect to t is written with an ordinary d (d) rather than a curly d (∂), since w is a function of the single variable t. ■

PROOF Let t change from t to $t + \Delta t$. This produces a change

$$\Delta x = g(t + \Delta t) - g(t)$$

in x from x to $x + \Delta x$ and a change

$$\Delta y = h(t + \Delta t) - h(t)$$

in y from y to $y + \Delta y$. Since g and h are differentiable, they are continuous at t, so both Δx and Δy approach zero as Δt approaches zero.

Next, observe that the changes of Δx in x and Δy in y in turn produce a change Δw in w from w to $w + \Delta w$. Since f is differentiable, we have

$$\Delta w = \frac{\partial w}{\partial x}\Delta x + \frac{\partial w}{\partial y}\Delta y + \varepsilon_1 \Delta x + \varepsilon_2 \Delta y$$

where $\varepsilon_1 \to 0$ and $\varepsilon_2 \to 0$ as $(\Delta x, \Delta y) \to (0, 0)$. Dividing both sides of this equation by Δt, we have

$$\frac{\Delta w}{\Delta t} = \frac{\partial w}{\partial x}\frac{\Delta x}{\Delta t} + \frac{\partial w}{\partial y}\frac{\Delta y}{\Delta t} + \varepsilon_1 \frac{\Delta x}{\Delta t} + \varepsilon_2 \frac{\Delta y}{\Delta t}$$

Letting $\Delta t \to 0$, we have

$$\frac{dw}{dt} = \lim_{\Delta t \to 0} \frac{\Delta w}{\Delta t}$$

$$= \frac{\partial w}{\partial x}\lim_{\Delta t \to 0}\frac{\Delta x}{\Delta t} + \frac{\partial w}{\partial y}\lim_{\Delta t \to 0}\frac{\Delta y}{\Delta t} + \lim_{\Delta t \to 0}\varepsilon_1 \lim_{\Delta t \to 0}\frac{\Delta x}{\Delta t} + \lim_{\Delta t \to 0}\varepsilon_2 \lim_{\Delta t \to 0}\frac{\Delta y}{\Delta t}$$

$$= \frac{\partial w}{\partial x}\frac{dx}{dt} + \frac{\partial w}{\partial y}\frac{dy}{dt} + 0 \cdot \frac{dx}{dt} + 0 \cdot \frac{dy}{dt}$$

$$= \frac{\partial w}{\partial x}\frac{dx}{dt} + \frac{\partial w}{\partial y}\frac{dy}{dt}$$ ■

The tree diagram in Figure 2 will help you recall this version of the Chain Rule. There are two "limbs" on this tree leading from w to t. To find dw/dt, multiply the partial derivatives along each limb, and then add the products of these partial derivatives.

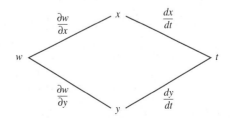

FIGURE 2
w depends on t via x and y.

EXAMPLE 1 Let $w = x^2y - xy^3$, where $x = \cos t$ and $y = e^t$. Find dw/dt and its value when $t = 0$.

Solution Observe that w is a function of x and y and that both these variables are functions of t. Thus, we have the situation depicted in the schematic in Figure 2. Using the Chain Rule, we have

$$\frac{dw}{dt} = \frac{\partial w}{\partial x}\frac{dx}{dt} + \frac{\partial w}{\partial y}\frac{dy}{dt}$$

$$= (2xy - y^3)(-\sin t) + (x^2 - 3xy^2)e^t$$

$$= y(y^2 - 2x)\sin t + x(x - 3y^2)e^t$$

To find the value of dw/dt when $t = 0$, we first observe that if $t = 0$, then $x = \cos 0 = 1$ and $y = e^0 = 1$. So

$$\left.\frac{dw}{dt}\right|_{t=0} = 0 + 1(1 - 3)e^0 = -2 \qquad\blacksquare$$

The Chain Rule in Theorem 1 can be extended to the case involving a function of any finite number of intermediate variables. For example, if $w = f(x_1, x_2, \ldots, x_n)$, where f is a differentiable function of $x_1, x_2, \ldots, x_n$ and $x_1 = f_1(t), x_2 = f_2(t), \ldots, x_n = f_n(t)$, where $f_1, f_2, \ldots, f_n$ are differentiable functions of t, then

$$\frac{dw}{dt} = \frac{\partial w}{\partial x_1}\frac{dx_1}{dt} + \frac{\partial w}{\partial x_2}\frac{dx_2}{dt} + \cdots + \frac{\partial w}{\partial x_n}\frac{dx_n}{dt}$$

This is easier to recall if you look at Figure 3, which shows the dependency of the variables involved: Multiply the derivatives along each limb leading from w to t, and add the products of these derivatives.

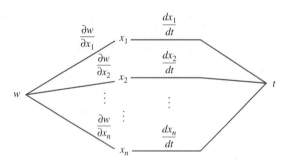

FIGURE 3
w depends on t via $x_1, x_2, \ldots, x_n$.

EXAMPLE 2 **Tracking a Missile Cruiser** Figure 4 depicts an AWACS (Airborne Warning and Control System) aircraft tracking a missile cruiser. The flight path of the plane is described by the parametric equations

$$x = 20 \cos 12t, \qquad y = 20 \sin 12t, \qquad z = 3$$

and the course of the missile cruiser is given by

$$x = 30 + 20t, \qquad y = 40 + 10t^2, \qquad z = 0$$

where $0 \le t \le 1$, and x, y, and z are measured in miles and t in hours. How fast is the distance between the AWACS plane and the missile cruiser changing when $t = 0$?

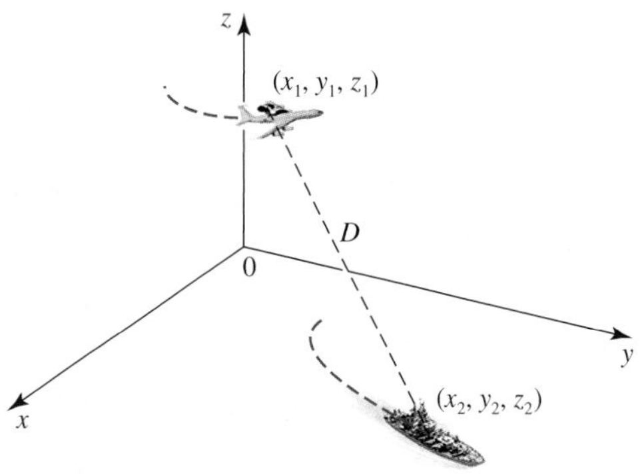

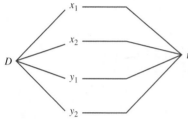

FIGURE 4
An AWACS aircraft tracking a missile cruiser.

FIGURE 5
D depends on t via the variables x_1, x_2, y_1, and y_2.

Solution At time t the position of the AWACS plane is given by the point (x_1, y_1, z_1) and the position of the missile cruiser is given by the point (x_2, y_2, z_2), so the distance D between the plane and the cruiser is

$$D = \sqrt{(x_2 - x_1)^2 + (y_2 - y_1)^2 + (z_2 - z_1)^2}$$
$$= \sqrt{(x_2 - x_1)^2 + (y_2 - y_1)^2 + 9}$$

We want to compute dD/dt when $t = 0$. To find dD/dt, we note that D is a function of the four variables x_1, x_2, y_1, and y_2—all of which are functions of the single variable t (Figure 5). By the Chain Rule we have

$$\frac{dD}{dt} = \frac{\partial D}{\partial x_1} \frac{dx_1}{dt} + \frac{\partial D}{\partial x_2} \frac{dx_2}{dt} + \frac{\partial D}{\partial y_1} \frac{dy_1}{dt} + \frac{\partial D}{\partial y_2} \frac{dy_2}{dt}$$

But

$$\frac{\partial D}{\partial x_1} = \frac{-(x_2 - x_1)}{\sqrt{(x_2 - x_1)^2 + (y_2 - y_1)^2 + 9}}, \qquad \frac{\partial D}{\partial x_2} = \frac{x_2 - x_1}{\sqrt{(x_2 - x_1)^2 + (y_2 - y_1)^2 + 9}}$$

$$\frac{\partial D}{\partial y_1} = \frac{-(y_2 - y_1)}{\sqrt{(x_2 - x_1)^2 + (y_2 - y_1)^2 + 9}}, \qquad \frac{\partial D}{\partial y_2} = \frac{y_2 - y_1}{\sqrt{(x_2 - x_1)^2 + (y_2 - y_1)^2 + 9}}$$

$$\frac{dx_1}{dt} = -240 \sin 12t, \qquad \frac{dx_2}{dt} = 20, \qquad \frac{dy_1}{dt} = 240 \cos 12t, \qquad \text{and} \qquad \frac{dy_2}{dt} = 20t$$

If $t = 0$, $x_1 = 20$, $y_1 = 0$, $x_2 = 30$, and $y_2 = 40$, then

$$\sqrt{(x_2 - x_1)^2 + (y_2 - y_1)^2 + 9} = \sqrt{(30 - 20)^2 + (40 - 0)^2 + 9} = \sqrt{1709}$$

Thus,

$$\frac{\partial D}{\partial x_1} = -\frac{10}{\sqrt{1709}} \approx -0.24, \qquad \frac{\partial D}{\partial x_2} = \frac{10}{\sqrt{1709}} \approx 0.24,$$

$$\frac{\partial D}{\partial y_1} = -\frac{40}{\sqrt{1709}} \approx -0.97, \qquad \frac{\partial D}{\partial y_2} = \frac{40}{\sqrt{1709}} \approx 0.97$$

and

$$\frac{dx_1}{dt} = 0, \qquad \frac{dx_2}{dt} = 20, \qquad \frac{dy_1}{dt} = 240, \qquad \frac{dy_2}{dt} = 0$$

Therefore, when $t = 0$,

$$\frac{dD}{dt} = (-0.24)(0) + (0.24)(20) + (-0.97)(240) + (0.97)(0)$$

$$= -228$$

that is, the distance between the AWACS aircraft and the missile cruiser is decreasing at the rate of 228 mph at that instant of time. ■

■ The Chain Rule for Functions Involving Two Independent Variables

We now look at the Chain Rule for the case in which a variable w depends on two intermediate variables x and y, each of which in turn depends on two variables u and v (so that w is a function of two independent variables u and v). More specifically, we have the following theorem.

THEOREM 2   The Chain Rule for Functions Involving Two Independent Variables

Let $w = f(x, y)$, where f is a differentiable function of x and y. Suppose that $x = g(u, v)$ and $y = h(u, v)$ and the partial derivatives $\partial g/\partial u$, $\partial g/\partial v$, $\partial h/\partial u$, and $\partial h/\partial v$ exist. Then

$$\frac{\partial w}{\partial u} = \frac{\partial w}{\partial x}\frac{\partial x}{\partial u} + \frac{\partial w}{\partial y}\frac{\partial y}{\partial u}$$

and

$$\frac{\partial w}{\partial v} = \frac{\partial w}{\partial x}\frac{\partial x}{\partial v} + \frac{\partial w}{\partial y}\frac{\partial y}{\partial v}$$

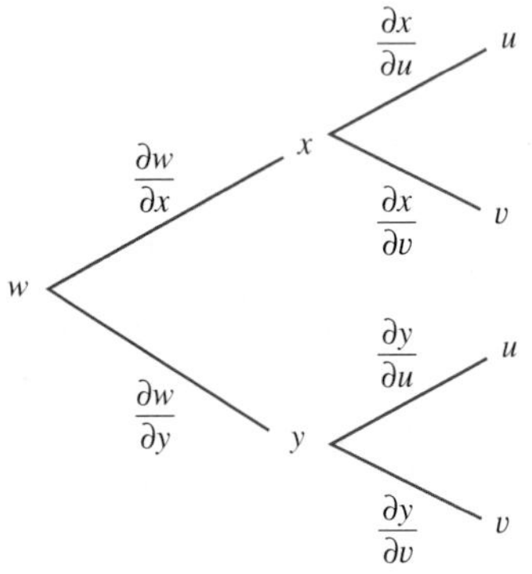

FIGURE 6
w depends on u and v via x and y.

PROOF   For $\partial w/\partial u$ we think of v as a constant, so g and h are differentiable functions of u. Then the result follows from Theorem 1. The expression $\partial w/\partial v$ is derived in a similar manner. ■

The tree diagram shown in Figure 6 will help you to recall the Chain Rule given in Theorem 2.

To obtain $\partial w/\partial u$, observe that w is connected to u by two "limbs," one from w to u via x and the other from w to u via y. Multiply the partial derivatives along each of these limbs, and add the product of these partial derivatives together to get $\partial w/\partial u$. The expression for $\partial w/\partial v$ is found in a similar manner.

EXAMPLE 3   Let $w = 2x^2y$, where $x = u^2 + v^2$ and $y = u^2 - v^2$. Find $\partial w/\partial u$ and $\partial w/\partial v$.

Solution   Observe that w is a function of x and y and that both of these variables are functions of u and v. Thus, we have the situation depicted in Figure 6. Using the Chain Rule (Theorem 2), we have

$$\frac{\partial w}{\partial u} = \frac{\partial w}{\partial x}\frac{\partial x}{\partial u} + \frac{\partial w}{\partial y}\frac{\partial y}{\partial u}$$

$$= 4xy(2u) + 2x^2(2u) = 4xu(2y + x)$$

and

$$\frac{\partial w}{\partial v} = \frac{\partial w}{\partial x}\frac{\partial x}{\partial v} + \frac{\partial w}{\partial y}\frac{\partial y}{\partial v}$$

$$= 4xy(2v) + 2x^2(-2v) = 4xv(2y - x) \qquad \blacksquare$$

■ The General Chain Rule

The Chain Rule in Theorem 2 can be extended to the case involving any finite number of intermediate variables and any finite number of independent variables. For example, if $w = f(x_1, x_2, \ldots, x_n)$, where f is a differentiable function of n intermediate variables, $x_1, x_2, \ldots, x_n$, and $x_1 = f_1(t_1, t_2, \ldots, t_m)$, $x_2 = f_2(t_1, t_2, \ldots, t_m), \ldots,$ $x_n = f_n(t_1, t_2, \ldots, t_m)$, where $f_1, f_2, \ldots, f_n$ are differentiable functions of m variables, $t_1, t_2, \ldots, t_m$, then

$$\frac{\partial w}{\partial t_1} = \frac{\partial w}{\partial x_1}\frac{\partial x_1}{\partial t_1} + \frac{\partial w}{\partial x_2}\frac{\partial x_2}{\partial t_1} + \cdots + \frac{\partial w}{\partial x_n}\frac{\partial x_n}{\partial t_1}$$

$$\frac{\partial w}{\partial t_2} = \frac{\partial w}{\partial x_1}\frac{\partial x_1}{\partial t_2} + \frac{\partial w}{\partial x_2}\frac{\partial x_2}{\partial t_2} + \cdots + \frac{\partial w}{\partial x_n}\frac{\partial x_n}{\partial t_2}$$

$$\vdots$$

$$\frac{\partial w}{\partial t_m} = \frac{\partial w}{\partial x_1}\frac{\partial x_1}{\partial t_m} + \frac{\partial w}{\partial x_2}\frac{\partial x_2}{\partial t_m} + \cdots + \frac{\partial w}{\partial x_n}\frac{\partial x_n}{\partial t_m}$$

(See Figure 7.)

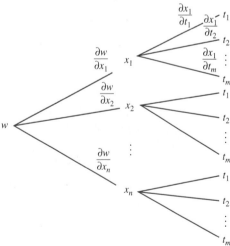

FIGURE 7

w depends on $t_1, t_2, \ldots, t_m$ via $x_1, x_2, \ldots, x_n$.

EXAMPLE 4 Let $w = x^2y + y^2z^3$, where $x = r\cos s$, $y = r\sin s$, and $z = re^s$. Find the value of $\partial w/\partial s$ when $r = 1$ and $s = 0$.

Solution Observe that w is a function of x, y, and z, which in turn are functions of r and s (Figure 8).

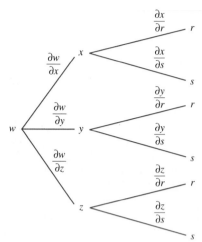

FIGURE 8
w depends on r and s via x, y, and z.

Multiplying the partial derivatives on the limbs that connect w to s on the tree diagram and adding the products of these derivatives, we obtain

$$\frac{\partial w}{\partial s} = \frac{\partial w}{\partial x}\frac{\partial x}{\partial s} + \frac{\partial w}{\partial y}\frac{\partial y}{\partial s} + \frac{\partial w}{\partial z}\frac{\partial z}{\partial s}$$

$$= 2xy(-r\sin s) + (x^2 + 2yz^3)(r\cos s) + 3y^2z^2(re^s)$$

When $r = 1$ and $s = 0$, we have $x = 1$, $y = 0$, and $z = 1$, so

$$\frac{\partial w}{\partial s} = 2(1)(0)(0) + (1)(1) + 3(0)(1)(1) = 1 \qquad \blacksquare$$

EXAMPLE 5 If $w = f(x^2 - y^2, y^2 - x^2)$ and f is differentiable, show that w satisfies the equation

$$y\frac{\partial w}{\partial x} + x\frac{\partial w}{\partial y} = 0$$

Solution Introduce the intermediate variables $u = x^2 - y^2$ and $v = y^2 - x^2$. Then $w = g(x, y) = f(u, v)$ (Figure 9).

Using the Chain Rule, we have

$$\frac{\partial w}{\partial x} = \frac{\partial w}{\partial u}\frac{\partial u}{\partial x} + \frac{\partial w}{\partial v}\frac{\partial v}{\partial x} = \frac{\partial w}{\partial u}(2x) + \frac{\partial w}{\partial v}(-2x)$$

and

$$\frac{\partial w}{\partial y} = \frac{\partial w}{\partial u}\frac{\partial u}{\partial y} + \frac{\partial w}{\partial v}\frac{\partial v}{\partial y} = \frac{\partial w}{\partial u}(-2y) + \frac{\partial w}{\partial v}(2y)$$

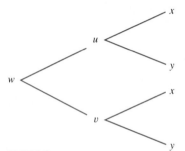

FIGURE 9
w depends on x and y via the intermediate variables u and v.

Therefore,

$$y\frac{\partial w}{\partial x} + x\frac{\partial w}{\partial y} = \left(2xy\frac{\partial w}{\partial u} - 2xy\frac{\partial w}{\partial v}\right) + \left(-2xy\frac{\partial w}{\partial u} + 2xy\frac{\partial w}{\partial v}\right) = 0 \qquad \blacksquare$$

EXAMPLE 6 Let $w = f(x, y)$, where f has continuous second-order partial derivatives, and let $x = r^2 + s^2$ and $y = 2rs$. Find $\partial^2 w/\partial r^2$.

Solution We begin by calculating $\partial w/\partial r$. Using the Chain Rule, we have

$$\frac{\partial w}{\partial r} = \frac{\partial w}{\partial x}\frac{\partial x}{\partial r} + \frac{\partial w}{\partial y}\frac{\partial y}{\partial r} = \frac{\partial w}{\partial x}(2r) + \frac{\partial w}{\partial y}(2s)$$

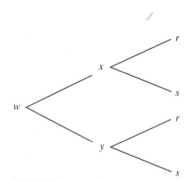

FIGURE 10
w depends on r and s via the intermediate variables x and y.

(See Figure 10.) Next, we apply the Product Rule to $\partial w/\partial r$ to obtain

$$\frac{\partial^2 w}{\partial r^2} = \frac{\partial}{\partial r}\left(2r\frac{\partial w}{\partial x} + 2s\frac{\partial w}{\partial y}\right)$$

$$= 2\frac{\partial w}{\partial x} + 2r\frac{\partial}{\partial r}\left(\frac{\partial w}{\partial x}\right) + 2s\frac{\partial}{\partial r}\left(\frac{\partial w}{\partial y}\right) \qquad \textbf{(1)}$$

To compute the partial derivatives appearing in the last two terms of Equation (1), we observe that since w is a function of r and s via the intermediate variables x and y, the same is true of $\partial w/\partial x$ and $\partial w/\partial y$ (Figure 11).

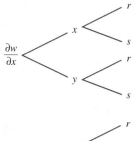

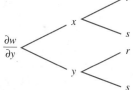

FIGURE 11
Both $\partial w / \partial x$ and $\partial w / \partial y$ depend on r and s via the intermediate variables x and y.

Using the Chain Rule once again, we have

$$\frac{\partial}{\partial r}\left(\frac{\partial w}{\partial x}\right) = \frac{\partial}{\partial x}\left(\frac{\partial w}{\partial x}\right)\frac{\partial x}{\partial r} + \frac{\partial}{\partial y}\left(\frac{\partial w}{\partial x}\right)\frac{\partial y}{\partial r}$$

$$= \frac{\partial^2 w}{\partial x^2}(2r) + \frac{\partial^2 w}{\partial y \, \partial x}(2s)$$

and

$$\frac{\partial}{\partial r}\left(\frac{\partial w}{\partial y}\right) = \frac{\partial}{\partial x}\left(\frac{\partial w}{\partial y}\right)\frac{\partial x}{\partial r} + \frac{\partial}{\partial y}\left(\frac{\partial w}{\partial y}\right)\frac{\partial y}{\partial r}$$

$$= \frac{\partial^2 w}{\partial x \, \partial y}(2r) + \frac{\partial^2 w}{\partial y^2}(2s)$$

Substituting these expressions into Equation (1) and observing that $f_{xy} = f_{yx}$ because they are continuous, we have

$$\frac{\partial^2 w}{\partial r^2} = 2\frac{\partial w}{\partial x} + 2r\left(2r\frac{\partial^2 w}{\partial x^2} + 2s\frac{\partial^2 w}{\partial y \, \partial x}\right) + 2s\left(2r\frac{\partial^2 w}{\partial x \, \partial y} + 2s\frac{\partial^2 w}{\partial y^2}\right)$$

$$= 2\frac{\partial w}{\partial x} + 4r^2\frac{\partial^2 w}{\partial x^2} + 8rs\frac{\partial^2 w}{\partial x \, \partial y} + 4s^2\frac{\partial^2 w}{\partial y^2} \qquad \blacksquare$$

■ Implicit Differentiation

The Chain Rule for a function of several variables can be used to find the derivative of a function implicitly. We will consider two situations.

First, suppose that the equation $F(x, y) = 0$, where F is a differentiable function, defines a differentiable function f of x via the equation $y = f(x)$. If we differentiate both sides of $w = F(x, y) = 0$ with respect to x, we obtain

$$\frac{\partial w}{\partial x} = \frac{\partial F}{\partial x} + \frac{\partial F}{\partial y}\frac{dy}{dx} = 0$$

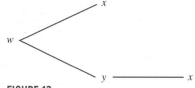

FIGURE 12
Tree diagram showing dependency of w on x directly and via y

(see Figure 12) which implies that

$$\frac{dy}{dx} = -\frac{\dfrac{\partial F}{\partial x}}{\dfrac{\partial F}{\partial y}} = -\frac{F_x}{F_y} \quad \text{if } F_y \neq 0$$

Let's summarize this result.

THEOREM 3   Implicit Differentiation: One Independent Variable

Suppose that the equation $F(x, y) = 0$, where F is differentiable, defines y implicitly as a differentiable function of x. Then

$$\frac{dy}{dx} = -\frac{F_x(x, y)}{F_y(x, y)} \qquad \text{if } F_y(x, y) \neq 0 \tag{2}$$

EXAMPLE 7 Find $\dfrac{dy}{dx}$ if $x^3 + xy + y^2 = 4$.

Solution The given equation can be rewritten as

$$F(x, y) = x^3 + xy + y^2 - 4 = 0$$

Then Equation (2) immediately gives

$$\frac{dy}{dx} = -\frac{F_x}{F_y} = -\frac{3x^2 + y}{x + 2y}$$

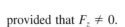

As a second application of the Chain Rule to implicit differentiation, suppose that the equation $F(x, y, z) = 0$, where F is a differentiable function, defines a differentiable function f of x and y via the equation $z = f(x, y)$. Differentiating both sides of $w = F(x, y, z) = 0$ with respect to x, we obtain

$$\frac{\partial w}{\partial x} = \frac{\partial F}{\partial x} + \frac{\partial F}{\partial z}\frac{\partial z}{\partial x} = F_x + F_z \frac{\partial z}{\partial x} = 0$$

(see Figure 13) which gives

$$\frac{\partial z}{\partial x} = -\frac{F_x}{F_z}$$

provided that $F_z \neq 0$.
Similarly, we see that

$$\frac{\partial z}{\partial y} = -\frac{F_y}{F_z} \quad \text{if } F_z \neq 0$$

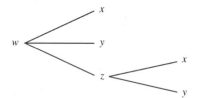

FIGURE 13
w depends on x and y directly and via z.

THEOREM 4   Implicit Differentiation: Two Independent Variables

Suppose the equation $F(x, y, z) = 0$, where F is differentiable, defines z implicitly as a differentiable function of x and y. Then

$$\frac{\partial z}{\partial x} = -\frac{F_x(x, y, z)}{F_z(x, y, z)} \quad \text{and} \quad \frac{\partial z}{\partial y} = -\frac{F_y(x, y, z)}{F_z(x, y, z)} \quad \text{if } F_z(x, y, z) \neq 0 \ \textbf{(3)}$$

EXAMPLE 8 Find $\dfrac{\partial z}{\partial x}$ and $\dfrac{\partial z}{\partial y}$ if $2x^2z - 3xy^2 + yz - 8 = 0$.

Solution Here, $F(x, y, z) = 2x^2z - 3xy^2 + yz - 8 = 0$, and Equation (3) gives

$$\frac{\partial z}{\partial x} = -\frac{F_x(x, y, z)}{F_z(x, y, z)} = -\frac{4xz - 3y^2}{2x^2 + y} = \frac{3y^2 - 4xz}{2x^2 + y}$$

and

$$\frac{\partial z}{\partial y} = -\frac{F_y(x, y, z)}{F_z(x, y, z)} = -\frac{-6xy + z}{2x^2 + y} = \frac{6xy - z}{2x^2 + y}$$

13.5 CONCEPT QUESTIONS

1. Suppose that $w = f(x, y)$, $x = g(t)$, and $y = h(t)$, where f, g, and h are differentiable functions. Write an expression for dw/dt. Illustrate with a tree diagram.

2. Suppose that $w = f(x, y)$, $x = g(u, v)$, and $y = h(u, v)$, where f, g, and h are differentiable functions. Write an expression for $\partial w/\partial v$. Illustrate with a tree diagram.

3. Suppose that $w = f(x_1, x_2, \ldots, x_n)$, $x_1 = f_1(t_1, t_2, \ldots, t_m)$, $x_2 = f_2(t_1, t_2, \ldots, t_m) \ldots$, $x_n = f_n(t_1, t_2, \ldots, t_m)$, where $f, f_1,$

$f_2, \ldots, f_n$ are differentiable functions. Write an expression for $\partial w/\partial t_i$, where $1 \le i \le m$. Illustrate with a tree diagram.

4. **a.** Suppose that $F(x, y) = 0$ defines y implicitly as a function of x and F is differentiable. Write an expression for dy/dx. Illustrate with a tree diagram.

  b. Suppose that $F(x, y, z) = 0$ defines z implicitly as a function of x and y and F is differentiable. Write an expression for $\partial z/\partial x$. Illustrate with a tree diagram.

13.5 EXERCISES

In Exercises 1–8, use the Chain Rule to find dw/dt.

1. $w = x^2 - y^2$, $x = t^2 + 1$, $y = t^3 + t$

2. $w = \sqrt{x^2 + 2y^2}$, $x = \sqrt{t}$, $y = \sqrt{2t + 1}$

3. $w = r \cos s + s \sin r$, $r = e^{-2t}$, $s = t^3 - 2t$

4. $w = \ln(x + y^2)$, $x = \tan t$, $y = \sec t$

5. $w = 2x^3 y^2 z$, $x = t$, $y = \cos t$, $z = t \sin t$

6. $w = pe^{qr}$, $p = \sqrt{t}$, $q = \sin 2t$, $r = \dfrac{t}{t^2 + 1}$

7. $w = \tan^{-1} xz + \dfrac{z}{y}$, $x = t$, $y = t^2$, $z = \sinh t$

8. $w = x\sqrt{y^2 + z^2}$, $x = \dfrac{1}{t}$, $y = e^{-t} \cos t$, $z = e^{-t} \sin t$

In Exercises 9–14, use the Chain Rule to find $\partial w/\partial u$ and $\partial w/\partial v$.

9. $w = x^3 + y^3$, $x = u^2 + v^2$, $y = 2uv$

10. $w = \sin xy$, $x = (u + v)^3$, $y = \sqrt{v}$

11. $w = e^x \cos y$, $x = \ln(u^2 + v^2)$, $y = \sqrt{uv}$

12. $w = x \ln y + 2^y$, $x = \ln u$, $y = ue^v$

13. $w = x \tan^{-1} yz$, $x = \sqrt{u}$, $y = e^{-2v}$, $z = v \cos u$

14. $w = x \cosh y + y \sinh z$, $x = u^2 - v^2$, $y = \ln(u + 1)$, $z = \dfrac{u}{v}$

In Exercises 15–18, write the Chain Rule for finding the indicated derivative with the aid of a tree diagram.

15. $w = f(r, s, u, v)$, $r = g(t)$, $s = h(t)$, $u = p(t)$, $v = q(t)$; $\dfrac{dw}{dt}$

16. $w = f(x, y)$, $x = g(u, v, t)$, $y = h(u, v, t)$; $\dfrac{\partial w}{\partial v}$

17. $w = f(x, y, z)$, $x = g(r, s, t)$, $y = h(r, s, t)$, $z = p(r, s, t)$; $\dfrac{\partial w}{\partial t}$

18. $w = f(x, y)$, $x = g(u, v, r, s)$, $y = h(u, v, r, s)$; $\dfrac{\partial w}{\partial r}$

In Exercises 19–26, use the Chain Rule to find the indicated derivative.

19. $w = x^2 + xy + y^2 + z^3$, $x = 2t$, $y = e^t$, $z = \cos 2t$; $\dfrac{dw}{dt}$

20. $z = x\sqrt{y} + \sqrt{x}$, $x = 2s + t$, $y = s^2 - 7t$; $\dfrac{\partial z}{\partial t}$ if $s = 4$ and $t = 1$

21. $u = \dfrac{x}{x^2 + y^2}$, $x = \sec 2t$, $y = \tan t$; $\dfrac{du}{dt}\Big|_{t=0}$

22. $w = \dfrac{u}{\sqrt{u^2 + v^2}}$, $u = x + 2y + 3z$, $v = x \cos \pi(y + z)$; $\dfrac{\partial w}{\partial x}$ and $\dfrac{\partial w}{\partial z}$ if $x = 0$, $y = 1$, and $z = 1$

23. $u = x \csc yz$, $x = rs$, $y = s^2 t$, $z = \dfrac{s}{t^2}$; $\dfrac{\partial u}{\partial s}$ and $\dfrac{\partial u}{\partial t}$

24. $w = \cos(2x + 3y)$, $x = r^2 st$, $y = s^2 tu$; $\dfrac{\partial w}{\partial r}$ and $\dfrac{\partial w}{\partial u}$

25. $w = \dfrac{x^2 y}{z^2}$, $x = re^{st}$, $y = se^{rt}$, $z = e^{rst}$; $\dfrac{\partial w}{\partial r}$ and $\dfrac{\partial w}{\partial t}$ if $r = 1$, $s = 2$, and $t = 0$

26. $w = \dfrac{x + y}{x + z}$, $x = r \cos s$, $y = r \sin s$, $z = s \tan t$; $\dfrac{\partial w}{\partial r}$ and $\dfrac{\partial w}{\partial t}$

27. Given the system
$$\begin{cases} x = u^2 + v^2 \\ y = u^2 - v^2 \end{cases}$$
find $\partial u/\partial x$, $\partial u/\partial y$, $\partial v/\partial x$ and $\partial v/\partial y$.

28. Given the system
$$\begin{cases} x = \dfrac{1}{2}(u^2 - v^2) \\ y = uv \end{cases}$$
find $\partial u/\partial x$, $\partial u/\partial y$, $\partial v/\partial x$ and $\partial v/\partial y$.

In Exercises 29–32, use Equation (2) to find dy/dx.

29. $x^3 - 2xy + y^3 = 4$ **30.** $x^4 + 2x^2y^2 - 3xy - x = 5$

31. $2x^2 + 3\sqrt{xy} - 2y = 4$ **32.** $x \sec y + y \cos x = 1$

In Exercises 33–36, use Equation (3) to find $\partial z/\partial x$ and $\partial z/\partial y$.

33. $x^2 + xy - x^2z + yz^2 = 0$

34. $x^2 + y^2 + z^2 - xy - yz - xz = 1$

35. $xe^y + ye^{xz} + x^2 e^{x/y} = 10$

36. $\ln(x^2 + y^2) + x \ln z - \cos(xyz) = 0$

37. Find dy/dx if $x^3 + y^3 - 3axy = 0$, $a > 0$.

38. The radius of a right circular cylinder is increasing at the rate of 0.1 cm/sec while its height is decreasing at the rate of 0.2 cm/sec. Find the rate at which the volume of the cylinder is changing when its radius is 60 cm and its height is 130 cm.

39. The radius of a right circular cone is decreasing at the rate of 0.2 in./min while its height is increasing at the rate of 0.1 in./min. Find the rate at which the area of its lateral surface is changing when its radius is 10 in. and its height is 18 in.

40. The pressure P (in pascals), the volume V (in liters), and the temperature T (in kelvins) of 1 mole of an ideal gas are related by the equation $PV = 8.314T$. Find the rate at which the pressure of the gas is changing when its volume is 20 L and is increasing at the rate of 0.2 L/sec and its temperature is 300 K and is increasing at the rate of 0.3 K/sec.

41. Car A is approaching an intersection from the north, and car B is approaching the same intersection from the east. At a certain instant of time car A is 0.4 mile from the intersection and approaching it at 45 mph, while car B is 0.3 mile from the intersection and approaching it at a speed of 30 mph. How fast is the distance between the two cars changing?

42. The position of boat A at time t is given by the parametric equations

$$x_1 = -5t, \qquad y_1 = 5t$$

and the position of boat B at time t is given by

$$x_2 = 5t, \qquad y_2 = 2t + t^2$$

where $0 < t < 15$, and x_1, y_1, x_2, y_2 are measured in feet and t is measured in seconds. How fast is the distance between the two boats changing when $t = 10$?

43. The total resistance R (in ohms) of n resistors with resistances $R_1, R_2, \ldots, R_n$ ohms connected in parallel is given by the formula

$$\frac{1}{R} = \frac{1}{R_1} + \frac{1}{R_2} + \cdots + \frac{1}{R_n}$$

Show that

$$\frac{\partial R}{\partial R_k} = \left(\frac{R}{R_k}\right)^2$$

44. **The Doppler Effect** Suppose a sound with frequency f is emitted by an object moving along a straight line with speed u and a listener is traveling along the same line in the opposite direction with speed v. Then the frequency F heard by the listener is given by

$$F = \left(\frac{c - v}{c + u}\right)f$$

where c is the speed of sound in still air—about 1100 ft/sec. (This phenomenon is called the **Doppler effect**.) Suppose that a railroad train is traveling at 100 ft/sec in still air and accelerating at the rate of 3 ft/sec^2 and that a note emitted by the locomotive whistle is 500 Hz. If a passenger is on a train that is moving at 50 ft/sec in the direction opposite to that of the first train and accelerating at the rate of 5 ft/sec^2, how fast is the frequency of the note he hears changing?

45. **Rate of Change in Temperature** The temperature at a point (x, y, z) is given by

$$T(x, y, z) = \frac{60}{1 + x^2 + y^2 + z^2}$$

where T is measured in degrees Fahrenheit and x, y, and z are measured in feet. Suppose the position of a flying insect is

$$\mathbf{r}(t) = 2t\mathbf{i} + t^2\mathbf{j} + t^3\mathbf{k} \qquad 0 \le t \le 5$$

where t is measured in seconds and the distance is measured in feet. Find the rate of change in temperature that the insect experiences at $t = 2$.

46. If $z = f(x, y)$, where $x = r \cos \theta$ and $y = r \sin \theta$, show that

$$\left(\frac{\partial z}{\partial x}\right)^2 + \left(\frac{\partial z}{\partial y}\right)^2 = \left(\frac{\partial z}{\partial r}\right)^2 + \frac{1}{r^2}\left(\frac{\partial z}{\partial \theta}\right)^2$$

47. If $u = f(x, y)$, where $x = e^r \cos \theta$ and $y = e^r \sin \theta$, show that

$$\left(\frac{\partial u}{\partial x}\right)^2 + \left(\frac{\partial u}{\partial y}\right)^2 = e^{-2r}\left[\left(\frac{\partial u}{\partial r}\right)^2 + \left(\frac{\partial u}{\partial \theta}\right)^2\right]$$

48. If $u = f(x, y)$, where $x = e^r \cos \theta$ and $y = e^r \sin \theta$, show that

$$\frac{\partial^2 u}{\partial x^2} + \frac{\partial^2 u}{\partial y^2} = e^{-2r}\left[\frac{\partial^2 u}{\partial r^2} + \frac{\partial^2 u}{\partial \theta^2}\right]$$

49. If $z = f(x, y)$, where $x = u - v$ and $y = v - u$, show that

$$\frac{\partial z}{\partial u} + \frac{\partial z}{\partial v} = 0$$

50. If $z = f(x, y)$, where $x = u + v$ and $y = u - v$, show that

$$\left(\frac{\partial z}{\partial x}\right)^2 - \left(\frac{\partial z}{\partial y}\right)^2 = \frac{\partial z}{\partial u}\frac{\partial z}{\partial v}$$

51. If $z = f(x + at) + g(x - at)$, show that z satisfies the wave equation

$$\frac{\partial^2 z}{\partial t^2} = a^2 \frac{\partial^2 z}{\partial x^2}$$

Hint: Let $u = x + at$ and $v = x - at$.

52. If $z = f(x^2 + y^2)$, show that

$$y\left(\frac{\partial z}{\partial x}\right) - x\left(\frac{\partial z}{\partial y}\right) = 0$$

Hint: Let $u = x^2 + y^2$.

53. If $z = f(u, v)$, where $u = g(x, y)$ and $v = h(x, y)$, show that

$$\frac{\partial^2 z}{\partial x^2} = \frac{\partial^2 z}{\partial u^2}\left(\frac{\partial u}{\partial x}\right)^2 + \left(\frac{\partial^2 z}{\partial v \, \partial u} + \frac{\partial^2 z}{\partial u \, \partial v}\right)\frac{\partial u}{\partial x}\frac{\partial v}{\partial x}$$
$$+ \frac{\partial^2 z}{\partial v^2}\left(\frac{\partial v}{\partial x}\right)^2 + \frac{\partial z}{\partial u}\frac{\partial^2 u}{\partial x^2} + \frac{\partial z}{\partial v}\frac{\partial^2 v}{\partial x^2}$$

Assume that all second-order partial derivatives are continuous.

54. If $z = f(u, v)$, where $u = g(x, y)$ and $v = h(x, y)$, show that

$$\frac{\partial^2 z}{\partial y \, \partial x} = \frac{\partial^2 z}{\partial u^2}\frac{\partial u}{\partial x}\frac{\partial u}{\partial y} + \frac{\partial^2 z}{\partial v \, \partial u}\frac{\partial u}{\partial x}\frac{\partial v}{\partial y} + \frac{\partial^2 z}{\partial u \, \partial v}\frac{\partial u}{\partial y}\frac{\partial v}{\partial x}$$
$$+ \frac{\partial^2 z}{\partial v^2}\frac{\partial v}{\partial x}\frac{\partial v}{\partial y} + \frac{\partial z}{\partial u}\frac{\partial^2 u}{\partial y \, \partial x} + \frac{\partial z}{\partial v}\frac{\partial^2 v}{\partial y \, \partial x}$$

Assume that all second-order partial derivatives are continuous.

55. A function f is *homogeneous of degree n* if $f(tx, ty) = t^n f(x, y)$ for every t, where n is an integer. Show that if f is homogeneous of degree n, then

$$x\frac{\partial f}{\partial x} + y\frac{\partial f}{\partial y} = nf$$

Hint: Differentiate both sides of the given equation with respect to t.

In Exercises 56–59, find the degree of homogeneity of f and show that f satisfies the equation

$$x\frac{\partial f}{\partial x} + y\frac{\partial f}{\partial y} = nf \qquad \text{See Exercise 55.}$$

56. $f(x, y) = 2x^3 + 4x^2 y + y^3$

57. $f(x, y) = \dfrac{xy^2}{\sqrt{x^2 + y^2}}$ **58.** $f(x, y) = \tan^{-1}\left(\dfrac{y}{x}\right)$

59. $f(x, y) = e^{x/y}$

60. Suppose that the functions $u = f(x, y)$ and $v = g(x, y)$ satisfy the *Cauchy-Riemann equations*

$$\frac{\partial u}{\partial x} = \frac{\partial v}{\partial y} \quad \text{and} \quad \frac{\partial u}{\partial y} = -\frac{\partial v}{\partial x}$$

If $x = r\cos\theta$ and $y = r\sin\theta$, show that u and v satisfy

$$\frac{\partial u}{\partial r} = \frac{1}{r}\frac{\partial v}{\partial \theta} \quad \text{and} \quad \frac{\partial v}{\partial r} = -\frac{1}{r}\frac{\partial u}{\partial \theta}$$

the polar coordinate form of the Cauchy-Riemann equations.

61. Show that the functions $u = \ln\sqrt{x^2 + y^2}$ and $v = \tan^{-1}\left(\dfrac{y}{x}\right)$ satisfy the Cauchy-Riemann equations (see Exercise 60).

62. a. Let $P(a, b)$ be a point on the curve defined by the equation $f(x, y) = 0$. Show that if the curve has a tangent line at $P(a, b)$, then an equation of the tangent line can be written in the form

$$f_x(a, b)(x - a) + f_y(a, b)(y - b) = 0$$

b. Find an equation of the tangent line to the ellipse

$$\frac{x^2}{4} + \frac{y^2}{9} = 1$$

at the point $\left(1, \frac{3\sqrt{3}}{2}\right)$.

63. a. Use implicit differentiation to find an expression for d^2y/dx^2 given the implicit equation $f(x, y) = 0$. (Assume that f has continuous second partial derivatives.)

b. Use the result of part (a) to find d^2y/dx^2 if $x^3 + y^3 - 3xy = 0$. What is its domain?

64. Course Taken by a Yacht The following figure depicts a bird's-eye view of the course taken by a yacht during an outing. The pier is located at the origin and the course is described by the equation

$$x^3 + y^3 - 9xy = 0 \qquad x \geq 0, y \geq 0$$

where x and y are measured in miles. When the yacht was at the point $(2, 4)$, it was sailing in an easterly direction at the rate of 16 mph. How fast was it moving in the northerly direction at that instant of time?

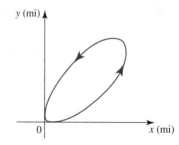

In Exercises 65–66, determine whether the statement is true or false. If it is true, explain why it is true. If it is false, give an example to show why it is false.

65. If $F(x, y) = 0$, where F is differentiable, then

$$\frac{dx}{dy} = -\frac{F_y(x, y)}{F_x(x, y)}$$

provided that $F_x(x, y) \neq 0$.

66. If $z = \cos xy$ for $x > 0$ and $y > 0$ and $xy \neq n\pi$, where n is an integer, then

$$\frac{\partial z}{\partial x} = \frac{1}{\partial x/\partial z}$$

provided that $\partial x/\partial z \neq 0$.

13.6   Directional Derivatives and Gradient Vectors

To study the heat conduction properties of a certain material, heat is applied to one corner of a thin rectangular sheet of that material. Suppose that the heated corner of the sheet is located at the origin of the xy-coordinate plane, as shown in Figure 1 and that the temperature at any point (x, y) on the sheet is given by $T = f(x, y)$.

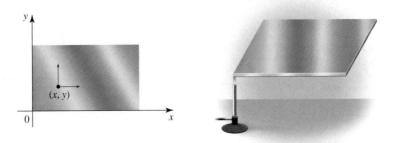

FIGURE 1
The temperature at the point
(x, y) is $T = f(x, y)$.

From our previous work we can find the rate at which the temperature is changing at the point (x, y) in the x-direction by computing $\partial f/\partial x$. Similarly, $\partial f/\partial y$ gives the rate of change of T in the y-direction. But how fast does the temperature change if we move in a direction other than those just mentioned?

In this section we will attempt to answer questions of this nature. More generally, we will be interested in the problem of finding the rate of change of a function f in a specified direction.

■ The Directional Derivative

Let's look at the problem from an intuitive point of view. Suppose that f is a function defined by the equation $z = f(x, y)$, and let $P(a, b)$ be a point in the domain D of f. Furthermore, let **u** be a unit (position) vector having a specified direction. Then the vertical plane containing the line L passing through $P(a, b)$ and having the same direction as **u** will intersect the surface $z = f(x, y)$ along a curve C (Figure 2). Intuitively, we see that the rate of change of z at the point $P(a, b)$ with respect to the distance measured along L is given by the slope of the tangent line T to the curve C at the point $P'(a, b, f(a, b))$.

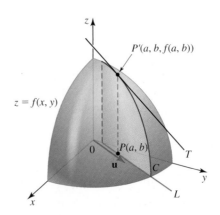

FIGURE 2
The rate of change of z at $P(a, b)$
with respect to the distance measured
along L is given by the slope of T.

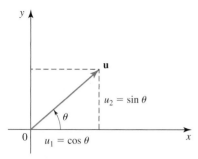

FIGURE 3
Any direction in the plane can be specified in terms of a unit vector **u**.

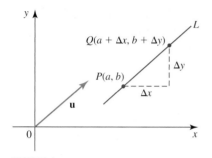

FIGURE 4
The point $Q(a + \Delta x, b + \Delta y)$ lies on L and is distinct from $P(a, b)$.

FIGURE 5
The secant line S passes through the points P' and Q' on the curve C.

Let's find the slope of T. First, observe that **u** may be specified by writing $\mathbf{u} = u_1 \mathbf{i} + u_2 \mathbf{j}$ for appropriate components u_1 and u_2. Equivalently, we may specify **u** by giving the angle θ that it makes with the positive x-axis, in which case $u_1 = \cos\theta$ and $u_2 = \sin\theta$ (Figure 3).

Next, let $Q(a + \Delta x, b + \Delta y)$ be any point distinct from $P(a, b)$ lying on the line L passing through P and having the same direction as **u** (Figure 4).

Since the vector $\overrightarrow{PQ}$ is parallel to **u**, it must be a scalar multiple of **u**. In other words, there exists a nonzero number h such that

$$\overrightarrow{PQ} = h\mathbf{u} = hu_1\mathbf{i} + hu_2\mathbf{j}$$

But $\overrightarrow{PQ}$ is also given by $\Delta x\mathbf{i} + \Delta y\mathbf{j}$, and therefore,

$$\Delta x = hu_1, \qquad \Delta y = hu_2, \qquad \text{and} \qquad h = \sqrt{(\Delta x)^2 + (\Delta y)^2}$$

So the point Q can be expressed as $Q(a + hu_1, b + hu_2)$. Therefore, the slope of the secant line S passing through the points P' and Q' (see Figure 5) is given by

$$\frac{\Delta z}{h} = \frac{f(a + hu_1, b + hu_2) - f(a, b)}{h} \tag{1}$$

Observe that Equation (1) also gives the average rate of change of $z = f(x, y)$ from $P(a, b)$ to $Q(a + \Delta x, b + \Delta y) = Q(a + hu_1, b + hu_2)$ in the direction of **u**.

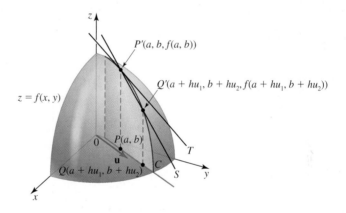

If we let h approach zero in Equation (1), we see that the slope of the secant line S approaches the slope of the tangent line at P'. Also, the average rate of change of z approaches the (instantaneous) rate of change of z at (a, b) in the direction of **u**. This limit, whenever it exists, is called the *directional derivative of f* at (a, b) in the direction of **u**. Since the point $P(a, b)$ is arbitrary, we can replace it by $P(x, y)$ and define the directional derivative of f at any point as follows.

DEFINITION **Directional Derivative**

Let f be a function of x and y and let $\mathbf{u} = u_1\mathbf{i} + u_2\mathbf{j}$ be a unit vector. Then the directional derivative of f at (x, y) in the direction of **u** is

$$D_{\mathbf{u}}f(x, y) = \lim_{h \to 0} \frac{f(x + hu_1, y + hu_2) - f(x, y)}{h} \tag{2}$$

if this limit exists.

Historical Biography

ADRIEN-MARIE LEGENDRE
(1752–1833)

Born to a wealthy family, Adrien-Marie Legendre was able to study at the College Mazarin in Paris, where he received instruction from highly regarded mathematicians of the time. In 1782 Legendre won a prize offered by the Berlin Academy to "determine the curve described by cannonballs and bombs" and to "give rules for obtaining the ranges corresponding to different initial velocities and to different angles of projection." This work was noticed by the famous mathematicians Pierre Lagrange and Simon Laplace, which led to the beginning of Legendre's research career. Legendre went on to produce important results in celestial mechanics, number theory, and the theory of elliptic functions. In 1794 Legendre published *Elements de geometrie*, which was a simplification of Euclid's *Elements*. This book became the standard textbook on elementary geometry for the next 100 years in both Europe and the United States. Legendre was strongly committed to Euclidean geometry and refused to accept non-Euclidean geometries. For nearly thirty years, he attempted to prove Euclid's parallel postulate. Legendre's research met many obstacles, including the French Revolution, Laplace's jealousy, and Legendre's arguments with Carl Friedrich Gauss over priority. In 1824 Legendre refused to vote for the government's candidate for the Institut National des Sciences et des Arts; as a result his government pension was stopped, and he died in poverty in 1833.

Note   If $\mathbf{u} = \mathbf{i}$ ($u_1 = 1$ and $u_2 = 0$), then Equation (2) gives

$$D_{\mathbf{i}}f(x, y) = \lim_{h \to 0} \frac{f(x + h, y) - f(x, y)}{h} = f_x(x, y)$$

That is, the directional derivative of f in the x-direction is the partial derivative of f in the x-direction, as expected. Similarly, you can show that $D_{\mathbf{j}}f(x, y) = f_y(x, y)$. ▪

The following theorem helps us to compute the directional derivatives of functions without appealing directly to the definition of the directional derivative. More specifically, it gives the directional derivative of f in terms of its partial derivatives f_x and f_y.

THEOREM 1 If f is a differentiable function of x and y, then f has a directional derivative in the direction of any unit vector $\mathbf{u} = u_1\mathbf{i} + u_2\mathbf{j}$ and

$$D_{\mathbf{u}}f(x, y) = f_x(x, y)u_1 + f_y(x, y)u_2 \tag{3}$$

PROOF Fix the point (a, b). Then the function g defined by

$$g(h) = f(a + hu_1, b + hu_2)$$

is a function of the single variable h. By the definition of the derivative,

$$g'(0) = \lim_{h \to 0} \frac{g(h) - g(0)}{h} = \lim_{h \to 0} \frac{f(a + hu_1, b + hu_2) - f(a, b)}{h}$$

$$= D_{\mathbf{u}}f(a, b)$$

Next, observe that g may be written as $g(h) = f(x, y)$ where $x = a + hu_1$ and $y = b + hu_2$. Therefore, by the Chain Rule we have

$$g'(h) = \frac{\partial f}{\partial x}\frac{dx}{dh} + \frac{\partial f}{\partial y}\frac{dy}{dh} = f_x(x, y)u_1 + f_y(x, y)u_2$$

In particular, when $h = 0$, we have $x = a$, $y = b$, so

$$g'(0) = f_x(a, b)u_1 + f_y(a, b)u_2$$

Comparing this expression for $g'(0)$ with the one obtained earlier, we conclude that

$$D_{\mathbf{u}}f(a, b) = f_x(a, b)u_1 + f_y(a, b)u_2$$

Finally, since (a, b) is arbitrary, we may replace it by (x, y) and the result follows.

▪

EXAMPLE 1 Find the directional derivative of $f(x, y) = 4 - 2x^2 - y^2$ at the point $(1, 1)$ in the direction of the unit vector $\mathbf{u}$ that makes an angle of $\pi/3$ radians with the positive x-axis.

Solution Here

$$\mathbf{u} = \cos\left(\frac{\pi}{3}\right)\mathbf{i} + \sin\left(\frac{\pi}{3}\right)\mathbf{j} = \frac{1}{2}\mathbf{i} + \frac{\sqrt{3}}{2}\mathbf{j}$$

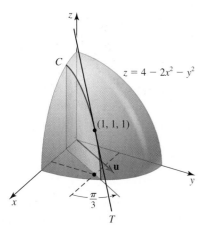

FIGURE 6
The slope of the tangent line to the curve C at $(1, 1, 1)$ is ≈ -3.732.

so $u_1 = \frac{1}{2}$ and $u_2 = \frac{\sqrt{3}}{2}$. Using Equation (3), we find that

$$D_{\mathbf{u}}f(x, y) = f_x(x, y)u_1 + f_y(x, y)u_2$$

$$= (-4x)\left(\frac{1}{2}\right) + (-2y)\left(\frac{\sqrt{3}}{2}\right) = -(2x + \sqrt{3}y)$$

In particular,

$$D_{\mathbf{u}}f(1, 1) = -(2 + \sqrt{3}) \approx -3.732$$

(See Figure 6.) ▪

EXAMPLE 2 Find the directional derivative of $f(x, y) = e^x \cos 2y$ at the point $\left(0, \frac{\pi}{4}\right)$ in the direction of $\mathbf{v} = 2\mathbf{i} + 3\mathbf{j}$.

Solution The unit vector $\mathbf{u}$ that has the same direction as $\mathbf{v}$ is

$$\mathbf{u} = \frac{\mathbf{v}}{|\mathbf{v}|} = \frac{2}{\sqrt{13}}\mathbf{i} + \frac{3}{\sqrt{13}}\mathbf{j}$$

Using Equation (3) with $u_1 = 2/\sqrt{13}$ and $u_2 = 3/\sqrt{13}$, we have

$$D_{\mathbf{u}}f(x, y) = f_x(x, y)u_1 + f_y(x, y)u_2$$

$$= (e^x \cos 2y)\left(\frac{2}{\sqrt{13}}\right) + (-2e^x \sin 2y)\left(\frac{3}{\sqrt{13}}\right)$$

In particular,

$$D_{\mathbf{u}}f\left(0, \frac{\pi}{4}\right) = \left(e^0 \cos \frac{\pi}{2}\right)\left(\frac{2}{\sqrt{13}}\right) - 2(e^0)\left(\sin \frac{\pi}{2}\right)\left(\frac{3}{\sqrt{13}}\right) = -\frac{6}{\sqrt{13}} = -\frac{6\sqrt{13}}{13}$$ ▪

■ The Gradient of a Function of Two Variables

The directional derivative $D_{\mathbf{u}}f(x, y)$ can be written as the dot product of the unit vector

$$\mathbf{u} = u_1\mathbf{i} + u_2\mathbf{j}$$

and the vector

$$f_x(x, y)\mathbf{i} + f_y(x, y)\mathbf{j}$$

Thus,

$$D_{\mathbf{u}}f(x, y) = (u_1\mathbf{i} + u_2\mathbf{j}) \cdot [f_x(x, y)\mathbf{i} + f_y(x, y)\mathbf{j}] = f_x(x, y)u_1 + f_y(x, y)u_2$$

The vector $f_x(x, y)\mathbf{i} + f_y(x, y)\mathbf{j}$ plays an important role in many other computations and is given a special name.

DEFINITION **Gradient of a Function of Two Variables**

Let f be a function of two variables x and y. The **gradient** of f is the vector function

$$\nabla f(x, y) = f_x(x, y)\mathbf{i} + f_y(x, y)\mathbf{j}$$

Notes
1. ∇f is read "del f."
2. $\nabla f(x, y)$ is sometimes written **grad** $f(x, y)$.

EXAMPLE 3   Find the gradient of $f(x, y) = x \sin y + y \ln x$ at the point (e, π).

Solution   Since

$$f_x(x, y) = \sin y + \frac{y}{x} \text{ and } f_y(x, y) = x \cos y + \ln x$$

we have

$$\nabla f(x, y) = f_x(x, y)\mathbf{i} + f_y(x, y)\mathbf{j}$$
$$= \left(\sin y + \frac{y}{x} \right) \mathbf{i} + (x \cos y + \ln x)\mathbf{j}$$

So the gradient of f at (e, π) is

$$\nabla f(e, \pi) = \left(\sin \pi + \frac{\pi}{e} \right) \mathbf{i} + (e \cos \pi + \ln e)\mathbf{j}$$
$$= \frac{\pi}{e}\mathbf{i} + (1 - e)\mathbf{j}$$

Theorem 1 can be rewritten in terms of the gradient of f as follows.

THEOREM 2

If f is a differentiable function of x and y, then f has a directional derivative in the direction of any unit vector $\mathbf{u}$, and

$$D_{\mathbf{u}}f(x, y) = \nabla f(x, y) \cdot \mathbf{u} \tag{4}$$

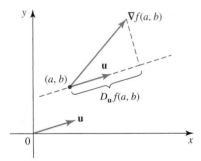

FIGURE 7
The directional derivative of f at (a, b) in the direction of $\mathbf{u}$ is the scalar component of the gradient of f at (a, b) along $\mathbf{u}$.

To give a geometric interpretation of Equation (4), suppose that (a, b) is a fixed point in the xy-plane. Then

$$D_{\mathbf{u}}f(a, b) = \nabla f(a, b) \cdot \mathbf{u} = \frac{\nabla f(a, b) \cdot \mathbf{u}}{|\mathbf{u}|} \qquad \text{since } |\mathbf{u}| = 1$$

so by Equation (6) of Section 11.3 we see that $D_{\mathbf{u}}f(a, b)$ can be viewed as the scalar component of $\nabla f(a, b)$ along $\mathbf{u}$ (Figure 7).

EXAMPLE 4   Let $f(x, y) = x^2 - 2xy$.

a. Find the gradient of f at the point $(1, -2)$.
b. Use the result of (a) to find the directional derivative of f at $(1, -2)$ in the direction from $P(-1, 2)$ to $Q(2, 3)$.

Solution
a. The gradient of f at any point (x, y) is

$$\nabla f(x, y) = (2x - 2y)\mathbf{i} - 2x\mathbf{j}$$

b. The gradient of f at the point $(1, -2)$ is

$$\nabla f(1, -2) = (2 + 4)\mathbf{i} - 2\mathbf{j} = 6\mathbf{i} - 2\mathbf{j}$$

The desired direction is given by the direction of the vector $\overrightarrow{PQ} = 3\mathbf{i} + \mathbf{j}$. A unit vector that has the same direction as $\overrightarrow{PQ}$ is

$$\mathbf{u} = \frac{3}{\sqrt{10}}\mathbf{i} + \frac{1}{\sqrt{10}}\mathbf{j}$$

Using Equation (4), we obtain

$$D_{\mathbf{u}}f(1, -2) = \nabla f(1, -2) \cdot \mathbf{u}$$

$$= (6\mathbf{i} - 2\mathbf{j}) \cdot \left(\frac{3}{\sqrt{10}}\mathbf{i} + \frac{1}{\sqrt{10}}\mathbf{j}\right)$$

$$= \frac{18}{\sqrt{10}} - \frac{2}{\sqrt{10}} = \frac{16}{\sqrt{10}} \approx 5.1$$

or a change in f of 5.1 per unit change in the direction of the vector $\mathbf{u}$. The gradient vector $\nabla f(1, -2)$, the unit vector $\mathbf{u}$, and the geometrical interpretation of $D_{\mathbf{u}}f(1, -2)$ are shown in Figure 8.

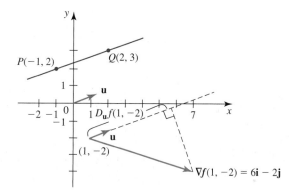

FIGURE 8
$D_{\mathbf{u}}f(1, -2)$ viewed as the scalar component of $\nabla f(1, -2)$ along $\mathbf{u}$.

Properties of the Gradient

The following theorem gives some important properties of the gradient of a function.

THEOREM 3 Properties of the Gradient

Suppose f is differentiable at the point (x, y).

1. If $\nabla f(x, y) = \mathbf{0}$, then $D_{\mathbf{u}}f(x, y) = 0$ for every $\mathbf{u}$.
2. The maximum value of $D_{\mathbf{u}}f(x, y)$ is $|\nabla f(x, y)|$, and this occurs when $\mathbf{u}$ has the same direction as $\nabla f(x, y)$.
3. The minimum value of $D_{\mathbf{u}}f(x, y)$ is $-|\nabla f(x, y)|$, and this occurs when $\mathbf{u}$ has the direction of $-\nabla f(x, y)$.

PROOF Suppose $\nabla f(x, y) = \mathbf{0}$. Then for any $\mathbf{u} = u_1\mathbf{i} + u_2\mathbf{j}$, we have

$$D_{\mathbf{u}}f(x, y) = \nabla f(x, y) \cdot \mathbf{u} = (0\mathbf{i} + 0\mathbf{j}) \cdot (u_1\mathbf{i} + u_2\mathbf{j}) = 0$$

Next, if $\nabla f(x, y) \neq \mathbf{0}$, then

$$D_{\mathbf{u}}f(x, y) = \nabla f(x, y) \cdot \mathbf{u} = |\nabla f(x, y)||\mathbf{u}|\cos\theta = |\nabla f(x, y)|\cos\theta$$

where θ is the angle between $\nabla f(x, y)$ and $\mathbf{u}$. Since the maximum value of $\cos\theta$ is 1 and this occurs when $\theta = 0$, we see that the maximum value of $D_{\mathbf{u}}f(x, y)$ is $|\nabla f(x, y)|$

and this occurs when both $\nabla f(x, y)$ and $\mathbf{u}$ have the same direction. Similarly, Property (3) is proved by observing that $\cos \theta$ has a minimum value of -1 when $\theta = \pi$. ∎

Notes

1. Property (2) of Theorem 3 tells us that f *increases* most rapidly in the direction of $\nabla f(x, y)$. This direction is called the direction of steepest ascent.
2. Property (3) of Theorem 3 says that f *decreases* most rapidly in the direction of $-\nabla f(x, y)$. This direction is called the direction of steepest descent. ∎

EXAMPLE 5 **Quickest Descent**   Suppose a hill is described mathematically by using the model $z = f(x, y) = 300 - 0.01x^2 - 0.005y^2$, where x, y, and z are measured in feet. If you are at the point $(50, 100, 225)$ on the hill, in what direction should you aim your toboggan if you want to achieve the quickest descent? What is the maximum rate of decrease of the height of the hill at this point?

Solution   The gradient of the "height" function is

$$\nabla f(x, y) = f_x(x, y)\mathbf{i} + f_y(x, y)\mathbf{j} = -0.02x\mathbf{i} - 0.01y\mathbf{j}$$

Therefore, the direction of greatest *increase* in z when you are at the point $(50, 100, 225)$ is given by the direction of

$$\nabla f(50, 100) = -\mathbf{i} - \mathbf{j}$$

So by pointing the toboggan in the direction of the vector

$$-\nabla f(50, 100) = -(-\mathbf{i} - \mathbf{j}) = \mathbf{i} + \mathbf{j}$$

you will achieve the quickest descent.

The maximum rate of *decrease* of the height of the hill at the point $(50, 100, 225)$ is

$$|\nabla f(50, 100)| = |-\mathbf{i} - \mathbf{j}| = \sqrt{2}$$

or approximately 1.41 ft/ft. The graph of f and the direction of greatest descent are shown in Figure 9. ∎

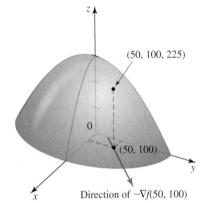

FIGURE 9
The direction of greatest descent is in the direction of $-\nabla f(50, 100)$.

EXAMPLE 6 **Path of a Heat-Seeking Object**   A heat-seeking object is located at the point $(2, 3)$ on a metal plate whose temperature at a point (x, y) is $T(x, y) = 30 - 8x^2 - 2y^2$. Find the path of the object if it moves continuously in the direction of maximum increase in temperature at each point.

Solution   Let the path of the object be described by the position function

$$\mathbf{r}(t) = x(t)\mathbf{i} + y(t)\mathbf{j}$$

where

$$\mathbf{r}(0) = 2\mathbf{i} + 3\mathbf{j}$$

Since the object moves in the direction of maximum increase in temperature, its velocity vector at time t has the same direction as the gradient of T at time t. Therefore, there exists a scalar function of t, k, such that $\mathbf{v}(t) = k\nabla T(x, y)$. But

$$\mathbf{v}(t) = \mathbf{r}'(t) = \frac{dx}{dt}\mathbf{i} + \frac{dy}{dt}\mathbf{j}$$

and $\nabla T = -16x\mathbf{i} - 4y\mathbf{j}$. So we have

$$\frac{dx}{dt}\mathbf{i} + \frac{dy}{dt}\mathbf{j} = -16kx\mathbf{i} - 4ky\mathbf{j}$$

or, equivalently, the system

$$\frac{dx}{dt} = -16kx \qquad \frac{dy}{dt} = -4ky$$

Therefore

$$\frac{dy}{dx} = \frac{\dfrac{dy}{dt}}{\dfrac{dx}{dt}} = \frac{-4ky}{-16kx} \qquad \text{or} \qquad \frac{dy}{dx} = \frac{y}{4x}$$

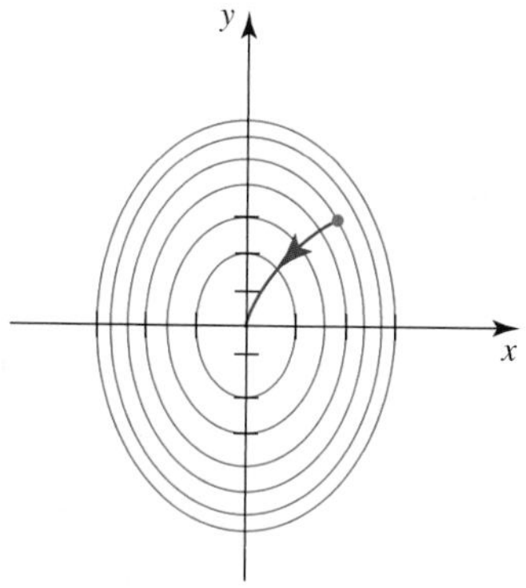

This is a first-order separable differential equation. The solution of this equation is $x = Cy^4$, where C is a constant. (See Section 8.1.) Using the initial condition $y(2) = 3$, we have $2 = C(3^4)$, or $C = 2/(81)$. So

$$x = \frac{2y^4}{81}$$

The path of the heat-seeking object is shown in Figure 10. ■

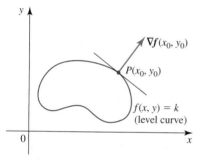

FIGURE 10
The path of the heat-seeking object

FIGURE 11
$\nabla f(x_0, y_0)$ is perpendicular to the level curve $f(x, y) = k$ at $P(x_0, y_0)$.

In Figure 10 observe that at each point where the path intersects a level curve that is part of the contour map of T, the gradient vector ∇T is perpendicular to the level curve at that point. To see why this makes sense, refer to Figure 11, where we show the level curve $f(x, y) = k$ of a function f for some k and a point $P(x_0, y_0)$ lying on the curve. If we move away from $P(x_0, y_0)$ along the level curve then the values of f remain constant (at k). It seems reasonable to conjecture that by moving away in a direction that is perpendicular to the tangent line to the level curve at $P(x_0, y_0)$, f will increase at the fastest rate. But this direction is given by the direction of $\nabla f(x_0, y_0)$. This will be demonstrated in Section 13.7.

■ Functions of Three Variables

The definitions of the directional derivative and the gradient of a function of three or more variables are similar to those for a function of two variables. Also, the algebraic results that are obtained for the case of a function of two variables carry over to the higher-dimensional case and are summarized in the following theorem.

THEOREM 4   Directional Derivative and Gradient of a Function of Three Variables

Let f be a differentiable function of x, y, and z, and let $\mathbf{u} = u_1\mathbf{i} + u_2\mathbf{j} + u_3\mathbf{k}$ be a unit vector. The directional derivative of f in the direction of $\mathbf{u}$ is given by

$$D_{\mathbf{u}}f(x, y, z) = f_x(x, y, z)u_1 + f_y(x, y, z)u_2 + f_z(x, y, z)u_3$$

The **gradient of** f is

$$\nabla f(x, y, z) = f_x(x, y, z)\mathbf{i} + f_y(x, y, z)\mathbf{j} + f_z(x, y, z)\mathbf{k}$$

We also write

$$D_{\mathbf{u}}f(x, y, z) = \nabla f(x, y, z) \cdot \mathbf{u}$$

The properties of the gradient given in Theorem 3 for a function of two variables are also valid for a function of three or more variables. For example, the direction of greatest increase of f coincides with that of the gradient of f and has magnitude $|\nabla f(x, y, z)|$.

EXAMPLE 7 **Electric Potential** Suppose a point charge Q (in coulombs) is located at the origin of a three-dimensional coordinate system. This charge produces an electric potential V (in volts) given by

$$V(x, y, z) = \frac{kQ}{\sqrt{x^2 + y^2 + z^2}}$$

where k is a positive constant and x, y, and z are measured in meters.

a. Find the rate of change of the potential at the point $P(1, 2, 3)$ in the direction of the vector $\mathbf{v} = 2\mathbf{i} + \mathbf{j} - 2\mathbf{k}$.

b. In which direction does the potential increase most rapidly at P, and what is the rate of increase?

Solution

a. We begin by computing the gradient of V. Since

$$V_x = \frac{\partial}{\partial x} [kQ(x^2 + y^2 + z^2)^{-1/2}] = kQ\left(-\tfrac{1}{2}\right)(x^2 + y^2 + z^2)^{-3/2}(2x)$$

$$= -\frac{kQx}{(x^2 + y^2 + z^2)^{3/2}}$$

and by symmetry

$$V_y = -\frac{kQy}{(x^2 + y^2 + z^2)^{3/2}} \quad \text{and} \quad V_z = -\frac{kQz}{(x^2 + y^2 + z^2)^{3/2}}$$

we obtain

$$\nabla V(x, y, z) = V_x\mathbf{i} + V_y\mathbf{j} + V_z\mathbf{k}$$

$$= -\frac{kQ}{(x^2 + y^2 + z^2)^{3/2}} (x\mathbf{i} + y\mathbf{j} + z\mathbf{k})$$

In particular,

$$\nabla V(1, 2, 3) = -\frac{kQ}{14^{3/2}} (\mathbf{i} + 2\mathbf{j} + 3\mathbf{k})$$

A unit vector $\mathbf{u}$ that has the same direction as $\mathbf{v} = 2\mathbf{i} + \mathbf{j} - 2\mathbf{k}$ is

$$\mathbf{u} = \tfrac{1}{3}(2\mathbf{i} + \mathbf{j} - 2\mathbf{k})$$

By Theorem 4 the rate of change of V at $P(1, 2, 3)$ in the direction of $\mathbf{v}$ is

$$D_{\mathbf{u}}V(1, 2, 3) = \nabla V(1, 2, 3) \cdot \mathbf{u} = -\frac{kQ}{14^{3/2}} (\mathbf{i} + 2\mathbf{j} + 3\mathbf{k}) \cdot \frac{(2\mathbf{i} + \mathbf{j} - 2\mathbf{k})}{3}$$

$$= -\frac{kQ}{(3)(14)\sqrt{14}} (2 + 2 - 6) = \frac{kQ}{21\sqrt{14}} = \frac{\sqrt{14}kQ}{294}$$

In other words, the potential is increasing at the rate of $\sqrt{14}kQ/294$ volts/m.

b. The maximum rate of change of V occurs in the direction of the gradient of V, that is, in the direction of the vector $-(\mathbf{i} + 2\mathbf{j} + 3\mathbf{k})$. Observe that this vector points toward the origin from $P(1, 2, 3)$. The maximum rate of change of V at $P(1, 2, 3)$ is given by

$$|\nabla V(1, 2, 3)| = \left| -\frac{kQ}{14^{3/2}} (\mathbf{i} + 2\mathbf{j} + 3\mathbf{k}) \right|$$

$$= \frac{kQ}{14^{3/2}} \sqrt{1 + 4 + 9} = \frac{kQ}{14}$$

or $kQ/14$ volts/m.

13.6 CONCEPT QUESTIONS

1. a. Let f be a function of x and y, and let $\mathbf{u} = u_1\mathbf{i} + u_2\mathbf{j}$ be a unit vector. Define the directional derivative of f in the direction of $\mathbf{u}$. Why is it necessary to use a unit vector to indicate the direction?

b. If f is a differentiable function of x, y, and z and $\mathbf{u} = u_1\mathbf{i} + u_2\mathbf{j} + u_3\mathbf{k}$ is a unit vector, express $D_\mathbf{u} f(x, y, z)$ in terms of the partial derivatives of f and the components of $\mathbf{u}$.

2. a. What is the gradient of a function $f(x, y)$ of two variables x and y?

b. What is the gradient of a function $f(x, y, z)$ of three variables x, y, and z?

c. If f is a differentiable function of x and y and $\mathbf{u}$ is a unit vector, write $D_\mathbf{u} f(x, y)$ in terms of f and $\mathbf{u}$.

d. If f is a differentiable function of x, y, and z and $\mathbf{u}$ is a unit vector, write $D_\mathbf{u} f(x, y, z)$ in terms of f and $\mathbf{u}$.

3. a. If f is a differentiable at (x, y), what can you say about $D_\mathbf{u} f(x, y)$ if $\nabla f(x, y) = \mathbf{0}$?

b. What is the maximum (minimum) value of $D_\mathbf{u} f(x, y)$, and when does it occur?

13.6 EXERCISES

In Exercises 1–4, find the directional derivative of the function f at the point P in the direction of the unit vector that makes the angle θ with the positive x-axis.

1. $f(x, y) = x^3 - 2x^2 + y^3$; $P(1, 2)$, $\theta = \frac{\pi}{6}$

2. $f(x, y) = \sqrt{y^2 - x^2}$; $P(4, 5)$, $\theta = \frac{3\pi}{4}$

3. $f(x, y) = (x + 1)e^y$; $P(3, 0)$, $\theta = \frac{\pi}{2}$

4. $f(x, y) = \sin xy$; $P(1, 0)$, $\theta = -\frac{\pi}{4}$

In Exercises 5–10, find the gradient of f at the point P.

5. $f(x, y) = 2x + 3xy - 3y + 4$; $P(2, 1)$

6. $f(x, y) = \dfrac{1}{x^2 + y^2}$; $P(1, 2)$

7. $f(x, y) = x \sin y + y \cos x$; $P\left(\frac{\pi}{4}, \frac{\pi}{2}\right)$

8. $f(x, y, z) = \dfrac{x + y}{x + z}$; $P(1, 2, 3)$

9. $f(x, y, z) = xe^{yz}$; $P(1, 0, 2)$

10. $f(x, y, z) = \ln(x^2 + y^2 + z^2)$; $P(1, 1, 1)$

In Exercises 11–28, find the directional derivative of the function f at the point P in the direction of the vector $\mathbf{v}$.

11. $f(x, y) = x^3 - x^2y^2 + xy + y^2$; $P(1, -1)$, $\mathbf{v} = \mathbf{i} - 2\mathbf{j}$

12. $f(x, y) = x^3 - y^3$; $P(2, 1)$, $\mathbf{v} = \dfrac{1}{\sqrt{2}}(\mathbf{i} + \mathbf{j})$

13. $f(x, y) = \dfrac{y}{x}$; $P(3, 1)$, $\mathbf{v} = -\mathbf{i}$

14. $f(x, y) = \sqrt{x^2 + y^2 + 1}$; $P(2, 2)$, $\mathbf{v} = 3\mathbf{i} + 4\mathbf{j}$

15. $f(x, y) = \dfrac{x + y}{x - y}$; $P(2, 1)$, $\mathbf{v} = -\mathbf{i} + 3\mathbf{j}$

16. $f(x, y) = xe^{xy}$; $P(2, 0)$, $\mathbf{v} = 2\mathbf{i} - \mathbf{j}$

17. $f(x, y) = x \sin^2 y$; $P\left(-1, \frac{\pi}{4}\right)$, $\mathbf{v} = -2\mathbf{i} + 3\mathbf{j}$

18. $f(x, y) = \tan^{-1}\dfrac{y}{x}$; $P(1, 1)$, $\mathbf{v} = \mathbf{i} - \mathbf{j}$

19. $f(x, y, z) = x^2y^3z^4$; $P(3, -2, 1)$, $\mathbf{v} = \mathbf{i} + \mathbf{j} + \mathbf{k}$

20. $f(x, y, z) = x^2 + 2xy^2 + 2yz^3$; $P(2, 1, -1)$, $\mathbf{v} = \mathbf{i} + 2\mathbf{j} + 2\mathbf{k}$

21. $f(x, y, z) = \sqrt{xyz}$; $P(4, 2, 2)$, $\mathbf{v} = 2\mathbf{i} - 4\mathbf{j} + 4\mathbf{k}$

22. $f(x, y, z) = \sqrt{xy^2 + 6y^2z^2}$; $P(2, 3, -1)$, $\mathbf{v} = 2\mathbf{i} - \mathbf{k}$

23. $f(x, y, z) = x^2 e^{yz}$; $\quad P(2, 3, 0)$, $\quad \mathbf{v} = \mathbf{i} - 2\mathbf{j} + 3\mathbf{k}$

24. $f(x, y, z) = \ln(x^2 + y^2 + z^2)$; $\quad P(1, 2, -1)$,
$\mathbf{v} = -3\mathbf{i} + 2\mathbf{j} + \mathbf{k}$

25. $f(x, y, z) = x^2 y \cos 2z$; $\quad P\left(-1, 2, \frac{\pi}{4}\right)$, $\quad \mathbf{v} = \mathbf{i} - \mathbf{j} + \mathbf{k}$

26. $f(x, y, z) = e^x(2 \cos y + 3 \sin z)$; $\quad P\left(1, \frac{\pi}{6}, \frac{\pi}{6}\right)$,
$\mathbf{v} = 2\mathbf{i} - \mathbf{j} + 3\mathbf{k}$

27. $f(x, y, z) = x \tan^{-1}\left(\dfrac{y}{z}\right)$; $\quad P(3, -2, 2)$, $\quad \mathbf{v} = \mathbf{i} + 2\mathbf{j} - \mathbf{k}$

28. $f(x, y, z) = x^2 \sin^{-1} yz$; $\quad P(2, 1, 0)$, $\quad \mathbf{v} = \dfrac{1}{\sqrt{3}} (\mathbf{i} - \mathbf{j} + \mathbf{k})$

In Exercises 29–32, find the directional derivative of the function f at the point P in the direction from P to the point Q.

29. $f(x, y) = x^3 + y^3$; $\quad P(1, 2)$, $\quad Q(2, 5)$

30. $f(x, y) = xe^{-y}$; $\quad P(2, 0)$, $\quad Q(-1, 2)$

31. $f(x, y, z) = x \sin(2y + 3z)$; $\quad P\left(1, \frac{\pi}{4}, -\frac{\pi}{12}\right)$, $\quad Q\left(3, \frac{\pi}{2}, -\frac{\pi}{4}\right)$

32. $f(x, y, z) = \dfrac{x + y}{y + z}$; $\quad P(2, 1, 1)$, $\quad Q(3, 2, -2)$

In Exercises 33–36, find a vector giving the direction in which the function f increases most rapidly at the point P. What is the maximum rate of increase?

33. $f(x, y) = \sqrt{2x + 3y^2}$; $\quad P(3, 2)$

34. $f(x, y) = e^{-2x} \cos y$; $\quad P\left(0, \frac{\pi}{4}\right)$

35. $f(x, y, z) = x^3 + 2xz + 2yz^2 + z^3$; $\quad P(-1, 3, 2)$

36. $f(x, y, z) = \ln(x^2 + 2y^2 + 3z^2)$; $\quad P(1, 2, -1)$

In Exercises 37–40, find a vector giving the direction in which the function f decreases most rapidly at the point P. What is the maximum rate of decrease?

37. $f(x, y) = \tan^{-1}(2x + y)$; $\quad P(0, 0)$

38. $f(x, y) = xe^{-y^2}$; $\quad P(1, 0)$

39. $f(x, y, z) = \dfrac{x}{y} + \dfrac{y}{z}$; $\quad P(1, -1, 2)$

40. $f(x, y, z) = \sqrt{xy} \cos z$; $\quad P\left(4, 1, \frac{\pi}{4}\right)$

41. The height of a hill (in feet) is given by

$$h(x, y) = 20(16 - 4x^2 - 3y^2 + 2xy + 28x - 18y)$$

where x is the distance (in miles) east and y the distance (in miles) north of Bolton. In what direction is the slope of the hill steepest at the point 1 mile north and 1 mile east of Bolton? What is the steepest slope at that point?

42. Path of Steepest Ascent The following figure shows the contour map of a hill with its summit denoted by S. Draw the curve from P to S that is associated with the path you will take to reach the summit by ascending the direction of the greatest increase in altitude.
Hint: Study Figure 10.

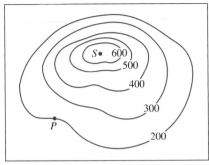

Note: This path is called the *path of steepest ascent.*

43. Path of Steepest Descent The figure shows a topographical map of a 620-ft hill with contours at 100-ft intervals.

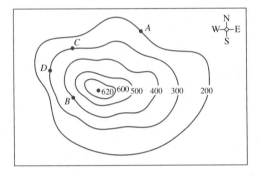

a. If you start from A and proceed in a southwesterly direction, will you be ascending, descending, or neither ascending nor descending? What if you start from B?

b. If you start from C and proceed in a westerly direction, will you be ascending, descending, or neither ascending nor descending?

c. If you start from D, in what direction should you proceed to have the steepest ascent?

d. If you want to climb to the summit of the hill using the gentlest ascent, would you start from the east or the west?

44. Steady-State Temperature Consider the upper half-disk $H = \{(x, y) \mid x^2 + y^2 \le 1, y \ge 0\}$ (see the figure). If the temperature at points on the upper boundary is kept at $100°C$ and the temperature at points on the lower boundary is kept at $50°C$, then the steady-state temperature at any point (x, y) inside the half-disk is given by

$$T(x, y) = 100 - \frac{100}{\pi} \tan^{-1} \frac{1 - x^2 - y^2}{2y}$$

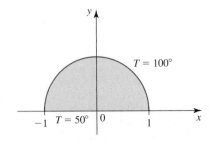

Find the rate of change of the temperature at the point $P\left(\frac{1}{2}, \frac{1}{2}\right)$ in the direction of the vector $\mathbf{v} = 2\mathbf{i} + 3\mathbf{j}$.

45. Steady-State Temperature Consider the upper half-disk $H = \{(x, y) \mid x^2 + y^2 \le 1, y \ge 0\}$ (see the figure). If the temperature at points on the upper boundary is kept at 100°C and the temperature at points on the lower boundary is kept at 0°C, then the steady-state temperature at any point (x, y) inside the half-disk is given by

$$T(x, y) = \frac{200}{\pi} \tan^{-1} \frac{2y}{1 - x^2 - y^2}$$

a. Find the gradient of T at the point $\left(\frac{\sqrt{7}}{4}, \frac{1}{4}\right)$, and interpret your result.

b. Sketch the isothermal curve of T passing through the point $\left(\frac{\sqrt{7}}{4}, \frac{1}{4}\right)$ and the gradient vector ∇T at that point on the same coordinate system.

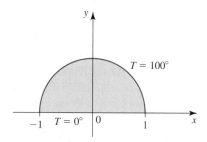

46. The temperature at a point $P(x, y, z)$ of a solid ball of radius 4 with center at the origin is given by $T(x, y, z) = xy + yz + xz$. Find the direction in which T is increasing most rapidly at $P(1, 1, 2)$.

47. Let $T(x, y, z)$ represent the temperature at a point $P(x, y, z)$ of a region R in space. If the isotherms of T are concentric spheres, show that the temperature gradient ∇T points either toward or away from the center of the spheres.
Hint: Recall that the isotherms of T are the sets on which T is constant.

48. Suppose the temperature at the point (x, y) on a thin sheet of metal is given by

$$T(x, y) = \frac{100(1 + 3x + 2y)}{1 + 2x^2 + 3y^2}$$

degrees Fahrenheit. In what direction will the temperature be increasing most rapidly at the point $(1, 2)$? In what direction will it be decreasing most rapidly?

49. The temperature (in degrees Fahrenheit) at a point (x, y) on a metal plate is

$$T(x, y) = 90 - 6x^2 - 2y^2$$

An insect located at the point $(1, 1)$ crawls in the direction in which the temperature drops most rapidly.
a. Find the path of the insect.
b. Sketch a few level curves of T and the path found in part (a).

50. Cobb-Douglas Production Function The output of a finished product is given by the production function

$$f(x, y) = 100x^{0.6}y^{0.4}$$

where x stands for the number of units of labor and y stands for the number of units of capital. Currently, the amount being spent on labor is 500 units, and the amount being spent on capital is 250 units. If the manufacturer wishes to expand production by injecting an additional 10 units into labor, how much more should be put into capital to maximize the increase in output?

51. The figure shows the contour map of a function f of two variables x and y. Use it to estimate the directional derivative of f at P_0 in the indicated direction.

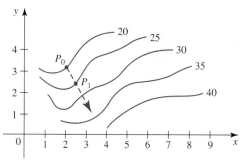

Hint: If $\mathbf{u}$ is a unit vector having indicated direction, then

$$D_\mathbf{u} f(P_0) \approx \frac{f(P_1) - f(P_0)}{d(P_0, P_1)}$$

when $f(P_i)$ is the value of f at P_i ($i = 0, 1$) and $d(P_0, P_1)$ is the distance between P_0 and P_1.

52. A rectangular metal plate of dimensions 8 in. × 4 in. is placed on a rectangular coordinate system with one corner at the origin and the longer side along the positive x-axis. The figure shows the contour map of the function f describing the temperature of the plate in degrees Fahrenheit. Use the contour map to estimate the rate of change of the temperature at the point $(3, 1)$ in the direction from the point $(3, 1)$ toward the point $(5, 4)$.

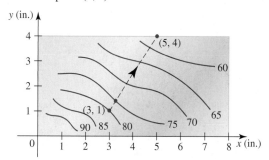

Hint: See Exercise 51.

53. Suppose that f is differentiable, and suppose that the directional derivative of f at the origin attains a maximum value of 5 in the direction of the vector from the origin to the point $(-3, 4)$. Find $\nabla f(0, 0)$.

54. Find unit vector(s) $\mathbf{u} = \langle u_1, u_2 \rangle$ such that the directional derivative of $f(x, y) = x^2 + e^{-xy}$ at the point $(1, 0)$ in the direction of $\mathbf{u}$ has value 1.

[cas] *In Exercises 55 and 56, (a) Plot several level curves of each pair of functions f and g using the same viewing window, and (b) Show analytically that each level curve of f intersects all level curves of g at right angles.*

55. $f(x, y) = x^2 - y^2, \quad g(x, y) = xy$

56. $f(x, y) = e^x \cos y, \quad g(x, y) = e^x \sin y$

57. Let $f(x, y) = x^2 + y^2$, and let $g(x, y) = x^2 - y^2$. Find the direction in which f increases most rapidly and the direction in which g increases most rapidly at $(0, 0)$. Is Theorem 3 applicable here?

In Exercises 58–62, determine whether the statement is true or false. If it is true, explain why it is true. If it is false, give an example to show why it is false.

58. If f is differentiable at each point, then the directional derivative of f exists in all directions.

59. If f is differentiable at each point, then the value of the directional derivative at any point in a given direction depends only on the direction and the partial derivatives f_x and f_y at that point.

60. If $f_x(a, b) = 0$ and $f_y(a, b) = 0$, then $\nabla f(a, b) = \mathbf{0}$.

61. The maximum value of $D_{\mathbf{u}} f(x, y)$ is $\sqrt{f_x^2(x, y) + f_y^2(x, y)}$.

62. If ∇f is known, then we can determine f completely.

13.7 Tangent Planes and Normal Lines

One compelling reason for studying the tangent line to a curve is that the curve may be approximated by its tangent line near a point of tangency (Figure 1). Answers to questions about the curve near a point of tangency may be obtained indirectly by analyzing the tangent line, a relatively simple task, rather than by studying the curve itself. As you might recall, both the approximation of the change in a function using its differential and Newton's method for finding the zeros of a function are based on this observation.

Our motivation for studying tangent planes to a surface in space is the same as that for studying tangent lines to a curve: Near a point of tangency, a surface may be approximated by its tangent plane (Figure 1). We will show later that approximating the change in $z = f(x, y)$ using the differential is tantamount to approximating this change by the change in z on the tangent plane.

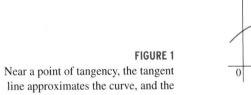

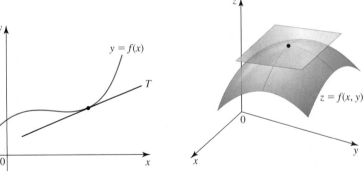

FIGURE 1
Near a point of tangency, the tangent line approximates the curve, and the tangent plane approximates the surface.

(a) The tangent line to a curve

(b) The tangent plane to a surface

■ Geometric Interpretation of the Gradient

We begin by looking at the geometric interpretation of the gradient of a function. This vector will play a central role in our effort to find the tangent plane to a surface.

Suppose that the temperature T at any point (x, y) in the plane is given by the function f; that is, $T = f(x, y)$. Then the level curve $f(x, y) = c$, where c is a constant, gives the set of points in the plane that have temperature c (Figure 2). Recall that such a curve is called an isothermal curve.

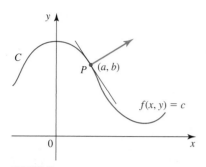

FIGURE 2
The level curve C defined by
$f(x, y) = c$ is an isothermal curve.

If we are at the point $P(a, b)$, in what direction should we move if we want to experience the greatest increase in temperature? Since the temperature remains constant if we move along C, it seems reasonable to conjecture that proceeding in the direction perpendicular to the tangent line to C at P will result in the greatest increase in temperature. But as we saw in Section 13.6, the function f (and hence the temperature) increases most rapidly in the direction given by its gradient $\nabla f(a, b)$. These observations suggest that the gradient $\nabla f(a, b)$ is perpendicular to the tangent line to the level curve $f(x, y) = c$ at P. That this is indeed the case can be demonstrated as follows:

Suppose that the curve C is represented by the vector function

$$\mathbf{r}(t) = g(t)\mathbf{i} + h(t)\mathbf{j}$$

where g and h are differentiable functions, $a = g(t_0)$ and $b = h(t_0)$, and t_0 lies in the parameter interval (Figure 3). Since the point $(x, y) = (g(t), h(t))$ lies on C, we have

$$f(g(t), h(t)) = c$$

for all t in the parameter interval.

FIGURE 3
The curve C may be represented by
$\mathbf{r}(t) = x\mathbf{i} + y\mathbf{j} = g(t)\mathbf{i} + h(t)\mathbf{j}$.

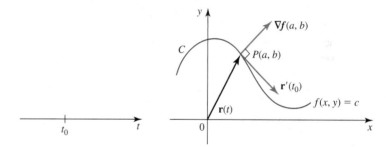

Differentiating both sides of this equation with respect to t and using the Chain Rule for a function of two variables, we obtain

$$\frac{\partial f}{\partial x}\frac{dx}{dt} + \frac{\partial f}{\partial y}\frac{dy}{dt} = 0$$

Recalling that

$$\nabla f(x, y) = \frac{\partial f}{\partial x}\mathbf{i} + \frac{\partial f}{\partial y}\mathbf{j} \quad \text{and} \quad \mathbf{r}'(t) = \frac{dx}{dt}\mathbf{i} + \frac{dy}{dt}\mathbf{j}$$

we can write this last equation in the form

$$\nabla f(x, y) \cdot \mathbf{r}'(t) = 0$$

In particular, when $t = t_0$, we have

$$\nabla f(a, b) \cdot \mathbf{r}'(t_0) = 0$$

Thus, if $\mathbf{r}'(t_0) \neq \mathbf{0}$, the vector $\nabla f(a, b)$ is orthogonal to the tangent vector $\mathbf{r}'(t_0)$ at $P(a, b)$. Loosely speaking, we have demonstrated the following:

∇f is orthogonal to the level curve $f(x, y) = c$ at P.   See Figure 3.

EXAMPLE 1 Let $f(x, y) = x^2 - y^2$. Find the level curve of f passing through the point $(5, 3)$. Also, find the gradient of f at that point, and make a sketch of both the level curve and the gradient vector.

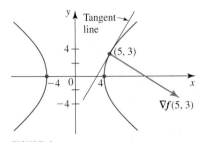

FIGURE 4
The gradient $\nabla f(5, 3)$ is orthogonal to the level curve $x^2 - y^2 = 16$ at $(5, 3)$.

Solution Since $f(5, 3) = 25 - 9 = 16$, the required level curve is the hyperbola $x^2 - y^2 = 16$. The gradient of f at any point (x, y) is

$$\nabla f(x, y) = 2x\mathbf{i} - 2y\mathbf{j}$$

and, in particular, the gradient of f at $(5, 3)$ is

$$\nabla f(5, 3) = 10\mathbf{i} - 6\mathbf{j}$$

The level curve and $\nabla f(5, 3)$ are shown in Figure 4.   ■

EXAMPLE 2 Refer to Example 1. Find equations of the normal line and the tangent line to the curve $x^2 - y^2 = 16$ at the point $(5, 3)$.

Solution We think of the curve $x^2 - y^2 = 16$ as the level curve $f(x, y) = k$ of the function $f(x, y) = x^2 - y^2$ for $k = 16$. From Example 1 we see that $\nabla f(5, 3) = 10\mathbf{i} - 6\mathbf{j}$. Since this gradient is normal to the curve $x^2 - y^2 = 16$ at $(5, 3)$ (see Figure 4), we see that the slope of the required normal line is

$$m_1 = -\frac{6}{10} = -\frac{3}{5}$$

Therefore, an equation of the normal line is

$$y - 3 = -\frac{3}{5}(x - 5) \qquad \text{or} \qquad y = -\frac{3}{5}x + 6$$

The slope of the required tangent line is

$$m_2 = -\frac{1}{m_1} = \frac{5}{3}$$

so an equation of the tangent line is

$$y - 3 = \frac{5}{3}(x - 5) \qquad \text{or} \qquad y = \frac{5}{3}x - \frac{16}{3} \qquad ■$$

Next, suppose that $F(x, y, z) = k$ is the level surface S of a differentiable function F defined by $T = F(x, y, z)$. You may think of the function F as giving the temperature at any point (x, y, z) in space and interpret the following argument in terms of this application.

Suppose that $P(a, b, c)$ is a point on S and let C be a smooth curve on S passing through P. Then C can be described by the vector function

$$\mathbf{r}(t) = f(t)\mathbf{i} + g(t)\mathbf{j} + h(t)\mathbf{k}$$

where $f(t_0) = a$, $g(t_0) = b$, $h(t_0) = c$, and t_0 is a point in the parameter interval (Figure 5).

FIGURE 5
The curve C is described by $\mathbf{r}(t) = f(t)\mathbf{i} + g(t)\mathbf{j} + h(t)\mathbf{k}$ with $P(a, b, c)$ corresponding to t_0.

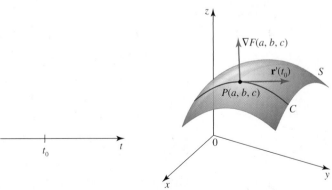

Since the point $(x, y, z) = (f(t), g(t), h(t))$ lies on S, we have

$$F(f(t), g(t), h(t)) = k$$

for all t in the parameter interval. If $\mathbf{r}$ is differentiable, then we can use the Chain Rule to differentiate both sides of this equation to obtain

$$\frac{\partial F}{\partial x}\frac{dx}{dt} + \frac{\partial F}{\partial y}\frac{dy}{dt} + \frac{\partial F}{\partial z}\frac{dz}{dt} = 0$$

This is the same as

$$[F_x(x, y, z)\mathbf{i} + F_y(x, y, z)\mathbf{j} + F_z(x, y, z)\mathbf{k}] \cdot \left[\frac{dx}{dt}\mathbf{i} + \frac{dy}{dt}\mathbf{j} + \frac{dz}{dt}\mathbf{k}\right] = 0$$

or, in an even more abbreviated form,

$$\nabla F(x, y, z) \cdot \mathbf{r}'(t) = 0$$

In particular, at $t = t_0$ we have

$$\nabla F(a, b, c) \cdot \mathbf{r}'(t_0) = 0$$

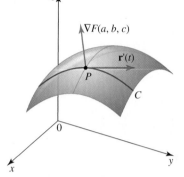

FIGURE 6

The gradient $\nabla F(a, b, c)$ is orthogonal to the tangent vector of *every* curve on S passing through $P(a, b, c)$.

This shows that if $\mathbf{r}'(t_0) \neq \mathbf{0}$, then the gradient vector $\nabla F(a, b, c)$ is orthogonal to the tangent vector $\mathbf{r}'(t_0)$ to C at P (Figure 6). Since this argument holds for any differentiable curve passing through $P(a, b, c)$ on S, we have shown that $\nabla F(a, b, c)$ is orthogonal to the tangent vector of *every* curve on S passing through P. Thus, loosely speaking, we have demonstrated the following result.

∇F is orthogonal to the level surface $F(x, y, z) = 0$ at P.

Note Interpreting the function F as giving the temperature at any point (x, y, z) in space as was suggested earlier, we see that the level surface $F(x, y, z) = k$ gives all points (x, y, z) in space whose temperature is k. The result that was just derived simply states that if you are at any point on this surface, then moving away from it in a direction of ∇F (perpendicular to the surface at that point) will result in the greatest increase in temperature. ∎

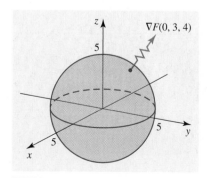

FIGURE 7

The gradient $\nabla F(0, 3, 4)$ is orthogonal to the level surface $x^2 + y^2 + z^2 = 25$ at $(0, 3, 4)$.

EXAMPLE 3 Let $F(x, y, z) = x^2 + y^2 + z^2$. Find the level surface that contains the point $(0, 3, 4)$. Also, find the gradient of F at that point, and make a sketch of both the level surface and the gradient vector.

Solution Since $F(0, 3, 4) = 0 + 9 + 16 = 25$, the required level surface is the sphere $x^2 + y^2 + z^2 = 25$ with center at the origin and radius 5. The gradient of F at any point (x, y, z) is

$$\nabla F(x, y, z) = 2x\mathbf{i} + 2y\mathbf{j} + 2z\mathbf{k}$$

so the gradient of F at $(0, 3, 4)$ is

$$\nabla F(0, 3, 4) = 6\mathbf{j} + 8\mathbf{k}$$

The level surface and $\nabla F(0, 3, 4)$ are sketched in Figure 7. ∎

■ Tangent Planes and Normal Lines

We are now in a position to define a tangent plane to a surface in space. But before doing so, let's digress a little to talk about the representation of surfaces in space. Up to now, we have assumed that a surface in space is described by a function f with explicit representation $z = f(x, y)$.

Another way of describing a surface in space is via a function that is represented implicitly by the equation

$$F(x, y, z) = 0 \tag{1}$$

Here, F is a function of the three variables x, y, and z described by the equation $w = F(x, y, z)$. Thus, we can think of Equation (1) as representing a *level surface* of F.

For a surface S that is given explicitly by $z = f(x, y)$, we define

$$F(x, y, z) = z - f(x, y)$$

This shows that we can also view S as the level surface of F given by Equation (1). For example, the surface described by $z = x^2 + 2y^2 + 1$ can be viewed as the level surface of F defined by $F(x, y, z) = 0$, where $F(x, y, z) = z - x^2 - 2y^2 - 1$.

To define a tangent plane, let S be a surface described by $F(x, y, z) = 0$, and let $P(a, b, c)$ be a point on S. Then, as we saw earlier, the gradient $\nabla F(a, b, c)$ at P is orthogonal to the tangent vector of *every* curve on S passing through P (Figure 8). This suggests that we define the *tangent plane* to S at P to be the plane passing through P and containing all these tangent vectors. Equivalently, the tangent plane should have $\nabla F(a, b, c)$ as a normal vector.

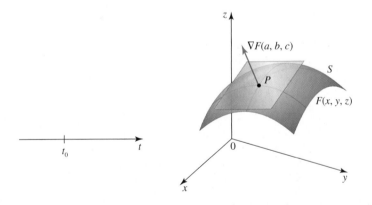

FIGURE 8
The tangent plane to S at P contains the tangent vectors to all curves on S passing through P.

DEFINITIONS   Tangent Plane and Normal Line

Let $P(a, b, c)$ be a point on the surface S described by $F(x, y, z) = 0$, where F is differentiable at P, and suppose that $\nabla F(a, b, c) \neq \mathbf{0}$. Then the **tangent plane** to S at P is the plane that passes through P and has normal vector $\nabla F(a, b, c)$. The **normal line** to S at P is the line that passes through P and has the same direction as $\nabla F(a, b, c)$.

Using Equation (4) from Section 11.5, we see that an equation of the tangent plane is

$$F_x(a, b, c)(x - a) + F_y(a, b, c)(y - b) + F_z(a, b, c)(z - c) = 0 \tag{2}$$

and using Equation (2) of Section 11.5, we see that the equations of the normal line (in symmetric form) are

$$\frac{x - a}{F_x(a, b, c)} = \frac{y - b}{F_y(a, b, c)} = \frac{z - c}{F_z(a, b, c)} \tag{3}$$

EXAMPLE 4 Find equations of the tangent plane and normal line to the ellipsoid with equation $4x^2 + y^2 + 4z^2 = 16$ at the point $(1, 2, \sqrt{2})$.

Solution The given equation can be written in the form $F(x, y, z) = 0$, where $F(x, y, z) = 4x^2 + y^2 + 4z^2 - 16$. The partial derivatives of F are

$$F_x(x, y, z) = 8x, \qquad F_y(x, y, z) = 2y, \qquad \text{and} \qquad F_z(x, y, z) = 8z$$

In particular, at the point $(1, 2, \sqrt{2})$

$$F_x(1, 2, \sqrt{2}) = 8, \qquad F_y(1, 2, \sqrt{2}) = 4, \qquad \text{and} \qquad F_z(1, 2, \sqrt{2}) = 8\sqrt{2}$$

Then, using Equation (2), we find that an equation of the tangent plane to the ellipsoid at $(1, 2, \sqrt{2})$ is

$$8(x - 1) + 4(y - 2) + 8\sqrt{2}(z - \sqrt{2}) = 0$$

or $2x + y + 2\sqrt{2}z = 8$. Next, using Equation (3), we obtain the following parametric equations of the normal line:

$$\frac{x - 1}{8} = \frac{y - 2}{4} = \frac{z - \sqrt{2}}{8\sqrt{2}} \qquad \text{or} \qquad \frac{x - 1}{2} = y - 2 = \frac{z - \sqrt{2}}{2\sqrt{2}}$$

The tangent plane and normal line are shown in Figure 9. ∎

EXAMPLE 5 Find equations of the tangent plane and normal line to the graph of the function f defined by $f(x, y) = 4x^2 + y^2 + 2$ at the point where $x = 1$ and $y = 1$.

Solution Here, the surface is defined by

$$z = f(x, y) = 4x^2 + y^2 + 2$$

and we recognize it to be a paraboloid. This equation can be rewritten in the form

$$F(x, y, z) = z - f(x, y) = 0$$

where $F(x, y, z) = z - 4x^2 - y^2 - 2$. The partial derivatives of F are

$$F_x(x, y, z) = -8x, \qquad F_y(x, y, z) = -2y, \qquad \text{and} \qquad F_z(x, y, z) = 1$$

If $x = 1$ and $y = 1$, then $z = f(1, 1) = 4 + 1 + 2 = 7$. At the point $(1, 1, 7)$ we have

$$F_x(1, 1, 7) = -8, \qquad F_y(1, 1, 7) = -2, \qquad \text{and} \qquad F_z(1, 1, 7) = 1$$

Then, using Equation (2), we find an equation of the tangent plane to the paraboloid at $(1, 1, 7)$ to be

$$-8(x - 1) - 2(y - 1) + 1(z - 7) = 0$$

or $8x + 2y - z = 3$. Next, using Equation (3), we find that the parametric equations of the normal line at $(1, 1, 7)$ are

$$\frac{x - 1}{-8} = \frac{y - 1}{-2} = \frac{z - 7}{1}$$

The tangent plane and normal line are shown in Figure 10. ∎

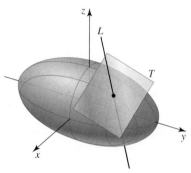

FIGURE 9
The tangent plane and normal line to the ellipsoid $4x^2 + y^2 + 4z^2 = 16$ at $(1, 2, \sqrt{2})$.

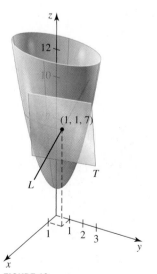

FIGURE 10
The tangent plane and normal line to the paraboloid $z = 4x^2 + y^2 + 2$ at $(1, 1, 7)$.

■ Using the Tangent Plane of f to Approximate the Surface $z = f(x, y)$

We conclude this section by showing that in approximating the change Δz in $z = f(x, y)$ as (x, y) changes from (a, b) to $(a + \Delta x, b + \Delta y)$ by the differential $dz = f_x(a, b)\,\Delta x + f_y(a, b)\,\Delta y$, we are in effect using the tangent plane of f near $P(a, b)$ to approximate the surface $z = f(x, y)$ near $P(a, b)$.

We begin by finding an expression for the tangent plane to the surface $z = f(x, y)$ at (a, b). Writing $z = f(x, y)$ in the form $F(x, y, z) = z - f(x, y) = 0$ we see that

$$F_x(a, b, c) = -f_x(a, b), \qquad F_y(a, b, c) = -f_y(a, b), \qquad \text{and} \qquad F_z(a, b, c) = 1$$

Using Equation (2), we find that the required equation is

$$-f_x(a, b)(x - a) - f_y(a, b)(y - b) + (z - c) = 0$$

or

$$z - f(a, b) = f_x(a, b)\,\Delta x + f_y(a, b)\,\Delta y \qquad c = f(a, b) \tag{4}$$

But the expression on the right is just the differential of f at (a, b). So Equation (4) implies that $dz = z - f(a, b)$; that is, dz represents the change in height of the tangent plane (Figure 11).

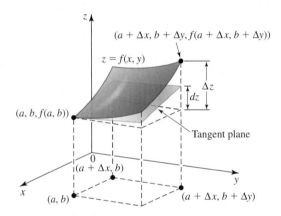

FIGURE 11
The relationship between Δz and dz

By Theorem 1 of Section 13.4 we have

$$\Delta z = f_x(a, b)\,\Delta x + f_y(a, b)\,\Delta y + \varepsilon_1\,\Delta x + \varepsilon_2\,\Delta y$$

or

$$\Delta z - dz = \varepsilon_1\,\Delta x + \varepsilon_2\,\Delta y$$

where ε_1 and ε_2 are functions of Δx and Δy that approach 0 as Δx and Δy approach 0. Therefore, as was pointed out in Section 13.4, $\Delta z \approx dz$ if Δx and Δy are small. Recalling the meaning of Δz (see Figure 11), we see that we are using the tangent plane at (a, b) to approximate the surface $z = f(x, y)$ when (x, y) is close to (a, b).

13.7 CONCEPT QUESTIONS

1. a. Consider the level curve $f(x, y) = c$, where f is differentiable and c is a constant. What can you say about ∇f at a point P on the level curve? Illustrate with a figure.
  b. Repeat part (a) for a level surface $F(x, y, z) = c$.

2. a. Define the tangent plane to a surface S described by $F(x, y, z) = 0$ at the point $P(a, b, c)$ on S. Illustrate with a figure and give an equation of the tangent plane.
  b. Repeat part (a) for the normal line to S at P.

13.7 EXERCISES

In Exercises 1–4, sketch (a) *the level curve of the function f that passes through the point P and* (b) *the gradient of f at P.*

1. $f(x, y) = y^2 - x^2$; $P(1, 2)$

2. $f(x, y) = 4x^2 + y^2$; $P\left(\frac{\sqrt{3}}{2}, 1\right)$

3. $f(x, y) = x^2 + y$; $P(1, 3)$

4. $f(x, y) = 2x + 3y$; $P(-3, 4)$

In Exercises 5–8, find equations of the normal and tangent lines to the curve at the given point.

5. $\dfrac{x^2}{9} + \dfrac{y^2}{16} = 1$; $\left(\frac{3\sqrt{3}}{2}, 2\right)$

6. $x^4 - x^2 + y^2 = 0$; $\left(\frac{1}{2}, \frac{\sqrt{3}}{4}\right)$

7. $x^4 + 2x^2y^2 + y^4 - 9x^2 + 9y^2 = 0$; $(\sqrt{5}, -1)$

8. $2x + y - e^{x-y} = 2$; $(1, 1)$

In Exercises 9–14, sketch (a) *the level surface of the function F that passes through the point P and* (b) *the gradient of F at P.*

9. $F(x, y, z) = x^2 + y^2 + z^2$; $P(1, 2, 2)$

10. $F(x, y, z) = z - x^2 - y^2$; $P(1, 1, 2)$

11. $F(x, y, z) = x^2 + y^2$; $P(0, 2, 4)$

12. $F(x, y, z) = 2x + 3y + z$; $P(2, 3, 1)$

13. $F(x, y, z) = -x^2 + y^2 - z^2$; $P(1, 3, 2)$

14. $F(x, y, z) = xy$; $P\left(2, \frac{1}{2}, 0\right)$

In Exercises 15–32, find equations for the tangent plane and the normal line to the surface with the equation at the given point.

15. $x^2 + 4y^2 + 9z^2 = 17$; $P(2, 1, 1)$

16. $2x^2 - y^2 + 3z^2 = 2$; $P(2, -3, 1)$

17. $x^2 - 2y^2 - 4z^2 = 4$; $P(4, -2, -1)$

18. $x^2 + y^2 + z^2 - 2xy + 4xz - x + y = 12$; $P(1, 0, 2)$

19. $xy + yz + xz = 11$; $P(1, 2, 3)$

20. $xyz = -4$; $P(2, -1, 2)$

21. $z = 9x^2 + 4y^2$; $P(-1, 2, 25)$

22. $z = y^2 - 2x^2$; $P(2, 4, 8)$

23. $xz^2 + yx^2 + y^2 - 2x + 3y + 6 = 0$; $P(-2, 1, 3)$

24. $x^3 - xy^2 + z^3 - 2x + 6 = 0$; $P(1, 2, -1)$

25. $z = xe^y$; $P(2, 0, 2)$

26. $z = e^x \sin \pi y$; $P(0, 1, 0)$

27. $z = \ln(xy + 1)$; $P(3, 0, 0)$

28. $z = \ln \dfrac{x}{y}$; $P(2, 2, 0)$

29. $z = \tan^{-1}\left(\dfrac{y}{x}\right)$; $P\left(1, 1, \dfrac{\pi}{4}\right)$

30. $z - x \cos y = 0$; $P\left(2, \dfrac{\pi}{3}, 1\right)$

31. $\sin xy + 3z = 3$; $P(0, 3, 1)$

32. $e^x(\cos y + 1) - 2z = -2$; $P(0, 0, 2)$

33. Show that an equation of the tangent plane to the ellipsoid

$$\frac{x^2}{a^2} + \frac{y^2}{b^2} + \frac{z^2}{c^2} = 1$$

at the point (x_0, y_0, z_0) can be written as

$$\frac{xx_0}{a^2} + \frac{yy_0}{b^2} + \frac{zz_0}{c^2} = 1$$

34. Show that an equation of the tangent plane to the hyperboloid

$$\frac{x^2}{a^2} + \frac{y^2}{b^2} - \frac{z^2}{c^2} = 1$$

at the point (x_0, y_0, z_0) can be written as

$$\frac{xx_0}{a^2} + \frac{yy_0}{b^2} - \frac{zz_0}{c^2} = 1$$

35. Find an equation of the tangent plane to the hyperboloid of two sheets

$$\frac{x^2}{a^2} - \frac{y^2}{b^2} - \frac{z^2}{c^2} = 1$$

at the point (x_0, y_0, z_0), and express it in a form similar to that of Exercise 34.

36. Show that an equation of the tangent plane to the elliptic paraboloid

$$\frac{x^2}{a^2} + \frac{y^2}{b^2} = cz$$

at the point (x_0, y_0, z_0) can be written as

$$\frac{2xx_0}{a^2} + \frac{2yy_0}{b^2} = c(z + z_0)$$

37. Find the points on the sphere $x^2 + y^2 + z^2 = 14$ at which the tangent plane is parallel to the plane $x + 2y + 3z = 12$.

38. Find the points on the hyperboloid of two sheets $-x^2 - 2y^2 + z^2 = 4$ at which the tangent plane is parallel to the plane $2x + 2y + 4z = 1$.

39. Find the points on the hyperboloid of one sheet $2x^2 - y^2 + z^2 = 1$ at which the normal line is parallel to the line passing through the points $(-1, 1, 2)$ and $(3, 3, 3)$.

40. Find the points on the surface $x^2 + 4y^2 + 3z^2 - 2xy = 16$ at which the tangent plane is horizontal.

41. Two surfaces are *tangent* to each other at a point P if and only if they have a common tangent plane at that point. Show that the elliptic paraboloid $2x^2 + y^2 - z - 5 = 0$ and the sphere $x^2 + y^2 + z^2 - 6x - 8y - z + 17 = 0$ are tangent to each other at the point $(1, 2, 1)$.

42. Two surfaces are *orthogonal* to each other at a point of intersection P if and only if their normal lines at P are orthogonal. Show that the sphere $x^2 + y^2 + z^2 - 17 = 0$ and the elliptic paraboloid $2x^2 - y + 2z^2 + 2 = 0$ are orthogonal to each other at the point $(1, 4, 0)$.

43. Show that any line that is tangent to the ellipse $(x^2/a^2) + (y^2/b^2) = 1$ has the equation

$$(b \cos \theta)x + (a \sin \theta)y = ab$$

where θ lies in the interval $[0, 2\pi)$.

44. Suppose that two surfaces $F(x, y, z) = 0$ and $G(x, y, z) = 0$ intersect along a curve C and that $P(x_0, y_0, z_0)$ is a point on C. Show that the vector $\nabla F(x_0, y_0, z_0) \times \nabla G(x_0, y_0, z_0)$ is parallel to the tangent line to C at P. Illustrate with a sketch.

45. Refer to Exercise 44. Let C be the intersection of the sphere $x^2 + y^2 + z^2 = 2$ and the paraboloid $z = x^2 + y^2$. Find the

parametric equations of the tangent line to C at the point $\left(-\frac{\sqrt{2}}{2}, \frac{\sqrt{2}}{2}, 1\right)$.

In Exercises 46–49, determine whether the statement is true or false. If it is true, explain why it is true. If it is false, give an example to show why it is false.

46. The tangent line at a point P on the level curve $f(x, y) = c$ is orthogonal to ∇f at P.

47. The line with equations

$$\frac{x - 2}{4} = \frac{y + 1}{6} = -\frac{z}{2}$$

is perpendicular to the plane with equation $2x + 3y - z = 4$.

48. If an equation of the tangent plane at the point $P_0(x_0, y_0, z_0)$ on the surface described by $F(x, y, z) = 0$ is $ax + by + cz = d$, then $\nabla F(x_0, y_0, z_0) = k\langle a, b, c \rangle$ for some scalar k.

49. The vector equation of the normal line passing through the point $P_0(x_0, y_0, z_0)$ on the surface with equation $F(x, y, z) = 0$ is $\mathbf{r}(t) = \langle x_0, y_0, z_0 \rangle + t \nabla F(x_0, y_0, z_0)$.

13.8 Extrema of Functions of Two Variables

■ Relative and Absolute Extrema

In Chapter 3 we saw that the solution of a problem often reduces to finding the extreme values of a function of one variable. A similar situation arises when we solve problems involving a function of two or more variables.

For example, suppose that the Scandi Company manufactures computer desks in both assembled and unassembled versions. Then its weekly profit P is a function of the number of assembled units, x, and the number of unassembled units, y, sold per week; that is, $P = f(x, y)$. A question of paramount importance to the manufacturer is: How many assembled desks and how many unassembled desks should the company manufacture per week to maximize its weekly profit? Mathematically, the problem is solved by finding the values of x and y that will make $f(x, y)$ a maximum.

In this section and the next section we will focus our attention on finding the extrema of a function of two variables. As in the case of a function of one variable, we distinguish between the relative (or local) extrema and the absolute extrema of a function of two variables.

> **DEFINITION** **Relative Extrema of a Function of Two Variables**
>
> Let f be a function defined on a region R containing the point (a, b). Then f has a **relative maximum** at (a, b) if $f(x, y) \leq f(a, b)$ for all points (x, y) in an open disk containing (a, b). The number $f(a, b)$ is called a **relative maximum value.**
>
> Similarly, f has a **relative minimum** at (a, b) with **relative minimum value** $f(a, b)$ if $f(x, y) \geq f(a, b)$ for all points (x, y) in an open disk containing (a, b).

Loosely speaking, f has a relative maximum at (a, b) if the point $(a, b, f(a, b))$ is the highest point on the graph of f when compared to all nearby points. A similar interpretation holds for a relative minimum.

If the inequalities in this last definition hold for all points (x, y) in the domain of f, then f has an **absolute maximum (absolute minimum)** at (a, b) with **absolute maximum value (absolute minimum value)** $f(a, b)$. Figure 1 shows the graph of a function defined on a domain D with relative maxima at (a, b) and (e, g) and a relative minimum at (c, d). The absolute maximum of f occurs at (e, g), and the absolute minimum of f occurs at (h, i).

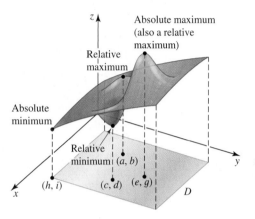

FIGURE 1

The relative and absolute extrema of the function f over the domain D

■ Critical Points—Candidates for Relative Extrema

Figure 2a shows the graph of a function f with a relative maximum at a point (a, b) lying inside the domain of f. As you can see, the tangent plane to the surface $z = f(x, y)$ at the point $(a, b, f(a, b))$ is horizontal. This means that all the directional derivatives of f at (a, b), if they exist, must be zero. In particular, $f_x(a, b) = 0$ and $f_y(a, b) = 0$. Next, Figure 2b shows the graph of a function with a relative maximum at a point (a, b). Note that both $f_x(a, b)$ and $f_y(a, b)$ do not exist because the surface $z = f(x, y)$ has a point $(a, b, f(a, b))$ that looks like a jagged mountain peak.

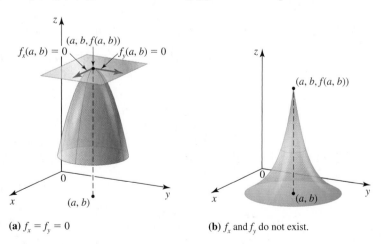

FIGURE 2

At a relative extremum of f, either $f_x = f_y = 0$ or one or both partial derivatives do not exist.

(a) $f_x = f_y = 0$

(b) f_x and f_y do not exist.

You are encouraged to draw similar graphs of functions having relative minima at points lying inside the domain of the functions. All of these points are critical points of a function of two variables.

> **DEFINITION   Critical Points of a Function**
>
> Let f be defined on an open region R containing the point (a, b). We call (a, b) a **critical point** of f if
>
> **a.** f_x and/or f_y do not exist at (a, b) or
> **b.** both $f_x(a, b) = 0$ and $f_y(a, b) = 0$.

The next theorem tells us that the relative extremum of a function f defined on an open region can occur only at a critical point of f.

> **THEOREM 1   The Critical Points of f Are Candidates for Relative Extrema**
>
> If f has a relative extremum (relative maximum or relative minimum) at a point (a, b) in the domain of f, then (a, b) must be a critical point of f.

PROOF   If either f_x or f_y does not exist at (a, b), then (a, b) is a critical point of f. So suppose that both $f_x(a, b)$ and $f_y(a, b)$ exist. Let $g(x) = f(x, b)$. If f has a relative extremum at (a, b), then g has a relative extremum at a, so by Theorem 1 of Section 3.1, $g'(a) = 0$. But

$$g'(a) = \lim_{h \to 0} \frac{f(a + h, b) - f(a, b)}{h} = f_x(a, b)$$

Therefore, $f_x(a, b) = 0$. Similarly, by considering the function $f(y) = f(a, y)$, we obtain $f_y(a, b) = 0$. Thus, (a, b) is a critical point of f. ∎

EXAMPLE 1   Let $f(x, y) = x^2 + y^2 - 4x - 6y + 17$. Find the critical point of f, and show that f has a relative minimum at that point.

Solution   To find the critical point of f, we compute

$$f_x(x, y) = 2x - 4 = 2(x - 2) \qquad \text{and} \qquad f_y(x, y) = 2y - 6 = 2(y - 3)$$

Observe that both f_x and f_y are continuous for all values of x and y. Setting f_x and f_y equal to zero, we find that $x = 2$ and $y = 3$, so $(2, 3)$ is the only critical point of f. Next, to show that f has a relative minimum at this point, we complete the squares in x and y and write $f(x, y)$ in the form

$$f(x, y) = (x - 2)^2 + (y - 3)^2 + 4$$

Notice that $(x - 2)^2 \geq 0$ and $(y - 3)^2 \geq 0$, so $f(x, y) \geq 4$ for all (x, y) in the domain of f. Therefore, $f(2, 3) = 4$ is a relative minimum value of f. In fact, we have shown that 4 is the absolute minimum value of f. The graph of f shown in Figure 3 confirms this result. ∎

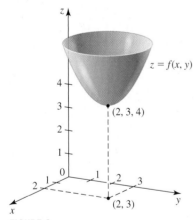

$z = f(x, y)$

$(2, 3, 4)$

$(2, 3)$

FIGURE 3
The function f has a relative minimum at $(2, 3)$.

EXAMPLE 2   Let $f(x, y) = 3 - \sqrt{x^2 + y^2}$. Show that $(0, 0)$ is the only critical point of f and that $f(0, 0) = 3$ is a relative maximum value of f.

Solution   The partial derivatives of f are

$$f_x(x, y) = -\frac{x}{\sqrt{x^2 + y^2}} \qquad \text{and} \qquad f_y(x, y) = -\frac{y}{\sqrt{x^2 + y^2}}$$

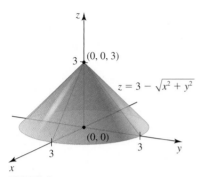

FIGURE 4
The function f has a relative maximum at $(0, 0)$.

Since both $f_x(x, y)$ and $f_y(x, y)$ are undefined at $(0, 0)$, we conclude that $(0, 0)$ is a critical point of f. Also, $f_x(x, y)$ and $f_y(x, y)$ are not both equal to zero at any point. This tells us that $(0, 0)$ is the only critical point of f. Finally, since $\sqrt{x^2 + y^2} \geq 0$ for all values of x and y, we see that $f(x, y) \leq 3$ for all points (x, y). We conclude that $f(0, 0) = 3$ is a relative (indeed, the absolute) maximum of f. The graph of f shown in Figure 4 confirms this result. ■

As in the case of a function of one variable, a critical point of a function of two variables is only a candidate for a relative extremum of the function. A critical point need not give rise to a relative extremum, as the following example shows.

EXAMPLE 3 Show that the point $(0, 0)$ is a critical point of $f(x, y) = y^2 - x^2$ but that it does not give rise to a relative extremum of f.

Solution The partial derivatives of f,

$$f_x(x, y) = -2x \qquad \text{and} \qquad f_y(x, y) = 2y$$

are continuous everywhere. Since f_x and f_y are both equal to zero at $(0, 0)$, we conclude that $(0, 0)$ is a critical point of f and that it is the only candidate for a relative extremum of f. But notice that for points on the x-axis we have $y = 0$, so $f(x, y) = -x^2 < 0$ if $x \neq 0$; and for points on the y-axis we have $x = 0$, so $f(x, y) = y^2 > 0$ if $y \neq 0$. Therefore, every open disk containing $(0, 0)$ has points where f takes on positive values as well as points where f takes on negative values. This shows that $f(0, 0) = 0$ cannot be a relative extremum of f. The graph of f is shown in Figure 5. The point $(0, 0, 0)$ is called a *saddle point*.

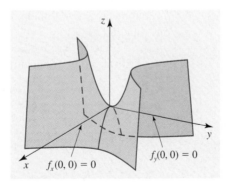

FIGURE 5
The point $(0, 0)$ is a critical point of $f(x, y) = y^2 - x^2$, but it does not give rise to a relative extremum of f.

■

As we have observed, the critical point $(0, 0)$ in Example 3 does not yield a relative maximum or minimum. In general, a critical point of a differentiable function of two variables that does not give rise to a relative extremum is called a **saddle point.** A saddle point is the analog of an inflection point for the case of a function of one variable.

■ The Second Derivative Test for Relative Extrema

In Examples 1 and 3 we were able to determine, either by inspection or with the help of simple algebraic manipulations, whether f did or did not possess a relative extremum at a critical point. For more complicated functions, the following test may be used. This test is the analog of the Second Derivative Test for a function of one variable. Its proof will be omitted.

> **THEOREM 2   The Second Derivative Test for a Function of Two Variables**
>
> Suppose that f has continuous second-order partial derivatives on an open region containing a critical point (a, b) of f. Let
>
> $$D(x, y) = f_{xx}(x, y)f_{yy}(x, y) - f_{xy}^2(x, y)$$
>
> **a.** If $D(a, b) > 0$ and $f_{xx}(a, b) < 0$, then $f(a, b)$ is a **relative maximum value.**
> **b.** If $D(a, b) > 0$ and $f_{xx}(a, b) > 0$, then $f(a, b)$ is a **relative minimum value.**
> **c.** If $D(a, b) < 0$, then $(a, b, f(a, b))$ is a **saddle point.**
> **d.** If $D(a, b) = 0$, then the test is inconclusive.

EXAMPLE 4   Find the relative extrema of $f(x, y) = x^3 + y^2 - 2xy + 7x - 8y + 2$.

Solution   First, we find the critical points of f. Since

$$f_x(x, y) = 3x^2 - 2y + 7 \quad \text{and} \quad f_y(x, y) = 2y - 2x - 8$$

are both continuous for all values of x and y, the only critical points of f, if any, are found by solving the system of equations $f_x(x, y) = 0$ and $f_y(x, y) = 0$, that is, by solving

$$3x^2 - 2y + 7 = 0 \quad \text{and} \quad 2y - 2x - 8 = 0$$

From the second equation we obtain $y = x + 4$, which upon substitution into the first equation yields

$$3x^2 - 2x - 1 = 0 \quad \text{or} \quad (3x + 1)(x - 1) = 0$$

Therefore, $x = -\frac{1}{3}$ or $x = 1$. Substituting each of these values of x into the expression for y gives $y = \frac{11}{3}$ and $y = 5$, respectively. Therefore, the critical points of f are $\left(-\frac{1}{3}, \frac{11}{3}\right)$ and $(1, 5)$.

Next, we use the Second Derivative Test to determine the nature of each of these critical points. We begin by computing $f_{xx}(x, y) = 6x, f_{yy}(x, y) = 2, f_{xy}(x, y) = -2$, and

$$D(x, y) = f_{xx}(x, y)f_{yy}(x, y) - f_{xy}^2(x, y)$$

$$= (6x)(2) - (-2)^2 = 4(3x - 1)$$

To test the point $\left(-\frac{1}{3}, \frac{11}{3}\right)$, we compute

$$D\left(-\frac{1}{3}, \frac{11}{3}\right) = 4(-1 - 1) = -8 < 0$$

from which we deduce that $\left(-\frac{1}{3}, \frac{11}{3}\right)$ gives rise to the saddle point $\left(-\frac{1}{3}, \frac{11}{3}, -\frac{373}{27}\right)$ of f. Next, to test the critical point $(1, 5)$, we compute

$$D(1, 5) = 4(3 - 1) = 8 > 0$$

which indicates that $(1, 5)$ gives a relative extremum of f. Since

$$f_{xx}(1, 5) = 6(1) = 6 > 0$$

we see that $(1, 5)$ yields a relative minimum of f. Its value is

$$f(1, 5) = (1)^3 + (5)^2 - 2(1)(5) + 7(1) - 8(5) + 2$$

$$= -15$$

The graph and contour map of f are shown in Figures 6a and 6b.

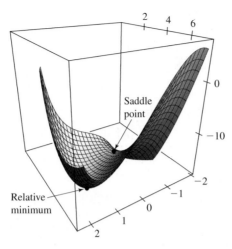

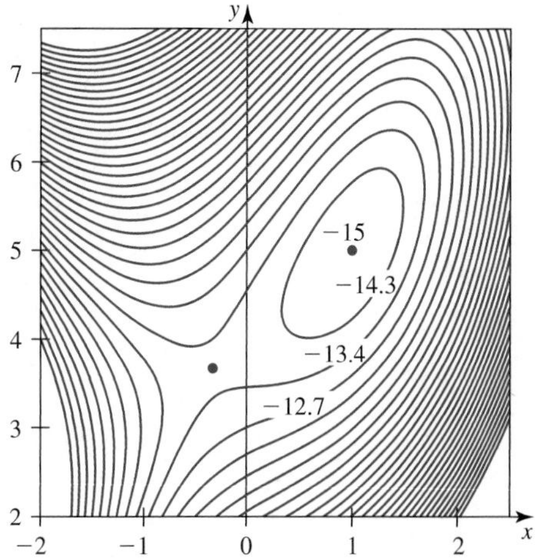

FIGURE 6   **(a)** The graph of $f(x, y) = x^3 + y^2 - 2xy + 7x - 8y + 2$ **(b)** The contour plot of f

EXAMPLE 5   **Priority Mail Regulations**   Postal regulations specify that the combined length and girth of a package sent by priority mail may not exceed 108 in. Find the dimensions of a rectangular package with the greatest possible volume satisfying these regulations.

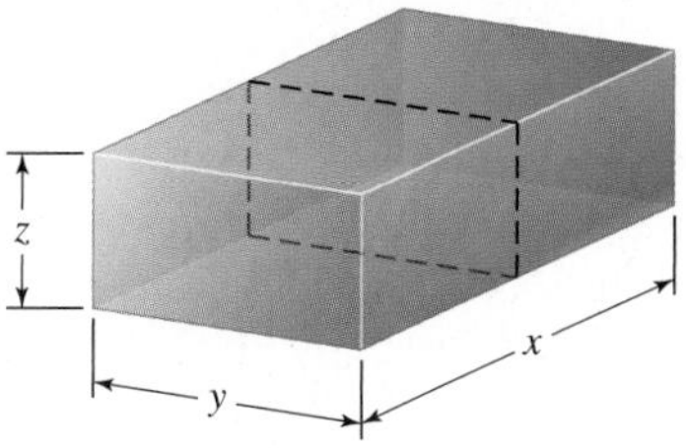

FIGURE 7
The combined length and girth of the package is $x + 2y + 2z$ inches.

Solution   Let the length, width, and height of the package be x, y, and z inches respectively, as shown in Figure 7. Then the volume of the package is $V = xyz$. Observe that the combined length and girth of the package is $(x + 2y + 2z)$ inches. Clearly, we should let this quantity be as large as possible, that is, we should let

$$x + 2y + 2z = 108$$

With the help of this equation we can express V as a function of two variables. For example, solving the equation for x in terms of y and z, we obtain

$$x = 108 - 2y - 2z$$

which, upon substitution into the expression for V, gives

$$V = f(y, z) = (108 - 2y - 2z)yz = 108yz - 2y^2z - 2yz^2$$

To find the critical points of f, we set

$$f_y = 108z - 4yz - 2z^2 = 2z(54 - 2y - z) = 0$$

and

$$f_z = 108y - 2y^2 - 4yz = 2y(54 - y - 2z) = 0$$

Since y and z are both nonzero (otherwise, V would be zero), we are led to the system

$$54 - 2y - z = 0$$

$$54 - y - 2z = 0$$

Multiplying the second equation by 2 gives $108 - 2y - 4z = 0$. Then subtracting this equation from the first equation gives $-54 + 3z = 0$, or $z = 18$. Substituting this value of z into either equation in the system then yields $y = 18$. Therefore, the only critical point of f is $(18, 18)$.

We could use the Second Derivative Test to show that the point $(18, 18)$ gives a relative maximum of V, or, as in this situation, we can simply argue from physical considerations that V must attain an absolute maximum at $(18, 18)$. Finally, from the equation

$$x = 108 - 2y - 2z$$

found earlier, we see that when $y = z = 18$,

$$x = 108 - 2(18) - 2(18) = 36$$

Therefore, the dimensions of the required package are 18 in. $\times$ 18 in. $\times$ 36 in.   ■

■ Finding the Absolute Extremum Values of a Continuous Function on a Closed Set

Recall that if f is a continuous function of one variable on a closed interval $[a, b]$, then the Extreme Value Theorem guarantees that f has an absolute maximum value and an absolute minimum value. The analog of this theorem for a function of two variables follows.

THEOREM 3   The Extreme Value Theorem for Functions of Two Variables

If f is continuous on a closed, bounded set D in the plane, then f attains an absolute maximum value $f(a, b)$ at some point (a, b) in D and an absolute minimum value $f(c, d)$ at some point (c, d) in D.

The following procedure for finding the extreme values of a function of two variables is the analog of the one for finding the extreme values of a function of one variable discussed in Section 3.1.

Finding the Absolute Extremum Values of f on a Closed, Bounded Set D

1. Find the values of f at the critical points of f in D.
2. Find the extreme values of f on the boundary of D.
3. The absolute maximum value of f and the absolute minimum value of f are precisely the largest and the smallest numbers found in Steps 1 and 2.

The justification for this procedure is similar to that for a function of one variable on a closed interval $[a, b]$: If an absolute extremum of f occurs in the interior of D, then it must also be a relative extremum of f, and hence it must occur at a critical point of f. Otherwise, the absolute extremum of f must occur at a boundary point of D.

EXAMPLE 6   Find the absolute maximum and the absolute minimum values of the function $f(x, y) = 2x^2 + y^2 - 4x - 2y + 3$ on the rectangle

$$D = \{(x, y) \,|\, 0 \le x \le 3, 0 \le y \le 2\}$$

Solution   Since f is a polynomial, it is continuous on the closed, bounded set D. Therefore, Theorem 3 guarantees the existence of an absolute maximum and an absolute minimum value of f on D.

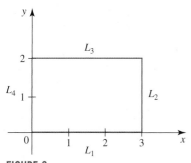

FIGURE 8
The boundary of D consists of the four line segments $L_1, L_2, L_3,$ and L_4.

Step 1 To find the critical point of f in D, we set $f_x = 4x - 4 = 0$ and $f_y = 2y - 2 = 0$. Solving this system of equations gives $(1, 1)$ as the only critical point of f. The value of f at this point is $f(1, 1) = 0$.

Step 2 Next, we look for extreme values of f on the boundary of D. We can think of this boundary as being made up of four line segments $L_1, L_2, L_3,$ and L_4, as shown in Figure 8.

On L_1: Here $y = 0$, so we have

$$f(x, 0) = 2x^2 - 4x + 3 \qquad 0 \le x \le 3$$

To find the extreme values of the continuous function $f(x, 0)$ of *one* variable on the closed bounded interval $[0, 3]$, we use the method of Section 3.1. Setting

$$f'(x, 0) = 4x - 4 = 0$$

gives $x = 1$ as the only critical number of $f(x, 0)$ in $(0, 3)$. Evaluating $f(x, 0)$ at $x = 1$, as well as at the endpoints of the interval $[0, 3]$, gives $f(0, 0) = 3$, $f(1, 0) = 1$, and $f(3, 0) = 9$. Thus, f has the absolute minimum value of 1 and the absolute maximum value of 9 on L_1.

On L_2: Here $x = 3$, so we have

$$f(3, y) = y^2 - 2y + 9 \qquad 0 \le y \le 2$$

Setting $f'(3, y) = 2y - 2 = 0$ yields $y = 1$ as the only critical number of $f(3, y)$ in $(0, 2)$. Evaluating $f(3, y)$ at the endpoints of $[0, 2]$ and at the critical number $y = 1$ gives $f(3, 0) = 9, f(3, 1) = 8,$ and $f(3, 2) = 9$. We see that f has the absolute minimum value of 8 and the absolute maximum value of 9 on L_2.

On L_3: Here $y = 2$, so we have

$$f(x, 2) = 2x^2 - 4x + 3 \qquad 0 \le x \le 3$$

Setting $f'(x, 2) = 4x - 4 = 0$ gives $x = 1$ as the only critical number of $f(x, 2)$ in $(0, 3)$. Since $f(0, 2) = 3, f(1, 2) = 1,$ and $f(3, 2) = 9$, we see that $f(x, 2)$ has the absolute minimum value of 1 and the absolute maximum value of 9 on L_3.

On L_4: Here $x = 0$, so we have

$$f(0, y) = y^2 - 2y + 3 \qquad 0 \le y \le 2$$

Setting $f'(0, y) = 2y - 2 = 0$ gives $y = 1$ as the only critical number of $f(0, y)$ in $(0, 2)$. Since $f(0, 0) = 3, f(0, 1) = 2,$ and $f(0, 2) = 3$, we see that $f(0, y)$ has the absolute minimum value of 2 and the absolute maximum value of 3 on L_4.

Step 3 Table 1 summarizes the results of our computations. Comparing the value of f obtained at the various points, we conclude that the absolute minimum value of f on D is 0 attained at the critical point $(1, 1)$ of f and that the absolute maximum value of f on D is 9 attained at the boundary points $(3, 0)$ and $(3, 2)$.

TABLE 1

	Critical point	Boundary point on L_1		Boundary point on L_2			Boundary point on L_3		Boundary point on L_4		
(x, y)	$(1, 1)$	$(1, 0)$	$(3, 0)$	$(3, 1)$	$(3, 0)$	$(3, 2)$	$(1, 2)$	$(3, 2)$	$(0, 1)$	$(0, 0)$	$(0, 2)$
Extreme value: $f(x, y)$	0	1	9	8	9	9	1	9	2	3	3

13.8 CONCEPT QUESTIONS

1. **a.** What does it mean when one says that f has a relative maximum (relative minimum) at a point (a, b) in the domain of f? What is $f(a, b)$ called in each case?
   b. What does it mean when one says that f has an absolute maximum (absolute minimum) at (a, b)? What is $f(a, b)$ called in each case?

2. **a.** What is a critical point of a function $f(x, y)$?
   b. What role does a critical point of a function play in the determination of the relative extrema of the function?
   c. State the Second Derivative Test for a function of two variables.

3. **a.** What can you say about the existence of a maximum value and a minimum value of a continuous function of two variables defined on a closed, bounded set on the plane?
   b. Describe a strategy for finding the absolute extreme values of a continuous function on a closed, bounded set in the plane.

13.8 EXERCISES

In Exercises 1–22, find and classify the relative extrema and saddle points of the function.

1. $f(x, y) = x^2 + y^2 - 2x + 4y$

2. $f(x, y) = 2x^2 + y^2 - 6x + 2y + 1$

3. $f(x, y) = -x^2 - 3y^2 + 4x - 6y + 8$

4. $f(x, y) = -2x^2 - 3y^2 + 6x - 4y - 6$

5. $f(x, y) = x^2 + 3xy + 3y^2$

6. $f(x, y) = x^2 + 3xy + 2y^2 + 1$

7. $f(x, y) = 2x^2 + y^2 - 2xy - 8x - 2y + 2$

8. $f(x, y) = x^2 + 3y^2 - 6xy - 2x + 4y$

9. $f(x, y) = x^2 + 2y^2 + x^2y + 3$

10. $f(x, y) = x^2 - y^2 + 2xy^2 + 1$

11. $f(x, y) = x^2 + 5y^2 + x^2y + 2y^3$

12. $f(x, y) = x^3 - 3xy + y^3 + 3$

13. $f(x, y) = x^2 - 6x - x\sqrt{y} + y$

14. $f(x, y) = xy(3 - x - y)$

15. $f(x, y) = \dfrac{x^2y^2 - 2y - 4x}{xy}$

16. $f(x, y) = -\dfrac{4y}{x^2 + y^2 + 1}$

17. $f(x, y) = e^{-x^2 - y^2}$

18. $f(x, y) = e^x \sin y, \quad x \geq 0, \quad 0 \leq y \leq 2\pi$

19. $f(x, y) = x \sin y, \quad x \geq 0, \quad 0 \leq y \leq 2\pi$

20. $f(x, y) = xe^x \sin y, \quad x \geq 0, \quad 0 \leq y \leq 2\pi$

21. $f(x, y) = e^{-x} \cos y, \quad x \geq 0, \quad 0 \leq y \leq 2\pi$

22. $f(x, y) = \sin x + \sin y, \quad 0 \leq x \leq 2\pi, \quad 0 \leq y \leq 2\pi$

In Exercises 23 and 24, (a) use the graph and the contour map of f to estimate the relative extrema and saddle point(s) of f, and (b) verify your guess analytically.

23. $f(x, y) = x^3 - 3xy^2 + y^4$

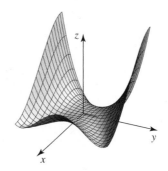
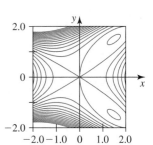

24. $f(x, y) = 3xy^2 - x^3$

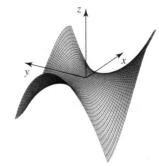
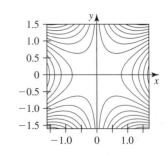

In Exercises 25–28, plot the graph and the contour map of f, and use them to estimate the relative maximum and minimum values and the saddle point(s) of f. Then find these values and saddle point(s) analytically.

25. $f(x, y) = \dfrac{1}{2}x^4 - 2x^3 + 4xy + y^2$

26. $f(x, y) = (x^2 + y^2)e^{-y}$

27. $f(x, y) = xy - \dfrac{2}{x} - \dfrac{4}{y} + 8$

28. $f(x, y) = 6xy^2 - 2x^3 - 3y^4$ ("Monkey Saddle")

In Exercises 29–32, use a graphing calculator or computer to find the critical points of f correct to three decimal places. Then use these results to find the relative extrema of f. Plot the graph of f.

29. $f(x, y) = 2x^4 - 8x^2 + y^2 + 4x - 2y - 5$

30. $f(x, y) = 2 - 2x^2 + 5xy + 2y - y^4$

31. $f(x, y) = -x^4 - y^4 + 2x^2y + x^2 + y - 2$

32. $f(x, y) = x^4 - 2x^2 + x + y^2 + e^{-y}$

In Exercises 33–40, find the absolute extrema of the function on the set D.

33. $f(x, y) = 2x + 3y - 6$;
$D = \{(x, y) \mid 0 \le x \le 2, -2 \le y \le 3\}$

34. $f(x, y) = x^2 + xy + y^2$;
$D = \{(x, y) \mid -2 \le x \le 2, -1 \le y \le 1\}$

35. $f(x, y) = 3x + 4y - 12$;   *D is the closed triangular region with vertices* $(0, 0)$, $(3, 0)$, and $(3, 4)$.

36. $f(x, y) = 3x^2 + 2xy + y^2$;   *D is the closed triangular region with vertices* $(-2, -1)$, $(1, -1)$, and $(1, 2)$.

37. $f(x, y) = xy - x^2$;   *D is the region bounded by the parabola* $y = x^2$ *and the line* $y = 4$.

38. $f(x, y) = 4x^2 + y^2$;   *D is the region bounded by the parabola* $y = 4 - x^2$ *and the x-axis.*

39. $f(x, y) = x^2 + 4y^2 + 3x - 1$; $D = \{(x, y) \mid x^2 + y^2 \le 4\}$

40. $f(x, y) = 4x^2 + y^2 + 2x - y$; $D = \{(x, y) \mid 4x^2 + y^2 \le 1\}$

41. Find the shortest distance from the origin to the plane $x + 2y + z = 4$.
Hint: The square of the distance from the origin to any point (x, y, z) on the plane is $d^2 = x^2 + y^2 + z^2 = x^2 + y^2 + (4 - x - 2y)^2$. Minimize $d^2 = f(x, y) = x^2 + y^2 + (4 - x - 2y)^2$.

42. Find the point on the plane $x + 2y - z = 5$ that is closest to the point $(2, 3, -1)$.
Hint: Study the hint of Exercise 41.

43. Find the points on the surface $z^2 = xy - x + 4y + 21$ that are closest to the origin. What is the shortest distance from the origin to the surface?

44. Find the points on the surface $xy^2z = 4$ that are closest to the origin. What is the shortest distance from the origin to the surface?

45. Find three positive real numbers whose sum is 500 and whose product is as large as possible.

46. Find the dimensions of an open rectangular box of maximum volume that can be constructed from 48 ft² of cardboard.

47. Find the dimensions of a closed rectangular box of maximum volume that can be constructed from 48 ft² of cardboard.

48. An open rectangular box having a volume of 108 in³. is to be constructed from cardboard. Find the dimensions of such a box if the amount of cardboard used in its construction is to be minimized.

49. Find the dimensions of the rectangular box of maximum volume with faces parallel to the coordinate planes that can be inscribed in the ellipsoid

$$\frac{x^2}{4} + \frac{y^2}{9} + \frac{z^2}{16} = 1$$

50. Solve the problem posed in Exercise 49 for the general case of an ellipsoid with equation

$$\frac{x^2}{a^2} + \frac{y^2}{b^2} + \frac{z^2}{c^2} = 1$$

where a, b, and c are positive real numbers.

51. Find the dimensions of the rectangular box of maximum volume lying in the first octant with three of its faces lying in the coordinate planes and one vertex lying in the plane $2x + 3y + z = 6$. What is the volume of such a box?

52. Solve the problem posed in Exercise 51 for the general case of a plane with equation

$$\frac{x}{a} + \frac{y}{b} + \frac{z}{c} = 1$$

where a, b, and c are positive real numbers.

53. An open rectangular box is to have a volume of 12 ft³. If the material for its base costs three times as much (per square foot) as the material for its sides, what are the dimensions of the box that can be constructed at a minimum cost?

54. A closed rectangular box is to have a volume of 16 ft³. If the material for its base costs twice as much (per square foot) as the material for its top and sides, find the dimensions of the box that can be constructed at a minimum cost.

55. Locating a Radio Station The following figure shows the locations of three neighboring communities. The operators of a newly proposed radio station have decided that the site $P(x, y)$ for the station should be chosen so that the sum of the squares of the distances from the site to each community is minimized. Find the location of the proposed radio station.

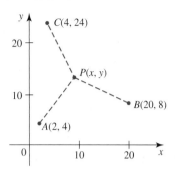

56. **Parcel Post Regulations** Postal regulations specify that a parcel sent by parcel post may have a combined length and girth of no more than 130 inches. Find the dimensions of a cylindrical package of greatest volume that can be sent through the mail. What is the volume of such a package?
Hint: The length plus the girth is $2\pi r + l$.

57. Suppose a relationship exists between two quantities x and y and that we have obtained the following data relating y to x:

x	x_1	x_2	$\cdots$	x_n
y	y_1	y_2	$\cdots$	y_n

The figure below shows the points (x_1, y_1), (x_2, y_2), ..., (x_n, y_n) plotted in the xy-plane. (This figure is called a **scatter diagram.**) If the data points are scattered about a straight line, as in this illustration, then it is reasonable to describe the relationship between x and y in terms of a linear equation $y = mx + b$.

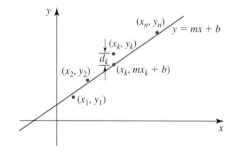

One criterion for determining the straight line that "best" fits the data calls for minimizing the sum of the squares of the deviations $\sum_{k=1}^{n} d_k^2$, where $d_k = y_k - (mx_k + b) = y_k - mx_k - b$. This sum is a function of m and b; that is,

$$g(m, b) = \sum_{k=1}^{n} (y_k - mx_k - b)^2$$

and it is minimized with respect to the variables m and b by solving the system comprising the equations $g_m(a, b) = 0$ and $g_b(a, b) = 0$ for m and b. Show that this leads to the system

$$\left(\sum_{k=1}^{n} x_k\right)m + nb = \sum_{k=1}^{n} y_k$$

$$\left(\sum_{k=1}^{n} x_k^2\right)m + \left(\sum_{k=1}^{n} x_k\right)b = \sum_{k=1}^{n} x_k y_k$$

This method of determining the equation $y = mx + b$ is called the **method of least squares,** and the line with equation $y = mx + b$ is called a **least squares** or **regression line.**

58. Use the method of least squares (Exercise 57) to find the straight line $y = mx + b$ that best fits the data points $(1, 3)$, $(2, 5)$, $(3, 5)$, $(4, 7)$, and $(5, 8)$. Plot the scatter diagram, and sketch the graph of the regression line on the same set of axes.

59. **Information Security Software Sales** Refer to Exercise 57. As online attacks persist, spending on information security software continues to rise. The following table gives the forecast for the worldwide sales (in billions of dollars) of information security software through 2007 ($x = 0$ corresponds to 2002).

Year, x	0	1	2	3	4	5
Spending, y	6.8	8.3	9.8	11.3	12.8	14.9

a. Find an equation of the least-squares line for these data.
b. Use the result of part (a) to forecast the spending on information security software in 2010, assuming that this trend continues.
Source: International Data Corp.

60. **Male Life Expectancy At 65** Refer to Exercise 57. The projections of male life expectancy at age 65 in the United States are summarized in the following table ($x = 0$ corresponds to 2000).

Year, x	0	10	20	30	40	50
Years beyond 65, y	15.9	16.8	17.6	18.5	19.3	20.3

a. Find an equation of the least-squares line for these data.

b. Use the result of (a) to estimate the life expectancy at 65 of a male in 2040. How does this result compare with the given data for that year?

c. Use the result of (a) to estimate the life expectancy at 65 of a male in 2030.

Source: U.S. Census Bureau.

61. Operations Management Consulting Spending Refer to Exercise 57. The following table gives the projected operations management consulting spending (in billions of dollars) from 2005 through 2010. Here, $x = 5$ corresponds to 2005.

Year, x	5	6	7	8	9	10
Spending, y	40	43.2	47.4	50.5	53.7	56.8

a. Find an equation of the least-squares line for these data.

b. Use the results of part (a) to estimate the average rate of change of operations management consulting spending from 2005 through 2010.

c. Use the results of part (a) to estimate the amount of spending on operations management consulting in 2011, assuming that the trend continues.

Source: Kennedy Information.

62. Let $f(x, y) = x^2 - y^2 + 2xy + 2$.

a. Show that f has no maximum or minimum values.

cas **b.** Find the maximum and minimum values of f in the region $D = \{(x, y) \mid x^2 + 4y^2 \leq 4\}$.

Hint: On the boundary of D, let $x = 2 \cos t$, $y = \sin t$ for $0 \leq t \leq 2\pi$.

63. Let $f(x, y) = -3x^2 + 6x - 4y^2 - 4y - 3$.

a. Show that f has no minimum value.

cas **b.** Find the maximum and minimum values of f in the region $D = \{(x, y) \mid x^2 + y^2 \leq 1\}$.

Hint: On the boundary of D, let $x = \cos t$, $y = \sin t$ for $0 \leq t \leq 2\pi$.

64. Let $f(x, y) = Ax^2 + 2Bxy + Cy^2$, where $B^2 - 4AC \neq 0$. Find conditions in terms of A, B, and C such that f has a relative minimum at $(0, 0)$; a relative maximum at $(0, 0)$; and a saddle point at $(0, 0)$.

In Exercises 65–68, determine whether the statement is true or false. If it is true, explain why it is true. If it is false, give an example to show why it is false.

65. If $f(x, y)$ has a relative maximum at (a, b), then $f_x(a, b) = 0$ and $f_y(a, b) = 0$.

66. Let $h(x, y) = f(x) + g(y)$, where f and g have second-order derivatives. If the graph of f is concave upward on $(-\infty, \infty)$ and the graph of g is concave downward on $(-\infty, \infty)$, then h cannot have a relative maximum or a relative minimum at any point.

67. If $\nabla f(a, b) = \mathbf{0}$, then f has a relative extremum at (a, b).

68. If $f(x, y)$ has continuous second-order partial derivatives and $f_{xx}(x, y) + f_{yy}(x, y) = 0$ and $f_{xy}(x, y) \neq 0$ for all (x, y), then f cannot have a relative extremum.

13.9 Lagrange Multipliers

Constrained Maxima and Minima

Many practical optimization problems involve maximizing or minimizing an objective function subject to one or more constraints, or side conditions. In Example 5 of Section 13.8 we discussed the problem of maximizing the (volume) function

$$V = f(x, y, z) = xyz$$

subject to the constraint

$$g(x, y, z) = x + 2y + 2z = 108$$

In this case the constraint expresses the condition that the combined length plus girth of a package is 108 in. (the maximum allowed by postal regulations).

As another example, consider a problem encountered in the construction of an AC transformer. Here, we are required to find the cross-shaped iron core of largest surface area that can be inserted into a coil of radius a (Figure 1). In terms of x and y we see that the surface area of the iron core is

$$S = 4xy + 4y(x - y) = 8xy - 4y^2$$

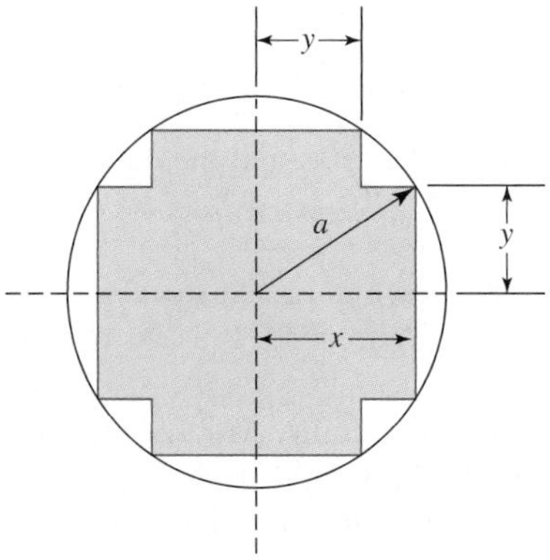

FIGURE 1
We want to find the core of largest surface area that can be inserted into a coil of radius a.

Next, observe that x and y must satisfy the equation $x^2 + y^2 = a^2$. Therefore, the problem is equivalent to one of maximizing the objective function

$$f(x, y) = 8xy - 4y^2$$

subject to the constraint

$$g(x, y) = x^2 + y^2 = a^2$$

We will complete the solution of this problem in Example 2.

Figure 2a shows the graph of a function f defined by the equation $z = f(x, y)$. Observe that f has an absolute minimum at $(0, 0)$ and an absolute minimum value of 0. However, if the independent variables x and y are subjected to a constraint of the form $g(x, y) = k$, then the points (x, y, z) that satisfy both $z = f(x, y)$ and $g(x, y) = k$ lie on the curve C, the intersection of the surface $z = f(x, y)$ and the cylinder $g(x, y) = k$ (Figure 2b). From the figure you can see that the absolute minimum of f subject to the constraint $g(x, y) = k$ occurs at the point (a, b). Furthermore, f has the **constrained** absolute minimum value $f(a, b)$ rather than the unconstrained absolute minimum value of 0 at $(0, 0)$.

FIGURE 2
The function f has an unconstrained minimum value of 0, but it has a constrained minimum value of $f(a, b)$ when subjected to the constraint $g(x, y) = k$.

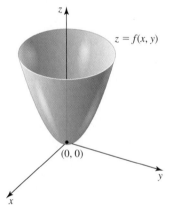

(a) f is not subject to any constraints.

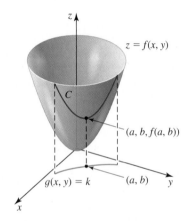

(b) f is subject to a constraint.

The problem that we discussed at the beginning of this section (maximizing the volume of a box subject to a given constraint) was first solved in Section 13.8. Recall the method of solution that we used:

First, we solved the constraint equation

$$g(x, y, z) = x + 2y + 2z = 108$$

for x in terms of y and z. We then substituted this expression for x into the equation

$$V = f(x, y, z) = xyz$$

thereby obtaining an expression for V involving the variables y and z and satisfying the constraint equation. Next, we found the maximum of V by treating V as an unconstrained function of y and z.

The major drawback of this method is that it relies on our ability to solve the constraint equation $g(x, y) = k$ for one variable explicitly in terms of the other (or $g(x, y, z) = k$ for one variable explicitly in terms of the other two variables in the case of a constraint involving three variables). This might not always be possible or convenient. Moreover, even when we are able to solve the constraint equation $g(x, y) = k$ for y explicitly in terms of x, the resulting function of one variable that is obtained by substituting this expression for y into the objective function $f(x, y)$ might turn out to be unnecessarily complicated.

■ The Method of Lagrange Multipliers

We will now consider a method, called the **method of Lagrange multipliers** (named after the French mathematician Joseph Lagrange, 1736–1813), which obviates the need to solve the constraint equation for one variable in terms of the other variables. To see how this method works, let's reexamine the problem of finding the absolute minimum of the objective function f subject to the constraint $g(x, y) = k$ that we considered earlier. Figure 3a shows the level curves of f drawn in the xyz-coordinate system. These level curves are reproduced in the xy-plane in Figure 3b.

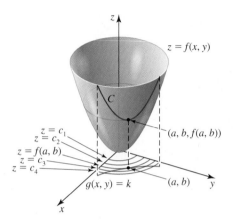

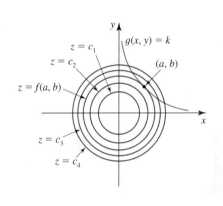

FIGURE 3   (a) The level curves of f in the xyz-plane   (b) The level curves of f in the xy-plane

Observe that the level curves of f with equations $f(x, y) = c$, where $c < f(a, b)$, have no points in common with the graph of the constraint equation $g(x, y) = k$ (for example, the level curves $f(x, y) = c_1$ and $f(x, y) = c_2$ shown in Figure 3). Thus, points lying on these curves are not candidates for the constrained minimum of f.

On the other hand, the level curves of f with equation $f(x, y) = c$, where $c \geq f(a, b)$, do intersect the graph of the constraint equation $g(x, y) = k$ (such as the level curves of $f(x, y) = c_3$ and $f(x, y) = c_4$). These points of intersection are candidates for the constrained minimum of f.

Finally, observe that the larger c is for $c \geq f(a, b)$, the larger the value $f(x, y)$ is for (x, y) lying on the level curve $g(x, y) = k$. This observation suggests that we can find the constrained minimum of f by choosing the smallest value of c so that the level curve $f(x, y) = c$ still intersects the curve $g(x, y) = k$. At such a point (a, b) the level curve of f just touches the graph of the constraint equation $g(x, y) = k$. That is, the two curves have a common tangent at (a, b) (see Figure 3b). Equivalently, their normal lines at this point coincide. Putting it yet another way, the gradient vectors $\nabla f(a, b)$ and $\nabla g(a, b)$ have the same direction, so $\nabla f(a, b) = \lambda \nabla g(a, b)$ for some scalar λ (lambda).

A similar result holds for the problem of maximizing or minimizing a function f of three variables defined by $w = f(x, y, z)$ and subject to the constraint $g(x, y, z) = k$. In this situation, f has a constrained maximum or constrained minimum at a point (a, b, c) where the level surface $f(x, y, z) = f(a, b, c)$ is tangent to the level surface $g(x, y, z) = k$. But this means that the normals of these surfaces, and therefore their gradient vectors, at the point (a, b, c) must be parallel to each other. Thus, there is a scalar λ such that $\nabla f(a, b, c) = \lambda \nabla g(a, b, c)$.

These geometric arguments suggest the following theorem.

> **THEOREM 1   Lagrange's Theorem**
>
> Let f and g have continuous first partial derivatives in some region D in the plane. If f has an extremum at a point (a, b) on the smooth constraint curve $g(x, y) = c$ lying in D and $\nabla g(a, b) \neq 0$, then there is a real number λ such that
>
> $$\nabla f(a, b) = \lambda \nabla g(a, b)$$

The number λ in Theorem 1 is called a **Lagrange multiplier.**

PROOF   Suppose that the smooth curve C described by $g(x, y) = c$ is represented by the vector function

$$\mathbf{r}(t) = x(t)\mathbf{i} + y(t)\mathbf{j}, \qquad \mathbf{r}'(t) \neq \mathbf{0}$$

where x' and y' are continuous on an open interval I (Figure 4). Then the values assumed by f on C are given by

$$h(t) = f(x(t), y(t))$$

for t in I. Suppose that f has an extreme value at (a, b). If t_0 is the point in I corresponding to the point (a, b), then h has an extreme value at t_0. Therefore, $h'(t_0) = 0$. Using the Chain Rule, we have

$$\begin{aligned} h'(t_0) &= f_x(x(t_0), y(t_0))x'(t_0) + f_y(x(t_0), y(t_0))y'(t_0) \\ &= f_x(a, b)x'(t_0) + f_y(a, b)y'(t_0) \\ &= \nabla f(a, b) \cdot \mathbf{r}'(t_0) = 0 \end{aligned}$$

This shows that $\nabla f(a, b)$ is orthogonal to $\mathbf{r}'(t_0)$. But as we demonstrated in Section 13.7, $\nabla g(a, b)$ is orthogonal to $\mathbf{r}'(t_0)$. Therefore, the gradient vectors $\nabla f(a, b)$ and $\nabla g(a, b)$ are parallel, so there is a number λ such that $\nabla f(a, b) = \lambda \nabla g(a, b)$.

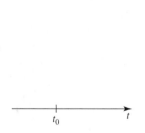

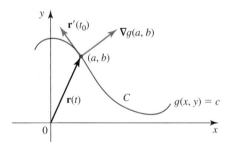

FIGURE 4

(a) The parameter interval I

(b) The smooth curve C is represented by the vector function $\mathbf{r}(t)$.

The proof of Lagrange's Theorem for functions of three variables is similar to that for functions of two variables. In the case involving three variables, level surfaces rather than level curves are involved. Lagrange's Theorem leads to the following procedure for finding the constrained extremum values of functions. We state it for the case of functions of three variables.

The Method of Lagrange Multipliers

Suppose f and g have continuous first partial derivatives. To find the maximum and minimum values of f subject to the constraint $g(x, y, z) = k$ (assuming that these extreme values exist and that $\nabla g \neq \mathbf{0}$ on $g(x, y, z) = k$):

1. Solve the equations

$$\nabla f(x, y, z) = \lambda \nabla g(x, y, z) \qquad \text{and} \qquad g(x, y, z) = k$$

for x, y, z, and λ.

2. Evaluate f at each solution point found in Step 1. The largest value yields the constrained maximum of f, and the smallest value yields the constrained minimum of f.

Note   Since

$$\nabla f(x, y, z) = f_x(x, y, z)\mathbf{i} + f_y(x, y, z)\mathbf{j} + f_z(x, y, z)\mathbf{k}$$

and

$$\nabla g(x, y, z) = g_x(x, y, z)\mathbf{i} + g_y(x, y, z)\mathbf{j} + g_z(x, y, z)\mathbf{k}$$

we see, by equating like components, that the vector equation

$$\nabla f(x, y, z) = \lambda \nabla g(x, y, z)$$

is equivalent to the three scalar equations

$$f_x(x, y, z) = \lambda g_x(x, y, z), \quad f_y(x, y, z) = \lambda g_y(x, y, z), \quad \text{and} \quad f_z(x, y, z) = \lambda g_z(x, y, z)$$

These scalar equations together with the constraint equation $g(x, y, z) = k$ give a system of four equations to be solved for the four unknowns x, y, z, and λ.   ■

EXAMPLE 1   Find the maximum and minimum values of the function $f(x, y) = x^2 - 2y$ subject to $x^2 + y^2 = 9$.

Solution   The constraint equation is $g(x, y) = x^2 + y^2 = 9$. Since

$$\nabla f(x, y) = 2x\mathbf{i} - 2\mathbf{j} \qquad \text{and} \qquad \nabla g(x, y) = 2x\mathbf{i} + 2y\mathbf{j}$$

the equation $\nabla f(x, y) = \lambda \nabla g(x, y)$ becomes

$$2x\mathbf{i} - 2\mathbf{j} = \lambda(2x\mathbf{i} + 2y\mathbf{j}) = 2\lambda x\mathbf{i} + 2\lambda y\mathbf{j}$$

Equating like components and rewriting the constraint equation lead to the following system of three equations in the three variables x, y, and λ:

$$2x = 2\lambda x \tag{1a}$$

$$-2 = 2\lambda y \tag{1b}$$

$$x^2 + y^2 = 9 \tag{1c}$$

From Equation (1a) we have

$$2x(1 - \lambda) = 0$$

so $x = 0$, or $\lambda = 1$. If $x = 0$, then Equation (1c) gives $y = \pm 3$. If $\lambda = 1$, then Equation (1b) gives $y = -1$, which upon substitution into Equation (1c) yields $x = \pm 2\sqrt{2}$. Therefore, f has possible extreme values at the points $(0, -3)$, $(0, 3)$, $(-2\sqrt{2}, -1)$, and $(2\sqrt{2}, -1)$. Evaluating f at each of these points gives

$$f(0, -3) = 6, \quad f(0, 3) = -6, \quad f(-2\sqrt{2}, -1) = 10, \quad \text{and} \quad f(2\sqrt{2}, -1) = 10$$

We conclude that the maximum value of f on the circle $x^2 + y^2 = 9$ is 10, attained at the points $(-2\sqrt{2}, -1)$ and $(2\sqrt{2}, -1)$, and that the minimum value of f on the circle is -6, attained at the point $(0, 3)$.

Figure 5 shows the graph of the constraint equation $x^2 + y^2 = 9$ and some level curves of the objective function f. Observe that the extreme values of f are attained at the points where the level curves of f are tangent to the graph of the constraint equation.

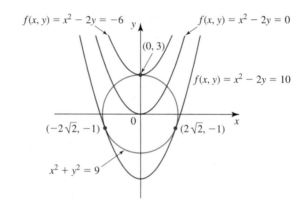

$f(x, y) = x^2 - 2y = -6$ $f(x, y) = x^2 - 2y = 0$

$(0, 3)$

$f(x, y) = x^2 - 2y = 10$

$(-2\sqrt{2}, -1)$ $(2\sqrt{2}, -1)$

$x^2 + y^2 = 9$

FIGURE 5
The extreme values of f occur at the points where the level curves of f are tangent to the graph of the constraint equation (the circle).

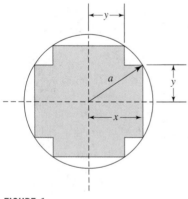

FIGURE 6
A cross-shaped iron core of largest surface area is to be inserted into the coil.

EXAMPLE 2 Complete the solution to the problem posed at the beginning of this section: Find the cross-shaped iron core of largest surface area that can be inserted into a coil of radius a (Figure 6).

Solution Recall that the problem reduces to one of finding the largest value of the objective function $f(x, y) = 8xy - 4y^2$ subject to the constraint $g(x, y) = x^2 + y^2 = a^2$. Since

$$\nabla f(x, y) = 8y\mathbf{i} + (8x - 8y)\mathbf{j} \quad \text{and} \quad \nabla g(x, y) = 2x\mathbf{i} + 2y\mathbf{j}$$

the equation $\nabla f(x, y) = \lambda \nabla g(x, y)$ becomes

$$8y\mathbf{i} + (8x - 8y)\mathbf{j} = \lambda(2x\mathbf{i} + 2y\mathbf{j}) = 2\lambda x\mathbf{i} + 2\lambda y\mathbf{j}$$

Equating like components and rewriting the constraint equation, we get the following system of three equations in the three variables x, y, and λ:

$$8y = 2\lambda x \tag{2a}$$

$$8x - 8y = 2\lambda y \tag{2b}$$

$$x^2 + y^2 = a^2 \tag{2c}$$

Solving Equation (2a) for y, we obtain $y = \frac{1}{4}\lambda x$. Substituting this expression for y into Equation (2b) gives

$$8x - 2\lambda x = \frac{1}{2}\lambda^2 x$$

or

$$x(\lambda^2 + 4\lambda - 16) = 0$$

Observe that $x \neq 0$; otherwise, Equation (2a) implies that $y = 0$, so Equation (2c) becomes $0 = a^2$, which is impossible. So we have $\lambda^2 + 4\lambda - 16 = 0$. Using the quadratic formula, we obtain

$$\lambda = \frac{-4 \pm \sqrt{16 + 64}}{2} = -2 \pm 2\sqrt{5}$$

Observe that λ must be positive; otherwise, Equation (2a) implies that x or y must be negative. So we choose $\lambda = -2 + 2\sqrt{5} \approx 2.4721$. Next, substituting $y = \frac{1}{4}\lambda x$ into Equation (2c) gives

$$x^2 + \frac{1}{16}\lambda^2 x^2 = a^2$$

$$x^2\left(1 + \frac{\lambda^2}{16}\right) = a^2$$

$$x^2\left(\frac{\lambda^2 + 16}{16}\right) = a^2$$

or

$$x = \frac{4a}{\sqrt{\lambda^2 + 16}} \approx \frac{4}{\sqrt{(2.4721)^2 + 16}}\, a \qquad \text{Recall that } \lambda \approx 2.4721.$$

$$\approx 0.8507a$$

Finally,

$$y = \frac{1}{4}\lambda x \approx \frac{1}{4}(2.4721)(0.8507a) \approx 0.5258a$$

Therefore, the core will have the largest surface area if $x \approx 0.8507a$ and $y \approx 0.5258a$, where a is the radius of the coil.

Figure 7 shows the graph of the constraint equation $x^2 + y^2 = a^2$ (the circle of radius a centered at the origin) and several level curves of the objective function f. Once again, observe that the maximum value of f, $f(0.8507a, 0.5258a) \approx 2.4725a^2$, occurs at the point $(0.8507a, 0.5258a)$, where the level curve of f is tangent to the graph of the constraint equation. ∎

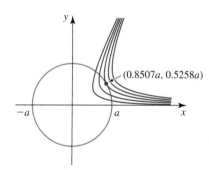

FIGURE 7
The maximum value of f occurs at the point where the level curve of f is tangent to the level curve of the constraint equation.

EXAMPLE 3 Find the dimensions of a rectangular package having the greatest possible volume and satisfying the postal regulation that specifies that the combined length and girth of the package may not exceed 108 inches. (See Example 5 in Section 13.8.)

Solution Recall that to solve this problem, we need to find the largest value of the volume function $f(x, y, z) = xyz$ subject to the constraint $g(x, y, z) = x + 2y + 2z = 108$. To solve this problem using the method of Lagrange multipliers, observe that

$$\nabla f(x, y, z) = yz\mathbf{i} + xz\mathbf{j} + xy\mathbf{k} \qquad \text{and} \qquad \nabla g(x, y, z) = \mathbf{i} + 2\mathbf{j} + 2\mathbf{k}$$

so the equation $\nabla f(x, y, z) = \lambda \nabla g(x, y, z)$ becomes

$$yz\mathbf{i} + xz\mathbf{j} + xy\mathbf{k} = \lambda(\mathbf{i} + 2\mathbf{j} + 2\mathbf{k})$$

Equating components and rewriting the constraint equation give the following system of four equations in the four variables x, y, z, and λ:

$$yz = \lambda \tag{3a}$$

$$xz = 2\lambda \tag{3b}$$

$$xy = 2\lambda \tag{3c}$$

$$x + 2y + 2z = 108 \tag{3d}$$

Substituting Equation (3a) into Equation (3b) yields

$$xz = 2yz \quad\text{or}\quad z(x - 2y) = 0$$

Since $z \neq 0$, we have $x = 2y$. Next, substituting Equation (3a) into Equation (3c) gives

$$xy = 2yz \quad\text{or}\quad y(x - 2z) = 0$$

Since $y \neq 0$, we have $x = 2z$. Equating the two expressions for x just obtained gives

$$2y = 2z \quad\text{or}\quad y = z$$

Finally, substituting the expressions for x and y into Equation (3d) gives

$$2z + 2z + 2z = 108 \quad\text{or}\quad z = 18$$

So $y = 18$ and $x = 2(18) = 36$. Therefore, the dimensions of the package are 18 in. $\times$ 18 in. $\times$ 36 in., as was obtained before.   ■

Interpreting Our Results

Geometrically, this problem is one of finding the point on the plane $x + 2y + 2z = 108$ at which $f(x, y, z) = xyz$ has the largest value. The point $(36, 18, 18)$ is precisely the point at which the level surface $xyz = f(36, 18, 18) = 11{,}664$ is tangent to the plane $x + 2y + 2z = 108$.

EXAMPLE 4   Find the dimensions of the open rectangular box of maximum volume that can be constructed from a rectangular piece of cardboard having an area of 48 ft^2. What is the volume of the box?

Solution   Let the length, width, and height of the box (in feet) be x, y, and z, as shown in Figure 8. Then the volume of the box is $V = xyz$. The area of the bottom of the box plus the area of the four sides is

$$xy + 2xz + 2yz$$

square feet, and this is equal to the area of the cardboard; that is,

$$xy + 2xz + 2yz = 48$$

Thus, the problem is one of maximizing the objective function

$$f(x, y, z) = xyz$$

subject to the constraint

$$g(x, y, z) = xy + 2xz + 2yz = 48$$

Since

$$\nabla f(x, y, z) = yz\mathbf{i} + xz\mathbf{j} + xy\mathbf{k}$$

FIGURE 8

An open rectangular box of maximum volume is to be constructed from a piece of cardboard. What are the dimensions of the box?

and

$$\nabla g(x, y, z) = (y + 2z)\mathbf{i} + (x + 2z)\mathbf{j} + (2x + 2y)\mathbf{k}$$

the equation $\nabla f(x, y, z) = \lambda \nabla g(x, y, z)$ becomes

$$yz\mathbf{i} + xz\mathbf{j} + xy\mathbf{k} = \lambda[(y + 2z)\mathbf{i} + (x + 2z)\mathbf{j} + (2x + 2y)\mathbf{k}]$$

Equating like components and rewriting the constraint equation give the following system of four equations in the unknowns x, y, z, and λ:

$$yz = \lambda(y + 2z) \tag{4a}$$

$$xz = \lambda(x + 2z) \tag{4b}$$

$$xy = \lambda(2x + 2y) \tag{4c}$$

$$xy + 2xz + 2yz = 48 \tag{4d}$$

Multiplying Equations (4a), (4b), and (4c) by x, y, and z, respectively, gives

$$xyz = \lambda(xy + 2xz) \tag{5a}$$

$$xyz = \lambda(xy + 2yz) \tag{5b}$$

$$xyz = \lambda(2xz + 2yz) \tag{5c}$$

From Equations (5a) and (5b), we obtain

$$\lambda(xy + 2xz) = \lambda(xy + 2yz) \tag{6}$$

Observe that $\lambda \neq 0$; otherwise, Equations (4a), (4b), and (4c) would imply that $yz = xz = xy = 0$, thus contradicting Equation (4d). Dividing both sides of (6) by λ and simplifying give

$$2xz = 2yz \qquad \text{or} \qquad 2z(x - y) = 0$$

Now $z \neq 0$; otherwise, Equation (4a) would imply that $\lambda = 0$, which, as was observed earlier, is impossible. Therefore, $x = y$.

Next, from Equations (5b) and (5c), we have

$$\lambda(xy + 2yz) = \lambda(2xz + 2yz) \tag{7}$$

Dividing both sides of Equation (7) by λ and simplifying, we get

$$xy = 2xz \qquad \text{or} \qquad x(y - 2z) = 0$$

Since $x \neq 0$, we have $y = 2z$. Finally, substituting $x = y = 2z$ into Equation (4d) gives

$$4z^2 + 4z^2 + 4z^2 = 48$$

or $z = 2$ (we reject the negative root, since z must be positive). Therefore, $x = y = 4$, and the dimensions of the box are 4 ft $\times$ 4 ft $\times$ 2 ft. Its volume is 32 ft^3. ∎

Interpreting Our Results

Geometrically, this problem is one of finding the point on the surface $xy + 2xz + 2yz = 48$ at which $f(x, y, z) = xyz$ has the largest value. The point $(4, 4, 2)$ is precisely the point at which the level surface $xyz = f(4, 4, 2) = 32$ is tangent to the surface $xy + 2xz + 2yz = 48$.

The next example shows how the method of Lagrange multipliers can be used to help find the absolute extreme values of a function on a closed, bounded set.

EXAMPLE 5 Find the absolute extreme values of $f(x, y) = 2x^2 + y^2 - 2y + 1$ subject to the constraint $x^2 + y^2 \le 4$.

Solution   The inequality $x^2 + y^2 \le 4$ defines the disk D, which is a closed, bounded set with boundary given by the circle $x^2 + y^2 = 4$. So, following the procedure given in Section 13.8, we first find the critical number(s) of f inside D. Setting

$$f_x(x, y) = 4x = 0$$

$$f_y(x, y) = 2y - 2 = 2(y - 1) = 0$$

simultaneously gives $(0, 1)$ as the only critical point of f in D.

Next, we find the critical numbers of f on the boundary of D using the method of Lagrange multipliers. Writing $g(x, y) = x^2 + y^2 = 4$, we have

$$\nabla f(x, y) = 4x\mathbf{i} + 2(y - 1)\mathbf{j} \quad \text{and} \quad \nabla g(x, y) = 2x\mathbf{i} + 2y\mathbf{j}$$

The equation $\nabla f(x, y) = \lambda g(x, y)$ and the constraint equation give the system

$$4x = 2\lambda x \tag{8a}$$

$$2(y - 1) = 2\lambda y \tag{8b}$$

$$x^2 + y^2 = 4 \tag{8c}$$

Equation (8a) gives

$$2x(\lambda - 2) = 0$$

that is, $x = 0$ or $\lambda = 2$. If $x = 0$, then Equation (8c) gives $y = \pm 2$. Next, if $\lambda = 2$, then Equation (8b) gives

$$2(y - 1) = 4y \quad \text{or} \quad y = -1$$

in which case $x = \pm\sqrt{3}$. So f has the critical points $(0, -2)$, $(0, 2)$, $(-\sqrt{3}, -1)$ and $(\sqrt{3}, -1)$ on the boundary of D.

Finally, we construct the following table.

(x, y)	$f(x, y) = 2x^2 + y^2 - 2y + 1$
$(0, 1)$	0
$(-\sqrt{3}, -1)$	10
$(\sqrt{3}, -1)$	10
$(0, -2)$	9
$(0, 2)$	1

From the table we see that f has an absolute minimum value of 0 attained at $(0, 1)$ and an absolute maximum value of 10 attained at $(-\sqrt{3}, -1)$ and $(\sqrt{3}, -1)$.   ∎

■ Optimizing a Function Subject to Two Constraints

Some applications involve maximizing or minimizing an objective function f subject to two or more constraints. Consider, for example, the problem of finding the extreme values of $f(x, y, z)$ subject to the two constraints

$$g(x, y, z) = k \quad \text{and} \quad h(x, y, z) = l$$

It can be shown that if f has an extremum at (a, b, c) subject to these constraints, then there are real numbers (Lagrange multipliers) λ and μ such that

$$\nabla f(a, b, c) = \lambda \nabla g(a, b, c) + \mu \nabla h(a, b, c) \tag{9}$$

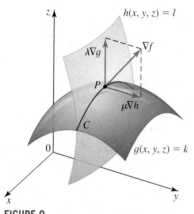

FIGURE 9
If f has an extreme value at $P(a, b, c)$, then $\nabla f(a, b, c) = \lambda \nabla g(a, b, c) + \mu \nabla h(a, b, c)$.

Geometrically, we are looking for the extreme values of $f(x, y, z)$ on the curve of intersection of the level surfaces $g(x, y, z) = k$ and $h(x, y, z) = l$. Condition (9) is a statement that at an extremum point (a, b, c), the gradient of f must lie in the plane determined by the gradient of g and the gradient of h. (See Figure 9.) The vector equation (9) is equivalent to three scalar equations. When combined with the two constraint equations, this leads to a system of five equations that can be solved for the five unknowns x, y, z, λ, and μ.

EXAMPLE 6 Find the maximum and minimum values of the function $f(x, y, z) = 3x + 2y + 4z$ subject to the constraints $x - y + 2z = 1$ and $x^2 + y^2 = 4$.

Solution Write the constraint equations in the form

$$g(x, y, z) = x - y + 2z = 1 \quad \text{and} \quad h(x, y, z) = x^2 + y^2 = 4$$

Then the equation $\nabla f(x, y, z) = \lambda \nabla g(x, y, z) + \mu \nabla h(x, y, z)$ becomes

$$3\mathbf{i} + 2\mathbf{j} + 4\mathbf{k} = \lambda(\mathbf{i} - \mathbf{j} + 2\mathbf{k}) + \mu(2x\mathbf{i} + 2y\mathbf{j})$$

$$= (\lambda + 2\mu x)\mathbf{i} + (-\lambda + 2\mu y)\mathbf{j} + 2\lambda\mathbf{k}$$

Equating like components and rewriting the constraint equations lead to the following system of five equations in the five variables, x, y, z, λ, and μ:

$$3 = \lambda + 2\mu x \tag{10a}$$

$$2 = -\lambda + 2\mu y \tag{10b}$$

$$4 = 2\lambda \tag{10c}$$

$$x - y + 2z = 1 \tag{10d}$$

$$x^2 + y^2 = 4 \tag{10e}$$

From Equation (10c) we have $\lambda = 2$. Next, substituting this value of λ into Equations (10a) and (10b) gives

$$3 = 2 + 2\mu x \quad \text{or} \quad 1 = 2\mu x \tag{11a}$$

and

$$2 = -2 + 2\mu y \quad \text{or} \quad 4 = 2\mu y \tag{11b}$$

Solving Equations (11a) and (11b) for x and y gives $x = 1/(2\mu)$ and $y = 2/\mu$. Substituting these values of x and y into Equation (10e) yields

$$\left(\frac{1}{2\mu}\right)^2 + \left(\frac{2}{\mu}\right)^2 = 4$$

$$1 + 16 = 16\mu^2 \quad \text{or} \quad \mu^2 = \frac{17}{16}$$

Therefore, $\mu = \pm\sqrt{17}/4$, so $x = \pm 2/\sqrt{17}$ and $y = \pm 8/\sqrt{17}$. Using Equation (10d), we have

$$z = \frac{1}{2}(1 - x + y) = \frac{1}{2}\left(1 \mp \frac{2}{\sqrt{17}} \pm \frac{8}{\sqrt{17}}\right)$$

$$= \frac{1}{2}\left(1 \pm \frac{6}{\sqrt{17}}\right)$$

The value of f at the point $\left(\frac{2}{\sqrt{17}}, \frac{8}{\sqrt{17}}, \frac{1}{2} + \frac{3}{\sqrt{17}}\right)$ is

$$3\left(\frac{2}{\sqrt{17}}\right) + 2\left(\frac{8}{\sqrt{17}}\right) + 4\left(\frac{1}{2} + \frac{3}{\sqrt{17}}\right) = 2 + \frac{34}{\sqrt{17}} = 2(1 + \sqrt{17})$$

and the value of f at the point $\left(-\frac{2}{\sqrt{17}}, -\frac{8}{\sqrt{17}}, \frac{1}{2} - \frac{3}{\sqrt{17}}\right)$ is

$$3\left(-\frac{2}{\sqrt{17}}\right) + 2\left(-\frac{8}{\sqrt{17}}\right) + 4\left(\frac{1}{2} - \frac{3}{\sqrt{17}}\right) = 2 - \frac{34}{\sqrt{17}} = 2(1 - \sqrt{17})$$

Therefore, the maximum value of f is $2(1 + \sqrt{17})$, and the minimum value of f is $2(1 - \sqrt{17})$. ■

13.9 CONCEPT QUESTIONS

1. What is a constrained maximum (minimum) value problem? Illustrate with examples.
2. Describe the method of Lagrange multipliers for finding the extrema of $f(x, y)$ subject to the constraint $g(x, y) = c$. State the method for the case in which f and g are functions of three variables.
3. The figure at the right shows the contour map of a function f and the curve of the equation $g(x, y) = 4$. Use the figure to obtain estimates of the maximum and minimum values of f subject to the constraint $g(x, y) = 4$.

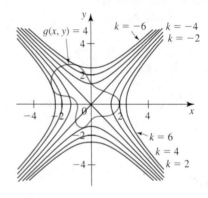

13.9 EXERCISES

In Exercises 1–4, use the method of Lagrange multipliers to find the extrema of the function f subject to the given constraint. Sketch the graph of the constraint equation and several level curves of f. Include the level curves that touch the graph of the constraint equation at the points where the extrema occur.

1. $f(x, y) = 3x + 4y; \quad x^2 + y^2 = 1$
2. $f(x, y) = x^2 + y^2; \quad 2x + 4y = 5$
3. $f(x, y) = x^2 + y^2; \quad xy = 1$
4. $f(x, y) = xy; \quad x^2 + y^2 = 4$

In Exercises 5–16, use the method of Lagrange multipliers to find the extrema of the function f subject to the given constraint.

5. $f(x, y) = xy; \quad 2x + 3y = 6$
6. $f(x, y) = x^2 - y^2; \quad x^2 + y^2 = 1$
7. $f(x, y) = xy; \quad x^2 + 4y^2 = 1$
8. $f(x, y) = 8x + 9y; \quad 4x^2 + 9y^2 = 36$
9. $f(x, y) = x^2 + xy + y^2; \quad x^2 + y^2 = 8$
10. $f(x, y) = x^2 + y^2; \quad x^4 + y^4 = 1$
11. $f(x, y, z) = x + 2y + z; \quad x^2 + 4y^2 - z = 0$

12. $f(x, y, z) = x + y + z; \quad x^2 + y^2 + z^2 = 1$
13. $f(x, y, z) = x + 2y - 2z; \quad x^2 + 2y^2 + 4z^2 = 1$
14. $f(x, y, z) = x^2 + y^2 + z^2; \quad y - x = 1$
15. $f(x, y, z) = xyz; \quad x^2 + 2y^2 + \frac{1}{2}z^2 = 6$
16. $f(x, y, z) = xy + xz; \quad x^2 + y^2 + z^2 = 8$

In Exercises 17–20, use the method of Lagrange multipliers to find the extrema of the function subject to the given constraints.

17. $f(x, y, z) = 2x + y; \quad x + y + z = 1, \quad y^2 + z^2 = 9$
18. $f(x, y, z) = x + y + z; \quad x^2 + y^2 = 1, \quad x + z = 2$
19. $f(x, y, z) = yz + xz; \quad xz = 1, \quad y^2 + z^2 = 1$
20. $f(x, y, z) = x^2 + y^2 + z^2; \quad 2x + y + z = 2, \quad x - 2y + 3z = -4$

In Exercises 21–22, use the method of Lagrange multipliers to find the extrema of the function subject to the inequality constraint.

21. $f(x, y) = 3x^2 + 2y^2 - 2x - 1; \quad x^2 + y^2 \leq 9$
22. $f(x, y) = x^2 y; \quad 4x^2 + y^2 \leq 4$

23. Find the point on the plane $x + 2y + z = 4$ that is closest to the origin.

24. Find the maximum and minimum distances from the origin to the curve $5x^2 + 6xy + 5y^2 - 10 = 0$.

25. Find the point on the plane $x + 2y - z = 5$ that is closest to the point $(2, 3, -1)$.

26. Find the points on the surface $z^2 = xy - x + 4y + 21$ that are closest to the origin. What is the shortest distance from the origin to the surface?

27. Find the points on the surface $xy^2z = 4$ that are closest to the origin. What is the shortest distance from the origin to the surface?

28. Find three positive real numbers whose sum is 500 and whose product is as large as possible.

29. Find the dimensions of a closed rectangular box of maximum volume that can be constructed from 48 ft^2 of cardboard.

30. Find the dimensions of an open rectangular box of maximum volume that can be constructed from 12 ft^2 of cardboard.

31. An open rectangular box having a volume of 108 in^3. is to be constructed from cardboard. Find the dimensions of such a box if the amount of cardboard used in its construction is to be minimized.

32. Find the dimensions of the rectangular box of maximum volume with faces parallel to the coordinate planes that can be inscribed in the ellipsoid

$$\frac{x^2}{4} + \frac{y^2}{9} + \frac{z^2}{16} = 1$$

33. Solve the problem posed in Exercise 32 for the general case of an ellipsoid with equation

$$\frac{x^2}{a^2} + \frac{y^2}{b^2} + \frac{z^2}{c^2} = 1$$

where a, b, and c are positive real numbers.

34. Find the dimensions of the rectangular box of maximum volume lying in the first octant with three of its faces lying in the coordinate planes and one vertex lying in the plane $2x + 3y + z = 6$. What is the volume of the box?

35. Solve the problem posed in Exercise 34 for the general case of a plane with equation

$$\frac{x}{a} + \frac{y}{b} + \frac{z}{c} = 1$$

where a, b, and c are positive real numbers.

36. An open rectangular box is to have a volume of 12 ft^3. If the material for its base costs three times as much (per square foot) as the material for its sides, what are the dimensions of the box that can be constructed at the minimum cost?

37. A rectangular box is to have a volume of 16 ft^3. If the material for its base costs twice as much (per square foot) as the material for its top and sides, find the dimensions of the box that can be constructed at the minimum cost.

38. Maximizing Profit The total daily profit (in dollars) realized by Weston Publishing in publishing and selling its dictionaries is given by the profit function

$$P(x, y) = -0.005x^2 - 0.003y^2 - 0.002xy + 14x + 12y - 200$$

where x stands for the number of deluxe editions and y denotes the number of standard editions sold daily. Weston's management decides that publication of these dictionaries should be restricted to a total of exactly 400 copies per day. How many deluxe copies and how many standard copies should be published each day to maximize Weston's daily profit?

39. Cobb-Douglas Production Function Suppose x units of labor and y units of capital are required to produce

$$f(x, y) = 100x^{3/4}y^{1/4}$$

units of a certain product. If each unit of labor costs \$200 and each unit of capital costs \$300 and a total of \$60,000 is available for production, determine how many units of labor and how many units of capital should be used to maximize production.

40. a. Find the distance between the point $P(x_1, y_1)$ and the line $ax + by + c = 0$ using the method of Lagrange multipliers.
   b. Use the result of part (a) to find the distance between the point $(2, -1)$ and the line $2x + 3y - 6 = 0$.

41. Let $f(x, y) = x - y$ and $g(x, y) = x + x^5 - y$.
   a. Use the method of Lagrange multipliers to find the point(s) where f may have a relative maximum or relative minimum subject to the constraint $g(x, y) = 1$.
   b. Plot the graph of g and the level curves of $f(x, y) = k$ for $k = -2, -1, 0, 1, 2$, using the viewing window $[-4, 4] \times [-4, 4]$. Then use this to explain why the point(s) found in part (a) does not give rise to a relative maximum or a relative minimum of f.
   c. Verify the observation made in part (b) analytically.

42. Let $f(x, y) = x^2 - y^2$, and let $g(x, y) = x + y$.
   a. Show that f has no maximum or minimum values when subjected to the constraint $g(x, y) = 1$.
   b. What happens when you try to use the method of Lagrange multipliers to find the extrema of f subject to $g(x, y) = 1$? Does this contradict Theorem 1?

43. Find the point on the line of intersection of the planes $x + 2y - 3z = 9$ and $2x - 3y + z = 4$ that is closest to the origin.

44. Find the shortest distance from the origin to the curve with equation $y = (x - 1)^{3/2}$. Explain why the method of Lagrange multipliers fails to give the solution.

45. a. Find the maximum distance from the origin to the Folium of Descartes, $x^3 + y^3 - 3axy = 0$, where $a > 0$, $x \geq 0$ and $y \geq 0$, using symmetry.
   b. Verify the result of part (a), using the method of Lagrange multipliers.

In Exercises 46 and 47, use the fact that a vector in n-space has the form $v = \langle v_1, v_2, \ldots, v_n \rangle$ and the gradient of a function of n variables, $f(x_1, x_2, \ldots, x_n)$, is defined by $\nabla f = \langle f_{x_1}, f_{x_2}, \ldots, f_{x_n} \rangle$. Also, assume that Theorem 1 holds for the n-dimensional case.

46. a. Find the maximum value of

$$f(x_1, x_2, \ldots, x_n, y_1, y_2, \ldots, y_n)$$
$$= \sum_{i=1}^{n} x_i y_i = x_1 y_1 + x_2 y_2 + \cdots + x_n y_n$$

subject to the constraints

$$\sum_{i=1}^{n} x_i^2 = x_1^2 + x_2^2 + \cdots + x_n^2 = 1$$

and

$$\sum_{i=1}^{n} y_i^2 = y_1^2 + y_2^2 + \cdots + y_n^2 = 1$$

b. Use the result of part (a) to show that if $a_1, a_2, \ldots, a_n$, $b_1, b_2, \ldots, b_n$ are any numbers, then

$$\sum_{i=1}^{n} a_i b_i \le \sqrt{\sum_{i=1}^{n} a_i^2} \sqrt{\sum_{i=1}^{n} b_i^2}$$

Hint: Put $x_i = \dfrac{a_i}{\sqrt{\sum_{i=1}^{n} a_i^2}}$ and $y_i = \dfrac{b_i}{\sqrt{\sum_{i=1}^{n} b_i^2}}$.

Note: This inequality is called the Cauchy-Schwarz Inequality. (Compare this with Exercise 9 in the Challenge Problems for Chapter 4.)

47. a. Let p and q be positive numbers satisfying $(1/p) + (1/q) = 1$. Find the minimum value of

$$f(x, y) = \frac{x^p}{p} + \frac{y^q}{q} \qquad x > 0, \quad y > 0$$

subject to the constraint $xy = c$, where c is a constant.

b. Use the result of part (a) to show that if x and y are positive numbers, then

$$\frac{x^p}{p} + \frac{y^q}{q} \ge xy$$

where $p > 0$ and $q > 0$ and $(1/p) + (1/q) = 1$.

48. a. Let $x_1, x_2, \ldots, x_n$ be positive numbers. Find the maximum value of

$$f(x_1, x_2, \ldots, x_n) = \sqrt[n]{x_1 x_2 \cdots x_n}$$

subject to the constraint $x_1 + x_2 + \cdots + x_n = c$, where c is a constant.

b. Use the result of part (a) to show that if $x_1, x_2, \ldots, x_n$ are positive numbers, then

$$\sqrt[n]{x_1 x_2 \cdots x_n} \le \frac{x_1 + x_2 + \cdots + x_n}{n}$$

This shows that the geometric mean of n positive numbers cannot exceed the arithmetic mean of the numbers.

49. Snell's Law of Refraction According to Fermat's Principle in optics, the path POQ taken by a ray of light (see the figure below) in traveling across the plane separating two optical media is such that the time taken is minimal. Using this principle, derive Snell's Law of Refraction:

$$\frac{v_1}{\sin \theta_1} = \frac{v_2}{\sin \theta_2}$$

where θ_1 is the angle of incidence, θ_2 is the angle of refraction, and v_1 and v_2 are the speeds of light in the two media.

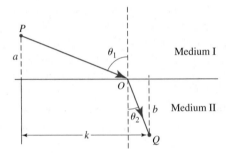

Hint: Show that the time taken by the ray of light in traveling from P to Q is

$$t = \frac{a}{v_1 \cos \theta_1} + \frac{b}{v_2 \cos \theta_2}$$

Then minimize $t = f(\cos \theta_1, \cos \theta_2)$ subject to $a \tan \theta_1 + b \tan \theta_2 = k$, where k is a constant.

In Exercises 50–52, determine whether the statement is true or false. If it is true, explain why it is true. If it is false, give an example to show why it is false.

50. Suppose f and g have continuous first partial derivatives in some region D in the plane. If f has an extremum at a point (a, b) subject to the constraint $g(x, y) = c$, then there exists a constant λ such that (a, b) is a critical point of $F = f + \lambda g$; that is, $F_x(a, b) = 0$, $F_y(a, b) = 0$, and $F_\lambda(a, b) = 0$.

51. If (a, b) gives rise to a (constrained) extremum of f subject to the constraint $g(x, y) = 0$, then (a, b) also gives rise to an unconstrained extremum of f.

52. If (a, b) gives rise to a (constrained) extremum of f subject to the constraint $g(x, y) = 0$, then $f_x(a, b) = 0$ and $f_y(a, b) = 0$ simultaneously.

CHAPTER 13 REVIEW

CONCEPT REVIEW

In Exercises 1–17, fill in the blanks.

1. a. A function f of two variables, x and y, is a _____ that assigns to each ordered pair _____ in the domain of f, exactly one real number $f(x, y)$.

   b. The number $z = f(x, y)$ is called a _____ variable, and x and y are _____ variables. The totality of the numbers z is called the _____ of the function f.

   c. The graph of f is the set $S =$ _____.

2. a. The curves in the xy-plane with equation $f(x, y) = k$, where k is a constant in the range of f, are called the _____ _____ of f.

   b. A level surface of a function f of three variables is the graph of the equation _____, where k is a constant in the range of _____.

3. $\lim_{(x, y) \to (a, b)} f(x, y) = L$ means there exists a number _____ such that $f(x, y)$ can be made as close to _____ as we please by restricting (x, y) to be sufficiently close to _____.

4. If $f(x, y)$ approaches L_1 as (x, y) approaches (a, b) along one path, and $f(x, y)$ approaches L_2 as (x, y) approaches (a, b) along another path with $L_1 \neq L_2$, then $\lim_{(x, y) \to (a, b)} f(x, y)$ _____ _____ exist.

5. a. $f(x, y)$ is continuous at (a, b) if $\lim_{(x, y) \to (a, b)} f(x, y) =$ _____.

   b. $f(x, y)$ is continuous on a region R if f is continuous at every point (x, y) in _____.

6. a. A polynomial function is continuous _____; a rational function is continuous at all points in its _____.

   b. If f is continuous at (a, b) and g is continuous at $f(a, b)$, then the composite function $h = g \circ f$ is continuous at _____.

7. a. The partial derivative of $f(x, y)$ with respect to x is _____ if the limit exists. The partial derivative $(\partial f / \partial x)(a, b)$ gives the slope of the tangent line to the curve obtained by the intersection of the plane _____ and the graph of $z = f(x, y)$ at _____; it also measures the rate of change of $f(x, y)$ in the _____-direction with y held _____ at _____.

   b. To compute $\partial f / \partial x$ where f is a function of x and y, treat _____ as a constant and differentiate with respect to _____ in the usual manner.

8. If $f(x, y)$ and its partial derivatives f_x, f_y, f_{xy}, and f_{yx} are continuous on an open region R, then $f_{xy}(x, y) =$ _____ for all (x, y) in R.

9. a. The total differential dz of $z = f(x, y)$ is $dz =$ _____.

   b. If $\Delta z = f(x + \Delta x, y + \Delta y) - f(x, y)$, then $\Delta z \approx$ _____.

   c. $\Delta z = f_x(x, y) \, \Delta x + f_y(x, y) \, \Delta y + \varepsilon_1 \, \Delta x + \varepsilon_2 \, \Delta y$, where ε_1 and ε_2 are functions of _____ and _____ such that $\lim_{(\Delta x, \Delta y) \to (0, 0)} \varepsilon_1 =$ _____ and $\lim_{(\Delta x, \Delta y) \to (0, 0)} \varepsilon_2 =$ _____.

   d. The function $z = f(x, y)$ is differentiable at (a, b) if Δz can be expressed in the form $\Delta z =$ _____, where _____ and _____ as $(\Delta x, \Delta y) \to$ _____.

10. a. If f is a function of x and y, and f_x and f_y are continuous on an open region R, then f is _____ in R.

   b. If f is differentiable at (a, b), then f is _____ at (a, b).

11. a. If $w = f(x, y)$, $x = g(t)$, and $y = h(t)$, then under suitable conditions the Chain Rule gives $dw/dt =$ _____.

   b. If $w = f(x, y)$, $x = g(u, v)$, and $y = h(u, v)$, then $\partial w / \partial u =$ _____.

   c. If $F(x, y) = 0$, where F is differentiable, then $dy/dx =$ _____, provided that _____.

   d. If $F(x, y, z) = 0$, where F is differentiable, and F defines z implicitly as a function of x and y, then $\partial z / \partial x =$ _____ and $\partial z / \partial y =$ _____, provided that _____.

12. a. If f is a function of x and y and $\mathbf{u} = u_1 \mathbf{i} + u_2 \mathbf{j}$ is a unit vector, then the directional derivative of f in the direction of $\mathbf{u}$ is $D_{\mathbf{u}} f(x, y) =$ _____ if the limit exists.

   b. The directional derivative $D_{\mathbf{u}} f(a, b)$ measures the rate of change of f at _____ in the direction of _____.

   c. If f is differentiable, then $D_{\mathbf{u}} f(x, y) =$ _____.

   d. The gradient of $f(x, y)$ is $\nabla f(x, y) =$ _____.

   e. In terms of the gradient, $D_{\mathbf{u}} f(x, y) =$ _____.

13. a. The maximum value of $D_{\mathbf{u}} f(x, y)$ is _____, and this occurs when $\mathbf{u}$ has the same direction as _____.

   b. The minimum value of $D_{\mathbf{u}} f(x, y)$ is _____, and this occurs when $\mathbf{u}$ has the direction of _____.

14. a. ∇f is _____ to the level curve $f(x, y) = c$ at P.

   b. ∇F is _____ to the level surface $F(x, y, z) = 0$ at P.

   c. The tangent plane to the surface $F(x, y, z) = 0$ at the point $P(a, b, c)$ is _____; the normal line passing through $P(a, b, c)$ has symmetric equations _____.

15. a. If $f(x, y) \leq f(a, b)$ for all points in an open disk containing (a, b), then f has a _____ _____ at (a, b).

   b. If $f(x, y) \geq f(a, b)$ for all points in the domain of f, then f has an _____ _____ at (a, b).

c. If f is defined on an open region R containing the point (a, b), then (a, b) is a critical point of f if (1) f_x and/or f_y _____ _____ _____ at (a, b) or (2) both $f_x(a, b)$ and $f_y(a, b)$ equal _____.

d. If f has a relative extremum at (a, b), then (a, b) must be a _____ _____ of f.

e. To determine whether a critical point of f does give rise to a relative extremum, we use the _____ _____ _____.

16. a. If f is continuous on a closed, bounded set D in the plane, then f has an absolute maximum value _____ at some point _____ in D, and f has an absolute minimum value _____ at some point _____ in D.

b. To find the absolute extreme values of f on a closed, bounded set D, (1) find the values of f at the _____ _____ _____ _____ in D, (2) find the extreme values of f on the _____ of D. Then the largest and smallest values found in (1) and (2) give the _____ value of f and the _____ value of f on D.

17. a. If $f(x, y)$ has an extremum at a point (a, b) lying on the curve with equation $g(x, y) = c$, then the extremum is called a _____ extremum.

b. If f has an extremum at (a, b) subject to the constraint $g(x, y) = c$, then $\nabla f(a, b) =$ _____, where λ is a real number called a Lagrange _____.

c. To find the maximum and minimum values of f subject to the constraint $g(x, y) = c$, we solve the system of equations $\nabla f(x, y) =$ _____ and $g(x, y) = c$ for x, y, and λ. We then evaluate _____ at each of the _____ _____ found in the last step. The largest value yields the constrained _____ of f, and the smallest value yields the constrained _____ of f.

REVIEW EXERCISES

In Exercises 1–4, find and sketch the domain of the function.

1. $f(x, y) = \dfrac{\sqrt{9 - x^2 - y^2}}{x^2 + y^2}$

2. $f(x, y) = \dfrac{\ln(x - 2y - 4)}{y + x}$

3. $f(x, y) = \sin^{-1} x + \tan^{-1} y$

4. $f(x, y) = \ln(xy - 1)$

In Exercises 5 and 6, sketch the graph of the function.

5. $f(x, y) = 4 - x^2 - y^2$

6. $f(x, y) = \sqrt{1 - x^2 - y^2}$

In Exercises 7–10, sketch several level curves for the function.

7. $f(x, y) = x^2 + 2y$

8. $f(x, y) = y^2 - x^2$

9. $f(x, y) = e^{x^2 + y^2}$

10. $f(x, y) = \ln xy$

In Exercises 11–14, find the limit or show that it does not exist.

11. $\displaystyle\lim_{(x, y) \to (0, 0)} \dfrac{\sqrt{xy + 4}}{2y + 3}$

12. $\displaystyle\lim_{(x, y) \to (0, 0)} \dfrac{x^2 y^2}{x^4 + 3y^4}$

13. $\displaystyle\lim_{(x, y) \to (1, 0^+)} \dfrac{x^2 y + x^3}{\sqrt{x} + \sqrt{y}}$

14. $\displaystyle\lim_{(x, y, z) \to (0, 0, 0)} \dfrac{x^2 - 2y^2 + 3z^2}{x^2 + y^2 + z^2}$

In Exercises 15 and 16, determine where the function is continuous.

15. $f(x, y) = \dfrac{\ln(x - y)}{(x^2 + y^2)^{3/2}}$

16. $f(x, y, z) = e^{x/y} \cos z + \sqrt{x - y}$

In Exercises 17–22, find the first partial derivatives of the function.

17. $f(x, y) = 2x^2 y - \sqrt{x}$

18. $f(x, y) = \dfrac{xy^2}{x^2 + y^2}$

19. $f(r, s) = re^{-(r^2 + s^2)}$

20. $f(u, v) = e^{2u} \cos(u^2 + v^2)$

21. $f(x, y, z) = \dfrac{x^2 - y^2}{z^2 - x^2}$

22. $f(r, s, t) = r \cos st + s \sin\left(\dfrac{s}{t}\right)$

In Exercises 23–26, find the second partial derivatives of the function.

23. $f(x, y) = x^4 - 2x^2 y^3 + y^2 - 2$

24. $f(x, y) = e^{-xy} \cos(2x + 3y)$

25. $f(x, y, z) = x^2 y z^3$

26. $f(u, v, w) = ue^{-v} \sin w$

27. If $u = \sqrt{x^2 + y^2 + z^2}$, show that
$$\frac{\partial^2 u}{\partial x^2} + \frac{\partial^2 u}{\partial y^2} + \frac{\partial^2 u}{\partial z^2} = \frac{2}{u}$$

28. Show that the function $u = e^{-t} \cos(x/c)$ satisfies the one-dimensional heat equation $u_t = c^2 u_{xx}$.

In Exercises 29 and 30, show that the function satisfies Laplace's equation $u_{xx} + u_{yy} + u_{zz} = 0$.

29. $u = 2z^2 - x^2 - y^2$

30. $u = z \tan^{-1} \dfrac{y}{x}$

31. Find dz if $z = x^2 \tan^{-1} y^3$.

32. Use differentials to approximate the change in $f(x, y) = x^2 - 3xy + y^2$ if (x, y) changes from $(2, -1)$ to $(1.9, -0.8)$.

33. Use differentials to approximate $(2.01)^2\sqrt{(1.98)^2 + (3.02)^3}$.

34. Estimating Changes in Profit The total daily profit function (in dollars) of Weston Publishing Company realized in publishing and selling its English language dictionaries is given by

$$P(x, y) = -0.0005x^2 - 0.003y^2 - 0.002xy + 14x + 12y - 200$$

where x denotes the number of deluxe copies and y denotes the number of standard copies published and sold daily. Currently, the number of deluxe and standard copies of the dictionaries published and sold daily are 1000 and 1700, respectively. Determine the approximate daily change in the total daily profit if the number of deluxe copies is increased to 1050 and the number of standard copies is decreased to 1650 per day.

35. Does a function f such that $\nabla f = -y\mathbf{i} + x\mathbf{j}$ exist? Explain.

36. According to Ohm's Law, $R = V/I$, where R is the resistance in ohms, V is the electromotive force in volts, and I is the current in amperes. If the errors in the measurements made in a certain experiment in V and I are 2% and 1%, respectively, use differentials to estimate the maximum percentage error in the calculated value of R.

37. Let $z = x^2y - \sqrt{y}$, where $x = e^{2t}$ and $y = \cos t$. Use the Chain Rule to find dz/dt.

38. Let $w = e^x \cos y + y \sin e^x$, where $x = u^2 - v^2$ and $y = \sqrt{uv}$. Use the Chain Rule to find $\partial w/\partial u$ and $\partial w/\partial v$.

39. Use partial differentiation to find dy/dx if $x^3 - 3x^2y + 2xy^2 + 2y^3 = 9$.

40. Find $\partial z/\partial x$ and $\partial z/\partial y$ if $x^3z^2 + yz^3 = \cos xz$.

In Exercises 41–44, find the gradient of the function f at the indicated point.

41. $f(x, y) = \sqrt{x^2 + y^2}$; $P(1, 2)$

42. $f(x, y) = e^{-x} \tan y$; $P\left(0, \frac{\pi}{4}\right)$

43. $f(x, y, z) = xy^2 - yz^2 + zx^2$; $P(2, 1, -3)$

44. $f(x, y, z) = x \ln y + y \ln z$; $P(2, 1, 1)$

In Exercises 45–48, find the directional derivative of the function f at the point P in the indicated direction.

45. $f(x, y) = x^3y^2 - xy^3$; $P(2, -1)$, in the direction of $\mathbf{v} = 3\mathbf{i} - 4\mathbf{j}$.

46. $f(x, y) = e^{-x^2} \cos y$; $P\left(0, \frac{\pi}{2}\right)$, in the direction from $P(1, 3)$ to $Q(3, 1)$.

47. $f(x, y, z) = x\sqrt{y^2 + z^2}$; $P(2, 3, 4)$ in the direction of $\mathbf{v} = \mathbf{i} - 2\mathbf{j} + 2\mathbf{k}$.

48. $f(x, y, z) = x^2 \ln y + xy^2e^z$; $P(2, 1, 0)$ in the direction of $\mathbf{v} = \langle 3, -1, 2 \rangle$.

49. Find the direction in which $f(x, y) = \sqrt{x} + xy^2$ increases most rapidly at the point $(4, 1)$. What is the maximum rate of increase?

50. Find the direction in which $f(x, y, z) = xe^{yz}$ decreases most rapidly at the point $(4, 3, 0)$. What is the greatest rate of decrease?

In Exercises 51–54, find equations for the tangent plane and the normal line to the surface with the equation at the given point.

51. $2x^2 + 4y^2 + 9z^2 = 27$; $P(1, 2, 1)$

52. $x^2 + 2y^2 - 3z^2 = 19$; $P(2, 3, 1)$

53. $z = x^2 + 3xy^2$; $P(3, 1, 18)$

54. $z = xe^{-y}$; $P(1, 0, 1)$

55. Let $f(x, y) = x^2 + y^2$, and let $g(x, y) = y^2/x^2$.

cas   **a.** Plot several level curves of f and g using the same viewing window.

   b. Show analytically that each level curve of f intersects all level curves of g at right angles.

56. Show that if $\nabla f(x_0, y_0) \neq 0$, then an equation of the tangent line to the level curve $f(x, y) = f(x_0, y_0)$ at the point (x_0, y_0) is

$$f_x(x_0, y_0)(x - x_0) + f_y(x_0, y_0)(y - y_0) = 0$$

In Exercises 57–60, find the relative extrema and saddle points of the function.

57. $f(x, y) = x^2 + xy + y^2 - 5x + 8y + 5$

58. $f(x, y) = 8x^3 - 6xy + y^3$

59. $f(x, y) = x^3 - 3xy + y^2$

60. $f(x, y) = \dfrac{2}{x} + \dfrac{4}{y} - xy$

In Exercises 61 and 62, find the absolute extrema of the function on the set D.

61. $f(x, y) = x^2 + xy^2 - y^3$; $D = \{(x, y) \mid -1 \le x \le 1,\ 0 \le y \le 2\}$

62. $f(x, y) = (x^2 + 3y^2)e^{-x}$; $D = \{(x, y) \mid x^2 + y^2 \le 9\}$

In Exercises 63–66, use the method of Lagrange multipliers to find the extrema of the function f subject to the constraints.

63. $f(x, y) = xy^2$; $x^2 + y^2 = 4$

64. $f(x, y) = \dfrac{1}{x} + \dfrac{1}{y}$; $\dfrac{1}{x^2} + \dfrac{1}{y^2} = 9$

65. $f(x, y, z) = xy + yz + xz$; $x + 2y + 3z = 1$

66. $f(x, y, z) = 3x^2 + 2y^2 + z^2$; $x + y + z = 1$, $2x - y + z = 2$

67. Let $f(x, y) = Ax^2 + Bxy + Cy^2 + Dx + Ey + F$. Show that if f has a relative maximum or a relative minimum at a point (x_0, y_0), then x_0 and y_0 must satisfy the system of equations

$$2Ax + By + D = 0$$

$$Bx + 2Cy + E = 0$$

simultaneously.

68. Let $f(x, y) = x^2 + 2Bxy + y^2$, where $B > 0$. For what value of B does f have a relative minimum at $(0, 0)$? A saddle point at $(0, 0)$? Are there any values of B such that f has a relative maximum at $(0, 0)$?

69. Find the point on the paraboloid

$$z = \frac{x^2}{4} + \frac{y^2}{25}$$

that is closest to the point $(3, 0, 0)$.

70. Isothermal Curves Consider the upper half-disk $H = \{(x, y) \mid x^2 + y^2 \leq 1, y \geq 0\}$ (see the figure). If the temperature at points on the upper boundary is kept at $100°C$ and the temperature at points on the lower boundary is kept at $50°C$, then the steady-state temperature $T(x, y)$ at any point inside the disk is given by

$$T(x, y) = 100 - \frac{100}{\pi} \tan^{-1} \frac{1 - x^2 - y^2}{2y}$$

Show that the isothermal curves $T(x, y) = k$ are arcs of circles that pass through the points $x = \pm 1$. Sketch the isothermal curve corresponding to a temperature of $75°C$.

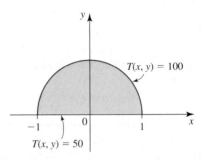

In Exercises 71 and 72, determine whether the statement is true or false. If it is true, explain why it is true. If it is false, give an example to show why it is false.

71. The directional derivative of $f(x, y)$ at the point (a, b) in the positive x-direction is $f_x(a, b)$.

72. If we know the gradient of $f(x, y, z)$ at the point $P(a, b, c)$, then we can compute the directional derivative of f in any direction at P.

CHALLENGE PROBLEMS

1. Find and sketch the domain of

$$f(x, y) = \sqrt{36 - 4x^2 - 9y^2}$$
$$+ \ln(x^2 - 2x + y^2)$$
$$+ \frac{1}{\sqrt{4x^2 + 16x + 4y^2 + 15}}$$

2. Describe the domain of

$$H(x, y, z) = \sqrt{x - a} + \sqrt{b - x} + \sqrt{y - c}$$
$$+ \sqrt{d - y} + \sqrt{z - e} + \sqrt{f - z}$$

where $a \leq b$, $c \leq d$, and $e \leq f$.

3. Suppose f has continuous second partial derivatives in x and y. Then the *second-order directional derivative* of f in the direction of the unit vector $\mathbf{u} = u_1\mathbf{i} + u_2\mathbf{j}$ is defined to be

$$D_{\mathbf{u}}^2 f(x, y) = D_{\mathbf{u}}(D_{\mathbf{u}}f)$$

a. Find an expression in terms of the partial derivatives of f for $D_{\mathbf{u}}^2 f$.
b. Find $D_{\mathbf{u}}^2 f(1, 0)$ if $f(x, y) = xy^2 + e^{xy}$ and $\mathbf{u}$ has the same direction as $\mathbf{v} = 2\mathbf{i} - 3\mathbf{j}$.

4. Consider the quadratic polynomial function

$$f(x, y) = Ax^2 + 2Bxy + Cy^2 + 2Dx + 2Ey + F$$

Find conditions on the coefficients of f such that f has (a) a relative maximum and (b) a relative minimum. What are the coordinates of the point in terms of the coefficients of f?

5. Let a, b, and c denote the sides of a triangle of area A, and let α, β, and γ denote the angles opposite them. If $A = f(a, b, c)$, show that

$$\frac{\partial f}{\partial a} = \frac{1}{2}R \cos \alpha$$

where R is the radius of the circumscribing circle.

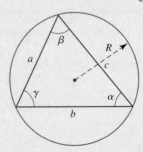

6. Linda has 24 feet of fencing with which to enclose a triangular flower garden. What should the lengths of the sides of the garden be if the area is to be as large as possible?

Hint: Heron's formula states that the area of a triangle with sides a, b, and c is given by $A = \sqrt{s(s-a)(s-b)(s-c)}$, where $s = \frac{1}{2}(a + b + c)$ is the semiperimeter.

7. Find the directional derivative at $\left(1, 2, \frac{\pi}{4}\right)$ of the function $f(x, y, z) = x^2 + y \cos z$ in the direction of increasing t along the curve in three-dimensional space described by the position vector $\mathbf{r}(t) = \langle t, t^2, t^3 \rangle$ at $\mathbf{r}(1)$.

8. Let

$$f(x, y) = \begin{cases} \dfrac{xy(x^2 - y^2)}{x^2 + y^2} & (x, y) \neq (0, 0) \\ 0 & (x, y) = (0, 0) \end{cases}$$

Use the definition of partial derivatives to show that $f_{xy}(0, 0) = -1$ and $f_{yx}(0, 0) = 1$.

9. Show that Laplace's equation $\dfrac{\partial^2 u}{\partial x^2} + \dfrac{\partial^2 u}{\partial y^2} + \dfrac{\partial^2 u}{\partial z^2} = 0$ in cylindrical coordinates takes the form

$$\frac{\partial^2 u}{\partial r^2} + \frac{1}{r} \cdot \frac{\partial u}{\partial r} + \frac{1}{r^2} \cdot \frac{\partial^2 u}{\partial \theta^2} + \frac{\partial^2 u}{\partial z^2} = 0$$

10. Consider the problem of determining the maximum and the minimum distances from the point (x_0, y_0, z_0) to the ellipsoid

$$\frac{x^2}{a^2} + \frac{y^2}{b^2} + \frac{z^2}{c^2} = 1$$

a. Show that the solutions are

$$x = \frac{a^2 x_0}{a^2 - \lambda}, \qquad y = \frac{b^2 y_0}{b^2 - \lambda}, \qquad z = \frac{c^2 z_0}{c^2 - \lambda}$$

where λ satisfies

$$\frac{a^2 x_0^2}{(a^2 - \lambda)^2} + \frac{b^2 y_0^2}{(b^2 - \lambda)^2} + \frac{c^2 z_0^2}{(c^2 - \lambda)^2} = 1$$

b. Use the result of part (a) to solve the problem with $a = 2$, $b = 3$, $c = 1$, and $(x_0, y_0, z_0) = (3, 2, 4)$.

A lawn sprinkler sprays water in a circular pattern. If we know the amount of the water per hour that the sprinkler delivers to any point within the circular region, can we find the total amount of water accumulated per hour in that part of the lawn? We will consider such problems in this chapter.

Photodisc/Getty Images

14 Multiple Integrals

IN THIS CHAPTER we extend the notion of the integral of a function of one variable to the integral of a function of two or three variables. Applications of double and triple integrals include finding the area of a surface, finding the center of mass of a planar object, and finding the centroid of a solid. We end the chapter by looking at how certain multiple integrals can be more easily evaluated by a change of variables.

V This symbol indicates that one of the following video types is available for enhanced student learning at **www.academic.cengage.com/login:**
- Chapter lecture videos
- Solutions to selected exercises

■ An Introductory Example

Suppose a piece of straight, thin wire of length $(b - a)$ is placed on the x-axis of a coordinate system, as shown in Figure 1. Further suppose that the wire has linear mass density given by $f(x)$ at x for $a \leq x \leq b$, where f is a nonnegative continuous function on $[a, b]$. Let $P = \{x_0, x_1, x_2, \ldots, x_n\}$, where $a = x_0$ and $b = x_n$, be a regular partition of $[a, b]$. Then the continuity of f tells us that, $f(x) \approx f(c_k)$ for each x in the kth subinterval $[x_{k-1}, x_k]$, where c_k is an evaluation point in $[x_{k-1}, x_k]$, provided n is large enough. Therefore, the mass of the piece of wire lying on $[x_{k-1}, x_k]$ is

$$\Delta m_k \approx f(c_k) \, \Delta x \qquad \Delta x = \frac{b - a}{n}$$

This leads to the definition of the mass of the wire as

$$m = \lim_{n \to \infty} \sum_{k=1}^{n} \Delta m_k = \lim_{n \to \infty} \sum_{k=1}^{n} f(c_k) \, \Delta x = \int_a^b f(x) \, dx$$

Thus, the mass of the curve has the same numerical value as that of the area under the graph of the (nonnegative) density function f shown in Figure 1.

Now let's consider a thin rectangular plate occupying the region

$$R = \{(x, y) \mid a \leq x \leq b, c \leq y \leq d\}$$

(See Figure 2.) If the plate is homogeneous (having a constant mass density of k g/cm^2), then its mass is given by

$$m = k(b - a)(d - c) \qquad \text{mass density} \cdot \text{area}$$

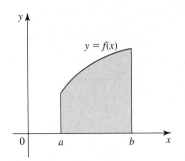

FIGURE 1
The mass of a straight wire of length $(b - a)$ is given by $\int_a^b f(x) \, dx$, where $f(x)$ is the density of the wire at any point x for $a \leq x \leq b$.

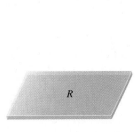

FIGURE 2

(a) A thin rectangular plate

(b) The plate placed in the xy-plane

Observe that m has the same numerical value as that of the volume of the rectangular box bounded above by the graph of the constant function $f(x, y) = k$ and below by R. (See Figure 3a.) Next, instead of being constant, suppose that the mass density of the

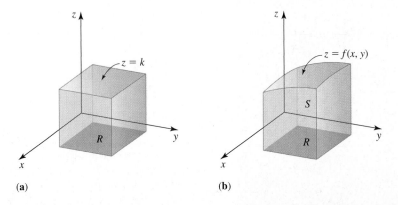

FIGURE 3
The mass of the plate R is numerically equal to that of the volume of the solid region lying directly above R and below the surface $z = f(x, y)$.

(a)

(b)

plate is given by the mass density function f. Then it seems reasonable to conjecture that the mass of the plate is given by the "volume" of the solid region S lying directly above R and below the graph of $z = f(x, y)$. (See Figure 3b.) We will show in Section 14.4 that this is indeed the case.

■ Volume of a Solid Between a Surface and a Rectangle

We will now show that the volume of a solid S can be defined as a limit of Riemann sums. Suppose that f is a nonnegative continuous function* of two variables that is defined on a rectangle

$$R = [a, b] \times [c, d] = \{(x, y) \mid a \le x \le b, c \le y \le d\}$$

and suppose that $f(x, y) \ge 0$ on R. Let

$$a = x_0 < x_1 < \cdots < x_{i-1} < x_i < \cdots < x_m = b$$

be a regular partition of the interval $[a, b]$ into m subintervals of length $\Delta x = (b - a)/m$, and let

$$c = y_0 < y_1 < \cdots < y_{j-1} < y_j < \cdots < y_n = d$$

be a regular partition of the interval $[c, d]$ into n subintervals of length $\Delta y = (d - c)/n$. The grid comprising segments of the vertical lines $x = x_i$ for $0 \le i \le m$ and the horizontal lines $y = y_j$ for $0 \le j \le n$ partition R into $N = mn$ subrectangles $R_{11}, R_{12}, \ldots, R_{ij}, \ldots, R_{mn}$, where $R_{ij} = [x_{i-1}, x_i] \times [y_{j-1}, y_j] = \{(x, y) \mid x_{i-1} \le x \le x_i, y_{j-1} \le y \le y_j\}$ as shown in Figure 4. The area of each subrectangle is $\Delta A = \Delta x \, \Delta y$. This partition is called a **regular partition** of R.

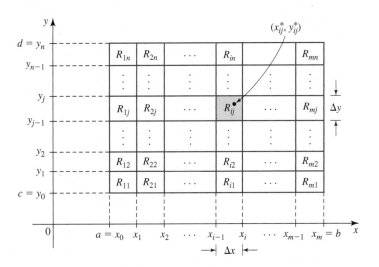

FIGURE 4
A partition $P = \{R_{ij}\}$ of R

The partition $P = \{R_{11}, R_{12}, \ldots, R_{ij}, \ldots, R_{mn}\}$ divides the solid S between the graph of $z = f(x, y)$ and R into $N = mn$ solids; the solid S_{ij} is bounded below by R_{ij} and bounded above by the part of the surface $z = f(x, y)$ that lies directly above R_{ij}. (See Figure 5.)

---

*As in the case of the integral of a function of one variable, these assumptions will simplify the discussion.

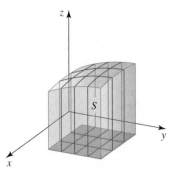

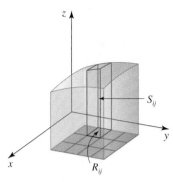

FIGURE 5

(**a**) The solid S is the union of $N = mn$ solids (shown here with $m = 3, n = 4$)

(**b**) A typical solid S_{ij}

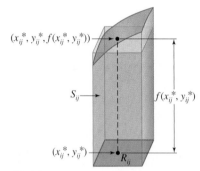

FIGURE 6
The volume of S_{ij} is approximated by the volume of the parallelepiped with base R_{ij} and height $f(x_{ij}^*, y_{ij}^*)$.

Let (x_{ij}^*, y_{ij}^*) be an evaluation point in R_{ij}. Then the parallelepiped with base R_{ij}, height $f(x_{ij}^*, y_{ij}^*)$, and volume

$$f(x_{ij}^*, y_{ij}^*) \, \Delta A$$

gives an approximation of the volume of S_{ij}. (See Figure 6.)

Therefore, the volume V of S is approximated by the volume of the sum of $N = mn$ parallelepipeds; that is,

$$V \approx \sum_{i=1}^{m} \sum_{j=1}^{n} f(x_{ij}^*, y_{ij}^*) \, \Delta A \tag{1}$$

If we take m and n to be larger and larger, then, intuitively, we can expect the approximation (1) to improve. This suggests the following definition.

DEFINITION Volume Under the Graph of $z = f(x, y)$

Let f be defined on the rectangle R and suppose $f(x, y) \geq 0$ on R. Then the volume V of the solid S that lies directly above R and below the surface $z = f(x, y)$ is

$$V = \lim_{m, n \to \infty} \sum_{i=1}^{m} \sum_{j=1}^{n} f(x_{ij}^*, y_{ij}^*) \, \Delta A \tag{2}$$

if the limit exists.

Because of the assumption that f be continuous, it can be shown that the limit in Equation (2) always exists regardless of how the evaluation points (x_{ij}^*, y_{ij}^*) in R_{ij}, for $1 \leq i \leq m$ and $1 \leq j \leq n$, are chosen.

EXAMPLE 1 Approximate the volume V of the solid lying under the graph of the elliptic paraboloid $z = 8 - 2x^2 - y^2$ and above the rectangle $R = \{(x, y) \,|\, 0 \leq x \leq 1, 0 \leq y \leq 2\}$. Use the partition P of R that is obtained by dividing R into four subrectangles with the lines $x = \frac{1}{2}$ and $y = 1$, and choose the evaluation point (x_{ij}^*, y_{ij}^*) to be the upper right-hand corner of R_{ij}. (See Figure 7.)

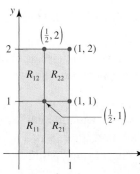

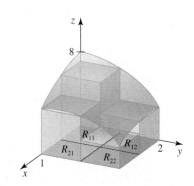

FIGURE 7

(**a**) The region R is divided into four subrectangles

(**b**) The solid lying under the graph of $z = 8 - 2x^2 - y^2$ and above R

Solution   Here,

$$\Delta x = \frac{1 - 0}{2} = \frac{1}{2} \quad \text{and} \quad \Delta y = \frac{2 - 0}{2} = 1$$

so $\Delta A = \left(\frac{1}{2}\right)(1) = \frac{1}{2}$. Also, $x_0 = 0$, $x_1 = \frac{1}{2}$, and $x_2 = 1$, and $y_0 = 0$, $y_1 = 1$, and $y_2 = 2$. Taking $(x_{11}^*, y_{11}^*) = (x_1, y_1) = \left(\frac{1}{2}, 1\right)$, $(x_{12}^*, y_{12}^*) = (x_1, y_2) = \left(\frac{1}{2}, 2\right)$, $(x_{21}^*, y_{21}^*) = (x_2, y_1) = (1, 1)$, and $(x_{22}^*, y_{22}^*) = (x_2, y_2) = (1, 2)$, we have

$$V \approx \sum_{i=1}^{2} \sum_{j=1}^{2} f(x_{ij}^*, y_{ij}^*) \, \Delta A = f(x_{11}^*, y_{11}^*) \, \Delta A + f(x_{12}^*, y_{12}^*) \, \Delta A + f(x_{21}^*, y_{21}^*) \, \Delta A + f(x_{22}^*, y_{22}^*) \, \Delta A$$

$$= f\left(\frac{1}{2}, 1\right) \Delta A + f\left(\frac{1}{2}, 2\right) \Delta A + f(1, 1) \, \Delta A + f(1, 2) \, \Delta A$$

$$= \left(\frac{13}{2}\right)\left(\frac{1}{2}\right) + \left(\frac{7}{2}\right)\left(\frac{1}{2}\right) + (5)\left(\frac{1}{2}\right) + (2)\left(\frac{1}{2}\right) = \frac{17}{2} \qquad ∎$$

The approximations to the volume in Example 1 get better and better as m and n increase, as shown in Figure 8.

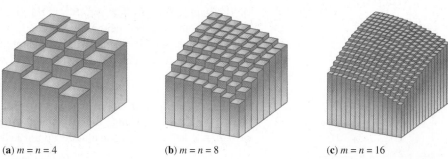

(**a**) $m = n = 4$ (**b**) $m = n = 8$ (**c**) $m = n = 16$

FIGURE 8

The approximation of V using the sum of the volumes of 16 parallelepipeds in (a), 64 parallelepipeds in (b), and 256 parallelepipeds in (c).

Note   Suppose that the mass density of a rectangular plate $R = \{(x, y) \,|\, 0 \le x \le 1,\ 0 \le y \le 2\}$ is $f(x, y) = 8 - 2x^2 - y^2$ g/cm^2. Then the result of Example 1 tells us that the mass of the plate is approximately $\frac{17}{2}$ g. ∎

■ The Double Integral Over a Rectangular Region

Thus far, we have assumed that $f(x, y) \geq 0$ on the rectangle R. This condition was imposed so that we could give a simple geometric interpretation for the limit in Equation (2). In the general situation we have the following.

DEFINITION   Riemann Sum

Let f be a continuous function of two variables defined on a rectangle R, and let $P = \{R_{ij}\}$ be a regular partition of R. A **Riemann sum of f** over R with respect to the partition P is a sum of the form

$$\sum_{i=1}^{m} \sum_{j=1}^{n} f(x_{ij}^*, y_{ij}^*) \, \Delta A \qquad (3)$$

where (x_{ij}^*, y_{ij}^*) is an evaluation point in R_{ij}.

DEFINITION   Double Integral of f Over a Rectangle R

Let f be a continuous function of two variables defined on a rectangle R. The **double integral of f** over R is

$$\iint\limits_{R} f(x, y) \, dA = \lim_{m, n \to \infty} \sum_{i=1}^{m} \sum_{j=1}^{n} f(x_{ij}^*, y_{ij}^*) \, \Delta A \qquad (4)$$

if this limit exists for all choices of the evaluation point (x_{ij}^*, y_{ij}^*) in R_{ij}.

Notes

1. If the double integral of f over R exists, then f is said to be **integrable over R.** It can be shown, although we will not do so here, that if f is continuous on R, then f is integrable over R.
2. If f is integrable, then the Riemann sum (3) is an approximation of the double integral (4).
3. If $f(x, y) \geq 0$ on R, then $\iint_R f(x, y) \, dA$ gives the volume of the solid lying directly above R and below the surface $z = f(x, y)$.
4. We use the (double) integral sign to denote the limit, whenever it exists, because it is related to the definite integral, as you will see in Section 14.2.   ■

EXAMPLE 2 Find an approximation for $\iint_R (x - 4y) \, dA$, where $R = \{(x, y) \mid 0 \leq x \leq 2, 0 \leq y \leq 1\}$, using the Riemann sum of $f(x, y) = x - 4y$ over R with $m = n = 2$ and taking the evaluation point (x_{ij}^*, y_{ij}^*) to be the center of R_{ij}.

Solution Here,

$$\Delta x = \frac{2 - 0}{2} = 1 \qquad \Delta y = \frac{1 - 0}{2} = \frac{1}{2}$$

and $x_0 = 0$, $x_1 = 1$, $x_2 = 2$, $y_0 = 0$, $y_1 = \frac{1}{2}$, $y_2 = 1$. The partition P is shown in Figure 9. Using Equation (3) with $f(x, y) = x - 4y$, $\Delta A = \Delta x \, \Delta y = (1)\left(\frac{1}{2}\right) = \frac{1}{2}$, and

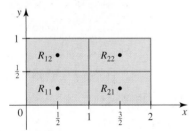

FIGURE 9
The partition
$P = \{R_{11}, R_{12}, R_{21}, R_{22}\}$ of R

$(x_{11}^*, y_{11}^*) = \left(\frac{1}{2}, \frac{1}{4}\right)$, $(x_{12}^*, y_{12}^*) = \left(\frac{1}{2}, \frac{3}{4}\right)$, $(x_{21}^*, y_{21}^*) = \left(\frac{3}{2}, \frac{1}{4}\right)$, and $(x_{22}^*, y_{22}^*) = \left(\frac{3}{2}, \frac{3}{4}\right)$, we obtain

$$\iint\limits_R f(x, y)\, dA \approx \sum_{i=1}^{2} \sum_{j=1}^{2} f(x_{ij}^*, y_{ij}^*)\, \Delta A$$

$$= f(x_{11}^*, y_{11}^*)\, \Delta A + f(x_{12}^*, y_{12}^*)\, \Delta A + f(x_{21}^*, y_{21}^*)\, \Delta A + f(x_{22}^*, y_{22}^*)\, \Delta A$$

$$= f\left(\frac{1}{2}, \frac{1}{4}\right)\frac{1}{2} + f\left(\frac{1}{2}, \frac{3}{4}\right)\frac{1}{2} + f\left(\frac{3}{2}, \frac{1}{4}\right)\frac{1}{2} + f\left(\frac{3}{2}, \frac{3}{4}\right)\frac{1}{2}$$

$$= \left(-\frac{1}{2}\right)\left(\frac{1}{2}\right) + \left(-\frac{5}{2}\right)\left(\frac{1}{2}\right) + \left(\frac{1}{2}\right)\left(\frac{1}{2}\right) + \left(-\frac{3}{2}\right)\left(\frac{1}{2}\right) = -2 \quad \blacksquare$$

■ Double Integrals Over General Regions

Next we will extend the definition of the double integral to more general functions and regions. Suppose that f is a bounded function defined on a bounded plane region D. If you like, you can think of f as a mass density function for a thin plate occupying a nonrectangular region D (in which case $f(x, y) \geq 0$ on D) and of what follows as a way of finding the mass of the plate. Since D is bounded, it can be enclosed in a rectangle R. Let Q be a regular partition of R into subrectangles $R_{11}, R_{12}, \dots, R_{ij}, \dots, R_{mn}$. (See Figure 10.)

Let's define the function

$$f_D(x, y) = \begin{cases} f(x, y) & \text{if } (x, y) \text{ is in } D \\ 0 & \text{if } (x, y) \text{ is in } R \text{ but not in } D \end{cases}$$

Note that f_D takes on the same value as f if (x, y) is in D, but it takes on the value zero if (x, y) lies outside D. (See Figure 11.)

Now let (x_{ij}^*, y_{ij}^*) be an evaluation point in the subrectangle R_{ij} of Q for $1 \leq i \leq m$ and $1 \leq j \leq n$. Then the sum

$$\sum_{i=1}^{m} \sum_{j=1}^{n} f_D(x_{ij}^*, y_{ij}^*)\, \Delta A$$

is a **Riemann sum of f over D** with respect to the partition Q. Taking the limit of these sums as $m, n \to \infty$ gives the **double integral of f over D**. Thus,

$$\iint\limits_D f(x, y)\, dA = \lim_{m, n \to \infty} \sum_{i=1}^{m} \sum_{j=1}^{n} f_D(x_{ij}^*, y_{ij}^*)\, \Delta A \tag{5}$$

if the limit exists. Again, it can be shown that if f is continuous, then the limit (5) always exists regardless of how the evaluation points (x_{ij}^*, y_{ij}^*) in R_{ij} are chosen.

Notes

1. If $f(x, y) \geq 0$ on D, then $\iint_D f(x, y)\, dA$ gives the volume of the solid lying directly above D and below the surface $z = f(x, y)$.
2. If $\rho(x, y) \geq 0$ on D, where ρ is a mass density function, then $\iint_D \rho(x, y)\, dA$ gives the mass of the thin plate occupying the plane region D in the xy-plane. This will be demonstrated in Section 14.4. ■

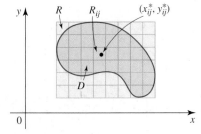

FIGURE 10
A partition Q of R

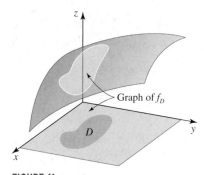

FIGURE 11
$f_D(x, y) = f(x, y)$ if (x, y) lies in D, but $f_D(x, y) = 0$ if (x, y) lies outside D.

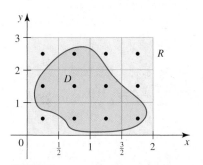

FIGURE 12
$f_D(x, y) = f(x, y)$ if (x, y) lies in D, but $f_D(x, y) = 0$ if (x, y) lies outside D.

EXAMPLE 3 Find an approximation for $\iint_D (x + 2y)\, dA$, where D is the region shown in Figure 12, using the Riemann sum of $f(x, y) = x + 2y$ over D with respect to the partition Q obtained by dividing the rectangle $\{(x, y) \mid 0 \leq x \leq 2, 0 \leq y \leq 3\}$

into 12 subrectangles by taking $m = 4$ and $n = 3$ and taking the evaluation points to be the center of R_{ij}.

Solution Here,

$$\Delta A = (\Delta x)(\Delta y) = \left(\frac{2 - 0}{4}\right)\left(\frac{3 - 0}{3}\right) = \frac{1}{2}$$

Next, define

$$f_D(x, y) = \begin{cases} f(x, y) & \text{if } (x, y) \text{ is in } D \\ 0 & \text{if } (x, y) \text{ is not in } D \end{cases}$$

Then

$$\iint\limits_D (x + 2y)\, dA \approx \sum_{i=1}^{4} \sum_{j=1}^{3} f_D(x_{ij}^*, y_{ij}^*)\, \Delta A \qquad f(x, y) = x + 2y$$

$$= \left[f_D\left(\frac{1}{4}, \frac{1}{2}\right) + f_D\left(\frac{1}{4}, \frac{3}{2}\right) + f_D\left(\frac{1}{4}, \frac{5}{2}\right) + f_D\left(\frac{3}{4}, \frac{1}{2}\right) + f_D\left(\frac{3}{4}, \frac{3}{2}\right) + f_D\left(\frac{3}{4}, \frac{5}{2}\right) \right.$$

$$\left. + f_D\left(\frac{5}{4}, \frac{1}{2}\right) + f_D\left(\frac{5}{4}, \frac{3}{2}\right) + f_D\left(\frac{5}{4}, \frac{5}{2}\right) + f_D\left(\frac{7}{4}, \frac{1}{2}\right) + f_D\left(\frac{7}{4}, \frac{3}{2}\right) + f_D\left(\frac{7}{4}, \frac{5}{2}\right) \right] \Delta A$$

$$= \frac{1}{2}\left[f\left(\frac{1}{4}, \frac{3}{2}\right) + f\left(\frac{3}{4}, \frac{1}{2}\right) + f\left(\frac{3}{4}, \frac{3}{2}\right) + f\left(\frac{3}{4}, \frac{5}{2}\right) + \left(\frac{5}{4}, \frac{1}{2}\right) + f\left(\frac{5}{4}, \frac{3}{2}\right) + f\left(\frac{7}{4}, \frac{1}{2}\right) \right]$$

$$= \frac{1}{2}\left\{ \left[\frac{1}{4} + 2\left(\frac{3}{2}\right)\right] + \left[\frac{3}{4} + 2\left(\frac{1}{2}\right)\right] + \left[\frac{3}{4} + 2\left(\frac{3}{2}\right)\right] + \left[\frac{3}{4} + 2\left(\frac{5}{2}\right)\right] + \left[\frac{5}{4} + 2\left(\frac{1}{2}\right)\right] \right.$$

$$\left. + \left[\frac{5}{4} + 2\left(\frac{3}{2}\right)\right] + \left[\frac{7}{4} + 2\left(\frac{1}{2}\right)\right] \right\} = 11.875 \qquad \blacksquare$$

■ Properties of Double Integrals

Double integrals have many of the properties that single integrals enjoy. We list some of them in the following theorem, the proof of which will be omitted.

THEOREM 1 **Properties of the Definite Integral**

Let f and g be defined on a suitably restricted region D, so that both $\iint_D f(x, y)\, dA$ and $\iint_D g(x, y)\, dA$ exist, and let c be a constant. Then

1. $\displaystyle\iint\limits_D cf(x, y)\, dA = c\iint\limits_D f(x, y)\, dA$

2. $\displaystyle\iint\limits_D [f(x, y) \pm g(x, y)]\, dA = \iint\limits_D f(x, y)\, dA \pm \iint\limits_D g(x, y)\, dA$

3. If $f(x, y) \geq 0$ on D, then $\displaystyle\iint\limits_D f(x, y)\, dA \geq 0$

4. If $f(x, y) \geq g(x, y)$ on D, then $\displaystyle\iint\limits_D f(x, y)\, dA \geq \iint\limits_D g(x, y)\, dA$

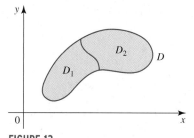

FIGURE 13
$D = D_1 \cup D_2$ where $D_1 \cap D_2 = \varnothing$.

5. If $D = D_1 \cup D_2$, where D_1 and D_2 are two nonoverlapping subregions with the possible exception of their common boundaries, then

$$\iint\limits_D f(x, y) \, dA = \iint\limits_{D_1} f(x, y) \, dA + \iint\limits_{D_2} f(x, y) \, dA$$

(See Figure 13.)

14.1 CONCEPT QUESTIONS

1. Let $f(x, y) = x + 2y$.
   a. Complete the table of values for $f(x, y)$ in the following table.

y x	1	$\frac{3}{2}$	2	$\frac{5}{2}$	3	$\frac{7}{2}$	4
0							
$\frac{1}{4}$							
$\frac{1}{2}$							
$\frac{3}{4}$							
1							

   b. Use the table of values from part (a) to estimate the volume of the solid lying under the graph of $z = x + 2y$ and above the rectangular region $R = [0, 1] \times [1, 4]$ using a regular partition with $m = 2$ and $n = 3$ and choosing the evaluation point (x_{ij}^*, y_{ij}^*) to be the lower left-hand corner of R_{ij}.
   c. Repeat part (b), this time choosing the evaluation point (x_{ij}^*, y_{ij}^*) to be the center of R_{ij}.

2. a. Let f be a continuous function defined on the rectangular region $[a, b] \times [c, d]$. Define $\iint_R f(x, y) \, dA$.
   b. Suppose that $f(x, y) = k$, where k is a constant. Find $\iint_R k \, dA$ for $R = [a, b] \times [c, d]$ using your definition from part (a).

14.1 EXERCISES

In Exercises 1–4, find an approximation for the volume V of the solid lying under the graph of the elliptic paraboloid $z = 8 - 2x^2 - y^2$ and above the rectangular region $R = \{(x, y) \mid 0 \le x \le 1, 0 \le y \le 2\}$. Use a regular partition P of R with $m = n = 2$, and choose the evaluation point (x_{ij}^*, y_{ij}^*) as indicated in each exercise.

1. The lower left-hand corner of R_{ij}

2. The upper left-hand corner of R_{ij}

3. The lower right-hand corner of R_{ij}

4. The center of R_{ij}

In Exercises 5–8, find the Riemann sum $\sum_{i=1}^{m} \sum_{j=1}^{n} f(x_{ij}^*, y_{ij}^*) \, \Delta A$ of f over the region R with respect to the regular partition P with the indicated values of m and n.

5. $f(x, y) = 2x + 3y$; $R = [0, 1] \times [0, 3]$; $m = 2, n = 3$; (x_{ij}^*, y_{ij}^*) is the lower left-hand corner of R_{ij}

6. $f(x, y) = x^2 - 2y$; $R = [1, 5] \times [1, 3]$; $m = 4, n = 2$; (x_{ij}^*, y_{ij}^*) is the upper right-hand corner of R_{ij}

7. $f(x, y) = x^2 + 2y^2$; $R = [-1, 3] \times [0, 4]$; $m = 4, n = 4$; (x_{ij}^*, y_{ij}^*) is the center of R_{ij}

8. $f(x, y) = 2xy$; $R = [-1, 1] \times [-2, 2]$; $m = 4, n = 4$; (x_{ij}^*, y_{ij}^*) is the center of R_{ij}

9. The figure on the following page shows a region D enclosed by a rectangular region R and a partition Q of R into subrectangles with $m = 5$ and $n = 3$. Suppose that f is continuous on D and the values of f at the evaluation points of Q that lie in D are as shown in the figure (next to the evaluation points). Define

$$f_D(x, y) = \begin{cases} f(x, y) & \text{if } (x, y) \text{ is in } D \\ 0 & \text{if } (x, y) \text{ is in } R \text{ but not in } D \end{cases}$$

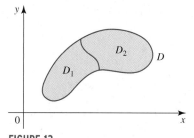 **V** Videos for selected exercises are available online at **www.academic.cengage.com/login**.

Compute $\sum_{i=1}^{5} \sum_{j=1}^{3} f_D(x_{ij}^*, y_{ij}^*) \, \Delta A$.

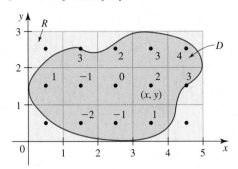

10. Volume of Water in a Pond The following figure depicts a pond that is 40 ft long and 20 ft wide. The depth of the pond is measured at the center of each subrectangle in the imaginary partition of the rectangle that is superimposed over the aerial view of the pond. These measurements (in feet) are shown in the figure. Estimate the volume of water in the pond.
Hint: See Exercise 9.

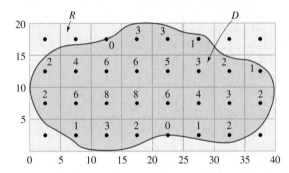

11. The figure shows the contour map of a function f on the set $R = \{(x, y) \mid 0 \le x \le 2, 0 \le y \le 2\}$. Estimate $\iint_R f(x, y) \, dA$ using a Riemann sum with $m = n = 2$ and choosing the evalution point (x_{ij}^*, y_{ij}^*) to be the center of R_{ij}.

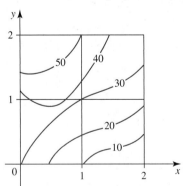

12. Room Temperature The figure represents a part of a room with a fireplace located at the origin. The curves shown are the level curves of the temperature function T, and are called *isothermals* because the temperature is the same at all points on an isothermal. Estimate the average temperature in this

part of the room using a regular partition with $m = n = 3$ and choosing the evaluation point (x_{ij}^*, y_{ij}^*) to be the center of R_{ij}.

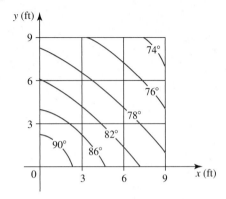

In Exercises 13–16, find the double integral by interpreting it as the volume of a solid.

13. $\iint_R 2 \, dA$, where $R = [-1, 3] \times [2, 5]$

14. $\iint_R 2x \, dA$, where $R = \{(x, y) \mid 0 \le x \le 2, 0 \le y \le 1\}$

15. $\iint_R (6 - 2y) \, dA$, where $R = \{(x, y) \mid 0 \le x \le 4, 0 \le y \le 2\}$

16. $\iint_R \sqrt{9 - x^2 - y^2} \, dA$, where

$R = \{(x, y) \mid x^2 + y^2 \le 9, x \ge 0, y \ge 0\}$

In Exercises 17 and 18, the double integral gives the volume of a solid. Describe the solid.

17. $\iint_R (4 - x^2) \, dA$, where $R = \{(x, y) \mid 0 \le y \le x, 0 \le x \le 2\}$

18. $\iint_R \left(3 - \dfrac{1}{2} x - \dfrac{3}{4} y\right) dA$, where

$R = \{(x, y) \mid 2x + 3y \le 12, x \ge 0, y \ge 0\}$

In Exercises 19 and 20, the expression is the limit of a Riemann sum of a function f over a rectangle R. Write this expression as a double integral over R.

19. $\displaystyle\lim_{m, n \to \infty} \sum_{i=1}^{m} \sum_{j=1}^{n} (3 - 2x_{ij}^* + y_{ij}^*) \, \Delta A$, $R = [-1, 2] \times [1, 3]$

20. $\displaystyle\lim_{m, n \to \infty} \sum_{i=1}^{m} \sum_{j=1}^{n} \sqrt{(x_{ij}^*)^2 + 2(y_{ij}^*)^2} \, \Delta A$, $R = [0, 1] \times [0, 2]$

cas *In Exercises 21 and 22, use a computer algebra system (CAS) to obtain an approximate value of the double integral using a regu-*

lar partition with the given value of m and n and choosing the evaluation point (x_{ij}^*, y_{ij}^*) to be the center of R_{ij}.

21. $\iint\limits_R \sqrt{1 + x^2 + y^2}\, dA$, where $R = [0, 1] \times [0, 1]$;
$m = 10, n = 10$

22. $\iint\limits_R \dfrac{1}{1 + e^{xy}}\, dA$, where $R = [0, 1] \times [1, 3]$;
$m = 10, n = 20$

23. Use Property 4 (in Theorem 1) of the double integral to show that if f and $|f|$ are integrable over D, then

$$\left| \iint\limits_D f(x, y)\, dA \right| \le \iint\limits_D |f(x, y)|\, dA$$

24. Use a geometric argument and Theorem 1 to show that if $f(x, y) = k$, where k is a constant, then $\iint_R k\, dA = k \cdot \text{area of } R$.

25. Let $R = \{(x, y) \mid 0 \le x \le 1, 0 \le y \le 1\}$. Show that $0 \le \iint_R e^{-x} \cos y\, dA \le 1$.

26. Let $R = \left[0, \tfrac{1}{2}\right] \times \left[0, \tfrac{1}{2}\right]$. Show that $0 \le \iint_R \sin(2x + 3y)\, dA \le \tfrac{1}{4}$.

In Exercises 27–30, determine whether the statement is true or false. If it is true, explain why. If it is false, explain why or give an example that shows it is false.

27. If f and g are continuous on D, then

$$\iint\limits_D [2f(x, y) - 3g(x, y)]\, dA = 2\iint\limits_D f(x, y)\, dA - 3\iint\limits_D g(x, y)\, dA$$

28. If f and g are continuous on D, then

$$\iint\limits_D [f(x, y)g(x, y)]\, dA = \left[\iint\limits_D f(x, y)\, dA\right]\left[\iint\limits_D g(x, y)\, dA\right]$$

29. $\iint\limits_R \dfrac{\sqrt{x^2 + xy + y^2 + 1}}{\cos(x^2 + y^2)}\, dA \ge \pi$, where
$R = \{(x, y) \mid x^2 + y^2 \le 1\}$

30. If f is nonnegative and continuous on
$D = \{(x, y) \mid 0 \le x \le 1, 0 \le y \le 1\}$ and
$E = \left\{(x, y) \mid 0 \le x \le 1, \tfrac{1}{2} \le y \le 1\right\}$, then

$$\iint\limits_D f(x, y)\, dA \ge \iint\limits_E f(x, y)\, dA$$

14.2 Iterated Integrals

■ Iterated Integrals Over Rectangular Regions

Just as it is difficult to find the value of an integral of a function of one variable directly from its definition, the task is even harder in the case of double integrals. Fortunately, as you will see, the value of a double integral can be found by evaluating two single integrals.

We begin by looking at the simple case in which f is a continuous function defined on the rectangular region $R = \{(x, y) \mid a \le x \le b, c \le y \le b\}$ shown in Figure 1b.

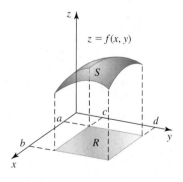
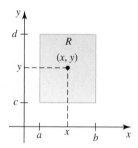

FIGURE 1   (a) The graph of f (b) The domain R of f

If we fix x, then $f(x, y)$ is a function of the single variable y for $c \le y \le d$. As such, we can integrate the function with respect to y over the interval $[c, d]$. This operation is called *partial integration with respect to y* and is the reverse of the operation of partial differentiation studied in Chapter 13. The result is the number

$$\int_c^d f(x, y)\, dy$$

that depends on the value of x in $[a, b]$. In other words, the rule

$$A(x) = \int_c^d f(x, y)\, dy \qquad a \le x \le b \tag{1}$$

defines a function A of x on $[a, b]$. If we integrate the function A with respect to x over $[a, b]$, we obtain

$$\int_a^b A(x)\, dx = \int_a^b \left[\int_c^d f(x, y)\, dy \right] dx \tag{2}$$

The integral on the right-hand side of Equation (2) is usually written in the form

$$\int_a^b \int_c^d f(x, y)\, dy\, dx \tag{3}$$

without the brackets and is called an *iterated* or *repeated integral.*

Similarly, by holding y fixed and integrating the resulting function with respect to x over $[a, b]$, we obtain a function of y on the interval $[c, d]$. If this function is then integrated with respect to y over $[c, d]$, we obtain the iterated integral

$$\int_c^d \int_a^b f(x, y)\, dx\, dy = \int_c^d \left[\int_a^b f(x, y)\, dx \right] dy \tag{4}$$

Observe that when we evaluate an iterated integral, we work *from the inside out.*

EXAMPLE 1 Evaluate the iterated integrals:

a. $\displaystyle\int_1^2 \int_0^1 3x^2 y\, dx\, dy$ **b.** $\displaystyle\int_0^1 \int_1^2 3x^2 y\, dy\, dx$

Solution
a. By definition,

$$\int_1^2 \int_0^1 3x^2 y\, dx\, dy = \int_1^2 \left[\int_0^1 3x^2 y\, dx \right] dy$$

Now the integral inside the brackets is found by integrating with respect to x while treating y as a constant. This gives

$$\int_0^1 3x^2 y\, dx = \left[x^3 y \right]_{x=0}^{x=1} = y$$

Therefore,

$$\int_1^2 \int_0^1 3x^2 y\, dx\, dy = \int_1^2 y\, dy$$

$$= \left[\frac{1}{2} y^2 \right]_1^2 = \frac{3}{2}$$

b. Here, we first integrate with respect to y and then with respect to x, obtaining

$$\int_0^1 \int_1^2 3x^2 y \, dy \, dx = \int_0^1 \left[\int_1^2 3x^2 y \, dy \right] dx$$

$$= \int_0^1 \left[\frac{3}{2} x^2 y^2 \right]_{y=1}^{y=2} dx$$

$$= \int_0^1 \frac{9}{2} x^2 \, dx = \left[\frac{3}{2} x^3 \right]_0^1 = \frac{3}{2}$$ ∎

Fubini's Theorem for Rectangular Regions

Observe that the two iterated integrals in Example 1 are equal. Thus, the example seems to suggest that the order of integration of the iterated integrals does not matter. To see why this might be true for continuous functions, consider the special case in which f is nonnegative. Let's calculate the volume V of the solid S lying under the graph of $z = f(x, y)$ and above the rectangular region $R = \{(x, y) \mid a \le x \le b, c \le y \le d\}$.

Using the method of cross sections of Section 5.2, we see that

$$V = \int_a^b A(x) \, dx$$

where $A(x)$ is the area of the cross section of S in the plane perpendicular to the x-axis at x. (See Figure 2a.) But from the figure, you can see that $A(x)$ is the area under the graph C of the function defined by $g(y) = f(x, y)$ for $c \le y \le d$, where x is fixed. So

$$A(x) = \int_c^d g(y) \, dy = \int_c^d f(x, y) \, dy \qquad x \text{ fixed}$$

Therefore,

$$V = \int_a^b A(x) \, dx = \int_a^b \left[\int_c^d f(x, y) \, dy \right] dx$$

Similarly, using cross sections perpendicular to the y-axis (Figure 2b), you can show that

$$V = \int_c^d \left[\int_a^b f(x, y) \, dx \right] dy$$

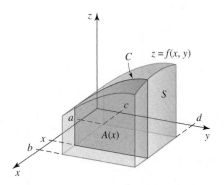

(**a**) $A(x)$ is the area of a cross section of S in the plane perpendicular to the x-axis.

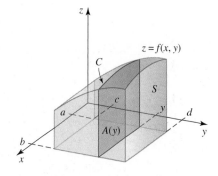

(**b**) $A(y)$ is the area of a cross section of S in the plane perpendicular to the y-axis.

FIGURE 2

Now, by definition,

$$V = \iint\limits_{R} f(x, y)\, dA$$

Therefore, we have shown that

$$\iint\limits_{R} f(x, y)\, dA = \int_{a}^{b} \int_{c}^{d} f(x, y)\, dy\, dx = \int_{c}^{d} \int_{a}^{b} f(x, y)\, dx\, dy$$

This discussion suggests the following theorem, which is named after the Italian mathematician Guido Fubini (1879–1943). Its proof lies outside the scope of this book and will be omitted.

THEOREM 1   Fubini's Theorem for Rectangular Regions

Let f be continuous over the rectangle $R = \{(x, y) \mid a \le x \le b, c \le y \le d\}$. Then

$$\iint\limits_{R} f(x, y)\, dA = \int_{a}^{b} \int_{c}^{d} f(x, y)\, dy\, dx = \int_{c}^{d} \int_{a}^{b} f(x, y)\, dx\, dy$$

Fubini's Theorem provides us with a practical method for finding double integrals by expressing them in terms of iterated integrals that we can evaluate by integrating with respect to one variable at a time. It also states that the order in which the integration is carried out does not matter, an important option, as you will see later on. Finally, observe that Fubini's Theorem holds for *any* continuous function; $f(x, y)$ may assume negative as well as positive values on R.

EXAMPLE 2   Evaluate $\iint_{R}(1 - 2xy^2)\, dA$, where

$$R = \{(x, y) \mid 0 \le x \le 2, -1 \le y \le 1\}$$

Solution   Using Fubini's Theorem, we obtain

$$\iint\limits_{R}(1 - 2xy^2)\, dA = \int_{-1}^{1} \int_{0}^{2} (1 - 2xy^2)\, dx\, dy$$

$$= \int_{-1}^{1} \left[x - x^2 y^2 \right]_{x=0}^{x=2} dy$$

$$= \int_{-1}^{1} (2 - 4y^2)\, dy = \left[2y - \frac{4}{3} y^3 \right]_{-1}^{1}$$

$$= \left(2 - \frac{4}{3} \right) - \left(-2 + \frac{4}{3} \right) = \frac{4}{3}$$

We leave it for you to verify that

$$\iint\limits_{R}(1 - 2xy^2)\, dA = \int_{0}^{2} \int_{-1}^{1} (1 - 2xy^2)\, dy\, dx = \frac{4}{3}$$

as well.

EXAMPLE 3 Find the volume of the solid lying under the elliptic paraboloid $z = 8 - 2x^2 - y^2$ and above the rectangular region $R = \{(x, y) \mid 0 \le x \le 1, 0 \le y \le 2\}$. (See Figure 3.) Compare with Example 1 in Section 14.1.

Solution Using Fubini's Theorem, we see that the required volume is

$$V = \iint_R (8 - 2x^2 - y^2) \, dA = \int_0^2 \int_0^1 (8 - 2x^2 - y^2) \, dx \, dy$$

$$= \int_0^2 \left[8x - \frac{2}{3}x^3 - xy^2 \right]_{x=0}^{x=1} dy$$

$$= \int_0^2 \left(\frac{22}{3} - y^2 \right) dy = \left[\frac{22}{3}y - \frac{1}{3}y^3 \right]_0^2 = 12$$

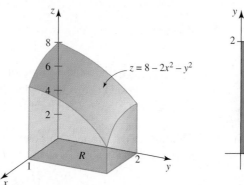

(a) The solid between the graph of $z = 8 - 2x^2 - y^2$ and the rectangular region R

(b) The region R

FIGURE 3

Iterated Integrals Over Nonrectangular Regions

Fubini's Theorem is valid for regions that are more general than rectangular regions. More specifically, it is valid for the two types of regions that we will now describe. A plane region R is said to be **y-simple** if it lies between two functions of x; that is,

$$R = \{(x, y) \mid a \le x \le b, g_1(x) \le y \le g_2(x)\}$$

where g_1 and g_2 are continuous on $[a, b]$. (See Figure 4.)

An **x-simple** region R is one that lies between two functions of y; that is,

$$R = \{(x, y) \mid c \le y \le d, h_1(y) \le x \le h_2(y)\}$$

where h_1 and h_2 are continuous on $[c, d]$. (See Figure 5.)

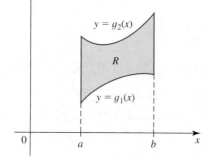

FIGURE 4
A y-simple region

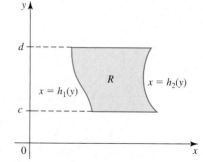

FIGURE 5
An x-simple region

The following theorem tells us that a double integral over a y-simple or an x-simple region can be found by evaluating an iterated integral.

THEOREM 2   Fubini's Theorem for General Regions

Let f be continuous on a region R.

1. If R is a y-simple region, then

$$\iint_R f(x, y)\, dA = \int_a^b \int_{g_1(x)}^{g_2(x)} f(x, y)\, dy\, dx$$

2. If R is an x-simple region, then

$$\iint_R f(x, y)\, dA = \int_c^d \int_{h_1(y)}^{h_2(y)} f(x, y)\, dx\, dy$$

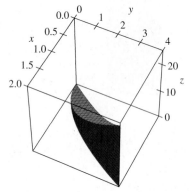

FIGURE 6

The graph of the solid S

EXAMPLE 4 Find the volume of the solid S lying under the graph of the surface $z = x^3 + 4y$ and above the region R in the xy-plane bounded by the line $y = 2x$ and the parabola $y = x^2$. (See Figure 6.)

Solution   First, we make a sketch of the region R. (See Figure 7a). We see that R can be viewed as a y-simple region; that is,

$$R = \{(x, y)\,|\,0 \le x \le 2,\, x^2 \le y \le 2x\}$$

where $g_1(x) = x^2$ and $g_2(x) = 2x$. Observe that if we integrate over a y-simple region, we integrate with respect to y first. The appropriate limits of integration can be found by drawing a vertical arrow as shown in Figure 7a. The arrow begins at the lower boundary of the region described by $y = g_1(x) = x^2$, giving the lower limit of integration as $g_1(x) = x^2$, and terminates at the upper boundary of the region described by $y = g_2(x) = 2x$, giving the upper limit of integration as $g_2(x) = 2x$. To find the limits for integrating with respect to x, observe that a vertical line sweeping from left to right meets the extreme left point of R when $x = 0$ (the lower limit of integration) and meets the extreme right point of R when $x = 2$ (the upper limit of integration). Using Fubini's Theorem for general regions, we obtain

$$V = \iint_R f(x, y)\, dA = \int_0^2 \int_{x^2}^{2x} (x^3 + 4y)\, dy\, dx$$

$$= \int_0^2 \left[x^3 y + 2y^2 \right]_{y=x^2}^{y=2x} dx = \int_0^2 \left[(2x^4 + 8x^2) - (x^5 + 2x^4) \right] dx$$

$$= \int_0^2 (8x^2 - x^5)\, dx = \left[\frac{8}{3} x^3 - \frac{1}{6} x^6 \right]_0^2 = \frac{32}{3}$$

Alternative Solution   We can view the region R as an x-simple region

$$R = \left\{ (x, y)\,|\,0 \le y \le 4,\, \tfrac{y}{2} \le x \le \sqrt{y} \right\}$$

where $h_1(y) = y/2$ and $h_2(y) = \sqrt{y}$ are obtained by solving $y = 2x$ and $y = x^2$ for x in terms of y, respectively. (See Figure 7b.) If we integrate over an x-simple region, we

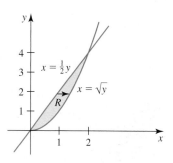

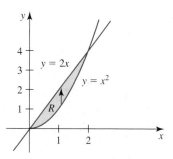

FIGURE 7   **(a)** The region R viewed as a y-simple region   **(b)** The region R viewed as an x-simple region

integrate with respect to x first. A horizontal arrow starting from the left boundary of R described by $h_1(y) = y/2$ and terminating at the right boundary of R described by $h_2(y) = \sqrt{y}$ gives the lower and upper limits of integration with respect to x. The limits of integration with respect to y are found by letting a horizontal line sweep through the region. This line meets the lowest point of R when $y = 0$ (the lower limit of integration) and the highest point of R when $y = 4$ (the upper limit of integration). Once again using Fubini's Theorem, we obtain

$$V = \iint\limits_{R} f(x, y) \, dA = \int_0^4 \int_{y/2}^{\sqrt{y}} (x^3 + 4y) \, dx \, dy$$

$$= \int_0^4 \left[\frac{1}{4} x^4 + 4xy \right]_{x=y/2}^{x=\sqrt{y}} dy = \int_0^4 \left[\left(\frac{1}{4} y^2 + 4y^{3/2} \right) - \left(\frac{1}{64} y^4 + 2y^2 \right) \right] dy$$

$$= \int_0^4 \left(-\frac{7}{4} y^2 + 4y^{3/2} - \frac{1}{64} y^4 \right) dy = \left[-\frac{7}{12} y^3 + \frac{8}{5} y^{5/2} - \frac{1}{320} y^5 \right]_0^4 = \frac{32}{3}$$

as before.   ■

EXAMPLE 5   Evaluate $\iint_R (2x - y) \, dA$, where R is the region bounded by the parabola $x = y^2$ and the straight line $x - y = 2$.

Solution   The region R is shown in Figure 8. It is both y-simple and x-simple. But observe that it is more convenient to view it as an x-simple region because the lower boundary of R consists of two curves when viewed as a y-simple region. In fact, viewing R as a y-simple region (Figure 8a) and using Fubini's Theorem, we find

$$\iint\limits_{R} (2x - y) \, dA = \int_0^1 \int_{-\sqrt{x}}^{\sqrt{x}} (2x - y) \, dy \, dx + \int_1^4 \int_{x-2}^{\sqrt{x}} (2x - y) \, dy \, dx$$

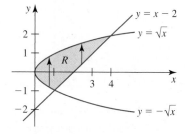

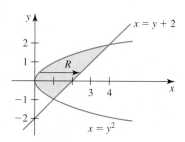

FIGURE 8   **(a)** R viewed as a y-simple region   **(b)** R viewed as an x-simple region

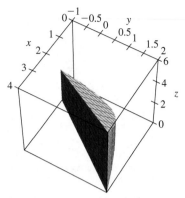

FIGURE 9
The solid S

On the other hand, viewing R as an x-simple region (Figure 8b), we have

$$\iint_R (2x - y)\, dA = \int_{-1}^{2} \int_{y^2}^{y+2} (2x - y)\, dx\, dy$$

$$= \int_{-1}^{2} \left[x^2 - xy \right]_{x=y^2}^{x=y+2} dy = \int_{-1}^{2} \left\{ \left[(y+2)^2 - y(y+2) \right] - \left[y^4 - y^3 \right] \right\} dy$$

$$= \int_{-1}^{2} (4 + 2y + y^3 - y^4)\, dy = \left[4y + y^2 + \frac{1}{4}y^4 - \frac{1}{5}y^5 \right]_{-1}^{2} = \frac{243}{20}$$

which is easier to evaluate.

The double integral $\iint_R (2x - y)\, dA$ gives the volume of the solid S shown in Figure 9. ∎

Example 5 shows that it is sometimes easier to integrate in one order rather than the other because of the shape of R. In certain instances the nature of the function dictates the order of integration, as the next example shows.

EXAMPLE 6 Evaluate $\displaystyle\int_0^1 \int_y^1 \frac{\sin x}{x}\, dx\, dy$.

Solution Because

$$\int \frac{\sin x}{x}\, dx$$

cannot be expressed in terms of elementary functions, the given integral cannot be evaluated as it stands. So let's attempt to evaluate it by reversing the order of integration. We begin by using Fubini's Theorem to express the iterated integral as a double integral. The order of integration of the given integral suggests that

$$\int_0^1 \int_y^1 \frac{\sin x}{x}\, dx\, dy = \iint_R \frac{\sin x}{x}\, dA$$

where $R = \{(x, y) \mid 0 \le y \le 1,\, y \le x \le 1\}$ is viewed as an x-simple region (see Figure 10a).

FIGURE 10

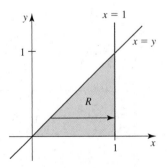

(a) R viewed as an x-simple region

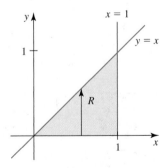

(b) R viewed as a y-simple region

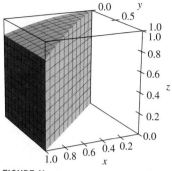

FIGURE 11
The solid S represented by the
double integral $\displaystyle\int_0^1\int_y^1 \frac{\sin x}{x}\,dx\,dy$

Viewing R as a y-simple region (Figure 10b), we find, again by Fubini's Theorem, that

$$\int_0^1\int_y^1 \frac{\sin x}{x}\,dx\,dy = \iint_R \frac{\sin x}{x}\,dA$$

$$= \int_0^1\int_0^x \frac{\sin x}{x}\,dy\,dx = \int_0^1\left[\frac{y\sin x}{x}\right]_{y=0}^{y=x}\,dx$$

$$= \int_0^1 \sin x\,dx = \left[-\cos x\right]_0^1 = -\cos 1 + 1 \approx 0.46$$

The double integral $\displaystyle\int_0^1\int_y^1 \frac{\sin x}{x}\,dx\,dy$ gives the volume of the solid S shown in Figure 11. ∎

14.2 CONCEPT QUESTIONS

1. Suppose that f is continuous on the rectangular region $R = [a, b] \times [c, d]$.
   a. Explain the difference between the iterated integrals

   $$\int_a^b\left[\int_c^d f(x, y)\,dy\right]dx \quad \text{and} \quad \int_c^d\left[\int_a^b f(x, y)\,dx\right]dy$$

   b. Give a geometric interpretation of each of the iterated integrals in part (a), where f is nonnegative.
   c. What does Fubini's Theorem say about the two iterated integrals in part (a)?

2. a. What is a y-simple region, and what is an x-simple region?
   b. Express $\iint_R f(x, y)\,dA$ as an iterated integral if R is a y-simple region. As an x-simple region.
   c. Explain why it is sometimes advantageous to reverse the order of integration of an iterated integral.

14.2 EXERCISES

In Exercises 1–12, evaluate the iterated integral.

1. $\displaystyle\int_0^1\int_0^2 (x + 2y)\,dy\,dx$

2. $\displaystyle\int_{-1}^1\int_0^3 (3x^2 + y)\,dx\,dy$

3. $\displaystyle\int_0^2\int_1^4 y\sqrt{x}\,dy\,dx$

4. $\displaystyle\int_0^1\int_0^1 \frac{x}{1 + xy}\,dy\,dx$

5. $\displaystyle\int_0^\pi\int_0^\pi \cos(x + y)\,dy\,dx$

6. $\displaystyle\int_0^{\pi/2}\int_0^{\ln 2} e^{-x}\sin y\,dx\,dy$

7. $\displaystyle\int_0^4\int_0^{\sqrt{x}} 2xy\,dy\,dx$

8. $\displaystyle\int_0^{1/2}\int_0^{\sqrt{1-x}} 2xy\,dy\,dx$

9. $\displaystyle\int_0^1\int_0^{\sqrt{1-y^2}} x\,dx\,dy$

10. $\displaystyle\int_0^1\int_0^{\sqrt{1-x^2}} (x + y)\,dy\,dx$

11. $\displaystyle\int_{-1}^1\int_x^{2x} e^{x+y}\,dy\,dx$

12. $\displaystyle\int_0^\pi\int_{e^{-2x}}^{e^{\cos x}} \frac{\ln y}{y}\,dy\,dx$

In Exercises 13–32, evaluate the double integral.

13. $\displaystyle\iint_R (x + y^2)\,dA$, where

$R = \{(x, y)\,|\,0 \le x \le 1, -1 \le y \le 2\}$

14. $\displaystyle\iint_R (3x^2 + 2xy^3)\,dA$, where

$R = \{(x, y)\,|\,-1 \le x \le 2, 0 \le y \le 2\}$

15. $\displaystyle\iint_R (x\cos y + y\sin x)\,dA$, where

$R = \left\{(x, y)\,\middle|\,0 \le x \le \frac{\pi}{2}, 0 \le y \le \frac{\pi}{4}\right\}$

16. $\displaystyle\iint_R ye^{xy}\,dA$, where $R = \{(x, y)\,|\,0 \le x \le 1, 0 \le y \le 1\}$

17. $\displaystyle\iint_R (x + 2y)\, dA$, where $R = \{(x, y) \,|\, 0 \le x \le 1, 0 \le y \le x\}$

18. $\displaystyle\iint_R \sqrt{1 - x^2}\, dA$, where $R = \{(x, y) \,|\, 0 \le x \le 1, 0 \le y \le x\}$

19. $\displaystyle\iint_R (x^3 + 2y)\, dA$, where

$R = \{(x, y) \,|\, 0 \le x \le 2, x^2 \le y \le 2x\}$

20. $\displaystyle\iint_R xy\, dA$, where

$R = \{(x, y) \,|\, -1 \le x \le 2, -x^2 \le y \le 1 + x^2\}$

21. $\displaystyle\iint_R (1 + 2x + 2y)\, dA$, where

$R = \{(x, y) \,|\, 0 \le y \le 1, y \le x \le 2y\}$

22. $\displaystyle\iint_R (x^2 + y^2)\, dA$, where

$R = \{(x, y) \,|\, 0 \le y \le 1, -y - 1 \le x \le y - 1\}$

23. $\displaystyle\iint_R x \cos y\, dA$, where

$R = \left\{(x, y) \,|\, 0 \le y \le \frac{\pi}{2}, 0 \le x \le \sin y\right\}$

24. $\displaystyle\iint_R \frac{1}{xy}\, dA$, where $R = \{(x, y) \,|\, 1 \le y \le e, y \le x \le y^2\}$

25. $\displaystyle\iint_R x^2 y\, dA$, where R is the region bounded by the graphs of

$y = x$, $y = 2x$, $x = 1$, and $x = 2$

26. $\displaystyle\iint_R xy\, dA$, where R is the region bounded by the graphs of

$y = x^3$, $y = 1$, and $x = 0$

27. $\displaystyle\iint_R (\sin x - y)\, dA$, where R is the region bounded by the

graphs of $y = \cos x$, $y = 0$, $x = 0$, and $x = \pi/2$

28. $\displaystyle\iint_R (x^2 + y)\, dA$, where R is the region bounded by the

graphs of $y = x^2 + 2$, $x = 0$, $x = 1$ and $y = 0$

29. $\displaystyle\iint_R 4x^3\, dA$, where R is the region bounded by the graphs of

$y = (x - 1)^2$ and $y = -x + 3$

30. $\displaystyle\iint_R 2xy^2\, dA$, where R is the region bounded by the graphs of

$x = y^2$ and $x = 3 - 2y^2$

31. $\displaystyle\iint_R ye^x\, dA$, where R is the triangular region with vertices

$(0, 0)$, $(4, 4)$, and $(6, 0)$

32. $\displaystyle\iint_R y\, dA$, where R is the half-disk defined by the inequalities

$x^2 + y^2 \le 1$ and $y \ge 0$

In Exercises 33–38, find the volume of the solid shown in the figure.

33.

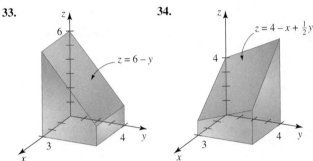

34.

35.

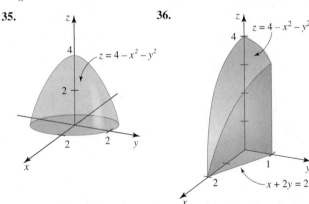

36.

37.

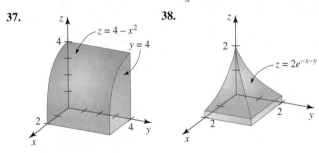

38.

In Exercises 39–46, find the volume of the solid.

39. The solid under the plane $z = 4 - 2x - y$ and above the region $R = \{(x, y) \,|\, 0 \le x \le 1, 0 \le y \le 2\}$ lying in the xy-plane

40. The solid under the plane $z = x + 2y$ and above the triangular region in the xy-plane bounded by the lines $y = 2x$, $y = 0$, and $x = 2$

41. The solid under the surface $z = xy$ and above the triangular region in the xy-plane bounded by the lines $y = 2x$, $y = -x + 6$, and $y = 0$

42. The solid under the surface $z = x^2 + y$ and above the region in the xy-plane bounded by the parabolas $y = x^2$ and $y = 2 - x^2$

43. The solid under the paraboloid $z = x^2 + y^2$ and above the region in the xy-plane bounded by the line $y = x$ and the parabola $y = x^2$

44. The solid under the paraboloid $z = x^2 + 3y^2$ and above the region in the xy-plane bounded by the graphs of $y = \sqrt{x}$, $y = 0$, and $x = 4$

45. The solid bounded by the cylinder $y^2 + z^2 = 9$ and the planes $x = 0$, $y = 0$, $z = 0$, and $2x + y = 2$

46. The solid bounded by the cylinder $x^2 + y^2 = 4$ and the planes $z = 4 - y$ and $z = 0$

In Exercises 47–54, sketch the region of integration for the iterated integral, and reverse the order of integration.

47. $\displaystyle\int_0^1 \int_0^{1-x} f(x, y)\, dy\, dx$

48. $\displaystyle\int_0^1 \int_{2x}^2 f(x, y)\, dy\, dx$

49. $\displaystyle\int_0^1 \int_{y^2}^{\sqrt{y}} f(x, y)\, dx\, dy$

50. $\displaystyle\int_{-2}^2 \int_{-\sqrt{4-y^2}}^{4-y^2} f(x, y)\, dx\, dy$

51. $\displaystyle\int_{-1}^{5/2} \int_{y^2-4}^{(3/2)y-3/2} f(x, y)\, dx\, dy$

52. $\displaystyle\int_{-1}^1 \int_{x^2}^{3-2x^2} f(x, y)\, dy\, dx$

53. $\displaystyle\int_1^e \int_0^{\ln x} f(x, y)\, dy\, dx$

54. $\displaystyle\int_0^{\pi/4} \int_0^{\tan x} f(x, y)\, dy\, dx$

In Exercises 55–60, evaluate the integral by reversing the order of integration.

55. $\displaystyle\int_0^1 \int_{2y}^2 e^{-x^2}\, dx\, dy$

56. $\displaystyle\int_0^2 \int_{y/2}^1 e^{y/x}\, dx\, dy$

57. $\displaystyle\int_0^4 \int_{\sqrt{x}}^2 \sin y^3\, dy\, dx$

58. $\displaystyle\int_0^2 \int_{x^2}^4 x \cos y^2\, dy\, dx$

59. $\displaystyle\int_0^4 \int_{\sqrt{y}}^2 \frac{1}{\sqrt{x^3 + 1}}\, dx\, dy$

60. $\displaystyle\int_0^1 \int_{\tan^{-1} y}^{\pi/4} \sec^2 x \sqrt{1 + \sec^2 x}\, dx\, dy$

61. Suppose that $f(x, y) = g(x)h(y)$ and let $R = \{(x, y) \mid a \le x \le b, c \le y \le d\}$. Show that

$$\iint_R f(x, y)\, dA = \left[\int_a^b g(x)\, dx\right]\left[\int_c^d h(y)\, dy\right]$$

62. Suppose that $f(x, y)$ has continuous second-order partial derivatives. Find

$$\iint_R f_{xy}(x, y)\, dA$$

where $R = \{(x, y) \mid a \le x \le b, c \le y \le d\}$.

63. The following figure depicts a semicircular metal plate whose density at the point (x, y) is $(1 + y)$ slugs/ft^2. What is the mass of the plate?

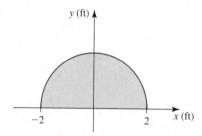

64. Population Density The population density (number of people per square mile) of a coastal town is described by the function

$$f(x, y) = \frac{10{,}000 e^y}{1 + 0.5|x|} \qquad -10 \le x \le 10, \qquad -4 \le y \le 0$$

where x and y are measured in miles. Find the population inside the rectangular area described by

$$R = \{(x, y) \mid -5 \le x \le 5, -2 \le y \le 0\}$$

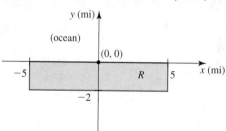

65. Population Density Refer to Exercise 64. Find the average population density inside the rectangular area R.

66. Population Density The population density (number of people per square mile) of a certain city is given by the function

$$f(x, y) = \frac{50{,}000 |xy|}{(x^2 + 20)(y^2 + 36)}$$

where the origin $(0, 0)$ gives the location of the government center. Find the population inside the rectangular area described by $R = \{(x, y) \mid -15 \le x \le 15, -20 \le y \le 20\}$.

cas 67. a. Plot the region R bounded by the graphs of $y = \cos x$ and $y = x^2 + x$ and the y-axis.

    b. Find the x-coordinate of the point of intersection of the graphs of $y = \cos x$ and $y = x^2 + x$ for $x > 0$ accurate to three decimal places.

    c. Estimate $\iint_R x\, dA$.

cas 68. a. Plot the region R bounded by the graphs of $y = e^{-2x}$ and $y = x\sqrt{1 - x^2}$. Then find the x-coordinates of the points of intersection of the two graphs accurate to three decimal places.

   b. Estimate $\iint_R x^{1/3} y^{2/3} \, dA$.

cas *In Exercises 69–72, use a calculator or computer to compute the iterated integral accurate to four decimal places.*

69. $\displaystyle\int_0^1 \int_0^2 x^2 y^3 \cos(x + y) \, dy \, dx$

70. $\displaystyle\int_0^1 \int_1^2 \frac{xy}{\sqrt{x^2 + y^2}} \, dy \, dx$

71. $\displaystyle\int_0^1 \int_0^{1-x} \sqrt{1 + x^2 + y^3} \, dy \, dx$

72. $\displaystyle\int_0^2 \int_{-\sqrt{4-x^2}}^{4-x^2} \frac{e^{xy}}{1 + x^2 + y^2} \, dy \, dx$

In Exercises 73–78, determine whether the statement is true or false. If it is true, explain why. If it is false, explain why or give an example that shows it is false.

73. If f is continuous on $R = [a, b] \times [c, d]$, then
$$\iint_R f(x, y) \, dA = \int_a^b \left[\int_c^d f(x, y) \, dy \right] dx = \int_c^d \left[\int_a^b f(x, y) \, dx \right] dy$$

74. If f is continuous on $R = [a, b] \times [c, d]$, then
$$\int_c^d \left[\int_a^b f(x, y) \, dx \right] dy = \int_d^c \left[\int_b^a f(x, y) \, dx \right] dy$$

75. If f is a nonnegative continuous function on the interval $[a, b]$, then the area under the graph of f on $[a, b]$ is $\int_a^b \left[\int_0^{f(x)} dy \right] dx$.

76. If f is continuous on $R = [0, 1] \times [0, 1]$, then
$$\int_0^1 \left[\int_0^y f(x, y) \, dx \right] dy = \int_0^1 \left[\int_0^x f(x, y) \, dy \right] dx$$

77. $\displaystyle\int_0^2 \int_{-1}^1 x \cos(y^2) \, dx \, dy \neq 0$

78. $\displaystyle\int_0^1 \int_0^1 (\sqrt{x} + y)\cos(\sqrt{xy}) \, dx \, dy \leq 1.2$

14.3 Double Integrals in Polar Coordinates

▉ Polar Rectangles

Some double integrals are easier to evaluate if they are expressed in terms of polar coordinates. This is especially true when the region of integration is a **polar rectangle,**

$$R = \{(r, \theta) \mid a \leq r \leq b, \alpha \leq \theta \leq \beta\}$$

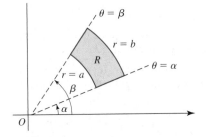

FIGURE 1
A polar rectangle is bounded by circular arcs and rays.

(See Figure 1.) Observe that R is a part of an annular ring with inner radius $r = a$ and outer radius $r = b$. Therefore, its area is the difference between the area of the circular sector of radius b and central angle $\Delta\theta = \beta - \alpha$, and the area of the circular sector of radius a and the same central angle $\Delta\theta$. Since the area of a circular sector of radius r and central angle θ is $\frac{1}{2} r^2 \theta$, we see that the area of R is

$$A = \frac{1}{2} b^2 \, \Delta\theta - \frac{1}{2} a^2 \, \Delta\theta = \frac{1}{2} (b^2 - a^2) \, \Delta\theta \tag{1}$$

$$= \frac{1}{2} (b + a)(b - a) \, \Delta\theta = \bar{r} \, \Delta r \, \Delta\theta$$

where $\Delta r = b - a$ and $\bar{r} = \frac{1}{2}(b + a)$ is the *average radius* of the polar rectangle.

▉ Double Integrals Over Polar Rectangles

To define a double integral over a polar rectangle R, suppose f is a continuous function on R. We start by taking a regular partition

$$a = r_0 < r_1 < r_2 < \cdots < r_{i-1} < r_i < \cdots < r_m = b$$

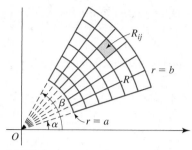

FIGURE 2

A polar partition of the polar region R with $m = 6$ and $n = 6$

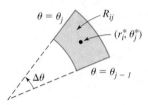

FIGURE 3

A polar subrectangle R_{ij} and its center (r_i^*, θ_j^*)

of $[a, b]$ into m subintervals of equal length $\Delta r = (b - a)/m$, and a regular partition

$$\alpha = \theta_0 < \theta_1 < \theta_2 < \cdots < \theta_{j-1} < \theta_j < \cdots < \theta_n = \beta$$

of $[\alpha, \beta]$ into n subintervals of equal length $\Delta \theta = (\beta - \alpha)/n$. Then the circles $r = r_i$ and the rays $\theta = \theta_j$ determine a **polar partition** P of R into $N = mn$ polar rectangles $R_{11}, R_{12}, \ldots, R_{ij}, \ldots, R_{mn}$, where $R_{ij} = \{(r, \theta) \mid r_{i-1} \le r \le r_i, \theta_{j-1} \le \theta \le \theta_j\}$, as shown in Figure 2. Figure 3 shows a typical polar subrectangle R_{ij} enlarged for the sake of clarity. The center of R_{ij} is the point (r_i^*, θ_j^*), where r_i^* is the average radius of R_{ij}, and θ_j^* is the average angle of R_{ij}. In other words, $r_i^* = \frac{1}{2}(r_{i-1} + r_i)$ and $\theta_j^* = \frac{1}{2}(\theta_{j-1} + \theta_j)$. Observe that the center of R_{ij}, when expressed in terms of rectangular coordinates, takes the form $(r_i^* \cos \theta_j^*, r_i^* \sin \theta_j^*)$. Also, from Equation (1) we see that the area of R_{ij} is $\Delta A_i = r_i^* \Delta r \Delta \theta$. Therefore, the Riemann sum of f over the polar partition P is

$$\sum_{i=1}^{m} \sum_{j=1}^{n} f(r_i^* \cos \theta_j^*, r_i^* \sin \theta_j^*) \Delta A_i = \sum_{i=1}^{m} \sum_{j=1}^{n} f(r_i^* \cos \theta_j^*, r_i^* \sin \theta_j^*) \, r_i^* \Delta r \Delta \theta$$

$$= \sum_{i=1}^{m} \sum_{j=1}^{n} g(r_i^*, \theta_j^*) \Delta r \Delta \theta$$

where $g(r, \theta) = rf(r \cos \theta, r \sin \theta)$. But the last sum is just a Riemann sum associated with the double integral

$$\int_{\alpha}^{\beta} \int_{a}^{b} g(r, \theta) \, dr \, d\theta$$

Therefore, we have

$$\iint_R f(x, y) \, dA = \lim_{m, n \to \infty} \sum_{i=1}^{m} \sum_{j=1}^{n} f(r_i^* \cos \theta_j^*, r_i^* \sin \theta_j^*) \Delta A$$

$$= \lim_{m, n \to \infty} \sum_{i=1}^{m} \sum_{j=1}^{n} g(r_i^*, \theta_j^*) \Delta r \Delta \theta$$

$$= \int_{\alpha}^{\beta} \int_{a}^{b} g(r, \theta) \, dr \, d\theta = \int_{\alpha}^{\beta} \int_{a}^{b} f(r \cos \theta, r \sin \theta) \, r \, dr \, d\theta$$

Transforming a Double Integral Over a Polar Rectangle to Polar Coordinates

Let f be continuous on a polar rectangle $R = \{(r, \theta) \mid 0 \le a \le r \le b, \alpha \le \theta \le \beta\}$, where $0 \le \beta - \alpha \le 2\pi$. Then

$$\iint_R f(x, y) \, dA = \int_{\alpha}^{\beta} \int_{a}^{b} f(r \cos \theta, r \sin \theta) \, r \, dr \, d\theta \tag{2}$$

Thus, we formally transform a double integral over a polar rectangle from rectangular to polar coordinates by substituting

$$x = r \cos \theta, \qquad y = r \sin \theta, \qquad dA = r \, dr \, d\theta$$

and inserting the appropriate limits.

⚠ Do not forget the factor r on the right-hand side of Equation (2). You can remember the expression for dA by making a sketch of the "infinitesimal polar rectangle" shown in Figure 4. The polar rectangle is similar to an ordinary rectangle with sides of length $r \, d\theta$ and dr, and therefore, it has "area" $dA = (r \, d\theta) \, dr = r \, dr \, d\theta$.

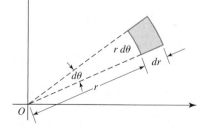

FIGURE 4

The infinitesimal polar rectangle has "area" $dA = r \, dr \, d\theta$.

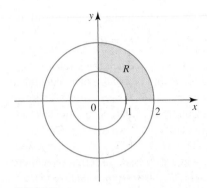

FIGURE 5
The region
$R = \left\{(r, \theta) \,|\, 1 \le r \le 2, 0 \le \theta \le \frac{\pi}{2}\right\}$

EXAMPLE 1 Evaluate $\iint_R (2x + 3y) \, dA$, where R is the region in the first quadrant bounded by the circles $x^2 + y^2 = 1$ and $x^2 + y^2 = 4$.

Solution The region R is a polar rectangle that can also be described in terms of polar coordinates by

$$R = \left\{(r, \theta) \,|\, 1 \le r \le 2, 0 \le \theta \le \frac{\pi}{2}\right\}$$

(See Figure 5.) Using Equation (2), we obtain

$$\iint_R (2x + 3y) \, dA = \int_0^{\pi/2} \int_1^2 (2r \cos \theta + 3r \sin \theta) \, r \, dr \, d\theta$$

$$= \int_0^{\pi/2} \int_1^2 (2r^2 \cos \theta + 3r^2 \sin \theta) \, dr \, d\theta$$

$$= \int_0^{\pi/2} \left[\frac{2}{3} r^3 \cos \theta + r^3 \sin \theta\right]_{r=1}^{r=2} d\theta$$

$$= \int_0^{\pi/2} \left(\frac{14}{3} \cos \theta + 7 \sin \theta\right) d\theta$$

$$= \left[\frac{14}{3} \sin \theta - 7 \cos \theta\right]_0^{\pi/2} = \frac{35}{3}$$

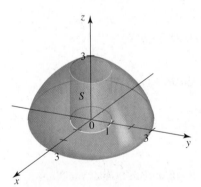

FIGURE 6
The solid S lies above the disk
$x^2 + y^2 \le 1$ and under the
hemisphere $z = \sqrt{9 - x^2 - y^2}$.

EXAMPLE 2 Find the volume of the solid S that lies below the hemisphere $z = \sqrt{9 - x^2 - y^2}$, above the xy-plane, and inside the cylinder $x^2 + y^2 = 1$.

Solution The solid S is shown in Figure 6. It lies between the hemisphere $z = \sqrt{9 - x^2 - y^2}$ and the circular disk centered at the origin with radius 1. A polar representation of R is

$$R = \{(r, \theta) \,|\, 0 \le r \le 1, 0 \le \theta \le 2\pi\}$$

Also, in polar coordinates we can write $z = \sqrt{9 - x^2 - y^2} = \sqrt{9 - r^2}$. Therefore, the required volume is given by

$$V = \iint_R f(x, y) \, dA = \int_0^{2\pi} \int_0^1 \sqrt{9 - r^2} \, r \, dr \, d\theta$$

$$= \int_0^{2\pi} \left[-\frac{1}{3} (9 - r^2)^{3/2}\right]_{r=0}^{r=1} d\theta$$

$$= \frac{1}{3} (27 - 16\sqrt{2}) \int_0^{2\pi} d\theta = \frac{2\pi}{3} (27 - 16\sqrt{2})$$

or approximately 9.16.

Note You can appreciate the role played by polar coordinates in Example 2 by observing that in rectangular coordinates,

$$V = \int_{-1}^1 \int_{-\sqrt{1-y^2}}^{\sqrt{1-y^2}} \sqrt{9 - x^2 - y^2} \, dx \, dy$$

which is not easy to evaluate.

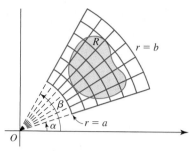

FIGURE 7
An inner polar partition of the region R

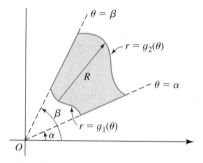

FIGURE 8
The polar region
$R = \{(r, \theta) \,|\, \alpha \leq \theta \leq \beta,$
$g_1(\theta) \leq r \leq g_2(\theta)\}$. Observe
that r runs from the curve
$r = g_1(\theta)$ to the curve $r = g_2(\theta)$
as indicated by the arrow.

■ Double Integrals Over General Regions

The results obtained thus far can be extended to more general regions. If the bounded region R is such a region, then we can transform the double integral $\iint_R f(x, y)\, dA$ into one involving polar coordinates by expressing it as a limit of Riemann sums associated with the function

$$f_R(x, y) = \begin{cases} f(x, y) & \text{if } (x, y) \text{ is in } R \\ 0 & \text{if } (x, y) \text{ is outside } R \end{cases}$$

(See Figure 7.)

We will not pursue the details. Instead, we will state the result for the type of region that occurs most frequently in practice: A region is **r-simple** if it is bounded by the graphs of two functions of θ. The r-simple region described by

$$R = \{(r, \theta) \,|\, \alpha \leq \theta \leq \beta, g_1(\theta) \leq r \leq g_2(\theta)\}$$

where g_1 and g_2 are continuous on $[\alpha, \beta]$, is shown in Figure 8.

Transforming a Double Integral Over a Polar Region to Polar Coordinates

Let f be continuous on a polar region of the form

$$R = \{(r, \theta) \,|\, \alpha \leq \theta \leq \beta, g_1(\theta) \leq r \leq g_2(\theta)\}$$

where $0 \leq \beta - \alpha \leq 2\pi$. Then

$$\iint_R f(x, y)\, dA = \int_\alpha^\beta \int_{g_1(\theta)}^{g_2(\theta)} f(r \cos \theta, r \sin \theta)\, r\, dr\, d\theta \tag{3}$$

Note θ-simple regions (regions that are bounded by the graphs of functions of r) will be considered in Exercise 44. ■

EXAMPLE 3 Use a double integral to find the area enclosed by one loop of the three-leaved rose $r = \sin 3\theta$.

Solution The graph of $r = \sin 3\theta$ is shown in Figure 9. Observe that a loop of the rose is described by the region

$$R = \left\{(r, \theta) \,\middle|\, 0 \leq \theta \leq \tfrac{\pi}{3}, 0 \leq r \leq \sin 3\theta\right\}$$

and may be viewed as being r-simple, where $g_1(\theta) = 0$ and $g_2(\theta) = \sin 3\theta$. Taking $f(x, y) = 1$ in Equation (3), we see that the required area is given by

$$A = \iint_R dA = \int_0^{\pi/3} \int_0^{\sin 3\theta} r\, dr\, d\theta$$

$$= \int_0^{\pi/3} \left[\frac{1}{2} r^2 \right]_{r=0}^{r=\sin 3\theta} d\theta$$

$$= \frac{1}{2} \int_0^{\pi/3} \sin^2 3\theta\, d\theta = \frac{1}{4} \int_0^{\pi/3} (1 - \cos 6\theta)\, d\theta \quad \sin^2 \theta = \frac{1 - \cos 2\theta}{2}$$

$$= \frac{1}{4} \left[\theta - \frac{1}{6} \sin 6\theta \right]_{\theta=0}^{\theta=\pi/3} = \frac{\pi}{12}$$

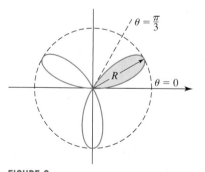

FIGURE 9
The region R viewed as an r-simple region

or approximately 0.26. ■

EXAMPLE 4 Evaluate $\iint_R y \, dA$, where R is the region in the first quadrant that is outside the circle $r = 2$ and inside the cardioid $r = 2(1 + \cos \theta)$.

Solution The required region

$$R = \left\{ (r, \theta) \,\middle|\, 0 \leq \theta \leq \frac{\pi}{2}, 2 \leq r \leq 2(1 + \cos \theta) \right\}$$

is sketched in Figure 10 and may be viewed as being r-simple. Recalling that $y = r \sin \theta$ and using Equation (3), we obtain

$$\iint_R y \, dA = \int_0^{\pi/2} \int_2^{2(1+\cos \theta)} r(\sin \theta) \, r \, dr \, d\theta = \int_0^{\pi/2} \int_2^{2(1+\cos \theta)} r^2 (\sin \theta) \, dr \, d\theta$$

$$= \int_0^{\pi/2} \left[\frac{1}{3} r^3 \sin \theta \right]_{r=2}^{r=2(1+\cos \theta)} d\theta$$

$$= \frac{8}{3} \int_0^{\pi/2} \left[(1 + \cos \theta)^3 \sin \theta - \sin \theta \right] d\theta$$

$$= \frac{8}{3} \left[-\frac{1}{4} (1 + \cos \theta)^4 + \cos \theta \right]_0^{\pi/2} = \frac{22}{3}$$

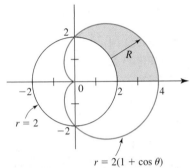

FIGURE 10
The polar region
$R = \left\{ (r, \theta) \,\middle|\, 0 \leq \theta \leq \frac{\pi}{2}, \right.$
$\left. 2 \leq r \leq 2(1 + \cos \theta) \right\}$

EXAMPLE 5 Find the volume of the solid that lies below the paraboloid $z = 4 - x^2 - y^2$, above the xy-plane, and inside the cylinder $(x - 1)^2 + y^2 = 1$.

Solution The solid S under consideration is shown in Figure 11a. It lies above the disk R bounded by the circle with center $(1, 0)$ and radius 1 shown in Figure 11b. This unit circle has polar equation $r = 2 \cos \theta$, as you can verify by replacing x and y in the rectangular equation of the circle by $x = r \cos \theta$ and $y = r \sin \theta$. Therefore,

$$R = \left\{ (r, \theta) \,\middle|\, -\frac{\pi}{2} \leq \theta \leq \frac{\pi}{2}, 0 \leq r \leq 2 \cos \theta \right\}$$

and may be viewed as being r-simple, where $g_1(\theta) = 0$ and $g_2(\theta) = 2 \cos \theta$. Using the relationship $x^2 + y^2 = r^2$ and taking advantage of symmetry, we see that the required volume is

$$V = \iint_R (4 - x^2 - y^2) \, dA = \int_{-\pi/2}^{\pi/2} \int_0^{2 \cos \theta} (4 - r^2) \, r \, dr \, d\theta = 2 \int_0^{\pi/2} \int_0^{2 \cos \theta} (4r - r^3) \, dr \, d\theta$$

$$= 2 \int_0^{\pi/2} \left[2r^2 - \frac{1}{4} r^4 \right]_{r=0}^{r=2 \cos \theta} d\theta = 8 \int_0^{\pi/2} (2 \cos^2 \theta - \cos^4 \theta) \, d\theta$$

$$= 8 \int_0^{\pi/2} \left[1 + \cos 2\theta - \left(\frac{1 + \cos 2\theta}{2} \right)^2 \right] d\theta \qquad \cos^2 \theta = \frac{1 + \cos 2\theta}{2}$$

$$= 8 \int_0^{\pi/2} \left[\frac{3}{4} + \frac{1}{2} \cos 2\theta - \frac{1 + \cos 4\theta}{8} \right] d\theta$$

$$= 8 \left[\frac{5}{8} \theta + \frac{1}{4} \sin 2\theta - \frac{1}{32} \sin 4\theta \right]_0^{\pi/2} = \frac{5\pi}{2}$$

or approximately 7.85.

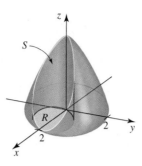

FIGURE 11   (a) The solid S (b) The region R is r-simple.

14.3 CONCEPT QUESTIONS

1. a. What is a polar rectangle? Illustrate with a sketch.
   b. Suppose f is continuous on a polar rectangle
   $R = \{(r, \theta) \mid a \leq r \leq b, \alpha \leq \theta \leq \beta\}$, where
   $0 \leq \beta - \alpha \leq 2\pi$. Write $\iint_R f(x, y) \, dA$ in terms of polar
   coordinates.

2. a. What is an r-simple region? Illustrate with a sketch.
   b. Suppose that f is continuous on a region of the form
   $R = \{(r, \theta) \mid \alpha \leq \theta \leq \beta, g_1(\theta) \leq r \leq g_2(\theta)\}$ where
   $0 \leq \beta - \alpha \leq 2\pi$. Write $\iint_R f(x, y) \, dA$ in terms of polar
   coordinates.

14.3 EXERCISES

*In Exercises 1–4, determine whether to use polar coordinates or
rectangular coordinates to evaluate the integral $\iint_R f(x, y) \, dA$,
where f is a continuous function. Then write an expression for
the (iterated) integral.*

1.

2.

3.

4.

*In Exercises 5–8, sketch the region of integration associated with
the integral.*

5. $\displaystyle\int_0^\pi \int_1^4 f(r \cos \theta, r \sin \theta) \, r \, dr \, d\theta$

6. $\displaystyle\int_0^\pi \int_0^{4 \sin \theta} f(r \cos \theta, r \sin \theta) \, r \, dr \, d\theta$

7. $\displaystyle\int_{\pi/4}^{\pi/2} \int_0^{2\sqrt{2}} f(r \cos \theta, r \sin \theta) \, r \, dr \, d\theta$

8. $\displaystyle\int_0^{2\pi} \int_0^{1 + \cos \theta} f(r \cos \theta, r \sin \theta) \, r \, dr \, d\theta$

*In Exercises 9–16, evaluate the integral by changing to polar
coordinates.*

9. $\displaystyle\iint_R 3y \, dA$, where R is the disk of radius 2 centered at the
origin

10. $\displaystyle\iint_R (x + 2y) \, dA$, where R is the region in the first quadrant
bounded by the circle $x^2 + y^2 = 9$

11. $\displaystyle\iint_R xy \, dA$, where R is the region in the first quadrant
bounded by the circle $x^2 + y^2 = 4$ and the lines $x = 0$ and
$x = y$

12. $\displaystyle\iint_R \sqrt{x^2 + y^2} \, dA$, where R is the region in the first quadrant
bounded by the circle $x^2 + y^2 = 4$ and the lines $y = 0$ and
$y = \sqrt{3}x$

13. $\displaystyle\iint_R \frac{y^2}{x^2 + y^2} \, dA$, where R is the annular region bounded by
the circles $x^2 + y^2 = 1$ and $x^2 + y^2 = 2$

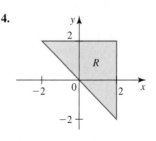

14. $\iint\limits_{R} \sin(x^2 + y^2)\, dA$, where R is the region in the first

quadrant bounded by the circles $x^2 + y^2 = 1$ and
$x^2 + y^2 = 9$

15. $\iint\limits_{R} y\, dA$, where R is the smaller of the two regions bounded

by the circle $x^2 + y^2 = 2x$ and the line $y = x$

16. $\iint\limits_{R} (x + y)\, dA$, where R is the region in the first quadrant

bounded by the circles $x^2 + y^2 = 4$ and $x^2 + y^2 = 2y$

*In Exercises 17–26, use polar coordinates to find the volume of
the solid region T.*

17. T lies below the paraboloid $z = x^2 + y^2$, above the xy-plane,
and inside the cylinder $x^2 + y^2 = 4$.

18. T lies below the paraboloid $z = 9 - x^2 - y^2$, above the
xy-plane, and inside the cylinder $x^2 + y^2 = 1$.

19. T lies below the cone $z = \sqrt{x^2 + y^2}$, above the xy-plane,
and inside the cylinder $x^2 + y^2 = 4$.

20. T lies below the cone $z = \sqrt{x^2 + y^2}$, above the xy-plane,
inside the cylinder $x^2 + y^2 = 4$, and outside the cylinder
$x^2 + y^2 = 1$.

21. T lies under the plane $3x + 4y + z = 12$, above the xy-
plane, and inside the cylinder $x^2 + y^2 = 2x$.

22. T lies under the paraboloid $z = x^2 + y^2$, above the xy-plane,
and inside the cylinder $x^2 + y^2 = 2y$.

23. T is bounded by the paraboloid $z = 9 - 2x^2 - 2y^2$ and the
plane $z = 1$.

24. T is bounded by the paraboloids $z = 5x^2 + 5y^2$ and
$z = 12 - x^2 - y^2$.

25. T is below the sphere $x^2 + y^2 + z^2 = 2$ and above the cone
$z = \sqrt{x^2 + y^2}$.

26. T is inside the sphere $x^2 + y^2 + z^2 = 4$ and inside the
cylinder $x^2 + y^2 = 2y$.

*In Exercises 27–32, use a double integral to find the area of the
region R.*

27. R is bounded by the circle $r = 3 \cos \theta$.

28. R is bounded by one loop of the four-leaved rose $r = \cos 2\theta$.

29. R is bounded by the cardioid $r = 3 - 3 \sin \theta$.

30. R is bounded by the lemniscate $r^2 = 4 \cos 2\theta$.

31. R is outside the circle $r = a$ and inside the circle
$r = 2a \sin \theta$.

32. R is inside the circle $r = 3 \sin \theta$ and outside the cardioid
$r = 1 + \sin \theta$.

*In Exercises 33–40, evaluate the integral by changing to polar
coordinates.*

33. $\displaystyle\int_{-2}^{2} \int_{0}^{\sqrt{4-x^2}} \sqrt{x^2 + y^2}\, dy\, dx$

34. $\displaystyle\int_{0}^{3} \int_{0}^{\sqrt{9-x^2}} (x^2 + y^2)^{3/2}\, dy\, dx$

35. $\displaystyle\int_{-1}^{1} \int_{0}^{\sqrt{1-y^2}} \frac{1}{1 + x^2 + y^2}\, dx\, dy$

36. $\displaystyle\int_{1}^{3} \int_{0}^{x} \frac{1}{\sqrt{x^2 + y^2}}\, dy\, dx$

37. $\displaystyle\int_{-2}^{2} \int_{0}^{\sqrt{4-x^2}} e^{x^2+y^2}\, dy\, dx$

38. $\displaystyle\int_{0}^{1} \int_{0}^{\sqrt{1-y^2}} \cos(x^2 + y^2)\, dx\, dy$

39. $\displaystyle\int_{0}^{2} \int_{-\sqrt{2x-x^2}}^{\sqrt{2x-x^2}} x\, dy\, dx$

40. $\displaystyle\int_{0}^{1} \int_{0}^{\sqrt{1-x^2}} \tan^{-1}\left(\frac{y}{x}\right) dy\, dx$

*In Exercises 41 and 42, write the sum of the double integrals as
a simple double integral using polar coordinates. Then evaluate
the resulting integral.*

41. $\displaystyle\int_{0}^{\sqrt{2}} \int_{0}^{x} xy\, dy\, dx + \int_{\sqrt{2}}^{2} \int_{0}^{\sqrt{4-x^2}} xy\, dy\, dx$

42. $\displaystyle\int_{0}^{1} \int_{\sqrt{1-x^2}}^{\sqrt{4-x^2}} \sqrt{x^2 + y^2}\, dy\, dx + \int_{1}^{2} \int_{0}^{\sqrt{4-x^2}} \sqrt{x^2 + y^2}\, dy\, dx$

43. a. Suppose that f is continuous on the region R bounded by
the lines $y = x$, $y = -x$, and $y = 1$. Show that

$$\iint\limits_{R} f(x, y)\, dA = \int_{\pi/4}^{3\pi/4} \int_{0}^{\csc \theta} f(r \cos \theta, r \sin \theta)\, r\, dr\, d\theta$$

b. Use the result of part (a) to evaluate

$$\int_{0}^{1} \int_{-y}^{y} \sqrt{x^2 + y^2}\, dx\, dy$$

44. A region is **θ-simple** if it is bounded by the graphs of two
functions of r. A θ-simple region is described by

$$R = \{(r, \theta) \mid a \leq r \leq b, g_1(r) \leq \theta \leq g_2(r)\}$$

where g_1 and g_2 are continuous on $[a, b]$. It can be shown
that if f is continuous on R, then

$$\iint\limits_{R} f(x, y)\, dA = \int_{a}^{b} \int_{g_1(r)}^{g_2(r)} f(r \cos \theta, r \sin \theta)\, r\, d\theta\, dr$$

Use this formula to find the area of the smaller region bounded by the spiral $r\theta = 1$, the circles $r = 1$ and $r = 2$, and the polar axis.

45. The integral $I = \int_{-\infty}^{\infty} e^{-x^2/2} \, dx$ occurs in the study of probability and statistics. Show that $I = \sqrt{2\pi}$ by verifying the following steps.

   a. Sketch the regions $R_1 = \{(x, y) \mid x^2 + y^2 \le a^2,$
   $x \ge 0, y \ge 0\}$, $R_2 = \{(x, y) \mid 0 \le x \le a, 0 \le y \le a\}$,
   and $R_3 = \{(x, y) \mid x^2 + y^2 \le 2a^2, x \ge 0, y \ge 0\}$ on the
   same plane. Observe that R_1 lies inside R_2 and that R_2
   lies inside R_3.

   b. Show that

   $$\iint_{R_1} f(r, \theta) \, dA = \frac{\pi}{4}\left(1 - e^{-a^2}\right)$$

   where $f(r, \theta) = e^{-r^2}$, and that

   $$\iint_{R_3} f(r, \theta) \, dA = \frac{\pi}{4}\left(1 - e^{-2a^2}\right)$$

   c. By considering $\iint_{R_2} f(x, y) \, dA$, where $f(x, y) = e^{-x^2 - y^2}$,
   and using the results of part (b), show that

   $$\frac{\pi}{4}\left(1 - e^{-a^2}\right) < \left(\int_0^a e^{-x^2} \, dx\right)^2 < \frac{\pi}{4}\left(1 - e^{-2a^2}\right)$$

   d. Show that $\int_0^{\infty} e^{-x^2} \, dx = \lim_{a \to \infty} \int_0^a e^{-x^2} \, dx = \sqrt{\pi}/2$, and hence deduce the result $I = \sqrt{2\pi}$.

Use the results of Exercise 45 to evaluate the integrals in Exercises 46 and 47.

46. $\displaystyle\int_0^{\infty} x^2 e^{-x^2} \, dx$

47. $\displaystyle\int_0^{\infty} \frac{e^{-x}}{\sqrt{x}} \, dx$

48. **Water Delivered by a Water Sprinkler** A lawn sprinkler sprays water in a circular pattern. It delivers water to a depth of $f(r) = 0.1re^{-0.1r}$ ft/hr at a distance of r ft from the sprinkler.

   a. Find the total amount of water that is accumulated in an hour in a circular region of radius 50 ft centered at the sprinkler.

   b. What is the average amount of water that is delivered to the region in part (a) in an hour?
   Hint: The average value of f over a region

   $$D = \frac{1}{A(D)} \iint_D f(x, y) \, dA$$

   where $A(D)$, is the area of D.

In Exercises 49 and 50, determine whether the statement is true or false. If it is true, explain why. If it is false, explain why or give an example that shows it is false.

49. If $R = \{(r, \theta) \mid \alpha \le \theta \le \beta, 0 \le r \le g(\theta)\}$, where
   $0 \le \beta - \alpha \le 2\pi$ and $f(r \cos \theta, r \sin \theta) = 1$ for all (r, θ) in
   R, then $\int_{\alpha}^{\beta} \int_0^{g(\theta)} f(r \cos \theta, r \sin \theta) \, r \, dr \, d\theta$ gives the area of R.

50. If R is the triangular region, whose vertices in rectangular coordinates are $(0, 0)$, $(1, 0)$, and $(1, 1)$, then
   $\iint_R f(x, y) \, dA = \int_0^{\pi/4} \int_0^{\sec \theta} f(r \cos \theta, r \sin \theta) \, r \, dr \, d\theta$, where
   r and θ are polar coordinates.

14.4 Applications of Double Integrals

▪ Mass of a Lamina

We mentioned in Section 14.1 that the mass of a thin rectangular plate R lying in the xy-plane and having mass density $\rho(x, y)$ at a point (x, y) in R is given by the volume of the solid region T lying directly above R and bounded above by $z = \rho(x, y)$. (See Figure 1.) We will now show that this is the case. In fact, we will demonstrate that the mass of a lamina occupying a region R in the xy-plane and having mass density $\rho(x, y)$ at a point (x, y), where ρ is a nonnegative continuous function, is given by $\iint_R \rho(x, y) \, dA$. The double integral also gives the volume of the solid region lying directly above R and bounded above by the surface $z = \rho(x, y)$. (See Figure 2.)

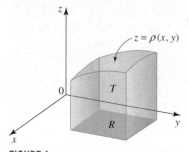

FIGURE 1
The mass of the plate R is numerically equal to the volume of the solid T.

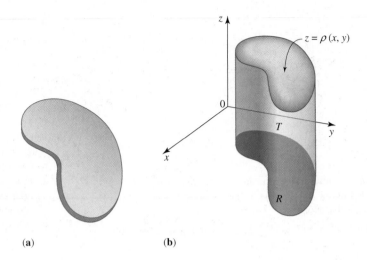

FIGURE 2
The mass of the lamina in part (a)
is numerically equal to the
volume of the solid T in part (b).

(a)   **(b)**

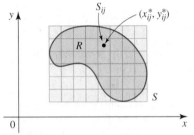

FIGURE 3
$P = \{S_{11}, S_{12}, \ldots, S_{ij}, \ldots, S_{mn}\}$ is a
partition of S.

Let S be a rectangle containing R, and let $P = \{S_{11}, S_{12}, \ldots, S_{ij}, \ldots, S_{mn}\}$ be a regular partition of S. (See Figure 3.) Define

$$\rho_R(x, y) = \begin{cases} \rho(x, y) & \text{if } (x, y) \text{ is in } R \\ 0 & \text{if } (x, y) \text{ is inside } S \text{ but outside } R \end{cases}$$

Let (x_{ij}^*, y_{ij}^*) be a point in S_{ij} that also lies in R. If both m and n are large (so that the dimensions of S_{ij} are small), then the continuity of ρ implies that $\rho(x, y)$ is approximately equal to $\rho(x_{ij}^*, y_{ij}^*)$ for all points (x, y) in S_{ij}. Therefore, the mass of that piece of R lying in S_{ij} with area ΔA is approximately

$$\rho(x_{ij}^*, y_{ij}^*)\, \Delta A \qquad \text{constant density} \cdot \text{area}$$

Summing the masses of all such pieces gives an approximation of the mass of R:

$$\sum_{i=1}^{m} \sum_{j=1}^{n} \rho(x_{ij}^*, y_{ij}^*)\, \Delta A$$

We can expect the approximation to improve as both m and n get larger and larger. Therefore, it is reasonable to define the mass of the lamina as the limiting value of the sums of this form. But each of these sums is just the Riemann sum of ρ_R over S. This leads to the following definition.

DEFINITION   **Mass of a Lamina**
Suppose that a lamina occupies a region R in the plane and the mass density of the lamina at a point (x, y) in R is $\rho(x, y)$, where ρ is a continuous density function. Then the mass of the lamina is given by

$$m = \iint\limits_{R} \rho(x, y)\, dA \tag{1}$$

Note   We obtain other physical interpretations of the double integral $\iint_R f(x, y)\, dA$ by letting f represent various types of densities. For example, if an electric charge is spread over a plane surface R and the charge density (charge per unit area) at a point (x, y) in R is $\sigma(x, y)$, then the total charge on the surface is given by

$$Q = \iint\limits_R \sigma(x, y) \, dA \tag{2}$$

For another example, suppose that the population density (number of people per unit area) at a point (x, y) in a plane region R is $\delta(x, y)$; then the total population in the region is given by

$$N = \iint\limits_R \delta(x, y) \, dA \tag{3}$$

∎

EXAMPLE 1 Find the mass of a lamina occupying a triangular region R with vertices $(0, 0)$, $(2, 0)$, and $(0, 2)$ if its mass density at a point (x, y) in R is $\rho(x, y) = x + 2y$.

Solution The region R is shown in Figure 4. Viewing R as a y-simple region and using Equation (1), we see that the required mass is given by

$$m = \iint\limits_R \rho(x, y) \, dA = \int_0^2 \int_0^{2-x} (x + 2y) \, dy \, dx$$

$$= \int_0^2 \left[xy + y^2 \right]_{y=0}^{y=2-x} dx = \int_0^2 \left[x(2 - x) + (2 - x)^2 \right] dx$$

$$= \int_0^2 (4 - 2x) \, dx = \left[4x - x^2 \right]_0^2 = 4$$

∎

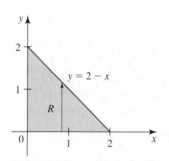

FIGURE 4
The region R is both x-simple and y-simple. Here, we view it as y-simple.

EXAMPLE 2 Electric Charge Over a Region An electric charge is spread over a region R lying in the first quadrant and inside the circle $x^2 + y^2 = 4$. Find the total charge on R if the charge density (measured in coulombs per square meter) at a point (x, y) in R is directly proportional to the square of the distance between the point and the origin.

Solution The region R is shown in Figure 5. The charge density function is given by $\sigma(x, y) = k(x^2 + y^2)$, where k is the constant of proportionality. Viewing R as a y-simple region and using Equation (2), we see that the total charge on R is given by

$$Q = \iint\limits_R \sigma(x, y) \, dA = \int_0^2 \int_0^{\sqrt{4-x^2}} k(x^2 + y^2) \, dy \, dx$$

or, changing to polar coordinates,

$$Q = \int_0^{\pi/2} \int_0^2 (kr^2) \, r \, dr \, d\theta = k \int_0^{\pi/2} \int_0^2 r^3 \, dr \, d\theta = k \int_0^{\pi/2} \left[\frac{1}{4} r^4 \right]_{r=0}^{r=2} d\theta$$

$$= 4k \int_0^{\pi/2} d\theta = 2\pi k$$

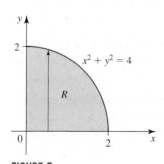

FIGURE 5
The region R is both x-simple and y-simple. Here, we view it as y-simple.

or $2\pi k$ coulombs.

∎

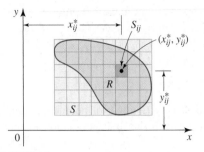

FIGURE 6
The region R contained in a rectangle S

Moments and Center of Mass of a Lamina

We considered the moments and the center of mass of a homogeneous lamina in Section 5.7. Using double integrals, we can now find the moments and center of mass of a lamina with *variable* density. Suppose that a lamina with continuous mass density function ρ occupies a region R in the xy-plane. (See Figure 6.)

Let S be a rectangle containing R, and let $P = \{S_{11}, S_{12}, \ldots, S_{ij}, \ldots, S_{mn}\}$ be a regular partition of S. Choose (x_{ij}^*, y_{ij}^*) to be any evaluation point in S_{ij}. If m and n are large, then the mass of the part of the lamina occupying the subrectangle S_{ij} is approximately $\rho(x_{ij}^*, y_{ij}^*)\,\Delta A$. Consequently, the moment of this part of the lamina with respect to the x-axis is approximately

$$[\rho(x_{ij}^*, y_{ij}^*)\,\Delta A]y_{ij}^* \qquad \text{mass} \cdot \text{moment arm}$$

Adding up these mn moments and taking the limit of the resulting sum as m and n approach infinity, we obtain the moment of the lamina with respect to the x-axis. A similar argument gives the moment of the lamina about the y-axis. These formulas and the formula for the center of mass of a lamina follow.

> **DEFINITION   Moments and Center of Mass of a Lamina**
>
> Suppose that a lamina occupies a region R in the xy-plane and the mass density of the lamina at a point (x, y) in R is $\rho(x, y)$, where ρ is a continuous density function. Then the **moments of mass** of the lamina with respect to the x- and y-axes are
>
> $$M_x = \iint\limits_R y\rho(x, y)\,dA \qquad \text{and} \qquad M_y = \iint\limits_R x\rho(x, y)\,dA \qquad \textbf{(4a)}$$
>
> Furthermore, the **center of mass** of the lamina is located at the point $(\bar{x}, \bar{y})$, where
>
> $$\bar{x} = \frac{M_y}{m} = \frac{1}{m}\iint\limits_R x\rho(x, y)\,dA \qquad \bar{y} = \frac{M_x}{m} = \frac{1}{m}\iint\limits_R y\rho(x, y)\,dA \qquad \textbf{(4b)}$$
>
> where the mass of the lamina is given by
>
> $$m = \iint\limits_R \rho(x, y)\,dA$$

Note   If the density function ρ is constant on R, then the point $(\bar{x}, \bar{y})$ is also called the *centroid* of the region R. (See Section 5.7.) ∎

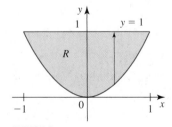

FIGURE 7
The region R occupied by the lamina viewed as being y-simple

EXAMPLE 3   A lamina occupies a region R in the xy-plane bounded by the parabola $y = x^2$ and the line $y = 1$. (See Figure 7.) Find the center of mass of the lamina if its mass density at a point (x, y) is directly proportional to the distance between the point and the x-axis.

Solution   The mass density of the lamina is given by $\rho(x, y) = ky$, where k is the constant of proportionality. Since R is symmetric with respect to the y-axis and the density of the lamina is directly proportional to the distance from the x-axis, we see that the center of mass is located on the y-axis. Thus, $\bar{x} = 0$. To find $\bar{y}$, we view R as being y-simple and first compute

$$m = \iint\limits_{R} \rho(x, y)\, dA = \int_{-1}^{1} \int_{x^2}^{1} ky \, dy \, dx = k \int_{-1}^{1} \left[\frac{1}{2} y^2 \right]_{y=x^2}^{y=1} dx$$

$$= \frac{k}{2} \int_{-1}^{1} (1 - x^4) \, dx = \frac{k}{2} \left[x - \frac{1}{5} x^5 \right]_{-1}^{1} = \frac{4k}{5}$$

Then using Equation (4b), we obtain

$$\bar{y} = \frac{1}{m} \iint\limits_{R} y\rho(x, y)\, dA = \frac{5}{4k} \int_{-1}^{1} \int_{x^2}^{1} y(ky) \, dy \, dx$$

$$= \frac{5}{4} \int_{-1}^{1} \int_{x^2}^{1} y^2 \, dy \, dx = \frac{5}{4} \int_{-1}^{1} \left[\frac{1}{3} y^3 \right]_{y=x^2}^{y=1} dx$$

$$= \frac{5}{12} \int_{-1}^{1} (1 - x^6) \, dx = \frac{5}{12} \left[x - \frac{1}{7} x^7 \right]_{-1}^{1} = \frac{5}{7}$$

Therefore, the center of mass of the lamina is located at $\left(0, \frac{5}{7} \right)$. ∎

Moments of Inertia

The moments of mass of a lamina, M_x and M_y, are called the **first moments** of the lamina with respect to the x- and y-axes. We can also consider the **second moment** or **moment of inertia** of a lamina about an axis. We begin by recalling that the moment of inertia of a particle of mass m with respect to an axis is defined to be

$$I = mr^2 \qquad \text{mass} \cdot \text{the square of the distance of the moment arm}$$

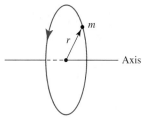

FIGURE 8

A particle of mass m rotating about a stationary axis

To understand the physical significance of the moment of inertia of a particle, suppose that a particle of mass m rotates with constant angular velocity ω about a stationary axis. (See Figure 8.) The velocity of the particle is $v = r\omega$, where r is the distance of the particle from the axis. The kinetic energy of the particle is

$$\frac{1}{2} mv^2 = \frac{1}{2} mr^2 \omega^2 = \frac{1}{2} I\omega^2 \qquad I = mr^2$$

This tells us that the moment of inertia I of the particle with respect to the axis plays the same role in rotational motion that the mass m of a particle plays in rectilinear motion. Since the mass m is a measure of the inertia or resistance to rectilinear motion (the larger m is, the greater the energy needed), we see that the moment of inertia I is a measure of the resistance of the particle to rotational motion.

To define the moment of inertia of a lamina occupying a region R in the xy-plane and having mass density described by a continuous function ρ, we proceed as before by enclosing R with a rectangle and partitioning the latter using a regular partition. The moment of inertia of the piece of the lamina occupying the subrectangle R_{ij} about the x-axis is approximately $[\rho(x_{ij}^*, y_{ij}^*) \Delta A](y_{ij}^*)^2$, where (x_{ij}^*, y_{ij}^*) is a point in R_{ij}. Taking the limit of the sum of the second moments as m and n approach infinity, we obtain the **moment of inertia** of the lamina **with respect to the x-axis.** In a similar manner we obtain the **moment of inertia** of a lamina **with respect to the y-axis.**

The formulas for these quantities and the formulas for the moment of inertia of a lamina with **respect to the origin** (the sum of the moments with respect to x and with respect to y) follow.

> **DEFINITION**   **Moments of Inertia of a Lamina**
>
> The **moment of inertia** of a lamina with respect to the *x*-axis, the *y*-axis, and the **origin** are, respectively, as follows:
>
> $$I_x = \lim_{m,\,n\to\infty} \sum_{i=1}^{m} \sum_{j=1}^{n} (y_{ij}^*)^2 \rho(x_{ij}^*, y_{ij}^*)\,\Delta A = \iint\limits_{R} y^2 \rho(x, y)\,dA \qquad \text{(5a)}$$
>
> $$I_y = \lim_{m,\,n\to\infty} \sum_{i=1}^{m} \sum_{j=1}^{n} (x_{ij}^*)^2 \rho(x_{ij}^*, y_{ij}^*)\,\Delta A = \iint\limits_{R} x^2 \rho(x, y)\,dA \qquad \text{(5b)}$$
>
> $$I_0 = \lim_{m,\,n\to\infty} \sum_{i=1}^{m} \sum_{j=1}^{n} [(x_{ij}^*)^2 + (y_{ij}^*)^2]\rho(x_{ij}^*, y_{ij}^*)\,\Delta A$$
>
> $$\text{(5c)}$$
>
> $$= \iint\limits_{R} (x^2 + y^2)\rho(x, y)\,dA = I_x + I_y$$

EXAMPLE 4   Find the moments of inertia with respect to the *x*-axis, the *y*-axis, and the origin of a thin homogeneous disk of mass *m* and radius *a*, centered at the origin.

Solution   Since the disk is homogeneous, its density is constant and given by $\rho(x, y) = m/(\pi a)^2$. Using Equation (5a), we see that the moment of inertia of the disk about the *x*-axis is given by

$$I_x = \iint\limits_{R} y^2 \rho(x, y)\,dA = \frac{m}{\pi a^2} \int_0^{2\pi} \int_0^a (r \sin \theta)^2\, r\, dr\, d\theta$$

$$= \frac{m}{\pi a^2} \int_0^{2\pi} \int_0^a r^3 \sin^2 \theta\, dr\, d\theta = \frac{m}{\pi a^2} \int_0^{2\pi} \left[\frac{1}{4} r^4 \sin^2 \theta \right]_{r=0}^{r=a} d\theta$$

$$= \frac{ma^2}{4\pi} \int_0^{2\pi} \sin^2 \theta\, d\theta = \frac{ma^2}{8\pi} \int_0^{2\pi} (1 - \cos 2\theta)\, d\theta$$

$$= \frac{ma^2}{8\pi} \left[\theta - \frac{1}{2} \sin 2\theta \right]_0^{2\pi} = \frac{1}{4} ma^2$$

By symmetry we see that $I_y = I_x = \frac{1}{4}ma^2$. Finally, using Equation (5c), we see that the moment of inertia of the disk about the origin is given by

$$I_0 = I_x + I_y = \frac{1}{4} ma^2 + \frac{1}{4} ma^2 = \frac{1}{2} ma^2 \qquad \blacksquare$$

■ Radius of Gyration of a Lamina

If we imagine that the mass of a lamina is concentrated at a point at a distance *R* from the axis, then the moment of inertia of this "point mass" would be the same as the moment of inertia of the lamina. (See Figure 9.) The distance *R* is called the **radius of gyration of the lamina with respect to the axis.** Thus, if the mass of the lamina is *m* and its moment of inertia with respect to the axis is *I*, then

$$mR^2 = I$$

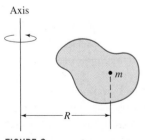

Axis

FIGURE 9
R is the radius of gyration of the lamina with respect to the axis.

from which we see that

$$R = \sqrt{\frac{I}{m}} \tag{6}$$

EXAMPLE 5 Find the radius of gyration of the disk of Example 4 with respect to the y-axis.

Solution Using the result of Example 4, we have $I_y = \frac{1}{4}ma^2$. Therefore, using Equation (6), we see that the radius of gyration of the disk about the y-axis is

$$\bar{\bar{x}} = \sqrt{\frac{I_y}{m}} = \sqrt{\frac{\frac{1}{4}ma^2}{m}} = \frac{1}{2}a$$ ∎

Note In Example 5 we have used the customary notation $\bar{\bar{x}}$ for the radius of gyration of a lamina with respect to the y-axis. The radius of gyration of a lamina with respect to the x-axis is denoted by $\bar{\bar{y}}$. ∎

14.4 CONCEPT QUESTIONS

1. A lamina occupies a region R in the plane. If the mass density of the lamina is $\rho(x, y)$, write an integral giving (a) the mass of the lamina, (b) the moments of mass of the lamina with respect to the x- and y-axes, and (c) the center of mass of the lamina.

2. A lamina occupies a region R in the plane.
   a. Write an integral giving the moment of inertia of the lamina with respect to the x-axis, the y-axis, and the origin.
   b. Write an integral giving the moment of inertia of the lamina with respect to a line L.
      Hint: Let $d(x, y)$ denote the distance between a point (x, y) in R and the line L.

3. What is the radius of gyration of a lamina with respect to an axis? Illustrate with a sketch.

14.4 EXERCISES

In Exercises 1–12, find the mass and the center of mass of the lamina occupying the region R and having the given mass density.

1. R is the rectangular region with vertices $(0, 0)$, $(3, 0)$, $(3, 2)$, and $(0, 2)$; $\rho(x, y) = y$

2. R is the rectangular region with vertices $(0, 0)$, $(3, 0)$, $(3, 1)$, and $(0, 1)$; $\rho(x, y) = x^2 + y^2$

3. R is the triangular region with vertices $(0, 0)$, $(2, 1)$, and $(4, 0)$; $\rho(x, y) = x$

4. R is the triangular region with vertices $(1, 0)$, $(1, 1)$, and $(0, 1)$; $\rho(x, y) = x + y$

5. R is the region bounded by the graphs of the equations $y = \sqrt{x}$, $y = 0$, and $x = 4$; $\rho(x, y) = xy$

6. R is the region bounded by the parabola $y = 4 - x^2$ and the x-axis; $\rho(x, y) = y$

7. R is the region bounded by the graphs of $y = e^x$, $y = 0$, $x = 0$, and $x = 1$; $\rho(x, y) = 2xy$

8. R is the region bounded by the graphs of $y = \ln x$, $y = 0$, and $x = e$; $\rho(x, y) = y/x$

9. R is the region bounded by the graphs of $y = \sin x$, $y = 0$, $x = 0$, and $x = \pi$; $\rho(x, y) = y$

10. R is the region in the first quadrant bounded by the circle $x^2 + y^2 = 1$; $\rho(x, y) = x + y$

11. R is the region bounded by the circle $r = 2\cos\theta$; $\rho(r, \theta) = r$

12. R is the region bounded by the cardioid $r = 1 + \cos\theta$; $\rho(r, \theta) = 3$

13. An electric charge is spread over a rectangular region $R = \{(x, y) \mid 0 \le x \le 3, 0 \le y \le 1\}$. Find the total charge on R if the charge density at a point (x, y) in R (measured in coulombs per square meter) is $\sigma(x, y) = 2x^2 + y^3$.

14. **Electric Charge on a Disk** An electric charge is spread over the half-disk H described by $x^2 + y^2 = 4$, $y \ge 0$. Find the total charge on H if the charge density at any point (x, y) in H (measured in coulombs per square meter) is $\sigma(x, y) = \sqrt{x^2 + y^2}$.

15. **Temperature of a Hot Plate** An 8-in. hot plate is described by the set $S = \{(x, y) \mid x^2 + y^2 \le 16\}$. The temperature at the

point (x, y) is $T(x, y) = 400 \cos(0.1\sqrt{x^2 + y^2})$, measured in degrees Fahrenheit. What is the average temperature of the hot plate?

16. **Population Density of a City** The population density (number of people per square mile) of a certain city is

$$\sigma(x, y) = 3000e^{-(x^2+y^2)}$$

where x and y are measured in miles. Find the population within a 1-mi radius of the town hall, located at the origin.

In Exercises 17–20, find the moments of inertia I_x, I_y, and I_0 and the radii of gyration $\bar{\bar{x}}$ and $\bar{\bar{y}}$ for the lamina occupying the region R and having uniform density ρ.

17. R is the rectangular region with vertices $(0, 0)$, $(a, 0)$, (a, b), and $(0, b)$.

18. R is the triangular region with vertices $(0, 0)$, $(a, 0)$, and $(0, b)$.

19. D is the half-disk $H = \{(x, y) \mid x^2 + y^2 \le R^2, y \ge 0\}$.

20. R is the region bounded by the ellipse $\dfrac{x^2}{a^2} + \dfrac{y^2}{b^2} = 1$.

In Exercises 21–24, find the moments of inertia I_x, I_y, and I_0 and the radii of gyration $\bar{\bar{x}}$ and $\bar{\bar{y}}$ for the lamina.

21. The lamina of Exercise 1

22. The lamina of Exercise 3

23. The lamina of Exercise 5

24. The lamina of Exercise 10

25. A thin metal plate has the shape of the region R inside the circle $x^2 + y^2 = 4$, below the line $y = x$, to the right of the line $x = 1$, and above the x-axis. Its density is $\rho(x, y) = y/x$ for (x, y) in R. Find the mass of the plate.

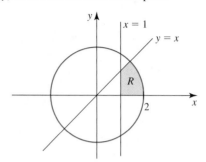

26. Find the rectangular coordinates of the centroid of the region lying between the circles $r = 2 \cos \theta$ and $r = 4 \cos \theta$.

In Exercises 27–29, determine whether the statement is true or false. If it is true, explain why. If it is false, explain why or give an example that shows it is false.

27. A piece of metal is laminated from two thin sheets of metal with mass density $\rho_1(x, y)$ and $\rho_2(x, y)$. If it occupies a region R in the plane, then the mass of the laminate is $\iint_R \rho_1(x, y) \, dA + \iint_R \rho_2(x, y) \, dA$.

28. If the region occupied by a lamina is symmetric with respect to both the x- and y-axes, then the center of mass of the lamina must be located at the origin.

29. If a lamina occupies a region R in the plane, then its center of mass must be located in R.

14.5   Surface Area

In Section 5.4 we saw that the area of a surface of revolution can be found by evaluating a simple integral. We now turn our attention to the problem of finding the area of more general surfaces. More specifically, we will consider surfaces that are graphs of functions of two variables. As you will see, the area of these surfaces can be found by using double integrals.

■ Area of a Surface $z = f(x, y)$

For simplicity we will consider the case in which f is defined in an open set containing a rectangular region $R = [a, b] \times [c, d] = \{(x, y) \mid a \le x \le b, c \le y \le d\}$ and $f(x, y) \ge 0$ on R. Furthermore, we assume that f has continuous first-order partial derivatives in that region. We wish first to define what we mean by the *area* of the surface S with equation $z = f(x, y)$ (Figure 1) and then to find a formula that will enable us to calculate this area.

Let P be a regular partition of R into $N = mn$ subrectangles $R_{11}, R_{12}, \ldots, R_{mn}$. Corresponding to the subrectangle R_{ij}, there is the part S_{ij} of S (called a **patch**) that lies

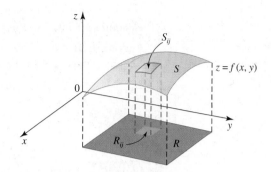

FIGURE 1

The surface S is the graph of
$z = f(x, y)$ for (x, y) in R.

directly above R_{ij} with area denoted by ΔS_{ij}. Since the subrectangles R_{ij} are nonover-lapping except for their common boundaries, so are the patches S_{ij} of S, so the area of S is given by

$$A = \sum_{i=1}^{m} \sum_{j=1}^{n} \Delta S_{ij} \qquad \textbf{(1)}$$

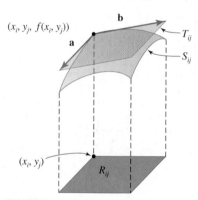

FIGURE 2

The tangent plane determined by **a** and **b** approximates S well if R_{ij} is small.

Next, let's find an approximation of ΔS_{ij}. Let (x_i, y_j) be the corner of R_{ij} closest to the origin, and let $(x_i, y_j, f(x_i, y_j))$ be the point directly above it. If you refer to Figure 2, you can see that ΔS_{ij} is approximated by the area of ΔT_{ij} of the parallelogram T_{ij} that is part of the tangent plane to S at the point $(x_i, y_j, f(x_i, y_j))$ and lying directly above R_{ij}. To find a formula for ΔT_{ij}, let **a** and **b** be vectors that have initial point $(x_i, y_j, f(x_i, y_j))$ and lie along the sides of the approximating parallelogram. Now from Section 13.3 we see that the slopes of the tangent lines passing through $(x_i, y_j, f(x_i, y_j))$ and having the directions of **a** and **b** are given by $f_x(x_i, y_j)$ and $f_y(x_i, y_j)$, respectively. Therefore,

$$\mathbf{a} = \Delta x \mathbf{i} + f_x(x_i, y_j) \, \Delta x \mathbf{k} \qquad \text{and} \qquad \mathbf{b} = \Delta y \mathbf{j} + f_y(x_i, y_j) \, \Delta y \mathbf{k}$$

From Section 11.4 we have $\Delta T_{ij} = |\mathbf{a} \times \mathbf{b}|$. But

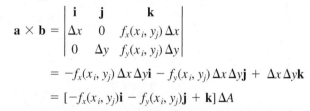

$$\mathbf{a} \times \mathbf{b} = \begin{vmatrix} \mathbf{i} & \mathbf{j} & \mathbf{k} \\ \Delta x & 0 & f_x(x_i, y_j) \, \Delta x \\ 0 & \Delta y & f_y(x_i, y_j) \, \Delta y \end{vmatrix}$$

$$= -f_x(x_i, y_j) \, \Delta x \, \Delta y \mathbf{i} - f_y(x_i, y_j) \, \Delta x \, \Delta y \mathbf{j} + \Delta x \, \Delta y \mathbf{k}$$

$$= [-f_x(x_i, y_j) \mathbf{i} - f_y(x_i, y_j) \mathbf{j} + \mathbf{k}] \Delta A$$

where $\Delta A = \Delta x \, \Delta y$ is the area of R_{ij}. Therefore,

$$\Delta T_{ij} = |\mathbf{a} \times \mathbf{b}| = \sqrt{[f_x(x_i, y_j)]^2 + [f_y(x_i, y_j)]^2 + 1} \, \Delta A \qquad \textbf{(2)}$$

If we approximate ΔS_{ij} by ΔT_{ij}, then Equation (1) becomes

$$A \approx \sum_{i=1}^{m} \sum_{j=1}^{n} \Delta T_{ij}$$

Intuitively, we see that the approximation should get better and better as both m and n get larger and larger. This suggests that we define

$$A = \lim_{m, n \to \infty} \sum_{i=1}^{m} \sum_{j=1}^{n} \sqrt{[f_x(x_i, y_j)]^2 + [f_y(x_i, y_j)]^2 + 1} \, \Delta A$$

Using the definition of the double integral, we obtain the following result, which is stated for the general case in which R is not necessarily rectangular and $f(x, y)$ is not necessarily positive.

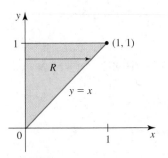

FIGURE 3
The region
$R = \{(x, y)\,|\,0 \le x \le y, 0 \le y \le 1\}$
viewed as an *x*-simple region

Formula for Finding the Area of a Surface $z = f(x, y)$

Let f be defined on a region R in the xy-plane and suppose that f_x and f_y are continuous. The area A of the surface $z = f(x, y)$ is

$$A = \iint\limits_{R} \sqrt{[f_x(x, y)]^2 + [f_y(x, y)]^2 + 1}\; dA \qquad (3)$$

EXAMPLE 1   Find the area of the part of the surface with equation $z = 2x + y^2$ that lies directly above the triangular region R in the xy-plane with vertices $(0, 0)$, $(1, 1)$, and $(0, 1)$.

Solution   The region R is shown in Figure 3. It is both a y-simple and an x-simple region. Viewed as an x-simple region

$$R = \{(x, y)\,|\,0 \le x \le y, 0 \le y \le 1\}$$

Using Equation (3) with $f(x, y) = 2x + y^2$, we see that the required area is

$$A = \iint\limits_{R} \sqrt{[f_x(x, y)]^2 + [f_y(x, y)]^2 + 1}\; dA$$

$$= \iint\limits_{R} \sqrt{2^2 + (2y)^2 + 1}\; dA = \int_0^1 \int_0^y \sqrt{4y^2 + 5}\; dx\, dy$$

$$= \int_0^1 \left[x\sqrt{4y^2 + 5}\, \right]_{x=0}^{x=y} dy = \int_0^1 y\sqrt{4y^2 + 5}\; dy$$

$$= \left[\frac{1}{8} \cdot \frac{2}{3} (4y^2 + 5)^{3/2} \right]_0^1 = \frac{1}{12}(27 - 5\sqrt{5})$$

or approximately 1.32.   ∎

EXAMPLE 2   Find the surface area of the part of the paraboloid $z = 9 - x^2 - y^2$ that lies above the plane $z = 5$.

Solution   The paraboloid is sketched in Figure 4a. The paraboloid intersects the plane $z = 5$ along the circle $x^2 + y^2 = 4$. Therefore, the surface of interest lies directly above the disk $R = \{(x, y)\,|\,x^2 + y^2 \le 4\}$ shown in Figure 4b. Using Equation (3) with $f(x, y) = 9 - x^2 - y^2$, we find the required area to be

$$A = \iint\limits_{R} \sqrt{[f_x(x, y)]^2 + [f_y(x, y)]^2 + 1}\; dA$$

$$= \iint\limits_{R} \sqrt{(-2x)^2 + (-2y)^2 + 1}\; dA$$

$$= \iint\limits_{R} \sqrt{4x^2 + 4y^2 + 1}\; dA$$

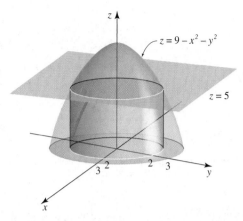

(a) The part of the paraboloid that lies above the plane $z = 5$

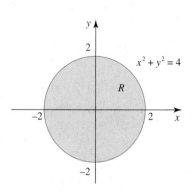

(b) The disk $R = \{(x,y) \mid x^2 + y^2 \le 4\}$

FIGURE 4

Changing to polar coordinates, we have

$$A = \int_0^{2\pi} \int_0^2 \sqrt{4r^2 + 1}\, r\, dr\, d\theta$$

$$= \int_0^{2\pi} \left[\frac{1}{8} \cdot \frac{2}{3} (4r^2 + 1)^{3/2} \right]_{r=0}^{r=2} d\theta$$

$$= \int_0^{2\pi} \left[\frac{1}{12} (17^{3/2} - 1) \right] d\theta = 2\pi\left(\frac{1}{12}\right)(17\sqrt{17} - 1) = \frac{1}{6}\pi(17\sqrt{17} - 1)$$

or approximately 36.2.

Area of Surfaces with Equations $y = g(x, z)$ and $x = h(y, z)$

Formulas for finding the area of surfaces that are graphs of $y = g(x, z)$ and $x = h(y, z)$ are developed in a similar manner.

> **Formulas for Finding the Area of Surfaces in the Form $y = g(x, z)$ and $x = h(y, z)$.**
>
> Let g be defined on a region R in the xz-plane, and suppose that g_x and g_z are continuous. The area A of the surface $y = g(x, z)$ is
>
> $$A = \iint_R \sqrt{[g_x(x, z)]^2 + [g_z(x, z)]^2 + 1}\, dA \tag{4}$$
>
> Let h be defined on a region R in the yz-plane, and suppose that h_y and h_z are continuous. The area A of the surface $x = h(y, z)$ is
>
> $$A = \iint_R \sqrt{[h_y(y, z)]^2 + [h_z(y, z)]^2 + 1}\, dA \tag{5}$$

These situations are depicted in Figure 5.

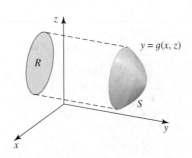
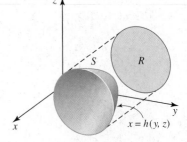

FIGURE 5

(a) The surface S has equation $y = g(x, z)$ and projection R onto the xz-plane.

(b) The surface S has equation $x = h(y, z)$ and projection R onto the yz-plane.

EXAMPLE 3 Find the area of that part of the plane $y + z = 2$ inside the cylinder $x^2 + z^2 = 1$.

Solution The surface S of interest is sketched in Figure 6a. The projection of S onto the xz-plane is the disk $R = \{(x, z) \mid x^2 + z^2 \leq 1\}$ shown in Figure 6b. Using Equation (4) with $g(x, z) = 2 - z$, we see that the area of S is

$$A = \iint\limits_{R} \sqrt{[g_x(x, z)]^2 + [g_z(x, z)]^2 + 1} \, dA$$

$$= \iint\limits_{R} \sqrt{0^2 + (-1)^2 + 1} \, dA = \sqrt{2} \iint\limits_{R} 1 \, dA = \sqrt{2}\pi$$

upon observing that the area of R is π.

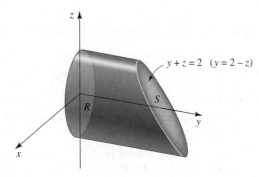
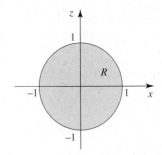

FIGURE 6   (a) The surface S

(b) The projection R of S onto the xz-plane

14.5   CONCEPT QUESTIONS

1. Write an integral giving the area of the surface $z = f(x, y)$ defined over a region R in the xy-plane.

2. Write an integral giving the area of the surface $x = f(y, z)$ defined over a region R in the yz-plane.

14.5 EXERCISES

In Exercises 1–14, find the area of the surface S.

1. S is the part of the plane $2x + 3y + z = 12$ that lies above the rectangular region $R = \{(x, y) \mid 0 \le x \le 2, 0 \le y \le 1\}$.

2. S is the part of the plane $3x + 2y + z = 6$ that lies above the triangular region with vertices $(0, 0)$, $(1, 3)$, and $(0, 3)$.

3. S is the part of the surface $z = \frac{1}{2}x^2 + y$ that lies above the triangular region with vertices $(0, 0)$, $(1, 0)$, and $(1, 1)$.

4. S is the part of the surface $z = 2 - x^2 + y$ that lies above the triangular region with vertices $(0, -1)$, $(1, 0)$, and $(0, 1)$.

5. S is the part of the paraboloid $z = 9 - x^2 - y^2$ that lies above the xy-plane.

6. S is the part of the paraboloid $y = 9 - x^2 - z^2$ that lies between the planes $y = 0$ and $y = 5$.

7. S is the part of the sphere $x^2 + y^2 + z^2 = 9$ that lies above the plane $z = 2$.

8. S is the part of the hyperbolic paraboloid $z = y^2 - x^2$ that lies above the annular region $A = \{(x, y) \mid 1 \le x^2 + y^2 \le 4\}$.

9. S is the part of the surface $x = yz$ that lies inside the cylinder $y^2 + z^2 = 16$.

10. S is the part of the sphere $x^2 + y^2 + z^2 = 9$ that lies to the right of the xz-plane and inside the cylinder $x^2 + z^2 = 4$.

11. S is the part of the sphere $x^2 + y^2 + z^2 = 8$ that lies inside the cone $z^2 = x^2 + y^2$.

12. S is the part of the hyperbolic paraboloid $y = x^2 - z^2$ that lies in the first octant and inside the cylinder $x^2 + z^2 = 4$.

13. S is the part of the sphere $x^2 + y^2 + z^2 = a^2$ that lies inside the cylinder $x^2 - ax + y^2 = 0$.

14. S comprises the parts of the cylinder $x^2 + z^2 = 1$ that lie within the cylinder $y^2 + z^2 = 1$.

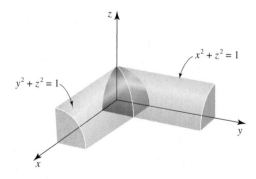

$y^2 + z^2 = 1$

$x^2 + z^2 = 1$

Hint: The figure shows the intersection of the two cylinders in the first octant. Use symmetry.

15. Let S be the part of the plane $ax + by + cz = d$ lying in the first octant whose projection onto the xy-plane is a region R. Prove that the area of S is $(1/c)\sqrt{a^2 + b^2 + c^2}\,A(R)$, where $A(R)$ is the area of R.

16. a. Let S be the part of the sphere $x^2 + y^2 + z^2 = a^2$ that lies above the region $R = \{(x, y) \mid x^2 + y^2 \le b^2, 0 \le b \le a\}$ in the xy-plane. Show that the area of S is $2\pi a(a - \sqrt{a^2 - b^2})$.

    b. Use the result of part (a) to deduce that the area of a sphere of radius a is $4\pi a^2$.

cas *In Exercises 17–20, use a calculator or a computer to approximate the area of the surface S, accurate to four decimal places.*

17. S is the part of the paraboloid $z = x^2 + y^2$ that lies above the square region $R = \{(x, y) \mid 0 \le x \le 2, 0 \le y \le 2\}$.

18. S is the part of the paraboloid $z = 9 - x^2 - y^2$ that lies above the square region $R = \{(x, y) \mid -2 \le x \le 2, -2 \le y \le 2\}$.

19. S is the part of the surface $z = e^{-x^2 - y^2}$ that lies inside the cylinder $x^2 + y^2 \le 4$.

20. S is the part of the surface $z = \sin(x^2 + y^2)$ that lies above the disk $x^2 + y^2 \le 1$.

In Exercises 21–24, write a double integral that gives the surface area of the part of the graph of f that lies above the region R. Do not evaluate the integral.

21. $f(x, y) = 3x^2y^2$; $R = \{(x, y) \mid -1 \le x \le 1, -1 \le y \le 1\}$

22. $f(x, y) = x^2 - 3xy + y^2$; R is the triangular region with vertices $(0, 0)$, $(1, 1)$, and $(0, 1)$

23. $f(x, y) = \dfrac{1}{2x + 3y}$; $R = \{(x, y) \mid 0 \le x \le 2, 0 \le y \le x\}$

24. $f(x, y) = e^{-xy}$; $R = \{(x, y) \mid 0 \le x \le 1, 0 \le y \le 2\}$

In Exercises 25 and 26, determine whether the statement is true or false. If it is true, explain why. If it is false, explain why or give an example that shows it is false.

25. If $f(x, y) = \sqrt{4 - x^2 - y^2}$, then $\iint_R \sqrt{f_x^2 + f_y^2 + 1}\,dA = 8\pi$, where $R = \{(x, y) \mid 0 \le x^2 + y^2 \le 4\}$.

26. If $z = f(x, y)$ is defined over a region R in the xy-plane, then $\iint_R \sqrt{f_x^2 + f_y^2 + 1}\,dA \ge A(R)$, where $A(R)$ denotes the area of R. (Assume that f_x and f_y exist.)

14.6   Triple Integrals

▪ Triple Integrals Over a Rectangular Box

Just as the mass of a piece of straight, thin wire of linear mass density $\delta(x)$, where $a \leq x \leq b$, is given by the single integral $\int_a^b \delta(x)\, dx$, and the mass of a thin plate D of mass density $\sigma(x, y)$ is given by the double integral $\iint_D \sigma(x, y)\, dA$, we will now see that the mass of a solid object T with mass density $\rho(x, y, z)$ is given by a *triple integral*.

Let's consider the simplest case in which the solid takes the form of a rectangular box:

$$B = [a, b] \times [c, d] \times [p, q] = \{(x, y, z) \mid a \leq x \leq b, c \leq y \leq d, p \leq z \leq q\}$$

Suppose that the mass density of the solid is $\rho(x, y, z)$ g/m^3, where ρ is a positive continuous function defined on B. Let

$$a = x_0 < x_1 < \cdots < x_{i-1} < x_i < \cdots < x_l = b$$

$$c = y_0 < y_1 < \cdots < y_{j-1} < y_j < \cdots < y_m = d$$

$$p = z_0 < z_1 < \cdots < z_{k-1} < z_k < \cdots < z_n = q$$

be regular partitions of the intervals $[a, b]$, $[c, d]$, and $[p, q]$ of length $\Delta x = (b - a)/l$, $\Delta y = (d - c)/m$, and $\Delta z = (q - p)/n$, respectively. The planes $x = x_i$, for $1 \leq i \leq l$, $y = y_j$, for $1 \leq j \leq m$, and $z = z_k$, for $1 \leq k \leq n$, parallel to the yz-, xz-, and xy-coordinate planes divide the box B into $N = lmn$ boxes $B_{111}, B_{112}, \ldots, B_{ijk}, \ldots, B_{lmn}$, as shown in Figure 1. The volume of B_{ijk} is $\Delta V = \Delta x\, \Delta y\, \Delta z$.

Let $(x_{ijk}^*, y_{ijk}^*, z_{ijk}^*)$ be an arbitrary point in B_{ijk}. If l, m, and n are large (so that the dimensions of B_{ijk} are small), then the continuity of ρ implies that $\rho(x, y, z)$ does not vary appreciably from $\rho(x_{ijk}^*, y_{ijk}^*, z_{ijk}^*)$, whenever (x, y, z) is in B_{ijk}. Therefore, we can approximate the mass of B_{ijk} by

$$\rho(x_{ijk}^*, y_{ijk}^*, z_{ijk}^*)\, \Delta V \qquad \text{constant mass density} \cdot \text{volume}$$

where $\Delta V = \Delta x\, \Delta y\, \Delta z$. Adding up the masses of the N boxes, we see that the mass of the box B is approximately

$$\sum_{i=1}^{l} \sum_{j=1}^{m} \sum_{k=1}^{n} \rho(x_{ijk}^*, y_{ijk}^*, z_{ijk}^*)\, \Delta V \tag{1}$$

We expect the approximation to improve as l, m, and n get larger and larger. Therefore, it is reasonable to define the mass of the box B as

$$\lim_{l, m, n \to \infty} \sum_{i=1}^{l} \sum_{j=1}^{m} \sum_{k=1}^{n} \rho(x_{ijk}^*, y_{ijk}^*, z_{ijk}^*)\, \Delta V \tag{2}$$

The expression in (1) is an example of a *Riemann sum* of a function of three variables over a box and the corresponding limit in (2) is the *triple integral* of f over B. More generally, we have the following definitions. Notice that no assumption regarding the sign of $f(x, y, z)$ is made in these definitions.

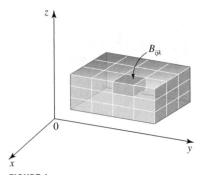

FIGURE 1
A partition $P = \{B_{ijk}\}$ of B

> **DEFINITION** **Triple Integral of f Over a Rectangular Box B**
>
> Let f be a continuous function of three variables defined on a rectangular box B, and let $P = \{B_{ijk}\}$ be a partition of B.
>
> **1.** A **Riemann sum of f over B** with respect to the partition P is a sum of the form
>
> $$\sum_{i=1}^{l} \sum_{j=1}^{m} \sum_{k=1}^{n} f(x_{ijk}^{*}, y_{ijk}^{*}, z_{ijk}^{*}) \, \Delta V$$
>
> where $(x_{ijk}^{*}, y_{ijk}^{*}, z_{ijk}^{*})$ is a point in B_{ijk}.
>
> **2.** The **triple integral of f over B** is
>
> $$\iiint\limits_{B} f(x, y, z) \, dV = \lim_{l,\, m,\, n \to \infty} \sum_{i=1}^{l} \sum_{j=1}^{m} \sum_{k=1}^{n} f(x_{ijk}^{*}, y_{ijk}^{*}, z_{ijk}^{*}) \, \Delta V$$
>
> if the limit exists for all choices of $(x_{ijk}^{*}, y_{ijk}^{*}, z_{ijk}^{*})$ in B_{ijk}.

As in the case of double integrals, a triple integral may be found by evaluating an appropriate iterated integral.

> **THEOREM 1**
>
> Let f be continuous on the rectangular box
>
> $$B = \{(x, y, z) \mid a \leq x \leq b, c \leq y \leq d, p \leq z \leq q\}$$
>
> Then
>
> $$\iiint\limits_{B} f(x, y, z) \, dV = \int_{p}^{q} \int_{c}^{d} \int_{a}^{b} f(x, y, z) \, dx \, dy \, dz \qquad (3)$$

The iterated integral in Equation (3) is evaluated by first integrating with respect to x while holding y and z constant, then integrating with respect to y while holding z constant, and finally integrating with respect to z. The triple integral in Equation (3) can also be expressed as any one of five other iterated integrals, each with a different order of integration. For example, we can write

$$\iiint\limits_{B} f(x, y, z) \, dV = \int_{a}^{b} \int_{p}^{q} \int_{c}^{d} f(x, y, z) \, dy \, dz \, dx$$

where the iterated integral is evaluated by successively integrating with respect to y, z, and then x. (Remember, we work "from the inside out.")

EXAMPLE 1 Evaluate $\iiint_{B} (x^2 y + y z^2) \, dV$, where

$$B = \{(x, y, z) \mid -1 \leq x \leq 1, 0 \leq y \leq 3, 1 \leq z \leq 2\}$$

Solution   We can express the given integral as one of six integrals. For example, if we choose to integrate with respect to x, y, and z, in that order, then we obtain

$$\iiint\limits_{B} (x^2y + yz^2)\, dV = \int_1^2 \int_0^3 \int_{-1}^1 (x^2y + yz^2)\, dx\, dy\, dz$$

$$= \int_1^2 \int_0^3 \left[\frac{1}{3}x^3y + xyz^2\right]_{x=-1}^{x=1} dy\, dz$$

$$= \int_1^2 \int_0^3 \left[\frac{2}{3}y + 2yz^2\right] dy\, dz$$

$$= \int_1^2 \left[\frac{1}{3}y^2 + y^2z^2\right]_{y=0}^{y=3} dz$$

$$= \int_1^2 (3 + 9z^2)\, dz = \left[3z + 3z^3\right]_1^2 = 24 \quad \blacksquare$$

■ Triple Integrals Over General Bounded Regions in Space

We can extend the definition of the triple integral to more general regions using the same technique that we used for double integrals. Suppose that T is a bounded solid region in space. Then it can be enclosed in a rectangular box $B = [a, b] \times [c, d] \times [p, q]$. Let P be a regular partition of B into $N = lmn$ boxes with sides of length $\Delta x = (b - a)/l$, $\Delta y = (d - c)/m$, $\Delta z = (q - p)/n$, and volume $\Delta V = \Delta x\, \Delta y\, \Delta z$. Thus, $P = \{B_{111}, B_{112}, \ldots, B_{ijk}, \ldots, B_{lmn}\}$. (See Figure 2.)

Define

$$F(x, y, z) = \begin{cases} f(x, y, z) & \text{if } (x, y, z) \text{ is in } T \\ 0 & \text{if } (x, y, z) \text{ is in } B \text{ but not in } T \end{cases}$$

Then a **Riemann sum of f over T** with respect to the partition P is given by

$$\sum_{i=1}^{l} \sum_{j=1}^{m} \sum_{k=1}^{n} F(x_{ijk}^*, y_{ijk}^*, z_{ijk}^*)\, \Delta V$$

where $(x_{ijk}^*, y_{ijk}^*, z_{ijk}^*)$ is an arbitrary point in B_{ijk} and ΔV is the volume of B_{ijk}. If we take the limit of these sums as l, m, n approach infinity, we obtain the **triple integral of f over T**. Thus,

$$\iiint\limits_{T} f(x, y, z)\, dV = \lim_{l, m, n \to \infty} \sum_{i=1}^{l} \sum_{j=1}^{m} \sum_{k=1}^{n} F(x_{ijk}^*, y_{ijk}^*, z_{ijk}^*)\, \Delta V$$

provided that the limit exists for all choices of $(x_{ijk}^*, y_{ijk}^*, z_{ijk}^*)$ in T.

Notes

1. If f is continuous and the surface bounding T is "sufficiently smooth," it can be shown that f is integrable over T.
2. The properties of double integrals that are listed in Theorem 1, Section 14.1, with the necessary modifications are also enjoyed by triple integrals.   ■

■ Evaluating Triple Integrals Over General Regions

We will now restrict our attention to certain types of regions. A region T is **z-simple** if it lies between the graphs of two continuous functions of x and y, that is, if

$$T = \{(x, y, z) \mid (x, y) \in R, k_1(x, y) \le z \le k_2(x, y)\}$$

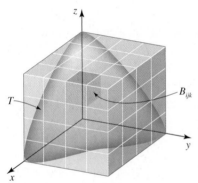

FIGURE 2
The box B_{ijk} is a typical element of the partition of B.

where R is the projection of T onto the xy-plane. (See Figure 3.) If f is continuous on T, then

$$\iiint_T f(x, y, z)\, dV = \iint_R \left[\int_{k_1(x, y)}^{k_2(x, y)} f(x, y, z)\, dz \right] dA \qquad (4)$$

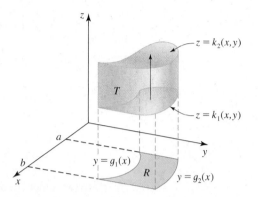

FIGURE 3
A z-simple region T is bounded by the surfaces $z = k_1(x, y)$ and $z = k_2(x, y)$.

The iterated integral on the right-hand side of Equation (4) is evaluated by first integrating with respect to z while holding x and y constant. The resulting double integral is then evaluated by using the method of Section 14.2. For example, if R is y-simple, as shown in Figure 3, then

$$R = \{ (x, y) \mid a \le x \le b, g_1(x) \le y \le g_2(x) \}$$

in which case Equation (4) becomes

$$\iiint_T f(x, y, z)\, dV = \int_a^b \int_{g_1(x)}^{g_2(x)} \int_{k_1(x, y)}^{k_2(x, y)} f(x, y, z)\, dz\, dy\, dx$$

To determine the "limits of integration" with respect to z, notice that z runs from the lower surface $z = k_1(x, y)$ to the upper surface $z = k_2(x, y)$ as indicated by the arrow in Figure 3.

EXAMPLE 2 Evaluate $\iiint_T z\, dV$ where T is the solid in the first octant bounded by the graphs of $z = 1 - x^2$ and $y = x$.

Solution The solid T is shown in Figure 4a. The solid is z-simple because it is bounded below by the graph of $z = k_1(x, y) = 0$ and above by $z = k_2(x, y) = 1 - x^2$.

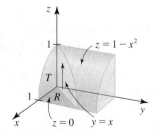

(a) The solid T is z-simple.

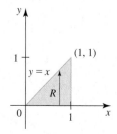

(b) The projection of the solid T onto R in the xy-plane is y-simple.

FIGURE 4

The projection of T onto the xy-plane is the set R that is sketched in Figure 4b. Regarding R as a y-simple region, we obtain

$$\iiint_T z \, dV = \iint_R \left[\int_{z=k_1(x,y)}^{z=k_2(x,y)} z \, dz \right] dA = \int_0^1 \int_0^x \int_0^{1-x^2} z \, dz \, dy \, dx$$

$$= \int_0^1 \int_0^x \left[\frac{1}{2} z^2 \right]_{z=0}^{z=1-x^2} dy \, dx$$

$$= \frac{1}{2} \int_0^1 \int_0^x (1 - x^2)^2 \, dy \, dx$$

$$= \frac{1}{2} \int_0^1 \left[(1 - x^2)^2 y \right]_{y=0}^{y=x} dx$$

$$= \frac{1}{2} \int_0^1 x(1 - x^2)^2 \, dx = \left[\left(\frac{1}{2} \right) \left(-\frac{1}{2} \right) \left(\frac{1}{3} \right) (1 - x^2)^3 \right]_0^1 = \frac{1}{12} \quad \blacksquare$$

There are two other simple regions besides the z-simple region just considered. An **x-simple region T** is one that lies between the graphs of two continuous functions of y and z. In other words, T may be described as

$$T = \{(x, y, z) \mid (y, z) \in R, k_1(y, z) \le x \le k_2(y, z)\}$$

where R is the projection of T onto the yz-plane. (See Figure 5.) Here, we have

$$\iiint_T f(x, y, z) \, dV = \iint_R \left[\int_{k_1(y,z)}^{k_2(y,z)} f(x, y, z) \, dx \right] dA \tag{5}$$

The (double) integral over the plane region R is evaluated by integrating with respect to y or z first depending on whether R is y-simple or z-simple.

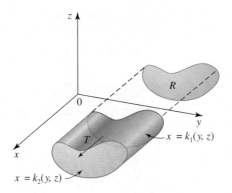

FIGURE 5
An x-simple region T is bounded by the surfaces $x = k_1(y, z)$ and $x = k_2(y, z)$.

A **y-simple region T** lies between the graphs of two continuous functions of x and z. In other words, T may be described as

$$T = \{(x, y, z) \mid (x, z) \in R, k_1(x, z) \le y \le k_2(x, z)\}$$

where R is the projection of T onto the xz-plane. (See Figure 6.) In this case we have

$$\iiint_T f(x, y, z) \, dV = \iint_R \left[\int_{k_1(x,z)}^{k_2(x,z)} f(x, y, z) \, dy \right] dA \tag{6}$$

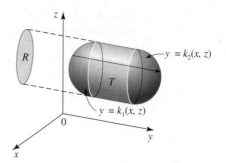

FIGURE 6
A y-simple region T is bounded by the surfaces $y = k_1(x, z)$ and $y = k_2(x, z)$.

Again depending on whether R is an x-simple or z-simple plane region, the double integration is carried out first with respect to x or z.

EXAMPLE 3 Evaluate $\iiint_T \sqrt{x^2 + z^2}\, dV$, where T is the region bounded by the cylinder $x^2 + z^2 = 1$ and the planes $y + z = 2$ and $y = 0$.

Solution The solid T is shown in Figure 7a. Although T can be viewed as an x-simple or z-simple region, it is easier to view it as a y-simple region. (Try It!) In this case we see that T is bounded to the left by the graph of the function $y = k_1(x, z) = 0$ and to the right by the graph of the function $y = k_2(x, z) = 2 - z$. The projection of T onto the xz-plane is the set R, which is sketched in Figure 7b. We have

$$\iiint_T \sqrt{x^2 + z^2}\, dV = \iint_R \left[\int_{k_1(x,z)}^{k_2(x,z)} \sqrt{x^2 + z^2}\, dy \right] dA$$

$$= \iint_R \left[\int_0^{2-z} \sqrt{x^2 + z^2}\, dy \right] dA$$

$$= \iint_R \left[\sqrt{x^2 + z^2}\, y \right]_{y=0}^{y=2-z} dA$$

$$= \iint_R \sqrt{x^2 + z^2}\,(2 - z)\, dA$$

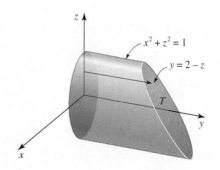

(a) The solid T is viewed as being y-simple.

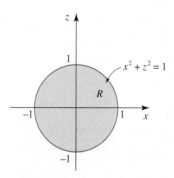

(b) The projection of the solid T onto R in the xz-plane

FIGURE 7

Since R is a circular region, it is more convenient to use polar coordinates when integrating over R. So letting $x = r \cos \theta$ and $z = r \sin \theta$, we have

$$
\iint_R \sqrt{x^2 + z^2}\,(2 - z)\,dA = \int_0^{2\pi} \int_0^1 r(2 - r \sin \theta)\,r\,dr\,d\theta
$$

$$
= \int_0^{2\pi} \int_0^1 (2r^2 - r^3 \sin \theta)\,dr\,d\theta
$$

$$
= \int_0^{2\pi} \left[\frac{2}{3} r^3 - \frac{1}{4} r^4 \sin \theta \right]_{r=0}^{r=1} d\theta
$$

$$
= \int_0^{2\pi} \left(\frac{2}{3} - \frac{1}{4} \sin \theta \right) d\theta
$$

$$
= \left[\frac{2}{3} \theta + \frac{1}{4} \cos \theta \right]_0^{2\pi} = \frac{4\pi}{3}
$$

Therefore,

$$
\iiint_T \sqrt{x^2 + z^2}\,dV = \frac{4\pi}{3}
$$

■

■ Volume, Mass, Center of Mass, and Moments of Inertia

Before looking at other examples, let's list some applications of triple integrals. Let $f(x, y, z) = 1$ for all points in a solid T. Then the triple integral of f over T gives the **volume** V of T; that is,

$$
V = \iiint_T dV \tag{7}
$$

We also have the following.

DEFINITIONS   Mass, Center of Mass, and Moments of Inertia for Solids in Space

Suppose that $\rho(x, y, z)$ gives the mass density at the point (x, y, z) of a solid T. Then the **mass** m of T is

$$
m = \iiint_T \rho(x, y, z)\,dV \tag{8}
$$

The **moments** of T about the three coordinate planes are

$$
M_{yz} = \iiint_T x \rho(x, y, z)\,dV \tag{9a}
$$

$$
M_{xz} = \iiint_T y \rho(x, y, z)\,dV \tag{9b}
$$

$$
M_{xy} = \iiint_T z \rho(x, y, z)\,dV \tag{9c}
$$

The **center of mass** of T is located at the point $(\bar{x}, \bar{y}, \bar{z})$, where

$$\bar{x} = \frac{M_{yz}}{m}, \qquad \bar{y} = \frac{M_{xz}}{m}, \qquad \bar{z} = \frac{M_{xy}}{m} \tag{10}$$

and the **moments of inertia** of T about the three coordinate axes are

$$I_x = \iiint\limits_T (y^2 + z^2)\rho(x, y, z)\, dV \tag{11a}$$

$$I_y = \iiint\limits_T (x^2 + z^2)\rho(x, y, z)\, dV \tag{11b}$$

$$I_z = \iiint\limits_T (x^2 + y^2)\rho(x, y, z)\, dV \tag{11c}$$

If the mass density is constant, then the center of mass of a solid is called the **centroid** of T.

EXAMPLE 4 Let T be the solid tetrahedron bounded by the plane $x + y + z = 1$ and the three coordinate planes $x = 0$, $y = 0$, and $z = 0$. Find the mass of T if the mass density of T is directly proportional to the distance between a base of T and a point on T.

Solution The solid T is shown in Figure 8a. It is x-, y-, and z-simple. For example, it can be viewed as being x-simple if you observe that it is bounded by the surface $x = k_1(y, z) = 0$ and the surface $x = k_2(y, z) = 1 - y - z$. (Solve the equation $x + y + z = 1$ for x.)

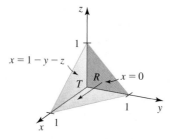

(**a**) The solid T viewed as an x-simple region

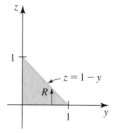

(**b**) The projection of the solid T onto the yz-plane viewed as a z-simple region

FIGURE 8

The projection of T onto the yz-plane is the set R shown in Figure 8b. Observe that the upper boundary of R lies along the line that is the intersection of $x + y + z = 1$ and the plane $x = 0$ and hence has equation $y + z = 1$ or $z = 1 - y$. If we take the base of T as the face of the tetrahedron lying on the xy-plane (actually, by symmetry, any face will do), then the mass density function for T is $\rho(x, y, z) = kz$, where k is the constant of proportionality. Using Equation (8), we see that the required mass is

$$m = \iiint_T \rho(x, y, z) \, dV = \iiint_T kz \, dV$$

$$= k \int_0^1 \int_0^{1-y} \int_0^{1-y-z} z \, dx \, dz \, dy \qquad \text{View } T \text{ as } x\text{-simple.}$$

$$= k \int_0^1 \int_0^{1-y} \left[zx \right]_{x=0}^{x=1-y-z} dz \, dy = k \int_0^1 \int_0^{1-y} \left[(1-y)z - z^2 \right] dz \, dy \qquad \text{View } R \text{ as } z\text{-simple.}$$

$$= k \int_0^1 \left[\frac{1}{2} (1-y)z^2 - \frac{1}{3} z^3 \right]_{z=0}^{z=1-y} dy$$

$$= k \int_0^1 \frac{1}{6} (1-y)^3 \, dy = k \left[\left(\frac{1}{6} \right) \left(-\frac{1}{4} \right) (1-y)^4 \right]_0^1 = \frac{k}{24} \qquad \blacksquare$$

EXAMPLE 5 Let T be the solid that is bounded by the parabolic cylinder $y = x^2$ and the planes $z = 0$ and $y + z = 1$. Find the center of mass of T, given that it has uniform density $\rho(x, y, z) = 1$.

Solution The solid T is shown in Figure 9a. It is x-, y-, and z-simple. Let's choose to view it as being z-simple. (You are also encouraged to solve the problem by viewing T as x-simple and y-simple.) In this case we see that T lies between the xy-plane $z = k_1(x, y) = 0$ and the plane $z = k_2(x, y) = 1 - y$. The projection of T onto the xy-plane is the region R shown in Figure 9b. As a first step toward finding the center of mass of T, let's find the mass of T. Using Equation (8), we have

$$m = \iiint_T \rho(x, y, z) \, dV = \iiint_T dV$$

$$= \int_{-1}^1 \int_{x^2}^1 \int_0^{1-y} dz \, dy \, dx = \int_{-1}^1 \int_{x^2}^1 \left[z \right]_{z=0}^{z=1-y} dy \, dx$$

$$= \int_{-1}^1 \int_{x^2}^1 (1-y) \, dy \, dx = \int_{-1}^1 \left[y - \frac{1}{2} y^2 \right]_{y=x^2}^{y=1} dx$$

$$= \int_{-1}^1 \left(\frac{1}{2} - x^2 + \frac{1}{2} x^4 \right) dx = \left[\frac{1}{2} x - \frac{1}{3} x^3 + \frac{1}{10} x^5 \right]_{-1}^1 = \frac{8}{15}$$

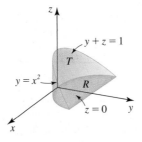

(a) The solid T is viewed as a z-simple region.

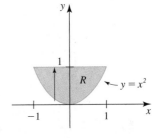

(b) The projection R of T onto the xy-plane viewed as being y-simple

FIGURE 9

By symmetry we see that $\bar{x} = 0$. Next, using Equations (9b) and (10), we have

$$\bar{y} = \frac{1}{m} \iiint_T y\rho(x, y, z) \, dV = \frac{15}{8} \iiint_T y \, dV$$

$$= \frac{15}{8} \int_{-1}^{1} \int_{x^2}^{1} \int_{0}^{1-y} y \, dz \, dy \, dx = \frac{15}{8} \int_{-1}^{1} \int_{x^2}^{1} \left[yz \right]_{z=0}^{z=1-y} dy \, dx$$

$$= \frac{15}{8} \int_{-1}^{1} \int_{x^2}^{1} (y - y^2) \, dy \, dx = \frac{15}{8} \int_{-1}^{1} \left[\frac{1}{2} y^2 - \frac{1}{3} y^3 \right]_{y=x^2}^{y=1} dx$$

$$= \frac{15}{8} \int_{-1}^{1} \left(\frac{1}{6} - \frac{1}{2} x^4 + \frac{1}{3} x^6 \right) dx = 2 \left(\frac{15}{8} \right) \int_{0}^{1} \left(\frac{1}{6} - \frac{1}{2} x^4 + \frac{1}{3} x^6 \right) dx$$

The integrand is an even function.

$$= \frac{15}{4} \left[\frac{1}{6} x - \frac{1}{10} x^5 + \frac{1}{21} x^7 \right]_{0}^{1} = \frac{3}{7}$$

Similarly, you can verify that

$$\bar{z} = \frac{1}{m} \iiint_T z\rho(x, y, z) \, dV = \frac{15}{8} \iiint_T z \, dV \qquad \text{Use Equation (9c).}$$

$$= \frac{15}{8} \int_{-1}^{1} \int_{x^2}^{1} \int_{0}^{1-y} z \, dz \, dy \, dx = \frac{2}{7}$$

Therefore, the center of mass of T is located at the point $\left(0, \frac{3}{7}, \frac{2}{7} \right)$. ■

EXAMPLE 6 Find the moments of inertia about the three coordinate axes for the solid rectangular parallelepiped of constant density k shown in Figure 10.

Solution Using Equation (11a) with $\rho(x, y, z) = k$, we obtain

$$I_x = \iiint_T (y^2 + z^2)k \, dV$$

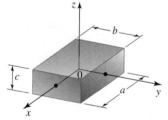

FIGURE 10
The center of the solid is placed at the origin.

$$= \int_{-c/2}^{c/2} \int_{-b/2}^{b/2} \int_{-a/2}^{a/2} k(y^2 + z^2) \, dx \, dy \, dz$$

Observe that the integrand is an even function of x, y, and z. Taking advantage of symmetry, we can write

$$I_x = 8k \int_{0}^{c/2} \int_{0}^{b/2} \int_{0}^{a/2} (y^2 + z^2) \, dx \, dy \, dz = 8k \int_{0}^{c/2} \int_{0}^{b/2} \left[(y^2 + z^2)x \right]_{x=0}^{x=a/2} dy \, dz$$

$$= 4ka \int_{0}^{c/2} \int_{0}^{b/2} (y^2 + z^2) \, dy \, dz = 4ka \int_{0}^{c/2} \left[\frac{1}{3} y^3 + z^2 y \right]_{y=0}^{y=b/2} dz$$

$$= 4ka \int_{0}^{c/2} \left(\frac{b^3}{24} + \frac{bz^2}{2} \right) dz = 4ka \left(\frac{b^3}{24} z + \frac{b}{6} z^3 \right) \Big|_{z=0}^{z=c/2}$$

$$= 4ka \left(\frac{b^3 c}{48} + \frac{bc^3}{48} \right) = \frac{kabc}{12} (b^2 + c^2)$$

$$= \frac{1}{12} m(b^2 + c^2) \qquad m = kabc = \text{mass of the solid}$$

Similarly, we find

$$I_y = \frac{1}{12} m(a^2 + c^2) \qquad \text{and} \qquad I_z = \frac{1}{12} m(a^2 + b^2)$$ ∎

14.6   CONCEPT QUESTIONS

1. **a.** Define the Riemann sum of f over a rectangular box B.
   b. Define the triple integral of f over B.
2. Suppose that f is continuous on the rectangular box $B = [a, b] \times [c, d] \times [p, q]$.
   a. Explain how you would evaluate $\iiint_B f(x, y, z)\, dV$.
   b. Write all iterated integrals that are associated with the triple integral of part (a).

3. **a.** What is a z-simple region in space? An x-simple region? A y-simple region?
   b. Write the integral $\iiint_T f(x, y, z)\, dV$, where T is a z-simple region. An x-simple region. A y-simple region.

14.6   EXERCISES

In Exercises 1–4, evaluate the integral $\iiint_B f(x, y, z)\, dV$ using the indicated order of integration.

1. $f(x, y, z) = x + y + z$; $B = \{(x, y, z) \,|\, 0 \le x \le 2, 0 \le y \le 1, 0 \le z \le 3\}$. Integrate (a) with respect to x, y, and z, in that order, and (b) with respect to z, y, and x, in that order.

2. $f(x, y, z) = xyz$; $B = \{(x, y, z) \,|\, -1 \le x \le 1, 0 \le y \le 2, -2 \le z \le 6\}$. Integrate (a) with respect to y, x, and z, in that order, and (b) with respect to x, z, and y, in that order.

3. $f(x, y, z) = xy^2 + yz^2$; $B = \{(x, y, z) \,|\, 0 \le x \le 2, -1 \le y \le 1, 0 \le z \le 3\}$. Integrate (a) with respect to z, y, and x, in that order, and (b) with respect to x, y, and z, in that order.

4. $f(x, y, z) = xy^2 \cos z$; $B = \left\{(x, y, z) \,|\, 0 \le x \le 2, 0 \le y \le 3, 0 \le z \le \frac{\pi}{2}\right\}$. Integrate (a) with respect to y, z, and x, in that order, and (b) with respect to y, x, and z, in that order.

In Exercises 5–10, evaluate the iterated integral.

5. $\displaystyle\int_0^1 \int_0^x \int_0^{x+y} x\, dz\, dy\, dx$

6. $\displaystyle\int_0^1 \int_0^z \int_0^y 2xz\, dx\, dy\, dz$

7. $\displaystyle\int_0^{\pi/2} \int_1^2 \int_0^{\sqrt{1-z}} y \cos x\, dy\, dz\, dx$

8. $\displaystyle\int_{-1}^1 \int_0^2 \int_0^{\sqrt{4-z^2}} y^2 z\, dx\, dz\, dy$

9. $\displaystyle\int_0^4 \int_0^1 \int_0^x 2\sqrt{y}\, e^{-x^2}\, dz\, dx\, dy$

10. $\displaystyle\int_1^e \int_1^x \int_0^{1/(xy)} 2 \ln y\, dz\, dy\, dx$

In Exercises 11–14, the figure shows the region of integration for $\iiint_T f(x, y, z)\, dV$. Express the triple integral as an iterated integral in six different ways using different orders of integration.

11.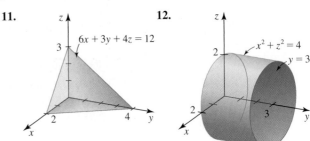

12.

13.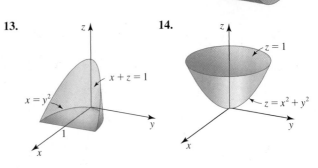

14.

In Exercises 15–22, evaluate the integral $\iiint_T f(x, y, z)\, dV$.

15. $f(x, y, z) = x$; T is the tetrahedron bounded by the planes $x = 0$, $y = 0$, $z = 0$, and $x + y + z = 1$

16. $f(x, y, z) = y$; T is the region bounded by the planes $x = 0$, $y = 0$, $z = 0$, and $2x + 3y + z = 6$

17. $f(x, y, z) = 2z$; T is the region bounded by the cylinder $y = x^3$ and the planes $y = x$, $z = 2x$, and $z = 0$

18. $f(x, y, z) = x + 2y$; T is the region bounded by the cylinder $y = \sqrt{x}$ and the planes $y = x$, $z = 2x$, and $z = 0$

19. $f(x, y, z) = y$; T is the region bounded by the paraboloid $y = x^2 + z^2$ and the plane $y = 4$

20. $f(x, y, z) = z$; T is the region bounded by the parabolic cylinder $y = x^2$ and the planes $y + z = 1$ and $z = 0$

21. $f(x, y, z) = z$; T is the region bounded by the cylinder $x^2 + z^2 = 4$ and the planes $x = 2y$, $y = 0$, and $z = 0$

22. $f(x, y, z) = \sqrt{x^2 + z^2}$; T is the region bounded by the paraboloids $y = x^2 + z^2$ and $y = 8 - x^2 - z^2$

In Exercises 23–28, sketch the solid bounded by the graphs of the equations, and then use a triple integral to find the volume of the solid.

23. $3x + 2y + z = 6$, $x = 0$, $y = 0$, $z = 0$

24. $y = 2z$, $y = x^2$, $y = 4$, $z = 0$

25. $x = 4 - y^2$, $x + z = 4$, $x = 0$, $z = 0$

26. $z = 1 - x^2$, $y = x$, $y = 2 - x$, $z = 0$

27. $z = x^2 + y^2$, $z = 8 - x^2 - y^2$

28. $x^2 + z^2 = 4$, $y^2 + z^2 = 4$

29. Find the volume of the tetrahedron with vertices $(0, 0, 0)$, $(1, 0, 0)$, $(0, 3, 0)$, and $(0, 0, 2)$.

30. Find the volume of the tetrahedron with vertices $(0, 0, 0)$, $(1, 0, 0)$, $(1, 0, 1)$, and $(1, 1, 0)$.

In Exercises 31–34, sketch the solid whose volume is given by the iterated integral.

31. $\displaystyle\int_0^1 \int_0^{1-y} \int_0^{1-x-y} dz\, dx\, dy$ **32.** $\displaystyle\int_0^1 \int_0^{1-y} \int_0^{2-2z} dx\, dz\, dy$

33. $\displaystyle\int_{-2}^{2} \int_0^{4-y^2} \int_0^{y+2} dz\, dx\, dy$ **34.** $\displaystyle\int_0^1 \int_{-\sqrt{1-y}}^{\sqrt{1-y}} \int_0^y dz\, dx\, dy$

In Exercises 35–38, express the triple integral $\iiint_T f(x, y, z)\, dV$ as an iterated integral in six different ways using different orders of integration.

35. T is the solid bounded by the planes $x + 2y + 3z = 6$, $x = 0$, $y = 0$, and $z = 0$.

36. T is the tetrahedron bounded by the planes $z = 0$, $x = 0$, $y = 0$, $y = 2 - 2z$, and $z = 1 - x$.

37. T is the solid bounded by the circular cylinder $x^2 + y^2 = 1$ and the planes $z = 0$ and $z = 2$.

38. T is the solid bounded by the parabolic cylinder $y = x^2$ and the planes $z = 0$ and $z = 4 - y$.

39. Let $f(x, y, z) = x + y + z$ and let $B = \{(x, y, z) \mid 0 \le x \le 4, 0 \le y \le 4, 0 \le z \le 4\}$.

 a. Use a Riemann sum with $m = n = p = 2$, and choose the evaluation point $(x_{ijk}^*, y_{ijk}^*, z_{ijk}^*)$ to be the midpoint

of the subrectangles $R_{ijk}(1 \le i, j, k \le 2)$ to estimate $\iiint_B f(x, y, z)\, dV$.

 b. Find the exact value of $\iiint_B f(x, y, z)\, dV$.

40. Let $f(x, y, z) = \sqrt{x^2 + y^2 + z^2}$ and let $B = \{(x, y, z) \mid 0 \le x \le 4, 0 \le y \le 2, 0 \le z \le 1\}$.

 a. Use a Riemann sum with $m = n = p = 2$, and choose the evaluation point $(x_{ijk}^*, y_{ijk}^*, z_{ijk}^*)$ to be the midpoint of the subrectangles $R_{ijk}(1 \le i, j, k \le 2)$ to estimate $\iiint_B f(x, y, z)\, dV$.

 cas **b.** Use a computer algebra system to estimate $\iiint_B f(x, y, z)\, dV$ accurate to four decimal places.

cas *In Exercises 41 and 42, use a computer algebra system to estimate the triple integral accurate to four decimal places.*

41. $\displaystyle\int_{-1}^{1} \int_0^2 \int_1^2 \frac{\cos xy}{\sqrt{1 + xyz^2}}\, dx\, dy\, dz$

42. $\displaystyle\int_0^1 \int_0^{1-x} \int_0^{1-x^2} xe^{yz}\, dz\, dy\, dx$

In Exercises 43–46, find the center of mass of the solid T having the given mass density.

43. T is the tetrahedron bounded by the planes $x = 0$, $y = 0$, $z = 0$, and $x + y + z = 1$. The mass density at a point P of T is directly proportional to the distance between P and the yz-plane.

44. T is the wedge bounded by the planes $x = 0$, $y = 0$, $z = 0$, $z = -\frac{2}{3}y + 2$ and $x = 1$. The mass density at a point P of T is directly proportional to the distance between P and the xy-plane.

45. T is the solid bounded by the cylinder $y^2 + z^2 = 4$ and the planes $x = 0$ and $x = 3$. The mass density at a point P of T is directly proportional to the distance between P and the yz-plane.

46. T is the solid bounded by the parabolic cylinder $z = 1 - x^2$ and the planes $y + z = 1$, $y = 0$, and $z = 0$. T has uniform mass density $\rho(x, y, z) = k$, where k is a constant.

In Exercises 47–50, set up, but do not evaluate, the iterated integral giving the mass of the solid T having mass density given by the function ρ.

47. T is the solid bounded by the cylinder $x^2 + z^2 = 1$ in the first octant and the plane $z + y = 1$; $\rho(x, y, z) = xy + z^2$

48. T is the solid bounded by the ellipsoid $36x^2 + 9y^2 + 4z^2 = 36$ and the planes $y = 0$ and $z = 0$; $\rho(x, y, z) = \sqrt{yz}$

49. T is the solid bounded by the parabolic cylinder $z = 1 - y^2$ and the planes $2x + y = 2$, $y = 0$, and $z = 0$; $\rho(x, y, z) = \sqrt{x^2 + y^2 + z^2}$

50. T is the upper hemisphere bounded by the sphere $x^2 + y^2 + z^2 = 1$ and the plane $z = 0$; $\rho(x, y, z) = \sqrt{1 + x^2 + y^2}$

51. Let T be a cube bounded by the planes $x = 0$, $x = 1$, $y = 0$, $y = 1$, $z = 0$, and $z = 1$. Find the moments of inertia of T with respect to the coordinate axes if T has constant mass density k.

52. Let T be a rectangular box bounded by the planes $x = 0$, $x = a$, $y = 0$, $y = b$, $z = 0$, and $z = c$. Find the moments of inertia of T with respect to the coordinate axes if T has constant mass density k.

53. Let T be the solid bounded by the planes $x + y + z = 1$, $x = 0$, $y = 0$, and $z = 0$. Find the moments of inertia of T with respect to the x-, y-, and z-axes if T has mass density given by $\rho(x, y, z) = x$.

54. Let T be the solid bounded by the cylinder $y = x^2$ and the planes $y = x$, $z = 0$, and $z = x$. Find the moments of inertia of T with respect to the coordinate axes if T has mass density given by $\rho(x, y, z) = z$.

The average value of a function f of three variables over a solid region T is defined to be

$$f_{av} = \frac{1}{V(T)} \iiint_T f(x, y, z)\, dV$$

where $V(T)$ is the volume of T. Use this definition in Exercises 55–58.

55. Find the average value of $f(x, y, z) = x + y + z$ over the rectangular box T bounded by the planes $x = 0$, $x = 1$, $y = 0$, $y = 2$, $z = 0$, and $z = 3$.

56. Find the average value of $f(x, y, z) = x^2 + y^2 + z^2$ over the tetrahedron bounded by the planes $x + y + z = 1$, $x = 0$, $y = 0$, and $z = 0$.

57. Find the average value of $f(x, y, z) = xyz$ over the solid region lying inside the spherical ball of radius 2 with center at the origin and in the first octant.

58. **Average Temperature in a Room** A rectangular room can be described by the set $B = \{(x, y, z) \mid 0 \le x \le 20,\ 0 \le y \le 40,\ 0 \le z \le 9\}$. If the temperature (in degrees Fahrenheit) at a point (x, y, z) in the room is given by $f(x, y, z) = 60 + 0.2x + 0.1y + 0.2z$, what is the average temperature in the room?

59. Find the region T that will make the value of $\iiint_T (1 - 2x^2 - 3y^2 - z^2)^{1/3}\, dV$ as large as possible.

60. Find the values of a and b that will maximize $\iiint_T (4 - x^2 - y^2 - z^2)\, dV$, where $T = \{(x, y, z) \mid 1 \le a \le x^2 + y^2 + z^2 \le b \le 2\}$.

In Exercises 61–64, determine whether the statement is true or false. If it is true, explain why. If it is false, explain why or give an example that shows it is false.

61. If $B = [-1, 1] \times [-2, 2] \times [-3, 3]$, then $\iiint_B \sqrt{x^2 + y^2 + z^2}\, dV > 0$.

62. If T is a solid sphere of radius a centered at the origin, then $\iiint_T x\, dV = 0$.

63. $12 \le \displaystyle\int_1^2 \int_1^3 \int_1^4 \sqrt{1 + x^2 + y^2 + z^2}\, dz\, dy\, dx \le 6\sqrt{30}$

64. $\displaystyle\iiint_T k\, dV = \frac{28\pi k}{3}$, where $T = \{(x, y, z) \mid 1 \le (x - 1)^2 + (y - 2)^2 + (z + 1)^2 \le 4\}$ and k is a constant

---

14.7   Triple Integrals in Cylindrical and Spherical Coordinates

Just as some double integrals are easier to evaluate by using polar coordinates, we will see that some triple integrals are easier to evaluate by using cylindrical or spherical coordinates.

■ Cylindrical Coordinates

Let T be a *z*-**simple region** described by

$$T = \{(x, y, z) \mid (x, y) \in R,\ h_1(x, y) \le z \le h_2(x, y)\}$$

where R is the projection of T onto the *xy*-plane. (See Figure 1.) As we saw in Section 14.6, if f is continuous on T, then

$$\iiint_T f(x, y, z)\, dV = \iint_R \left[\int_{h_1(x, y)}^{h_2(x, y)} f(x, y, z)\, dz\right] dA \qquad (1)$$

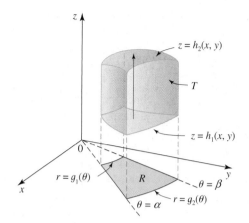

FIGURE 1
T viewed as a *z*-simple region

Now suppose that the region *R* can be described in polar coordinates by

$$R = \{(r, \theta) \mid \alpha \leq \theta \leq \beta, g_1(\theta) \leq r \leq g_2(\theta)\}$$

Then, since $x = r \cos \theta$, $y = r \sin \theta$, and $z = z$ in cylindrical coordinates, we use Equation (2) in Section 14.3 to obtain the following formula.

Triple Integral in Cylindrical Coordinates

$$\iiint_T f(x, y, z) \, dV = \int_\alpha^\beta \int_{g_1(\theta)}^{g_2(\theta)} \int_{h_1(r \cos \theta, r \sin \theta)}^{h_2(r \cos \theta, r \sin \theta)} f(r \cos \theta, r \sin \theta, z) \, r \, dz \, dr \, d\theta \quad \textbf{(2)}$$

Note   As an aid to remembering Equation (2), observe that the element of volume in cylindrical coordinates is $dV = r \, dz \, dr \, d\theta$, as is suggested by Figure 2.

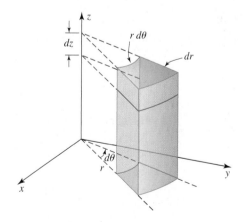

FIGURE 2
The element of volume in cylindrical
coordinates is $dV = r \, dz \, dr \, d\theta$.

EXAMPLE 1   A solid *T* is bounded by the cone $z = \sqrt{x^2 + y^2}$ and the plane $z = 2$. (See Figure 3.) The mass density at any point of the solid is proportional to the distance between the axis of the cone and the point. Find the mass of *T*.

Solution   The solid *T* is described by

$$T = \left\{(x, y, z) \mid (x, y) \in R, \sqrt{x^2 + y^2} \leq z \leq 2\right\}$$

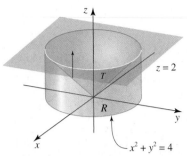

FIGURE 3
The arrow runs from the lower surface $z = h_1(x, y) = \sqrt{x^2 + y^2}$ to the upper surface $z = h_2(x, y) = 2$ of T.

where $R = \{(x, y) \mid 0 \le x^2 + y^2 \le 4\}$. In cylindrical coordinates,

$$T = \{(r, \theta, z) \mid 0 \le \theta \le 2\pi, 0 \le r \le 2, r \le z \le 2\}$$

and

$$R = \{(r, \theta) \mid 0 \le \theta \le 2\pi, 0 \le r \le 2\}$$

Since the density of the solid at (x, y, z) is proportional to the distance from the z-axis to the point in question, we see that the density function is

$$\rho(x, y, z) = k\sqrt{x^2 + y^2} = kr$$

where k is the constant of proportionality. Therefore, if we use Equation (8) in Section 14.6, the mass of T is

$$m = \iiint_T \rho(x, y, z)\, dV = \iiint_T k\sqrt{x^2 + y^2}\, dV$$

$$= \int_0^{2\pi} \int_0^2 \int_r^2 (kr)\, r\, dz\, dr\, d\theta$$

$$= k \int_0^{2\pi} \int_0^2 \left[r^2 z \right]_{z=r}^{z=2} dr\, d\theta = k \int_0^{2\pi} \int_0^2 (2r^2 - r^3)\, dr\, d\theta$$

$$= k \int_0^{2\pi} \left[\frac{2}{3} r^3 - \frac{1}{4} r^4 \right]_{r=0}^{r=2} d\theta = \frac{4}{3} k \int_0^{2\pi} d\theta = \frac{8}{3} \pi k \qquad \blacksquare$$

EXAMPLE 2 Find the centroid of a homogeneous solid hemisphere of radius a.

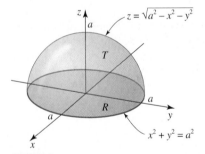

FIGURE 4
A homogeneous solid hemisphere of radius a

Solution   The solid T is shown in Figure 4. In rectangular coordinates we can write

$$T = \left\{ (x, y, z) \mid (x, y) \in R, 0 \le z \le \sqrt{a^2 - x^2 - y^2} \right\}$$

where

$$R = \{(x, y) \mid 0 \le x^2 + y^2 \le a^2\}$$

In cylindrical coordinates we have

$$T = \left\{ (r, \theta, z) \mid 0 \le \theta \le 2\pi, 0 \le r \le a, 0 \le z \le \sqrt{a^2 - r^2} \right\}$$

and

$$R = \{(r, \theta) \mid 0 \le \theta \le 2\pi, 0 \le r \le a\}$$

By symmetry the centroid lies on the z-axis. Therefore, it suffices to find $\bar{z} = M_{xy}/V$, where V, the volume of T, is $\frac{1}{2} \cdot \frac{4}{3}\pi a^3$, or $\frac{2}{3}\pi a^3$. Using Equation (9c) in Section 14.6, with $\rho(x, y, z) = 1$, we obtain

$$M_{xy} = \iiint_T z\, dV = \int_0^{2\pi} \int_0^a \int_0^{\sqrt{a^2 - r^2}} z\, r\, dz\, dr\, d\theta$$

$$= \int_0^{2\pi} \int_0^a \left[\frac{1}{2} z^2 \right]_{z=0}^{z=\sqrt{a^2 - r^2}} r\, dr\, d\theta = \frac{1}{2} \int_0^{2\pi} \int_0^a (a^2 - r^2)\, r\, dr\, d\theta$$

$$= \frac{1}{2} \int_0^{2\pi} \left[\frac{1}{2} a^2 r^2 - \frac{1}{4} r^4 \right]_{r=0}^{r=a} d\theta$$

$$= \frac{1}{2} \left(\frac{1}{4} a^4 \right) \int_0^{2\pi} d\theta = \frac{1}{8} a^4 (2\pi) = \frac{1}{4} \pi a^4$$

Therefore,

$$\bar{z} = \frac{M_{xy}}{V} = \frac{\pi a^4}{4} \cdot \frac{3}{2\pi a^3} = \frac{3}{8}a$$

so the centroid is located at the point $\left(0, 0, \frac{3a}{8}\right)$. ■

■ Spherical Coordinates

When the region of integration is bounded by portions of spheres and cones, a triple integral is generally easier to evaluate if it is expressed in terms of spherical coordinates. Recall from Section 11.7 that the relationship between spherical coordinates ρ, ϕ, θ and rectangular coordinates x, y, z is given by

$$x = \rho \sin \phi \cos \theta, \qquad y = \rho \sin \phi \sin \theta, \qquad z = \rho \cos \phi \qquad (3)$$

(See Figure 5.)

To see the role played by spherical coordinates in integration, let's consider the simplest case in which the region of integration is a **spherical wedge** (the analog of a rectangular box)

$$T = \{(\rho, \phi, \theta) \,|\, a \le \rho \le b, c \le \phi \le d, \alpha \le \theta \le \beta\}$$

where $a \ge 0$, $0 \le d - c \le \pi$, and $0 \le \beta - \alpha \le 2\pi$. To integrate over such a region, let

$$a = \rho_0 < \rho_1 < \cdots < \rho_{i-1} < \rho_i < \cdots < \rho_l = b$$

$$c = \phi_0 < \phi_1 < \cdots < \phi_{j-1} < \phi_j < \cdots < \phi_m = d$$

$$\alpha = \theta_0 < \theta_1 < \cdots < \theta_{k-1} < \theta_k < \cdots < \theta_n = \beta$$

be regular partitions of the intervals $[a, b]$, $[c, d]$, and $[\alpha, \beta]$, respectively, where $\Delta\rho = (b - a)/l$, $\Delta\phi = (d - c)/m$ and $\Delta\theta = (\beta - \alpha)/n$. The concentric spheres ρ_i, where $1 \le i \le l$, half-cones $\phi = \phi_j$, where $1 \le j \le m$, and the half-planes $\theta = \theta_k$, where $1 \le k \le n$, divide the spherical wedge T into $N = lmn$ spherical wedges $T_{111}, T_{112}, \ldots, T_{lmn}$. A typical wedge T_{ijk} comprising the spherical partition $P = \{T_{ijk}\}$ is shown in Figure 6.

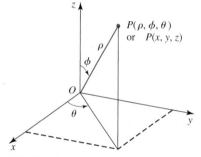

FIGURE 5
The point P has representation (ρ, ϕ, θ) in spherical coordinates and (x, y, z) in rectangular coordinates.

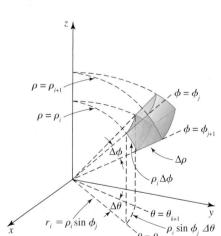

FIGURE 6
A typical spherical wedge in the partition P of the solid T

If you refer to Figure 6, you will see that T_{ijk} is approximately a rectangular box with dimensions $\Delta\rho$, $\rho_i \Delta\phi$ (the arc of a circle with radius ρ_i that subtends an angle of

$\Delta\phi$) and $\rho_i \sin \phi_j \Delta\theta$ (the arc of a circle with radius $\rho_i \sin \phi_j$ and subtending an angle of $\Delta\theta$). Thus, its volume ΔV is

$$\Delta V = \rho_i^2 \sin \phi_j \, \Delta\rho \, \Delta\phi \, \Delta\theta$$

Therefore, an approximation to a Riemann sum of f over T is

$$\sum_{i=1}^{l} \sum_{j=1}^{m} \sum_{k=1}^{n} f(\rho_i^* \sin \phi_j^* \cos \theta_k^*, \rho_i^* \sin \phi_j^* \sin \theta_k^*, \rho_i^* \cos \phi_j^*) \rho_i^{*2} \sin \phi_j^* \, \Delta\rho \, \Delta\phi \, \Delta\theta$$

But this is a Riemann sum of the function

$$F(\rho, \phi, \theta) = f(\rho \sin \phi \cos \theta, \rho \sin \phi \sin \theta, \rho \cos \phi)\rho^2 \sin \phi$$

and its limit is the triple integral

$$\int_\alpha^\beta \int_c^d \int_a^b F(\rho, \phi, \theta) \, \rho^2 \sin \phi \, d\rho \, d\phi \, d\theta$$

Therefore, we have the following formula for transforming a triple integral in rectangular coordinates into one involving spherical coordinates.

Triple Integral in Spherical Coordinates

$$\iiint_T f(x, y, z) \, dV = \int_\alpha^\beta \int_c^d \int_a^b f(\rho \sin \phi \cos \theta, \rho \sin \phi \sin \theta, \rho \cos \phi)\rho^2 \sin \phi \, d\rho \, d\phi \, d\theta \qquad \textbf{(4)}$$

where T is the spherical wedge

$$T = \{(\rho, \phi, \theta) \,|\, a \le \rho \le b, c \le \phi \le d, \alpha \le \theta \le \beta\}$$

Equation (4) states that to transform a triple integral in rectangular coordinates to one in spherical coordinates, make the substitutions

$$x = \rho \sin \phi \cos \theta, \quad y = \rho \sin \phi \sin \theta, \quad z = \rho \cos \phi, \quad \text{and} \quad x^2 + y^2 + z^2 = \rho^2$$

then make the appropriate change in the limits of integration, and replace dV by $\rho^2 \sin \phi \, d\rho \, d\phi \, d\theta$. This element of volume can be recalled with the help of Figure 7.

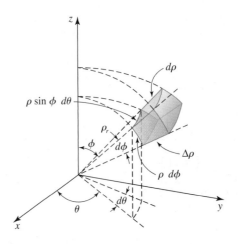

FIGURE 7
The element of volume in spherical coordinates is $dV = \rho^2 \sin \phi \, d\rho \, d\phi \, d\theta$.

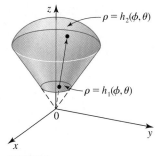

FIGURE 8

A ρ-simple region is bounded by the surfaces $\rho = h_1(\phi, \theta)$ and $\rho = h_2(\phi, \theta)$

Equation (4) can be extended to include more general regions. For example, if T is **ρ-simple,** that is, if the region T can be described by

$$T = \{(\rho, \phi, \theta) \mid h_1(\phi, \theta) \le \rho \le h_2(\phi, \theta), c \le \phi \le d, \alpha \le \theta \le \beta\}$$

then

$$\iiint_T f(x, y, z) \, dV$$

$$= \int_\alpha^\beta \int_c^d \int_{h_1(\phi, \theta)}^{h_2(\phi, \theta)} f(\rho \sin\phi \cos\theta, \rho \sin\phi \sin\theta, \rho \cos\phi) \rho^2 \sin\phi \, d\rho \, d\phi \, d\theta \qquad \textbf{(5)}$$

Observe that ρ-simple regions are precisely those regions that lie between two surfaces $\rho = h_1(\phi, \theta)$ and $\rho = h_2(\phi, \theta)$, as shown in Figure 8. To find the limits of integration with respect to ρ, we draw a radial line emanating from the origin. The line first intersects the surface, $\rho = h_1(\phi, \theta)$, giving the lower limit of integration, and then intersects the surface $\rho = h_2(\phi, \theta)$, giving the upper limit of integration.

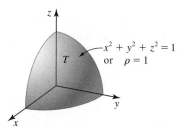

FIGURE 9

T is the part of the ball $x^2 + y^2 + z^2 \le 1$ lying in the first octant.

EXAMPLE 3 Evaluate $\iiint_T x \, dV$, where T is the part of the region in the first octant lying inside the sphere $x^2 + y^2 + z^2 = 1$.

Solution The solid T is shown in Figure 9. Since the boundary of T is part of a sphere, let's use spherical coordinates. In terms of spherical coordinates we can write

$$T = \left\{(\rho, \phi, \theta) \mid 0 \le \rho \le 1, 0 \le \phi \le \frac{\pi}{2}, 0 \le \theta \le \frac{\pi}{2}\right\}$$

Furthermore, $x = \rho \sin\phi \cos\theta$. Therefore, using Equation (4), we obtain

$$\iiint_T x \, dV = \int_0^{\pi/2} \int_0^{\pi/2} \int_0^1 (\rho \sin\phi \cos\theta)\rho^2 \sin\phi \, d\rho \, d\phi \, d\theta$$

$$= \int_0^{\pi/2} \int_0^{\pi/2} \int_0^1 \rho^3 \sin^2\phi \cos\theta \, d\rho \, d\phi \, d\theta$$

$$= \int_0^{\pi/2} \int_0^{\pi/2} \left[\frac{1}{4}\rho^4 \sin^2\phi \cos\theta\right]_{\rho=0}^{\rho=1} d\phi \, d\theta = \frac{1}{4}\int_0^{\pi/2} \int_0^{\pi/2} \sin^2\phi \cos\theta \, d\phi \, d\theta$$

$$= \frac{1}{8}\int_0^{\pi/2} \int_0^{\pi/2} (1 - \cos 2\phi)\cos\theta \, d\phi \, d\theta = \frac{1}{8}\int_0^{\pi/2} \cos\theta\left[\phi - \frac{1}{2}\sin 2\phi\right]_{\phi=0}^{\phi=\pi/2} d\theta$$

$$= \frac{\pi}{16}\int_0^{\pi/2} \cos\theta \, d\theta = \frac{\pi}{16}\sin\theta\Big|_0^{\pi/2} = \frac{\pi}{16} \qquad \blacksquare$$

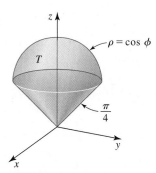

FIGURE 10

The solid T is bounded below by part of a cone and above by part of a sphere.

EXAMPLE 4 Find the center of mass of the solid T of uniform density bounded by the cone $z = \sqrt{x^2 + y^2}$ and the sphere $x^2 + y^2 + z^2 = z$. (See Figure 10.)

Solution We first express the given equations in terms of spherical coordinates. The equation of the cone is

$$\rho \cos\phi = \sqrt{\rho^2 \sin^2\phi \cos^2\theta + \rho^2 \sin^2\phi \sin^2\theta} = \rho \sin\phi$$

which simplifies to $\cos\phi = \sin\phi$, $\tan\phi = 1$, or $\phi = \pi/4$. Next, we see that the equation of the sphere is

$$\rho^2 = \rho\cos\phi \qquad \text{or} \qquad \rho = \cos\phi$$

Therefore, the solid under consideration can be described by

$$T = \left\{(\rho, \phi, \theta) \,\middle|\, 0 \le \rho \le \cos\phi, 0 \le \phi \le \tfrac{\pi}{4}, 0 \le \theta \le 2\pi\right\}$$

Let the uniform density of T be k. Then the mass of T is

$$m = k\iiint_T dV = k\int_0^{2\pi}\int_0^{\pi/4}\int_0^{\cos\phi} \rho^2\sin\phi\, d\rho\, d\phi\, d\theta \qquad h_1(\phi,\theta) = 0,\ h_2(\phi,\theta) = \cos\phi$$

$$= k\int_0^{2\pi}\int_0^{\pi/4}\left[\frac{1}{3}\rho^3\sin\phi\right]_{\rho=0}^{\rho=\cos\phi} d\phi\, d\theta$$

$$= \frac{k}{3}\int_0^{2\pi}\int_0^{\pi/4}\cos^3\phi\sin\phi\, d\phi\, d\theta = \frac{k}{3}\int_0^{2\pi}\left[-\frac{1}{4}\cos^4\phi\right]_{\phi=0}^{\phi=\pi/4} d\theta$$

$$= \frac{k}{16}\int_0^{2\pi} d\theta = \frac{\pi k}{8}$$

By symmetry the center of mass lies on the z-axis, so it suffices to find $\bar{z} = M_{xy}/m$. Using Equation (9c) in Section 14.6, with $\rho(x, y, z) = 1$, we obtain

$$M_{xy} = \iiint_T kz\, dV = k\int_0^{2\pi}\int_0^{\pi/4}\int_0^{\cos\phi} (\rho\cos\phi)\rho^2\sin\phi\, d\rho\, d\phi\, d\theta$$

$$= k\int_0^{2\pi}\int_0^{\pi/4}\left[\frac{1}{4}\rho^4\cos\phi\sin\phi\right]_{\rho=0}^{\rho=\cos\phi} d\phi\, d\theta$$

$$= \frac{k}{4}\int_0^{2\pi}\int_0^{\pi/4}\cos^5\phi\sin\phi\, d\phi\, d\theta = \frac{k}{4}\int_0^{2\pi}\left[-\frac{1}{6}\cos^6\phi\right]_{\phi=0}^{\phi=\pi/4} d\theta$$

$$= -\frac{k}{24}\left(\frac{(\sqrt{2})^6}{2^6} - 1\right)\int_0^{2\pi} d\theta = \frac{7k}{192}\int_0^{2\pi} d\theta = \frac{7k\pi}{96}$$

Therefore,

$$\bar{z} = \frac{M_{xy}}{m} = \frac{7k\pi}{96}\cdot\frac{8}{\pi k} = \frac{7}{12}$$

so the center of mass is located at $\left(0, 0, \tfrac{7}{12}\right)$. ∎

14.7 CONCEPT QUESTIONS

1. Write the triple integral $\iiint_T f(x, y, z)\, dV$ in cylindrical coordinates if

$$T = \{(r, \theta, z) \mid \alpha \le \theta \le \beta, g_1(\theta) \le r \le g_2(\theta),$$

$$h_1(r\cos\theta, r\sin\theta) \le z \le h_2(r\cos\theta, r\sin\theta)\}$$

2. Write the triple integral $\iiint_T f(x, y, z)\, dV$ in spherical coordinates if

$$T = \{(\rho, \phi, \theta) \mid h_1(\phi, \theta) \le \rho \le h_2(\phi, \theta), c \le \phi \le d, \alpha \le \theta \le \beta\}$$

3. Write the element of volume dV in (a) cylindrical coordinates and (b) spherical coordinates.

14.7 EXERCISES

In Exercises 1–4, sketch the solid whose volume is given by the integral, and evaluate the integral.

1. $\int_0^{\pi/2} \int_0^3 \int_0^{r^2} r \, dz \, dr \, d\theta$

2. $\int_0^{2\pi} \int_1^2 \int_0^{2-r} r \, dz \, dr \, d\theta$

3. $\int_0^{2\pi} \int_0^{\pi/2} \int_0^2 \rho^2 \sin \phi \, d\rho \, d\phi \, d\theta$

4. $\int_0^{2\pi} \int_0^{\pi/4} \int_0^{2\sec\phi} \rho^2 \sin \phi \, d\rho \, d\phi \, d\theta$

In Exercises 5–18, solve the problem using cylindrical coordinates.

5. Evaluate $\iiint_T \sqrt{x^2 + y^2} \, dV$, where T is the solid bounded by the cylinder $x^2 + y^2 = 1$ and the planes $z = 1$ and $z = 3$.

6. Evaluate $\iiint_T e^{x^2+y^2} \, dV$, where T is the solid bounded by the cylinder $x^2 + y^2 = 4$ and the planes $z = 0$ and $z = 4$.

7. Evaluate $\iiint_T y \, dV$, where T is the part of the solid in the first octant lying inside the paraboloid $z = 4 - x^2 - y^2$.

8. Evaluate $\iiint_T x \, dV$, where T is the part of the solid in the first octant bounded by the paraboloid $z = x^2 + y^2$ and the plane $z = 4$.

9. Evaluate $\iiint_T (x^2 + y^2) \, dV$, where T is the solid bounded by the cone $z = 4 - \sqrt{x^2 + y^2}$ and the xy-plane.

10. Evaluate $\iiint_T y^2 \, dV$, where T is the solid that lies within the cylinder $x^2 + y^2 = 1$ and between the xy-plane and the paraboloid $z = 2x^2 + 2y^2$.

11. Find the volume of the solid bounded above by the sphere $x^2 + y^2 + z^2 = 9$ and below by the paraboloid $8z = x^2 + y^2$.

12. Find the volume of the solid bounded by the paraboloids $z = x^2 + y^2$ and $z = 12 - 2x^2 - 2y^2$.

13. A solid is bounded by the cylinder $x^2 + y^2 = 4$ and the planes $z = 0$ and $z = 3$. Find the center of mass of the solid if the mass density at any point is directly proportional to its distance from the xy-plane.

14. A solid is bounded by the cone $z = \sqrt{x^2 + y^2}$ and the plane $z = 4$. Find its center of mass if the mass density at $P(x, y, z)$ is directly proportional to the distance between P and the z-axis.

15. Find the center of mass of a homogeneous solid bounded by the paraboloid $z = 4 - x^2 - y^2$ and $z = 0$.

16. Find the center of mass of a homogeneous solid bounded by the paraboloids $z = x^2 + y^2$ and $z = 36 - 3x^2 - 3y^2$.

17. Find the moment of inertia about the z-axis of a homogeneous solid bounded by the cone $z = \sqrt{x^2 + y^2}$ and the paraboloid $z = x^2 + y^2$.

18. Find the moment of inertia about the z-axis of a solid bounded by the cylinder $x^2 + y^2 = 4$ and the planes $z = 0$ and $z = 3$ if the mass density at any point on the solid is directly proportional to its distance from the xy-plane.

In Exercises 19–24, solve the problem by using spherical coordinates.

19. Evaluate $\iiint_B \sqrt{x^2 + y^2 + z^2} \, dV$, where B is the unit ball $x^2 + y^2 + z^2 \le 1$.

20. Evaluate $\iiint_B e^{(x^2+y^2+z^2)^{3/2}} dV$, where B is the part of the unit ball $x^2 + y^2 + z^2 \le 1$ lying in the first octant.

21. Evaluate $\iiint_T y \, dV$, where T is the solid bounded by the hemisphere $z = \sqrt{1 - x^2 - y^2}$ and the xy-plane.

22. Evaluate $\iiint_T x^2 \, dV$, where T is the part of the unit ball $x^2 + y^2 + z^2 \le 1$ lying in the first octant.

23. Evaluate $\iiint_T xz \, dV$, where T is the solid bounded above by the sphere $x^2 + y^2 + z^2 = 4$ and below by the cone $z = \sqrt{x^2 + y^2}$.

24. Evaluate $\iiint_T z \, dV$, where T is the solid bounded above by the sphere $x^2 + y^2 + z^2 = 4$ and below by the cone $z = \sqrt{x^2 + y^2}$.

25. Find the volume of the solid that is bounded above by the plane $z = 1$ and below by the cone $z = \sqrt{x^2 + y^2}$.

26. Find the volume of the solid bounded by the cone $z = \sqrt{x^2 + y^2}$, the cylinder $x^2 + y^2 = 4$, and the plane $z = 0$.

27. Find the volume of the solid lying outside the cone $z = \sqrt{x^2 + y^2}$ and inside the upper hemisphere $x^2 + y^2 + z^2 \le 1$.

28. Find the volume of the solid lying above the cone $\phi = \pi/6$ and below the sphere $\rho = 4 \cos \phi$.

29. Find the centroid of a homogeneous solid hemisphere of radius a.

30. Find the centroid of the solid of Exercise 28.

31. Find the mass of a solid hemisphere of radius a if the mass density at any point on the solid is directly proportional to its distance from the base of the solid.

32. Find the center of mass of the solid of Exercise 31.

33. Find the mass of the solid bounded by the cone $z = \sqrt{x^2 + y^2}$ and the plane $z = 2$ if the mass density at any point on the solid is directly proportional to the square of its distance from the origin.

Videos for selected exercises are available online at **www.academic.cengage.com/login**.

34. Find the center of mass of the solid of Exercise 33.

35. Find the moment of inertia about the z-axis of the solid of Exercise 28, assuming that it has constant mass density.

36. Find the moment of inertia with respect to the axis of symmetry for a solid hemisphere of radius a if the density at a point is directly proportional to its distance from the center of the base.

37. Find the moment of inertia with respect to a diameter of the base of a homogeneous solid hemisphere of radius a.

38. Show that the average distance from the center of a circle of radius a to other points of the circle is $2a/3$ and that the average distance from the center of a sphere of radius a to other points of the sphere is $3a/4$.

39. Let T be a uniform solid of mass m bounded by the spheres $\rho = a$ and $\rho = b$, where $0 < a < b$. Show that the moment of inertia of T about a diameter of T is

$$I = \frac{2m}{5}\left(\frac{b^5 - a^5}{b^3 - a^3}\right)$$

40. a. Use the result of Exercise 39 to find the moment of inertia of a uniform solid ball of mass m and radius b about a diameter of the ball.

 b. Use the result of Exercise 39 to find the moment of inertia of a hollow spherical shell of mass m and radius b about a diameter of the shell.
 Hint: Find $\lim_{a \to b^-} I$.

In Exercises 41 and 42, evaluate the integral by using cylindrical coordinates.

41. $\displaystyle\int_{-1}^{1}\int_{0}^{\sqrt{1-x^2}}\int_{0}^{\sqrt{4-x^2-y^2}} z \, dz \, dy \, dx$

42. $\displaystyle\int_{-1}^{1}\int_{-\sqrt{1-x^2}}^{\sqrt{1-x^2}}\int_{\sqrt{x^2+y^2}}^{2-x^2-y^2} (x^2 + y^2)^{3/2} \, dz \, dy \, dx$

In Exercises 43 and 44, evaluate the integral by using spherical coordinates.

43. $\displaystyle\int_{0}^{1}\int_{0}^{\sqrt{1-x^2}}\int_{\sqrt{x^2+y^2}}^{\sqrt{2-x^2-y^2}} (x^2 + y^2 + z^2)^{3/2} \, dz \, dy \, dx$

44. $\displaystyle\int_{-3}^{3}\int_{-\sqrt{9-x^2}}^{\sqrt{9-x^2}}\int_{4}^{\sqrt{25-x^2-y^2}} (x^2 + y^2 + z^2)^{-1/2} \, dz \, dy \, dx$

45. The temperature (in degrees Fahrenheit) at a point (x, y, z) of a solid ball of radius 3 in. centered at the origin is given by $T(x, y, z) = 20(x^2 + y^2 + z^2)$. What is the average temperature of the ball?

In Exercises 46–50, determine whether the statement is true or false. If it is true, explain why. If it is false, explain why or give an example that shows it is false.

46. The volume of the solid bounded above by the paraboloid $z = 4 - x^2 - y^2$ and below by the xy-plane in cylindrical coordinates is $\int_{0}^{2\pi}\int_{0}^{2}\int_{0}^{4-r^2} dz \, dr \, d\theta$.

47. $\int_{0}^{\pi/2}\int_{0}^{2\pi}\int_{0}^{2} \rho^2 \sin \phi \, d\rho \, d\theta \, d\phi = \frac{16\pi}{3}$

48. If $T = \left\{(\rho, \phi, \theta) \,|\, a < \rho < b, 0 \le \phi \le \frac{\pi}{2}, 0 \le \theta \le \frac{\pi}{2}\right\}$, then $\iiint_T dV = \frac{\pi}{6}(b^3 - a^3)$.

49. If T is a solid with constant density k, then its moment of inertia about the z-axis is given by $I_z = k\iiint_T \rho^2 \sin^2 \phi \, dV$.

50. If $T = \left\{(\rho, \phi, \theta) \,|\, 0 < \rho < a, 0 \le \phi \le \frac{\pi}{2}, 0 \le \theta \le 2\pi\right\}$, then $\iiint_T \rho \cos \theta \, dV = 0$.

14.8 Change of Variables in Multiple Integrals

We often use a change of variable (a substitution) when we integrate a function of one variable to transform the given integral into one that is easier to evaluate. For example, using the substitution $x = \sin \theta$, we find

$$\int_{0}^{1} \sqrt{1 - x^2} \, dx = \int_{0}^{\pi/2} \cos^2 \theta \, d\theta = \frac{1}{2}\int_{0}^{\pi/2} (1 + \cos 2\theta) \, d\theta$$

$$= \frac{\pi}{4}$$

Observe that the interval of integration is $[0, 1]$ if we integrate with respect to x, and it changes to $\left[0, \frac{\pi}{2}\right]$ if we integrate with respect to θ. More generally, the substitution $x = g(u)$ [so $dx = g'(u) \, du$] enables us to write

$$\int_{a}^{b} f(x) \, dx = \int_{c}^{d} f(g(u))g'(u) \, du \tag{1}$$

where $a = g(c)$ and $b = g(d)$.

As you have also seen on many occasions, a change of variables can be used to help us to evaluate integrals involving a function of two or more variables. For example, in evaluating a double integral $\iint_R f(x, y)\, dA$, where R is a circular region, it is often helpful to use the substitution

$$x = r\cos\theta \qquad y = r\sin\theta$$

to transform the original integral into one involving polar coordinates. In this instance we have

$$\iint\limits_R f(x, y)\, dA = \iint\limits_D f(r\cos\theta, r\sin\theta)\, r\, dr\, d\theta$$

where D is in the region in the $r\theta$-plane that corresponds to the region R in the xy-plane.

These examples raise the following questions:

1. If an integral $\iint f(x, y)\, dA$ cannot be readily found when we are integrating with respect to the variables x and y, can we find a substitution $x = g(u, v)$, $y = h(u, v)$ that transforms this integral into one involving the variables u and v that is more convenient to evaluate?
2. What form does the latter integral take?

■ Transformations

The substitutions that are used to change an integral involving the variables x and y into one involving the variables u and v are determined by a **transformation** or function T from the uv-plane to the xy-plane. This function associates with each point (u, v) in a region S in the uv-plane exactly one point (x, y) in the xy-plane. (See Figure 1.) The point (x, y), called the **image** of the point (u, v) under the transformation T, is written $(x, y) = T(u, v)$ and is defined by the equations

$$x = g(u, v) \qquad y = h(u, v) \tag{2}$$

where g and h are functions of two variables. The totality of all points in the xy-plane that are images of all points in S is called the **image of S** and denoted by $T(S)$. Figure 1 gives a geometric visualization of a transformation T that maps a region S in the uv-plane onto a region R in the xy-plane.

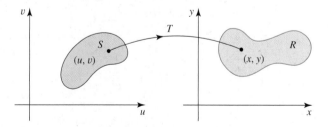

FIGURE 1

T maps the region S in the uv-plane onto the region R in the xy-plane.

A transformation T is **one-to-one** if no two distinct points in the uv-plane have the same image. In this case it may be possible to solve Equation (2) for u and v in terms of x and y to obtain the equations

$$u = G(x, y) \qquad v = H(x, y)$$

which defines the **inverse transformation** T^{-1} from the xy-plane to the uv-plane.

EXAMPLE 1 Let T be a transformation defined by the equations

$$x = u + v \qquad y = v$$

Find the image of the rectangular region $S = \{(u, v) \mid 0 \le u \le 2, 0 \le v \le 1\}$ under the transformation T.

Solution Let's see how the sides of the rectangle S are transformed by T. Referring to Figure 2a, observe $0 \le u \le 2$ and $v = 0$ on S_1. Using the given equations describing T, we see that $x = u$ and $y = 0$. This shows that S_1 is mapped onto the line segment $0 \le x \le 2$ and $y = 0$ (labeled $T(S_1)$ in Figure 2b). On S_2, $u = 2$ and $0 \le v \le 1$, so $x = 2 + y$, for $0 \le y \le 1$. This gives the image of S_2 under T as the line segment $T(S_2)$. On S_3, $0 \le u \le 2$ and $v = 1$, so $x = u + 1$ and $y = 1$, which means that S_3 is mapped onto the line segment $T(S_3)$ described by $1 \le x \le 3$, $y = 1$. Finally, on S_4, $u = 0$ and $0 \le v \le 1$, and this gives the image of S_4 as the line segment $x = y$, for $0 \le y \le 1$. Observe that as the perimeter of S is traced in a counterclockwise direction, so too is the boundary of the image $R = T(S)$ of S. The image of S under T is the region inside and on the parallelogram R.

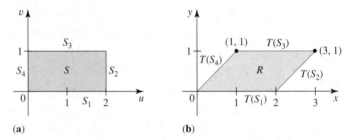

FIGURE 2
The region S in part (a) is transformed onto the region R in part (b) by T.

(a)   (b)

■ Change of Variables in Double Integrals

To see how a double integral is changed under the transformation T defined by Equation (2), let's consider the effect that T has on the area of a small rectangular region S in the uv-plane with vertices (u_0, v_0), $(u_0 + \Delta u, v_0)$, $(u_0 + \Delta u, v_0 + \Delta v)$, and $(u_0, v_0 + \Delta v)$ as shown in Figure 3a. The image of S is the region $R = T(S)$ in the xy-plane shown in Figure 3b. The lower left-hand corner point of S, (u_0, v_0), is mapped onto the point $(x_0, y_0) = T(u_0, v_0) = (g(u_0, v_0), h(u_0, v_0))$ by T. On the side L_1 of S, $u_0 \le u \le u_0 + \Delta u$ and $v = v_0$. Therefore, the image $T(L_1)$ of L_1 under T is the curve with equations

$$x = g(u, v_0) \qquad y = h(u, v_0)$$

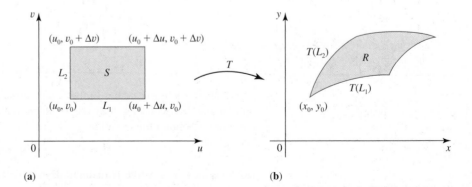

FIGURE 3
The transformation T maps S onto R.

(a)   (b)

or, in vector form,

$$\mathbf{r}(u, v_0) = g(u, v_0)\mathbf{i} + h(u, v_0)\mathbf{j}$$

with parameter interval $[u_0, u_0 + \Delta u]$. As you can see from Figure 4, the vector

$$\mathbf{a} = \mathbf{r}(u_0 + \Delta u, v_0) - \mathbf{r}(u_0, v_0)$$

provides us with an approximation of $\mathbf{T}(L_1)$. Similarly, we see that the vector

$$\mathbf{b} = \mathbf{r}(u_0, v_0 + \Delta v) - \mathbf{r}(u_0, v_0)$$

provides us with an approximation of $\mathbf{T}(L_2)$.

But we can write

$$\mathbf{a} = \left[\frac{\mathbf{r}(u_0 + \Delta u, v_0) - \mathbf{r}(u_0, v_0)}{\Delta u} \right] \Delta u$$

If Δu is small, as we have assumed, then the term inside the brackets is approximately equal to $\mathbf{r}_u(u_0, v_0)$. So

$$\mathbf{a} \approx \Delta u \, \mathbf{r}_u(u_0, v_0)$$

Similarly, we see that

$$\mathbf{b} \approx \Delta v \, \mathbf{r}_v(u_0, v_0)$$

This suggests that we can approximate R by the parallelogram having $\Delta u \, \mathbf{r}_u(u_0, v_0)$ and $\Delta v \, \mathbf{r}_v(u_0, v_0)$ as adjacent sides. (See Figure 5.) The area of this parallelogram is $|\mathbf{a} \times \mathbf{b}|$, or

$$|(\Delta u \, \mathbf{r}_u) \times (\Delta v \, \mathbf{r}_v)| = |\mathbf{r}_u \times \mathbf{r}_v| \, \Delta u \, \Delta v$$

where the partial derivatives are evaluated at (u_0, v_0). But

$$\mathbf{r}_u = g_u \mathbf{i} + h_u \mathbf{j} = \frac{\partial x}{\partial u} \mathbf{i} + \frac{\partial y}{\partial u} \mathbf{j}$$

where the partial derivatives are evaluated at (u_0, v_0). Similarly,

$$\mathbf{r}_v = g_v \mathbf{i} + h_v \mathbf{j} = \frac{\partial x}{\partial v} \mathbf{i} + \frac{\partial y}{\partial v} \mathbf{j}$$

So

$$\mathbf{r}_u \times \mathbf{r}_v = \begin{vmatrix} \mathbf{i} & \mathbf{j} & \mathbf{k} \\ \dfrac{\partial x}{\partial u} & \dfrac{\partial y}{\partial u} & 0 \\ \dfrac{\partial x}{\partial v} & \dfrac{\partial y}{\partial v} & 0 \end{vmatrix} = \begin{vmatrix} \dfrac{\partial x}{\partial u} & \dfrac{\partial y}{\partial u} \\ \dfrac{\partial x}{\partial v} & \dfrac{\partial y}{\partial v} \end{vmatrix} \mathbf{k} = \begin{vmatrix} \dfrac{\partial x}{\partial u} & \dfrac{\partial x}{\partial v} \\ \dfrac{\partial y}{\partial u} & \dfrac{\partial y}{\partial v} \end{vmatrix} \mathbf{k}$$

Before proceeding, let's define the following determinant, which is named after the German mathematician Carl Jacobi (1804–1851).

DEFINITION   The Jacobian

The Jacobian of the transformation T defined by $x = g(u, v)$ and $y = h(u, v)$ is

$$\frac{\partial(x, y)}{\partial(u, v)} = \begin{vmatrix} \dfrac{\partial x}{\partial u} & \dfrac{\partial x}{\partial v} \\ \dfrac{\partial y}{\partial u} & \dfrac{\partial y}{\partial v} \end{vmatrix} = \frac{\partial x}{\partial u} \frac{\partial y}{\partial v} - \frac{\partial y}{\partial u} \frac{\partial x}{\partial v}$$

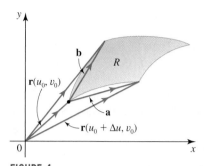

FIGURE 4
The vector
$\mathbf{a} = \mathbf{r}(u_0 + \Delta u, v_0) - \mathbf{r}(u_0, v_0)$

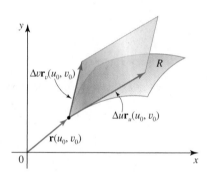

FIGURE 5
The image region R is approximated by the parallelogram with sides $\Delta u \, \mathbf{r}_u(u_0, v_0)$ and $\Delta v \, \mathbf{r}_v(u_0, v_0)$.

In terms of the Jacobian we can write the approximation of the area ΔA of R as

$$\Delta A \approx |\mathbf{r}_u \times \mathbf{r}_v| \Delta u \, \Delta v = \left| \frac{\partial(x, y)}{\partial(u, v)} \right| \Delta u \, \Delta v \tag{3}$$

where the Jacobian is evaluated at (u_0, v_0).

Now let R be the image (in the xy-plane) under T of the region S in the uv-plane; that is, let $R = T(S)$ as shown in Figure 6. Enclose S by a rectangle, and partition the latter into mn rectangles S_{ij}, where $1 \le i \le m$, $1 \le j \le n$. The images S_{ij} are transformed onto images R_{ij} in the xy-plane, as shown in Figure 6.

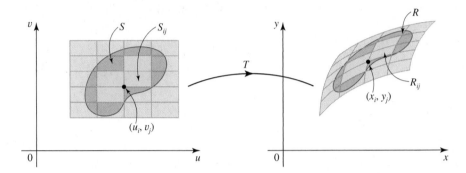

FIGURE 6
The images S_{ij} in the uv-plane are transformed onto the images R_{ij} in the xy-plane.

Suppose that f is continuous on R, and define F by

$$F_R(x, y) = \begin{cases} f(x, y) & \text{if } (x, y) \in R \\ 0 & \text{if } (x, y) \notin R \end{cases}$$

Using the approximation in Equation (3) on each subrectangle R_{ij}, we can write the double integral of f over R as

$$\iint_R f(x, y) \, dA = \lim_{m, n \to \infty} \sum_{i=1}^{m} \sum_{j=1}^{n} F_R(x_i, y_j) \, \Delta A$$

$$= \lim_{m, n \to \infty} \sum_{i=1}^{m} \sum_{j=1}^{n} F_R(g(u_i, v_j), h(u_i, v_j)) \left| \frac{\partial(x, y)}{\partial(u, v)} \right| \Delta u \, \Delta v$$

where the Jacobian is evaluated at (u_i, v_j). But the sum on the right is the Riemann sum associated with the integral

$$\iint_S f(g(u, v), h(u, v)) \left| \frac{\partial(x, y)}{\partial(u, v)} \right| du \, dv$$

This discussion suggests the following result. Its proof can be found in books on advanced calculus.

THEOREM 1   Change of Variables in Double Integrals

Let T be a one-to-one transformation defined by $x = g(u, v)$, $y = h(u, v)$ that maps a region S in the uv-plane onto a region R in the xy-plane. Suppose that the boundaries of both R and S consist of finitely many piecewise smooth, simple, closed curves. Furthermore, suppose that the first-order partial derivatives of g and h are continuous functions. If f is continuous on R and the Jacobian of T is nonzero, then

$$\iint_R f(x, y) \, dA = \iint_S f(g(u, v), h(u, v)) \left| \frac{\partial(x, y)}{\partial(u, v)} \right| du \, dv \tag{4}$$

Note Theorem 1 tells us that we can formally transform an integral $\iint_R f(x, y)\, dA$ involving the variables x and y into an integral involving the variables u and v by replacing x by $g(u, v)$ and y by $h(u, v)$ and the area element dA in x and y by the area element

$$dA = \left| \frac{\partial(x, y)}{\partial(u, v)} \right| du\, dv$$

in u and v. If you compare Equation (4) with Equation (1), you will see that the absolute value of the Jacobian of T plays the same role as the derivative $g'(u)$ of the "transformation" g defined by $x = g(u)$ in the one-dimensional case. ■

EXAMPLE 2 Use the transformation T defined by the equations $x = u + v$, $y = v$ to evaluate $\iint_R (x + y)\, dA$, where R is the parallelogram shown in Figure 2b. (See Example 1.)

Solution Recall that the transformation T maps the much simpler rectangular region $S = \{(u, v) \mid 0 \le u \le 2, 0 \le v \le 1\}$ onto R and that this is precisely the reason for choosing this transformation. The Jacobian of T is

$$\frac{\partial(x, y)}{\partial(u, v)} = \begin{vmatrix} \dfrac{\partial x}{\partial u} & \dfrac{\partial x}{\partial v} \\ \dfrac{\partial y}{\partial u} & \dfrac{\partial y}{\partial v} \end{vmatrix} = \begin{vmatrix} 1 & 1 \\ 0 & 1 \end{vmatrix} = 1$$

Using Theorem 1, we obtain

$$\iint_R (x + y)\, dA = \iint_S [(u + v) + v](1)\, du\, dv$$

$$= \int_0^1 \int_0^2 (u + 2v)\, du\, dv = \int_0^1 \left[\frac{1}{2} u^2 + 2uv \right]_{u=0}^{u=2} dv$$

$$= \int_0^1 (2 + 4v)\, dv = \left[2v + 2v^2 \right]_0^1 = 4$$ ■

In Example 2 the transformation T was chosen so that the region S in the uv-plane corresponding to the region R could be described more simply. This made it easier to evaluate the transformed integral. In other instances the transformation is chosen so that the corresponding integrand in u and v is easier to integrate than the original integrand in the variables x and y, as the following example shows.

EXAMPLE 3 Evaluate

$$\iint_R \cos\left(\frac{x - y}{x + y} \right) dA$$

where R is the trapezoidal region with vertices $(1, 0)$, $(2, 0)$, $(0, 2)$, and $(0, 1)$.

Solution As it stands, this integral is difficult to evaluate. But observe that the form of the integrand suggests that we make the substitution

$$u = x - y \qquad v = x + y$$

These equations define a transformation T^{-1} from the xy-plane to the uv-plane. If we solve these equations for x and y in terms of u and v, we obtain the transformation T from the uv-plane to the xy-plane defined by

$$x = \frac{1}{2}(u + v) \qquad y = \frac{1}{2}(v - u)$$

The given region R is shown in Figure 7.

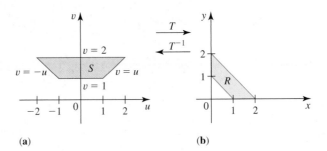

FIGURE 7
T maps S onto R, and
T^{-1} maps R onto S.

To find the region S in the uv-plane that is mapped onto R under the transformation T, observe that the sides of R lie on the lines

$$y = 0, \qquad y + x = 2, \qquad x = 0, \qquad \text{and} \qquad y + x = 1$$

Using the equations defining T^{-1}, we see that the sides of S corresponding to these sides of R are

$$v = u, \qquad v = 2, \qquad v = -u, \qquad \text{and} \qquad v = 1$$

The region S is shown in Figure 7a.

The Jacobian of T is

$$\frac{\partial(x, y)}{\partial(u, v)} = \begin{vmatrix} \dfrac{\partial x}{\partial u} & \dfrac{\partial x}{\partial v} \\[2mm] \dfrac{\partial y}{\partial u} & \dfrac{\partial y}{\partial v} \end{vmatrix} = \begin{vmatrix} \dfrac{1}{2} & \dfrac{1}{2} \\[2mm] -\dfrac{1}{2} & \dfrac{1}{2} \end{vmatrix} = \frac{1}{2}$$

If we use Theorem 1 while viewing S as a u-simple region, we find

$$\iint_R \cos\left(\frac{x - y}{x + y}\right) dA = \iint_S \cos\left(\frac{u}{v}\right)\left|\frac{\partial(x, y)}{\partial(u, v)}\right| du\, dv$$

$$= \int_1^2 \int_{-v}^{v} \cos\left(\frac{u}{v}\right) \cdot \left(\frac{1}{2}\right) du\, dv = \frac{1}{2}\int_1^2 \left[v \sin\left(\frac{u}{v}\right)\right]_{u=-v}^{u=v} dv$$

$$= \sin 1 \int_1^2 v\, dv = \frac{3}{2}\sin 1 \qquad \blacksquare$$

The next example shows how the formula for integration in polar coordinates can be derived with the help of Theorem 1.

EXAMPLE 4 Suppose that f is continuous on a polar rectangle

$$R = \{(r, \theta) \mid a \le r \le b, \alpha \le \theta \le \beta\}$$

in the xy-plane. Show that

$$\iint\limits_R f(x, y)\, dA = \iint\limits_S f(r \cos \theta, r \sin \theta)\, r\, dr\, d\theta$$

where S is the region in the $r\theta$-plane mapped onto R under the transformation T defined by

$$x = g(r, \theta) = r \cos \theta \qquad y = h(r, \theta) = r \sin \theta$$

Solution Observe that T maps the r-simple region

$$S = \{(r, \theta) \mid a \le r \le b,\, \alpha \le \theta \le \beta\}$$

onto the polar rectangle R as shown in Figure 8. The Jacobian of T is

$$\frac{\partial(x, y)}{\partial(r, \theta)} = \begin{vmatrix} \dfrac{\partial x}{\partial r} & \dfrac{\partial x}{\partial \theta} \\[2mm] \dfrac{\partial y}{\partial r} & \dfrac{\partial y}{\partial \theta} \end{vmatrix} = \begin{vmatrix} \cos \theta & -r \sin \theta \\ \sin \theta & r \cos \theta \end{vmatrix}$$

$$= r \cos^2 \theta + r \sin^2 \theta = r > 0$$

Using Theorem 1, we obtain

$$\iint\limits_R f(x, y)\, dA = \iint\limits_S f(g(r, \theta), h(r, \theta)) \left| \frac{\partial(x, y)}{\partial(r, \theta)} \right| dr\, d\theta$$

$$= \int_\alpha^\beta \int_{g_1(\theta)}^{g_2(\theta)} f(r \cos \theta, r \sin \theta)\, r\, dr\, d\theta$$

as was to be shown.

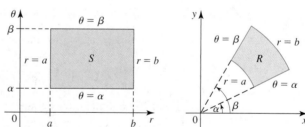

FIGURE 8
T maps the region S onto the polar rectangle R.

Change of Variables in Triple Integrals

The results for a change of variables for double integrals can be extended to the case involving triple integrals. Let T be a transformation from the uvw-space to the xyz-space defined by the equations

$$x = g(u, v, w), \qquad y = h(u, v, w), \qquad z = k(u, v, w)$$

and suppose that T maps a region S in uvw-space onto a region R in xyz-space. The Jacobian of T is

$$\frac{\partial(x, y, z)}{\partial(u, v, w)} = \begin{vmatrix} \dfrac{\partial x}{\partial u} & \dfrac{\partial x}{\partial v} & \dfrac{\partial x}{\partial w} \\[2mm] \dfrac{\partial y}{\partial u} & \dfrac{\partial y}{\partial v} & \dfrac{\partial y}{\partial w} \\[2mm] \dfrac{\partial z}{\partial u} & \dfrac{\partial z}{\partial v} & \dfrac{\partial z}{\partial w} \end{vmatrix}$$

The following is the analog of Equation (4) for triple integrals.

Change of Variables in Triple Integrals

$$\iiint\limits_{R} f(x, y, z)\, dV = \iiint\limits_{S} f(g(u, v, w), h(u, v, w), k(u, v, w)) \left| \frac{\partial(x, y, z)}{\partial(u, v, w)} \right| du\, dv\, dw \qquad \textbf{(5)}$$

EXAMPLE 5 Use Equation (5) to derive the formula for changing a triple integral in rectangular coordinates to one in spherical coordinates.

Solution The required transformation is defined by the equations

$$x = \rho \sin \phi \cos \theta, \qquad y = \rho \sin \phi \sin \theta, \qquad z = \rho \cos \phi$$

where ρ, ϕ, and θ are spherical coordinates. The Jacobian of T is

$$\frac{\partial(x, y, z)}{\partial(\rho, \phi, \theta)} = \begin{vmatrix} \sin \phi \cos \theta & \rho \cos \phi \cos \theta & -\rho \sin \phi \sin \theta \\ \sin \phi \sin \theta & \rho \cos \phi \sin \theta & \rho \sin \phi \cos \theta \\ \cos \phi & -\rho \sin \phi & 0 \end{vmatrix}$$

Expanding the determinant by the third row, we find

$$\frac{\partial(x, y, z)}{\partial(\rho, \phi, \theta)} = \cos \phi \begin{vmatrix} \rho \cos \phi \cos \theta & -\rho \sin \phi \sin \theta \\ \rho \cos \phi \sin \theta & \rho \sin \phi \cos \theta \end{vmatrix} + \rho \sin \phi \begin{vmatrix} \sin \phi \cos \theta & -\rho \sin \phi \sin \theta \\ \sin \phi \sin \theta & \rho \sin \phi \cos \theta \end{vmatrix}$$

$$= \cos \phi (\rho^2 \cos \phi \sin \phi \cos^2 \theta + \rho^2 \cos \phi \sin \phi \sin^2 \theta) + \rho \sin \phi (\rho \sin^2 \phi \cos^2 \theta + \rho \sin^2 \phi \sin^2 \theta)$$

$$= \rho^2 \cos^2 \phi \sin \phi + \rho^2 \sin^3 \phi = \rho^2 \sin \phi$$

Since $0 \le \phi \le \pi$, we see that $\sin \phi \ge 0$, so

$$\left| \frac{\partial(x, y, z)}{\partial(\rho, \phi, \theta)} \right| = |\rho^2 \sin \phi| = \rho^2 \sin \phi$$

Using Equation (5), we obtain

$$\iiint\limits_{R} f(x, y, z)\, dV = \iiint\limits_{S} f(\rho \sin \phi \cos \theta, \rho \sin \phi \sin \theta, \rho \cos \phi)\rho^2 \sin \phi\, d\rho\, d\phi\, d\theta$$

which is Equation (4) in Section 14.7, the formula for integrating a triple integral in spherical coordinates. ∎

14.8 CONCEPT QUESTIONS

1. a. Let T be a transformation defined by $x = g(u, v)$ and $y = h(u, v)$. What is the Jacobian of T?

  b. Write the Jacobian of the transformation T given by $x = g(u, v, w)$, $y = h(u, v, w)$, and $z = k(u, v, w)$.

2. a. Let T be the one-to-one transformation defined by $x = g(u, v)$ and $y = h(u, v)$ that maps a region S in the uv-plane onto a region R in the xy-plane. Write the formula for transforming the integral $\iint_R f(x, y)\, dA$ into an integral involving u and v over the region S.

  b. Repeat part (a) for the case of a triple integral.

14.8 EXERCISES

In Exercises 1–6, sketch the image $R = T(S)$ of the set S under the transformation T defined by the equations $x = g(u, v)$, $y = h(u, v)$.

1. $S = \{(u, v) \mid 0 \le u \le 2, 0 \le v \le 1\}$; $x = u - v$, $y = v$

2. $S = \{(u, v) \mid 0 \le u \le 1, 0 \le v \le 2\}$; $x = u + v$, $y = u - v$

3. S is the triangular region with vertices $(0, 0)$, $(1, 1)$, $(0, 1)$; $x = u + 2v$, $y = 2v$.

4. S is the trapezoidal region with vertices $(-2, 0)$, $(-1, 0)$, $(0, 1)$, $(0, 2)$; $x = u + v$, $y = u - v$

5. $S = \{(u, v) \mid u^2 + v^2 \le 1, u \ge 0, v \ge 0\}$; $x = u^2 - v^2$; $y = 2uv$

6. $S = \{(u, v) \mid 1 \le u \le 2, 0 \le v \le \frac{\pi}{2}\}$; $x = u \cos v$, $y = u \sin v$

In Exercises 7–12, find the Jacobian of the transformation T defined by the equations.

7. $x = 2u + v$, $y = u^2 - v$

8. $x = u^2 - v^2$, $y = 2uv$

9. $x = e^u \cos 2v$, $y = e^u \sin 2v$

10. $x = u \ln v$, $y = v \ln u$

11. $x = u + v + w$, $y = u - v + w$, $z = u - 2v + 3w$

12. $x = 2u + w$, $y = u^2 - v^2$, $z = u + v^2 - 2w^2$

In Exercises 13–20, evaluate the integral using the transformation T.

13. $\iint\limits_R (x + y)\, dA$, where R is the parallelogram bounded by the lines with equations $y = -2x$, $y = \frac{1}{2}x - \frac{15}{2}$, $y = -2x + 10$, and $y = \frac{1}{2}x$; T is defined by $x = u + 2v$ and $y = v - 2u$

14. $\iint\limits_R (2x + 3y)\, dA$, where R is the parallelogram bounded by the lines with equations $y = 2x$, $y = \frac{1}{2}x + 3$, $y = 2x + 3$, and $y = \frac{1}{2}x$; T is defined by $x = u - 2v$ and $y = 2u - v$

15. $\iint\limits_R 2xy\, dA$, where R is the region in the first quadrant bounded by the ellipse $4x^2 + 9y^2 = 36$; T is defined by $x = 3u$ and $y = 2v$

16. $\iint\limits_R \cos(x^2 - xy + y^2)\, dA$, where R is the region bounded by the ellipse $x^2 - xy + y^2 = 2$; T is defined by $x = \sqrt{2}u - \sqrt{2/3}v$ and $y = \sqrt{2}u + \sqrt{2/3}v$

17. $\iint\limits_R \sqrt{1 - \dfrac{x^2}{4} - \dfrac{y^2}{9}}\, dA$, where R is the region bounded by

the ellipse $\dfrac{x^2}{4} + \dfrac{y^2}{9} = 1$; T is defined by $x = 2u$ and $y = 3v$

18. $\iint\limits_R xy^2\, dA$, where R is the region in the first quadrant bounded by the hyperbolas $xy = 1$ and $xy = 2$ and the lines $y = x$ and $y = 2x$; T is defined by $x = \dfrac{u}{v}$ and $y = v$

19. $\iint\limits_R \dfrac{1}{\sqrt{x^2 + y^2}}\, dA$, where $R = \{(x, y) \mid x^2 + y^2 \le 1, y \ge 0\}$; T is defined by $x = u^2 - v^2$ and $y = 2uv$, where $u, v \ge 0$.

20. $\iint\limits_R y \sin x\, dA$, where R is the region bounded by the graphs of $x = y^2$, $x = 0$, and $y = 1$; T is defined by $x = u^2$ and $y = v$

In Exercises 21–26, evaluate the integral by making a suitable change of variables.

21. $\iint\limits_R (2x + y)\, dA$, where R is the parallelogram bounded by the lines $x + y = -1$, $x + y = 3$, $2x - y = 0$, and $2x - y = 4$

22. $\iint\limits_R (x + y) \sin(2x - y)\, dA$, where R is the parallelogram bounded by the lines $y = -x$, $y = -x + 1$, $y = 2x$, and $y = 2x - 2$

23. $\iint\limits_R e^{(x-y)/(x+y)}\, dA$, where R is the triangular region bounded by the lines $x = 0$, $y = 0$, and $x + y = 1$

24. $\iint\limits_R e^{(x+y)/(x-y)}\, dA$, where R is the trapezoidal region with vertices $(-2, 0)$, $(-1, 0)$, $(0, 1)$, and $(0, 2)$

25. $\iint\limits_R xy\, dA$, where R is the region in the first quadrant bounded by the ellipse $\dfrac{x^2}{a^2} + \dfrac{y^2}{b^2} = 1$

26. $\iint\limits_R \ln(4x^2 + 25y^2 + 1)\, dA$, where R is the region bounded by the ellipse $4x^2 + 25y^2 = 1$

27. Find the volume V of the solid E enclosed by the ellipsoid

$$\frac{x^2}{a^2} + \frac{y^2}{b^2} + \frac{z^2}{c^2} = 1$$

Hint: $V = \iiint_E dV$. Use the transformation $x = au$, $y = bv$, and $z = cw$.

V Videos for selected exercises are available online at **www.academic.cengage.com/login**.

28. Let E be the solid enclosed by the ellipsoid

$$\frac{x^2}{a^2} + \frac{y^2}{b^2} + \frac{z^2}{c^2} = 1$$

Find the mass of E if it has constant mass density δ.
Hint: Use the transformation of Exercise 27.

29. Find the moment of inertia, I_x, of the lamina that has constant mass density ρ and occupies the disk $x^2 + y^2 - ax \leq 0$ about the x-axis.

30. Show that the moment of inertia of the solid of Exercise 28 about the z-axis is $I_z = \frac{1}{5}m(a^2 + b^2)$, where $m = \frac{4}{3}\pi\delta abc$ is the mass of the solid.

31. Use Formula (5) to find the formula for changing a triple integral in rectangular coordinates to one in cylindrical coordinates.

In Exercises 32 and 33, determine whether the statement is true or false. If it is true, explain why. If it is false, explain why or give an example that shows it is false.

32. If T is defined by $x = g(u, v)$ and $y = h(u, v)$ and maps a region S in the uv-plane onto a region R in the xy-plane, then the area of R is the same as the area of S.

33. If T is defined by $x = g(u, v)$, $y = h(u, v)$ and maps a region S onto a region R, then

$$\iint_R (x^2 + y^2)\, dx\, dy = \iint_S (u^2 + v^2) \left| \frac{\partial(x, y)}{\partial(u, v)} \right| du\, dv$$

CHAPTER 14 REVIEW

CONCEPT QUESTIONS

In Exercises 1–12, fill in the blanks.

1. a. If f is a continuous function defined on a rectangle $R = [a, b] \times [c, d]$, then the Riemann sum of f over R with respect to a partition $P = \{R_{ij}\}$ is _____, where (x_{ij}^*, y_{ij}^*) is a point in R_{ij}.
  b. The double integral $\iint_R f(x, y)\, dA = $ _____ if the limit exists for all choices of _____ in R_{ij}.
  c. If $f(x, y) \geq 0$ on R, then $\iint_R f(x, y)\, dA$ gives the _____ of the solid lying directly above R and below the surface _____.
  d. If D is a bounded region that is not rectangular, then $\iint_D f(x, y)\, dA = $ _____, where $f_D(x, y) = $ _____ if (x, y) is in D and $f_D(x, y) = $ _____ if (x, y) is not in D.

2. The following properties hold for double integrals:
  a. $\iint_D cf(x, y)\, dA = $ _____
  b. $\iint_D [f(x, y) \pm g(x, y)]\, dA = $ _____
  c. If $f(x, y) \geq 0$ on D, then $\iint_D f(x, y)\, dA$ _____.
  d. If $f(x, y) \geq g(x, y)$ on D, then $\iint_D f(x, y)\, dA$ _____.
  e. If $D = D_1 \cup D_2$ and $D_1 \cap D_2 = \varnothing$, then $\iint_D f(x, y)\, dA = $ _____.

3. a. If $R = [a, b] \times [c, d]$, then the two iterated integrals of f over R are _____ and _____.
  b. Fubini's Theorem for a rectangular region $R = [a, b] \times [c, d]$ states that $\iint_R f(x, y)\, dA$ is equal to the _____ integrals in part (a).

4. a. A y-simple region has the form $R = $ _____, where g_1 and g_2 are continuous functions on $[a, b]$.

b. An x-simple region has the form $R = $ _____, where h_1 and h_2 are continuous functions on $[c, d]$.
  c. Fubini's Theorem for the y-simple region R of part (a), states that $\iint_R f(x, y)\, dA = $ _____. If R is the x-simple region R of part (b), then $\iint_R f(x, y)\, dA = $ _____.

5. a. A polar rectangle is a set of the form $R = $ _____.
  b. If f is continuous on a polar rectangle R, then $\iint_R f(x, y)\, dA = $ _____.
  c. An r-simple region is a set of the form $R = $ _____.
  d. If f is continuous on an r-simple region R, then $\iint_R f(x, y)\, dA = $ _____.

6. If a lamina occupies a region R in the plane and the mass density of the lamina is $\rho(x, y)$, then
  a. The mass of the lamina is given by $m = $ _____.
  b. The moments of the lamina with respect to the x- and y-axes are $M_x = $ _____ and $M_y = $ _____. The coordinates of the center of mass of the lamina are $\bar{x} = $ _____ and $\bar{y} = $ _____.
  c. The moments of inertia of the lamina with respect to the x-axis, the y-axis, and the origin are $I_x = $ _____, $I_y = $ _____, and $I_0 = $ _____, respectively.
  d. If the moment of inertia of a lamina with respect to an axis is I, then its radius of gyration with respect to the axis is $R = $ _____.

7. a. If f_x and f_y are continuous on a region R in the xy-plane, then the area of the surface $z = f(x, y)$ over R is $A = $ _____.

b. If g is defined in a region R in the xz-plane, then the area of the surface $y = g(x, z)$ is $A = $ _____.

c. If h is defined in a region R in the yz-plane, then the area of the surface $x = h(y, z)$ is $A =$ _____.

8. a. If f is a continuous function defined on a rectangular box $B = [a, b] \times [c, d] \times [p, q]$, then the Riemann sum of f over B with respect to a partition $P = \{B_{ijk}\}$ is _____, where $(x_{ijk}^*, y_{ijk}^*, z_{ijk}^*)$ is a point in B_{ijk}.

b. The triple integral $\iiint_B f(x, y, z)\, dV =$ _____ if the limit exists for all choices of $(x_{ijk}^*, y_{ijk}^*, z_{ijk}^*)$ in B_{ijk}.

c. If f is continuous on a bounded solid region T in space, B is a rectangular box that contains T, $Q = \{B_{111}, B_{112}, \dots, B_{ijk}, \dots, B_{lmn}\}$ is a partition of B, F is a function defined by

$$F(x, y, z) = \begin{cases} f(x, y, z) & \text{if } (x, y, z) \text{ is in } T \\ 0 & \text{if } (x, y, z) \text{ is in } B \text{ but not in } T \end{cases}$$

then a Riemann sum of f over T is _____.

d. The triple integral of f over T is $\iiint_T f(x, y, z)\, dV =$ _____ provided that the limit exists for all choices of $(x_{ijk}^*, y_{ijk}^*, z_{ijk}^*)$ in T.

9. a. If f is continuous on $B = [a, b] \times [c, d] \times [p, q]$, then $\iiint_B f(x, y, z)\, dV$ is equal to any of six iterated integrals depending on the _____ of integration. If we integrate with respect to x, y, and z, in that order, then $\iiint_B f(x, y, z)\, dV =$ _____.

b. If f is continuous on a z-simple region $T = \{(x, y, z) \mid (x, y) \in R,\ k_1(x, y) \le z \le k_2(x, y)\}$, where R is the projection of T onto the xy-plane, then $\iiint_T f(x, y, z)\, dV =$ _____.

10. If $\rho(x, y, z)$ gives the density at the point (x, y, z) of a solid T, then

a. The mass of T is $m =$ _____.

b. The moment of T about the yz-plane is $M_{yz} =$ _____, the moment of T about the xz-plane is $M_{xz} =$ _____,

and the moment of T about the xy-plane is $M_{xy} =$ _____.

c. The _____ _____ _____ of T is located at the point $(\bar{x}, \bar{y}, \bar{z})$, where $\bar{x} =$ _____, $\bar{y} =$ _____, and $\bar{z} =$ _____.

d. The moments of inertia of T about the x-, y-, and z-axes are $I_x =$ _____, $I_y =$ _____, and $I_z =$ _____.

11. a. If T is a z-simple region described by $T = \{(x, y, z) \mid (x, y) \in R,\ h_1(x, y) \le z \le h_2(x, y)\}$, where $R = \{(r, \theta) \mid \alpha \le \theta \le \beta,\ g_1(\theta) \le r \le g_2(\theta)\}$, then in terms of cylindrical coordinates, $\iiint_T f(x, y, z)\, dV =$ _____.

b. If $T = \{(\rho, \phi, \theta) \mid a \le \rho \le b,\ c \le \phi \le d,\ \alpha \le \theta \le \beta\}$ is a spherical wedge, then in terms of spherical coordinates, $\iiint_T f(x, y, z)\, dV =$ _____.

c. If T is ρ-simple, $T = \{(\rho, \phi, \theta) \mid h_1(\phi, \theta) \le \rho \le h_2(\phi, \theta),\ c \le \phi \le d,\ \alpha \le \theta \le \beta\}$, then $\iiint_T f(x, y, z)\, dV =$ _____.

12. a. If T is a transformation defined by $x = g(u, v)$ and $y = h(u, v)$, then the Jacobian of T is $\dfrac{\partial(x, y)}{\partial(u, v)} =$ _____.

b. If T maps S in the uv-plane onto a region R in the xy-plane, then the formula for transforming the integral $\iint_R f(x, y)\, dx\, dy$ into one involving u and v is $\iint_R f(x, y)\, dx\, dy =$ _____.

c. If T maps S in uvw-space onto R in xyz-space and is defined by $x = g(u, v, w)$, $y = h(u, v, w)$, and $z = k(u, v, w)$, then the Jacobian T is $\dfrac{\partial(x, y, z)}{\partial(u, v, w)} =$ _____, and the change of variable formula for triple integrals is $\iiint_R f(x, y, z)\, dx\, dy\, dz =$ _____.

REVIEW EXERCISES

In Exercises 1–8 evaluate the iterated integral.

1. $\displaystyle\int_0^2 \int_{-1}^2 (2x + 3xy^2)\, dx\, dy$

2. $\displaystyle\int_0^\pi \int_0^1 x \sin xy\, dy\, dx$

3. $\displaystyle\int_0^1 \int_x^{\sqrt{x}} (2x + 3y)\, dy\, dx$

4. $\displaystyle\int_0^1 \int_0^{\sqrt{1-y^2}} 2y\, dx\, dy$

5. $\displaystyle\int_0^2 \int_y^2 \frac{1}{4 + y^2}\, dx\, dy$

6. $\displaystyle\int_1^e \int_0^{1/x} \sqrt{\ln x}\, dy\, dx$

7. $\displaystyle\int_0^2 \int_0^{\sqrt{z}} \int_0^x (x + 2z)\, dy\, dx\, dz$

8. $\displaystyle\int_1^2 \int_x^3 \int_0^y \frac{y}{y + z}\, dz\, dy\, dx$

In Exercises 9–12, sketch the region of integration for the iterated integral.

9. $\displaystyle\int_1^2 \int_{\ln x}^{\sqrt[3]{x}} f(x, y)\, dy\, dx$

10. $\displaystyle\int_0^1 \int_0^{\sin^{-1} y} f(x, y)\, dx\, dy$

11. $\displaystyle\int_0^\pi \int_0^{1 + \cos \theta} f(r, \theta)\, r\, dr\, d\theta$

12. $\displaystyle\int_{-\sqrt{2}}^{\sqrt{2}} \int_{y^2}^2 \int_0^{2-x} f(x, y, z)\, dz\, dx\, dy$

In Exercises 13 and 14, reverse the order of integration, and evaluate the resulting integral.

13. $\displaystyle\int_0^1 \int_y^1 \sin x^2\, dx\, dy$

14. $\displaystyle\int_0^1 \int_y^{\sqrt{y}} \frac{\cos x}{x}\, dx\, dy$

In Exercises 15–26, evaluate the multiple integral.

15. $\iint\limits_{R} (x^2 + 3y^2) \, dA$, where

$R = \{(x, y) \mid -1 \le x \le 1, 0 \le y \le 2\}$

16. $\iint\limits_{R} (x + y) \, dA$, where

$R = \{(x, y) \mid 0 \le x \le 1, 0 \le y \le \sqrt{1 - x^2}\}$

17. $\iint\limits_{R} y \, dA$, where R is the region bounded by the parabola

$x = y^2$ and the line $x - 2y = 3$

18. $\iint\limits_{R} (x + 2y) \, dA$, where R is the region bounded by the

graphs of $x = 4 - y^2$, $x = 0$, and $y = 0$

19. $\iint\limits_{R} x \, dA$, where R is the region in the first quadrant bounded

by the ellipse $4x^2 + 9y^2 = 36$

20. $\iint\limits_{R} \ln x \, dy \, dx$, where R is the region bounded by the graphs

of $y = 1/x$, $y = x$, and $x = e$

21. $\iiint\limits_{T} xy \, dV$, where

$T = \{(x, y, z) \mid 0 \le x \le 1, 0 \le y \le x^2, 0 \le z \le x + y\}$

22. $\iiint\limits_{T} z \, dV$, where R is the tetrahedron bounded by the planes

$x + 2y + z = 6$, $x = 0$, $y = 0$, and $z = 0$

23. $\iiint\limits_{T} xyz \, dV$, where T is the region bounded by the hemi-

sphere $z = \sqrt{1 - x^2 - y^2}$ and the plane $z = 0$

24. $\iiint\limits_{T} z \, dV$, where T is the region bounded by the cylinder

$x^2 + z^2 = 1$ and the planes $y = x$, $y = 2x$, and $z = 0$

25. $\iiint\limits_{T} x^2 z \, dV$, where T is the region bounded above by the

paraboloid $y = 1 - x^2 - z^2$, above the plane $z = 0$, and to
the left by the plane $y = 0$

26. $\iiint\limits_{T} \dfrac{1}{\sqrt{x^2 + y^2 + z^2}} \, dV$, where T is the region bounded

above by the hemisphere $z = \sqrt{1 - x^2 - y^2}$ and below by
the plane $z = 0$

In Exercises 27–32, find the volume of the solid.

27. The solid under the surface $z = xy^2$ and above the rectangular region $R = \{(x, y) \mid 0 \le x \le 1, 1 \le y \le 2\}$

28. The solid under the paraboloid $z = 4 - x^2 - y^2$ and above
the triangular region in the xy-plane with vertices $(0, 0)$,
$(1, 1)$, and $(0, 1)$

29. The solid bounded by the paraboloid $z = x^2 + y^2$, the cylinder $x^2 + y^2 = 1$, and the plane $z = 0$

30. The solid under the paraboloid $z = 9 - x^2 - y^2$ and above
the circular region $x^2 + y^2 \le 4$ in the xy-plane

31. The solid under the surface $z = e^{-(x^2 + y^2)}$, within the cylinder
$x^2 + y^2 = 1$ and above the plane $z = 0$

32. The solid bounded above by the paraboloid
$z = 4 - x^2 - y^2$ and below by the cone $z = \sqrt{x^2 + y^2}$

*In Exercises 33–36, find the mass and the center of mass of
the lamina occupying the region D and having the given mass
density.*

33. D is the region in the first quadrant bounded by the graphs
of $y = x$ and $y = x^3$; $\rho(x, y) = y$

34. D is the region bounded by the parabola $y = x^2$ and the line
$y = 4$; $\rho(x, y) = x^2 y$

35. D is the region in the first quadrant bounded by the circle
$x^2 + y^2 = 1$; $\rho(x, y) = \sqrt{x^2 + y^2}$

36. D is the region bounded by the semicircle $y = \sqrt{4 - x^2}$
and the x-axis; $\rho(x, y) = x^2 y$

*In Exercises 37 and 38, find the moments of inertia I_x, I_y, and I_0
of the lamina occupying the region D and having the given mass
density.*

37. D is the region bounded by the triangle with vertices $(0, 0)$,
$(0, 1)$, and $(1, 1)$; $\rho(x, y) = x^2 + y^2$

38. D is the region bounded by the graphs of $y = x$ and $y = x^2$;
$\rho(x, y) = x$

In Exercises 39 and 40, find the area of the surface S.

39. S is the part of the plane $2x + 3y + z = 6$ in the first octant.

40. S is the part of the paraboloid $z = x^2 + y^2$ below the plane
$z = 4$.

*In Exercises 41 and 42, evaluate the integral by changing to
cylindrical or spherical coordinates.*

41. $\displaystyle\int_0^2 \int_0^{\sqrt{4-x^2}} \int_0^1 (x^2 + y^2)^{3/2} \, dz \, dy \, dx$

42. $\displaystyle\int_0^3 \int_0^{\sqrt{9-x^2}} \int_0^{\sqrt{9-x^2-y^2}} z\sqrt{x^2 + y^2 + z^2} \, dz \, dy \, dx$

43. Express the triple integral $\iiint_T f(x, y, z) \, dV$ as an iterated
integral in six different ways using different orders of integration, where T is the tetrahedron bounded by the planes
$2x + 3y + z = 6$, $x = 0$, $y = 0$, and $z = 0$.

44. Set up, but do not evaluate, the iterated integral giving the
mass of the solid bounded by the cone $z = \sqrt{x^2 + y^2}$ and

the sphere $x^2 + y^2 + z^2 = 8$ if the density of the solid at any point P is $\rho(x, y, z) = \sqrt{1 + xz}$.

45. Find the Jacobian of the transformation T defined by the equations $x = u + w^2$, $y = 2u^2 + v$, and $z = u^2 - v^2 + 2w$.

46. Use the transformation $x = u/v$ and $y = v$ to evaluate $\iint_R y \cos xy \, dy \, dx$, where R is the region bounded by the hyperbolas $xy = 1$ and $xy = 4$ and the lines $y = 1$ and $y = 4$.

47. Evaluate $\iint_R e^{(x-y)/(x+y)} \, dA$, where R is the triangular region bounded by the lines $y = x$, $x + y = 2$, and $y = 0$.

In Exercises 48–53, state whether the statement is true or false. Give a reason for your answer.

48. $\displaystyle\int_a^b \int_a^b f(x)f(y) \, dx \, dy = \left[\int_a^b f(x) \, dx \right]^2$

49. $\displaystyle\int_0^1 \int_{-2}^3 (x + \cos xy) \, dx \, dy = \int_{-2}^3 \int_0^1 (x + \cos xy) \, dy \, dx$

50. $\displaystyle\int_0^1 \int_0^y f(x, y) \, dx \, dy = \int_0^1 \int_0^x f(x, y) \, dy \, dx$

51. If $\iint_D f(x, y) \, dA \geq 0$, then $f(x, y) \geq 0$ for all (x, y) in D.

52. $\displaystyle\int_{-1}^1 \int_0^3 x^3 \sin y^2 \, dy \, dx = 0$

53. $\displaystyle\int_0^1 \int_1^3 [\sqrt{x} + \cos^2(xy)] \, dx \, dy \leq 6$

CHALLENGE PROBLEMS

1. a. Use the definition of the double integral as a limit of a Riemann sum to compute $\iint_R (3x^2 + 2y) \, dA$, where $R = \{(x, y) \,|\, 0 \leq x \leq 2, 0 \leq y \leq 1\}$.
   Hint: Take $\Delta x = 2/m$, and $\Delta y = 1/n$, so that $x_i = 2i/m$, where $1 \leq i \leq m$, and $y_j = j/n$, where $1 \leq j \leq n$.
   b. Verify the result of part (a) by evaluating an appropriate iterated integral.

2. The following figure shows a triangular lamina. Its mass density at (x, y) is $f(x, y) = \cos(y^2)$. Find its mass.

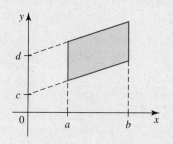

3. Show that the area of the parallelogram shown in the figure is $(b - a)(d - c)$, where $a \leq b$ and $c \leq d$.

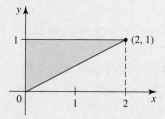

4. Using the result of Problem 3, show that the area of the parallelogram determined by the vectors $\mathbf{a} = \langle a_1, a_2 \rangle$ and $\mathbf{b} = \langle b_1, b_2 \rangle$ is $|a_1 b_2 - a_2 b_1|$.

5. Monte Carlo Integration This is a method that is used to find the area of complicated bounded regions in the xy-plane. To describe the method, suppose that D is such a region completely enclosed by a rectangle $R = \{(x, y) \,|\, a \leq x \leq b, c \leq y \leq d\}$, as shown in the figure. Using a random number generator, we then pick points in R. If $A(D)$ denotes the area of D, then

$$\frac{A(D)}{A(R)} \approx \frac{N(D)}{n}$$

where $N(D)$ denotes the number of points landing in D, $A(R) = (b - a)(d - c)$, and n is the number of points picked. Then

$$A(D) \approx \frac{(b - a)(d - c)}{n} N(D)$$

Use Monte Carlo integration with $n = 5000$ to estimate the area of the disk of radius 5.

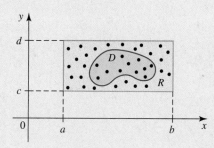

6. The expression $\sum_{i=1}^{m} \sum_{j=1}^{n} (x_i^2 + y_j^3) \, \Delta x \, \Delta y$, where $x_i = i/m$, $i = 1, 2, \ldots, m$, and $y_j = 1 + (j/n)$, $j = 1, 2, \ldots, n$, is the Riemann sum of a function $f(x, y)$ over a region associated with a regular pattern.

   a. Write a double integral corresponding to this Riemann sum.

   b. Write an iterated integral corresponding to this Riemann sum.

7. a. Suppose that $f(x, y)$ is continuous in the triangular region $R = \{(x, y) \mid x \le b, y \ge a, y \le x\}$. Show that

$$\int_a^b \left[\int_a^x f(x, y) \, dy \right] dx = \int_a^b \left[\int_y^b f(x, y) \, dx \right] dy$$

   b. Use the result of part (a) to evaluate

$$\int_0^1 \left[\int_y^1 \sin x^2 \, dx \right] dy$$

8. Let f be a continuous function of one variable. Show that

$$\int_a^x \int_a^y \int_a^z f(t) \, dt \, dz \, dy = \frac{1}{2} \int_a^x (x - t)^2 f(t) \, dt$$

   Hint: Use the result of Exercise 7.

9. a. Let R be a region in the xy-plane that is symmetric with respect to the y-axis, and let f be a function that satisfies the condition $f(-x, y) = -f(x, y)$. Show that $\iint_R f(x, y) \, dA = 0$.

   b. Use the result of part (a) to show that if a lamina with uniform density ρ that occupies a plane region that is symmetric with respect to a straight line L, then the centroid of the lamina lies on L.

10. In Exercise 6 in the Challenge Problems for Chapter 11, you were asked to show that the area of the portion of the plane $ax + by + cz = d$, where a, b, and c are positive constants, in the first octant is given by

$$\frac{d^2 \sqrt{a^2 + b^2 + c^2}}{2abc}$$

   Derive this formula again, this time using integration. Show that the result can also be written as

$$\frac{A(R)}{c} \sqrt{a^2 + b^2 + c^2}$$

where $A(R)$ is the area of the region R in the xy-plane.

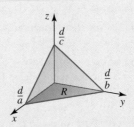

11. A thin rectangular metal plate has dimensions a ft by b ft and a constant density of k slugs/ft^2. The plate is placed in the xy-plane as shown in the figure and is allowed to rotate about the z-axis at a constant angular velocity of ω radians/sec.

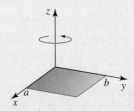

   a. Show that the kinetic energy of the plate is given by

$$E = \frac{k\omega^2}{2} \int_0^b \int_0^a (x^2 + y^2) \, dx \, dy = \frac{1}{3} (a^2 + b^2) m\omega^2$$

where $m = kab$.

   Hint: The kinetic energy of a particle of mass m slugs and velocity v ft/sec is $\frac{1}{2} m v^2$ ft-lb.

   b. Show that $E = \frac{1}{2} I \omega^2$, where $I = \frac{2}{3}(a^2 + b^2)m$.

12. The Schwartz inequality for functions of one variable holds for multiple integrals. (See Exercise 9 in the Challenge Problems for Chapter 4.) Thus,

$$\left| \iint_D f(x, y) g(x, y) \, dA \right| \le \sqrt{\iint_D [f(x, y)]^2 \, dA \iint_D [g(x, y)]^2 \, dA}$$

   a. Use Schwartz's inequality to prove that

$$\left| \iint_D \sqrt{4x^2 - y^2} \, dA \right| \le \frac{2\sqrt{3}}{3}$$

where D is the triangle with vertices $A(0, 0)$, $B(1, 2)$, and $C(1, 0)$.

   b. Find the exact value of the integral in part (a). How accurate is the estimate?

A vector field in a region in three-dimensional space is a vector-valued function that assigns a vector to each point in the region. Vector fields are used in aerodynamics to model the speed and direction of air flow around an airplane. The photograph shows the air flow from the wing of an agricultural plane. The air flow was made visible by a technique that uses colored smoke rising from the ground. The *wingtip vortex*, a tube of circulating air that is left behind by the wing as it generates lift, exerts a powerful influence on the flow field behind the plane. This is the reason that the Federal Aviation Administration (FAA) requires aircraft to maintain set distances behind each other when they land.

NASA Langley Research Center

15 Vector Analysis

A *VECTOR FIELD* is a function that assigns a vector to each point in a region. The study of vector fields is motivated by many physical fields such as force fields and velocity fields. Gravitational and electric fields are examples of force fields, and the flow of water through a channel and the flow of air around an airfoil are examples of velocity fields.

The calculus of vector fields enables us to calculate many quantities of interest associated with force fields and velocity fields. For example, using the notion of the line integral, which is a generalization of the definite integral, we are able to calculate the work done by a force field in moving a body from one point to another along a curve. Using *surface integrals*, which are generalizations of double integrals, we can calculate the flux (flow of fluids and gases) across a surface.

The calculations involving line integrals and surface integrals are facilitated by the theorems of Green and Stokes and the Divergence Theorem, all of which may be regarded as analogs of the Fundamental Theorem of Calculus in higher dimensions.

V This symbol indicates that one of the following video types is available for enhanced student learning at **www.academic.cengage.com/login:**
- Chapter lecture videos
- Solutions to selected exercises

15.1   Vector Fields

Figure 1 shows the airflow around an airfoil in a wind tunnel. The smooth curves, traced by the individual air particles and made visible by kerosene smoke, are called **stream-lines.**

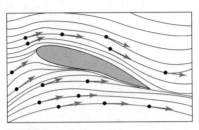

FIGURE 1
A vector field associated with the
airflow around an airfoil

FIGURE 2
A vector field associated with the flow
of blood in an artery

To facilitate the analysis of this flow, we can associate a tangent vector with each point on a streamline. The direction of the vector indicates the direction of flow of the air particle, and the length of the vector gives the speed of the particle. If we assign a tangent vector to each point on every streamline, we obtain what is called a *vector field* associated with this flow.

Another example of a vector field arises in the study of the flow of blood through an artery. Here, the vectors give the direction of flow and the speed of the blood cells (see Figure 2).

DEFINITION   Vector Field in Two-Dimensional Space

Let R be a region in the plane. A **vector field in R** is a vector-valued function **F** that associates with each point (x, y) in R a two-dimensional vector

$$\mathbf{F}(x, y) = P(x, y)\mathbf{i} + Q(x, y)\mathbf{j}$$

where P and Q are functions of two variables defined on R.

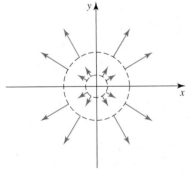

FIGURE 3
Some vectors representing the
vector field $\mathbf{F}(x, y) = x\mathbf{i} + y\mathbf{j}$

EXAMPLE 1   A vector field **F** in R^2 (two-dimensional space) is defined by $\mathbf{F}(x, y) = x\mathbf{i} + y\mathbf{j}$. Describe **F**, and sketch a few vectors representing the vector field.

Solution   The vector-valued function **F** associates with each point (x, y) in R^2 its position vector $\mathbf{r} = x\mathbf{i} + y\mathbf{j}$. This vector points directly away from the origin and has length

$$|\mathbf{F}(x, y)| = |\mathbf{r}| = \sqrt{x^2 + y^2} = r$$

which is equal to the distance of (x, y) from the origin. As an aid to sketching some vectors representing **F**, observe that each point on a circle of radius r centered at the origin is associated with a vector of length r. Figure 3 shows a few vectors representing this vector field.   ∎

EXAMPLE 2   A vector field **F** in R^2 is defined by $\mathbf{F}(x, y) = -y\mathbf{i} + x\mathbf{j}$. Describe **F**, and sketch a few vectors representing the vector field.

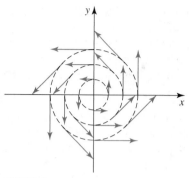

FIGURE 4
Some vectors representing the
vector field $\mathbf{F}(x, y) = -y\mathbf{i} + x\mathbf{j}$

Solution   Let $\mathbf{r} = x\mathbf{i} + y\mathbf{j}$ be the position vector of the point (x, y). Then

$$\mathbf{F} \cdot \mathbf{r} = (-y\mathbf{i} + x\mathbf{j}) \cdot (x\mathbf{i} + y\mathbf{j})$$

$$= -yx + xy = 0$$

and this shows that $\mathbf{F}$ is orthogonal to the vector $\mathbf{r}$. This means that $\mathbf{F}(x, y)$ is tangent to the circle of radius $r = |\mathbf{r}|$ with center at the origin. Furthermore,

$$|\mathbf{F}(x, y)| = \sqrt{(-y)^2 + x^2} = \sqrt{x^2 + y^2} = r$$

gives the length of the position vector. Therefore, $\mathbf{F}$ associates with each point (x, y) a vector of length equal to the distance between the origin and (x, y) and direction that is perpendicular to the position vector of (x, y). A few vectors representing this vector field are sketched in Figure 4. As in Example 1, this task is facilitated by first sketching a few concentric circles centered at the origin.   ■

The "spin" vector field of Example 2 is used to describe phenomena as diverse as whirlpools and the motion of a ferris wheel. It is called a **velocity field.**

The definition of vector fields in three-dimensional space is similar to that in two-dimensional vector fields.

DEFINITION   Vector Field in Three-Dimensional Space

Let T be a region in space. A **vector field in T** is a vector-valued function $\mathbf{F}$ that associates with each point (x, y, z) in T a three-dimensional vector

$$\mathbf{F}(x, y, z) = P(x, y, z)\mathbf{i} + Q(x, y, z)\mathbf{j} + R(x, y, z)\mathbf{k}$$

where P, Q, and R are functions of three variables defined on T.

Important applications of vector fields in three-dimensional space occur in the form of *gravitational* and *electric fields,* as described in the following examples.

EXAMPLE 3   **Gravitational Field**   Suppose that an object O of mass M is located at the origin of a three-dimensional coordinate system. We can think of this object as inducing a **force field F** in space. The effect of this **gravitational field** is to attract any object placed in the vicinity of O toward it with a force that is governed by Newton's Law of Gravitation. To find an expression for $\mathbf{F}$, suppose that an object of mass m is located at a point (x, y, z) with position vector $\mathbf{r} = x\mathbf{i} + y\mathbf{j} + z\mathbf{k}$. Then, according to Newton's Law of Gravitation, the force of attraction of the object O of mass M on the object of mass m has magnitude

$$\frac{GmM}{|\mathbf{r}|^2}$$

and direction given by the unit vector $-\mathbf{r}/|\mathbf{r}|$, where G is the gravitational constant. Therefore, we can write

$$\mathbf{F}(x, y, z) = -\frac{GM}{|\mathbf{r}|^3} \mathbf{r}$$

$$= -\frac{GMx}{(x^2 + y^2 + z^2)^{3/2}} \mathbf{i} - \frac{GMy}{(x^2 + y^2 + z^2)^{3/2}} \mathbf{j} - \frac{GMz}{(x^2 + y^2 + z^2)^{3/2}} \mathbf{k}$$

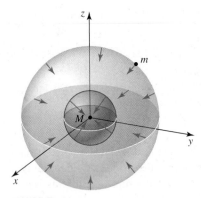

FIGURE 5
A gravitational force field

The force exerted by the gravitational field $\mathbf{F}$ on a particle of mass m with position vector $\mathbf{r}$ is $m\mathbf{F}$. The vector field $\mathbf{F}$ is sketched in Figure 5.

Observe that all the arrows point toward the origin and that the lengths of the arrows decrease as one moves farther away from the origin. Physically, $\mathbf{F}(x, y, z)$ is the force per unit mass that would be exerted on a test mass placed at the point $P(x, y, z)$. ∎

EXAMPLE 4   **Electric Field**   Suppose that a charge of Q coulombs is located at the origin of a three-dimensional coordinate system. Then, according to **Coulomb's Law,** the electric force exerted by this charge on a charge of q coulombs located at a point (x, y, z) with position vector $\mathbf{r} = x\mathbf{i} + y\mathbf{j} + z\mathbf{k}$ has magnitude

$$\frac{k|q\|Q|}{}$$

(where k, the electrical constant, depends on the units used) and direction given by the unit vector $\mathbf{r}/|\mathbf{r}|$ for like charges Q and q (repulsion). Therefore, we can write the **electric field $\mathbf{E}$** that is induced by Q as

$$\mathbf{E}(x, y, z) = \frac{kQ}{|\mathbf{r}|^3}\mathbf{r}$$

$$= \frac{kQx}{(x^2 + y^2 + z^2)^{3/2}}\mathbf{i} + \frac{kQy}{(x^2 + y^2 + z^2)^{3/2}}\mathbf{j} + \frac{kQz}{(x^2 + y^2 + z^2)^{3/2}}\mathbf{k}$$

The force exerted by the electric field $\mathbf{E}$ on a charge of q coulombs, located at (x, y, z), is $q\mathbf{E}$. Physically, $\mathbf{E}(x, y, z)$ is the force per unit charge that would be exerted on a test charge placed at the point $P(x, y, z)$. ∎

■ Conservative Vector Fields

Recall from our work in Section 13.6 that if f is a scalar function of three variables, then the *gradient* of f, written ∇f or grad f, is defined by

$$\nabla f(x, y, z) = f_x(x, y, z)\mathbf{i} + f_y(x, y, z)\mathbf{j} + f_z(x, y, z)\mathbf{k}$$

If f is a function of two variables, then

$$\nabla f(x, y) = f_x(x, y)\mathbf{i} + f_y(x, y)\mathbf{j}$$

Since ∇f assigns to each point (x, y, z) the vector $\nabla f(x, y, z)$, we see that ∇f is a vector field that associates with each point in its domain a vector giving the direction of greatest increase of f. (See Section 13.6.) The vector field ∇f is called the **gradient vector field** of f.

EXAMPLE 5   Find the gradient vector field of $f(x, y, z) = x^2 + xy + y^2z^3$.

Solution   The required gradient vector field is given by

$$\nabla f(x, y, z) = \frac{\partial f}{\partial x}\mathbf{i} + \frac{\partial f}{\partial y}\mathbf{j} + \frac{\partial f}{\partial z}\mathbf{k}$$

$$= \frac{\partial}{\partial x}(x^2 + xy + y^2z^3)\mathbf{i} + \frac{\partial}{\partial y}(x^2 + xy + y^2z^3)\mathbf{j} + \frac{\partial}{\partial z}(x^2 + xy + y^2z^3)\mathbf{k}$$

$$= (2x + y)\mathbf{i} + (x + 2yz^3)\mathbf{j} + 3y^2z^2\mathbf{k}$$

∎

Before we proceed further, it should be pointed out that vector fields in both two- and three-dimensional space can be plotted with the help of most computer algebra systems. The computer often scales the lengths of the vectors but still gives a good visual representation of the vector field. The vector fields of Examples 1 and 2 and two examples of vector fields in 3-space are shown in Figures 6a–6d.

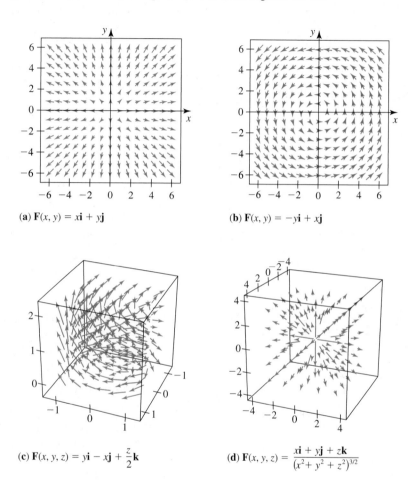

(a) $\mathbf{F}(x, y) = x\mathbf{i} + y\mathbf{j}$

(b) $\mathbf{F}(x, y) = -y\mathbf{i} + x\mathbf{j}$

(c) $\mathbf{F}(x, y, z) = y\mathbf{i} - x\mathbf{j} + \dfrac{z}{2}\mathbf{k}$

(d) $\mathbf{F}(x, y, z) = \dfrac{x\mathbf{i} + y\mathbf{j} + z\mathbf{k}}{(x^2 + y^2 + z^2)^{3/2}}$

FIGURE 6
Some computer-generated graphs of vector fields

Not all vector fields are gradients of scalar functions, but those that are play an important role in the physical sciences.

> **DEFINITION**   **Conservative Vector Field**
>
> A vector field $\mathbf{F}$ in a region R is **conservative** if there exists a scalar function f defined in R such that
>
> $$\mathbf{F} = \nabla f$$
>
> The function f is called a **potential function** for $\mathbf{F}$.

The reason for using the words *conservative* and *potential* in this definition will be apparent when we discuss the law of conservation of energy in Section 15.4.

Vector fields of the form

$$\mathbf{F}(x, y, z) = \frac{k}{|\mathbf{r}|^3} \mathbf{r}$$

are called **inverse square fields.** The gravitational and electric fields in Examples 3 and 4 are inverse square fields. The next example shows that these fields are conservative.

EXAMPLE 6   Find the gradient vector field of the function

$$f(x, y, z) = -\frac{k}{\sqrt{x^2 + y^2 + z^2}}$$

and hence deduce that the inverse square field **F** is conservative.

Solution   The gradient vector field of f is given by

$$\nabla f(x, y, z) = f_x(x, y, z)\mathbf{i} + f_y(x, y, z)\mathbf{j} + f_z(x, y, z)\mathbf{k}$$

$$= \frac{kx}{(x^2 + y^2 + z^2)^{3/2}}\mathbf{i} + \frac{ky}{(x^2 + y^2 + z^2)^{3/2}}\mathbf{j} + \frac{kz}{(x^2 + y^2 + z^2)^{3/2}}\mathbf{k}$$

$$= \frac{k}{|\mathbf{r}|^3} \mathbf{r}$$

where $\mathbf{r} = x\mathbf{i} + y\mathbf{j} + z\mathbf{k}$. This shows that the inverse square field

$$\mathbf{F}(x, y, z) = \frac{k}{|\mathbf{r}|^3} \mathbf{r}$$

is the gradient of the potential function f and is therefore conservative.   ■

Note   In Example 6 we were able to show that an inverse square field **F** is conservative because we were *given* a potential function f such that $\mathbf{F} = \nabla f$. In Section 15.4 we will learn how to find the potential function f for a conservative vector field. We will also learn how to determine whether a vector field is conservative without knowing its potential function.   ■

15.1   CONCEPT QUESTIONS

1. a. What is a vector field in the plane? In space? Give examples of each.
   b. Give three examples of vector fields with a physical interpretation.

2. a. What is a conservative vector field? Give an example.
   b. What is a potential function? Give an example.

15.1   EXERCISES

In Exercises 1–6, match the vector field with one of the plots labeled (a)–(f).

(a)

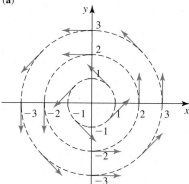

(b)

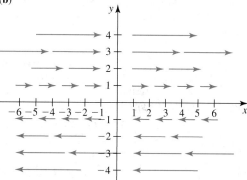

(c)

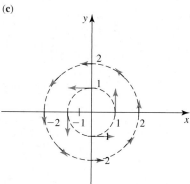

(d)

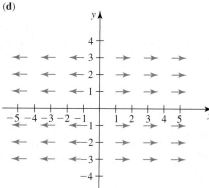

(e)

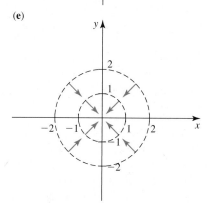

(f)

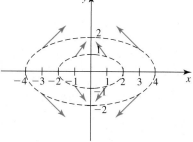

1. $\mathbf{F}(x, y) = y\mathbf{i}$

2. $\mathbf{F}(x, y) = \dfrac{x}{|x|}\mathbf{i}$

3. $\mathbf{F}(x, y) = -\dfrac{y}{x^2 + y^2}\mathbf{i} + \dfrac{x}{x^2 + y^2}\mathbf{j}$

4. $\mathbf{F}(x, y) = -\dfrac{y}{\sqrt{x^2 + y^2}}\mathbf{i} + \dfrac{x}{\sqrt{x^2 + y^2}}\mathbf{j}$

5. $\mathbf{F}(x, y) = -\dfrac{x}{\sqrt{x^2 + y^2}}\mathbf{i} - \dfrac{y}{\sqrt{x^2 + y^2}}\mathbf{j}$

6. $\mathbf{F}(x, y) = -\dfrac{1}{2}x\mathbf{i} + y\mathbf{j}$

In Exercises 7–18, sketch several vectors associated with the vector field $\mathbf{F}$.

7. $\mathbf{F}(x, y) = 2\mathbf{i}$ **8.** $\mathbf{F}(x, y) = \mathbf{i} + \mathbf{j}$

9. $\mathbf{F}(x, y) = -x\mathbf{i} - y\mathbf{j}$ **10.** $\mathbf{F}(x, y) = y\mathbf{i} - x\mathbf{j}$

11. $\mathbf{F}(x, y) = x\mathbf{i} - 2y\mathbf{j}$ **12.** $\mathbf{F}(x, y) = x\mathbf{i} + 3y\mathbf{j}$

13. $\mathbf{F}(x, y) = \dfrac{x}{\sqrt{x^2 + y^2}}\mathbf{i} + \dfrac{y}{\sqrt{x^2 + y^2}}\mathbf{j}$

14. $\mathbf{F}(x, y) = \dfrac{y}{\sqrt{x^2 + y^2}}\mathbf{i} - \dfrac{x}{\sqrt{x^2 + y^2}}\mathbf{j}$

15. $\mathbf{F}(x, y, z) = c\mathbf{j}$, c a constant

16. $\mathbf{F}(x, y, z) = z\mathbf{k}$ **17.** $\mathbf{F}(x, y, z) = \mathbf{i} + \mathbf{j} + \mathbf{k}$

18. $\mathbf{F}(x, y, z) = x\mathbf{i} + y\mathbf{j} + z\mathbf{k}$

In Exercises 19–22, match the vector field with one of the plots labeled (a)–(d).

19. $\mathbf{F}(x, y, z) = \mathbf{i} + \mathbf{j} + 2\mathbf{k}$ **20.** $\mathbf{F}(x, y, z) = x\mathbf{i} + y\mathbf{j} + 2\mathbf{k}$

21. $\mathbf{F}(x, y, z) = -x\mathbf{i} - y\mathbf{j} - z\mathbf{k}$

22. $\mathbf{F}(x, y, z) = \dfrac{x}{\sqrt{x^2 + y^2 + z^2}}\mathbf{i} + \dfrac{y}{\sqrt{x^2 + y^2 + z^2}}\mathbf{j}$
$\qquad + \dfrac{z}{\sqrt{x^2 + y^2 + z^2}}\mathbf{k}$

(a)   **(b)**

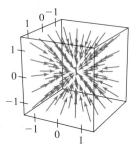

(c)   **(d)**

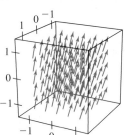

cas *In Exercises 23–26, use a computer algebra system to plot the vector field.*

23. $\mathbf{F}(x, y) = \dfrac{1}{10}(x + y)\mathbf{i} + \dfrac{1}{10}(x - y)\mathbf{j}$

24. $\mathbf{F}(x, y) = 2xy\mathbf{i} + 2x^2y\mathbf{j}$

25. $\mathbf{F}(x, y, z) = \dfrac{1}{5}(-y\mathbf{i} + x\mathbf{j} + z\mathbf{k})$

26. $\mathbf{F}(x, y, z) = -\dfrac{x\mathbf{i} + y\mathbf{j} + z\mathbf{k}}{\sqrt{x^2 + y^2 + z^2}}$

*In Exercises 27–32, find the gradient vector field of the scalar function f. (That is, find the conservative vector field **F** for the potential function f of **F**.)*

27. $f(x, y) = x^2y - y^3$ **28.** $f(x, y) = e^{-2x}\sin 3y$

29. $f(x, y, z) = xyz$ **30.** $f(x, y, z) = xy^2 - yz^3$

31. $f(x, y, z) = y\ln(x + z)$ **32.** $f(x, y, z) = \tan^{-1}(xyz)$

33. Velocity of a Particle A particle is moving in a velocity field

$$\mathbf{V}(x, y, z) = 2x\mathbf{i} + (x + 3y)\mathbf{j} + z^2\mathbf{k}$$

At time $t = 2$ the particle is located at the point $(1, 3, 2)$.
a. What is the velocity of the particle at $t = 2$?
b. What is the approximate location of the particle at $t = 2.01$?

34. Velocity of Flow The following figure shows a lateral section of a tube through which a liquid is flowing. The velocity of flow may vary from point to point, but it is independent of time.
   a. Assuming that the flow is from right to left, sketch vectors emanating from the indicated points representing the speed and direction of fluid flow. Give a reason for your answer. (The answer is not unique.)

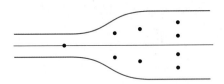

   b. Explain why it is a bad idea to seek shelter in a tunnel when a tornado is approaching.

35. Show that the vector field $\mathbf{F}(x, y) = y\mathbf{i}$ is not a gradient vector field of a scalar function f.
   Hint: If $\mathbf{F}$ is a gradient vector field of f, then $\partial f/\partial x = y$ and $\partial f/\partial y = 0$. Show that f cannot exist.

36. Is $\mathbf{F}(x, y) = -y\mathbf{i} + x\mathbf{j}$ a gradient vector field of a scalar function f? Explain your answer.

In Exercises 37–40, determine whether the statement is true or false. If it is true, explain why. If it is false, explain why or give an example that shows it is false.

37. If $\mathbf{F}$ is a vector field in the plane, then $\mathbf{G} = c\mathbf{F}$ defined by $\mathbf{G}(x, y) = c\mathbf{F}(x, y)$, where c is a constant, is also a vector field.

38. If $\mathbf{F}$ is a velocity field in space, then $|\mathbf{F}(x, y, z)|$ gives the speed of a particle at the point (x, y, z), and $\mathbf{F}(x, y, z)/|\mathbf{F}(x, y, z)|$, where $|\mathbf{F}(x, y, z)| \neq 0$, is a unit vector giving its direction.

39. A constant vector field $\mathbf{F}(x, y, z) = a\mathbf{i} + b\mathbf{j} + c\mathbf{k}$ is a gradient vector field.

40. All the vectors of the vector field $\mathbf{F}(x, y) = x^2\mathbf{i} + y^2\mathbf{j}$ point outward in a radial direction from the origin.

15.2   Divergence and Curl

In this section we will look at two ways of measuring the rate of change of a vector field **F**: the **divergence** of **F** at a point P and the **curl** of **F** at P. The divergence and curl of a vector field play a very important role in describing fluid flow, heat conduction, and electromagnetism.

■ Divergence

Suppose that **F** is a vector field in 2- or 3-space and P is a point in its domain. For the purpose of this discussion, let's also suppose that the vector field **F** describes the flow of a fluid in 2- or 3-space. Then the divergence of **F** at P, written div **F**(P), measures the rate per unit area (or volume) at which the fluid departs or accumulates at P. Let's consider several examples.

EXAMPLE 1

a. Figure 1a shows the vector field $\mathbf{F}(x, y) = x\mathbf{i} + y\mathbf{j}$ described in Example 1 of Section 15.1. Let P be a point in the plane, and let N be a neighborhood of P with center P. Referring to Figure 1b, observe that an arrow entering N along a streamline is matched by one that emerges from N and has a greater length (because it is located farther away from the origin). This shows that more fluid leaves than enters a neighborhood of P. We will show in Example 2a that the vector field **F** is "divergent" at P; that is, the divergence of **F** at P is positive.

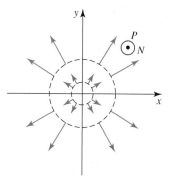

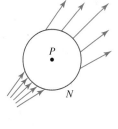

FIGURE 1

(a) The vector field $\mathbf{F}(x, y) = x\mathbf{i} + y\mathbf{i}$

(b) Flow through a neighborhood of P (enlarged and not to scale)

b. Figure 2a shows the vector field $\mathbf{F}(x, y) = y\mathbf{i}$ for $x \geq 0$ and $y \geq 0$. Observe that the streamlines are parallel to the x-axis and that the lengths of the arrows on each horizontal line are constant. We can think of **F** as describing the flow of a river near one side of a riverbank. The velocity of flow is near zero close to the bank (the x-axis) and increases as we move away from it. You can see from Figure 2b that the amount of fluid flowing into the neighborhood N of P is matched by the same amount that exits N. Consequently, we expect the "divergence" at P to be zero. We will show that this is the case in Example 2b.

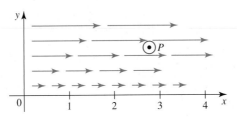
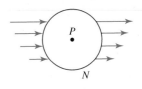

(a) The vector field $\mathbf{F}(x, y) = y\mathbf{i}$

(b) Flow through a neighborhood of P (enlarged and not to scale)

FIGURE 2

c. Figure 3a shows the vector field

$$\mathbf{F}(x, y) = \frac{1}{x + 1}\mathbf{i}$$

for $x \geq 0$ and $y \geq 0$. Observe that the streamlines are parallel to the x-axis and that the lengths of the arrows on each horizontal line get smaller as x increases. From Figure 3b you can see that the "flow" into a neighborhood N of P is greater than the flow that emerges from N. In this case, more fluid enters the neighborhood than leaves it, and the "divergence" is negative. We will show that our intuition is correct in Example 2.

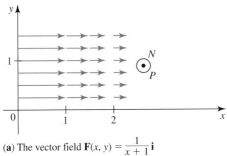
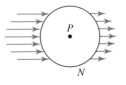

(a) The vector field $\mathbf{F}(x, y) = \frac{1}{x + 1}\mathbf{i}$

(b) Flow through a neighborhood of P (enlarged and not to scale)

FIGURE 3

Up to now, we have looked at the notion of "divergence" intuitively. The divergence of a vector field can be defined as follows.

DEFINITION **Divergence of a Vector Field**

Let $\mathbf{F}(x, y, z) = P\mathbf{i} + Q\mathbf{j} + R\mathbf{k}$ be a vector field in space, where P, Q, and R have first-order partial derivatives in some region T. The divergence of $\mathbf{F}$ is the scalar function defined by

$$\text{div } \mathbf{F} = \frac{\partial P}{\partial x} + \frac{\partial Q}{\partial y} + \frac{\partial R}{\partial z} \tag{1}$$

(We will justify this definition of divergence in Section 15.8.) In two-dimensional space,

$$\mathbf{F}(x, y) = P\mathbf{i} + Q\mathbf{j} \qquad \text{and} \qquad \text{div } \mathbf{F} = \frac{\partial P}{\partial x} + \frac{\partial Q}{\partial y}$$

As an aid to remembering Equation (1), let's introduce the vector differential operator ∇ (read "del") defined by

$$\nabla = \frac{\partial}{\partial x}\mathbf{i} + \frac{\partial}{\partial y}\mathbf{j} + \frac{\partial}{\partial z}\mathbf{k}$$

If we let ∇ operate on a scalar function $f(x, y, z)$, we obtain

$$\nabla f(x, y, z) = \left(\frac{\partial}{\partial x}\mathbf{i} + \frac{\partial}{\partial y}\mathbf{j} + \frac{\partial}{\partial z}\mathbf{k}\right)f(x, y, z)$$

$$= \frac{\partial}{\partial x}f(x, y, z)\mathbf{i} + \frac{\partial}{\partial y}f(x, y, z)\mathbf{j} + \frac{\partial}{\partial z}f(x, y, z)\mathbf{k}$$

$$= \frac{\partial f}{\partial x}(x, y, z)\mathbf{i} + \frac{\partial f}{\partial y}(x, y, z)\mathbf{j} + \frac{\partial f}{\partial z}(x, y, z)\mathbf{k}$$

which is the gradient of f. If we take the "dot product" of ∇ with the vector field $\mathbf{F}(x, y, z) = P\mathbf{i} + Q\mathbf{j} + R\mathbf{k}$, we obtain

$$\nabla \cdot \mathbf{F} = \left(\frac{\partial}{\partial x}\mathbf{i} + \frac{\partial}{\partial y}\mathbf{j} + \frac{\partial}{\partial z}\mathbf{k}\right) \cdot (P\mathbf{i} + Q\mathbf{j} + R\mathbf{k})$$

$$= \frac{\partial}{\partial x}P + \frac{\partial}{\partial y}Q + \frac{\partial}{\partial z}R = \frac{\partial P}{\partial x} + \frac{\partial Q}{\partial y} + \frac{\partial R}{\partial z}$$

which is the divergence of the vector field $\mathbf{F}$. Thus, we can write the divergence of $\mathbf{F}$ symbolically as

$$\operatorname{div}\mathbf{F} = \nabla \cdot \mathbf{F} \qquad (2)$$

Let's apply the definition of divergence to the vector fields that we discussed in Example 1.

EXAMPLE 2 Find the divergence of (a) $\mathbf{F}(x, y) = x\mathbf{i} + y\mathbf{j}$, (b) $\mathbf{F}(x, y) = y\mathbf{i}$, and (c) $\mathbf{F}(x, y) = \dfrac{1}{x + 1}\mathbf{i}$. Reconcile your results with the intuitive observations that were made in Example 1.

Solution

a. $\operatorname{div}\mathbf{F} = \dfrac{\partial}{\partial x}(x) + \dfrac{\partial}{\partial y}(y) = 1 + 1 = 2$. Here, $\operatorname{div}\mathbf{F} > 0$, as expected.

b. Here, $\mathbf{F} = y\mathbf{i} + 0\mathbf{j}$, so $\operatorname{div}\mathbf{F} = \dfrac{\partial}{\partial x}(y) + \dfrac{\partial}{\partial y}(0) = 0$. In this case, $\operatorname{div}\mathbf{F} = 0$, as was observed in Example 1b.

c. With $\mathbf{F} = (x + 1)^{-1}\mathbf{i} + 0\mathbf{j}$ we find

$$\operatorname{div}\mathbf{F} = \frac{\partial}{\partial x}(x + 1)^{-1} + \frac{\partial}{\partial y}(0) = -(x + 1)^{-2} = -\frac{1}{(x + 1)^2}$$

and $\operatorname{div}\mathbf{F} < 0$, as we concluded intuitively in Example 1c. ∎

We turn now to an example involving a vector field whose streamlines are not so easily visualized.

EXAMPLE 3 Find the divergence of $\mathbf{F}(x, y, z) = xyz\mathbf{i} + x^2y^2z\mathbf{j} + xy^2\mathbf{k}$ at the point $(1, -1, 2)$.

Solution

$$\operatorname{div} \mathbf{F} = \frac{\partial}{\partial x}(xyz) + \frac{\partial}{\partial y}(x^2y^2z) + \frac{\partial}{\partial z}(xy^2)$$

$$= yz + 2x^2yz$$

In particular, at the point $(1, -1, 2)$ we find

$$\operatorname{div} \mathbf{F}(1, -1, 2) = (-1)(2) + 2(1)^2(-1)(2) = -6 \qquad \blacksquare$$

The divergence of the vector field $\mathbf{F}(x, y) = y\mathbf{i}$ of Examples 1b and 2b is zero. In general, if $\operatorname{div} \mathbf{F} = 0$, then $\mathbf{F}$ is called **incompressible.** In electromagnetic theory a vector field $\mathbf{F}$ that satisfies $\nabla \cdot \mathbf{F} = 0$ is called **solenoidal.** For example, the electric field $\mathbf{E}$ in Example 4 is solenoidal. We will study the *divergence* of a vector field in greater detail in Section 15.8.

EXAMPLE 4 Show that the divergence of the electric field $\mathbf{E}(x, y, z) = \dfrac{kQ}{|\mathbf{r}|^3}\mathbf{r}$, where $\mathbf{r} = x\mathbf{i} + y\mathbf{j} + z\mathbf{k}$, is zero.

Solution   We first write

$$\mathbf{E}(x, y, z) = \frac{kQx}{(x^2 + y^2 + z^2)^{3/2}}\mathbf{i} + \frac{kQy}{(x^2 + y^2 + z^2)^{3/2}}\mathbf{j} + \frac{kQz}{(x^2 + y^2 + z^2)^{3/2}}\mathbf{k}$$

Then

$$\operatorname{div} \mathbf{E} = kQ\left\{\frac{\partial}{\partial x}\left[\frac{x}{(x^2 + y^2 + z^2)^{3/2}}\right] + \frac{\partial}{\partial y}\left[\frac{y}{(x^2 + y^2 + z^2)^{3/2}}\right] + \frac{\partial}{\partial z}\left[\frac{z}{(x^2 + y^2 + z^2)^{3/2}}\right]\right\}$$

But

$$\frac{\partial}{\partial x}\left[\frac{x}{(x^2 + y^2 + z^2)^{3/2}}\right] = \frac{\partial}{\partial x}[x(x^2 + y^2 + z^2)^{-3/2}]$$

$$= (x^2 + y^2 + z^2)^{-3/2} + x \cdot \left(-\frac{3}{2}\right)(x^2 + y^2 + z^2)^{-5/2}(2x)$$

$$= (x^2 + y^2 + z^2)^{-5/2}[(x^2 + y^2 + z^2) - 3x^2]$$

$$= \frac{-2x^2 + y^2 + z^2}{(x^2 + y^2 + z^2)^{5/2}}$$

Similarly, we find

$$\frac{\partial}{\partial y}\left[\frac{y}{(x^2 + y^2 + z^2)^{3/2}}\right] = \frac{x^2 - 2y^2 + z^2}{(x^2 + y^2 + z^2)^{5/2}}$$

and

$$\frac{\partial}{\partial z}\left[\frac{z}{(x^2 + y^2 + z^2)^{3/2}}\right] = \frac{x^2 + y^2 - 2z^2}{(x^2 + y^2 + z^2)^{5/2}}$$

Therefore,

$$\operatorname{div} \mathbf{E} = kQ\left[\frac{-2x^2 + y^2 + z^2}{(x^2 + y^2 + z^2)^{5/2}} + \frac{x^2 - 2y^2 + z^2}{(x^2 + y^2 + z^2)^{5/2}} + \frac{x^2 + y^2 - 2z^2}{(x^2 + y^2 + z^2)^{5/2}}\right] = 0 \qquad \blacksquare$$

FIGURE 4
A paddle wheel

■ Curl

We now turn our attention to the other measure of the rate of change of a vector field **F**. Let **F** be a vector field in 3-space, and let P be a point in its domain. Once again, let's think of the vector field as one that describes the flow of fluid. Suppose that a small paddle wheel, like the one shown in Figure 4, is immersed in the fluid at P. Then the curl of **F**, written curl **F**, is a measure of the tendency of the fluid to rotate the device about its vertical axis at P. Later, we will show that the paddle wheel will rotate most rapidly if its axis coincides with the direction of curl **F** at P and that its maximum rate of rotation at P is given by the length of curl **F** at P.

EXAMPLE 5

a. Consider the vector field $\mathbf{F}(x, y, z) = y\mathbf{i}$ for $x \geq 0$ similar to that of Example 1b. This field is shown in Figure 5a. Notice that the positive z-axis points vertically out of the page. Suppose that a paddle wheel is planted at a point P. Referring to Figure 5b, you can see that the arrows in the upper half of the circle with center at P are longer than those in the lower half. This shows that the net clockwise flow of the fluid is greater than the net counterclockwise flow. This will cause the paddle to rotate in a clockwise direction, as we will show in Example 6.

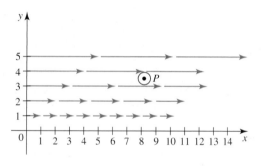

(**a**) The vector field $\mathbf{F}(x, y, z) = y\mathbf{i}$

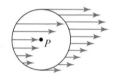

(**b**) Flow through a neighborhood of P at which a paddle wheel is located (enlarged and not to scale)

FIGURE 5

b. Consider the vector field $\mathbf{F}(x, y, z) = -y\mathbf{i} + x\mathbf{j}$ shown in Figure 6a. Observe that it is similar to the spin vector field of Example 2 in Section 15.1. Again, the positive z-axis points vertically out of the page. If a paddle wheel is placed at the origin, it is easy to see that it will rotate in a counterclockwise direction. Next, suppose that the paddle wheel is planted at a point P other than the origin. If you refer to Figure 6b, you can see that the circle with center at P is divided into two arcs by the points of tangency of the two half-lines starting from the origin. Notice that the arc farther from the origin is longer than the one closer to the origin and that the flow on the larger arc is counterclockwise, whereas the flow on the shorter arc is clockwise. Furthermore, the arrows emanating from the longer arc are longer than those emanating from the shorter arc. This shows that the amount of fluid flowing in the counterclockwise direction is greater than that flowing in the clockwise direction. Therefore, the paddle wheel will rotate in a counterclockwise direction, as we will show in Example 6.

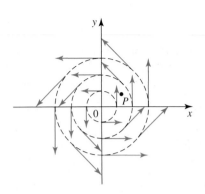

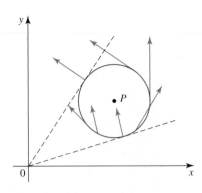

(**a**) The vector field $\mathbf{F}(x, y, z) = -y\mathbf{i} + x\mathbf{j}$

(**b**) Flow through a neighborhood of P at which a paddle wheel is located (enlarged and not to scale)

FIGURE 6

c. Consider the vector field $\mathbf{F}(x, y, z) = x\mathbf{i} + y\mathbf{j}$ shown in Figure 7a. Note that it is similar to the vector field in Example 1 in Section 15.1. Suppose that a paddle wheel is placed at a point P. Then referring to Figure 7(b) and using an argument involving symmetry, you can convince yourself that the paddle wheel will not rotate. Again, we will show in Example 6 that this is true.

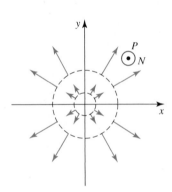

(**a**) The vector field $\mathbf{F}(x, y, z) = x\mathbf{i} + y\mathbf{j}$

(**b**) Flow through a neighborhood of P at which a paddle wheel is located (enlarged and not to scale)

FIGURE 7

The following definition provides us with an exact way to measure the curl of a vector field.

DEFINITION　Curl of a Vector Field

Let $\mathbf{F}(x, y, z) = P\mathbf{i} + Q\mathbf{j} + R\mathbf{k}$ be a vector field in space, where P, Q, and R have first-order partial derivatives in some region T. The curl of $\mathbf{F}$ is the vector field defined by

$$\text{curl } \mathbf{F} = \nabla \times \mathbf{F} = \left(\frac{\partial R}{\partial y} - \frac{\partial Q}{\partial z} \right)\mathbf{i} + \left(\frac{\partial P}{\partial z} - \frac{\partial R}{\partial x} \right)\mathbf{j} + \left(\frac{\partial Q}{\partial x} - \frac{\partial P}{\partial y} \right)\mathbf{k}$$

(We will justify this definition in Section 15.9.)

As in the case of the cross product of two vectors, we can remember the expression for the curl of a vector field by writing it (formally) in determinant form:

$$\text{curl } \mathbf{F} = \nabla \times \mathbf{F} = \begin{vmatrix} \mathbf{i} & \mathbf{j} & \mathbf{k} \\ \dfrac{\partial}{\partial x} & \dfrac{\partial}{\partial y} & \dfrac{\partial}{\partial z} \\ P & Q & R \end{vmatrix}$$

$$= \left(\frac{\partial R}{\partial y} - \frac{\partial Q}{\partial z} \right)\mathbf{i} + \left(\frac{\partial P}{\partial z} - \frac{\partial R}{\partial x} \right)\mathbf{j} + \left(\frac{\partial Q}{\partial x} - \frac{\partial P}{\partial y} \right)\mathbf{k}$$

Let's apply this definition to the vector fields that we discussed in Example 5.

EXAMPLE 6 Find the curl of (a) $\mathbf{F}(x, y, z) = y\mathbf{i}$ for $x \geq 0$, (b) $\mathbf{F}(x, y, z) = -y\mathbf{i} + x\mathbf{j}$, and (c) $\mathbf{F}(x, y, z) = x\mathbf{i} + y\mathbf{j}$. Reconcile your results with the intuitive observations that were made in Example 5.

Solution

a.

$$\text{curl } \mathbf{F} = \nabla \times \mathbf{F} = \begin{vmatrix} \mathbf{i} & \mathbf{j} & \mathbf{k} \\ \dfrac{\partial}{\partial x} & \dfrac{\partial}{\partial y} & \dfrac{\partial}{\partial z} \\ y & 0 & 0 \end{vmatrix}$$

$$= \left[\frac{\partial}{\partial y}(0) - \frac{\partial}{\partial z}(0) \right]\mathbf{i} - \left[\frac{\partial}{\partial x}(0) - \frac{\partial}{\partial z}(y) \right]\mathbf{j} + \left[\frac{\partial}{\partial x}(0) - \frac{\partial}{\partial y}(y) \right]\mathbf{k} = -\mathbf{k}$$

This shows that curl $\mathbf{F}$ is a (unit) vector that points vertically into the page. Applying the right-hand rule, we see that this result tells us that at any point in the vector field, the paddle wheel will rotate in a clockwise direction, as was observed earlier.

b.

$$\text{curl } \mathbf{F} = \nabla \times \mathbf{F} = \begin{vmatrix} \mathbf{i} & \mathbf{j} & \mathbf{k} \\ \dfrac{\partial}{\partial x} & \dfrac{\partial}{\partial y} & \dfrac{\partial}{\partial z} \\ -y & x & 0 \end{vmatrix}$$

$$= \left[\frac{\partial}{\partial y}(0) - \frac{\partial}{\partial z}(x) \right]\mathbf{i} - \left[\frac{\partial}{\partial x}(0) - \frac{\partial}{\partial z}(-y) \right]\mathbf{j} + \left[\frac{\partial}{\partial x}(x) - \frac{\partial}{\partial y}(-y) \right]\mathbf{k}$$

$$= 2\mathbf{k}$$

The result tells us that curl $\mathbf{F}$ points vertically out of the page, so the paddle wheel will rotate in a counterclockwise direction when placed at any point in the vector field $\mathbf{F}$.

c.

$$\text{curl } \mathbf{F} = \nabla \times \mathbf{F} = \begin{vmatrix} \mathbf{i} & \mathbf{j} & \mathbf{k} \\ \dfrac{\partial}{\partial x} & \dfrac{\partial}{\partial y} & \dfrac{\partial}{\partial z} \\ x & y & 0 \end{vmatrix}$$

$$= \left[\frac{\partial}{\partial y}(0) - \frac{\partial}{\partial z}(y) \right]\mathbf{i} - \left[\frac{\partial}{\partial x}(0) - \frac{\partial}{\partial z}(x) \right]\mathbf{j} + \left[\frac{\partial}{\partial x}(y) - \frac{\partial}{\partial y}(x) \right]\mathbf{k} = \mathbf{0}$$

This shows that a paddle wheel placed at any point in $\mathbf{F}$ will not rotate, as observed earlier.

The vector field **F** in Example 6c has the property that curl **F** = **0** at any point P. In general, if curl **F** = **0** at a point P, then **F** is said to be **irrotational** at P. This means that there are no vortices or whirlpools there.

EXAMPLE 7

a. Find curl **F** if $\mathbf{F}(x, y, z) = xy\mathbf{i} + xz\mathbf{j} + xyz^2\mathbf{k}$.
b. What is curl $\mathbf{F}(-1, 2, 1)$?

Solution
a. By definition,

$$
\text{curl } \mathbf{F} = \nabla \times \mathbf{F} = \begin{vmatrix} \mathbf{i} & \mathbf{j} & \mathbf{k} \\ \dfrac{\partial}{\partial x} & \dfrac{\partial}{\partial y} & \dfrac{\partial}{\partial z} \\ xy & xz & xyz^2 \end{vmatrix}
$$

$$
= \left[\frac{\partial}{\partial y}(xyz^2) - \frac{\partial}{\partial z}(xz) \right]\mathbf{i} - \left[\frac{\partial}{\partial x}(xyz^2) - \frac{\partial}{\partial z}(xy) \right]\mathbf{j} + \left[\frac{\partial}{\partial x}(xz) - \frac{\partial}{\partial y}(xy) \right]\mathbf{k}
$$

$$
= (xz^2 - x)\mathbf{i} - yz^2\mathbf{j} + (z - x)\mathbf{k}
$$

$$
= x(z^2 - 1)\mathbf{i} - yz^2\mathbf{j} + (z - x)\mathbf{k}
$$

b. curl $\mathbf{F}(-1, 2, 1) = (-1)(1^2 - 1)\mathbf{i} - (2)(1^2)\mathbf{j} + [1 - (-1)]\mathbf{k} = -2\mathbf{j} + 2\mathbf{k}$ ■

The *div* and *curl* of vector fields enjoy some algebraic properties as illustrated in the following examples. Other properties can be found in the exercises at the end of this section.

EXAMPLE 8 Let f be a scalar function, and let **F** be a vector field. If f and the components of **F** have first-order partial derivatives, show that

$$
\text{div}(f\mathbf{F}) = f\,\text{div } \mathbf{F} + \mathbf{F} \cdot \nabla f
$$

Solution Let's write $\mathbf{F} = P\mathbf{i} + Q\mathbf{j} + R\mathbf{k}$, where P, Q, and R are functions of x, y, and z. Then

$$
f\mathbf{F} = f(P\mathbf{i} + Q\mathbf{j} + R\mathbf{k}) = fP\mathbf{i} + fQ\mathbf{j} + fR\mathbf{k}
$$

so the left-hand side of the given equation reads

$$
\text{div}(f\mathbf{F}) = \nabla \cdot (f\mathbf{F}) = \left(\frac{\partial}{\partial x}\mathbf{i} + \frac{\partial}{\partial y}\mathbf{j} + \frac{\partial}{\partial z}\mathbf{k} \right) \cdot (fP\mathbf{i} + fQ\mathbf{j} + fR\mathbf{k})
$$

$$
= \frac{\partial}{\partial x}(fP) + \frac{\partial}{\partial y}(fQ) + \frac{\partial}{\partial z}(fR)
$$

$$
= f\frac{\partial P}{\partial x} + \frac{\partial f}{\partial x}P + f\frac{\partial Q}{\partial y} + \frac{\partial f}{\partial y}Q + f\frac{\partial R}{\partial z} + \frac{\partial f}{\partial z}R
$$

$$
= f\left(\frac{\partial P}{\partial x} + \frac{\partial Q}{\partial y} + \frac{\partial R}{\partial z} \right) + \left(\frac{\partial f}{\partial x}P + \frac{\partial f}{\partial y}Q + \frac{\partial f}{\partial z}R \right)
$$

$$
= f(\nabla \cdot \mathbf{F}) + (\nabla f) \cdot \mathbf{F}
$$

$$
= f\,\text{div } \mathbf{F} + \mathbf{F} \cdot \nabla f
$$

which is equal to the right-hand side. ■

EXAMPLE 9 Let $\mathbf{F} = P\mathbf{i} + Q\mathbf{j} + R\mathbf{k}$ be a vector field in space, and suppose that P, Q, and R have continuous second-order partial derivatives. Show that

$$\text{div curl } \mathbf{F} = 0$$

Solution Direct computation shows that

$$\text{div curl } \mathbf{F} = \nabla \cdot (\nabla \times \mathbf{F})$$

$$= \left(\frac{\partial}{\partial x}\mathbf{i} + \frac{\partial}{\partial y}\mathbf{j} + \frac{\partial}{\partial z}\mathbf{k} \right) \cdot \left[\left(\frac{\partial R}{\partial y} - \frac{\partial Q}{\partial z} \right)\mathbf{i} + \left(\frac{\partial P}{\partial z} - \frac{\partial R}{\partial x} \right)\mathbf{j} + \left(\frac{\partial Q}{\partial x} - \frac{\partial P}{\partial y} \right)\mathbf{k} \right]$$

$$= \frac{\partial}{\partial x}\left(\frac{\partial R}{\partial y} - \frac{\partial Q}{\partial z} \right) + \frac{\partial}{\partial y}\left(\frac{\partial P}{\partial z} - \frac{\partial R}{\partial x} \right) + \frac{\partial}{\partial z}\left(\frac{\partial Q}{\partial x} - \frac{\partial P}{\partial y} \right)$$

$$= \frac{\partial^2 R}{\partial x\, \partial y} - \frac{\partial^2 Q}{\partial x\, \partial z} + \frac{\partial^2 P}{\partial y\, \partial z} - \frac{\partial^2 R}{\partial y\, \partial x} + \frac{\partial^2 Q}{\partial z\, \partial x} - \frac{\partial^2 P}{\partial z\, \partial y}$$

$$= 0$$

Here we have used the fact that the mixed derivatives are equal because, by assumption, they are continuous.

15.2 CONCEPT QUESTIONS

1. **a.** Define the divergence of a vector field $\mathbf{F}$ and give a formula for finding it.
   b. Define the curl of a vector field $\mathbf{F}$, and give a formula for finding it.
   c. Suppose that $\mathbf{F}$ is the velocity vector field associated with the airflow around an airfoil. Give an interpretation of $\nabla \cdot \mathbf{F}$ and $\nabla \times \mathbf{F}$.

2. **a.** What is meant by a vector field $\mathbf{F}$ that is incompressible? Give a physical example of an (almost) incompressible field.
   b. Repeat part (a) for an irrotational vector field.

15.2 EXERCISES

In Exercises 1–4, you are given the vector field $\mathbf{F}$ and a plot of the vector field in the xy-plane. (The z-component of $\mathbf{F}$ is 0.) (a) By studying the plot of $\mathbf{F}$, determine whether div $\mathbf{F}$ is positive, negative, or zero. Justify your answer. (b) Find div $\mathbf{F}$, and reconcile your result with your answer in part (a). (c) By studying the plot of $\mathbf{F}$, determine whether a paddle wheel planted at a point in the field will rotate clockwise, rotate counterclockwise, or not rotate at all. Justify your answer. (d) Find curl $\mathbf{F}$, and reconcile your result with your answer in part (c).

1. $\mathbf{F}(x, y, z) = \dfrac{x}{|x|}\mathbf{i}, \quad x \neq 0$

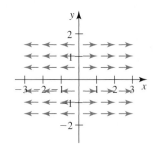

2. $\mathbf{F}(x, y, z) = -x\mathbf{j}$

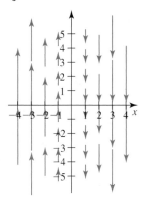

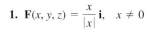

3. $\mathbf{F}(x, y, z) = \dfrac{x}{\sqrt{x^2 + y^2}}\mathbf{i} + \dfrac{y}{\sqrt{x^2 + y^2}}\mathbf{j}$

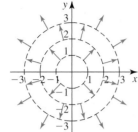

4. $\mathbf{F}(x, y, z) = -\dfrac{y}{\sqrt{x^2 + y^2}}\mathbf{i} + \dfrac{x}{\sqrt{x^2 + y^2}}\mathbf{j}$

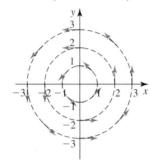

In Exercises 5–12, find (a) *the divergence and* (b) *the curl of the vector field* $\mathbf{F}$.

5. $\mathbf{F}(x, y, z) = yz\mathbf{i} + xz\mathbf{j} + xy\mathbf{k}$

6. $\mathbf{F}(x, y, z) = x^2 y\mathbf{i} - xy^2\mathbf{j} + xyz\mathbf{k}$

7. $\mathbf{F}(x, y, z) = x^2 y^3\mathbf{i} + xz^2\mathbf{k}$

8. $\mathbf{F}(x, y, z) = yz^2\mathbf{i} + x^2 z\mathbf{j}$

9. $\mathbf{F}(x, y, z) = \sin x\mathbf{i} + x\cos y\mathbf{j} + \sin z\mathbf{k}$

10. $\mathbf{F}(x, y, z) = x\cos y\mathbf{i} + y\tan x\mathbf{j} + \sec z\mathbf{k}$

11. $\mathbf{F}(x, y, z) = e^{-x}\cos y\mathbf{i} + e^{-x}\sin y\mathbf{j} + \ln z\mathbf{k}$

12. $\mathbf{F}(x, y, z) = e^{xyz}\mathbf{i} + \cos(x + y)\mathbf{j} - \ln(x + z)\mathbf{k}$

In Exercises 13–15, let $\mathbf{F}$ *be a vector field, and let* f *be a scalar field. Determine whether each expression is meaningful. If so, state whether the expression represents a scalar field or a vector field.*

13. a. $\nabla \times f$ **b.** $\nabla \cdot f$
    c. $\nabla \times (\nabla f)$ **d.** $\operatorname{grad} \mathbf{F}$

14. a. $\operatorname{div}(\nabla f)$ **b.** $\operatorname{grad}(\nabla f)$
    c. $\nabla \times (\operatorname{grad} f)$ **d.** $\operatorname{curl}(\operatorname{curl} \mathbf{F})$

15. a. $\nabla \times (\nabla \times \mathbf{F})$ **b.** $\nabla \cdot (\nabla f)$
    c. $\nabla \cdot (\nabla \cdot \mathbf{F})$ **d.** $\nabla \times [\nabla \times (\nabla f)]$

16. Find $\operatorname{div} \mathbf{F}$ if $\mathbf{F} = \operatorname{grad} f$, where $f(x, y, z) = 2xy^2 z^3$.

17. Show that the vector field $\mathbf{F}(x, y, z) = f(y, z)\mathbf{i} + g(x, z)\mathbf{j} + h(x, y)\mathbf{k}$, where f, g, and h are differentiable, is incompressible.

18. Show that the vector field $\mathbf{F}(x, y, z) = f(x)\mathbf{i} + g(y)\mathbf{j} + h(z)\mathbf{k}$, where f, g and h are differentiable, is irrotational.

In Exercises 19–26, prove the property for vector fields $\mathbf{F}$ *and* $\mathbf{G}$ *and scalar fields* f *and* g. *Assume that the appropriate partial derivatives exist and are continuous.*

19. $\operatorname{div}(\mathbf{F} + \mathbf{G}) = \operatorname{div} \mathbf{F} + \operatorname{div} \mathbf{G}$

20. $\operatorname{curl}(\mathbf{F} + \mathbf{G}) = \operatorname{curl} \mathbf{F} + \operatorname{curl} \mathbf{G}$

21. $\operatorname{curl}(f\mathbf{F}) = f\operatorname{curl} \mathbf{F} + (\nabla f) \times \mathbf{F}$

22. $\operatorname{curl}(\nabla f) = \mathbf{0}$

23. $\operatorname{div}(\mathbf{F} \times \mathbf{G}) = \mathbf{G} \cdot \operatorname{curl} \mathbf{F} - \mathbf{F} \cdot \operatorname{curl} \mathbf{G}$

24. $\operatorname{div}(\nabla f \times \nabla g) = 0$

25. $\nabla \times (\nabla \times \mathbf{F}) = \nabla (\nabla \cdot \mathbf{F}) - \nabla^2 \mathbf{F}$, where
$$\nabla^2 \mathbf{F} = \left(\frac{\partial^2}{\partial x^2} + \frac{\partial^2}{\partial y^2} + \frac{\partial^2}{\partial z^2} \right)\mathbf{F}$$

26. $\nabla \times [\nabla f + (\nabla \times \mathbf{F})] = \nabla \times (\nabla \times \mathbf{F})$
Hint: Use the results of Exercises 20 and 22.

27. Show that there is no vector field $\mathbf{F}$ in space such that $\operatorname{curl} \mathbf{F} = xy\mathbf{i} - yz\mathbf{j} + xy\mathbf{k}$.
Hint: See Example 9.

28. Find the value of the constant c such that the vector field
$$\mathbf{G}(x, y, z) = (2x + 3y + z^2)\mathbf{i} + (cy - z)\mathbf{j} + (x - y + 2z)\mathbf{k}$$
is the curl of some vector field $\mathbf{F}$.

29. Show that $\mathbf{F} = (\cos x)y\mathbf{i} + (\sin y)x\mathbf{j}$ is not a gradient vector field.
Hint: See Exercise 22.

30. Let f be a differentiable function, $\mathbf{r} = x\mathbf{i} + y\mathbf{j} + z\mathbf{k}$, and $r = |\mathbf{r}|$.
    a. Find $\operatorname{curl}[f(r)\mathbf{r}]$ by interpreting it geometrically.
    b. Verify your answer to part (a) analytically.

In Exercises 31–34, let $\mathbf{r} = x\mathbf{i} + y\mathbf{j} + z\mathbf{k}$ *and* $r = |\mathbf{r}|$.

31. Show that $\nabla r = \mathbf{r}/r$.

32. Show that $\nabla(1/r) = -\mathbf{r}/r^3$.

33. Show that $\nabla(\ln r) = \mathbf{r}/r^2$.

34. Show that $\nabla r^n = nr^{n-2}\mathbf{r}$.

In Exercises 35–38, the differential operator ∇^2 *(called the* **Laplacian**) *is defined by* $\nabla^2 = \nabla \cdot \nabla = \dfrac{\partial^2}{\partial x^2} + \dfrac{\partial^2}{\partial y^2} + \dfrac{\partial^2}{\partial z^2}$.

It acts on f *to produce the function* $\nabla^2 f = \dfrac{\partial^2 f}{\partial x^2} + \dfrac{\partial^2 f}{\partial y^2} + \dfrac{\partial^2 f}{\partial z^2}$.

Assume that f *and* g *have second-order partial derivatives.*

35. Show that $\nabla \cdot (\nabla f) = \nabla^2 f$.

36. Show that $\nabla^2(fg) = f\nabla^2 g + g\nabla^2 f + 2\nabla f \cdot \nabla g$.

37. Show that $\nabla^2 r^3 = 12r$, where $r = |\mathbf{r}|$ and $\mathbf{r} = x\mathbf{i} + y\mathbf{j} + z\mathbf{k}$.

38. Show that $\nabla^2\left(\dfrac{1}{r}\right) = 0$, where $r = |\mathbf{r}|$ and $\mathbf{r} = x\mathbf{i} + y\mathbf{j} + z\mathbf{k}$.

39. Angular Velocity of a Particle A particle located at the point P is rotating about the z-axis on a circle of radius R that lies in the plane $z = h$, as shown in the figure. Suppose that the angular speed of the particle is a constant ω. Then this rotational motion can be described by the vector $\mathbf{w} = \omega\mathbf{k}$, which gives the **angular velocity** of P.

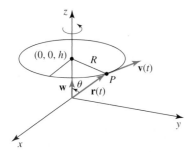

a. Show that the velocity $\mathbf{v}(t)$ of P is given by $\mathbf{v} = \mathbf{w} \times \mathbf{r}$.
  Hint: The position of P is $\mathbf{r}(t) = R \cos \omega t\mathbf{i} + R \sin \omega t\mathbf{j} + h\mathbf{k}$.
b. Show that $\mathbf{v} = -\omega y\mathbf{i} + \omega x\mathbf{j}$.
c. Show that curl $\mathbf{v} = 2\mathbf{w}$. This shows that the angular velocity of P is one half the curl of its tangential velocity.

40. Maxwell's Equations Maxwell's equations relating the electric field $\mathbf{E}$ and the magnetic field $\mathbf{H}$, where c is the speed of light, are given by

$$\nabla \cdot \mathbf{E} = 0, \qquad \nabla \cdot \mathbf{H} = 0,$$

$$\nabla \times \mathbf{E} = -\frac{1}{c}\frac{\partial \mathbf{H}}{\partial t}, \qquad \nabla \times \mathbf{H} = \frac{1}{c}\frac{\partial \mathbf{E}}{\partial t}$$

Show that

a. $\nabla^2\mathbf{E} = \dfrac{1}{c^2}\dfrac{\partial^2 \mathbf{E}}{\partial t^2}$

b. $\nabla^2\mathbf{H} = \dfrac{1}{c^2}\dfrac{\partial^2 \mathbf{H}}{\partial t^2}$

  Hint: Use Exercise 25.

In Exercises 41–48, determine whether the statement is true or false. If it is true, explain why. If it is false, explain why or give an example that shows it is false.

41. If $\mathbf{F}$ is a nonconstant vector field, then div $\mathbf{F} \neq 0$.

42. If $\mathbf{F}(x, y) \neq \mathbf{0}$ and div $\mathbf{F} = 0$ for all x and y, then the streamlines of $\mathbf{F}$ must be closed curves.

43. If the streamlines of a vector field $\mathbf{F}$ are straight lines, then div $\mathbf{F} = 0$.

44. If the streamlines of a vector field $\mathbf{F}$ are concentric circles, then curl $\mathbf{F} = \mathbf{0}$.

45. If the streamlines of a vector field $\mathbf{F}$ are straight lines, then curl $\mathbf{F} = \mathbf{0}$.

46. The curl of a "spin" field is never equal to $\mathbf{0}$.

47. There is no nonzero vector field $\mathbf{F}$ such that div $\mathbf{F} = 0$ and curl $\mathbf{F} = \mathbf{0}$, simultaneously.

48. If curl $\mathbf{F} = \mathbf{0}$, then $\mathbf{F}$ must be a constant vector field.

15.3 Line Integrals

Line Integrals

Once again recall that the mass of a thin, straight wire of length $(b - a)$ and linear mass density $f(x)$ is given by

$$m = \int_a^b f(x)\, dx$$

which has the same numerical value as the area under the graph of f on $[a, b]$. (See Figure 1.)

Instead of being straight, suppose that the wire takes the shape of a plane curve C described by the parametric equations $x = x(t)$ and $y = y(t)$, where $a \le t \le b$, or, equivalently, by the vector equation $\mathbf{r}(t) = x(t)\mathbf{i} + y(t)\mathbf{j}$ with parameter interval $[a, b]$. (See Figure 2a.) Furthermore, suppose that the linear mass density of the wire is given by a continuous function $f(x, y)$. Then one might conjecture that the mass of the curved wire should be numerically equal to the area of the region under the graph of $z = f(x, y)$ with (x, y) lying on C. (See Figure 2b.)

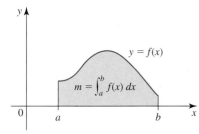

FIGURE 1
The mass of a wire of length $(b - a)$ and linear mass density $f(x)$ is $\int_a^b f(x)\, dx$.

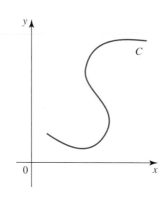
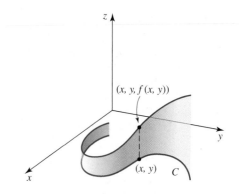

FIGURE 2

(a) The curve C gives the shape of a wire with linear density $f(x, y)$.

(b) The region under the graph of f along C

But how do we define this area, and how do we compute it? As we will now see, this area can be defined in terms of an integral called a *line integral*, even though the term "curve integral" would seem more appropriate.

Let C be a smooth plane curve defined by the parametric equations

$$x = x(t), \qquad y = y(t), \qquad a \le t \le b$$

or, equivalently, by the vector equation $\mathbf{r}(t) = x(t)\mathbf{i} + y(t)\mathbf{j}$, and let P be a regular partition of the parameter interval $[a, b]$ with partition points

$$a = t_0 < t_1 < t_2 \cdots < t_n = b$$

If $x_k = x(t_k)$ and $y_k = y(t_k)$, then the points $P_k(x_k, y_k)$ divide C into n subarcs $\widehat{P_0 P_1}$, $\widehat{P_1 P_2}, \ldots, \widehat{P_{n-1} P_n}$ of lengths $\Delta s_1, \Delta s_2, \ldots, \Delta s_n$, respectively. (See Figure 3.) Next, we pick any evaluation point t_k^* in the subinterval $[t_{k-1}, t_k]$. This point is mapped onto the point $P_k^*(x_k^*, y_k^*)$ lying in the subarc $\widehat{P_{k-1} P_k}$. If f is any function of two variables with domain that contains the curve C, then we can evaluate f at the point (x_k^*, y_k^*), obtaining $f(x_k^*, y_k^*)$. If f is positive, we can think of the product $f(x_k^*, y_k^*) \Delta s_k$ as representing the area of a curved panel with a curved base of length Δs_k and constant height $f(x_k^*, y_k^*)$. (See Figure 4.) This panel is an approximation of the area under the curve $z = f(x, y)$ on the subarc $\widehat{P_{k-1} P_k}$. Therefore, the sum

$$\sum_{k=1}^{n} f(x_k^*, y_k^*) \, \Delta s_k \tag{1}$$

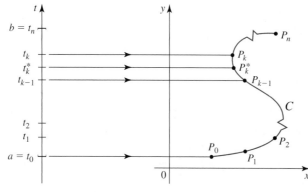

FIGURE 3

(a) A parameter interval

(b) The point $P_k(x_k, y_k)$ corresponds to the point t_k.

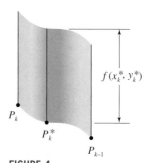

FIGURE 4

The product $f(x_k^*, y_k^*) \, \Delta s_k$ gives the area of a curved panel with a curved base of length Δs_k and with constant height.

gives an approximation of the area under the curve $z = f(x, y)$ and along the curve C. If we let $n \to \infty$, then it seems reasonable to expect that this sum will approach the area under the curve $z = f(x, y)$ along the curve C. This observation suggests the following definition.

DEFINITION **Line Integral**

If f is defined in a region containing a smooth curve C with parametric representation $\mathbf{r}(t)$, where $a \le t \le b$, then the **line integral of f along C** is

$$\int_C f(x, y) \, ds = \lim_{n \to \infty} \sum_{k=1}^{n} f(x_k^*, y_k^*) \Delta s_k \qquad (2)$$

provided that the limit exists.

Note   Observe that over a small piece of a curved wire represented by the segment $\overparen{P_{k-1}P_k}$, the linear density of the wire does not vary by much. Therefore, we may assume that the linear mass density of the wire in the segment $\overparen{P_{k-1}P_k}$ is approximately $f(x_k^*, y_k^*)$, so the mass of this segment is approximately $f(x_k^*, y_k^*) \Delta s_k$. (This is also the area of a typical panel.) Adding the masses of all the segments of the wire leads to the sum (1). Taking the limit as $n \to \infty$ in (1) then gives the mass of the wire.   ■

In general, it can be shown that if f is continuous, then the limit in Equation (2) always exists, and the line integral can be evaluated as an ordinary definite integral with respect to a single variable by using the following formula.

$$\int_C f(x, y) \, ds = \int_a^b f(x(t), y(t)) \sqrt{[x'(t)]^2 + [y'(t)]^2} \, dt \qquad (3)$$

Notes
1. Equation (3) is easier to remember by observing that the element of arc length is given by $ds = |\mathbf{r}'(t)| \, dt = \sqrt{[x'(t)]^2 + [y'(t)]^2} \, dt$.
2. If C is given by the interval $[a, b]$, then C is just the line segment joining $(a, 0)$ to $(b, 0)$. So C can be described by the parametric equations $x = t$ and $y = 0$, where $a \le t \le b$. In this case, Equation (3) becomes $\int_C f(x, y) \, ds = \int_a^b f(t, 0) \, dt = \int_a^b g(x) \, dx$, where $g(x) = f(x, 0)$. So the line integral reduces to an integral of a function defined on an interval $[a, b]$, as expected.   ■

EXAMPLE 1   Evaluate $\int_C (1 + xy) \, ds$, where C is the quarter-circle described by $\mathbf{r}(t) = \cos t \mathbf{i} + \sin t \mathbf{j}$, $0 \le t \le \frac{\pi}{2}$, as shown in Figure 5.

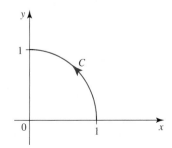

FIGURE 5
The curve C is described by $\mathbf{r}(t) = \cos t \mathbf{i} + \sin t \mathbf{j}$, where $0 \le t \le \frac{\pi}{2}$.

Solution Here, $x(t) = \cos t$ and $y(t) = \sin t$, so $x'(t) = -\sin t$ and $y'(t) = \cos t$. Therefore, using Equation (3), we obtain

$$\int_C (1 + xy)\, ds = \int_0^{\pi/2} (1 + \cos t \sin t)\sqrt{[x'(t)]^2 + [y'(t)]^2}\, dt$$

$$= \int_0^{\pi/2} (1 + \cos t \sin t)\sqrt{(-\sin t)^2 + (\cos t)^2}\, dt$$

$$= \int_0^{\pi/2} (1 + \cos t \sin t)\, dt = \left[t + \frac{1}{2}\sin^2 t \right]_0^{\pi/2}$$

$$= \frac{\pi}{2} + \frac{1}{2} = \frac{1}{2}(\pi + 1) \qquad\blacksquare$$

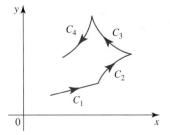

FIGURE 6

A piecewise-smooth curve composed of four smooth curves ($n = 4$)

A curve is **piecewise-smooth** if it is made up of a finite number of smooth curves $C_1, C_2, \dots, C_n$ connected at consecutive endpoints as shown in Figure 6. If f is continuous in a region containing C, then it can be shown that

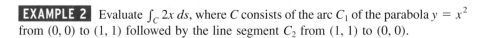

$$\int_C f(x, y)\, ds = \int_{C_1} f(x, y)\, ds + \int_{C_2} f(x, y)\, ds + \cdots + \int_{C_n} f(x, y)\, ds$$

EXAMPLE 2 Evaluate $\int_C 2x\, ds$, where C consists of the arc C_1 of the parabola $y = x^2$ from $(0, 0)$ to $(1, 1)$ followed by the line segment C_2 from $(1, 1)$ to $(0, 0)$.

Solution The curve C is shown in Figure 7. C_1 can be parametrized by taking $x = t$, where t is a parameter. Thus,

$$C_1: \quad x(t) = t, \qquad y(t) = t^2, \qquad 0 \leq t \leq 1$$

Therefore,

$$\int_{C_1} 2x\, ds = \int_0^1 2x\sqrt{[x'(t)]^2 + [y'(t)]^2}\, dt$$

$$= 2\int_0^1 t\sqrt{1 + 4t^2}\, dt$$

$$= \left[2\left(\frac{1}{8}\right)\left(\frac{2}{3}\right)(1 + 4t^2)^{3/2} \right]_0^1 = \frac{5\sqrt{5} - 1}{6}$$

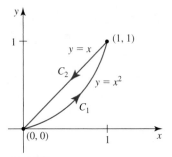

FIGURE 7

C is composed of two smooth curves C_1 and C_2.

C_2 can be parametrized by taking $x = 1 - t$. Thus,

$$C_2: \quad x(t) = 1 - t, \qquad y(t) = 1 - t, \qquad 0 \leq t \leq 1$$

Therefore,

$$\int_{C_2} 2x\, ds = \int_0^1 2x\sqrt{[x'(t)]^2 + [y'(t)]^2}\, dt$$

$$= 2\int_0^1 (1 - t)\sqrt{1 + 1}\, dt$$

$$= \left[2\sqrt{2}\left(t - \frac{1}{2}t^2 \right) \right]_0^1 = \sqrt{2}$$

Putting these results together, we have

$$\int_C 2x \, ds = \int_{C_1} 2x \, ds + \int_{C_2} 2x \, ds = \frac{5\sqrt{5} - 1}{6} + \sqrt{2}$$ ∎

As we saw earlier, the mass of a thin wire represented by C that has linear mass density $\rho(x, y)$ is given by

$$m = \int_C \rho(x, y) \, ds$$

The **center of mass** of the wire is located at the point $(\bar{x}, \bar{y})$, where

$$\bar{x} = \frac{1}{m} \int_C x\rho(x, y) \, ds \qquad \bar{y} = \frac{1}{m} \int_C y\rho(x, y) \, ds \qquad (4)$$

EXAMPLE 3   **The Mass and Center of Mass of a Wire**   A thin wire has the shape of a semicircle of radius a. The linear mass density of the wire is proportional to the distance from the diameter that joins the two endpoints of the wire. Find the mass of the wire and the location of its center of mass.

Solution   If the wire is placed on a coordinate system as shown in Figure 8, then it coincides with the curve C described by the parametric equations $x = a \cos t$ and $y = a \sin t$, where $0 \le t \le \pi$. Its linear mass density is given by $\rho(x, y) = ky$, where k is a positive constant. Since $x'(t) = -a \sin t$ and $y'(t) = a \cos t$, we see that the mass of the wire is

$$m = \int_C \rho(x, y) \, ds = \int_C ky \, ds = \int_0^\pi ky\sqrt{[x'(t)]^2 + [y'(t)]^2} \, dt$$

$$= \int_0^\pi ka \sin t \sqrt{(-a \sin t)^2 + (a \cos t)^2} \, dt$$

$$= ka^2 \int_0^\pi \sin t \, dt = \left[-ka^2 \cos t\right]_0^\pi = 2ka^2$$

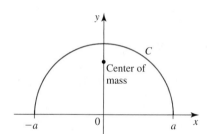

FIGURE 8
The curve C has parametric equations $x = a \cos t$ and $y = a \sin t$, where $0 \le t \le \pi$.

Next, we note that by symmetry, $\bar{x} = 0$. Using Equation (4), we obtain

$$\bar{y} = \frac{1}{m} \int_C y\rho(x, y) \, ds = \frac{1}{2ka^2} \int_0^\pi ky^2 \, ds = \frac{1}{2a^2} \int_0^\pi a(a \sin t)^2 \, dt$$

$$= \frac{a}{2} \int_0^\pi \sin^2 t \, dt = \frac{a}{4} \int_0^\pi (1 - \cos 2t) \, dt$$

$$= \frac{a}{4}\left[t - \frac{1}{2}\sin 2t\right]_0^\pi = \frac{1}{4}\pi a$$

Therefore, the center of mass of the curve is located at $\left(0, \frac{\pi a}{4}\right)$. (See Figure 8.)   ∎

■ Line Integrals with Respect to Coordinate Variables

The line integrals that we have dealt with up to now are taken *with respect to arc length*. Two other line integrals are obtained by replacing Δs_k in Equation (2) by $\Delta x_k = x(t_k) - x(t_{k-1})$ and $\Delta y_k = y(t_k) - y(t_{k-1})$. In the first instance we have the **line** integral of f along C with respect to x,

$$\int_C f(x, y)\, dx = \lim_{n \to \infty} \sum_{k=1}^{n} f(x_k^*, y_k^*)\, \Delta x_k$$

and in the second instance we have the **line integral of f along C with respect to y,**

$$\int_C f(x, y)\, dy = \lim_{n \to \infty} \sum_{k=1}^{n} f(x_k^*, y_k^*)\, \Delta y_k$$

Line integrals with respect to both coordinate variables can also be evaluated as ordinary definite integrals with respect to a single variable. In fact, since $x = x(t)$ and $y = y(t)$, we see that $dx = x'(t)\, dt$ and $dy = y'(t)\, dt$. This leads to the following formulas:

$$\int_C f(x, y)\, dx = \int_a^b f(x(t), y(t))x'(t)\, dt \tag{5a}$$

$$\int_C f(x, y)\, dy = \int_a^b f(x(t), y(t))y'(t)\, dt \tag{5b}$$

Thus, if P and Q are continuous functions of x and y, then

$$\int_C P(x, y)\, dx + Q(x, y)\, dy = \int_C P(x, y)\, dx + \int_C Q(x, y)\, dy$$

can be evaluated as an ordinary integral of a single variable using the formula

$$\int_C P(x, y)\, dx + Q(x, y)\, dy = \int_a^b [P(x(t), y(t))x'(t) + Q(x(t), y(t))y'(t)]\, dt \tag{6}$$

EXAMPLE 4 Evaluate $\int_C y\, dx + x^2\, dy$, where (a) C is the line segment C_1 from $(1, -1)$ to $(4, 2)$, (b) C is the arc C_2 of the parabola $x = y^2$ from $(1, -1)$ to $(4, 2)$, and (c) C is the arc C_3 of the parabola $x = y^2$ from $(4, 2)$ to $(1, -1)$. (See Figure 9.)

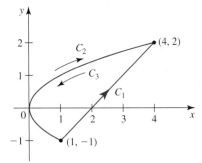

FIGURE 9
The curves C_1, C_2, and C_3

Solution

a. C_1 can be described by the parametric equations

$$x = 1 + 3t, \qquad y = -1 + 3t, \qquad 0 \le t \le 1$$

(See Section 10.5.) We have $dx = 3\, dt$ and $dy = 3\, dt$, so Equation (6) gives

$$\int_{C_1} y\, dx + x^2\, dy = \int_0^1 (-1 + 3t)(3\, dt) + (1 + 3t)^2(3\, dt)$$

$$= 27 \int_0^1 (t^2 + t)\, dt = 27\left[\frac{1}{3}t^3 + \frac{1}{2}t^2\right]_0^1 = \frac{45}{2}$$

b. A parametric representation of C_2 is obtained by letting $y = t$. Thus,

$$C_2\colon \quad x = t^2, \qquad y = t, \qquad -1 \le t \le 2$$

Then $dx = 2t\, dt$ and $dy = dt$, so Equation (6) gives

$$\int_{C_2} y\, dx + x^2\, dy = \int_{-1}^2 t(2t\, dt) + (t^2)^2\, dt$$

$$= \int_{-1}^2 (2t^2 + t^4)\, dt = \left[\frac{2}{3}t^3 + \frac{1}{5}t^5\right]_{-1}^2 = \frac{63}{5}$$

c. C_3 can be parametrized by taking $y = -t$. Thus,

$$C_3: \quad x = t^2, \qquad y = -t, \qquad -2 \le t \le 1$$

Then $dx = 2t\,dt$ and $dy = -dt$, so Equation (6) gives

$$\int_{C_3} y\,dx + x^2\,dy = \int_{-2}^{1} (-t)(2t\,dt) + (t^2)^2\,(-dt)$$

$$= -1\int_{-2}^{1} (2t^2 + t^4)\,dt = -\left[\frac{2}{3}t^3 + \frac{1}{5}t^5\right]_{-2}^{1} = -\frac{63}{5} \quad \blacksquare$$

Example 4 sheds some light on the nature of line integrals. First of all, the results of parts (a) and (b) suggest that the value of a line integral depends not only on the endpoints, but also on the curve joining these points. Second, the results of parts (b) and (c) seem to suggest that reversing the direction in which a curve is traced changes the sign of the value of the line integral.

This latter observation turns out to be true in the general case. For example, suppose that the **orientation** of the curve C (the direction in which it is traced as t increases) is reversed. Let $-C$ denote precisely the curve C with its orientation reversed (so that the curve is traced from B to A instead of from A to B as shown in Figure 10). Then

$$\int_{-C} P\,dx + Q\,dy = -\int_{C} P\,dx + Q\,dy$$

In contrast, note that the value of a line integral taken with respect to arc length does *not* change sign when C is reversed. These results follow because the terms $x'(t)$ and $y'(t)$ change sign but ds does not when the orientation of C is reversed.

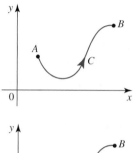

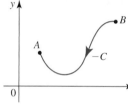

FIGURE 10
$-C$ is the curve consisting of the points of C but traversed in the opposite direction.

Line Integrals in Space

The line integrals in two-dimensional space that we have just considered can be extended to line integrals in three-dimensional space. Suppose that C is a smooth space curve described by the parametric equations

$$x = x(t), \qquad y = y(t), \qquad z = z(t), \qquad a \le t \le b$$

or, equivalently, by the vector equation $\mathbf{r}(t) = x(t)\mathbf{i} + y(t)\mathbf{j} + z(t)\mathbf{k}$, and let f be a function of three variables that is defined and continuous on some region containing C. We define the **line integral of f along C** (with respect to arc length) by

$$\int_{C} f(x, y, z)\,ds = \lim_{n\to\infty} \sum_{k=1}^{n} f(x_k^*, y_k^*, z_k^*)\,\Delta s_k$$

This integral can be evaluated as an ordinary integral by using the following formula, which is the analog of Equation (3) for the three-dimensional case.

$$\int_{C} f(x, y, z)\,ds = \int_{a}^{b} f(x(t), y(t), z(t))\sqrt{\left(\frac{dx}{dt}\right)^2 + \left(\frac{dy}{dt}\right)^2 + \left(\frac{dz}{dt}\right)^2}\,dt \qquad (7)$$

If we make use of vector notation, Equation (7) can be written in the equivalent form

$$\int_{C} f(x, y, z)\,ds = \int_{a}^{b} f(\mathbf{r}(t))|\mathbf{r}'(t)|\,dt$$

EXAMPLE 5 Evaluate $\int_C kz\,ds$, where k is a constant and C is the circular helix with parametric equations $x = \cos t$, $y = \sin t$, and $z = t$, where $0 \le t \le 2\pi$.

Solution   With $x'(t) = -\sin t$, $y'(t) = \cos t$, and $z'(t) = 1$, Equation (7) gives

$$\int_C kz\,ds = \int_0^{2\pi} kt\sqrt{[x'(t)]^2 + [y'(t)]^2 + [z'(t)]^2}\,dt$$

$$= \int_0^{2\pi} kt\sqrt{\sin^2 t + \cos^2 t + 1}\,dt$$

$$= \sqrt{2}k\int_0^{2\pi} t\,dt = \sqrt{2}k\left[\frac{1}{2}t^2\right]_0^{2\pi} = 2\sqrt{2}k\pi^2 \qquad \blacksquare$$

Note   In Example 5, suppose that C represents a thin wire whose linear mass density is directly proportional to its height. Then our calculations tell us that its mass is $2\sqrt{2}k\pi^2$ units. $\blacksquare$

Line integrals along a curve C in space with respect to x, y, and z are defined in much the same way as line integrals along a curve in two-dimensional space. For example, the **line integral of f along C with respect to x** is given by

$$\int_C f(x, y, z)\,dx = \lim_{n\to\infty} \sum_{k=1}^{n} f(x_k^*, y_k^*, z_k^*)\,\Delta x_k$$

so

$$\int_C f(x, y, z)\,dx = \int_a^b f(x(t), y(t), z(t))x'(t)\,dt \qquad (8)$$

If the line integrals with respect to x, y, and z occur together, we have

$$\int_C P(x, y, z)\,dx + Q(x, y, z)\,dy + R(x, y, z)\,dz$$

$$= \int_a^b \left[P(x(t), y(t), z(t))\frac{dx}{dt} + Q(x(t), y(t), z(t))\frac{dy}{dt} + R(x(t), y(t), z(t))\frac{dz}{dt} \right]dt \qquad (9)$$

EXAMPLE 6 Evaluate $\int_C y\,dx + z\,dy + x\,dz$, where C consists of part of the twisted cubic C_1 with parametric equations $x = t$, $y = t^2$, and $z = t^3$, where $0 \le t \le 1$, followed by the line segment C_2 from $(1, 1, 1)$ to $(0, 1, 0)$.

Solution   The curve C is shown in Figure 11. Integrating along C_1, we have $dx = dt$, $dy = 2t\,dt$, and $dz = 3t^2\,dt$. Therefore,

$$\int_{C_1} y\,dx + z\,dy + x\,dz = \int_0^1 t^2\,dt + t^3(2t\,dt) + t(3t^2)\,dt$$

$$= \int_0^1 (t^2 + 3t^3 + 2t^4)\,dt$$

$$= \left[\frac{1}{3}t^3 + \frac{3}{4}t^4 + \frac{2}{5}t^5\right]_0^1 = \frac{89}{60}$$

Next, we write the parametric equations of the line segment from $(1, 1, 1)$ to $(0, 1, 0)$.

On C_2: $x = 1 - t$, $y = 1$, $z = 1 - t$, $0 \le t \le 1$

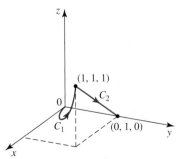

FIGURE 11
The curve C is composed of C_1 and C_2 traversed in the directions shown.

Then $dx = -dt$, $dy = 0$, and $dz = -dt$. Therefore,

$$\int_{C_2} y\,dx + z\,dy + x\,dz = \int_0^1 1(-dt) + (1 - t)(0) + (1 - t)(-dt)$$

$$= \int_0^1 (t - 2)\,dt = \left[\frac{1}{2}t^2 - 2t\right]_0^1 = -\frac{3}{2}$$

Finally, putting these results together, we have

$$\int_C y\,dx + z\,dy + x\,dz = \frac{89}{60} - \frac{3}{2} = -\frac{1}{60}$$

■

■ Line Integrals of Vector Fields

Up to now, we have considered line integrals involving a scalar function f. We now turn our attention to the study of line integrals of vector fields. Suppose that we want to find the work done by a continuous force field $\mathbf{F}$ in moving a particle from a point A to a point B along a smooth curve C in space. Let C be represented parametrically by

$$x = x(t), \qquad y = y(t), \qquad z = z(t), \qquad a \le t \le b$$

or, equivalently, by the vector equation $\mathbf{r}(t) = x(t)\mathbf{i} + y(t)\mathbf{j} + z(t)\mathbf{k}$ with parameter interval $[a, b]$. Take a regular partition P of the parameter interval $[a, b]$ with partition points

$$a = t_0 < t_1 < t_2 < \cdots < t_n = b$$

If $x_k = x(t_k)$, $y_k = y(t_k)$, and $z_k = z(t_k)$, then the points $P_k(x_k, y_k, z_k)$ divide C into n subarcs $\overparen{P_0 P_1}$, $\overparen{P_1 P_2}$, ..., $\overparen{P_{n-1} P_n}$ of lengths $\Delta s_1, \Delta s_2, \ldots, \Delta s_n$, respectively. (See Figure 12.) Furthermore, because $\mathbf{r}$ is smooth, the unit tangent vector $\mathbf{T}(t)$ at any point on the subarc $\overparen{P_{k-1} P_k}$ will not exhibit an appreciable change in direction and may be approximated by $\mathbf{T}(t_k^*)$. Also, because $\mathbf{F}$ is continuous, the force $\mathbf{F}(x(t), y(t), z(t))$ for $t_{k-1} \le t \le t_k$ is approximated by $\mathbf{F}(x_k^*, y_k^*, z_k^*)$. Therefore, we can approximate the work done by $\mathbf{F}$ in moving the particle along the curve from P_{k-1} to P_k by the work done by the component of the constant force $\mathbf{F}(x_k^*, y_k^*, z_k^*)$ in the direction of the line segment (approximated by $\mathbf{T}(t_k^*)$) from P_{k-1} to P_k, that is, by

$$\Delta W_k = \mathbf{F}(x_k^*, y_k^*, z_k^*) \cdot \mathbf{T}(t_k^*)\,\Delta s_k \qquad \text{Constant force in the direction of } \mathbf{T}(x_k^*) \text{ times displacement}$$

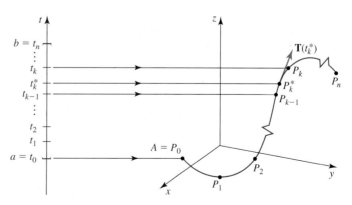

FIGURE 12

(a) A parameter interval

(b) The partition P of $[a, b]$ breaks the curve C into n subarcs.

Here, we have used the fact that the length of the line segment from P_{k-1} to P_k is approximately Δs_k. So the total work done by $\mathbf{F}$ in moving the particle from A to B is

$$W \approx \sum_{k=1}^{n} \mathbf{F}(x_k^*, y_k^*, z_k^*) \cdot \mathbf{T}(t_k^*) \, \Delta s_k$$

This approximation suggests that we define the **work** W done by the force field $\mathbf{F}$ as

$$W = \lim_{n \to \infty} \sum_{k=1}^{n} \mathbf{F}(x_k^*, y_k^*, z_k^*) \cdot \mathbf{T}(t_k^*) \, \Delta s_k = \int_C \mathbf{F} \cdot \mathbf{T} \, ds \qquad (10)$$

Since $\mathbf{T}(t) = \mathbf{r}'(t)/|\mathbf{r}'(t)|$, Equation (10) can also be written in the form

$$W = \int_a^b \left[\mathbf{F}(\mathbf{r}(t)) \cdot \frac{\mathbf{r}'(t)}{|\mathbf{r}'(t)|} \right] |\mathbf{r}'(t)| \, dt$$

$$= \int_a^b \mathbf{F}(\mathbf{r}(t)) \cdot \mathbf{r}'(t) \, dt$$

The last integral is usually written in the form $\int_C \mathbf{F} \cdot d\mathbf{r}$. In words, it says that the work done by a force is given by the line integral of the tangential component of the force with respect to arc length. Although this integral was defined in the context of work done by a force, integrals of this type occur frequently in many other areas of physics and engineering.

DEFINITION **Line Integral of Vector Fields**

Let $\mathbf{F}$ be a continuous vector field defined in a region that contains a smooth curve C described by a vector function $\mathbf{r}(t)$, $a \le t \le b$. Then the **line integral of F along C** is

$$\int_C \mathbf{F} \cdot d\mathbf{r} = \int_C \mathbf{F} \cdot \mathbf{T} \, ds = \int_a^b \mathbf{F}(\mathbf{r}(t)) \cdot \mathbf{r}'(t) \, dt \qquad (11)$$

Note We remind you that $d\mathbf{r}$ is an abbreviation for $\mathbf{r}'(t) \, dt$ and that $\mathbf{F}(\mathbf{r}(t))$ is an abbreviation for $\mathbf{F}(x(t), y(t), z(t))$. ■

EXAMPLE 7 Find the work done by the force field $\mathbf{F}(x, y, z) = -y\mathbf{i} + x\mathbf{j} + z\mathbf{k}$ in moving a particle along the helix C described by the parametric equations $x = \cos t$, $y = \sin t$, and $z = t$ from $(1, 0, 0)$ to $\left(0, 1, \frac{\pi}{2}\right)$. (See Figure 13.)

Solution Since $x(t) = \cos t$, $y(t) = \sin t$, and $z(t) = t$, we see that

$$\mathbf{F}(\mathbf{r}(t)) = \mathbf{F}(x(t), y(t), z(t)) = -y\mathbf{i} + x\mathbf{j} + z\mathbf{k}$$

$$= -\sin t \, \mathbf{i} + \cos t \, \mathbf{j} + t\mathbf{k}$$

Furthermore, observe that the vector equation of C is

$$\mathbf{r}(t) = x(t)\mathbf{i} + y(t)\mathbf{j} + z(t)\mathbf{k} = \cos t \, \mathbf{i} + \sin t \, \mathbf{j} + t\mathbf{k} \qquad 0 \le t \le \frac{\pi}{2}$$

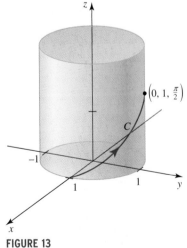

FIGURE 13
The curve C is described by
$\mathbf{r}(t) = \cos t \, \mathbf{i} + \sin t \, \mathbf{j} + t\mathbf{k}, \, 0 \le t \le \frac{\pi}{2}$.

from which we have

$$\mathbf{r}'(t) = -\sin t\mathbf{i} + \cos t\mathbf{j} + \mathbf{k}$$

Therefore, the work done by the force is

$$W = \int_C \mathbf{F} \cdot d\mathbf{r} = \int_0^{\pi/2} \mathbf{F}(\mathbf{r}(t)) \cdot \mathbf{r}'(t)\, dt$$

$$= \int_0^{\pi/2} (-\sin t\mathbf{i} + \cos t\mathbf{j} + t\mathbf{k}) \cdot (-\sin t\mathbf{i} + \cos t\mathbf{j} + \mathbf{k})\, dt$$

$$= \int_0^{\pi/2} (\sin^2 t + \cos^2 t + t)\, dt = \int_0^{\pi/2} (1 + t)\, dt = \left[t + \frac{1}{2}t^2 \right]_0^{\pi/2} = \frac{\pi}{2}\left(1 + \frac{\pi}{4} \right)$$

We close this section by pointing out the relationship between line integrals of vector fields and line integrals of scalar fields with respect to the coordinate variables. Suppose that a vector field $\mathbf{F}$ in space is defined by $\mathbf{F} = P(x, y, z)\mathbf{i} + Q(x, y, z)\mathbf{j} + R(x, y, z)\mathbf{k}$. Then by Equation (11) we have

$$\int_C \mathbf{F} \cdot d\mathbf{r} = \int_a^b \mathbf{F}(\mathbf{r}(t)) \cdot \mathbf{r}'(t)\, dt = \int_a^b (P\mathbf{i} + Q\mathbf{j} + R\mathbf{k}) \cdot (x'(t)\mathbf{i} + y'(t)\mathbf{j} + z'(t)\mathbf{k})\, dt$$

$$= \int_a^b [P(x(t), y(t), z(t))x'(t) + Q(x(t), y(t), z(t))y'(t) + R(x(t), y(t), z(t))z'(t)]\, dt$$

But the integral on the right is just the line integral of Equation (9). Therefore, we have shown that

$$\int_C \mathbf{F} \cdot d\mathbf{r} = \int_C P\, dx + Q\, dy + R\, dz \qquad \text{where} \quad \mathbf{F} = P\mathbf{i} + Q\mathbf{j} + R\mathbf{k} \qquad \textbf{(12)}$$

You are urged to rework Example 7 with the aid of Equation (12).

As a consequence of Equation (12), we have the result

$$\int_{-C} \mathbf{F} \cdot d\mathbf{r} = -\int_C \mathbf{F} \cdot d\mathbf{r}$$

(see page 1247). This result also follows from the equation

$$\int_{-C} \mathbf{F} \cdot d\mathbf{r} = -\int_C \mathbf{F} \cdot \mathbf{T}\, ds$$

and we observe that even though line integrals with respect to arc length do not change sign when the direction traversed is reversed, the unit vector $\mathbf{T}$ does change sign when C is replaced by $-C$.

EXAMPLE 8 Let $\mathbf{F}(x, y) = -\frac{1}{8}(x - y)\mathbf{i} - \frac{1}{8}(x + y)\mathbf{j}$ be the force field shown in Figure 14. Find the work done on a particle that moves along the quarter-circle of radius 1 centered at the origin (a) in a counterclockwise direction from $(1, 0)$ to $(0, 1)$ and (b) in a clockwise direction from $(0, 1)$ to $(1, 0)$.

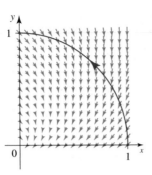

(a) The direction of the path goes against the direction of **F**.

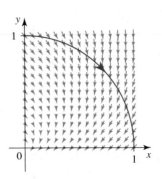

(b) The direction of the path has the same direction as the direction of **F**.

FIGURE 14
The force field
$$\mathbf{F}(x, y) = -\tfrac{1}{8}(x - y)\mathbf{i} - \tfrac{1}{8}(x + y)\mathbf{j}$$

Solution

a. The path of the particle may be represented by $\mathbf{r}(t) = \cos t\mathbf{i} + \sin t\mathbf{j}$ for $0 \le t \le \frac{\pi}{2}$. Since $x = \cos t$ and $y = \sin t$, we find

$$\mathbf{F}(\mathbf{r}(t)) = -\frac{1}{8}(\cos t - \sin t)\mathbf{i} - \frac{1}{8}(\cos t + \sin t)\mathbf{j}$$

and

$$\mathbf{r}'(t) = -\sin t\mathbf{i} + \cos t\mathbf{j}$$

Therefore, the work done by the force on the particle is

$$\int_C \mathbf{F} \cdot d\mathbf{r} = \int_0^{\pi/2} \mathbf{F}(\mathbf{r}(t)) \cdot \mathbf{r}'(t)\, dt = \frac{1}{8}\int_0^{\pi/2} (\cos t \sin t - \sin^2 t - \cos^2 t - \sin t \cos t)\, dt$$

$$= -\frac{1}{8}\int_0^{\pi/2} dt = -\frac{\pi}{16}$$

b. Here, we can represent the path by $\mathbf{r}(t) = \sin t\mathbf{i} + \cos t\mathbf{j}$ for $0 \le t \le \frac{\pi}{2}$. Then $x = \sin t$ and $y = \cos t$. So

$$\int_C \mathbf{F} \cdot d\mathbf{r} = \int_0^{\pi/2} \mathbf{F}(\mathbf{r}(t)) \cdot \mathbf{r}'(t)\, dt = \frac{1}{8}\int_0^{\pi/2} (-\sin t \cos t + \cos^2 t + \sin^2 t + \sin t \cos t)\, dt$$

$$= \frac{1}{8}\int_0^{\pi/2} dt = \frac{\pi}{16} \qquad \blacksquare$$

In Example 8, observe that the work done by **F** on the particle in part (a) is negative because the force field opposes the motion of the particle.

15.3 CONCEPT QUESTIONS

1. a. Define the line integral of a function $f(x, y, z)$ along a smooth curve C with parametric representation $\mathbf{r}(t)$, where $a \le t \le b$.
  b. Write a formula for evaluating the line integral of part (a).
2. a. Define the line integral of a function $f(x, y, z)$ along a smooth curve with respect to x, with respect to y, and with respect to z.

  b. Write formulas for evaluating the line integrals for the integrals of part (a).
  c. Write a formula for evaluating $\int_C P\, dx + Q\, dy + R\, dz$.
3. a. Define the line integral of a vector field **F** along a smooth curve C.
  b. If **F** is a force field, what does the line integral in part (a) represent?

15.3   EXERCISES

In Exercises 1–22, evaluate the line integral over the given curve C.

1. $\int_C (x + y)\, ds;$ $C: \mathbf{r}(t) = 3t\mathbf{i} + 4t\mathbf{j},$ $0 \le t \le 1$

2. $\int_C (x^2 + 2y)\, ds;$ $C: \mathbf{r}(t) = t\mathbf{i} + (t + 1)\mathbf{j},$ $0 \le t \le 2$

3. $\int_C y\, ds;$ $C: \mathbf{r}(t) = 2t\mathbf{i} + t^3\mathbf{j}, 0 \le t \le 1$

4. $\int_C (x + y^3)\, ds;$ $C: \mathbf{r}(t) = t^3\mathbf{i} + t\mathbf{j},$ $0 \le t \le 1$

5. $\int_C (xy^2 + yx^2)\, ds,$ where C is the upper semicircle $y = \sqrt{4 - x^2}$

6. $\int_C (x^2 + y^2)\, ds,$ where C is the right half of the circle $x^2 + y^2 = 9$

7. $\int_C 2xy\, ds,$ where C is the line segment joining $(-2, -1)$ to $(1, 3)$

8. $\int_C (x^2 + 2y)\, ds,$ where C is the line segment joining $(-1, 1)$ to $(0, 3)$

9. $\int_C (x + 3y^2)\, dx;$ $C: \mathbf{r}(t) = (-1 + 2t)\mathbf{i} + (1 + 3t)\mathbf{j},$ $0 \le t \le 1$

10. $\int_C (x + 3y^2)\, dy;$ $C: \mathbf{r}(t) = (-1 + 2t)\mathbf{i} + (1 + 3t)\mathbf{j},$ $0 \le t \le 1$

11. $\int_C xy\, dx + (x + y)\, dy,$ where C consists of the line segment from $(1, 2)$ to $(3, 4)$ and the line segment from $(3, 4)$ to $(4, 0)$

12. $\int_C (y - x)\, dx + y^2\, dy,$ where C consists of the line segment from $(0, 0)$ to $(1, 0)$, and the line segment from $(1, 0)$ to $(2, 4)$

13. $\int_C y\, dx + x\, dy,$ where C consists of the arc of the parabola $y = 4 - x^2$ from $(-2, 0)$ to $(0, 4)$ and the line segment from $(0, 4)$ to $(2, 0)$

14. $\int_C (2x + y)\, dx + 2y\, dy,$ where C consists of the elliptical path $9x^2 + 16y^2 = 144$ from $(4, 0)$ to $(0, 3)$ and the circular path $x^2 + y^2 = 9$ from $(0, 3)$ to $(-3, 0)$

15. $\int_C xyz\, ds;$ $C: \mathbf{r}(t) = (1 + t)\mathbf{i} + 2t\mathbf{j} + (1 - t)\mathbf{k},$ $0 \le t \le 1$

16. $\int_C xyz^2\, ds,$ where C is the line segment joining $(1, 1, 0)$ to $(2, 3, 1)$

17. $\int_C xy^2\, ds;$ $C: \mathbf{r}(t) = \cos 2t\mathbf{i} + \sin 2t\mathbf{j} + 3t\mathbf{k},$ $0 \le t \le \frac{\pi}{2}$

18. $\int_C (8x + 27z)\, ds;$ $C: \mathbf{r}(t) = t\mathbf{i} + 2t^2\mathbf{j} + 3t^3\mathbf{k},$ $0 \le t \le 1$

19. $\int_C (x + y)\, dx + xy\, dy + y\, dz;$ $C: \mathbf{r}(t) = e^t\mathbf{i} + e^{-t}\mathbf{j} + 2e^{2t}\mathbf{k},$ $0 \le t \le 1$

20. $\int_C x\, dx - y^2\, dy + yz\, dz;$ $C: \mathbf{r}(t) = t\mathbf{i} + \cos t\mathbf{j} + \sin t\mathbf{k},$ $0 \le t \le \frac{\pi}{4}$

21. $\int_C xy\, dx - yz\, dy + x^2\, dz,$ where C consists of the line segment from $(0, 0, 0)$ to $(1, 1, 0)$ and the line segment from $(1, 1, 0)$ to $(2, 3, 5)$

22. $\int_C (x + y + z)\, dx + (x - y)\, dy + xz\, dz,$ where C consists of the line segment from $(0, 0, 0)$ to $(1, 1, 1)$ and the line segment from $(1, 1, 1)$ to $(-1, -2, 3)$

23. A thin wire has the shape of a semicircle of radius a. Find the mass and the location of the center of mass of the wire if it has a constant linear mass density k.

24. A thin wire in the shape of a quarter-circle $\mathbf{r}(t) = a \cos t\mathbf{i} + a \sin t\mathbf{j}, 0 \le t \le \frac{\pi}{2}$, has linear mass density $\pi(x, y) = k(x + y)$, where k is a positive constant. Find the mass and the location of the center of mass of the wire.

25. A thin wire has the shape of a semicircle $x^2 + y^2 = a^2$, $y \ge 0$. Find the center of mass of the wire if the linear mass density of the wire at any point is proportional to its distance from the line $y = a$.

26. A thin wire of constant linear mass density k takes the shape of an arch of the cycloid $x = a(t - \sin t), y = a(1 - \cos t)$, $0 \le t \le 2\pi$. Determine the mass of the wire, and find the location of its center of mass.

27. A thin wire of constant linear mass density k has the shape of the astroid $x = \cos^3 t$, $y = \sin^3 t$, $0 \le t \le \frac{\pi}{2}$. Determine the location of its center of mass.

28. A thin wire has the shape of the helix $x = a \cos t$, $y = a \sin t$, $z = bt$, $0 \le t \le 3\pi$. Find the mass and the center of mass of the wire if it has constant linear mass density k.

    Hint: $\bar{x} = \dfrac{1}{m} \displaystyle\int_C x\rho(x, y, z) \, ds$, $\bar{y} = \dfrac{1}{m} \displaystyle\int_C y\rho(x, y, z) \, ds$,

    $\bar{z} = \dfrac{1}{m} \displaystyle\int_C z\rho(x, y, z) \, ds$, where $m = \displaystyle\int_C \rho(x, y, z) \, ds$

29. The vector field $\mathbf{F}(x, y) = (x - y)\mathbf{i} + (x + y)\mathbf{j}$ is shown in the figure. A particle is moved from the point $(-2, 0)$ to the point $(2, 0)$ along the upper semicircle of radius 2 with center at the origin.

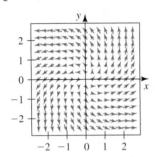

    a. By inspection, determine whether the work done by $\mathbf{F}$ on the particle is positive, zero, or negative.
    b. Find the work done by $\mathbf{F}$ on the particle.

30. The vector field $\mathbf{F}(x, y) = -\dfrac{y}{\sqrt{x^2 + y^2}}\mathbf{i} + \dfrac{x}{\sqrt{x^2 + y^2}}\mathbf{j}$

    is shown in the figure. A particle is moved once around the circle of radius 2 with center at the origin in the counterclockwise direction.

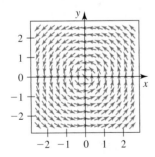

    a. By inspection, determine whether the work done by $\mathbf{F}$ on the particle is positive, zero, or negative.
    b. Find the work done by $\mathbf{F}$ on the particle.

In Exercises 31–36, find the work done by the force field $\mathbf{F}$ on a particle that moves along the curve C.

31. $\mathbf{F}(x, y) = (x^2 + y^2)\mathbf{i} + xy\mathbf{j}$; $C: \mathbf{r}(t) = t^2\mathbf{i} + t^3\mathbf{j}$,
    $0 \le t \le 1$

32. $\mathbf{F}(x, y) = \ln x\mathbf{i} + y^2\mathbf{j}$; $C: \mathbf{r}(t) = t\mathbf{i} + t^2\mathbf{j}$, $1 \le t \le 2$

33. $\mathbf{F}(x, y) = xe^y\mathbf{i} + y\mathbf{j}$, where C is the part of the parabola $y = x^2$ from $(-1, 1)$ to $(2, 4)$

34. $\mathbf{F}(x, y) = x\mathbf{i} + (y + 1)\mathbf{j}$, where C is an arch of the cycloid $x = t - \sin t$, $y = 1 - \cos t$, $0 \le t \le 2\pi$

35. $\mathbf{F}(x, y, z) = x^2\mathbf{i} + y^2\mathbf{j} + z^2\mathbf{k}$; $C: \mathbf{r}(t) = t\mathbf{i} + t^2\mathbf{j} + t^3\mathbf{k}$,
    $0 \le t \le 1$

36. $\mathbf{F}(x, y, z) = (x + 2y)\mathbf{i} + 2z\mathbf{j} + (x - y)\mathbf{k}$, where C is the line segment from $(-1, 3, 2)$ to $(1, -2, 4)$.

37. **Walking up a Spiral Staircase** A spiral staircase is described by the parametric equations

    $$x = 5 \cos t, \qquad y = 5 \sin t, \qquad z = \frac{16}{\pi}t, \qquad 0 \le t \le \frac{\pi}{2}$$

    where the distance is measured in feet. If a 90-lb girl walks up the staircase, what is the work done by her against gravity in walking to the top of the staircase?
    Note: You can also obtain the answer using elementary physics.

38. A particle is moved along a path from $(0, 0)$ to $(1, 2)$ by the force $\mathbf{F} = 2xy^2\mathbf{i} + 3yx^2\mathbf{j}$. Which of the following polygonal paths results in the least work?
    a. The path from $(0, 0)$ to $(1, 0)$ to $(1, 2)$
    b. The path from $(0, 0)$ to $(0, 2)$ to $(1, 2)$
    c. The path from $(0, 0)$ to $(1, 2)$

39. **Newton's Second Law of Motion** Suppose that the position of a particle of varying mass $m(t)$ in 3-space at time t is $\mathbf{r}(t)$. According to Newton's Second Law of Motion, the force acting on the particle at $\mathbf{r}(t)$ is

    $$\mathbf{F}(\mathbf{r}(t)) = \frac{d}{dt}[m(t)\mathbf{v}(t)]$$

    a. Show that $\mathbf{F}(\mathbf{r}(t)) \cdot \mathbf{r}'(t) = m'(t)v^2(t) + m(t)v(t)v'(t)$, where $v = |\mathbf{r}'|$ is the speed of the particle.
    b. Show that if m is constant, then the work done by the force in moving the particle along its path from $t = a$ to $t = b$ is

    $$W = \frac{m}{2}[v^2(b) - v^2(a)]$$

    Note: The function $W(t) = \frac{1}{2}mv^2(t)$ is the kinetic energy of the particle.

40. **Work Done by an Electric Field** Suppose that a charge of Q coulombs is located at the origin of a three-dimensional coordinate system. This charge induces an electric field

    $$\mathbf{E}(x, y, z) = \frac{cQ}{|\mathbf{r}|^3}\mathbf{r}$$

    where $\mathbf{r} = x\mathbf{i} + y\mathbf{j} + z\mathbf{k}$ and c is a constant (see Example 4 in Section 15.1). Find the work done by the electric field on a particle of charge q coulombs as it is moved along the path $C: \mathbf{r}(t) = t\mathbf{i} + 2t\mathbf{j} + (1 + 4t)\mathbf{k}$, where $0 \le t \le 1$.

41. Work Done by an Electric Field The electric field **E** at any point (x, y, z) induced by a point charge Q located at the origin is given by

$$\mathbf{E} = \frac{Q\mathbf{r}}{4\pi\varepsilon_0 |\mathbf{r}|^3}$$

where $\mathbf{r} = \langle x, y, z \rangle$ and ε_0 is a positive constant called the permittivity of free space.

a. Find the work done by the field when a particle of charge q coulombs is moved from $A(2, 1, 0)$ to $D(0, 5, 5)$ along the indicated paths.

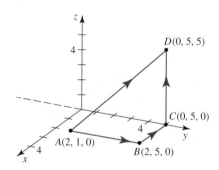

(i) The straight line segment from A to D.

(ii) The polygonal path from $A(2, 1, 0)$ to $B(2, 5, 0)$ to $C(0, 5, 0)$ and then to $D(0, 5, 5)$.

b. Is there any difference in the work done in part (a) and part (b)?

42. Magnitude of a Magnetic Field The following figure shows a long straight wire that is carrying a steady current I. This current induces a magnetic field **B** whose direction is *circumferential;* that is, it circles around the wire. Ampere's Law states that

$$\int_C \mathbf{B} \cdot d\mathbf{r} = \mu_0 I$$

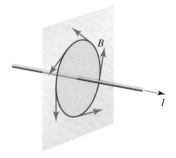

In words, the line integral of the tangential component of the magnetic field around a closed loop C is proportional to the current I passing through any surface bounded by the loop. The constant μ_0 is called the permeability of free space. By taking the loop to be a circle of radius r centered on the wire, show that the magnitude $B = |\mathbf{B}|$ of the magnetic field at a distance r from the center of the wire is

$$B = \frac{\mu_0 I}{2\pi r}$$

cas *In Exercises 43 and 44, plot the graph of the vector field **F** and the curve C on the same set of axes. Guess at whether the line integral of **F** over C is positive, negative, or zero. Verify your answer by evaluating the line integral.*

43. $\mathbf{F}(x, y) = \dfrac{1}{2}(x - y)\mathbf{i} + \dfrac{1}{2}(x + y)\mathbf{j}$; C is the curve

$\mathbf{r}(t) = 2\sin t\,\mathbf{i} + 2\cos t\,\mathbf{j},\quad 0 \le t \le \pi$

44. $\mathbf{F}(x, y) = \dfrac{1}{4}x\mathbf{i} + \dfrac{1}{2}y\mathbf{j}$; C is the curve $\mathbf{r}(t) = t\mathbf{i} + (1 - t^2)\mathbf{j}$,

$-1 \le t \le 1$

In Exercises 45–48, determine whether the statement is true or false. If it is true, explain why. If it is false, explain why or give an example that shows it is false.

45. If $\mathbf{F}(x, y) = x\mathbf{i} + y\mathbf{j}$, then $\int_C \mathbf{F} \cdot d\mathbf{r} = 0$, where C is any circular path centered at the origin.

46. If $f(x, y)$ is continuous and C is a smooth curve, then $\int_C f(x, y)\,ds = -\int_{-C} f(x, y)\,ds$.

47. If C is a smooth curve defined by $\mathbf{r}(t) = x(t)\mathbf{i} + y(t)\mathbf{j}$ with $a \le t \le b$, then $\int_C xy\,dy = \frac{1}{2}xy^2 \Big|_{t=a}^{t=b}$.

48. If $f(x, y)$ is continuous and C is a smooth curve defined by $\mathbf{r}(t) = x(t)\mathbf{i} + y(t)\mathbf{j}$ with $a \le t \le b$, then $\left[\int_C f(x, y)\,ds\right]^2 = \left[\int_C f(x, y)\,dx\right]^2 + \left[\int_C f(x, y)\,dy\right]^2$.

15.4   Independence of Path and Conservative Vector Fields

The gravitational field possesses an important property that we will demonstrate in the following example.

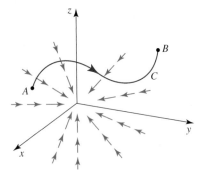

FIGURE 1
The particle moves from A to B along the path C in the gravitational field.

EXAMPLE 1   Work Done on a Particle by a Gravitational Field   Consider the gravitational field $\mathbf{F}$ induced by an object of mass M located at the origin (see Example 3, Section 15.1):

$$\mathbf{F}(x, y, z) = -\frac{GM}{|\mathbf{r}|^3}\mathbf{r}$$

$$= -\frac{GMx}{(x^2 + y^2 + z^2)^{3/2}}\mathbf{i} - \frac{GMy}{(x^2 + y^2 + z^2)^{3/2}}\mathbf{j} - \frac{GMz}{(x^2 + y^2 + z^2)^{3/2}}\mathbf{k}$$

Suppose that a particle with mass m moves in the gravitational field $\mathbf{F}$ from the point $A(x(a), y(a), z(a))$ to the point $B(x(b), y(b), z(b))$ along a smooth curve C defined by

$$\mathbf{r}(t) = x(t)\mathbf{i} + y(t)\mathbf{j} + z(t)\mathbf{k}$$

with parameter interval $[a, b]$. (See Figure 1.) What is the work W done by $\mathbf{F}$ on the particle?

Solution   To find W, we note that the particle moving in the gravitational field $\mathbf{F}$ is subjected to a force of $m\mathbf{F}$, so the work done by the force on the particle is

$$W = \int_C m\mathbf{F} \cdot d\mathbf{r} = \int_a^b m\mathbf{F}(\mathbf{r}(t)) \cdot \mathbf{r}'(t)\, dt$$

$$= -GMm\int_a^b \left[\frac{x}{(x^2 + y^2 + z^2)^{3/2}}\mathbf{i} + \frac{y}{(x^2 + y^2 + z^2)^{3/2}}\mathbf{j} + \frac{z}{(x^2 + y^2 + z^2)^{3/2}}\mathbf{k}\right]$$

$$\cdot \left[\frac{dx}{dt}\mathbf{i} + \frac{dy}{dt}\mathbf{j} + \frac{dz}{dt}\mathbf{k}\right] dt$$

$$= -GMm\int_a^b \left[\frac{x}{(x^2 + y^2 + z^2)^{3/2}}\frac{dx}{dt} + \frac{y}{(x^2 + y^2 + z^2)^{3/2}}\frac{dy}{dt} + \frac{z}{(x^2 + y^2 + z^2)^{3/2}}\frac{dz}{dt}\right] dt$$

But the expression inside the brackets can be written as

$$\frac{d}{dt}f(x, y, z) = \frac{\partial f}{\partial x}\frac{dx}{dt} + \frac{\partial f}{\partial y}\frac{dy}{dt} + \frac{\partial f}{\partial z}\frac{dz}{dt}$$

where

$$\mathrm{s}f(x, y, z) = \frac{1}{\sqrt{x^2 + y^2 + z^2}}$$

as you can verify. (Also, see Example 6 in Section 15.1.) Using this result, we can write

$$W = -GMm\int_a^b \frac{d}{dt}\left(\frac{1}{\sqrt{x^2 + y^2 + z^2}}\right) dt = -\frac{GMm}{\sqrt{x^2 + y^2 + z^2}}\Big|_{t=a}^{t=b}$$

$$= -GMm\,f(x, y, z)\Big|_{t=a}^{t=b} = -GMm[f(x(b), y(b), z(b)) - f(x(a), y(a), z(a))] \quad \blacksquare$$

Note Don't worry about finding the potential function

$$f(x, y, z) = \frac{1}{\sqrt{x^2 + y^2 + z^2}}$$

for the gravitational field **F**. We will develop a systematic method for finding potential functions f of gradient fields ∇f later in this section. ∎

Example 1 shows that the work done on a particle by a gravitational field **F** depends *only* on the initial point A and the endpoint B of a curve C and *not* on the curve itself. We say that the value of the line integral along the path C is *independent of the path*. (A path is a piecewise-smooth curve.)

More generally, we say that the line integral $\int_C \mathbf{F} \cdot d\mathbf{r}$ is **independent of path** if

$$\int_{C_1} \mathbf{F} \cdot d\mathbf{r} = \int_{C_2} \mathbf{F} \cdot d\mathbf{r}$$

for any two paths C_1 and C_2 that have the same initial and terminal points.

Observe that the gravitational field **F** happens to be a *conservative* vector field with potential function f; that is, $\mathbf{F} = \nabla f$. Also, Example 1 seems to suggest that if $\mathbf{F} = \nabla f$ is a gradient vector field with potential function f, then

$$\int_C \mathbf{F} \cdot d\mathbf{r} = \int_C \nabla f \cdot d\mathbf{r} = f(x(b), y(b), z(b)) - f(x(a), y(a), z(a)) \tag{1}$$

This expression reminds us of Part 2 of the Fundamental Theorem of Calculus which states that

$$\int_a^b F'(x)\, dx = F(b) - F(a)$$

where F is continuous on $[a, b]$. The Fundamental Theorem of Calculus, Part 2, tells us that if the derivative of F in the interior of the interval $[a, b]$ is known, then the integral of F' over $[a, b]$ is given by the difference of the values of F (an antiderivative of F') at the endpoints of $[a, b]$. If we think of ∇f as some kind of derivative of f, then Equation (1) says that if we know the "derivative" of f, then the line integral of ∇f is given by the difference of the values of the potential function f ("antiderivative" of ∇f) at the endpoints of the curve C.

We now show that Equation (1) is indeed true for all conservative vector fields. We state and prove the result for a function f of two variables and a curve C in the plane.

THEOREM 1 Fundamental Theorem for Line Integrals

Let $\mathbf{F}(x, y) = \nabla f(x, y)$ be a conservative vector field in an open region R, where f is a differentiable potential function for **F**. If C is any piecewise-smooth curve lying in R given by

$$\mathbf{r}(t) = x(t)\mathbf{i} + y(t)\mathbf{j} \qquad a \le t \le b$$

then

$$\int_C \mathbf{F} \cdot d\mathbf{r} = \int_C \nabla f \cdot d\mathbf{r} = f(x(b), y(b)) - f(x(a), y(a))$$

PROOF We will give the proof for a smooth curve C. Since $\mathbf{F}(x, y) = \nabla f = f_x(x, y)\mathbf{i} + f_y(x, y)\mathbf{j}$, we see that

$$\int_C \mathbf{F} \cdot d\mathbf{r} = \int_C \nabla f \cdot d\mathbf{r} = \int_a^b \nabla f \cdot \frac{d\mathbf{r}}{dt}\, dt$$

$$= \int_a^b \left[\frac{\partial f}{\partial x} \frac{dx}{dt} + \frac{\partial f}{\partial y} \frac{dy}{dt} \right] dt$$

$$= \int_a^b \frac{d}{dt}\, [f(x(t), y(t))]\, dt \qquad \text{Use the Chain Rule.}$$

$$= f(x(t), y(t)) \Big|_{t=a}^{t=b} = f(x(b), y(b)) - f(x(a), y(a)) \qquad \blacksquare$$

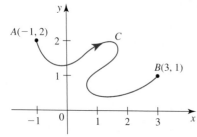

FIGURE 2
C is a piecewise smooth curve joining
A to B.

EXAMPLE 2 Let $\mathbf{F}(x, y) = 2xy\mathbf{i} + x^2\mathbf{j}$ be a force field.

a. Prove that $\mathbf{F}$ is conservative by showing that it is the gradient of the potential function $f(x, y) = x^2 y$.
b. Use the Fundamental Theorem for Line Integrals to evaluate $\int_C \mathbf{F} \cdot d\mathbf{r}$, where C is any piecewise-smooth curve joining the point $A(-1, 2)$ to the point $B(3, 1)$ (See Figure 2.)

Solution

a. Since $\nabla f(x, y) = \dfrac{\partial}{\partial x}(x^2 y)\mathbf{i} + \dfrac{\partial}{\partial y}(x^2 y)\mathbf{j} = 2xy\mathbf{i} + x^2\mathbf{j} = \mathbf{F}(x, y)$, we conclude that $\mathbf{F}$ is indeed conservative.

b. Thanks to the Fundamental Theorem for Line Integrals, we do not need to know the rule defining the curve C; the integral depends only on the coordinates of the endpoints A and B of the curve. We have

$$\int_C \mathbf{F} \cdot d\mathbf{r} = f(3, 1) - f(-1, 2) = x^2 y \Big|_{(-1, 2)}^{(3, 1)}$$

$$= (3)^2(1) - (-1)^2(2) = 7 \qquad \blacksquare$$

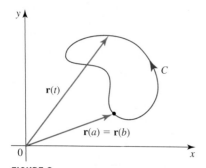

FIGURE 3
On the closed curve C, the tip of $\mathbf{r}(t)$
starts at $\mathbf{r}(a)$, traverses C, and ends up
back at $\mathbf{r}(b) = \mathbf{r}(a)$.

◼ Line Integrals Along Closed Paths

A path is **closed** if its terminal point coincides with its initial point. If a curve C has parametric representation $\mathbf{r}(t)$ with parameter interval $[a, b]$, then C is closed if $\mathbf{r}(a) = \mathbf{r}(b)$. (See Figure 3.)

The following theorem gives an alternative method for determining whether a line integral is independent of path.

THEOREM 2

Suppose that $\mathbf{F}$ is a continuous vector field in a region R. Then $\int_C \mathbf{F} \cdot d\mathbf{r}$ is independent of path if and only if $\int_C \mathbf{F} \cdot d\mathbf{r} = 0$ for every closed path C in R.

PROOF   Suppose that $\int_C \mathbf{F} \cdot d\mathbf{r}$ is independent of path in R, and let C be any closed path in R. We can pick any two points A and B on C and regard C as being made up of the path from A to B and the path from B to A. (See Figure 4a.) Then

$$\int_C \mathbf{F} \cdot d\mathbf{r} = \int_{C_1} \mathbf{F} \cdot d\mathbf{r} + \int_{C_2} \mathbf{F} \cdot d\mathbf{r} = \int_{C_1} \mathbf{F} \cdot d\mathbf{r} - \int_{-C_2} \mathbf{F} \cdot d\mathbf{r}$$

where $-C_2$ is the path C_2 traversed in the opposite direction. But both C_1 and $-C_2$ have the same initial point A and the same terminal point B. Since the line integral is assumed to be independent of path, we have

$$\int_{C_1} \mathbf{F} \cdot d\mathbf{r} = \int_{-C_2} \mathbf{F} \cdot d\mathbf{r}$$

and this implies that $\int_C \mathbf{F} \cdot d\mathbf{r} = 0$.

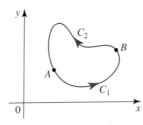

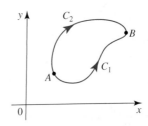

FIGURE 4
C is a closed path in an open region R.

(a) C is made up of C_1 and C_2.

(b) C is made up of C_1 and $-C_2$.

Conversely, suppose that $\int_C \mathbf{F} \cdot d\mathbf{r} = 0$ for every closed path C in R. Let A and B be any two points in R and let C_1 and C_2 be any two paths in R connecting A to B, respectively. (See Figure 4b.) Let C be the closed path composed of C_1 followed by $-C_2$. Then

$$0 = \int_C \mathbf{F} \cdot d\mathbf{r} = \int_{C_1} \mathbf{F} \cdot d\mathbf{r} + \int_{-C_2} \mathbf{F} \cdot d\mathbf{r} = \int_{C_1} \mathbf{F} \cdot d\mathbf{r} - \int_{C_2} \mathbf{F} \cdot d\mathbf{r}$$

so $\int_{C_1} \mathbf{F} \cdot d\mathbf{r} = \int_{C_2} \mathbf{F} \cdot d\mathbf{r}$, which shows that the line integral is independent of path.   ■

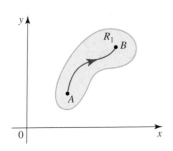

(a) The plane region R_1 is connected.

As a consequence of Theorem 1, we see that if a body moves along a closed path that ends where it began, then the work done by a conservative force field on the body is zero.

■ Independence of Path and Conservative Vector Fields

The Fundamental Theorem for Line Integrals tells us that the line integral of a *conservative* vector field is independent of path. A question that arises naturally is: Is a vector field whose integral is independent of path necessarily a conservative vector field? To answer this question, we need to consider regions that are both *open* and *connected*. A region is **open** if it doesn't contain any of its boundary points. It is **connected** if any two points in the region can be joined by a path that lies in the region. (See Figure 5.) The following theorem provides an answer to part of the first question that we raised.

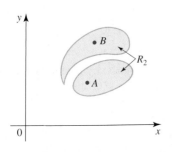

(b) The region R_2 is not connected, since it is impossible to find a path from A to B lying strictly within R_2.

FIGURE 5

> ### THEOREM 3   Independence of Path and Conservative Vector Fields
>
> Let $\mathbf{F}$ be a continuous vector field in an open, connected region R. The line integral $\int_C \mathbf{F} \cdot d\mathbf{r}$ is independent of path if and only if $\mathbf{F}$ is conservative, that is, if and only if $\mathbf{F} = \nabla f$ for some scalar function f.

PROOF   If $\mathbf{F}$ is conservative, then the Fundamental Theorem for Line Integrals implies that the line integral is independent of path. We will prove the converse for the case in which R is a plane region; the proof for the three-dimensional case is similar. Suppose that the integral is independent of path in R. Let (x_0, y_0) be a fixed point in R, and let (x, y) be any point in R. If C is any path from (x_0, y_0) to (x, y), we define the function f by

$$f(x, y) = \int_C \mathbf{F} \cdot d\mathbf{r} = \int_{(x_0, y_0)}^{(x, y)} \mathbf{F} \cdot d\mathbf{r}$$

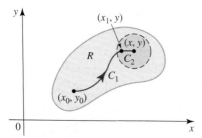

FIGURE 6

The path C consists of an arbitrary path C_1 from (x_0, y_0) to (x_1, y) followed by the horizontal line segment from (x_1, y) to (x, y).

Since R is open, there exists a disk contained in R with center (x, y). Pick any point (x_1, y) in the disk with $x_1 < x$. Now, by assumption, the line integral is independent of path, so we can choose C to be the path consisting of any path C_1 from (x_0, y_0) to (x_1, y) followed by the horizontal line segment C_2 from (x_1, y) to (x, y), as shown in Figure 6. Then

$$f(x, y) = \int_{C_1} \mathbf{F} \cdot d\mathbf{r} + \int_{C_2} \mathbf{F} \cdot d\mathbf{r} = \int_{(x_0, y_0)}^{(x_1, y)} \mathbf{F} \cdot d\mathbf{r} + \int_{C_2} \mathbf{F} \cdot d\mathbf{r}$$

Since the first of the two integrals on the right does not depend on x, we have

$$\frac{\partial}{\partial x} f(x, y) = 0 + \frac{\partial}{\partial x} \int_{C_2} \mathbf{F} \cdot d\mathbf{r}$$

If we write $\mathbf{F}(x, y) = P(x, y)\mathbf{i} + Q(x, y)\mathbf{j}$, then

$$\int_{C_2} \mathbf{F} \cdot d\mathbf{r} = \int_{C_2} P(x, y)\, dx + Q(x, y)\, dy$$

Now C_2 can be represented parametrically by $x(t) = t$, $y(t) = y$, where $x_1 \leq t \leq x$ and y is a constant. This gives $dx = x'(t)\, dt = dt$ and $dy = 0$ since y is constant on C_2. Therefore,

$$\frac{\partial}{\partial x} f(x, y) = \frac{\partial}{\partial x} \int_{C_2} P(x, y)\, dx + Q(x, y)\, dy$$

$$= \frac{\partial}{\partial x} \int_{x_1}^{x} P(t, y)\, dt = P(x, y)$$

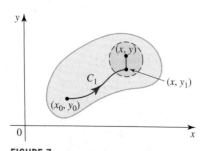

FIGURE 7

The path C consists of an arbitrary path from (x_0, y_0) to (x, y_1) followed by the vertical line segment from (x, y_1) to (x, y).

upon using the Fundamental Theorem of Calculus, Part 1. Similarly, by choosing C to be the path with a vertical line segment as shown in Figure 7, we can show that

$$\frac{\partial}{\partial y} f(x, y) = Q(x, y)$$

Therefore,

$$\mathbf{F} = P\mathbf{i} + Q\mathbf{j} = \frac{\partial f}{\partial x}\mathbf{i} + \frac{\partial f}{\partial y}\mathbf{j} = \nabla f$$

that is, $\mathbf{F}$ is conservative.

■ Determining Whether a Vector Field Is Conservative

Although Theorem 3 provides us with a good characterization of conservative vector fields, it does not help us to determine whether a vector field is conservative, since it is not practical to evaluate the line integral of **F** over all possible paths. Before stating a criterion for determining whether a vector field is conservative, we look at a condition that must be satisfied by a conservative vector field.

THEOREM 4

If $\mathbf{F}(x, y) = P(x, y)\mathbf{i} + Q(x, y)\mathbf{j}$ is a conservative vector field in an open region R and both P and Q have continuous first-order partial derivatives in R, then

$$\frac{\partial Q}{\partial x} = \frac{\partial P}{\partial y}$$

at each point (x, y) in R.

PROOF Because $\mathbf{F} = P\mathbf{i} + Q\mathbf{j}$ is conservative in R, there exists a function f such that $\mathbf{F} = \nabla f$, that is,

$$P\mathbf{i} + Q\mathbf{j} = f_x\mathbf{i} + f_y\mathbf{j}$$

This equation is equivalent to the two equations

$$P = f_x \qquad \text{and} \qquad Q = f_y$$

Since P_y and Q_x are continuous by assumption, it follows from Clairaut's Theorem in Section 13.3 that

$$\frac{\partial P}{\partial y} = f_{xy} = f_{yx} = \frac{\partial Q}{\partial x} \qquad \blacksquare$$

The converse of Theorem 4 holds only for a certain type of region. To describe this region, we need the notion of a *simple curve*. A plane curve described by $\mathbf{r} = \mathbf{r}(t)$ is a **simple curve** if it does not intersect itself anywhere except possibly at its endpoints; that is, $\mathbf{r}(t_1) \neq \mathbf{r}(t_2)$ if $a < t_1 < t_2 < b$. (See Figure 8.)

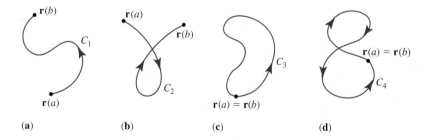

FIGURE 8
C_1 is simple, C_2 is not simple, C_3 is simple and closed, and C_4 is closed but not simple.

(a)   (b)   (c)   (d)

A connected region R in the plane is a **simply-connected region** if every simple closed curve C in R encloses only points that are in R. As is illustrated in Figure 9, a simply-connected region not only is connected, but also does not have any hole(s).

FIGURE 9
R_1 is simply-connected; R_2 is not simply-connected because the simple closed curve shown encloses points outside R_2; R_3 is not simply-connected because it is not connected.

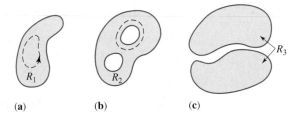

(a)   (b)   (c)

The following theorem, which is a partial converse of Theorem 4, gives us a test to determine whether a vector field on a simply-connected region in the plane is conservative.

THEOREM 5   Test for a Conservative Vector Field in the Plane

Let $\mathbf{F} = P\mathbf{i} + Q\mathbf{j}$ be a vector field in an open simply-connected region R in the plane. If P and Q have continuous first-order partial derivatives on R and

$$\frac{\partial Q}{\partial x} = \frac{\partial P}{\partial y} \tag{2}$$

for all (x, y) in R, then $\mathbf{F}$ is conservative in R.

The proof of this theorem can be found in advanced calculus books.

EXAMPLE 3 Determine whether the vector field $\mathbf{F}(x, y) = (x^2 - 2xy + 1)\mathbf{i} + (y^2 - x^2)\mathbf{j}$ is conservative.

Solution   Here, $P(x, y) = x^2 - 2xy + 1$ and $Q(x, y) = y^2 - x^2$. Since

$$\frac{\partial P}{\partial y} = -2x = \frac{\partial Q}{\partial x}$$

for all (x, y) in the plane, which is open and simply-connected, we conclude by Theorem 5 that $\mathbf{F}$ is conservative.   ∎

EXAMPLE 4 Determine whether the vector field $\mathbf{F}(x, y) = 2xy^2\mathbf{i} + x^2y\mathbf{j}$ is conservative.

Solution   Here, $P(x, y) = 2xy^2$ and $Q(x, y) = x^2y$. So

$$\frac{\partial P}{\partial y} = 4xy \qquad \text{and} \qquad \frac{\partial Q}{\partial x} = 2xy$$

Since $\partial P/\partial y \neq \partial Q/\partial x$ except along the x- or y-axis, we see that Equation (2) of Theorem 5 is not satisfied for all points (x, y) in any open simply-connected region in the plane. Therefore, $\mathbf{F}$ is not conservative.   ∎

■ Finding a Potential Function

Once we have ascertained that a vector field $\mathbf{F}$ is conservative, how do we go about finding a potential function f for $\mathbf{F}$? One such technique is utilized in the following example.

EXAMPLE 5 Let $\mathbf{F}(x, y) = 2xy\mathbf{i} + (1 + x^2 - y^2)\mathbf{j}$.

a. Show that $\mathbf{F}$ is conservative, and find a potential function f such that $\mathbf{F} = \nabla f$.
b. If $\mathbf{F}$ is a force field, find the work done by $\mathbf{F}$ in moving a particle along any path from $(1, 0)$ to $(2, 3)$.

Solution

a. Here, $P(x, y) = 2xy$ and $Q(x, y) = 1 + x^2 - y^2$. Since

$$\frac{\partial P}{\partial y} = 2x = \frac{\partial Q}{\partial x}$$

for all points in the plane, we see that $\mathbf{F}$ is conservative. Therefore, there exists a function f such that $\mathbf{F} = \nabla f$. In this case the equation reads

$$2xy\mathbf{i} + (1 + x^2 - y^2)\mathbf{j} = \frac{\partial f}{\partial x}\mathbf{i} + \frac{\partial f}{\partial y}\mathbf{j}$$

This vector equation is equivalent to the system of scalar equations

$$\frac{\partial f}{\partial x} = 2xy \tag{3}$$

$$\frac{\partial f}{\partial y} = 1 + x^2 - y^2 \tag{4}$$

Integrating Equation (3) with respect to x, (so that y is treated as a constant), we have

$$f(x, y) = x^2 y + g(y) \tag{5}$$

where $g(y)$ is the constant of integration. (Remember that y is treated as a constant, so the most general expression of a constant here involves a function of y.) To determine $g(y)$, we differentiate Equation (5) with respect to y, obtaining

$$\frac{\partial f}{\partial y} = x^2 + g'(y) \tag{6}$$

Comparing Equation (6) with Equation (4) leads to

$$x^2 + g'(y) = 1 + x^2 - y^2$$

or

$$g'(y) = 1 - y^2 \tag{7}$$

Integrating Equation (7) with respect to y gives

$$g(y) = y - \frac{1}{3}y^3 + C$$

where C is a constant. Finally, substituting $g(y)$ into Equation (5) gives

$$f(x, y) = x^2 y + y - \frac{1}{3}y^3 + C$$

the desired potential function.

b. Since $\mathbf{F}$ is conservative, we know that the work done by $\mathbf{F}$ in moving a particle from $(1, 0)$ to $(2, 3)$ is independent of the path connecting these two points.

Using Equation (1), we see that the work done by $\mathbf{F}$ is

$$W = \int_C \mathbf{F} \cdot d\mathbf{r} = \int_C \nabla f \cdot d\mathbf{r} = f(2, 3) - f(1, 0)$$

$$= \left[(2^2)(3) + 3 - \frac{1}{3}(3^3) \right] - \left[(1^2)(0) + 0 - \frac{1}{3}(0) \right] = 6 \quad \blacksquare$$

Note In Example 5a you may also integrate Equation (4) first with respect to y and proceed in a similar manner. $\quad\blacksquare$

The following theorem provides us with a test to determine whether a vector field in space is conservative. Theorem 6 is an extension of Theorem 5, and its proof will be omitted.

THEOREM 6   Test for a Conservative Vector Field in Space

Let $\mathbf{F} = P\mathbf{i} + Q\mathbf{j} + R\mathbf{k}$ be a vector field in an open, simply connected region D in space. If P, Q, and R have continuous first-order partial derivatives in space, then $\mathbf{F}$ is conservative if curl $\mathbf{F} = \mathbf{0}$ for all points in D. Equivalently, $\mathbf{F}$ is conservative if

$$\frac{\partial R}{\partial y} = \frac{\partial Q}{\partial z}, \qquad \frac{\partial R}{\partial x} = \frac{\partial P}{\partial z}, \qquad \text{and} \qquad \frac{\partial Q}{\partial x} = \frac{\partial P}{\partial y}$$

The following example illustrates how to find a potential function for a conservative vector field in space.

EXAMPLE 6 Let $\mathbf{F}(x, y, z) = 2xyz^2\mathbf{i} + x^2z^2\mathbf{j} + 2x^2yz\mathbf{k}$.

a. Show that $\mathbf{F}$ is conservative, and find a function f such that $\mathbf{F} = \nabla f$.
b. If $\mathbf{F}$ is a force field, find the work done by $\mathbf{F}$ in moving a particle along any path from $(0, 1, 0)$ to $(1, 2, -1)$.

Solution
a. We compute

$$\text{curl } \mathbf{F} = \begin{vmatrix} \mathbf{i} & \mathbf{j} & \mathbf{k} \\ \dfrac{\partial}{\partial x} & \dfrac{\partial}{\partial y} & \dfrac{\partial}{\partial z} \\ 2xyz^2 & x^2z^2 & 2x^2yz \end{vmatrix}$$

$$= (2x^2z - 2x^2z)\mathbf{i} - (4xyz - 4xyz)\mathbf{j} + (2xz^2 - 2xz^2)\mathbf{k}$$

$$= \mathbf{0}$$

Since curl $\mathbf{F} = \mathbf{0}$ for all points in R^3, we see that $\mathbf{F}$ is a conservative vector field by Theorem 6. Therefore, there exists a function f such that $\mathbf{F} = \nabla f$. In this case the equation reads

$$2xyz^2\mathbf{i} + x^2z^2\mathbf{j} + 2x^2yz\mathbf{k} = \frac{\partial f}{\partial x}\mathbf{i} + \frac{\partial f}{\partial y}\mathbf{j} + \frac{\partial f}{\partial z}\mathbf{k}$$

This vector equation is equivalent to the system of three scalar equations

$$\frac{\partial f}{\partial x} = 2xyz^2 \tag{8}$$

$$\frac{\partial f}{\partial y} = x^2z^2 \tag{9}$$

$$\frac{\partial f}{\partial z} = 2x^2yz \tag{10}$$

Integrating Equation (8) with respect to x (so that y and z, are treated as constants), we have

$$f(x, y, z) = x^2yz^2 + g(y, z) \tag{11}$$

where $g(y, z)$ is the constant of integration. To determine $g(y, z)$, we differentiate Equation (11) with respect to y, obtaining

$$\frac{\partial f}{\partial y} = x^2z^2 + \frac{\partial g}{\partial y} \tag{12}$$

Comparing Equation (12) with Equation (9) leads to

$$x^2z^2 + \frac{\partial g}{\partial y} = x^2z^2$$

or

$$\frac{\partial g}{\partial y} = 0 \tag{13}$$

Integrating Equation (13) with respect to y (so that z, is treated as a constant), we obtain $g(y, z) = h(z)$, so

$$f(x, y, z) = x^2yz^2 + h(z) \tag{14}$$

Differentiating Equation (14) with respect to z, and comparing the result with Equation (10), we have

$$\frac{\partial f}{\partial z} = 2x^2yz + h'(z) = 2x^2yz$$

Therefore, $h'(z) = 0$ and $h(z) = C$, where C is a constant. Finally, substituting the value of $h(z)$ into Equation (14) gives

$$f(x, y, z) = x^2yz^2 + C$$

as the desired potential function.

b. Since **F** is conservative, we know that the work done by **F** in moving a particle from $(0, 1, 0)$ to $(1, 2, -1)$ is independent of the path connecting these two points. Therefore, the work done by **F** is

$$W = \int_C \mathbf{F} \cdot d\mathbf{r} = \int_C \nabla f \cdot d\mathbf{r} = f(1, 2, -1) - f(0, 1, 0)$$

$$= (1)^2(2)(-1)^2 - 0 = 2 \qquad \blacksquare$$

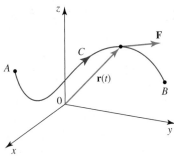

FIGURE 10
The path C of a body from $A = \mathbf{r}(a)$
to $B = \mathbf{r}(b)$

Conservation of Energy

The Fundamental Theorem for Line Integrals can be used to derive one of the most important laws of physics: the Law of Conservation of Energy. Suppose that a body of mass m is moved from A to B along a piecewise-smooth curve C such that its position at any time t is given by $\mathbf{r}(t)$, $a \leq t \leq b$, and suppose that the body is subjected to the action of a continuous conservative force field $\mathbf{F}$. (See Figure 10.) To find the work done by the force on the body, we use Newton's Second Law of Motion to write $\mathbf{F} = m\mathbf{a} = m\mathbf{v}'(t) = m\mathbf{r}''(t)$, where $\mathbf{v}(t) = \mathbf{r}'(t)$ and $\mathbf{a}(t) = \mathbf{r}''(t)$ are the velocity and acceleration of the body at any time t, respectively. The work done by the force $\mathbf{F}$ on the body as it is moved from A to B along C is

$$
\begin{aligned}
W &= \int_C \mathbf{F} \cdot d\mathbf{r} = \int_a^b \mathbf{F}(\mathbf{r}(t)) \cdot \mathbf{r}'(t)\, dt \\[2mm]
&= \int_a^b m\mathbf{r}''(t) \cdot \mathbf{r}'(t)\, dt \\[2mm]
&= \frac{m}{2} \int_a^b \frac{d}{dt} [\mathbf{r}'(t) \cdot \mathbf{r}'(t)]\, dt \qquad \text{Use Theorem 2 in Section 12.2.} \\[2mm]
&= \frac{m}{2} \int_a^b \frac{d}{dt} |\mathbf{r}'(t)|^2\, dt \\[2mm]
&= \frac{m}{2} \Big[|\mathbf{r}'(t)|^2 \Big]_a^b \qquad \text{Use the Fundamental Theorem for Line Integrals.} \\[2mm]
&= \frac{m}{2} \left(|\mathbf{r}'(b)|^2 - |\mathbf{r}'(a)|^2 \right) \\[2mm]
&= \frac{1}{2} m |\mathbf{v}(b)|^2 - \frac{1}{2} m |\mathbf{v}(a)|^2 \qquad \text{Since } \mathbf{v}(t) = \mathbf{r}'(t)
\end{aligned}
$$

Since the kinetic energy K of a particle of mass m and speed v is $\frac{1}{2}mv^2$, we can write

$$W = K(B) - K(A) \tag{15}$$

which says that the work done by the force field on the body as it moves from A to B along C is equal to the change in kinetic energy of the body at A and B.

Since $\mathbf{F}$ is conservative, there is a scalar function f such that $\mathbf{F} = \nabla f$. The **potential energy** P of a body at the point (x, y, z) in a conservative force field is defined to be $P(x, y, z) = -f(x, y, z)$, so we have $\mathbf{F} = -\nabla P$. Consequently, the work done by $\mathbf{F}$ on the body as it is moved from A to B along C is given by

$$
\begin{aligned}
W &= \int_C \mathbf{F} \cdot d\mathbf{r} = -\int_C \nabla P \cdot d\mathbf{r} \\[2mm]
&= \Big[-P(\mathbf{r}(t)) \Big]_a^b = -[P(\mathbf{r}(b)) - P(\mathbf{r}(a))] \\[2mm]
&= P(A) - P(B)
\end{aligned}
$$

Comparing this equation with Equation (15), we see that

$$P(A) + K(A) = P(B) + K(B)$$

which states that as the body moves from one point to another in a conservative force field, then the sum of its potential energy and kinetic energy remains constant. This is the **Law of Conservation of Energy** and is the reason why certain vector fields are called *conservative*.

15.4 CONCEPT QUESTIONS

1. State the Fundamental Theorem for Line Integrals.
2. **a.** Explain what it means for the line integral $\int_C \mathbf{F} \cdot d\mathbf{r}$ to be independent of path.
   b. If $\int_C \mathbf{F} \cdot d\mathbf{r}$ is independent of path for all paths C in an open, connected region R, what can you say about $\mathbf{F}$?

3. **a.** How do you determine whether a vector field $\mathbf{F} = P(x, y)\mathbf{i} + Q(x, y)\mathbf{j}$ is conservative?
   b. How do you determine whether a vector field $\mathbf{F} = P(x, y, z)\mathbf{i} + Q(x, y, z)\mathbf{j} + R(x, y, z)\mathbf{k}$ is conservative?

15.4 EXERCISES

In Exercises 1–10, determine whether $\mathbf{F}$ *is conservative. If so, find a function f such that* $\mathbf{F} = \nabla f$.

1. $\mathbf{F}(x, y) = (4x + 3y)\mathbf{i} + (3x - 2y)\mathbf{j}$
2. $\mathbf{F}(x, y) = (2x^2 + 4y)\mathbf{i} + (2x - 3y^2)\mathbf{j}$
3. $\mathbf{F}(x, y) = (2x + y^2)\mathbf{i} + (x^2 + y)\mathbf{j}$
4. $\mathbf{F}(x, y) = (x^2 + y^2)\mathbf{i} + 2xy\mathbf{j}$
5. $\mathbf{F}(x, y) = y^2 \cos x\mathbf{i} + (2y \sin x + 3)\mathbf{j}$
6. $\mathbf{F}(x, y) = (x \cos y + \sin y)\mathbf{i} + (\cos y - x \sin y)\mathbf{j}$
7. $\mathbf{F}(x, y) = (e^{-x} - 2y \cos 2x)\mathbf{i} + (\sin 2x + ye^{-x})\mathbf{j}$
8. $\mathbf{F}(x, y) = (\tan y + 2xy)\mathbf{i} + (x \sec^2 y + x^2)\mathbf{j}$
9. $\mathbf{F}(x, y) = \left(x^2 + \dfrac{y}{x}\right)\mathbf{i} + (y^2 + \ln x)\mathbf{j}$
10. $\mathbf{F}(x, y) = (e^x \cos y + y \sec^2 x)\mathbf{i} + (\tan x - e^x \cos y)\mathbf{j}$

In Exercises 11–18, (a) show that $\mathbf{F}$ *is conservative and find a function f such that* $\mathbf{F} = \nabla f$, *and (b) use the result of part* (a) *to evaluate* $\int_C \mathbf{F} \cdot d\mathbf{r}$, *where C is any path from* $A(x_0, y_0)$ *to* $B(x_1, y_1)$.

11. $\mathbf{F}(x, y) = (2y + 1)\mathbf{i} + (2x + 3)\mathbf{j}$; $A(0, 0)$ and $B(-1, 1)$
12. $\mathbf{F}(x, y) = (x - 2y)\mathbf{i} + (y - 2x)\mathbf{j}$; $A(0, 0)$ and $B(1, 1)$
13. $\mathbf{F}(x, y) = (2xy^2 + 2y)\mathbf{i} + (2x^2y + 2x)\mathbf{j}$; $A(-1, 1)$ and $B(1, 2)$
14. $\mathbf{F}(x, y) = 2xy^3\mathbf{i} + (3x^2y^2 + 1)\mathbf{j}$; $A(1, 1)$ and $B(2, 0)$
15. $\mathbf{F}(x, y) = xe^{2y}\mathbf{i} + x^2e^{2y}\mathbf{j}$; $A(0, 0)$ and $B(-1, 1)$
16. $\mathbf{F}(x, y) = 2x \sin y\mathbf{i} + x^2 \cos y\mathbf{j}$; $A(0, 0)$ and $B\left(1, \frac{\pi}{2}\right)$
17. $\mathbf{F}(x, y) = e^x \sin y\mathbf{i} + (e^x \cos y + y)\mathbf{j}$; $A(0, 0)$ and $B(0, \pi)$
18. $\mathbf{F}(x, y) = (x + \tan^{-1} y)\mathbf{i} + \dfrac{x + y}{1 + y^2}\mathbf{j}$; $A(0, 0)$ and $B(1, 1)$

In Exercises 19 and 20, evaluate $\int_C \mathbf{F} \cdot d\mathbf{r}$ *for the vector field* $\mathbf{F}$ *and the path C. (Hint: Show that* $\mathbf{F}$ *is conservative, and pick a simpler path.)*

19. $\mathbf{F}(x, y) = (2xy^2 + \cos y)\mathbf{i} + (2x^2y - x \sin y)\mathbf{j}$
    $C: \mathbf{r}(t) = (1 - \cos t)\mathbf{i} + \sin t\mathbf{j}$, $0 \le t \le \pi$
20. $\mathbf{F}(x, y) = (e^y - y^2 \sin x)\mathbf{i} + (xe^y + 2y \cos x)\mathbf{j}$
    $C: \mathbf{r}(t) = 4 \cos t\mathbf{i} + 3 \sin t\mathbf{j}$, $0 \le t \le \pi$

In Exercises 21 and 22, find the work done by the force field $\mathbf{F}$ *on a particle moving along a path from P to Q.*

21. $\mathbf{F}(x, y) = 2\sqrt{y}\mathbf{i} + \dfrac{x}{\sqrt{y}}\mathbf{j}$; $A(1, 1)$, $B(2, 9)$
22. $\mathbf{F}(x, y) = -e^{-x} \cos y\mathbf{i} - e^{-x} \sin y\mathbf{j}$; $A(0, 0)$, $B(1, \pi)$
23. Show that the line integral $\int_C yz\, dx + xz\, dy + xyz\, dz$ is not independent of path.
24. Show that the following line integral is not independent of path: $\int_C e^{-y} \sin z\, dx - xe^{-y} \sin z\, dy - xe^{-y} \cos z\, dz$

In Exercises 25–32, determine whether $\mathbf{F}$ *is conservative. If so, find a function f such that* $\mathbf{F} = \nabla f$.

25. $\mathbf{F}(x, y, z) = yz\mathbf{i} + xz\mathbf{j} + xy\mathbf{k}$
26. $\mathbf{F}(x, y, z) = 2xy^2z\mathbf{i} + 2x^2yz\mathbf{j} + x^2y^2\mathbf{k}$
27. $\mathbf{F}(x, y, z) = 2xy\mathbf{i} + (x^2 + z^2)\mathbf{j} + xy\mathbf{k}$
28. $\mathbf{F}(x, y, z) = \sin y\mathbf{i} + (x \cos y + \cos z)\mathbf{j} + \sin z\mathbf{k}$
29. $\mathbf{F}(x, y, z) = e^x \cos z\mathbf{i} + z \sinh y\mathbf{j} + (\cosh y - e^x \sin z)\mathbf{k}$
30. $\mathbf{F}(x, y, z) = ze^{xz}\mathbf{i} + \ln z\mathbf{j} + \left(xe^{xz} + \dfrac{y}{z}\right)\mathbf{k}$
31. $\mathbf{F}(x, y, z) = z \cos(x + y)\mathbf{i} + z \sin(x + y)\mathbf{j} + \cos(x + y)\mathbf{k}$
32. $\mathbf{F}(x, y, z) = \dfrac{1}{yz}\mathbf{i} - \dfrac{x}{y^2z}\mathbf{j} - \dfrac{x}{yz^2}\mathbf{k}$

In Exercises 33–36, (a) show that $\mathbf{F}$ *is conservative, and find a function f such that* $\mathbf{F} = \nabla f$, *and (b) use the result of part* (a) *to evaluate* $\int_C \mathbf{F} \cdot d\mathbf{r}$, *where C is any curve from* $A(x_0, y_0, z_0)$ *to* $B(x_1, y_1, z_1)$.

33. $\mathbf{F}(x, y, z) = yz^2\mathbf{i} + xz^2\mathbf{j} + 2xyz\mathbf{k}$; $A(0, 0, 1)$ and $B(1, 3, 2)$
34. $\mathbf{F}(x, y, z) = 2xy^2z^3\mathbf{i} + 2x^2yz^3\mathbf{j} + 3x^2y^2z^2\mathbf{k}$; $A(0, 0, 0)$ and $B(1, 1, 1)$
35. $\mathbf{F}(x, y, z) = \cos y\mathbf{i} + (z^2 - x \sin y)\mathbf{j} + 2yz\mathbf{k}$; $A(1, 0, 0)$ and $B(2, 2\pi, 1)$
36. $\mathbf{F}(x, y, z) = e^y\mathbf{i} + (xe^y + \ln z)\mathbf{j} + \left(\dfrac{y}{z}\right)\mathbf{k}$; $A(0, 1, 1)$ and $B(1, 0, 2)$
37. Evaluate $\int_C (2xy^2 - 3)\, dx + (2x^2y + 1)\, dy$, where C is the curve $x^4 - 6xy^3 - 4y^2 = 0$ from $(0, 0)$ to $(2, 1)$.

38. Evaluate $\int_C (3x^2y + e^y)\, dx + (x^3 + xe^y - 2y)\, dy$, where C is the curve of Exercise 37.

39. Let

$$\mathbf{E}(x, y, z) = \frac{kQ}{|\mathbf{r}|^3}\, \mathbf{r}$$

where k is a constant, and let $\mathbf{r} = x\mathbf{i} + y\mathbf{j} + z\mathbf{k}$ be the electric field induced by a charge Q located at the origin. (See Example 4 in Section 15.1.) Find the work done by $\mathbf{E}$ in moving a charge of q coulombs from the point $A(1, 3, 2)$ along any path to the point $B(2, 4, 1)$.

40. Find the work that is done by the force field $\mathbf{F}(x, y, z) = y^2z\mathbf{i} + 2xyz\mathbf{j} + xy^2\mathbf{k}$ on a particle moving along a path from $P(1, 1, 1)$ to $Q(2, 1, 3)$.

41. Let

$$\mathbf{F}(x, y) = \frac{y}{x^2 + y^2}\,\mathbf{i} - \frac{x}{x^2 + y^2}\,\mathbf{j}$$

a. Show that $\dfrac{\partial Q}{\partial x} = \dfrac{\partial P}{\partial y}$.

b. Show that $\int_C \mathbf{F} \cdot d\mathbf{r}$ is not independent of path by computing $\int_{C_1} \mathbf{F} \cdot d\mathbf{r}$ and $\int_{C_2} \mathbf{F} \cdot d\mathbf{r}$, where C_1 and C_2 are the upper and lower semicircles of radius 1, centered at the origin, from $(1, 0)$ to $(-1, 0)$.

c. Do your results contradict Theorem 5? Explain.

42. Let

$$\mathbf{F}(x, y, z) = \frac{y}{(y^2 + z^2)^2}\,\mathbf{j} + \frac{z}{(y^2 + z^2)^2}\,\mathbf{k}$$

a. Show that curl $\mathbf{F} = \mathbf{0}$.

b. Is $\mathbf{F}$ conservative? Explain.

In Exercises 43–48, determine whether the statement is true or false. If it is true, explain why. If it is false, explain why or give an example that shows it is false.

43. The region $R = \{(x, y) \mid 0 < x^2 + y^2 < 1\}$ is simply-connected.

44. If $\mathbf{F}$ is a nonconservative vector field, then $\int_C \mathbf{F} \cdot d\mathbf{r} \neq 0$ whenever C is a closed path.

45. If $\mathbf{F}$ has continuous first-order partial derivatives in space and C is any smooth curve, then $\int_C \nabla f \cdot d\mathbf{r}$ depends only on the endpoints of C.

46. If $\mathbf{F} = P\mathbf{i} + Q\mathbf{j}$ is in an open connected region R and $\dfrac{\partial Q}{\partial x} = \dfrac{\partial P}{\partial y}$ for all (x, y) in R, then $\int_C \mathbf{F} \cdot d\mathbf{r} = 0$ for any smooth curve C in R.

47. If $\mathbf{F}(x, y)$ is continuous and C is a smooth curve, then $\int_C \mathbf{F} \cdot d\mathbf{r} = -\int_{-C} \mathbf{F} \cdot d\mathbf{r}$.

48. If $\mathbf{F}$ has first-order partial derivatives in a simply-connected region R, then $\int_C \mathbf{F} \cdot d\mathbf{r} = 0$ for every closed path in R.

15.5 Green's Theorem

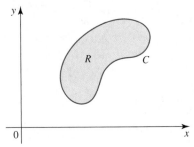

FIGURE 1

A plane region R bounded by a simple closed plane curve C

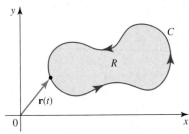

FIGURE 2

The curve C traversed in the positive or counterclockwise direction

■ Green's Theorem for Simple Regions

Green's Theorem, named after the English mathematical physicist George Green (1793–1841), relates a line integral around a simple closed plane curve C to a double integral over the plane region R bounded by C. (See Figure 1.)

Before stating Green's Theorem, however, we need to explain what is meant by the *orientation* of a simple closed curve. Suppose that C is defined by the vector function $\mathbf{r}(t)$, where $a \leq t \leq b$. Then C is traversed in the **positive** or **counterclockwise** direction if the region R is always on the left as the terminal point of $\mathbf{r}(t)$ traces the boundary curve C. (See Figure 2.)

> **THEOREM 1 Green's Theorem**
>
> Let C be a piecewise-smooth, simple closed curve that bounds a region R in the plane. If P and Q have continuous partial derivatives on an open set that contains R, then
>
> $$\oint_C P\, dx + Q\, dy = \iint_R \left[\frac{\partial Q}{\partial x} - \frac{\partial P}{\partial y} \right] dA \qquad (1)$$
>
> where the line integral over C is taken in the positive (counterclockwise) direction.

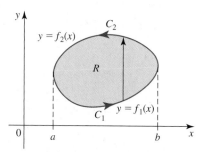

FIGURE 3
The simple region R viewed as a y-simple region

Note The notation

$$\oint_C P\,dx + Q\,dy \qquad \text{or} \qquad \oint_C P\,dx + Q\,dy$$

is sometimes used to indicate that the line integral over a simple closed curved C is taken in the positive, or counterclockwise, direction. ■

Since it is not easy to prove Green's Theorem for general regions, we will prove it only for the special case in which the region R is both a y-simple and an x-simple region. (See Section 14.2.) Such regions are called **simple** or **elementary regions.**

PROOF OF GREEN'S THEOREM FOR SIMPLE REGIONS Let R be a simple region with boundary C as shown in Figure 3. Since

$$\oint_C P\,dx + Q\,dy = \oint_C P\,dx + \oint_C Q\,dy$$

we can consider each integral on the right separately. Since R is a y-simple region, it can be described as

$$R = \{(x, y)\,|\,a \le x \le b,\, f_1(x) \le y \le f_2(x)\}$$

where f_1 and f_2 are continuous on $[a, b]$. Observe that the boundary C of R consists of the curves C_1 and C_2 that are the graphs of the functions f_1 and f_2 as shown in the figure. Therefore,

$$\oint_C P\,dx = \int_{C_1} P\,dx + \int_{C_2} P\,dx$$

where C_1 and C_2 are oriented as shown in Figure 3.

Observe that the point $(x, f_1(x))$ traces C_1 as x increases from a to b, whereas the point $(x, f_2(x))$ traces C_2 as x *decreases* from b to a. Therefore,

$$\oint_C P\,dx = \oint_{C_1} P\,dx + \oint_{C_2} P\,dx$$

$$= \int_a^b P(x, f_1(x))\,dx + \int_b^a P(x, f_2(x))\,dx$$

$$= \int_a^b P(x, f_1(x))\,dx - \int_a^b P(x, f_2(x))\,dx$$

$$= \int_a^b [P(x, f_1(x)) - P(x, f_2(x))]\,dx \tag{2}$$

Next, we find

$$\iint_R \frac{\partial P}{\partial y}\,dA = \int_a^b \int_{f_1(x)}^{f_2(x)} \frac{\partial P}{\partial y}(x, y)\,dy\,dx$$

$$= \int_a^b [P(x, f_2(x)) - P(x, f_1(x))]\,dx \tag{3}$$

where the last equality is obtained with the aid of the Fundamental Theorem of Calculus. Comparing Equation (3) with Equation (2), we see that

$$\oint_C P\,dx = -\iint_R \frac{\partial P}{\partial y}\,dA \tag{4}$$

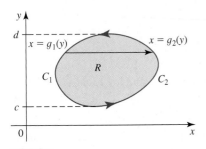

FIGURE 4
The simple region R viewed as an x-simple region

By viewing R as an x-simple region (Figure 4),

$$R = \{(x, y) \mid c \le y \le d, g_1(y) \le x \le g_2(y)\}$$

you can show in a similar manner that

$$\oint_C Q \, dy = \iint_R \frac{\partial Q}{\partial x} \, dA \tag{5}$$

(See Exercise 48.) Adding Equation (4) and Equation (5), we obtain Equation (1), the conclusion of Green's Theorem for the case of a simple region. ∎

EXAMPLE 1   Evaluate $\oint_C x^2 \, dx + (xy + y^2) \, dy$, where C is the boundary of the region R bounded by the graphs of $y = x$ and $y = x^2$ and is oriented in a positive direction.

Solution   The region R is shown in Figure 5. Observe that R is simple. Using Green's Theorem with $P(x, y) = x^2$ and $Q(x, y) = xy + y^2$, we have

$$\oint_C x^2 \, dx + (xy + y^2) \, dy = \iint_R \left[\frac{\partial Q}{\partial x} - \frac{\partial P}{\partial y} \right] dA = \int_0^1 \int_{x^2}^x (y - 0) \, dy \, dx$$

$$= \int_0^1 \left[\frac{1}{2} y^2 \right]_{y=x^2}^{y=x} dx = \frac{1}{2} \int_0^1 (x^2 - x^4) \, dx$$

$$= \frac{1}{2} \left(\frac{1}{3} x^3 - \frac{1}{5} x^5 \right) \bigg|_0^1 = \frac{1}{15}$$ ∎

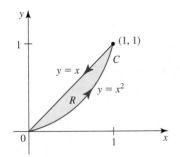

FIGURE 5
The curve C is the boundary of the region R.

EXAMPLE 2   Evaluate $\oint_C (y^2 + \tan x) \, dx + (x^3 + 2xy + \sqrt{y}) \, dy$, where C is the circle $x^2 + y^2 = 4$ and is oriented in a positive direction.

Solution   The simple region R bounded by C is the disk $R = \{(x, y) \mid x^2 + y^2 \le 4\}$ shown in Figure 6. Using Green's Theorem with $P(x, y) = y^2 + \tan x$ and $Q(x, y) = x^3 + 2xy + \sqrt{y}$, we find

$$\frac{\partial Q}{\partial x} = \frac{\partial}{\partial x} (x^3 + 2xy + \sqrt{y}) = 3x^2 + 2y \qquad \text{and} \qquad \frac{\partial P}{\partial y} = \frac{\partial}{\partial y} (y^2 + \tan x) = 2y$$

and so

$$\oint_C (y^2 + \tan x) \, dx + (x^3 + 2xy + \sqrt{y}) \, dy = \iint_R \left[\frac{\partial Q}{\partial x} - \frac{\partial P}{\partial y} \right] dA = \iint_R 3x^2 \, dA$$

$$= 3 \int_0^{2\pi} \int_0^2 (r \cos \theta)^2 r \, dr \, d\theta \qquad \text{Use polar coordinates.}$$

$$= 3 \int_0^{2\pi} \int_0^2 r^3 \cos^2 \theta \, dr \, d\theta$$

$$= 3 \int_0^{2\pi} \left[\frac{1}{4} r^4 \cos^2 \theta \right]_{r=0}^{r=2} d\theta$$

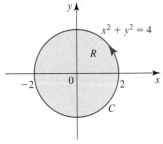

FIGURE 6
The region R is the disk bounded by the circle $x^2 + y^2 \le 4$.

$$= 12 \int_0^{2\pi} \cos^2 \theta \, d\theta$$

$$= 6 \int_0^{2\pi} (1 + \cos 2\theta) \, d\theta$$

$$= 6 \left[\theta + \frac{1}{2} \sin 2\theta \right]_0^{2\pi} = 12\pi \quad \blacksquare$$

The results obtained in Examples 1 and 2 can be verified by evaluating the given line integrals directly without the benefit of Green's Theorem, but this entails much more work than evaluating the corresponding double integrals. In certain situations, however, the opposite is true; that is, it is easier to evaluate a line integral than it is to evaluate the corresponding double integral. This fact is exploited in the following formulas based on Green's Theorem for finding the area of a plane region.

THEOREM 2   Finding Area Using Line Integrals

Let R be a plane region bounded by a piecewise-smooth simple closed curve C. Then the area of R is given by

$$A = \oint_C x \, dy = -\oint_C y \, dx = \frac{1}{2} \oint_C x \, dy - y \, dx \tag{6}$$

PROOF   Taking $P(x, y) = 0$ and $Q(x, y) = x$, Green's Theorem gives

$$\oint_C x \, dy = \iint_R \left[\frac{\partial Q}{\partial x} - \frac{\partial P}{\partial y} \right] dA = \iint_R 1 \, dA = A$$

Similarly, by taking $P(x, y) = -y$ and $Q(x, y) = 0$, we have

$$\oint_C -y \, dx = \iint_R \left[\frac{\partial Q}{\partial x} - \frac{\partial P}{\partial y} \right] dA = \iint_R 1 \, dA = A$$

Finally, with $P(x, y) = -\frac{1}{2}y$ and $Q(x, y) = \frac{1}{2}x$, we have

$$\oint_C -\frac{1}{2} y \, dx + \frac{1}{2} x \, dy = \oint_C P \, dx + Q \, dy = \iint_R \left[\frac{\partial Q}{\partial x} - \frac{\partial P}{\partial y} \right] dA = \iint_R \left(\frac{1}{2} + \frac{1}{2} \right) dA = A \quad \blacksquare$$

EXAMPLE 3   Find the area enclosed by the ellipse $\dfrac{x^2}{a^2} + \dfrac{y^2}{b^2} = 1$.

Solution   The ellipse C can be represented by the parametric equations $x = a \cos t$ and $y = b \sin t$, where $0 \le t \le 2\pi$. Also observe that the ellipse is traced in the counterclockwise direction as t increases from 0 to 2π. Using Equation (6), we have

$$A = \frac{1}{2} \oint_C x \, dy - y \, dx = \frac{1}{2} \int_0^{2\pi} (a \cos t)(b \cos t) \, dt - (b \sin t)(-a \sin t) \, dt$$

$$= \frac{ab}{2} \int_0^{2\pi} dt = \pi ab \quad \blacksquare$$

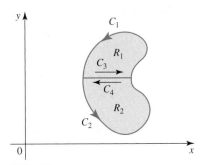

FIGURE 7
The region R is the union of two simple regions R_1 and R_2.

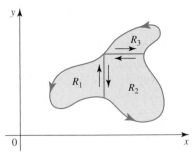

FIGURE 8
The region R is a union of three simple regions R_1, R_2, and R_3.

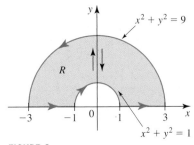

FIGURE 9
The region R is divided into two simple regions by the crosscut that lies on the y-axis.

■ Green's Theorem for More General Regions

So far, we have proved Green's Theorem for the case in which R is a simple region, but the theorem can be extended to the case in which the region R is a finite union of simple regions. For example, the region R shown in Figure 7 is not simple, but it can be written as $R = R_1 \cup R_2$, where R_1 and R_2 are both simple. The boundary of R_1 is $C_1 \cup C_3$, and the boundary of R_2 is $C_2 \cup C_4$, where C_3 and C_4 are paths along the crosscut traversed in the indicated directions.

Applying Green's Theorem to each of the regions R_1 and R_2 gives

$$\oint_{C_1 \cup C_3} P \, dx + Q \, dy = \iint_{R_1} \left[\frac{\partial Q}{\partial x} - \frac{\partial P}{\partial y} \right] dA$$

and

$$\oint_{C_2 \cup C_4} P \, dx + Q \, dy = \iint_{R_2} \left[\frac{\partial Q}{\partial x} - \frac{\partial P}{\partial y} \right] dA$$

Adding these two equations and observing that the line integrals along C_3 and C_4 cancel each other, we obtain

$$\oint_{C_1 \cup C_3} P \, dx + Q \, dy + \oint_{C_2 \cup C_4} P \, dx + Q \, dy = \oint_{C_1 \cup C_2} P \, dx + Q \, dy = \iint_{R} \left[\frac{\partial Q}{\partial x} - \frac{\partial P}{\partial y} \right] dA$$

which is Green's Theorem for the region $R = R_1 \cup R_2$ with boundary $C = C_1 \cup C_2$.

A similar argument enables us to establish Green's Theorem for the general case in which R is the union of any finite number of nonoverlapping, except perhaps for the common boundaries, simple regions (see Figure 8).

EXAMPLE 4   Evaluate $\oint_C (e^x + y^2) \, dx + (x^2 + 3xy) \, dy$, where C is the positively oriented closed curve lying on the boundary of the semiannular region R bounded by the upper semicircles $x^2 + y^2 = 1$ and $x^2 + y^2 = 9$ and the x-axis as shown in Figure 9.

Solution   The region R is not simple, but it can be divided into two simple regions by means of the crosscut that is the intersection of R and the y-axis. Also notice that in polar coordinates,

$$R = \{(r, \theta) \,|\, 1 \le r \le 3, 0 \le \theta \le \pi\}$$

Using Green's Theorem with $P(x, y) = e^x + y^2$ and $Q(x, y) = x^2 + 3xy$, we have

$$\frac{\partial Q}{\partial x} = \frac{\partial}{\partial x}(x^2 + 3xy) = 2x + 3y \qquad \text{and} \qquad \frac{\partial P}{\partial y} = \frac{\partial}{\partial y}(e^x + y^2) = 2y$$

and so

$$\oint_C (e^x + y^2) \, dx + (x^2 + 3xy) \, dy = \iint_R \left[\frac{\partial Q}{\partial x} - \frac{\partial P}{\partial y} \right] dA = \iint_R (2x + y) \, dA$$

$$= \int_0^\pi \int_1^3 (2r \cos \theta + r \sin \theta) r \, dr \, d\theta \qquad \text{Use polar coordinates.}$$

$$= \int_0^\pi (2 \cos \theta + \sin \theta) \left[\frac{1}{3} r^3 \right]_1^3 d\theta$$

$$= \frac{26}{3} \left[2 \sin \theta - \cos \theta \right]_0^\pi = \frac{52}{3} \qquad ■$$

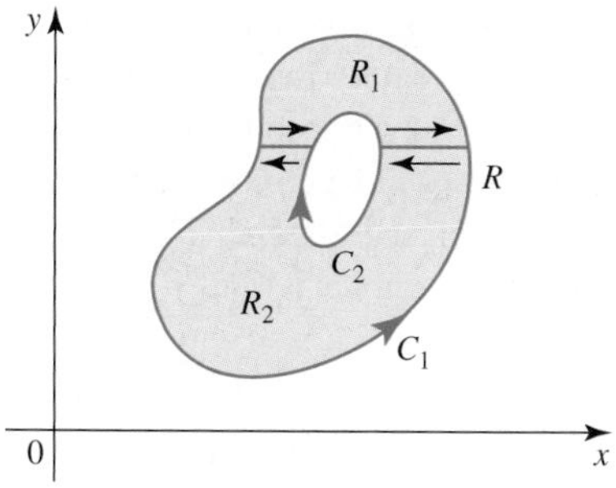

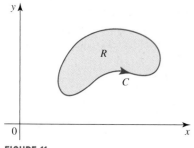

FIGURE 10

The annular region R can be divided into two simple regions using two crosscuts.

Green's Theorem can be extended to even more general regions. Recall that a region R is simply-connected if for every simple closed curve C that lies in R, the region bounded by C is also in R. Thus, as was noted earlier, a simply-connected region "has no holes." For example, a rectangle is simply-connected, but an annulus (a ring bounded by two concentric circles) is not. Also, **multiply-connected regions** may have one or more holes in them and also may have boundaries that consist of two or more simple closed curves. For example, the annular region R shown in Figure 10 has a boundary C consisting of two simple closed curves C_1 and C_2. Observe that C is traversed in the positive direction provided that C_1 is traversed in the counterclockwise direction and C_2 is traversed in the clockwise direction (so that the region R always lies to the left as the curve is traced).

The region R can be divided into two simple regions, R_1 and R_2, by means of two crosscuts, as shown in Figure 10. Applying Green's Theorem to each of these subregions of R, we obtain

$$\iint\limits_{R} \left[\frac{\partial Q}{\partial x} - \frac{\partial P}{\partial y} \right] dA = \iint\limits_{R_1} \left[\frac{\partial Q}{\partial x} - \frac{\partial P}{\partial y} \right] dA + \iint\limits_{R_2} \left[\frac{\partial Q}{\partial x} - \frac{\partial P}{\partial y} \right] dA$$

$$= \iint\limits_{\partial R_1} P \, dx + Q \, dy + \iint\limits_{\partial R_2} P \, dx + Q \, dy$$

where ∂R_1 and ∂R_2 denote the boundaries of R_1 and R_2, respectively. Since the line integrals along the crosscuts are traversed in opposite directions, they cancel out, and we have

$$\iint\limits_{R} \left[\frac{\partial Q}{\partial x} - \frac{\partial P}{\partial y} \right] dA = \oint_{C_1} P \, dx + Q \, dy + \oint_{C_2} P \, dx + Q \, dy = \oint_{C} P \, dx + Q \, dy$$

which is Green's Theorem for the region R. Observe that the second line integral above is traversed in the clockwise direction.

EXAMPLE 5   Let C be a smooth, simple, closed curve that does not pass through the origin. Show that

$$\oint_{C} -\frac{y}{x^2 + y^2} \, dx + \frac{x}{x^2 + y^2} \, dy$$

is equal to zero if C does not enclose the origin but is equal to 2π if C encloses the origin.

Solution   Suppose that C does not enclose the origin. (See Figure 11.) Using Green's Theorem with $P(x, y) = -y/(x^2 + y^2)$ and $Q(x, y) = x/(x^2 + y^2)$ so that

$$\frac{\partial Q}{\partial x} = \frac{(x^2 + y^2)(1) - x(2x)}{(x^2 + y^2)^2} = \frac{y^2 - x^2}{(x^2 + y^2)^2}$$

and

$$\frac{\partial P}{\partial y} = \frac{(x^2 + y^2)(-1) - (-y)(2y)}{(x^2 + y^2)^2} = \frac{y^2 - x^2}{(x^2 + y^2)^2} = \frac{\partial Q}{\partial x}$$

we obtain

$$\oint_{C} -\frac{y}{x^2 + y^2} \, dx + \frac{x}{x^2 + y^2} \, dy = \iint\limits_{R} \left[\frac{\partial Q}{\partial x} - \frac{\partial P}{\partial y} \right] dA = \iint\limits_{R} 0 \, dA = 0$$

Here, R denotes the region enclosed by C.

FIGURE 11

C does not enclose the origin.

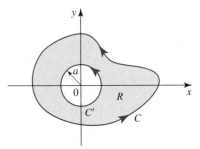

FIGURE 12
C encloses the origin.

Next, suppose that C *encloses* the origin. Since P and Q are *not* continuous in the region enclosed by C, Green's Theorem is not directly applicable. Let C' be a counterclockwise-oriented circle with center at the origin and radius a chosen small enough so that C' lies inside C. (See Figure 12.) Then both P and Q have continuous partial derivatives in the annular region bounded by C and C'. Applying Green's Theorem to the multiply-connected region R with its positively oriented boundary $C \cup (-C')$, we obtain

$$\oint_C P\,dx + Q\,dy + \oint_{-C'} P\,dx + Q\,dy = \iint_R \left[\frac{\partial Q}{\partial x} - \frac{\partial P}{\partial y}\right] dA = \iint_R 0\,dA = 0$$

or, upon reversing the direction of traversal of the second line integral,

$$\oint_C P\,dx + Q\,dy - \oint_{C'} P\,dx + Q\,dy = 0$$

Therefore,

$$\oint_C P\,dx + Q\,dy = \oint_{C'} P\,dx + Q\,dy$$

Up to this point, we have shown that the required line integral is equal to the line integral taken over the circle C' in the counterclockwise direction. To evaluate this integral, we represent the circle by the parametric equations $x = a\cos t$ and $y = a\sin t$, where $0 \le t \le 2\pi$. We obtain

$$\oint_{C'} -\frac{y}{x^2 + y^2}\,dx + \frac{x}{x^2 + y^2}\,dy = \int_0^{2\pi} -\frac{(a\sin t)(-a\sin t)}{(a\cos t)^2 + (a\sin t)^2}\,dt + \frac{(a\cos t)(a\cos t)}{(a\cos t)^2 + (a\sin t)^2}\,dt$$

$$= \int_0^{2\pi} 1\,dt = 2\pi$$

Therefore,

$$\oint_C -\frac{y}{x^2 + y^2}\,dx + \frac{x}{x^2 + y^2}\,dy = 2\pi \qquad \blacksquare$$

■ Vector Form of Green's Theorem

The vector form of Green's Theorem has two useful versions: one involving the curl of a vector field and another involving the divergence of a vector field.

Suppose that the curve C, the plane region R, and the functions P and Q satisfy the hypothesis of Green's Theorem. Let $\mathbf{F} = P\mathbf{i} + Q\mathbf{j}$ be a vector field. Then

$$\oint_C \mathbf{F} \cdot \mathbf{T}\,ds = \oint_C P\,dx + Q\,dy$$

Recalling that P and Q are functions of x and y, we have

$$\text{curl } \mathbf{F} = \nabla \times \mathbf{F} = \begin{vmatrix} \mathbf{i} & \mathbf{j} & \mathbf{k} \\ \dfrac{\partial}{\partial x} & \dfrac{\partial}{\partial y} & \dfrac{\partial}{\partial z} \\ P & Q & 0 \end{vmatrix} = \left(\frac{\partial Q}{\partial x} - \frac{\partial P}{\partial y}\right)\mathbf{k}$$

Remember that P and Q are functions of x and y.

so

$$(\text{curl } \mathbf{F}) \cdot \mathbf{k} = \left(\frac{\partial Q}{\partial x} - \frac{\partial P}{\partial y}\right)\mathbf{k} \cdot \mathbf{k} = \frac{\partial Q}{\partial x} - \frac{\partial P}{\partial y}$$

Therefore, Green's Theorem can be written in the vector form

$$\oint_C \mathbf{F} \cdot \mathbf{T}\, ds = \iint_R \text{curl } \mathbf{F} \cdot \mathbf{k}\, dA \qquad (7)$$

Equation (7) states that the line integral of the tangential component of $\mathbf{F}$ around a closed curve C is equal to the double integral of the normal component to R of curl $\mathbf{F}$ over the region R enclosed by C.

Next, let the curve C be represented by the vector equation $\mathbf{r}(t) = x(t)\mathbf{i} + y(t)\mathbf{j}$, $a \le t \le b$. Then the outer unit normal vector to C is

$$\mathbf{n}(t) = \frac{y'(t)}{|\mathbf{r}'(t)|}\mathbf{i} - \frac{x'(t)}{|\mathbf{r}'(t)|}\mathbf{j}$$

which you can verify by showing that $\mathbf{n}(t) \cdot \mathbf{T}(t) = 0$, where

$$\mathbf{T}(t) = \frac{x'(t)}{|\mathbf{r}'(t)|}\mathbf{i} + \frac{y'(t)}{|\mathbf{r}'(t)|}\mathbf{j}$$

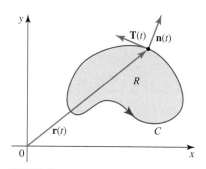

FIGURE 13
$\mathbf{n}(t)$ is the outer normal vector to C.

is the unit tangent vector to C. (See Figure 13.) We have

$$\oint_C \mathbf{F} \cdot \mathbf{n}\, ds = \int_a^b (\mathbf{F} \cdot \mathbf{n})(t)|\mathbf{r}'(t)|\, dt$$

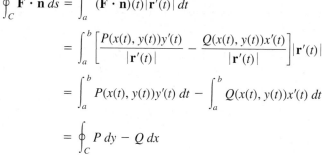

$$= \int_a^b \left[\frac{P(x(t), y(t))y'(t)}{|\mathbf{r}'(t)|} - \frac{Q(x(t), y(t))x'(t)}{|\mathbf{r}'(t)|}\right]|\mathbf{r}'(t)|\, dt$$

$$= \int_a^b P(x(t), y(t))y'(t)\, dt - \int_a^b Q(x(t), y(t))x'(t)\, dt$$

$$= \oint_C P\, dy - Q\, dx$$

But by Green's Theorem,

$$\oint_C P\, dy - Q\, dx = \iint_R \left[\frac{\partial}{\partial x}(P) - \frac{\partial}{\partial y}(-Q)\right] dA$$

$$= \iint_R \left(\frac{\partial P}{\partial x} + \frac{\partial Q}{\partial y}\right) dA$$

Observing that the integrand of the last integral is just the divergence of $\mathbf{F}$, we obtain the second vector form of Green's Theorem:

$$\oint_C \mathbf{F} \cdot \mathbf{n}\, ds = \iint_R \text{div } \mathbf{F}\, dA \qquad (8)$$

Equation (8) states that the line integral of the normal component of $\mathbf{F}$ around a closed curve C is equal to the double integral of the divergence of $\mathbf{F}$ over R.

15.5   CONCEPT QUESTIONS

1. State Green's Theorem.

2. Write three line integrals that give the area of a region bounded by a piecewise smooth curve C.

15.5   EXERCISES

In Exercises 1–4, evaluate the line integral (a) *directly and* (b) *by using Green's Theorem, where C is positively oriented.*

1. $\oint_C 2xy \, dx + 3xy^2 \, dy$, where C is the square with vertices $(0, 0)$, $(1, 0)$, $(1, 1)$, and $(0, 1)$

2. $\oint_C x^2 \, dx + xy \, dy$, where C is the triangle with vertices $(0, 0)$, $(1, 0)$, and $(0, 1)$

3. $\oint_C y^2 \, dx + (x^2 + 2xy) \, dy$, where C is the boundary of the region bounded by the graphs of $y = x$ and $y = x^3$ lying in the first quadrant

4. $\oint_C 2x \, dx - 3y \, dy$, where C is the circle $x^2 + y^2 = a^2$

In Exercises 5–16, use Green's Theorem to evaluate the line integral along the positively oriented closed curve C.

5. $\oint_C x^3 \, dx + xy \, dy$, where C is the triangle with vertices $(0, 0)$, $(1, 1)$, and $(0, 1)$

6. $\oint_C (x^2 + y^2) \, dx - 2xy \, dy$, where C is the square with vertices $(\pm 1, \pm 1)$

7. $\oint_C (x^2 y + x^3) \, dx + 2xy \, dy$, where C is the boundary of the region bounded by the graphs of $y = x$ and $y = x^2$

8. $\oint_C (-y^3 + \cos x) \, dx + e^{y^2} \, dy$, where C is the boundary of the region bounded by the parabolas $y = x^2$ and $x = y^2$

9. $\oint_C (y^2 + \cos x) \, dx + (x - \tan^{-1} y) \, dy$, where C is the boundary of the region bounded by the graphs of $y = 4 - x^2$ and $y = 0$

10. $\oint_C x^2 y \, dx + y^3 \, dy$, where C consists of the line segment from $(-1, 0)$ to $(1, 0)$ and the upper half of the circle $x^2 + y^2 = 1$

11. $\oint_C (x^2 - y) \, dx + \sqrt{1 + y^2} \, dy$, where C is the astroid $x^{2/3} + y^{2/3} = a^{2/3}$

12. $\oint_C 6xy \, dx + (3x^2 + \ln(1 + y)) \, dy$, where C is the cardioid $r = 1 + \cos \theta$

13. $\oint_C (x + e^x \sin y) \, dx + (x + e^x \cos y) \, dy$, where C is the ellipse $\dfrac{x^2}{9} + \dfrac{y^2}{4} = 1$

14. $\oint_C \dfrac{y}{1 + x^2} \, dx + (x + \tan^{-1} x) \, dy$, where C is the right-hand loop of the lemniscate $r^2 = \cos 2\theta$

15. $\oint_C (-y \, dx + x \, dy)$, where C is the boundary of the annular region formed by circles $x^2 + y^2 = 1$ and $x^2 + y^2 = 4$

16. $\oint_C 3x^2 y \, dx + (x^3 + x) \, dy$, where C is the boundary of the region lying between the ellipse $\dfrac{x^2}{4} + \dfrac{y^2}{9} = 1$ and the circle $x^2 + y^2 = 1$

17. Use Green's Theorem to find the work done by the force $\mathbf{F}(x, y) = (x^2 - y^2)\mathbf{i} + 2xy\mathbf{j}$ in moving a particle in the positive direction once around the triangle with vertices $(0, 0)$, $(1, 0)$, and $(0, 1)$.

18. Use Green's Theorem to find the work done by the force $\mathbf{F}(x, y) = 3y\mathbf{i} - 2x\mathbf{j}$ in moving a particle once around the ellipse $\dfrac{x^2}{4} + \dfrac{y^2}{9} = 1$ in the clockwise direction.

In Exercises 19–22, use one of the formulas on page 1271 to find the area of the indicated region.

19. The region enclosed by the astroid $x^{2/3} + y^{2/3} = a^{2/3}$

20. The region bounded by an arc of the cycloid $x = a(t - \sin t)$, $y = a(1 - \cos t)$, and the x-axis

21. The region enclosed by the curve $x = a \sin t$ and $y = b \sin 2t$

22. The region enclosed by the curve $x = \cos t$ and $y = 4 \sin^3 t$, where $0 \le t \le 2\pi$

23. a. Plot the curve C defined by $x = t(1 - t^2)$ and $y = t^2(1 - t^3)$, where $0 \le t \le 1$.

b. Find the area of the region enclosed by the curve C.

24. a. Plot the deltoid defined by $x = \frac{1}{4}(2 \cos t + \cos 2t)$ and $y = \frac{1}{4}(2 \sin t - \sin 2t)$, where $0 \le t \le 2\pi$.

b. Find the area of the region enclosed by the deltoid.

25. Swallowtail Catastrophe

a. Plot the swallowtail catastrophe defined by $x = 2t(1 - t^2)$ and $y = \frac{1}{2}t^2(3t^2 - 2)$, where $-1 \le t \le 1$.

b. Find the area of the region enclosed by the swallowtail catastrophe.

26. Refer to the following figure. Suppose that $\displaystyle\int_{C_2} \mathbf{F} \cdot d\mathbf{r} = 3\pi$, where $\mathbf{F}(x, y) = P(x, y)\mathbf{i} + Q(x, y)\mathbf{j}$, and that $\left(\dfrac{\partial Q}{\partial x} - \dfrac{\partial P}{\partial y} \right) = 6$ for all (x, y) in the region R bounded by the circles C_1 and C_2, and oriented in a counterclockwise direction. Use Green's Theorem to find $\displaystyle\int_{C_1} \mathbf{F} \cdot d\mathbf{r}$.

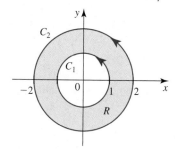

27. Refer to the figure below. Suppose that $\displaystyle\oint_{C_2} \mathbf{F} \cdot d\mathbf{r} = 2\pi$ and $\displaystyle\oint_{C_3} \mathbf{F} \cdot d\mathbf{r} = 3\pi$, where $\mathbf{F}(x, y) = P(x, y)\mathbf{i} + Q(x, y)\mathbf{j}$, and that $\left(\dfrac{\partial Q}{\partial x} - \dfrac{\partial P}{\partial y} \right) = 6$ for all (x, y) in the region R lying inside the curve C_1 and outside the curves C_2 and C_3. Use Green's Theorem to find $\displaystyle\oint_{C_1} \mathbf{F} \cdot d\mathbf{r}$.

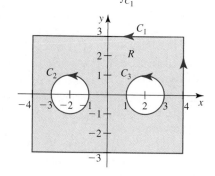

28. Evaluate $\int_{C_1} (x^2 + 2y)\, dx + \left(4x + e^{y^2} \right) dy$, where C_1 is the semi-elliptical path from A to B shown in the figure.

Hint: Use Green's Theorem, noting that $C_1 \cup C_2$, where C_2 is the straight path from $(-3, 0)$ to $(3, 0)$, is a closed path.

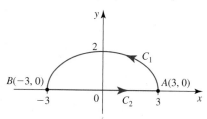

29. Evaluate $\int_{C_1}(x^2 + 2y)\, dx + (3x - \sinh y)\, dy$, where C_1 is the path $ABCDEF$ shown in the figure.

Hint: See the hint in Exercise 28.

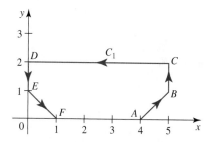

30. a. Let C be the line segment joining the points (x_1, y_1) and (x_2, y_2). Show that $\int_C -y\, dx + x\, dy = x_1 y_2 - x_2 y_1$.

b. Use the result of part (a) to show that the area of a polygon with vertices $(x_1, y_1), (x_2, y_2), \ldots, (x_n, y_n)$ (appearing in the counterclockwise order) is

$$A = \frac{1}{2}[(x_1 y_2 - x_2 y_1) + (x_2 y_3 - x_3 y_2) + \cdots$$
$$+ (x_{n-1} y_n - x_n y_{n-1}) + (x_n y_1 - x_1 y_n)]$$

In Exercises 31 and 32, use the result of Exercise 30 to find the area of the shaded region.

31.

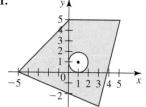

32.

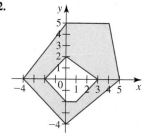

In Exercises 33 and 34, use the result of Exercise 30 to find the area of the polygon.

33. Pentagon with vertices $(0, 0)$, $(2, 0)$, $(3, 1)$, $(1, 3)$, and $(-1, 1)$.

34. Hexagon with vertices $(0, 0)$, $(3, 0)$, $(4, 1)$, $(2, 4)$, $(0, 3)$, and $(-2, 1)$.

35. Let R be a plane region of area A bounded by a piecewise-smooth simple closed curve C. Use Green's Theorem to show that the centroid of R is $(\bar{x}, \bar{y})$, where

$$\bar{x} = \frac{1}{2A} \oint_C x^2 \, dy \qquad \bar{y} = -\frac{1}{2A} \oint_C y^2 \, dx$$

In Exercises 36 and 37, use the result of Exercise 35 to find the centroid of the region.

36. The triangle with vertices $(0, 0)$, $(1, 0)$, and $(1, 1)$.

37. The region bounded by the graphs of $y = 0$ and $y = 9 - x^2$.

38. A plane lamina with constant density ρ has the shape of a region bounded by a piecewise-smooth simple closed curve C. Show that its moments of inertia about the axes are

$$I_x = -\frac{\rho}{3} \oint_C y^3 \, dx \qquad I_y = \frac{\rho}{3} \oint_C x^3 \, dy$$

39. Use the result of Exercise 38 to find the moment of inertia of a circular lamina of radius a and constant density ρ about a diameter.

40. Show that if f and g have continuous derivatives, then

$$\oint_C f(x) \, dx + g(y) \, dy = 0$$

for every piecewise-smooth simple closed curve C.

41. Let C be a piecewise-smooth simple closed curve that encloses a region R of area A. Show that

$$\oint_C (ay + b) \, dx + (cx + d) \, dy = (c - a)A$$

42. Let C be a piecewise-smooth simple closed curve that does not pass through the origin. Evaluate

$$\oint_C \frac{x}{x^2 + y^2} \, dx + \frac{y}{x^2 + y^2} \, dy$$

(a) where C does not enclose the origin and (b) where C encloses the origin.

43. Let $P(x, y) = -\dfrac{y}{x^2 + y^2}$ and $Q(x, y) = \dfrac{x}{x^2 + y^2}$.

   a. Show that $\oint_C (P \, dx + Q \, dy) \neq 0$, where C is the circle of radius 1 centered at the origin.

   b. Verify that $\dfrac{\partial P}{\partial y} = \dfrac{\partial Q}{\partial x}$.

   c. Do parts (a) and (b) contradict each other? Explain.

44. Let R be the region bounded by the circles of radius 1 and 3 centered at the origin, and let C be the circle of radius 2 centered at the origin described by $\mathbf{r}(t) = 2 \cos t \mathbf{i} + 2 \sin t \mathbf{j}$, where $0 \leq t \leq 2\pi$. Let

$$P(x, y) = -\frac{y}{x^2 + y^2} \qquad \text{and} \qquad Q(x, y) = \frac{x}{x^2 + y^2}$$

   a. Show that $\dfrac{\partial P}{\partial y} = \dfrac{\partial Q}{\partial x}$ in R but $\oint_C P \, dx + Q \, dy \neq 0$.

   b. Does this contradict Green's Theorem? Explain.

45. a. Use Green's Theorem to show that

$$\oint_C (\cos x + x^3 y) \, dx + (x^4 + e^y) \, dy = 0$$

   where C is the boundary of the square with vertices $(-1, -1)$, $(1, -1)$, $(1, 1)$, and $(-1, 1)$.

   b. Note that

$$\frac{\partial}{\partial x} (x^4 + e^y) \neq \frac{\partial}{\partial y} (\cos x + x^3 y)$$

   Does this contradict Theorem 4 of Section 15.4? Explain.

   c. Evaluate the line integral of part (a), taking C to be the boundary of the square with vertices $(0, 0)$, $(1, 0)$, $(1, 1)$, and $(0, 1)$.

46. Can Green's Theorem be applied to evaluate

$$\oint_C \frac{x}{\sqrt{(x - 2)^2 + y^2}} \, dx + \frac{y}{\sqrt{(x - 2)^2 + y^2}} \, dy$$

where C is the circle of radius 1 centered at the origin? Explain.

47. Show that if $P(y)$ and $Q(x)$ have continuous derivatives, then

$$\oint_C P(y) \, dx + Q(x) \, dy = 2\big[Q(t) + P(t)\big]_{t=-1}^{t=1}$$

where C is the rectangular path that is traced in a counterclockwise direction with vertices $(-1, -1)$, $(1, -1)$, $(1, 1)$, and $(-1, 1)$.

48. Refer to the proof of Green's Theorem. Show that by viewing R as an x-simple region, we have

$$\oint_C Q \, dy = \iint_R \frac{\partial Q}{\partial x} \, dA$$

In Exercises 49–51, determine whether the statement is true or false. If it is true, explain why. If it is false, explain why or give an example that shows it is false.

49. If a and b are constants, then $\oint_C a \, dx + b \, dy \neq 0$, where C is a simple closed curve.

50. If C is a piecewise-smooth simple closed curve that bounds a region R in the plane, then $\oint_C xy^2 \, dx + (x^2 y + x) \, dy$ is equal to the area of R.

51. The work done by the force field $\mathbf{F}(x, y) = -\frac{1}{2} y \mathbf{i} + \frac{1}{2} x \mathbf{j}$ on a particle that moves once around a piecewise-smooth simple closed curve in a counterclockwise direction is numerically equal to the area of the region bounded by the curve.

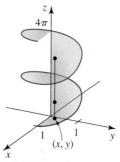

FIGURE 1
The helicoid shown here
is *not* the graph of a
function $z = f(x, y)$.

■ Why We Use Parametric Surfaces

In Chapter 13 we studied surfaces that are graphs of functions of two variables. However, not every surface is the graph of a function $z = f(x, y)$. Consider, for example, the *helicoid* shown in Figure 1. Observe that the point (x, y) in the xy-plane is associated with *more than one* point on the helicoid, so this surface cannot be the graph of a function $z = f(x, y)$.

Just as we found it useful to describe a curve in the plane (and in space) as the image of a line under a vector-valued function **r** rather than as the graph of a function, we will now see that a similar situation exists for surfaces. Instead of a single parameter, however, we will use two parameters and view a surface in space as the image of a plane region. More specifically, we have the following.

DEFINITION **Parametric Surface**

Let

$$\mathbf{r}(u, v) = x(u, v)\mathbf{i} + y(u, v)\mathbf{j} + z(u, v)\mathbf{k}$$

be a vector-valued function defined for all points (u, v) in a region D in the uv-plane. The set of all points (x, y, z) in R^3 satisfying the **parametric equations**

$$x = x(u, v), \qquad y = y(u, v), \qquad z = z(u, v)$$

as (u, v) ranges over D is called a **parametric surface S represented by r.** The region D is called the **parameter domain.**

Thus, as (u, v) ranges over D, the tip of the vector $\mathbf{r}(u, v)$ traces out the surface S (see Figure 2). Put another way, we can think of **r** as mapping each point (u, v) in D onto a point $(x(u, v), y(u, v), z(u, v))$ on S in such a way that the plane region D is bent, twisted, stretched, and/or shrunk to yield the surface S.

FIGURE 2
The function **r** maps D
onto the surface S.

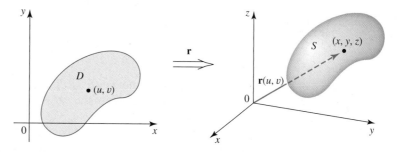

EXAMPLE 1 Identify and sketch the surface represented by

$$\mathbf{r}(u, v) = 2 \cos u \mathbf{i} + 2 \sin u \mathbf{j} + v \mathbf{k}$$

with parameter domain $D = \{(u, v) \mid 0 \le u \le 2\pi, 0 \le v \le 3\}$.

Solution The parametric equations for the surface are

$$x = 2 \cos u, \qquad y = 2 \sin u, \qquad z = v$$

Eliminating the parameters u and v in the first two equations, we obtain

$$x^2 + y^2 = 4 \cos^2 u + 4 \sin^2 u = 4$$

Observe that the variable z is missing in this equation, so it represents a cylinder with the z-axis as its axis. (See Section 11.6.) Furthermore, the trace in the xy-plane is a circle of radius 2, and we conclude that the cylinder is a circular cylinder. Finally, because $0 \leq v \leq 3$, the third equation $z = v$ tells us that $0 \leq z \leq 3$. Thus, the required surface is the truncated cylinder shown in Figure 3.

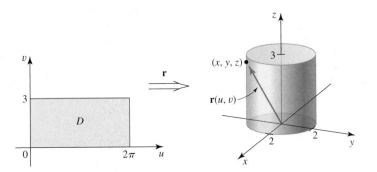

FIGURE 3
The function **r** "bends" the rectangular region D into a cylinder.

There is another way of visualizing the way **r** maps the domain D onto a surface S. If we fix u by setting $u = u_0$, where u_0 is a constant, and allow v to vary so that the points (u_0, v) lie in D, then we obtain a vertical line segment L_1 lying in D. When restricted to L_1, the function **r** becomes a function involving one parameter v whose domain is the parameter interval L_1. Therefore, $\mathbf{r}(u_0, v)$ maps L_1 onto a curve C_1 lying on S (see Figure 4).

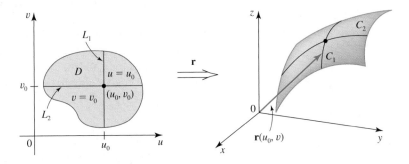

FIGURE 4
r maps L_1 onto C_1 and L_2 onto C_2.

Similarly, by holding v fixed, say, $v = v_0$, where v_0 is a constant, the tip of the resulting vector $\mathbf{r}(u, v_0)$ traces the curve C_2 as u is allowed to assume values in the parameter interval L_2. The curves C_1 and C_2 are called **grid curves.**

By way of illustration, if we set $u = u_0$ in Example 1, then both $x = 2 \cos u_0$ and $y = 2 \sin u_0$ are constant. So the vertical line $u = u_0$ is mapped onto the vertical line segment $(2 \cos u_0, 2 \sin u_0, v)$, $0 \leq v \leq 3$. Similarly, you can verify that a horizontal line segment $v = v_0$ in D is mapped onto a circle on the cylinder at a height of v_0 units from the xy-plane.

EXAMPLE 2 Use a computer algebra system (CAS) to generate the surface represented by

$$\mathbf{r}(u, v) = \sin u \cos v \mathbf{i} + \sin u \sin v \mathbf{j} + \cos u \mathbf{k}$$

with parameter domain $D = \{(u, v) \mid 0 \le u \le \pi, 0 \le v \le 2\pi\}$. Identify the curves on the surface that correspond to the curves with u held constant and those with v held constant.

Solution The required surface is the unit sphere centered at the origin. (See Figure 5a.) You can verify that this is the case by eliminating u and v in the parametric equations

$$x = \sin u \cos v, \qquad y = \sin u \sin v, \qquad z = \cos u$$

to obtain the rectangular equation $x^2 + y^2 + z^2 = 1$ for the sphere. Fixing $u = u_0$, where u_0 is a constant, leads to the equations

$$x = \sin u_0 \cos v, \qquad y = \sin u_0 \sin v, \qquad z = \cos u_0$$

We have

$$x^2 + y^2 = \sin^2 u_0 \cos^2 v + \sin^2 u_0 \sin^2 v$$
$$= \sin^2 u_0 (\cos^2 v + \sin^2 v) = \sin^2 u_0$$

The system of equations

$$\left. \begin{array}{r} x^2 + y^2 = \sin^2 u_0 \\ z = \cos u_0 \end{array} \right\}$$

for a fixed u_0 lying in $[0, \pi]$ or, equivalently, the vector-valued function

$$\mathbf{r}(u_0, v) = \sin u_0 \cos v \mathbf{i} + \sin u_0 \sin v \mathbf{j} + \cos u_0 \mathbf{k}$$

represents a circle of radius $\sin u_0$ on the sphere that is parallel to the xy-plane. Thus, if we think of the sphere as a globe then the horizontal line segments in the domain of $\mathbf{r}$ are mapped onto the latitudinal lines, or *parallels*. (See Figure 5b.) Similarly, we can show that the vertical line segments in the domain of $\mathbf{r}$ with $v = v_0$, where v_0 is a constant, are mapped by

$$\mathbf{r}(u, v_0) = \sin u \cos v_0 \mathbf{i} + \sin u \sin v_0 \mathbf{j} + \cos u \mathbf{k}$$

onto the *meridians of longitude*—great circles on the surface of the globe passing through the poles.

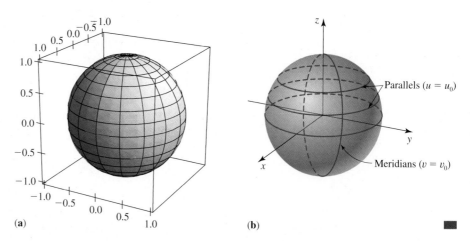

FIGURE 5 (a) (b)

Finding Parametric Representations of Surfaces

We now turn our attention to finding vector-valued function representations of surfaces. We begin by showing that if a surface is the graph of a function $f(x, y)$, then it has a simple parametric representation.

EXAMPLE 3

a. Find a parametric representation for the graph of a function $f(x, y)$.
b. Use the result of part (a) to find a parametric representation for the elliptic paraboloid $z = 4x^2 + y^2$.

Solution

a. Suppose that S is the graph of $z = f(x, y)$ defined on a domain D in the xy-plane. (See Figure 6.) We simply pick x and y to be the parameters; in other words, we write the desired parametric equations as

$$x = x(u, v) = u, \qquad y = y(u, v) = v, \qquad z = z(u, v) = f(u, v)$$

and take the domain of f to be the parameter domain. Equivalently, we obtain the vector-valued representation by writing

$$\mathbf{r}(u, v) = u\mathbf{i} + v\mathbf{j} + f(u, v)\mathbf{k}$$

b. The surface is the graph of the function $f(x, y) = 4x^2 + y^2$. So we can let x and y be the parameters. Thus, the required parametric equations are

$$x = u, \qquad y = v, \qquad z = 4u^2 + v^2$$

and the corresponding vector-valued function is

$$\mathbf{r}(u, v) = u\mathbf{i} + v\mathbf{j} + (4u^2 + v^2)\mathbf{k}$$

The parameter domain is $D = \{(u, v) \mid -\infty < u < \infty, -\infty < v < \infty\}$. ■

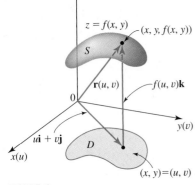

FIGURE 6
The vector $\mathbf{r}(u, v) = u\mathbf{i} + v\mathbf{j} + f(u, v)\mathbf{k}$ by the rule for vector addition.

EXAMPLE 4   Find a parametric representation for the cone $x = \sqrt{y^2 + z^2}$.

Solution   The surface is the graph of the function $f(y, z) = \sqrt{y^2 + z^2}$. So we can let y and z be the parameters. Thus, the required parametric equations are

$$x = \sqrt{u^2 + v^2}, \qquad y = u, \qquad z = v$$

and the corresponding vector-valued function is

$$\mathbf{r}(u, v) = \sqrt{u^2 + v^2}\,\mathbf{i} + u\mathbf{j} + v\mathbf{k}$$

The parameter domain is $D = \{(u, v) \mid -\infty < u < \infty, -\infty < v < \infty\}$. ■

EXAMPLE 5

a. Find a parametric representation of the plane that passes through the point P_0 with position vector $\mathbf{r_0}$ and contains two nonparallel vectors $\mathbf{a}$ and $\mathbf{b}$.
b. Using the result of part (a), find a parametric representation of the plane passing through the point $P_0(3, -1, 1)$ and containing the vectors $\mathbf{a} = -2\mathbf{i} + 5\mathbf{j} + \mathbf{k}$ and $\mathbf{b} = -3\mathbf{i} + 2\mathbf{j} + 3\mathbf{k}$. (This is the plane in Example 6 in Section 11.5.)

Solution

a. Let P be a point lying on the plane, and let $\mathbf{r} = \overline{OP}$. Since $\overline{P_0P}$ lies in the plane determined by $\mathbf{a}$ and $\mathbf{b}$, there exist real numbers u and v such that $\overline{P_0P} = u\mathbf{a} + v\mathbf{b}$. (See Figure 7.) Furthermore, we see that $\mathbf{r} = \mathbf{r_0} + \overline{P_0P} = \mathbf{r_0} + u\mathbf{a} + v\mathbf{b}$. Finally, since any point on the plane is located at the tip of $\mathbf{r}$ for an appropriate choice of u and v, we see that the required representation is

$$\mathbf{r}(u, v) = \mathbf{r_0} + u\mathbf{a} + v\mathbf{b}$$

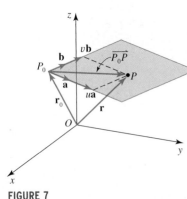

FIGURE 7
$\mathbf{r} = \mathbf{r_0} + \overline{P_0P} = \mathbf{r_0} + u\mathbf{a} + v\mathbf{b}$

The parameter domain is $D = \{(u, v) \mid -\infty < u < \infty, -\infty < v < \infty\}$.

b. The required representation is

$$\mathbf{r}(u, v) = (3\mathbf{i} - \mathbf{j} + \mathbf{k}) + u(-2\mathbf{i} + 5\mathbf{j} + \mathbf{k}) + v(-3\mathbf{i} + 2\mathbf{j} + 3\mathbf{k})$$

$$= (-2u - 3v + 3)\mathbf{i} + (5u + 2v - 1)\mathbf{j} + (u + 3v + 1)\mathbf{k}$$

with domain $D = \{(u, v) \mid -\infty < u < \infty, -\infty < v < \infty\}$.

Note The representation in Example 5b is by no means unique. For example, an equation of the plane in question is $13x + 3y + 11z = 47$. (See Example 6 in Section 11.5.) Solving this equation for z in terms of x and y, we obtain $z = f(x, y) = \frac{1}{11}(47 - 13x - 3y)$. Thus, the plane is the graph of the function f, and this observation leads us to the representation

$$\mathbf{r}(u, v) = u\mathbf{i} + v\mathbf{j} + \left(\frac{47 - 13u - 3v}{11}\right)\mathbf{k}$$

The next two examples involve surfaces that are not graphs of functions.

EXAMPLE 6 Find a parametric representation for the cone $x^2 + y^2 = z^2$.

Solution The cone has a simple representation $r^2 = z^2$ in cylindrical coordinates. This suggests that we choose r and θ as parameters. Writing u for r and v for θ, we have

$$x = u \cos v, \qquad y = u \sin v, \qquad z = u$$

as the required parametric equations. In vector form we have

$$\mathbf{r}(u, v) = u \cos v\mathbf{i} + u \sin v\mathbf{j} + u\mathbf{k}$$

EXAMPLE 7 Find a parametric representation for the helicoid shown in Figure 1.

Solution Recall that the parametric equations for a helix are

$$x = a \cos \theta, \qquad y = a \sin \theta, \qquad z = \theta$$

where θ and z are in cylindrical coordinates. This suggests that we let u denote r and v denote θ. Then the parametric equations for the helicoid are

$$x = u \cos v, \qquad y = u \sin v, \qquad z = v$$

with parameter domain $D = \{(u, v) \mid -1 \leq u \leq 1, 0 \leq v \leq 4\pi\}$. In vector form we have

$$\mathbf{r}(u, v) = u \cos v\mathbf{i} + u \sin v\mathbf{j} + v\mathbf{k}$$

We now turn our attention to finding the parametric representation for surfaces of revolution. Suppose that a surface S is obtained by revolving the graph of the function $y = f(x)$ for $a \leq x \leq b$ about the x-axis, where $f(x) \geq 0$. (See Figure 8.) Letting u denote x and v denote the angle shown in the figure, we see that if (x, y, z) is any point on S, then

$$x = u, \qquad y = f(u) \cos v, \qquad z = f(u) \sin v \qquad \text{(1)}$$

or, equivalently,

$$r(u, v) = u\mathbf{i} + f(u) \cos v\mathbf{j} + f(u) \sin v\mathbf{k}$$

The parameter domain is $D = \{(u, v) \mid a \leq u \leq b, 0 \leq v \leq 2\pi\}$.

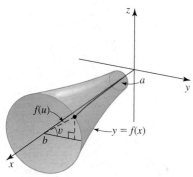

FIGURE 8
S is obtained by revolving the graph of f between $x = a$ and $x = b$ about the x-axis.

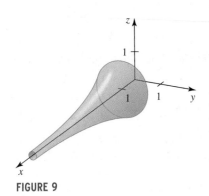

FIGURE 9
Gabriel's Horn

EXAMPLE 8 **Gabriel's Horn** Find a parametric representation for the surface obtained by revolving the graph of $f(x) = 1/x$, where $1 \leq x < \infty$, about the x-axis.

Solution Using Equation (1), we obtain the parametric equations

$$x = u, \qquad y = \frac{1}{u} \cos v, \qquad z = \frac{1}{u} \sin v$$

with parametric domain $D = \{(u, v) \mid 1 \leq u < \infty, 0 \leq v \leq 2\pi\}$. The resulting surface is a portion of Gabriel's Horn as shown in Figure 9. ∎

■ Tangent Planes to Parametric Surfaces

Suppose that S is a parametric surface represented by the vector function

$$\mathbf{r}(u, v) = x(u, v)\mathbf{i} + y(u, v)\mathbf{j} + z(u, v)\mathbf{k}$$

and P_0 is a point on the surface S represented by the vector $\mathbf{r}(u_0, v_0)$, where (u_0, v_0) is a point in the parameter domain D of $\mathbf{r}$. If we fix u by putting $u = u_0$ and allow v to vary, then the tip of $\mathbf{r}(u_0, v)$ traces the curve C_1 lying on S. (See Figure 10.) The tangent vector to C_1 at P_0 is given by

$$\mathbf{r}_v(u_0, v_0) = \frac{\partial x}{\partial v}(u_0, v_0)\mathbf{i} + \frac{\partial y}{\partial v}(u_0, v_0)\mathbf{j} + \frac{\partial z}{\partial v}(u_0, v_0)\mathbf{k}$$

Similarly, by holding v fast, $v = v_0$, and allowing u to vary, the tip of $\mathbf{r}(u, v_0)$ traces the curve C_2 lying on S, with tangent vector at P_0 given by

$$\mathbf{r}_u(u_0, v_0) = \frac{\partial x}{\partial u}(u_0, v_0)\mathbf{i} + \frac{\partial y}{\partial u}(u_0, v_0)\mathbf{j} + \frac{\partial z}{\partial u}(u_0, v_0)\mathbf{k}$$

If $\mathbf{r}_u(u, v) \times \mathbf{r}_v(u, v) \neq \mathbf{0}$ for each (u, v) in the parameter domain of $\mathbf{r}$, then the surface S is said to be **smooth**. For a smooth surface the **tangent plane** to S at P_0 is the plane that contains the tangent vectors $\mathbf{r}_u(u_0, v_0)$ and $\mathbf{r}_v(u_0, v_0)$ and thus has a normal vector given by

$$\mathbf{n} = \mathbf{r}_u(u_0, v_0) \times \mathbf{r}_v(u_0, v_0)$$

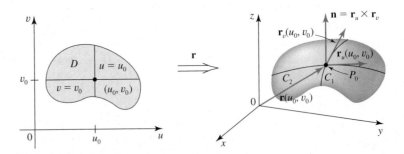

FIGURE 10

EXAMPLE 9 Find an equation of the tangent plane to the helicoid

$$\mathbf{r}(u, v) = u \cos v\mathbf{i} + u \sin v\mathbf{j} + v\mathbf{k}$$

of Example 7 at the point where $u = \frac{1}{2}$ and $v = \frac{\pi}{4}$.

Solution   We start by finding the partial derivatives of $\mathbf{r}$. Thus,

$$\mathbf{r}_u(u, v) = \cos v \mathbf{i} + \sin v \mathbf{j}$$

$$\mathbf{r}_v(u, v) = -u \sin v \mathbf{i} + u \cos v \mathbf{j} + \mathbf{k}$$

So

$$\mathbf{r}_u\left(\frac{1}{2}, \frac{\pi}{4}\right) = \frac{\sqrt{2}}{2} \mathbf{i} + \frac{\sqrt{2}}{2} \mathbf{j}$$

$$\mathbf{r}_v\left(\frac{1}{2}, \frac{\pi}{4}\right) = -\frac{1}{2} \cdot \frac{\sqrt{2}}{2} \mathbf{i} + \frac{1}{2} \cdot \frac{\sqrt{2}}{2} \mathbf{j} + \mathbf{k} = -\frac{\sqrt{2}}{4} \mathbf{i} + \frac{\sqrt{2}}{4} \mathbf{j} + \mathbf{k}$$

A normal vector to the tangent plane is

$$\mathbf{n} = \mathbf{r}_u\left(\frac{1}{2}, \frac{\pi}{4}\right) \times \mathbf{r}_v\left(\frac{1}{2}, \frac{\pi}{4}\right) = \begin{vmatrix} \mathbf{i} & \mathbf{j} & \mathbf{k} \\ \frac{\sqrt{2}}{2} & \frac{\sqrt{2}}{2} & 0 \\ -\frac{\sqrt{2}}{4} & \frac{\sqrt{2}}{4} & 1 \end{vmatrix} = \frac{\sqrt{2}}{2} \mathbf{i} - \frac{\sqrt{2}}{2} \mathbf{j} + \frac{1}{2} \mathbf{k}$$

Since any normal vector will do, let's take $\mathbf{n} = \sqrt{2}\mathbf{i} - \sqrt{2}\mathbf{j} + \mathbf{k}$.

Next note that the point $\left(\frac{1}{2}, \frac{\pi}{4}\right)$ in the parameter domain is mapped onto the point with coordinates

$$x_0 = \frac{1}{2} \cos \frac{\pi}{4} = \frac{1}{2} \cdot \frac{\sqrt{2}}{2} = \frac{\sqrt{2}}{4}, \qquad y_0 = \frac{1}{2} \sin \frac{\pi}{4} = \frac{\sqrt{2}}{4}, \qquad z_0 = \frac{\pi}{4}$$

Therefore, an equation of the required tangent plane at $\left(\frac{\sqrt{2}}{4}, \frac{\sqrt{2}}{4}, \frac{\pi}{4}\right)$ is

$$\sqrt{2}\left(x - \frac{\sqrt{2}}{4}\right) - \sqrt{2}\left(y - \frac{\sqrt{2}}{4}\right) + 1\left(z - \frac{\pi}{4}\right) = 0$$

$$\sqrt{2}x - \sqrt{2}y + z - \frac{\pi}{4} = 0$$

or

$$4\sqrt{2}x - 4\sqrt{2}y + 4z - \pi = 0$$

Area of a Parametric Surface

In Section 14.5, we learned how to find the area of a surface that is the graph of a function $z = f(x, y)$. We now take on the task of finding the areas of parametric surfaces, which are more general than the surfaces (graphs) defined by functions.

For simplicity, let's assume that the parametric surface S defined by

$$\mathbf{r}(u, v) = x(u, v)\mathbf{i} + y(u, v)\mathbf{j} + z(u, v)\mathbf{k}$$

has parameter domain R that is a rectangle. (See Figure 11.) Let P be a regular partition of R into $n = mn$ subrectangles $R_{11}, R_{12}, \ldots, R_{mn}$. The subrectangle R_{ij} is mapped by $\mathbf{r}$ onto the patch S_{ij} with area denoted by ΔS_{ij}. Since the subrectangles R_{ij} are nonoverlapping, except for their common boundaries, so are the patches S_{ij}, and so the area of S is given by

$$S = \sum_{i=1}^{m} \sum_{j=1}^{n} \Delta S_{ij}$$

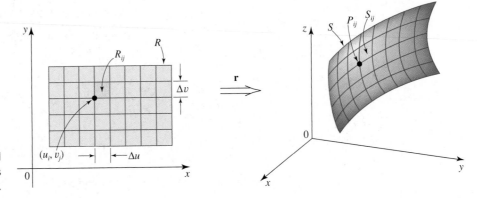

FIGURE 11
The subrectangle R_{ij} is
mapped onto the patch S_{ij}.

Next, let's find an approximation of ΔS_{ij}. Let (u_i, v_j) be the corner of R_{ij} closest to the origin with image the point P_{ij} represented by $\mathbf{r}(u_i, v_j)$ as shown in Figure 12. For the sake of clarity, both R_{ij} and S_{ij} are shown enlarged. The sides of S_{ij} with corner represented by $\mathbf{r}(u_i, v_j)$ are approximated by $\mathbf{a}$ and $\mathbf{b}$, where $\mathbf{a} = \mathbf{r}(u_i + \Delta u, v_j) - \mathbf{r}(u_i, v_j)$ and $\mathbf{b} = \mathbf{r}(u_i, v_j + \Delta v) - \mathbf{r}(u_i, v_j)$. So ΔS_{ij} may be approximated by the area of the parallelogram with $\mathbf{a}$ and $\mathbf{b}$ as adjacent sides, that is,

$$\Delta S_{ij} \approx |\mathbf{a} \times \mathbf{b}|$$

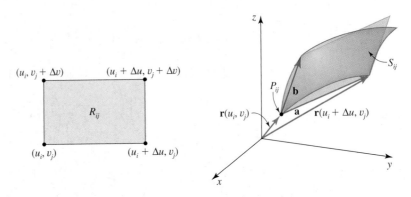

FIGURE 12

But we can write

$$\mathbf{a} = \left[\frac{\mathbf{r}(u_i + \Delta u, v_j) - \mathbf{r}(u_i, v_j)}{\Delta u}\right]\Delta u$$

If Δu is small, as we assume, then the term inside the brackets is approximately equal to $\mathbf{r}_u(u_i, v_j)$. So

$$\mathbf{a} \approx \Delta u \, \mathbf{r}_u(u_i, v_j)$$

Similarly, we see that

$$\mathbf{b} \approx \Delta v \, \mathbf{r}_v(u_i, v_j)$$

Therefore,

$$\Delta S_{ij} \approx |[\Delta u \, \mathbf{r}_u(u_i, v_j)] \times [\Delta v \, \mathbf{r}_v(u_i, v_j)]|$$

$$= |\mathbf{r}_u(u_i, v_j) \times \mathbf{r}_v(u_i, v_j)|\Delta u \, \Delta v$$

and the area of S may be approximated by

$$\sum_{i=1}^{m} \sum_{j=1}^{n} |\mathbf{r}_u(u_i, v_j) \times \mathbf{r}_v(u_i, v_j)|\Delta u \, \Delta v$$

Intuitively, the approximation gets better and better as m and n get larger and larger. But the double sum is the Riemann sum of $|\mathbf{r}_u \times \mathbf{r}_v|$, and we are led to define the area of S as

$$\lim_{m, n \to \infty} \sum_{i=1}^{m} \sum_{j=1}^{n} |\mathbf{r}_u(u_i, v_j) \times \mathbf{r}_v(u_i, v_j)| \, \Delta u \, \Delta v$$

Alternatively, we have the following definition.

DEFINITION   Surface Area (Parametric Form)

Let S be a smooth surface represented by the equation

$$\mathbf{r}(u, v) = x(u, v)\mathbf{i} + y(u, v)\mathbf{j} + z(u, v)\mathbf{k}$$

with parameter domain D. If S is covered just once as (u, v) varies throughout D, then the **surface area** of S is

$$A(S) = \iint_D |\mathbf{r}_u \times \mathbf{r}_v| \, dA \tag{2}$$

EXAMPLE 10 Find the surface area of a sphere of radius a.

Solution The sphere centered at the origin with radius a is represented by the equation

$$\mathbf{r}(u, v) = a \sin u \cos v \mathbf{i} + a \sin u \sin v \mathbf{j} + a \cos u \mathbf{k}$$

with parameter domain $D = \{(u, v) \mid 0 \le u \le \pi, 0 \le v \le 2\pi\}$. We find

$$\mathbf{r}_u(u, v) = a \cos u \cos v \mathbf{i} + a \cos u \sin v \mathbf{j} - a \sin u \mathbf{k}$$

$$\mathbf{r}_v(u, v) = -a \sin u \sin v \mathbf{i} + a \sin u \cos v \mathbf{j}$$

and so

$$\mathbf{r}_u \times \mathbf{r}_v = \begin{vmatrix} \mathbf{i} & \mathbf{j} & \mathbf{k} \\ a \cos u \cos v & a \cos u \sin v & -a \sin u \\ -a \sin u \sin v & a \sin u \cos v & 0 \end{vmatrix}$$

$$= a^2 \sin^2 u \cos v \mathbf{i} + a^2 \sin^2 u \sin v \mathbf{j} + a^2 \sin u \cos u \mathbf{k}$$

Therefore

$$|\mathbf{r}_u \times \mathbf{r}_v| = \sqrt{a^4 \sin^4 u \cos^2 v + a^4 \sin^4 u \sin^2 v + a^4 \sin^2 u \cos^2 u}$$

$$= \sqrt{a^4 \sin^4 u + a^4 \sin^2 u \cos^2 u} = \sqrt{a^4 \sin^2 u}$$

$$= a^2 \sin u$$

since $\sin u \ge 0$ for $0 \le u \le \pi$. Using Equation (2), the area of the sphere is

$$A = \iint_D |\mathbf{r}_u \times \mathbf{r}_v| \, dA = \int_0^{2\pi} \int_0^{\pi} a^2 \sin u \, du \, dv$$

$$= \int_0^{2\pi} \left[-a^2 \cos u \right]_{u=0}^{u=\pi} dv = \int_0^{2\pi} 2a^2 \, dv = 2a^2(2\pi) = 4\pi a^2 \qquad \blacksquare$$

EXAMPLE 11 Find the area of one complete turn of the helicoid of width one represented by the equation $\mathbf{r}(u, v) = u \cos v \mathbf{i} + u \sin v \mathbf{j} + v \mathbf{k}$ with parameter domain $D = \{(u, v) \mid 0 \le u \le 1, 0 \le v \le 2\pi\}$. (Refer to Figure 1.)

Solution   We first find

$$\mathbf{r}_u = \cos v \mathbf{i} + \sin v \mathbf{j}$$

$$\mathbf{r}_v = -u \sin v \mathbf{i} + u \cos v \mathbf{j} + \mathbf{k}$$

and so

$$\mathbf{r}_u \times \mathbf{r}_v = \begin{vmatrix} \mathbf{i} & \mathbf{j} & \mathbf{k} \\ \cos v & \sin v & 0 \\ -u \sin v & u \cos v & 1 \end{vmatrix}$$

$$= \sin v \mathbf{i} - \cos v \mathbf{j} + (u \cos^2 v + u \sin^2 v) \mathbf{k}$$

$$= \sin v \mathbf{i} - \cos v \mathbf{j} + u \mathbf{k}$$

Therefore,

$$|\mathbf{r}_u \times \mathbf{r}_v| = \sqrt{\sin^2 v + \cos^2 v + u^2} = \sqrt{1 + u^2}$$

So, the required area is

$$A = \iint_D |\mathbf{r}_u \times \mathbf{r}_v| \, dA = \int_0^{2\pi} \int_0^1 \sqrt{1 + u^2} \, du \, dv$$

$$= \int_0^{2\pi} \left[\frac{u}{2} \sqrt{1 + u^2} + \frac{1}{2} \ln(u + \sqrt{1 + u^2}) \right]_{u=0}^{u=1} dv$$

$$= \int_0^{2\pi} \left[\frac{1}{2} \sqrt{2} + \frac{1}{2} \ln(1 + \sqrt{2}) \right] dv \qquad \text{Use Formula 37 from the Table of Integrals.}$$

$$= \pi[\sqrt{2} + \ln(1 + \sqrt{2})] \approx 7.212 \qquad \blacksquare$$

15.6   CONCEPT QUESTIONS

1. **a.** Define a parametric surface.
    b. What are the grid curves of a parametric surface?
2. **a.** What is a smooth surface?
    b. Explain how you would find an equation of the tangent plane to a smooth surface with representation $\mathbf{r}(u, v)$ at the point represented by $\mathbf{r}(u_0, v_0)$.

3. Write a double integral giving the area of a surface S defined by a vector function $\mathbf{r}(u, v)$, where (u, v) lies in the parameter domain D of $\mathbf{r}$.

15.6 EXERCISES

In Exercises 1–4, match the equation with one of the graphs labeled (a)–(d). Give a reason for your choice.

1. $\mathbf{r}(u, v) = 2 \cos u \mathbf{i} + 2 \sin u \mathbf{j} + v \mathbf{k}$

2. $\mathbf{r}(u, v) = u \cos v \mathbf{i} + u \sin v \mathbf{j} + u \mathbf{k}$

3. $\mathbf{r}(u, v) = u \cos v \mathbf{i} + u \sin v \mathbf{j} + u^2 \mathbf{k}$

4. $\mathbf{r}(u, v) = u \cos v \mathbf{i} + u \sin v \mathbf{j} + v \mathbf{k}$

(a)    (b)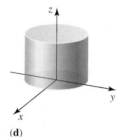

(c)    (d)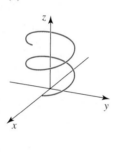

In Exercises 5–8, find an equation in rectangular coordinates, and then identify and sketch the surface.

5. $\mathbf{r}(u, v) = (u - v)\mathbf{i} + 3v\mathbf{j} + (u + v)\mathbf{k}$

6. $\mathbf{r}(u, v) = (u^2 + v^2)\mathbf{i} + u\mathbf{j} + v\mathbf{k}$

7. $\mathbf{r}(u, v) = 3 \sin u \mathbf{i} + 2 \cos u \mathbf{j} + v \mathbf{k}, \quad 0 \le v \le 2$

8. $\mathbf{r}(u, v) = 2 \cos v \cos u \mathbf{i} + 2 \cos v \sin u \mathbf{j} + 3 \sin v \mathbf{k}$

cas *In Exercises 9–14, use a computer algebra system (CAS) to graph the surface represented by the vector function.*

9. $\mathbf{r}(u, v) = (u + v)\mathbf{i} + (u - v)\mathbf{j} + (u^2 + v^2)\mathbf{k}; \quad -1 \le u \le 1,$
$-1 \le v \le 1$

10. $\mathbf{r}(u, v) = u\mathbf{i} + (v - 1)\mathbf{j} + (v^3 - v)\mathbf{k}; \quad 0 \le u \le 1,$
$-2 \le v \le 1$

11. $\mathbf{r}(u, v) = \cos u \sin v \mathbf{i} + \sin u \sin v \mathbf{j} + (1 + \cos v)\mathbf{k};$
$0 \le u \le 2\pi, \quad 0 \le v \le 2\pi$

12. $\mathbf{r}(u, v) = v \cos u \sin v \mathbf{i} + v \sin u \sin v \mathbf{j} + u \cos v \mathbf{k};$
$0 \le u \le 2\pi, \quad 0 \le v \le \pi$

13. $\mathbf{r}(u, v) = v \cos u \mathbf{i} + v \sin u \mathbf{j} + v^2 \mathbf{k}; \quad 0 \le u \le 2\pi,$
$0 \le v \le 1$

14. $\mathbf{r}(u, v) = \left[2 \cos u + v \cos\left(\dfrac{u}{2}\right) \right] \mathbf{i}$
$\qquad + \left[2 \sin u + v \cos\left(\dfrac{u}{2}\right) \right] \mathbf{j} + v \sin\left(\dfrac{u}{2}\right) \mathbf{k};$
$\qquad\qquad\qquad 0 \le u \le 2\pi, \quad -\tfrac{1}{2} \le v \le \tfrac{1}{2}$

Note: This is a representation for the **Möbius strip.**

In Exercises 15–22, find a vector representation for the surface.

15. The plane that passes through the point $(2, 1, -3)$ and contains the vectors $2\mathbf{i} + \mathbf{j} - \mathbf{k}$ and $\mathbf{i} - 2\mathbf{j} - \mathbf{k}$

16. The plane $2x + 3y + z = 6$

17. The lower half of the sphere $x^2 + y^2 + z^2 = 1$

18. The upper half of the ellipsoid $9x^2 + 4y^2 + 36z^2 = 36$

19. The part of the cylinder $x^2 + y^2 = 4$ between $z = -1$ and $z = 3$

20. The part of the cylinder $9y^2 + 4z^2 = 36$ between $x = 0$ and $x = 3$

21. The part of the paraboloid $z = 9 - x^2 - y^2$ inside the cylinder $x^2 + y^2 = 4$

22. The part of the plane $z = x + 2$ that lies inside the cylinder $x^2 + y^2 = 1$

In Exercises 23–26, find a vector equation for the surface obtained by revolving the graph of the function about the indicated axis. Graph the surface.

23. $y = \sqrt{x}, \quad 0 \le x \le 4; \quad x\text{-axis}$

24. $y = e^{-x}, \quad 0 \le x \le 1; \quad x\text{-axis}$

25. $x = 9 - y^2, \quad 0 \le y \le 3; \quad y\text{-axis}$

26. $y = \cos z, \quad -\pi \le z \le \pi; \quad z\text{-axis}$

In Exercises 27–32, find an equation of the tangent plane to the parametric surface represented by $\mathbf{r}$ at the specified point.

27. $\mathbf{r}(u, v) = (u + v)\mathbf{i} + (u - v)\mathbf{j} + v^2\mathbf{k}; \quad (2, 0, 1)$

28. $\mathbf{r}(u, v) = u\mathbf{i} + (u^2 + v^2)\mathbf{j} + v\mathbf{k}; \quad (2, 5, 1)$

29. $\mathbf{r}(u, v) = u \cos v \mathbf{i} + 2u \sin v \mathbf{j} + u^2\mathbf{k}; \quad u = 1, \quad v = \pi$

30. $\mathbf{r}(u, v) = \cos u \sin v \mathbf{i} + \sin u \sin v \mathbf{j} + \cos v \mathbf{k}; \quad u = \dfrac{\pi}{2},$
$v = \dfrac{\pi}{4}$

31. $\mathbf{r}(u, v) = ue^v\mathbf{i} + ve^u\mathbf{j} + uv\mathbf{k}; \quad u = 0, \quad v = \ln 2$

32. $\mathbf{r}(u, v) = uv\mathbf{i} + u \ln v \mathbf{j} + v\mathbf{k}; \quad u = 1, \quad v = 1$

In Exercises 33–40, find the area of the surface.

33. The part of the plane $\mathbf{r}(u, v) = (u + 2v - 1)\mathbf{i} + (2u + 3v + 1)\mathbf{j} + (u + v + 2)\mathbf{k}; \quad 0 \le u \le 1, \quad 0 \le v \le 2$

V Videos for selected exercises are available online at www.academic.cengage.com/login.

34. The part of the plane $2x + 3y - z = 1$ that lies above the rectangular region $[1, 2] \times [1, 3]$

35. The part of the plane $z = 8 - 2x - 3y$ that lies inside the cylinder $x^2 + y^2 = 4$

36. The part of the paraboloid $\mathbf{r}(u, v) = u \cos v\mathbf{i} + u \sin v\mathbf{j} + u^2\mathbf{k}$; $0 \leq u \leq 3$, $0 \leq v \leq 2\pi$

37. The part of the cone $\mathbf{r}(u, v) = u \cos v\mathbf{i} + u \sin v\mathbf{j} + u\mathbf{k}$; $1 \leq u \leq 2$, $0 \leq v \leq \frac{\pi}{2}$

38. The part of the sphere $\mathbf{r}(u, v) = a \sin u \cos v\mathbf{i} + a \sin u \sin v\mathbf{j} + a \cos u\mathbf{k}$ that lies in the first octant

39. The surface $\mathbf{r}(u, v) = \sin u \cos v\mathbf{i} + \sin u \sin v\mathbf{j} + u\mathbf{k}$; $0 \leq u \leq \pi$, $0 \leq v \leq 2\pi$

40. The part of the surface $\mathbf{r}(u, v) = u^2\mathbf{i} + uv\mathbf{j} + \frac{1}{2}v^2\mathbf{k}$; $0 \leq u \leq 1$, $0 \leq v \leq 2$

cas 41. a. Show that the vector equation $\mathbf{r}(u, v) = a \sin u \cos v\mathbf{i} + b \sin u \sin v\mathbf{j} + c \cos u\mathbf{k}$, where $0 \leq u \leq \pi$ and $0 \leq v \leq 2\pi$, represents the ellipsoid

$$\frac{x^2}{a^2} + \frac{y^2}{b^2} + \frac{z^2}{c^2} = 1$$

b. Use a CAS to graph the ellipsoid with $a = 3$, $b = 4$, and $c = 5$.

c. Use a CAS to find the approximate surface area of the ellipsoid of part (b).

cas 42. a. Show that the vector equation $\mathbf{r}(u, v) = a \sin^3 u \cos^3 v\mathbf{i} + a \sin^3 u \sin^3 v\mathbf{j} + a \cos^3 u\mathbf{k}$, where $0 \leq u \leq \pi$ and $0 \leq v \leq 2\pi$, represents the **astroidal sphere** $x^{2/3} + y^{2/3} + z^{2/3} = a^{2/3}$.

b. Use a CAS to graph the astroidal sphere with $a = 1$.

c. Find the area of the astroidal sphere with $a = 1$.

43. Find the area of the part of the cone $z = \sqrt{x^2 + y^2}$ that is cut off by the cylinder $x^2 + (y - 1)^2 = 1$.

44. In Section 13.7 we showed that the tangent plane to the graph S of a function $f(x, y)$ at the point $(a, b, f(a, b))$ is given by the equation

$$z - f(a, b) = f_x(a, b)(x - a) + f_y(a, b)(y - b)$$

(See Equation (4) in Section 13.7.) Show that parametrizing S by $\mathbf{r}(x, y) = x\mathbf{i} + y\mathbf{j} + f(x, y)\mathbf{k}$ yields the same tangent plane.

45. In Section 5.4 we defined the area of the surface of revolution obtained by revolving the graph of a nonnegative smooth function f on $[a, b]$ about the x-axis as

$$S = 2\pi \int_a^b f(x)\sqrt{1 + [f'(x)]^2} \, dx$$

Use Equation (1) to derive this formula.

46. If the circle with center at $(a, 0, 0)$ and radius b, where $0 < b < a$, in the xz-plane is revolved about the z-axis, we obtain a torus represented parametrically by

$$x = (a + b \cos v)\cos u$$
$$y = (a + b \cos v)\sin u$$
$$z = b \sin v$$

with parametric domain $D = \{(u, v) \mid 0 \leq u \leq 2\pi, 0 \leq v \leq 2\pi\}$. (See the figure below.) Find the surface area of the torus.

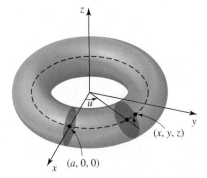

In Exercises 47 and 48, determine whether the statement is true or false. If it is true, explain why. If it is false, explain why or give an example that shows it is false.

47. The surface described by $\mathbf{r}(u, v) = u \cos v\mathbf{i} + u \sin v\mathbf{j} + u\mathbf{k}$, where $-2 \leq u \leq 2$ and $0 \leq v \leq 2\pi$, is smooth.

48. If $\mathbf{r}(u, v) = 2 \sin u \cos v\mathbf{i} + 2 \sin u \sin v\mathbf{j} + 2 \cos u\mathbf{k}$, where $0 \leq u \leq \frac{\pi}{2}$ and $0 \leq v \leq \frac{\pi}{2}$, then

$$\int_0^{\pi/2} \int_0^{\pi/2} |\mathbf{r}_u \times \mathbf{r}_v| \, du \, dv = 2\pi$$

15.7 Surface Integrals

■ Surface Integrals of Scalar Fields

As we saw in Section 14.1, the mass of a thin plate lying in a plane region can be found by evaluating the double integral $\iint_R \sigma(x, y) \, dA$, where $\sigma(x, y)$ is the mass density of the plate at any point (x, y) in R. Now, instead of a flat plate, let's suppose that we have a plate that takes the form of a curved surface. How do we determine the mass of this plate?

For simplicity, let's suppose that the thin plate has the shape of the surface S that is the graph of a continuous function g of two variables defined by $z = g(x, y)$. To further simplify our discussion, suppose that the domain of g is a rectangular region $R = \{(x, y) \mid a \le x \le b, c \le y \le d\}$. A typical surface is shown in Figure 1.

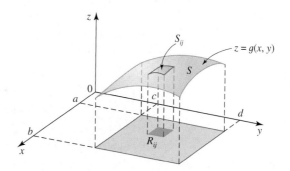

FIGURE 1
A thin plate that takes the shape of
a surface S defined by $z = g(x, y)$

Let the mass density of the plate at any point on S be $\sigma(x, y, z)$, where σ is a nonnegative continuous function defined on an open region containing S, and let $P = \{R_{ij}\}$ be a partition of R into $N = mn$ subrectangles. Corresponding to each subrectangle R_{ij}, there is a part of S, S_{ij}, that lies directly above R_{ij} with area ΔS_{ij}. Then

$$\Delta S_{ij} = \sqrt{[g_x(x_i, y_j)]^2 + [g_y(x_i, y_j)]^2 + 1}\; \Delta A \tag{1}$$

where (x_i, y_j) is the corner of R_{ij} closest to the origin and ΔA is the area of R_{ij}. If m and n are large so that the dimensions of R_{ij} are small, then the continuity of g and σ implies that $\sigma(x, y, z)$ does not differ appreciably from $\sigma(x_i, y_j, g(x_i, y_j))$. Therefore, the mass of the part of the plate that lies on S and directly above R_{ij} is

$$\Delta m_{ij} \approx \sigma(x_i, y_j, g(x_i, y_j))\, \Delta S_{ij} \qquad \text{Constant mass density} \cdot \text{surface area}$$

Using Equation (1), we see that the mass of the plate is approximately

$$\sum_{i=1}^{m} \sum_{j=1}^{n} \sigma(x_i, y_j, g(x_i, y_j)) \sqrt{[g_x(x_i, y_j)]^2 + [g_y(x_i, y_j)]^2 + 1}\; \Delta A$$

The approximation should improve as m and n approach infinity. This suggests that we define the mass of the plate to be

$$\lim_{n, m \to \infty} \sum_{i=1}^{m} \sum_{j=1}^{n} \sigma(x_i, y_j, g(x_i, y_j)) \sqrt{[g_x(x_i, y_j)]^2 + [g_y(x_i, y_j)]^2 + 1}\; \Delta A$$

Using the definition of the double integral, we see that the required mass, m, is

$$m = \iint\limits_{S} \sigma(x, y, z)\, dS = \iint\limits_{R} \sigma(x, y, g(x, y)) \sqrt{[g_x(x, y)]^2 + [g_y(x, y)]^2 + 1}\; dA \tag{2}$$

if we assume that both g_x and g_y are continuous on R.

The integral that appears in Equation (2) is a *surface integral*. More generally, we can define the surface integral of a function f over nonrectangular regions as follows.

> **DEFINITION**   **Surface Integral of a Scalar Function**
> Let f be a function of three variables defined in a region in space containing a surface S. The **surface integral of f over S** is
> $$\iint\limits_{S} f(x, y, z) \, dS = \lim_{n, \, m \to \infty} \sum_{i=1}^{m} \sum_{j=1}^{n} f(x_i, y_j, g(x_i, y_j)) \, \Delta S_{ij}$$

We also have the following formulas for evaluating a surface integral depending on the way S is defined.

> **THEOREM 1**   **Evaluation of Surface Integrals**
> (for Surfaces That Are Graphs)
> 1. If S is defined by $z = g(x, y)$ and the projection of S onto the xy-plane is R (Figure 2a), then
> $$\iint\limits_{S} f(x, y, z) \, dS = \iint\limits_{R} f(x, y, g(x, y))\sqrt{[g_x(x, y)]^2 + [g_y(x, y)]^2 + 1} \; dA \quad \textbf{(3)}$$
> 2. If S is defined by $y = g(x, z)$ and the projection of S onto the xz-plane is R (Figure 2b), then
> $$\iint\limits_{S} f(x, y, z) \, dS = \iint\limits_{R} f(x, g(x, z), z)\sqrt{[g_x(x, z)]^2 + [g_z(x, z)]^2 + 1} \; dA \quad \textbf{(4)}$$
> 3. If S is defined by $x = g(y, z)$ and the projection of S onto the yz-plane is R (Figure 2c), then
> $$\iint\limits_{S} f(x, y, z) \, dS = \iint\limits_{R} f(g(y, z), y, z)\sqrt{[g_y(y, z)]^2 + [g_z(y, z)]^2 + 1} \; dA \quad \textbf{(5)}$$

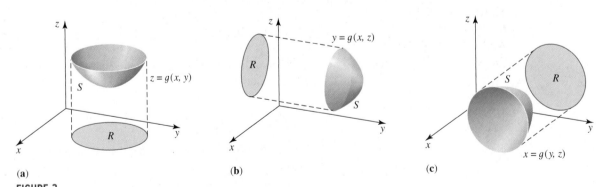

(a)   (b)   (c)

FIGURE 2
The surfaces S and their projections onto the coordinate planes

Note   If we take $f(x, y, z) = 1$, then each of the formulas gives the area of S.   ∎

EXAMPLE 1   Evaluate $\iint_S x \, dS$, where S is the part of the plane $2x + 3y + z = 6$ that lies in the first octant.

Solution   The plane S is shown in Figure 3a, and its projection onto the xy-plane is shown in Figure 3b. Using Equation (3) with $f(x, y, z) = x$ and $z = g(x, y) = 6 - 2x - 3y$, we have

$$\iint_S f(x, y, z) \, dS = \iint_S x \, dS = \iint_R x\sqrt{[g_x(x, y)]^2 + [g_y(x, y)]^2 + 1} \, dA$$

$$= \iint_R x\sqrt{(-2)^2 + (-3)^2 + 1} \, dA = \sqrt{14} \iint_R x \, dA$$

$$= \sqrt{14} \int_0^3 \int_0^{2-(2/3)x} x \, dy \, dx \qquad \text{View } R \text{ as } y\text{-simple.}$$

$$= \sqrt{14} \int_0^3 \left[xy \right]_{y=0}^{y=2-(2/3)x} dx$$

$$= \sqrt{14} \int_0^3 \left(2x - \frac{2}{3}x^2 \right) dx = \sqrt{14} \left[x^2 - \frac{2}{9}x^3 \right]_0^3 = 3\sqrt{14}$$

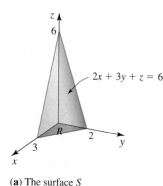

(a) The surface S

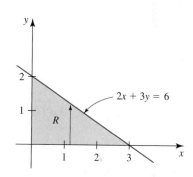

(b) The projection R of S onto the xy-plane viewed as y-simple

FIGURE 3

EXAMPLE 2   Find the mass of the surface S composed of the part of the paraboloid $y = x^2 + z^2$ between the planes $y = 1$ and $y = 4$ if the density at a point P on S is inversely proportional to the distance between P and the axis of symmetry of S.

Solution   The surface S is shown in Figure 4a, and its projection onto the xz-plane is shown in Figure 4b. Using Equation (4) with $f(x, y, z) = \sigma(x, y, z) = k(x^2 + z^2)^{-1/2}$, where k is the constant of proportionality and $y = g(x, z) = x^2 + z^2$, we have

$$m = \iint_S \sigma(x, y, z) \, dS = \iint_S k(x^2 + z^2)^{-1/2} \, dS$$

$$= k\iint_R (x^2 + z^2)^{-1/2} \sqrt{[g_x(x, z)]^2 + [g_z(x, z)]^2 + 1} \, dA$$

$$= k\iint_R (x^2 + z^2)^{-1/2} \sqrt{(2x)^2 + (2z)^2 + 1} \, dA$$

$$= k\iint_R (x^2 + z^2)^{-1/2} \sqrt{4x^2 + 4z^2 + 1} \, dA$$

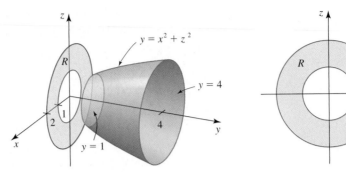

FIGURE 4   **(a)** The surface S **(b)** The projection R of S onto the xz-plane

Changing to polar coordinates, $x = r \cos \theta$ and $z = r \sin \theta$, we obtain

$$m = k \int_0^{2\pi} \int_1^2 \left(\frac{1}{r}\right) \sqrt{4r^2 + 1} \; r \, dr \, d\theta = 2k \int_0^{2\pi} \int_1^2 \sqrt{r^2 + \left(\tfrac{1}{2}\right)^2} \, dr \, d\theta$$

$$= 2k \int_0^{2\pi} \left[\frac{r}{2} \sqrt{r^2 + \tfrac{1}{4}} + \frac{1}{8} \ln \left| r + \sqrt{r^2 + \tfrac{1}{4}} \right| \right]_{r=1}^{r=2} d\theta \qquad \text{Use Formula 37 from the Table of Integrals.}$$

$$= k \left[\sqrt{17} - \frac{1}{2} \sqrt{5} + \frac{1}{4} \ln \left(\frac{4 + \sqrt{17}}{2 + \sqrt{5}} \right) \right] \int_0^{2\pi} d\theta$$

$$= k\pi \left[2\sqrt{17} - \sqrt{5} + \frac{1}{2} \ln \left(\frac{4 + \sqrt{17}}{2 + \sqrt{5}} \right) \right] \qquad \blacksquare$$

■ Parametric Surfaces

If a surface S is represented by a vector equation

$$\mathbf{r}(u, v) = x(u, v)\mathbf{i} + y(u, v)\mathbf{j} + z(u, v)\mathbf{k}$$

with parameter domain D, then an element of surface area is given by

$$|\mathbf{r}_u(u, v) \times \mathbf{r}_v(u, v)| \, dA$$

as we saw in Section 15.6. This leads to the following formula for evaluating a surface integral in which the surface is defined parametrically.

THEOREM 2   Evaluation of Surface Integrals (for Parametric Surfaces)

If f is a continuous function in a region that contains a smooth surface S with parametric representation

$$\mathbf{r}(u, v) = x(u, v)\mathbf{i} + y(u, v)\mathbf{j} + z(u, v)\mathbf{k} \qquad (u, v) \in D$$

then the **surface integral of f over S** is

$$\iint_S f(x, y, z) \, dS = \iint_D f(\mathbf{r}(u, v)) |\mathbf{r}_u \times \mathbf{r}_v| \, dA \qquad \textbf{(6)}$$

where $f(\mathbf{r}(u, v)) = f(x(u, v), y(u, v), z(u, v))$.

Note You can show that if S is the graph of a function $z = g(x, y)$, then Equation (3) follows from Equation (6) by putting $\mathbf{r}(u, v) = u\mathbf{i} + v\mathbf{j} + g(u, v)\mathbf{k}$. (See Exercise 46.) ∎

EXAMPLE 3 Evaluate $\displaystyle\iint_S \frac{x - y}{\sqrt{2z + 1}}\, dS$, where S is the surface represented by $\mathbf{r}(u, v) = (u + v)\mathbf{i} + (u - v)\mathbf{j} + (u^2 + v^2)\mathbf{k}$, where $0 \le u \le 1$ and $0 \le v \le 2$.

Solution We first find

$$\mathbf{r}_u(u, v) = \mathbf{i} + \mathbf{j} + 2u\mathbf{k}$$

$$\mathbf{r}_v(u, v) = \mathbf{i} - \mathbf{j} + 2v\mathbf{k}$$

$$\mathbf{r}_u \times \mathbf{r}_v = \begin{vmatrix} \mathbf{i} & \mathbf{j} & \mathbf{k} \\ 1 & 1 & 2u \\ 1 & -1 & 2v \end{vmatrix}$$

$$= 2[(u + v)\mathbf{i} + (u - v)\mathbf{j} - \mathbf{k}]$$

so

$$|\mathbf{r}_u \times \mathbf{r}_v| = 2\sqrt{(u + v)^2 + (u - v)^2 + 1} = 2\sqrt{2u^2 + 2v^2 + 1}$$

Therefore,

$$\iint_S \frac{x - y}{\sqrt{2z + 1}}\, dS = \int_0^2 \int_0^1 \frac{(u + v) - (u - v)}{\sqrt{2(u^2 + v^2) + 1}} \cdot 2\sqrt{2u^2 + 2v^2 + 1}\, du\, dv$$

$$= 4\int_0^2 \int_0^1 v\, du\, dv$$

$$= 4\int_0^2 \big[uv\big]_{u=0}^{u=1}\, dv$$

$$= 4\int_0^2 v\, dv = 2v^2 \Big|_0^2 = 8$$ ∎

■ Oriented Surfaces

One of the most important applications of surface integrals involves the computation of the flux of a vector field across an *oriented* surface. Before explaining the notion of *flux,* however, we need to elaborate on the meaning of *orientation.*

A surface S is **orientable** or **two-sided** if it has a unit normal vector $\mathbf{n}$ that varies *continuously* over S, that is, if the components of $\mathbf{n}$ are continuous at each point (x, y, z) on S. Closed surfaces (surfaces that are boundaries of solids) such as spheres are examples of orientable surfaces. There are two possible choices of $\mathbf{n}$ for orientable surfaces: the **unit inner normal** that points *inward* from S and the **unit outer normal** that points *outward* from S (see Figure 5). By convention, however, the **positive orientation** for a closed surface S is the one for which the unit normal vector points outward from S.

An example of a nonorientable surface is the Möbius strip, which can be constructed by taking a long, rectangular strip of paper, giving it a half-twist, and then taping the short edges together to produce the surface shown in Figure 6. If you take a unit normal $\mathbf{n}$ starting at P (see Figure 6), then you can move it along the surface in such a way that upon returning to the starting point (and without crossing any edges), it will point in a direction precisely opposite to its initial direction. This shows that $\mathbf{n}$ does not vary continuously on a Möbius strip, and accordingly, the strip is not orientable.

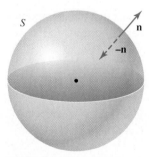

FIGURE 5
Unit inner and outer normals to an (orientable) closed surface S

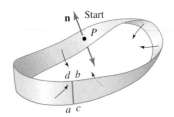

FIGURE 6

The Möbius strip can be constructed by using a rectangular strip of paper.

■ Surface Integrals of Vector Fields

Suppose that $\mathbf{F}$ is a continuous vector field defined in a region R in space. We can think of $\mathbf{F}(x, y, z)$ as giving the velocity of a fluid at a point (x, y, z) in R, and S as a smooth, oriented surface lying in R. If S is flat and $\mathbf{F}$ is a constant field, then the flux, or rate of flow (volume of fluid crossing S per unit time), is equal to

$$\mathbf{F} \cdot \mathbf{n}\, A(S)$$

(the *normal component* of $\mathbf{F}$ with respect to S times the area of S). Geometrically, the flux is given by the volume of fluid in the prism in Figure 7.

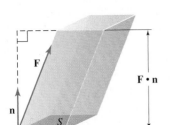

FIGURE 7

If S is flat and $\mathbf{F}$ is constant, then the flux is equal to the volume of the prism.

More generally, suppose that S is the graph of a function of two variables defined by $z = g(x, y)$, where, for simplicity, we assume that the domain of g is a rectangular region $R = \{(x, y) \mid a \le x \le b, c \le y \le d\}$. (See Figure 8.) Let $P = \{R_{ij}\}$ be a partition of R into $N = mn$ subrectangles. Corresponding to each subrectangle R_{ij} there is the part of S that lies directly above R_{ij} with area ΔS_{ij}. As in Section 14.5, let (x_i, y_j) be the corner of R_{ij} closest to the origin, and let $(x_i, y_j, g(x_i, y_j))$ be the point directly above it, as shown in Figure 9. Let $\mathbf{n}_{ij}$ denote the unit normal vector to S at $(x_i, y_j, g(x_i, y_j))$. If m and n are large so that the dimensions of R_{ij} are small, then the continuity of $\mathbf{F}$ implies that $\mathbf{F}(x, y, z)$ does not differ appreciably from $\mathbf{F}(x_i, y_j, g(x_i, y_j))$ on R_{ij}.

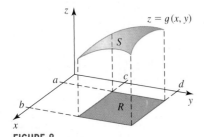

FIGURE 8

A smooth surface S defined by $z = g(x, y)$ for (x, y) in R

Furthermore, the continuity of g implies that S_{ij} may be approximated by T_{ij}, the parallelogram that is part of the tangent plane to S at the point $(x_i, y_j, g(x_i, y_j))$ lying directly above R_{ij}. But the flux of $\mathbf{F}$ across (the flat) T_{ij} is approximately

$$\mathbf{F}(x_i, y_j, g(x_i, y_j)) \cdot \mathbf{n}_{ij}(\Delta T_{ij})$$

where ΔT_{ij} is the area of T_{ij}. Since $\Delta T_{ij} \approx \Delta S_{ij}$, we see that the flux of $\mathbf{F}$ across S may be approximated by the sum

$$\sum_{i=1}^{m} \sum_{j=1}^{n} \mathbf{F}(x_i, y_j, g(x_i, y_j)) \cdot \mathbf{n}_{ij}\, \Delta S_{ij}$$

$$= \sum_{i=1}^{m} \sum_{j=1}^{n} \mathbf{F}(x_i, y_j, g(x_i, y_j)) \cdot \mathbf{n}_{ij} \sqrt{[g_x(x_i, y_j)]^2 + [g_y(x_i, y_j)]^2 + 1}\, \Delta A$$

This last equality follows upon using Equation (1). We can expect that the approximation will get better as the partition P becomes finer. This observation leads to the following definition.

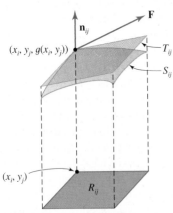

FIGURE 9

DEFINITION   Surface Integral of a Vector Field

Let $\mathbf{F}$ be a continuous vector field defined in a region containing an oriented surface S with unit normal vector $\mathbf{n}$. The **surface integral of F across S in the direction of n** is

$$\iint_S \mathbf{F} \cdot d\mathbf{S} = \iint_S \mathbf{F} \cdot \mathbf{n}\, dS = \lim_{m, n \to \infty} \sum_{i=1}^{m} \sum_{j=1}^{n} \mathbf{F}(x_i, y_j, f(x_i, y_j)) \cdot \mathbf{n}_{ij} \Delta S_{ij}$$

Thus, the surface integral (also called **flux integral**) of a vector field **F** across an oriented surface S is the integral of the *normal* component of **F** over S. If the fluid has density $\rho(x, y, z)$ at (x, y, z), then the flux integral

$$\iint_S \rho \mathbf{F} \cdot \mathbf{n} \, dS$$

gives the *mass* of the fluid flowing across S per unit time.

To obtain a formula for finding the flux integral in terms of $g(x, y)$, recall from Section 13.6 that the normal to the surface $z = g(x, y)$ is given by ∇G, where $G(x, y, z) = z - g(x, y)$. Therefore, the unit normal to S is

$$\mathbf{n} = \frac{\nabla G(x, y, z)}{|\nabla G(x, y, z)|} = \frac{-g_x(x, y)\mathbf{i} - g_y(x, y)\mathbf{j} + \mathbf{k}}{\sqrt{[g_x(x, y)]^2 + [g_y(x, y)]^2 + 1}}$$

Furthermore, in Section 14.5 we showed that the "element of area" dS is given by

$$dS = \sqrt{[g_x(x, y)]^2 + [g_y(x, y)]^2 + 1} \, dA$$

so

$$\iint_S \mathbf{F} \cdot \mathbf{n} \, dS = \iint_D \frac{\mathbf{F} \cdot [-g_x(x, y)\mathbf{i} - g_y(x, y)\mathbf{j} + \mathbf{k}]}{\sqrt{[g_x(x, y)]^2 + [g_y(x, y)]^2 + 1}} \cdot \sqrt{[g_x(x, y)]^2 + [g_y(x, y)]^2 + 1} \, dA$$

$$= \iint_D \mathbf{F} \cdot [-g_x(x, y)\mathbf{i} - g_y(x, y)\mathbf{j} + \mathbf{k}] \, dA$$

where D is the projection of S onto the xy-plane.

If $\mathbf{F}(x, y, z) = P(x, y, z)\mathbf{i} + Q(x, y, z)\mathbf{j} + R(x, y, z)\mathbf{k}$, then we can write

$$\iint_S \mathbf{F} \cdot \mathbf{n} \, dS = \iint_D (-Pg_x - Qg_y + R) \, dA$$

THEOREM 3 Evaluation of Surface Integrals (for Graphs)

If $\mathbf{F} = P\mathbf{i} + Q\mathbf{j} + R\mathbf{k}$ is a continuous vector field in a region that contains a smooth oriented surface S given by $z = g(x, y)$ and D is its projection onto the xy-plane, then

$$\iint_S \mathbf{F} \cdot d\mathbf{S} = \iint_D (-Pg_x - Qg_y + R) \, dA \tag{7}$$

Before looking at the next example, we note the following property of surface integrals: If $S = S_1 \cup S_2 \cup \cdots \cup S_n$, where each of the surfaces is smooth and intersect only along their boundaries, then

$$\iint_S \mathbf{F} \cdot d\mathbf{S} = \iint_{S_1} \mathbf{F} \cdot d\mathbf{S} + \cdots + \iint_{S_n} \mathbf{F} \cdot d\mathbf{S}$$

EXAMPLE 4 Evaluate $\iint_S \mathbf{F} \cdot d\mathbf{S}$, where $\mathbf{F}(x, y, z) = x\mathbf{i} + y\mathbf{j} + z\mathbf{k}$ and S is the surface that is composed of the part of the paraboloid $z = 1 - x^2 - y^2$ lying above the xy-plane and the disk $D = \{(x, y) \mid 0 \le x^2 + y^2 \le 1\}$.

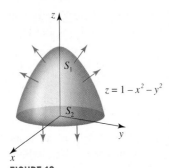

FIGURE 10
The part of the paraboloid
$z = 1 - x^2 - y^2$ that lies above
the xy-plane and is oriented so that
the unit normal vector **n** points
upward. The unit normal for the
disk $D = S_2$ points downward.

Solution The (closed) surface S together with a few vectors from the vector field **F** is shown in Figure 10. Writing the equation of the surface S_1 in the form $g(x, y) = 1 - x^2 - y^2$, we find that $g_x = -2x$ and $g_y = -2y$. Observe that the projection of S onto the xy-plane is $D = \{(x, y) \mid 0 \le x^2 + y^2 \le 1\}$. Also, $P(x, y, z) = x$, $Q(x, y, z) = y$, and $R(x, y, z) = z$. So using Equation (7), we obtain

$$\iint_{S_1} \mathbf{F} \cdot d\mathbf{S} = \iint_D (-Pg_x - Qg_y + R)\, dA$$

$$= \iint_D [-x(-2x) - y(-2y) + z]\, dA$$

$$= \iint_D (2x^2 + 2y^2 + z)\, dA$$

$$= \iint_D [2x^2 + 2y^2 + (1 - x^2 - y^2)]\, dA \qquad z = 1 - x^2 - y^2$$

$$= \iint_D (1 + x^2 + y^2)\, dA$$

$$= \int_0^{2\pi} \int_0^1 (1 + r^2) r\, dr\, d\theta \qquad \text{Use polar coordinates.}$$

$$= \int_0^{2\pi} \left[\frac{1}{2} r^2 + \frac{1}{4} r^4 \right]_{r=0}^{r=1} d\theta = \int_0^{2\pi} \frac{3}{4}\, d\theta = \frac{3}{2}\pi$$

Next, observe that the normal for the surface S_2 is $\mathbf{n} = -\mathbf{k}$. (Remember that the normal for a closed surface, by convention, points outward.) So we have

$$\iint_{S_2} \mathbf{F} \cdot d\mathbf{S} = \iint_{S_2} \mathbf{F} \cdot (-\mathbf{k})\, dS = \iint_D (-z)\, dA = \iint_D 0\, dA = 0$$

because $z = 0$ on S_2. Therefore,

$$\iint_S \mathbf{F} \cdot d\mathbf{S} = \iint_{S_1} \mathbf{F} \cdot d\mathbf{S} + \iint_{S_2} \mathbf{F} \cdot d\mathbf{S} = \frac{3}{2}\pi + 0 = \frac{3\pi}{2} \qquad \blacksquare$$

Notes
1. If the vector field **F** of Example 4 describes the velocity of a fluid flowing through the paraboloidal surface S, then the integral $\iint_S \mathbf{F} \cdot \mathbf{n}\, dS$ that we have just evaluated tells us that the fluid is flowing out through S at the rate of $3\pi/2$ cubic units per unit time.
2. In Example 4, if we had wanted the paraboloid to be oriented so that the normal pointed downward, then we would simply have picked the normal to be $-\mathbf{n}$. In this case the fluid flows into S at the rate of $3\pi/2$ cubic units per unit time. $\blacksquare$

■ Parametric Surfaces

If an oriented surface S is a smooth surface represented by a vector equation

$$\mathbf{r}(u, v) = x(u, v)\mathbf{i} + y(u, v)\mathbf{j} + z(u, v)\mathbf{k}$$

with parameter domain D, then the normal to S is

$$\mathbf{n} = \frac{\mathbf{r}_u \times \mathbf{r}_v}{|\mathbf{r}_u \times \mathbf{r}_v|}$$

Therefore,

$$\iint_S \mathbf{F} \cdot d\mathbf{S} = \iint_S \mathbf{F} \cdot \mathbf{n} \, dS = \iint_S \mathbf{F} \cdot \frac{\mathbf{r}_u \times \mathbf{r}_v}{|\mathbf{r}_u \times \mathbf{r}_v|} \, dS$$

$$= \iint_D \left[\mathbf{F}(\mathbf{r}(u, v)) \cdot \frac{\mathbf{r}_u \times \mathbf{r}_v}{|\mathbf{r}_u \times \mathbf{r}_v|} \right] |\mathbf{r}_u \times \mathbf{r}_v| \, dA$$

$$= \iint_D \mathbf{F}(\mathbf{r}(u, v)) \cdot (\mathbf{r}_u \times \mathbf{r}_v) \, dA$$

THEOREM 4   Evaluation of Surface Integrals of a Vector Field
(for Parametric Surfaces)

If $\mathbf{F}$ is a continuous vector field in a region that contains a smooth, oriented surface S with parametric representation

$$\mathbf{r}(u, v) = x(u, v)\mathbf{i} + y(u, v)\mathbf{j} + z(u, v)\mathbf{k} \qquad (u, v) \in D$$

then the surface integral of f over S is

$$\iint_S \mathbf{F} \cdot d\mathbf{S} = \iint_D \mathbf{F}(\mathbf{r}(u, v)) \cdot (\mathbf{r}_u \times \mathbf{r}_v) \, dA \qquad \text{(8)}$$

where

$$\mathbf{F}(\mathbf{r}(u, v)) = \mathbf{F}(x(u, v), y(u, v), z(u, v))$$

EXAMPLE 5   Find the flux of the vector field $\mathbf{F}(x, y, z) = y\mathbf{i} + x\mathbf{j} + 2z\mathbf{k}$ across the unit sphere $x^2 + y^2 + z^2 = 1$.

Solution   The unit sphere has parametric representation

$$\mathbf{r}(u, v) = \sin u \cos v \mathbf{i} + \sin u \sin v \mathbf{j} + \cos u \mathbf{k}$$

with parameter domain $D = \{(u, v) \,|\, 0 \leq u \leq \pi, 0 \leq v \leq 2\pi\}$. Proceeding as in Example 10 in Section 15.6 and taking $a = 1$, we find

$$\mathbf{r}_u \times \mathbf{r}_v = \sin^2 u \cos v \mathbf{i} + \sin^2 u \sin v \mathbf{j} + \sin u \cos u \mathbf{k}$$

Therefore,

$$\mathbf{F}(\mathbf{r}(u, v)) \cdot (\mathbf{r}_u \times \mathbf{r}_v) = (\sin u \sin v \mathbf{i} + \sin u \cos v \mathbf{j} + 2 \cos u \mathbf{k})$$

$$\cdot (\sin^2 u \cos v \mathbf{i} + \sin^2 u \sin v \mathbf{j} + \sin u \cos u \mathbf{k})$$

$$= \sin^3 u \sin v \cos v + \sin^3 u \cos v \sin v + 2 \cos^2 u \sin u$$

$$= 2(\sin^3 u \sin v \cos v + \cos^2 u \sin u)$$

Using Equation (8), the flux across the sphere is

$$\iint_S \mathbf{F} \cdot d\mathbf{S} = \iint_D \mathbf{F}(\mathbf{r}(u,v)) \cdot (\mathbf{r}_u \times \mathbf{r}_v) \, dA$$

$$= 2 \int_0^{2\pi} \int_0^{\pi} (\sin^3 u \sin v \cos v + \cos^2 u \sin u) \, du \, dv$$

$$= 2 \int_0^{\pi} \sin^3 u \, du \int_0^{2\pi} \sin v \cos v \, dv + 2 \int_0^{\pi} \cos^2 u \sin u \, du \int_0^{2\pi} dv$$

The first term on the right is equal to zero because

$$\int_0^{2\pi} \sin v \cos v \, dv = \frac{1}{2} \sin^2 v \Big|_0^{2\pi} = 0$$

so

$$\iint_S \mathbf{F} \cdot d\mathbf{S} = 2 \int_0^{\pi} \cos^2 u \sin u \, du \int_0^{2\pi} dv$$

$$= 2 \left(-\frac{1}{3} \cos^3 u \right) \Big|_0^{\pi} \cdot 2\pi$$

$$= \frac{8\pi}{3} \qquad \blacksquare$$

We have used an example involving fluid flow to illustrate the concept of the surface integral of a vector field. But these integrals have wider applications in the physical sciences. For example, if $\mathbf{E}$ is the electric field induced by an electric charge of q coulombs located at the origin of a three-dimensional coordinate system, then by Coulomb's Law (Example 4 in Section 15.1),

$$\mathbf{E} = \frac{q}{4\pi\varepsilon_0} \cdot \frac{\mathbf{r}}{|\mathbf{r}|^3}$$

where ε_0 is a constant called the *permittivity of free space*. If S is a sphere of radius r centered at the origin, then the surface integral

$$\iint_S \mathbf{E} \cdot \mathbf{n} \, dS$$

is the flux of $\mathbf{E}$ passing through S.

Yet another application of surface integrals can be found in the study of heat flow. Suppose that the temperature at a point (x, y, z) in a homogeneous body is $T(x, y, z)$. Empirical results suggest that heat will flow from points at higher temperatures to those at lower temperatures. Since the temperature gradient ∇T points in the direction of maximum increase of the temperature, we see that the flow of heat can be described by the vector field

$$\mathbf{q} = -k\nabla T$$

where k is a constant of proportionality known as the thermal conductivity of the body. The rate at which heat flows across a surface S in the body is given by the surface integral

$$\iint_S \mathbf{q} \cdot \mathbf{n} \, dS = -k \iint_S \nabla T \cdot \mathbf{n} \, dS$$

EXAMPLE 6 **Rate of Flow of Heat Across a Sphere** The temperature at a point $P(x, y, z)$ in a medium with thermal conductivity k is inversely proportional to the distance between P and the origin. Find the rate of flow of heat across a sphere S of radius a, centered at the origin.

Solution We have

$$T(x, y, z) = \frac{c}{\sqrt{x^2 + y^2 + z^2}}$$

where c is the constant of proportionality. Then the flow of heat is

$$\mathbf{q} = -k \nabla \mathbf{T} = -k \left[-\frac{cx}{(x^2 + y^2 + z^2)^{3/2}} \mathbf{i} - \frac{cy}{(x^2 + y^2 + z^2)^{3/2}} \mathbf{j} - \frac{cz}{(x^2 + y^2 + z^2)^{3/2}} \mathbf{k} \right]$$

$$= \frac{ck}{(x^2 + y^2 + z^2)^{3/2}} (x\mathbf{i} + y\mathbf{j} + z\mathbf{k})$$

The outward unit normal to the sphere $x^2 + y^2 + z^2 = a^2$ at the point (x, y, z) is

$$\mathbf{n} = \frac{1}{a} (x\mathbf{i} + y\mathbf{j} + z\mathbf{k})$$

So the rate at which heat flows across S is

$$\iint_S \mathbf{q} \cdot \mathbf{n} \, dS = \iint_S \frac{ck}{(x^2 + y^2 + z^2)^{3/2}} (x\mathbf{i} + y\mathbf{j} + z\mathbf{k}) \cdot \left[\frac{1}{a} (x\mathbf{i} + y\mathbf{j} + z\mathbf{k}) \right] dS$$

$$= \frac{ck}{a} \iint_S \frac{1}{\sqrt{x^2 + y^2 + z^2}} \, dS$$

$$= \frac{ck}{a^2} \iint_S dS \qquad \text{Since } x^2 + y^2 + z^2 = a^2 \text{ on } S$$

$$= \frac{ck}{a^2} A(S) = \frac{ck}{a^2} (4\pi a^2) = 4\pi ck$$

15.7 CONCEPT QUESTIONS

1. **a.** Define the surface integral of a scalar function f over a surface that is the graph of a function $z = f(x, y)$.
   b. How do you evaluate the integral of part (a)?
   c. How do you evaluate the surface integral if the surface is represented by a vector function $\mathbf{r}(u, v)$?

2. What is an orientable surface? Give an example of a surface that is not orientable.

3. **a.** Define the surface (flux) integral of a vector field $\mathbf{F}$ over an oriented surface S with a unit normal $\mathbf{n}$.
   b. How do you evaluate the surface integral if the surface is the graph of a function $z = g(x, y)$?
   c. How do you evaluate the surface integral if the surface is represented by the vector function $\mathbf{r}(u, v)$?

15.7 EXERCISES

In Exercises 1–14, evaluate $\iint_S f(x, y, z) \, dS$.

1. $f(x, y, z) = x + y$; S is the part of the plane $3x + 2y + z = 6$ in the first octant

2. $f(x, y, z) = xy$; S is the part of the plane $2x + 3y + z = 6$ in the first octant

V Videos for selected exercises are available online at **www.academic.cengage.com/login**.

3. ; is the part of the surface
above the rectangular region
$R = \{(x, y) \mid 0 \le x \le 2, 0 \le y \le 1\}$

4. $f(x, y, z) = xz^2$; S is the part of the surface
$y = x^2 + 2z$ to the right of the square region
$R = \{(x, z) \mid 0 \le x \le 1, 0 \le z \le 1\}$

5. $f(x, y, z) = x + 2y + z$; S is the part of the plane
$y + z = 4$ inside the cylinder $x^2 + y^2 = 1$

6. $f(x, y, z) = xz$; S is the part of the plane $y + z = 4$
inside the cylinder $x^2 + y^2 = 4$

7. $f(x, y, z) = x^2 z$; S is the part of the cone $z = \sqrt{x^2 + y^2}$
inside the cylinder $x^2 + y^2 = 1$

8. $f(x, y, z) = x + 2y + 3z$; S is the part of the cone
$x = \sqrt{y^2 + z^2}$ between the planes $x = 1$ and $x = 4$

9. $f(x, y, z) = xyz$; S is the part of the cylinder $x^2 + y^2 = 4$
in the first octant between $z = 0$ and $z = 4$

10. $f(x, y, z) = z^2$; S is the hemisphere $z = \sqrt{9 - x^2 - y^2}$

11. $f(x, y, z) = x + \dfrac{y}{\sqrt{4z + 5}}$; S is the surface with

vector representation $\mathbf{r}(u, v) = u\mathbf{i} + v\mathbf{j} + (v^2 - 1)\mathbf{k}$,
$0 \le u \le 1, -1 \le v \le 1$

12. $f(x, y, z) = x + z$; S is the surface with vector representation
$\mathbf{r}(u, v) = u \sin v\mathbf{i} + u \cos v\mathbf{j} + u^2\mathbf{k}$, $0 \le u \le 1, 0 \le v \le \frac{\pi}{2}$

13. $f(x, y, z) = z\sqrt{1 + x^2 + y^2}$; S is the helicoid with vector
representation $\mathbf{r}(u, v) = u \cos v\mathbf{i} + u \sin v\mathbf{j} + v\mathbf{k}$,
$0 \le u \le 1, 0 \le v \le 2\pi$

14. $f(x, y, z) = z$; S is the part of the torus with vector
representation $\mathbf{r}(u, v) = (a + b \cos v)\cos u\mathbf{i} +$
$(a + b \cos v)\sin u\mathbf{j} + b \sin v\mathbf{k}$, $0 \le u \le 2\pi, 0 \le v \le \frac{\pi}{2}$

*In Exercises 15–18, find the mass of the surface S having the
given mass density.*

15. S is the part of the plane $x + 2y + 3z = 6$ in the first
octant; the density at a point P on S is directly proportional
to the square of the distance between P and the yz-plane.

16. S is the part of the paraboloid $z = x^2 + y^2$ between the
planes $z = 1$ and $z = 4$; the density at a point P on S is
directly proportional to the distance between P and the axis
of symmetry of S.

17. S is the hemisphere $x^2 + y^2 + z^2 = 4$, $z \ge 0$; the density at
a point P on S is directly proportional to the distance
between P and the xy-plane.

18. S is the part of the sphere $x^2 + y^2 + z^2 = 1$ that lies above
the cone $z = \sqrt{x^2 + y^2}$; the density at a point P on S is
directly proportional to the distance between P and the
xy-plane.

*In Exercises 19–28, evaluate $\iint_S \mathbf{F} \cdot d\mathbf{S}$, that is, find the flux of
$\mathbf{F}$ across S. If S is closed, use the positive (outward) orientation.*

19. $\mathbf{F}(x, y, z) = 2x\mathbf{i} + 2y\mathbf{j} + z\mathbf{k}$; S is the part of the paraboloid
$z = 4 - x^2 - y^2$ above the xy-plane; $\mathbf{n}$ points upward

20. $\mathbf{F}(x, y, z) = 3x\mathbf{i} + 3y\mathbf{j} + 2z\mathbf{k}$; S is the part of the paraboloid $z = x^2 + y^2$ between the planes $z = 0$ and $z = 4$;
$\mathbf{n}$ points downward

21. $\mathbf{F}(x, y, z) = x\mathbf{i} + y\mathbf{j} + z\mathbf{k}$; S is the part of the plane
$3x + 2y + z = 6$ in the first octant; $\mathbf{n}$ points upward

22. $\mathbf{F}(x, y, z) = 6z\mathbf{i} + 2x\mathbf{j} + y\mathbf{k}$; S is the part of the plane
$2x + 3y + 6z = 12$ in the first octant; $\mathbf{n}$ points upward

23. $\mathbf{F}(x, y, z) = -y\mathbf{i} + x\mathbf{j} + 2z\mathbf{k}$; S is the hemisphere
$z = \sqrt{4 - x^2 - y^2}$; $\mathbf{n}$ points upward

24. $\mathbf{F}(x, y, z) = x\mathbf{i} + y\mathbf{j} + z\mathbf{k}$; S is the hemisphere
$z = \sqrt{9 - x^2 - y^2}$; $\mathbf{n}$ points upward
Hint: First evaluate $\iint_{\bar{S}} \mathbf{F} \cdot \mathbf{n} \, dS$, where $\bar{S}$ is the part of the hemisphere $z = \sqrt{9 - x^2 - y^2}$ inside the cylinder $x^2 + y^2 = a^2$, where
$0 < a < 3$. Then take the limit as $a \to 3^-$.

25. $\mathbf{F}(x, y, z) = 2\mathbf{i} + 3\mathbf{j} + \mathbf{k}$; S is the part of the cone
$z = \sqrt{x^2 + y^2}$ inside the cylinder $x^2 + y^2 = 1$; $\mathbf{n}$ points
upward

26. $\mathbf{F}(x, y, z) = x\mathbf{i} + 2y\mathbf{j} + z\mathbf{k}$; S is the part of the cone
$y = \sqrt{x^2 + z^2}$ inside the cylinder $x^2 + z^2 = 1$; $\mathbf{n}$ points
to the right

27. $\mathbf{F}(x, y, z) = y^3\mathbf{i} + x^2\mathbf{j} + z\mathbf{k}$; S is the boundary of the
cylindrical solid bounded by $x^2 + y^2 = 9$, $z = 0$, and $z = 3$

28. $\mathbf{F}(x, y, z) = x^2\mathbf{i} + xy\mathbf{j} + xz\mathbf{k}$; S is the surface of the tetrahedron with vertices $(0, 0, 0)$, $(1, 0, 0)$, $(0, 2, 0)$, and
$(0, 0, 3)$

*In Exercises 29 and 30, a thin sheet has the shape of the surface
S. If its density (mass per unit area) at the point (x, y, z) is
$\rho(x, y, z)$, then its **center of mass** is $(\bar{x}, \bar{y}, \bar{z})$, where*

$$\bar{x} = \frac{1}{m} \iint_S x\rho(x, y, z) \, dS, \qquad \bar{y} = \frac{1}{m} \iint_S y\rho(x, y, z) \, dS,$$

$$\bar{z} = \frac{1}{m} \iint_S z\rho(x, y, z) \, dS$$

*and m is the mass of the sheet. Find the center of mass of the
sheet.*

29. S is the upper hemisphere $x^2 + y^2 + z^2 = a^2$, $z \ge 0$,
$\rho(x, y, z) = k$, where k is a constant.

30. S is the part of the paraboloid $z = 4 - \frac{1}{2}x^2 - \frac{1}{2}y^2$, $z \ge 0$,
$\rho(x, y, z) = k$, where k is a constant.

In Exercises 31 and 32, a thin sheet has the shape of a surface S. If its density (mass per unit area) at the point (x, y, z) is $\rho(x, y, z)$, then its moment of inertia about the z-axis is $I_z = \iint_S (x^2 + y^2)\rho(x, y, z)\, dS$.

31. Show that the moment of inertia of a spherical shell of uniform density about its diameter is $\frac{2}{3}ma^2$, where m is its mass and a is its radius.

32. Find the moment of inertia of the conical shell $z^2 = x^2 + y^2$, where $0 \le z \le 2$, if it has constant density k.

In Exercises 33 and 34 the electric charge density at a point (x, y, z) on S is $\sigma(x, y, z)$, and the total charge on S is given by $Q = \iint_S \sigma(x, y, z)\, dS$.

33. Electric Charge Find the total charge on the part of the plane $2x + 3y + z = 6$ in the first octant if the charge density at a point P on the surface is directly proportional to the square of the distance between P and the xy-plane.

34. Electric Charge Find the total charge on the part of the hemisphere $z = \sqrt{25 - x^2 - y^2}$ that lies directly above the plane region $R = \{(x, y)\,|\,x^2 + y^2 \le 9\}$ if the charge density at a point P on the surface is directly proportional to the square of the distance between P and the xy-plane.

35. Flow of a Fluid The flow of a fluid is described by the vector field $\mathbf{F}(x, y, z) = 2x\mathbf{i} + 2y\mathbf{j} + 3z\mathbf{k}$. Find the rate of flow of the fluid upward through the surface S that is the part of the plane $2x + 3y + z = 6$ in the first octant.

36. Flow of a Liquid The flow of a liquid is described by the vector field $\mathbf{F}(x, y, z) = x\mathbf{i} + y\mathbf{j} + 3z\mathbf{k}$. If the mass density of the fluid is 1000 (in appropriate units), find the rate of flow (mass per unit time) upward of the liquid through the surface S that is part of the paraboloid $z = 9 - x^2 - y^2$ above the xy-plane.
Hint: The flux is $\iint_S \rho\mathbf{F} \cdot \mathbf{n}\, dS$, where ρ is the mass density function.

37. Rate of Flow of Heat The temperature at a point (x, y, z) in a homogeneous body with thermal conductivity $k = 5$ is $T(x, y, z) = x^2 + y^2$. Find the rate of flow of heat across the cylindrical surface $x^2 + y^2 = 1$ between the planes $z = 0$ and $z = 1$.

38. Rate of Flow of Heat The temperature at a point $P(x, y, z)$ in a medium with thermal conductivity k is proportional to the square of the distance between P and the origin. Find the rate of flow of heat across a sphere S of radius a, centered at the origin.

39. a. Suppose that $\mathbf{F} = P\mathbf{i} + Q\mathbf{j} + R\mathbf{k}$ is a continuous vector field in a region that contains a smooth oriented surface S given by $y = g(x, z)$ and D is its projection onto the xz-plane. Write a double integral similar to Equation (7) that gives $\iint_S \mathbf{F} \cdot d\mathbf{S}$.

b. Use the result of part (a) to evaluate $\iint_S \mathbf{F} \cdot d\mathbf{S}$, where $\mathbf{F}(x, y, z) = y\mathbf{i} + z\mathbf{j} - 3yz^2\mathbf{k}$ and S is the part of the cylinder $x^2 + y^2 = 4$ that lies in the first octant between $z = 0$ and $z = 3$ and oriented away from the origin.

40. Flux of an Electric Field Find the flux of the electric field
$$\mathbf{E} = \frac{q}{4\pi\varepsilon_0}\frac{\mathbf{r}}{|\mathbf{r}|^3}$$
across the sphere $x^2 + y^2 + z^2 = a^2$. Is the flux independent of the radius of the sphere? Give a physical interpretation.

41. Suppose that the density at each point of a thin spherical shell of radius R is proportional to the (linear) distance from the point to a fixed point on the sphere. Find the total mass of the shell.

42. Mass of a Ramp Suppose that the density at each point of a spiral ramp represented by the vector equation $\mathbf{r}(u, v) = u\cos v\mathbf{i} + u\sin v\mathbf{j} + v\mathbf{k}$, where $0 \le u \le 3$ and $0 \le v \le 6\pi$, is proportional to the distance of the point from the central axis of the ramp. What is the total mass of the ramp?

43. Suppose that f is a nonnegative real-valued function defined on the interval $[a, b]$ and f has a continuous derivative in (a, b). Show that the area of the surface of revolution S obtained by revolving the graph of f about the x-axis is given by
$$2\pi \int_a^b f(x)\sqrt{1 + [f'(x)]^2}\, dx$$

Hint: First find a parametric representation of S (see Section 15.6).

44. Find the flux of $\mathbf{F}(x, y, z) = 2xz\mathbf{i} + yz\mathbf{j} - z^2\mathbf{k}$ out of a unit cube $T = \{(x, y, z)\,|\,0 \le x \le 1, 0 \le y \le 1, 0 \le z \le 1\}$.
Hint: The flux out of the cube is the sum of the fluxes across the sides of the cube.

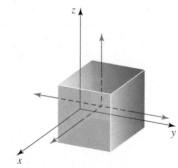

Four of the six unit normal vectors are shown.

45. a. Let f be a function of three variables defined on a region in space containing a surface S. Suppose that S is the graph of the function $z = g(x, y)$ that is represented implicitly by the equation $F(x, y, z) = 0$, where F is differentiable. Show that

$$\iint_S f(x, y, z) \, dS = \iint_D \frac{f\sqrt{F_x^2 + F_y^2 + F_z^2}}{|F_z|} \, dA$$

where D is the projection of S onto the xy-plane.

b. Re-solve Example 1 using the result of part (a).

46. Show that if the surface S is the graph of a function $z = g(x, y)$, then Equation (3) follows from Equation (6) by putting $\mathbf{r}(u, v) = u\mathbf{i} + v\mathbf{j} + g(u, v)\mathbf{k}$.

47. Let $\mathbf{F}$ and $\mathbf{G}$ be continuous vector fields defined on a smooth, oriented surface S. If a and b are constants, show that

$$\iint_S (a\mathbf{F} + b\mathbf{G}) \cdot d\mathbf{S} = a\iint_S \mathbf{F} \cdot d\mathbf{S} + b\iint_S \mathbf{F} \cdot d\mathbf{S}$$

In Exercises 48 and 49, determine whether the statement is true or false. If it is true, explain why. If it is false, explain why or give an example that shows it is false.

48. If $f(x, y, z) \geq 0$, then $\iint_S f \, dS \geq A(S)$, where $A(S)$ is the area of S.

49. If $\mathbf{F}$ is a constant vector field and S is a sphere, then $\iint_S \mathbf{F} \cdot d\mathbf{S} = 0$.

15.8 The Divergence Theorem

Recall that Green's Theorem can be written in the form

$$\oint_C \mathbf{F} \cdot \mathbf{n} \, ds = \iint_R \operatorname{div} \mathbf{F} \, dA \qquad \text{Equation (8) in Section 15.5}$$

where $\mathbf{F}$ is a vector field in the plane, C is an oriented, piecewise-smooth, simple closed curve that bounds a region R, and $\mathbf{n}$ is the outer normal vector to C. The theorem states that the line integral of the normal component of a vector field in two-dimensional space around a simple closed curve is equal to the double integral of the divergence of the vector field over the plane region bounded by the curve.

■ The Divergence Theorem

The *Divergence Theorem* generalizes this result to the case involving vector fields in three-dimensional space. This theorem, also called *Gauss's Theorem* in honor of the German mathematician Karl Friedrich Gauss (1777–1855), relates the surface integral of the normal component of a vector field $\mathbf{F}$ in three-dimensional space over a closed surface S to the triple integral of the divergence of $\mathbf{F}$ over the solid region T bounded by S.

Although the Divergence Theorem is true for very general surfaces, we will restrict our attention to the case in which the solid regions T are simultaneously x-, y-, and z-simple. These regions are called **simple solid regions.** Examples are regions bounded by spheres, ellipsoids, cubes, and tetrahedrons.

THEOREM 1 The Divergence Theorem

Let T be a simple solid region bounded by a closed piecewise-smooth surface S, and let $\mathbf{n}$ be the unit *outer* normal to S. If $\mathbf{F} = P\mathbf{i} + Q\mathbf{j} + R\mathbf{k}$ is a vector field, where P, Q, and R have continuous partial derivatives on an open region containing T, then

$$\iint_S \mathbf{F} \cdot d\mathbf{S} = \iint_S \mathbf{F} \cdot \mathbf{n} \, dS = \iiint_T \operatorname{div} \mathbf{F} \, dV \qquad (1)$$

In words, the surface integral of the normal component of **F** over a closed surface S is equal to the volume integral of the divergence of **F** over the solid T bounded by S.

PROOF Recall that if $\mathbf{F} = P\mathbf{i} + Q\mathbf{j} + R\mathbf{k}$, then

$$\nabla \cdot \mathbf{F} = \frac{\partial P}{\partial x} + \frac{\partial Q}{\partial y} + \frac{\partial R}{\partial z}$$

Therefore, the right-hand side of Equation (1) takes the form

$$\iiint_T \operatorname{div} \mathbf{F} \, dV = \iiint_T \frac{\partial P}{\partial x} \, dV + \iiint_T \frac{\partial Q}{\partial y} \, dV + \iiint_T \frac{\partial R}{\partial z} \, dV$$

Next, if **n** is the unit outer normal vector to S, then the left-hand side of Equation (1) assumes the form

$$\iint_S \mathbf{F} \cdot \mathbf{n} \, dS = \iint_S (P\mathbf{i} + Q\mathbf{j} + R\mathbf{k}) \cdot \mathbf{n} \, dS$$

$$= \iint_S P\mathbf{i} \cdot \mathbf{n} \, dS + \iint_S Q\mathbf{j} \cdot \mathbf{n} \, dS + \iint_S R\mathbf{k} \cdot \mathbf{n} \, dS$$

By equating the last two expressions, we see that the Divergence Theorem will be proved if we can show that

$$\iint_S P\mathbf{i} \cdot \mathbf{n} \, dS = \iiint_T \frac{\partial P}{\partial x} \, dV \tag{2}$$

$$\iint_S Q\mathbf{j} \cdot \mathbf{n} \, dS = \iiint_T \frac{\partial Q}{\partial y} \, dV \tag{3}$$

and

$$\iint_S R\mathbf{k} \cdot \mathbf{n} \, dS = \iiint_T \frac{\partial R}{\partial z} \, dV \tag{4}$$

PROOF OF EQUATION (4) Because T is z-simple, it can be described by the set

$$T = \{(x, y, z) \mid (x, y) \in D, \ k_1(x, y) \le z \le k_2(x, y)\}$$

where D is the projection of T onto the xy-plane, and k_1 and k_2 are continuous functions of x and y. (See Figure 1.) Using Equation (4) in Section 14.6, we obtain

$$\iiint_T \frac{\partial R}{\partial z} \, dV = \iint_D \left[\int_{k_1(x, y)}^{k_2(x, y)} \frac{\partial R}{\partial z} \, dz \right] dA$$

$$= \iint_D [R(x, y, k_2(x, y)) - R(x, y, k_1(x, y))] \, dA \tag{5}$$

that gives an alternative expression for the right-hand side of Equation (4).

Next, observe that the boundary of T may consist of up to six surfaces (Figure 1). On each of the vertical sides, $\mathbf{k} \cdot \mathbf{n} = 0$, so

$$\iint_{S_3} R\mathbf{k} \cdot \mathbf{n} \, dS = \iint_{S_4} R\mathbf{k} \cdot \mathbf{n} \, dS = \iint_{S_5} R\mathbf{k} \cdot \mathbf{n} \, dS = \iint_{S_6} R\mathbf{k} \cdot \mathbf{n} \, dS = 0$$

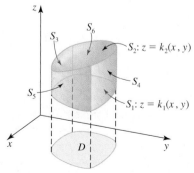

FIGURE 1
T viewed as a z-simple region

Therefore,

$$\iint\limits_S R\mathbf{k} \cdot \mathbf{n}\, dS = \iint\limits_{S_1} R\mathbf{k} \cdot \mathbf{n}\, dS + \iint\limits_{S_2} R\mathbf{k} \cdot \mathbf{n}\, dS \tag{6}$$

To evaluate the integrals on the right-hand side of Equation (6), observe that the outer normal $\mathbf{n}$ points downward on S_1. Writing $g_1(x, y, z) = z - k_1(x, y)$, we find

$$\mathbf{n} = -\frac{\nabla g_1(x, y, z)}{|\nabla g_1(x, y, z)|} = \frac{\dfrac{\partial k_1}{\partial x}\mathbf{i} + \dfrac{\partial k_1}{\partial y}\mathbf{j} - \mathbf{k}}{\sqrt{\left(\dfrac{\partial k_1}{\partial x}\right)^2 + \left(\dfrac{\partial k_1}{\partial y}\right)^2 + 1}}$$

Therefore, using Equation (6), we obtain

$$\iint\limits_{S_1} R\mathbf{k} \cdot \mathbf{n}\, dS = \iint\limits_D R(x, y, k_1(x, y)) \cdot \frac{-1}{\sqrt{\left(\dfrac{\partial k_1}{\partial x}\right)^2 + \left(\dfrac{\partial k_1}{\partial y}\right)^2 + 1}} \sqrt{\left(\dfrac{\partial k_1}{\partial x}\right)^2 + \left(\dfrac{\partial k_1}{\partial y}\right)^2 + 1}\, dA$$

$$= -\iint\limits_D R(x, y, k_1(x, y))\, dA$$

On S_2 the outer normal $\mathbf{n}$ points upward. Writing $g_2(x, y, z) = z - k_2(x, y)$, we see that

$$\mathbf{n} = \frac{\nabla g_2(x, y, z)}{|\nabla g_2(x, y, z)|} = \frac{-\dfrac{\partial k_2}{\partial x}\mathbf{i} - \dfrac{\partial k_2}{\partial y}\mathbf{j} + \mathbf{k}}{\sqrt{\left(\dfrac{\partial k_2}{\partial x}\right)^2 + \left(\dfrac{\partial k_2}{\partial y}\right)^2 + 1}}$$

so

$$\iint\limits_{S_2} R\mathbf{k} \cdot \mathbf{n}\, dS = \iint\limits_D R(x, y, k_2(x, y)) \cdot \frac{1}{\sqrt{\left(\dfrac{\partial k_2}{\partial x}\right)^2 + \left(\dfrac{\partial k_2}{\partial y}\right)^2 + 1}} \sqrt{\left(\dfrac{\partial k_2}{\partial x}\right)^2 + \left(\dfrac{\partial k_2}{\partial y}\right)^2 + 1}\, dA$$

$$= \iint\limits_D R(x, y, k_2(x, y))\, dA$$

Therefore, Equation (6) becomes

$$\iint\limits_S R\mathbf{k} \cdot \mathbf{n}\, dS = \iint\limits_D [R(x, y, k_2(x, y)) - R(x, y, k_1(x, y))]\, dA$$

Comparing this with Equation (5), we have

$$\iint\limits_S R\mathbf{k} \cdot \mathbf{n}\, dS = \iiint\limits_T \frac{\partial R}{\partial z}\, dV$$

so Equation (4) is established. Equations (2) and (3) are proved in a similar manner by viewing R as x-simple and y-simple, respectively. ∎

EXAMPLE 1 Compute $\iint_S \mathbf{F} \cdot \mathbf{n}\, dS$ given that

$$\mathbf{F}(x, y, z) = (x + \sin z)\mathbf{i} + (2y + \cos x)\mathbf{j} + (3z + \tan y)\mathbf{k}$$

and S is the unit sphere $x^2 + y^2 + z^2 = 1$.

Solution To evaluate the integral directly would be a difficult task. Applying the Divergence Theorem, we have

$$\iint_S \mathbf{F} \cdot \mathbf{n} \, dS = \iiint_T \text{div } \mathbf{F} \, dV$$

But

$$\nabla \cdot \mathbf{F} = \frac{\partial}{\partial x}(x + \sin z) + \frac{\partial}{\partial y}(2y + \cos x) + \frac{\partial}{\partial z}(3z + \tan y) = 1 + 2 + 3 = 6$$

and the solid T is the unit ball B bounded by the unit sphere $x^2 + y^2 + z^2 = 1$. Therefore,

$$\iint_S \mathbf{F} \cdot \mathbf{n} \, dS = \iiint_B \nabla \cdot \mathbf{F} \, dV = \iiint_B 6 \, dV = 6V(B) = 6\left[\frac{4}{3}\pi(1)^3\right] = 8\pi \quad \blacksquare$$

EXAMPLE 2 Let T be the solid bounded by the cylinder $x^2 + y^2 = 4$ and the planes $z = 0$ and $z = 3$, and let S be the surface of T. Calculate the outward flux of the vector field $\mathbf{F}(x, y, z) = xy^2\mathbf{i} + yz^2\mathbf{j} + zx^2\mathbf{k}$ over S.

Solution The surface S is shown in Figure 2. The flux of $\mathbf{F}$ over S is given by $\iint_S \mathbf{F} \cdot \mathbf{n} \, dS$, which by the Divergence Theorem can also be found by evaluating $\iiint_T \nabla \cdot \mathbf{F} \, dV$. Now

$$\text{div } \mathbf{F} = \frac{\partial}{\partial x}(xy^2) + \frac{\partial}{\partial y}(yz^2) + \frac{\partial}{\partial z}(zx^2) = y^2 + z^2 + x^2$$

Therefore,

$$\iint_S \mathbf{F} \cdot \mathbf{n} \, dS = \iiint_T \text{div } \mathbf{F} \, dV = \iiint_T (x^2 + y^2 + z^2) \, dV$$

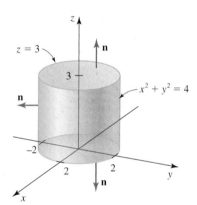

FIGURE 2
The surface S and some of the outer normals to S

Using cylindrical coordinates to evaluate the triple integral, we have

$$\iint_S \mathbf{F} \cdot \mathbf{n} \, dS = \int_0^{2\pi} \int_0^2 \int_0^3 (r^2 + z^2)r \, dz \, dr \, d\theta$$

$$= \int_0^{2\pi} \int_0^2 \left[r^3 z + \frac{1}{3}rz^3\right]_{z=0}^{z=3} dr \, d\theta$$

$$= \int_0^{2\pi} \int_0^2 (3r^3 + 9r) \, dr \, d\theta = \int_0^{2\pi} \left[\frac{3}{4}r^4 + \frac{9}{2}r^2\right]_{r=0}^{r=2} d\theta$$

$$= \int_0^{2\pi} 30 \, d\theta = 60\pi \quad \blacksquare$$

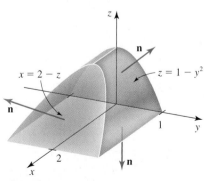

FIGURE 3
The surface S and some of the outer normals to S

EXAMPLE 3 Let T be the region bounded by the parabolic cylinder $z = 1 - y^2$ and the planes $z = 0$, $x = 0$, and $x + z = 2$, and let S be the surface of T. If $\mathbf{F}(x, y, z) = xy^2\mathbf{i} + \left(\frac{1}{3}y^3 - \cos xz\right)\mathbf{j} + xe^y\mathbf{k}$, find $\iint_S \mathbf{F} \cdot \mathbf{n} \, dS$.

Solution The surface S is shown in Figure 3. We first compute

$$\nabla \cdot \mathbf{F} = \frac{\partial}{\partial x}(xy^2) + \frac{\partial}{\partial y}\left(\frac{1}{3}y^3 - \cos xz\right) + \frac{\partial}{\partial z}(xe^y) = y^2 + y^2 = 2y^2$$

Viewing T as an x-simple region, we use the Divergence Theorem to obtain

$$\iint_S \mathbf{F} \cdot \mathbf{n} \, dS = \iiint_T \text{div } \mathbf{F} \, dV = \iiint_T 2y^2 \, dV$$

$$= \int_{-1}^{1} \int_0^{1-y^2} \int_0^{2-z} 2y^2 \, dx \, dz \, dy = \int_{-1}^{1} \int_0^{1-y^2} \left[2y^2 x \right]_{x=0}^{x=2-z} dz \, dy$$

$$= \int_{-1}^{1} \int_0^{1-y^2} 2y^2 (2 - z) \, dz \, dy = \int_{-1}^{1} \left[-y^2 (2 - z)^2 \right]_{z=0}^{z=1-y^2} dy$$

$$= \int_{-1}^{1} \left[-y^2 (1 + y^2)^2 + 4y^2 \right] dy = \int_{-1}^{1} (3y^2 - 2y^4 - y^6) \, dy$$

$$= 2 \int_0^1 (3y^2 - 2y^4 - y^6) \, dy$$

$$= 2 \left[y^3 - \frac{2}{5} y^5 - \frac{1}{7} y^7 \right]_0^1 = \frac{32}{35} \qquad \blacksquare$$

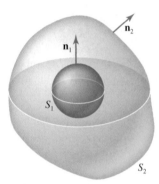

FIGURE 4
The region T lies between S_1 and S_2.

The Divergence Theorem was stated for simple solid regions. But it can be extended to include regions that are finite unions of simple solid regions. For example, let T be the region that lies between the closed surfaces S_1 and S_2 with S_1 lying inside S_2. Then the boundary of T is $S = S_1 \cup S_2$, as shown in Figure 4. If $\mathbf{n}_1$ and $\mathbf{n}_2$ denote the outward normals of S_1 and S_2, respectively, then the normal to S is given by $\mathbf{n} = -\mathbf{n}_1$ on S_1 and by $\mathbf{n} = \mathbf{n}_2$ on S_2. Applying the Divergence Theorem to T, we obtain

$$\iiint_T \text{div } \mathbf{F} \, dV = \iint_S \mathbf{F} \cdot \mathbf{n} \, dS = \iint_{S_1} \mathbf{F} \cdot \mathbf{n} \, dS + \iint_{S_2} \mathbf{F} \cdot \mathbf{n} \, dS$$

$$= \iint_{S_1} \mathbf{F} \cdot (-\mathbf{n}_1) \, dS + \iint_{S_2} \mathbf{F} \cdot \mathbf{n}_2 \, dS$$

$$= -\iint_{S_1} \mathbf{F} \cdot \mathbf{n}_1 \, dS + \iint_{S_2} \mathbf{F} \cdot \mathbf{n}_2 \, dS$$

$$= -\iint_{S_1} \mathbf{F} \cdot d\mathbf{S} + \iint_{S_2} \mathbf{F} \cdot d\mathbf{S} \qquad (7)$$

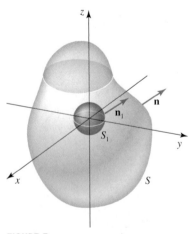

FIGURE 5
The region T lies between the sphere S_1 and the surface S.

EXAMPLE 4 **Flux of an Electric Field** Consider the electric field

$$\mathbf{E} = \frac{q}{4\pi\varepsilon_0} \frac{\mathbf{r}}{|\mathbf{r}|^3}$$

induced by a point charge q placed at the origin of a three-dimensional coordinate system, where $\mathbf{r} = x\mathbf{i} + y\mathbf{j} + z\mathbf{k}$. (See Example 4 in Section 15.1.) Find the flux of $\mathbf{E}$ across a smooth surface S that encloses the origin.

Solution The Divergence Theorem is not immediately applicable because $\mathbf{E}$ is not continuous at the origin. To avoid this difficulty, let's construct a sphere with center at the origin and a radius a that is small enough to ensure that the sphere lies completely inside S. (See Figure 5.) If we denote this sphere by S_1, then $\mathbf{E}$ satisfies the conditions of the Divergence Theorem for the solid T that lies between S_1 and S. Using Equation (7), we obtain

$$\iiint\limits_{T} \operatorname{div} \mathbf{E} \, dV = -\iint\limits_{S_1} \mathbf{E} \cdot d\mathbf{S} + \iint\limits_{S} \mathbf{E} \cdot d\mathbf{S}$$

But we showed that div $\mathbf{E} = 0$ in Example 4 in Section 15.2. So we have

$$\iint\limits_{S} \mathbf{E} \cdot d\mathbf{S} = \iint\limits_{S_1} \mathbf{E} \cdot d\mathbf{S} = \iint\limits_{S_1} \mathbf{E} \cdot \mathbf{n} \, dS$$

To evaluate the integral on the right, note that the unit normal to the sphere S_1 is $\mathbf{n} = \mathbf{r}/|\mathbf{r}|$. Therefore,

$$\mathbf{E} \cdot \mathbf{n} = \frac{q}{4\pi\varepsilon_0} \frac{\mathbf{r}}{|\mathbf{r}|^3} \cdot \left(\frac{\mathbf{r}}{|\mathbf{r}|}\right) = \frac{q}{4\pi\varepsilon_0} \frac{\mathbf{r} \cdot \mathbf{r}}{|\mathbf{r}|^4}$$

$$= \frac{q}{4\pi\varepsilon_0|\mathbf{r}|^2} \qquad \mathbf{r} \cdot \mathbf{r} = |\mathbf{r}|^2$$

$$= \frac{q}{4\pi\varepsilon_0 a^2}$$

because $|\mathbf{r}| = a$ on the sphere S_1. Therefore, we have

$$\iint\limits_{S} \mathbf{E} \cdot d\mathbf{S} = \iint\limits_{S_1} \mathbf{E} \cdot d\mathbf{S} = \frac{q}{4\pi\varepsilon_0 a^2} \iint\limits_{S_1} dS = \frac{q}{4\pi\varepsilon_0 a^2} A(S_1)$$

$$= \frac{q}{4\pi\varepsilon_0 a^2} (4\pi a^2) = \frac{q}{\varepsilon_0} \qquad \blacksquare$$

The result in Example 4 shows that the flux across *any* closed surface that contains the charge q is q/ε_0. This is intuitively clear, since any closed surface enclosing the charge q would trap the same number of field lines.

Furthermore, by using the principle of superposition (the field induced by several electric charges is the vector sum of the fields due to the individual charges), it can be shown that for any closed surface S,

$$\iint\limits_{S} \mathbf{E} \cdot \mathbf{n} \, dS = \frac{Q}{\varepsilon_0}$$

where Q is the total charge enclosed by S. This is one of the most important laws in electrostatics and is known as **Gauss's Law.**

■ Interpretation of Divergence

For a physical interpretation of the divergence of a vector field $\mathbf{F}$, we can think of $\mathbf{F}$ as representing the velocity field associated with the flow of a fluid. Let $P_0(x_0, y_0, z_0)$ be a point in the fluid, and let B_r be a ball with radius r, centered at P_0, and having the sphere S_r for its boundary, as shown in Figure 6. If r is small, then the continuity of

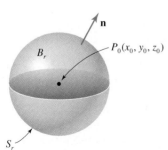

FIGURE 6
B_r is a ball of radius r centered at $P_0(x_0, y_0, z_0)$.

div $\mathbf{F}$ guarantees that $(\text{div }\mathbf{F})(P) \approx (\text{div }\mathbf{F})(P_0)$ for all points P in B_r. Therefore, using the Divergence Theorem, we have

$$\iint\limits_{S_r} \mathbf{F} \cdot \mathbf{n}\, dS = \iiint\limits_{B_r} \text{div} \cdot \mathbf{F}\, dV \approx \iiint\limits_{B_r} (\text{div }\mathbf{F})(P_0)\, dV$$

$$= (\text{div }\mathbf{F})(P_0)\iiint\limits_{B_r} dV = (\text{div }\mathbf{F})(P_0)V(B_r)$$

where $V(B_r) = \frac{4}{3}\pi r^3$. This approximation improves as $r \to 0$, and we have

$$(\text{div }\mathbf{F})(P_0) = \lim_{r \to 0} \frac{1}{V(B_r)} \iint\limits_{B_r} \mathbf{F} \cdot \mathbf{n}\, dS \qquad (8)$$

Equation (8) tells us that we can interpret div $\mathbf{F}(P_0)$ as the rate of flow outward of the fluid per unit volume at P_0—hence the term *divergence*. In general, if div $\mathbf{F}(P) > 0$, the net flow is outward near P, and P is called a **source**. If div $\mathbf{F}(P) < 0$, the net flow is inward near P, and P is called a **sink**. Finally, if the fluid is incompressible and there are no sources or sinks present, then no fluid exits or enters B_r and, accordingly, div $\mathbf{F}(P) = 0$ at every point P.

15.8   CONCEPT QUESTIONS

1. State the Divergence Theorem.
2. Suppose that the vector field $\mathbf{F}$ is associated with the flow of a fluid and P is a point in the domain of $\mathbf{F}$. Give a physical interpretation for div $\mathbf{F}(P)$. What happens to the flow at P if div $\mathbf{F}(P) > 0$? If div $\mathbf{F}(P) < 0$? If div $\mathbf{F}(P) = 0$?

15.8   EXERCISES

In Exercises 1–4, verify the Divergence Theorem for the given vector field $\mathbf{F}$ and region T.

1. $\mathbf{F}(x, y, z) = x\mathbf{i} + y\mathbf{j} + z\mathbf{k}$; T is the cube bounded by the planes $x = 0$, $x = 1$, $y = 0$, $y = 1$, $z = 0$, and $z = 1$

2. $\mathbf{F}(x, y, z) = 2xy\mathbf{i} - y^2\mathbf{j} + 3yz\mathbf{k}$; T is the cube bounded by the planes $x = 0$, $x = 2$, $y = 0$, $y = 2$, $z = 0$, and $z = 2$

3. $\mathbf{F}(x, y, z) = y\mathbf{i} + z\mathbf{j} - 3yz^2\mathbf{k}$; T is the region bounded by the cylinder $x^2 + y^2 = 4$ in the first octant between $z = 0$ and $z = 3$

4. $\mathbf{F}(x, y, z) = x\mathbf{i} + y\mathbf{j} + 2z^2\mathbf{k}$; T is the region bounded by the paraboloid $z = x^2 + y^2$ and the plane $z = 1$

In Exercises 5–18, use the Divergence Theorem to find the flux of $\mathbf{F}$ across S; that is, calculate $\iint_S \mathbf{F} \cdot \mathbf{n}\, dS$.

5. $\mathbf{F}(x, y, z) = xy^2\mathbf{i} + 2yz\mathbf{j} - 3x^2y^3\mathbf{k}$; S is the surface of the cube bounded by the planes $x = \pm 1$, $y = \pm 1$, and $z = \pm 1$

6. $\mathbf{F}(x, y, z) = 2xz\mathbf{i} - y^2\mathbf{j} + yz\mathbf{k}$; S is the surface of the rectangular box bounded by the planes $x = 0$, $x = 2$, $y = 0$, $y = 3$, $z = -1$, and $z = 1$

7. $\mathbf{F}(x, y, z) = (x^3 + \cos y)\mathbf{i} + (y^3 + \sin xz)\mathbf{j} + (z^3 + 2e^{-x})\mathbf{k}$; S is the surface of the region bounded by the cylinder $y^2 + z^2 = 1$ and the planes $x = 0$ and $x = 3$

8. $\mathbf{F}(x, y, z) = \sin y\mathbf{i} + (x^2y + e^z)\mathbf{j} + (2x^2z + e^{-x})\mathbf{k}$; S is the surface of the region bounded by the cylinder $z = 4 - x^2$ and the planes $z = 0$, $y = 0$ and $y + z = 5$

9. $\mathbf{F}(x, y, z) = 2xy\mathbf{i} + y^2\mathbf{j} + (x^2 + yz)\mathbf{k}$; S is the surface of the tetrahedron bounded by the planes $x + y + z = 1$, $x = 0$, $y = 0$, and $z = 0$

10. $\mathbf{F}(x, y, z) = x^2\mathbf{i} + xz^2\mathbf{j} + (2xz + \sin xy)\mathbf{k}$; S is the surface of the tetrahedron bounded by the planes $x + 2y + 3z = 6$, $x = 0$, $y = 0$, and $z = 0$

11. $\mathbf{F}(x, y, z) = x\mathbf{i} + 2y\mathbf{j} + 3z\mathbf{k}$; S is the sphere $x^2 + y^2 + z^2 = 9$

12. $\mathbf{F}(x, y, z) = (x + y^2)\mathbf{i} + (\sqrt{x^2 + z^2} + y)\mathbf{j} + \cos xy\, \mathbf{k}$; S is the surface of the region bounded by the cylinder $x^2 + z^2 = 1$ and the planes $y = 0$ and $y = 1$

13. $\mathbf{F}(x, y, z) = xz\mathbf{i} - yz\mathbf{j} + xy\mathbf{k}$; S is the ellipsoid $9x^2 + 4y^2 + 36z^2 = 36$

14. $\mathbf{F}(x, y, z) = (x + ye^z)\mathbf{i} - (y + \tan xz)\mathbf{j} + (x + \cosh y)\mathbf{k}$;
S is the surface of the region bounded by the cone
$y = \sqrt{x^2 + z^2}$ and the plane $y = 4$

15. $\mathbf{F}(x, y, z) = (x^3 + 1)\mathbf{i} + (yz^2 + \cos xz)\mathbf{j} + (2y^2z + e^{\tan x})\mathbf{k}$;
S is the sphere $x^2 + y^2 + z^2 = 1$

16. $\mathbf{F}(x, y, z) = x^3\mathbf{i} - 3yz^2\mathbf{j} + 2z^3\mathbf{k}$; S is the surface of the
region bounded by the paraboloid $y = 9 - x^2 - z^2$ and the
xz-plane

17. $\mathbf{F}(x, y, z) = xz\mathbf{i} + x^2y\mathbf{j} + (y^2z + 1)\mathbf{k}$; S is the surface of
the region that lies between the cylinders $x^2 + y^2 = 1$ and
$x^2 + y^2 = 4$ and between the planes $z = 1$ and $z = 3$

18. $\mathbf{F}(x, y, z) = yz^2\mathbf{i} + (y^3 + xz)\mathbf{j} - (y^2z + 10x)\mathbf{k}$; S is the
surface of the region between the spheres $x^2 + y^2 + z^2 = 1$
and $x^2 + y^2 + z^2 = 4$

*In Exercises 19–25, assume that S and T satisfy the conditions of
the Divergence Theorem.*

19. Show that the volume of T is given by $V(T) = \frac{1}{3}\iint_S \mathbf{r} \cdot \mathbf{n}\, dS$,
where $\mathbf{r} = x\mathbf{i} + y\mathbf{j} + z\mathbf{k}$.

20. Show that $\iint_S \mathbf{a} \cdot \mathbf{n}\, dS = 0$, where $\mathbf{a}$ is a constant vector.

21. Show that if $\mathbf{F}$ has continuous second-order derivatives, then

$$\iint_S \text{curl } \mathbf{F} \cdot \mathbf{n}\, dS = 0$$

22. Show that if f has continuous second-order partial derivatives, then

$$\iiint_T \nabla^2 f\, dV = \iint_S D_{\mathbf{n}} f\, dS$$

where $\nabla^2 f = \dfrac{\partial^2 f}{\partial x^2} + \dfrac{\partial^2 f}{\partial y^2} + \dfrac{\partial^2 f}{\partial z^2}$ and $D_{\mathbf{n}} f$ is the directional
derivative of f in the direction of an outer normal $\mathbf{n}$ of S.

23. Show that $\iiint_T \nabla f\, dV = \iint_S f\mathbf{n}\, dS$.
Hint: Apply the Divergence Theorem to $\mathbf{F} = f\mathbf{c}$, where $\mathbf{c}$ is a constant vector.

24. Show that if f and g have continuous second-order partial
derivatives, then

$$\iiint_T (f\nabla^2 g + \nabla f \cdot \nabla g)\, dV = \iint_S (f\nabla g) \cdot \mathbf{n}\, dS$$

25. Show that if f and g have continuous second-order partial
derivatives, then

$$\iiint_T (f\nabla^2 g - g\nabla^2 f)\, dV = \iint_S (f\nabla g - g\nabla f) \cdot \mathbf{n}\, dS$$

26. Find the flux of the vector field

$$\mathbf{F}(x, y, z) = \frac{x\mathbf{i} + y\mathbf{j} + z\mathbf{k}}{(x^2 + y^2 + z^2)^{3/2}}$$

across the ellipsoid $(x^2/9) + (y^2/16) + (z^2/4) = 1$.

*In Exercises 27–29, determine whether the statement is true or
false. If it is true, explain why. If it is false, explain why or give
an example that shows it is false.*

27. If div $\mathbf{F} = 0$, then $\iint_S \mathbf{F} \cdot d\mathbf{S} = 0$ for every closed surface S.

28. If $\mathbf{F}$ is a constant vector field and S is a cube, then
$\iint_S \mathbf{F} \cdot d\mathbf{S} = 0$.

29. If $|\mathbf{F}(x, y, z)| \leq 1$ for all points (x, y, z) in a solid region T
bounded by a closed surface S, then $\iiint_T \text{div } \mathbf{F}\, dV \leq A(S)$,
where $A(S)$ is the area of S.

15.9 Stokes' Theorem

■ Stokes' Theorem

In this section we consider another generalization of Green's Theorem to higher dimensions. We start with the following version of Green's Theorem:

$$\oint_C \mathbf{F} \cdot \mathbf{T}\, ds = \iint_R \text{curl } \mathbf{F} \cdot \mathbf{k}\, dA \qquad \text{Equation (7) in Section 15.5}$$

where the plane curve C is an oriented piecewise-smooth simply closed curve that
bounds a region R. The theorem states that the line integral of the tangential component of a vector field in two-dimensional space around a closed curve is equal to the
double integral of the normal component to R of the curl of the vector field over the
plane region bounded by the curve.

Stokes' Theorem generalizes this version of Green's Theorem to three-dimensional
space. Named after the English mathematical physicist George G. Stokes (1819–1903),

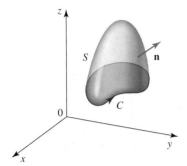

FIGURE 1
The curve C is a boundary of the surface S.

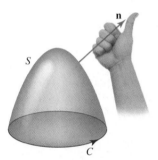

FIGURE 2
The orientation of S induces an orientation on C.

Stokes' Theorem relates the line integral of the tangential component of a vector field in three-dimensional space around a simple closed curve in space to the surface integral of the normal component of the curl of the vector field over any surface that has the closed curve as its boundary. (See Figure 1.)

The orientation of the surface induces an orientation on C that is determined by using the right-hand rule: Imagine grasping the normal vector $\mathbf{n}$ to S with your right hand in such a way that your thumb points in the direction of $\mathbf{n}$. Then your fingers will point toward the positive direction of C. (See Figure 2.)

THEOREM 1   Stokes' Theorem

Let S be an oriented piecewise-smooth surface that has a unit normal vector $\mathbf{n}$ and is bounded by a simple, closed, positively oriented curve C. If $\mathbf{F} = P\mathbf{i} + Q\mathbf{j} + R\mathbf{k}$ is a vector field, where P, Q, and R have continuous partial derivatives in an open region containing S, then

$$\oint_C \mathbf{F} \cdot d\mathbf{r} = \oint_C \mathbf{F} \cdot \mathbf{T}\,ds = \iint_S \text{curl } \mathbf{F} \cdot d\mathbf{S} = \iint_S \text{curl } \mathbf{F} \cdot \mathbf{n}\,dS \qquad \textbf{(1)}$$

In words, the line integral of the tangential component of $\mathbf{F}$ around a simple closed curve C is equal to the surface integral of the normal component of the curl of $\mathbf{F}$ over any surface S with C as its boundary.

Stokes' Theorem provides us with the following physical interpretation: If $\mathbf{F}$ is a force field, then the work done by $\mathbf{F}$ along C is equal to the flux of curl $\mathbf{F}$ across S. The proof of Stokes' Theorem can be found in more advanced textbooks.

EXAMPLE 1 Verify Stokes' Theorem for the case in which $\mathbf{F}(x, y, z) = 3z\mathbf{i} + 2x\mathbf{j} + y^2\mathbf{k}$, S is the part of the paraboloid $z = 4 - x^2 - y^2$ with $z \geq 0$, and C is the trace of S on the xy-plane.

Solution The surface S and the curve C are sketched in Figure 3. We begin by calculating

$$\text{curl } \mathbf{F} = \begin{vmatrix} \mathbf{i} & \mathbf{j} & \mathbf{k} \\ \dfrac{\partial}{\partial x} & \dfrac{\partial}{\partial y} & \dfrac{\partial}{\partial z} \\ 3z & 2x & y^2 \end{vmatrix} = 2y\mathbf{i} + 3\mathbf{j} + 2\mathbf{k}$$

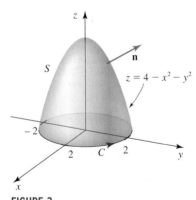

FIGURE 3
The outer normal $\mathbf{n}$ to S induces the positive direction for C as shown.

Next, writing $g(x, y) = 4 - x^2 - y^2$, we find that $g_x = -2x$ and $g_y = -2y$. Also, observe that the projection of S onto the xy-plane is $R = \{(x, y) \mid x^2 + y^2 \leq 4\}$. So using Equation (7) of Section 15.7 with $P(x, y) = 2y$, $Q(x, y) = 3$, and $R(x, y) = 2$, we obtain

$$\iint_S \text{curl } \mathbf{F} \cdot d\mathbf{S} = \iint_R (-Pg_x - Qg_y + R)\,dA$$

$$= \iint_R (4xy + 6y + 2)\,dA$$

Changing to polar coordinates, we find

$$\iint_S \text{curl } \mathbf{F} \cdot d\mathbf{S} = \int_0^{2\pi} \int_0^2 (4r^2 \cos \theta \sin \theta + 6r \sin \theta + 2)r \, dr \, d\theta$$

$$= \int_0^{2\pi} \int_0^2 (4r^3 \cos \theta \sin \theta + 6r^2 \sin \theta + 2r) \, dr \, d\theta$$

$$= \int_0^{2\pi} \left[r^4 \cos \theta \sin \theta + 2r^3 \sin \theta + r^2 \right]_{r=0}^{r=2} d\theta$$

$$= \int_0^{2\pi} [16 \cos \theta \sin \theta + 16 \sin \theta + 4) \, d\theta$$

$$= \left[8 \sin^2 \theta - 16 \cos \theta + 4\theta \right]_0^{2\pi} = 8\pi$$

To evaluate the line integral on the left-hand side of Equation (1), we observe that C can be parametrized by

$$\mathbf{r}(t) = 2 \cos t\mathbf{i} + 2 \sin t\mathbf{j} + 0\mathbf{k} \qquad 0 \le t \le 2\pi$$

Therefore, with $\mathbf{F}(\mathbf{r}(t)) = 0\mathbf{i} + 4 \cos t\mathbf{j} + 4 \sin^2 t\mathbf{k}$ we have

$$\oint_C \mathbf{F} \cdot d\mathbf{r} = \int_0^{2\pi} \mathbf{F}(\mathbf{r}(t)) \cdot \mathbf{r}'(t) \, dt$$

$$= \int_0^{2\pi} (4 \cos t\mathbf{j} + 4 \sin^2 t\mathbf{k}) \cdot (-2 \sin t\mathbf{i} + 2 \cos t\mathbf{j}) \, dt$$

$$= \int_0^{2\pi} 8 \cos^2 t \, dt = 4 \int_0^{2\pi} (1 + \cos 2t) \, dt$$

$$= 4 \left[t + \frac{1}{2} \sin 2t \right]_0^{2\pi} = 8\pi$$

which is the same as the surface integral. This verifies the solution. ▪

EXAMPLE 2 Evaluate $\oint_C \mathbf{F} \cdot d\mathbf{r}$, where $\mathbf{F}(x, y, z) = \cos z\mathbf{i} + x^2\mathbf{j} + 2y\mathbf{k}$ and C is the curve of intersection of the plane $x + z = 2$ and the cylinder $x^2 + y^2 = 1$.

Solution The curve C is an ellipse, as shown in Figure 4. We can evaluate $\oint_C \mathbf{F} \cdot d\mathbf{r}$ directly (see Exercise 29). But it is easier to use Stokes' Theorem. Thus,

$$\oint_C \mathbf{F} \cdot d\mathbf{r} = \oint_C \mathbf{F} \cdot \mathbf{T} \, ds = \iint_S \text{curl } \mathbf{F} \cdot d\mathbf{S}$$

where S is the elliptic plane region lying in the plane $x + z = 2$ and enclosed by C. Of all the surfaces that have C as their boundary, this choice of S is clearly the most convenient for our purpose.

We first find

$$\text{curl } \mathbf{F} = \begin{vmatrix} \mathbf{i} & \mathbf{j} & \mathbf{k} \\ \dfrac{\partial}{\partial x} & \dfrac{\partial}{\partial y} & \dfrac{\partial}{\partial z} \\ \cos z & x^2 & 2y \end{vmatrix} = 2\mathbf{i} - \sin z\mathbf{j} + 2x\mathbf{k}$$

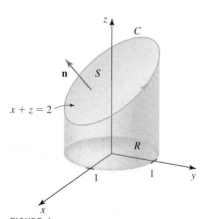

FIGURE 4
The surface S is enclosed by C and the disk R is the projection of S onto the xy-plane.

Then, writing $g(x, y) = -x + 2$, we find $g_x = -1$ and $g_y = 0$. Next, observe that the projection of S onto the xy-plane is $R = \{(x, y) \mid x^2 + y^2 \le 1\}$. So using Equation (7) of Section 15.7 with $P(x, y) = 2$, $Q(x, y) = -\sin z$, and $R(x, y) = 2x$, we obtain

$$
\iint_S \text{curl } \mathbf{F} \cdot d\mathbf{S} = \iint_R (-Pg_x - Qg_y + R) \, dA
$$

$$
= \iint_R (2 + 2x) \, dx
$$

$$
= 2 \iint_R (1 + x) \, dA
$$

Changing to polar coordinates, we obtain

$$
\iint_S \text{curl } \mathbf{F} \cdot d\mathbf{S} = 2 \int_0^{2\pi} \int_0^1 (1 + r \cos \theta) r \, dr \, d\theta = 2 \int_0^{2\pi} \int_0^1 (r + r^2 \cos \theta) \, dr \, d\theta
$$

$$
= 2 \int_0^{2\pi} \left[\frac{1}{2} r^2 + \frac{1}{3} r^3 \cos \theta \right]_{r=0}^{r=1} d\theta = 2 \int_0^{2\pi} \left(\frac{1}{2} + \frac{1}{3} \cos \theta \right) d\theta
$$

$$
= 2 \left[\frac{1}{2} \theta + \frac{1}{3} \sin \theta \right]_0^{2\pi} = 2\pi
$$

so by Stokes' Theorem we have

$$
\oint_C \mathbf{F} \cdot d\mathbf{r} = \iint_S \text{curl } \mathbf{F} \cdot d\mathbf{S} = 2\pi \qquad \blacksquare
$$

EXAMPLE 3 Evaluate $\iint_S \text{curl } \mathbf{F} \cdot d\mathbf{S}$, where $\mathbf{F}(x, y, z) = yz\mathbf{i} - xz\mathbf{j} + z^3\mathbf{k}$ and S is the part of the sphere $x^2 + y^2 + z^2 = 8$ lying inside the cone $z = \sqrt{x^2 + y^2}$. (See Figure 5.)

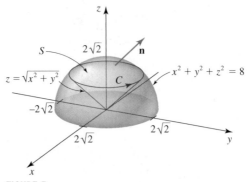

FIGURE 5
The surface S is enclosed by C; the outer normal $\mathbf{n}$ induces the positive direction of C shown.

Solution The boundary of S is the curve whose equation is obtained by solving the equations $x^2 + y^2 + z^2 = 8$ and $z = \sqrt{x^2 + y^2}$ simultaneously. Squaring the second equation and substituting this result into the first equation give $2z^2 = 8$, or $z = 2$ (since $z > 0$). Therefore, the boundary of S is the circle C with equations $x^2 + y^2 = 4$ and

$z = 2$. Since C is easily parametrized, we can use Stokes' Theorem to help us find the value of the given surface integral by evaluating the line integral $\oint_C \mathbf{F} \cdot d\mathbf{r}$. A vector equation of C is

$$\mathbf{r}(t) = 2\cos t\,\mathbf{i} + 2\sin t\,\mathbf{j} + 2\mathbf{k} \qquad 0 \le t \le 2\pi$$

so

$$\mathbf{r}'(t) = -2\sin t\,\mathbf{i} + 2\cos t\,\mathbf{j}$$

Furthermore, we have

$$\mathbf{F}(\mathbf{r}(t)) = (2\sin t)(2)\mathbf{i} - (2\cos t)(2)\mathbf{j} + (2^3)\mathbf{k} = 4\sin t\,\mathbf{i} - 4\cos t\,\mathbf{j} + 8\mathbf{k}$$

Therefore,

$$\iint_S \text{curl } \mathbf{F} \cdot d\mathbf{S} = \oint_C \mathbf{F} \cdot d\mathbf{r}$$

$$= \oint_C \mathbf{F}(\mathbf{r}(t)) \cdot \mathbf{r}'(t)\, dt$$

$$= \int_0^{2\pi} (4\sin t\,\mathbf{i} - 4\cos t\,\mathbf{j} + 8\mathbf{k}) \cdot (-2\sin t\,\mathbf{i} + 2\cos t\,\mathbf{j})\, dt$$

$$= \int_0^{2\pi} (-8\sin^2 t - 8\cos^2 t)\, dt$$

$$= -8\int_0^{2\pi} dt = -16\pi \qquad \blacksquare$$

■ Interpretation of Curl

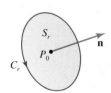

FIGURE 6
C_r is a circular disk of
radius r at $P_0(x_0, y_0, z_0)$.

Just as the divergence theorem can be used to give a physical interpretation of the divergence of a vector field, Stokes' Theorem can be used to give a physical interpretation of the curl vector. Once again, we think of $\mathbf{F}$ as representing the velocity field in fluid flow. Let $P_0(x_0, y_0, z_0)$ be a point in the fluid, and let S be a circular disk with radius r centered at P_0 and boundary C_r, as shown in Figure 6. Let $\mathbf{n}$ be a unit vector normal to S_r at P_0. Applying Stokes' Theorem to the vector field $\mathbf{F}$ on the surface S_r, we obtain

$$\oint_C \mathbf{F} \cdot d\mathbf{r} = \oint_C \mathbf{F} \cdot \mathbf{T}\, ds = \iint_{S_r} \text{curl } \mathbf{F} \cdot d\mathbf{S} \qquad (2)$$

Since $\mathbf{F} \cdot \mathbf{T}$ is the component of $\mathbf{F}$ tangent to C_r, we see that $\oint_{C_r} \mathbf{F} \cdot d\mathbf{r}$ is a measure of the tendency of the fluid to move around C_r, and accordingly, this line integral is called the **circulation of F around C_r.** By taking r small, we see that the circulation of $\mathbf{F}$ around C_r is a measure of the tendency of the field to rotate around the axis determined by $\mathbf{n}$.

Next, for small r the continuity of curl $\mathbf{F}$ implies that $(\text{curl } \mathbf{F})(P) \approx (\text{curl } \mathbf{F})(P_0)$ for all points P in S_r. Therefore, if r is small, we can write

$$\iint_{S_r} \text{curl } \mathbf{F} \cdot d\mathbf{S} = \iint_{S_r} (\text{curl } \mathbf{F})(P_0) \cdot \mathbf{n}\, dS$$

$$\approx (\text{curl } \mathbf{F})(P_0) \cdot \mathbf{n} \iint_{S_r} dS = (\text{curl } \mathbf{F})(P_0) \cdot \mathbf{n}(\pi r^2)$$

So for small r, Equation (2) gives

$$\oint_C \mathbf{F} \cdot d\mathbf{r} \approx (\text{curl } \mathbf{F})(P_0) \cdot \mathbf{n}(\pi r^2)$$

The approximation improves as $r \to 0$, and we have

$$(\text{curl } \mathbf{F})(P_0) \cdot \mathbf{n} = \lim_{r \to 0} \frac{1}{\pi r^2} \oint_C \mathbf{F} \cdot d\mathbf{r} \tag{3}$$

Equation (3) gives the relationship between the curl and the circulation. It tells us that we can think of $|\text{curl } \mathbf{F}(P_0)|$ as a measure of the magnitude of the tendency of the fluid to rotate about the axis determined by $\mathbf{n}$. It also tells us that we can think of curl $\mathbf{F}(P_0)$ as determining the axis about which the circulation of $\mathbf{F}$ is greatest near P_0.

We now summarize the types of line and surface integrals and the major theorems associated with these integrals.

■ Summary of Line and Surface Integrals

Line Integrals
a. Element of arc length: $ds = \|\mathbf{r}'(t)\| \, dt$ $\qquad\qquad\qquad\qquad\qquad = \sqrt{[x'(t)]^2 + [y'(t)]^2 + [z'(t)]^2} \, dt$
b. Line integral of a scalar function: $\displaystyle\int_C f(x, y, z) \, ds = \int_a^b f(x(t), y(t), z(t)) \|\mathbf{r}'(t)\| \, dt$
c. Line integral of a vector field: $\displaystyle\int_C \mathbf{F} \cdot d\mathbf{r} = \int_C \mathbf{F} \cdot \mathbf{T} \, ds = \int_a^b \mathbf{F}(\mathbf{r}(t)) \cdot \mathbf{r}'(t) \, dt$

Surface Integrals
a. Element of surface area: (i) If S is the graph of $z = g(x, y)$, then $dS = \sqrt{g_x^2 + g_y^2 + 1} \, dA$. (ii) If S is represented parametrically by $\mathbf{r}(u, v)$, then $dS = \|r_{\mathbf{u}} \times r_{\mathbf{v}}\| \, dA$.
b. Surface integral of a scalar function: (i) If S is the graph of $z = g(x, y)$, then $$\iint_S f(x, y, z) \, dS = \iint_R f(x, y, g(x, y)) \sqrt{g_x^2 + g_y^2 + 1} \, dA$$ (ii) If S is represented parametrically by $\mathbf{r}(u, v)$, then $$\iint_S f(\mathbf{r}(u, v)) \, dS = \iint_R f(x(u, v), y(u, v), z(u, v)) \|\mathbf{r}_u \times \mathbf{r}_v\| \, dA$$
c. Surface integral of a vector field: (i) If S is the graph of $z = g(x, y)$, then $$\iint_S \mathbf{F} \cdot d\mathbf{S} = \iint_S \mathbf{F} \cdot \mathbf{n} \, dS = \iint_R \mathbf{F} \cdot (-g_x \mathbf{i} - g_y \mathbf{j} + \mathbf{k}) \, dA$$ (ii) If S is represented parametrically by $\mathbf{r}(u, v)$, then $$\iint_S \mathbf{F} \cdot d\mathbf{S} = \iint_S \mathbf{F} \cdot \mathbf{n} \, dS = \iint_D \mathbf{F} \cdot (\mathbf{r}_u \times \mathbf{r}_v) \, dA$$

Summary of Major Theorems Involving Line Integrals and Surface Integrals

1. Fundamental Theorem for Line Integrals

$$\int_C \nabla f \cdot d\mathbf{r} = f(\mathbf{r}(b)) - f(\mathbf{r}(a))$$

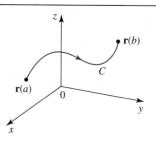

2. Green's Theorem

$$\oint_C P \, dx + Q \, dy = \iint_R \left(\frac{\partial Q}{\partial x} - \frac{\partial P}{\partial y} \right) dA$$

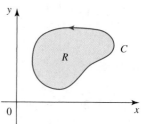

3. Divergence Theorem

$$\iint_S \mathbf{F} \cdot d\mathbf{S} = \iiint_T \operatorname{div} \mathbf{F} \, dV$$

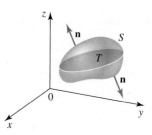

4. Stokes' Theorem

$$\oint_C \mathbf{F} \cdot d\mathbf{r} = \iint_S \operatorname{curl} \mathbf{F} \cdot d\mathbf{S}$$

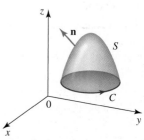

15.9 CONCEPT QUESTIONS

1. State Stokes' Theorem.
2. Suppose that $\mathbf{F} = P\mathbf{i} + Q\mathbf{j} + R\mathbf{k}$ is a vector field in three-dimensional space such that P, Q, and R have continuous partial derivatives. Let S_1 be the hemisphere $z = \sqrt{1 - x^2 - y^2}$, and let S_2 be the paraboloid $z = 1 - x^2 - y^2$. Explain why

$$\iint_{S_1} \operatorname{curl} \mathbf{F} \cdot d\mathbf{S} = \iint_{S_2} \operatorname{curl} \mathbf{F} \cdot d\mathbf{S}$$

15.9 EXERCISES

In Exercises 1–4, verify Stokes' Theorem for the given vector field F and the surface S, oriented with the normal pointing upward.

1. $\mathbf{F}(x, y, z) = 2z\mathbf{i} + 3x\mathbf{j} - 2y\mathbf{k}$; S is the part of the paraboloid $z = 9 - x^2 - y^2$ with $z \geq 0$

2. $\mathbf{F}(x, y, z) = 2y\mathbf{i} - 2x\mathbf{j} + z\mathbf{k}$; S is the part of the plane $z = 1$ lying within the cone $z = \sqrt{x^2 + y^2}$

3. $\mathbf{F}(x, y, z) = y\mathbf{i} + z\mathbf{j} + x\mathbf{k}$; S is the part of the plane $2x + 2y + z = 6$ lying in the first octant

4. $\mathbf{F}(x, y, z) = (x + 2y)\mathbf{i} + yz^2\mathbf{j} + y^2 z\mathbf{k}$; S is the hemisphere $z = \sqrt{1 - x^2 - y^2}$

In Exercises 5–10, use Stokes' Theorem to evaluate $\iint_S \text{curl } \mathbf{F} \cdot d\mathbf{S}$.

5. $\mathbf{F}(x, y, z) = 2y\mathbf{i} + xz^2\mathbf{j} + x^2 y e^z\mathbf{k}$; S is the hemisphere $z = \sqrt{4 - x^2 - y^2}$ oriented with normal pointing upward

6. $\mathbf{F}(x, y, z) = 5yz\mathbf{i} + 2xz\mathbf{j} - 3xz^2\mathbf{k}$; S is the part of the paraboloid $z = x^2 + y^2$ lying below the plane $z = 4$ and oriented with normal pointing downward

7. $\mathbf{F}(x, y, z) = xyz\mathbf{i} + 2x\mathbf{j} + \tan^{-1} y^2\mathbf{k}$; S is the part of the hemisphere $z = \sqrt{4 - x^2 - y^2}$ lying inside the cylinder $x^2 + y^2 = 1$ and oriented with normal pointing upward

8. $\mathbf{F}(x, y, z) = xy\mathbf{i} + yz\mathbf{j} + xz\mathbf{k}$; S is the part of the cylinder $z = \sqrt{1 - x^2}$ lying in the first octant between $y = 0$ and $y = 1$ and oriented with normal pointing in the positive x-direction

9. $\mathbf{F}(x, y, z) = z \sin x\mathbf{i} + 2x\mathbf{j} + e^x \cos z\mathbf{k}$; S is the part of the ellipsoid $9x^2 + 9y^2 + 4z^2 = 36$ lying above the xy-plane and oriented with normal pointing upward

10. $\mathbf{F}(x, y, z) = yz^2\mathbf{i} - xz\mathbf{j} + z^3\mathbf{k}$; S is the part of the cone $z = \sqrt{x^2 + y^2}$ between the planes $z = 1$ and $z = 3$ and oriented with normal pointing upward

In Exercises 11–16, use Stokes' Theorem to evaluate $\oint_C \mathbf{F} \cdot d\mathbf{r}$.

11. $\mathbf{F}(x, y, z) = (y - z)\mathbf{i} + (z - x)\mathbf{j} + (x - y)\mathbf{k}$; C is the boundary of the part of the plane $2x + 3y + z = 6$ in the first octant, oriented in a counterclockwise direction when viewed from above

12. $\mathbf{F}(x, y, z) = y\mathbf{i} + z\mathbf{j} + x\mathbf{k}$; C is the boundary of the triangle with vertices $(0, 0, 0)$, $(1, 0, 0)$, and $(0, 1, 1)$ oriented in a counterclockwise direction when viewed from above

13. $\mathbf{F}(x, y, z) = 3xz\mathbf{i} + e^{xz}\mathbf{j} + 2xy\mathbf{k}$; C is the circle obtained by intersecting the cylinder $x^2 + z^2 = 1$ with the plane $y = 3$ oriented in a counterclockwise direction when viewed from the right

14. $\mathbf{F}(x, y, z) = 2z\mathbf{i} + xy\mathbf{j} + 4y\mathbf{k}$; C is the ellipse obtained by intersecting the plane $y + z = 4$ with the cylinder $x^2 + y^2 = 4$, oriented in a counterclockwise direction when viewed from above

15. $\mathbf{F}(x, y, z) = xe^y\mathbf{i} + ye^x\mathbf{j} + (xyz)\mathbf{k}$; C is the path consisting of straight line segments joining the points $(0, 0, 0)$, $(0, 1, 0)$, $(2, 1, 0)$, $(2, 0, 0)$, $(2, 0, 1)$, $(0, 0, 1)$, and $(0, 0, 0)$ in that order

16. $\mathbf{F}(x, y, z) = \left(\dfrac{y}{1 + x^2}\right)\mathbf{i} + \tan^{-1} x\mathbf{j} + xy\mathbf{k}$; C is the curve obtained by intersecting the cylinder $x^2 + y^2 = 1$ with the hyperbolic paraboloid $z = y^2 - x^2$, oriented in a counterclockwise direction when viewed from above

17. Find the work done by the force field $\mathbf{F}(x, y, z) = (e^x + z)\mathbf{i} + (x^2 + \cosh y)\mathbf{j} + (y^2 + z^3)\mathbf{k}$ on a particle when it is moved along the triangular path that is obtained by intersecting the plane $2x + 2y + z = 2$ with the coordinate planes and oriented in a counterclockwise direction when viewed from above.

18. Find the work done by the force field $\mathbf{F}(x, y, z) = xy^2\mathbf{i} + (x/z)\mathbf{j} + (2x + y)\mathbf{k}$ on a particle when it is moved along the rectangular path with vertices $A(0, 0, 3)$, $B(2, 0, 3)$, $C(2, 4, 3)$, $D(0, 4, 3)$, and $A(0, 0, 3)$ in that order.

19. **Ampere's Law** A steady current in a long wire produces a magnetic field that is tangent to any circle that lies in the plane perpendicular to the wire and whose center lies on the wire. (See the figure below.)

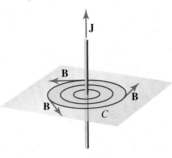

Let $\mathbf{J}$ denote the vector that points in the direction of the current and has magnitude measured in amperes per square meter. This vector is called the electric current density. One of Maxwell's equations states that curl $\mathbf{B} = \mu_0\mathbf{J}$, where $\mathbf{B}$ denotes the magnetic field intensity and μ_0 is a constant called the permeability of free space. Using Stokes' Theorem, show that $\oint_C \mathbf{B} \cdot d\mathbf{r} = \mu_0 I$, where C is any closed curve enclosing the curve and I is the net current that passes through any surface bounded by C. This is Ampere's Law. (See Exercise 42 in Section 15.3.)

20. Let S be a sphere, and suppose that $\mathbf{F}$ satisfies the conditions of Stokes' Theorem. Show that $\iint_S \text{curl } \mathbf{F} \cdot d\mathbf{S} = 0$.

21. Let f and g have continuous partial derivatives, and let C and S satisfy the conditions of Stokes' Theorem. Show that:

a. $\oint_C (f\nabla g) \cdot d\mathbf{r} = \iint_S (\nabla f \times \nabla g) \cdot d\mathbf{S}$

b. $\oint_C (f\nabla f) \cdot d\mathbf{r} = 0$ **c.** $\oint_C (f\nabla g + g\nabla f) \cdot d\mathbf{r} = 0$

22. Evaluate $\oint_C (2xy + z^2)\, dx + (x^2 - 1)\, dy + 2xz\, dz$, where C is the curve

$$\mathbf{r}(t) = (1 + \cos t)\mathbf{i} + (1 + \sin t)\mathbf{j} + (1 - \sin t - \cos t)\mathbf{k}$$
$$0 \le t \le 2\pi$$

23. Let $\mathbf{F}(x, y, z) = f(r)\mathbf{r}$, where $\mathbf{r} = x\mathbf{i} + y\mathbf{j} + z\mathbf{k}$, f is a differentiable function, and $r = |\mathbf{r}|$. Evaluate $\oint_C \mathbf{F} \cdot d\mathbf{r}$, where C is the boundary of the triangle with vertices $(4, 2, 0)$, $(1, 5, 2)$, and $(1, -1, 5)$, traced in a counterclockwise direction when viewed from above the plane.

24. Let $\mathbf{F}(x, y, z) = xy\mathbf{i} + (4x - yz)\mathbf{j} + (xy - \sqrt{z})\mathbf{k}$, and let C be a circle of radius r lying in the plane $x + y + z = 5$. (See the following figure.) If $\oint_C \mathbf{F} \cdot d\mathbf{r} = \sqrt{3}\pi$, where C is oriented in the counterclockwise direction when viewed from above the plane, what is the value of r?

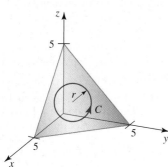

25. Use Stokes' Theorem to evaluate $\oint_C e^x \cos z\, dx + 2xy^2\, dy + \cot^{-1} y\, dz$, where C is the circle $x^2 + y^2 = 4$ and $z = 0$.
Hint: Find a surface S with C as its boundary and such that C is oriented counterclockwise when viewed from above.

26. Find $\iint_S \text{curl } \mathbf{F} \cdot d\mathbf{S}$ if $\mathbf{F}(x, y, z) = (x + y - z + 2)\mathbf{i} + (y \cos z + 4)\mathbf{j} + xz\mathbf{k}$, where S is the surface with the normal pointing outward and having the boundary $x^2 + y^2 = 1$, $z = 0$, shown in the figure.

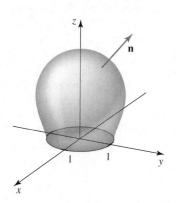

27. Let S be the oriented piecewise-smooth surface that is the boundary of a simple solid region T. If $\mathbf{F} = P\mathbf{i} + Q\mathbf{j} + R\mathbf{k}$ is a vector field, where P, Q, and R have continuous second-order partial derivatives in an open region containing T, show that

$$\iint_S \text{curl } \mathbf{F} \cdot d\mathbf{S} = 0$$

28. Suppose that f has continuous second-order partial derivatives in a simply connected set D. Use Stokes' Theorem to show that

$$\oint_C \nabla f \cdot d\mathbf{r} = 0$$

for any simple piecewise-smooth closed curve C lying in D.

29. Refer to Example 2. Evaluate $\oint_C \mathbf{F} \cdot d\mathbf{r}$ directly (that is, without using Stokes' Theorem), where $F(x, y, z) = \cos z\mathbf{i} + x^2\mathbf{j} + 2y\mathbf{k}$ and C is the curve of intersection of the plane $x + z = 2$ and the cylinder $x^2 + y^2 = 1$.

30. Refer to Exercise 16. Evaluate $\oint_C \mathbf{F} \cdot d\mathbf{r}$ directly (that is, without using Stokes' Theorem), where

$$F(x, y, z) = \left(\frac{y}{1 + x^2}\right)\mathbf{i} + \tan^{-1} x\mathbf{j} + xy\mathbf{k} \text{ and } C \text{ is}$$

the curve obtained by intersecting the cylinder $x^2 + y^2 = 1$ with the hyperbolic paraboloid $z = y^2 - x^2$, oriented in a counterclockwise direction when viewed from above.

In Exercises 31 and 32, determine whether the statement is true or false. If it is true, explain why. If it is false, explain why or give an example that shows it is false.

31. If $\mathbf{F} = P\mathbf{i} + Q\mathbf{j} + R\mathbf{k}$, where P, Q, and R have continuous partial derivatives in three-dimensional space and S_1 and S_2 are the upper and lower hemisphere $z = \sqrt{4 - x^2 - y^2}$ and $z = -\sqrt{4 - x^2 - y^2}$, respectively, then $\iint_{S_1} \text{curl } \mathbf{F} \cdot d\mathbf{S} = \iint_{S_2} \text{curl } \mathbf{F} \cdot d\mathbf{S}$.

32. If $\mathbf{F}$ has continuous partial derivatives on an open region containing an oriented surface S bounded by a piecewise-smooth simple closed curve C and curl $\mathbf{F}$ is tangent to S, then $\oint_C \mathbf{F} \cdot d\mathbf{r} = 0$.

CHAPTER 15 REVIEW

CONCEPT REVIEW

In Exercises 1–18, fill in the blanks.

1. A vector field on a region R is a function $\mathbf{F}$ that associates with each point (x, y) in two-dimensional space a two-dimensional _____; if R is a subset of three-dimensional space, then $\mathbf{F}$ associates each point (x, y, z) with a three-dimensional _____.

2. A vector field $\mathbf{F}$ is conservative if there exists a scalar function f such that $\mathbf{F} =$ _____. The function f is called a _____ function for $\mathbf{F}$.

3. **a.** The divergence of $\mathbf{F}$ (describing the flow of fluid) at a point P measures the rate of flow per unit area (or volume) at which the fluid _____ or _____ at P.
   b. The divergence of $\mathbf{F} = P\mathbf{i} + Q\mathbf{j} + R\mathbf{k}$ in three-dimensional space is defined by div $\mathbf{F} =$ _____; if div $\mathbf{F}(P) < 0$, more fluid _____ a neighborhood of P than _____ from it; if div $\mathbf{F}(P) = 0$, the amount of fluid entering a neighborhood of P _____ the amount departing from it; if div $\mathbf{F}(P) > 0$, more fluid _____ a neighborhood of P than _____ it.

4. **a.** If $\mathbf{F}$ describes fluid flow, then the curl of $\mathbf{F}$ measures the tendency of the fluid to _____ a paddle wheel.
   b. The curl of $\mathbf{F} = P\mathbf{i} + Q\mathbf{j} + R\mathbf{k}$ is defined by curl $\mathbf{F} =$ _____.

5. **a.** If C is a smooth curve, then the line integral of f along C is $\int_C f(x, y) \, ds =$ _____.
   b. The formula for evaluating the line integral of part (a) is $\int_C f(x, y) \, ds =$ _____.
   c. The line integral of f along C with respect to x is defined to be $\int_C f(x, y) \, dx =$ _____ with formula $\int_C f(x, y) \, dx =$ _____, where $\mathbf{r}(t) = x(t)\mathbf{i} + y(t)\mathbf{j}$, $a \le t \le b$.
   d. The line integral of f along C with respect to y is defined to be $\int_C f(x, y) \, dy =$ _____ with formula $\int_C f(x, y) \, dy =$ _____.

6. The formula for evaluating the line integral of f along a curve C (with respect to arc length) parametrized by $\mathbf{r}(t) = x(t)\mathbf{i} + y(t)\mathbf{j} + z(t)\mathbf{k}$, $a \le t \le b$, is $\int_C f(x, y, z) \, ds =$ _____.

7. **a.** If $\mathbf{F}$ is a continuous vector field and C is a smooth curve described by $\mathbf{r}(t)$, $a \le t \le b$, then the formula for evaluating the line integral of $\mathbf{F}$ along C is $\int_C \mathbf{F} \cdot d\mathbf{r} =$ _____.
   b. If $\mathbf{F}$ is a force field, then $\int_C \mathbf{F} \cdot d\mathbf{r}$ gives the _____ done by $\mathbf{F}$ on a particle as it moves along C from $t = a$ to $t = b$.

8. **a.** If $\int_{C_1} \mathbf{F} \cdot d\mathbf{r} = \int_{C_2} \mathbf{F} \cdot d\mathbf{r}$ for any two paths having the same initial and terminal points, then the line integral $\int_C \mathbf{F} \cdot d\mathbf{r}$ is _____ _____ _____.
   b. The Fundamental Theorem for Line Integrals states that if $\mathbf{F} = \nabla f$, where f is a _____ function for $\mathbf{F}$ and C is any piecewise-smooth curve described by $\mathbf{r}(t)$, $a \le t \le b$, then $\int_C \mathbf{F} \cdot d\mathbf{r} = \int_C \nabla f \cdot d\mathbf{r} =$ _____.

9. **a.** The line integral $\int_C \mathbf{F} \cdot d\mathbf{r}$ is independent of path if and only if $\int_C \mathbf{F} \cdot d\mathbf{r} = 0$ for every _____ path C.
   b. If $\mathbf{F}$ is continuous on an open, _____ region R, then the line integral $\int_C \mathbf{F} \cdot d\mathbf{r}$ is independent of path if and only if $\mathbf{F}$ is _____.

10. **a.** If $\mathbf{F}(x, y) = P(x, y)\mathbf{i} + Q(x, y)\mathbf{j}$ is a conservative vector field in an open region R and both P and Q have continuous first-order partial derivatives in R, then _____ at each point in R.
   b. If $\mathbf{F} = P\mathbf{i} + Q\mathbf{j}$ is defined on an open, _____ region R in the plane and P and Q have continuous first-order derivatives on R and _____, for all (x, y) in R, then $\mathbf{F}$ is conservative in R.

11. If $\mathbf{F} = P\mathbf{i} + Q\mathbf{j} + R\mathbf{k}$, where P, Q, and R have continuous first-order partial derivatives in space, then $\mathbf{F}$ is conservative if and only if _____ $= \mathbf{0}$, or, in terms of the partial derivatives of P, Q, and R, $\dfrac{\partial R}{\partial y} =$ _____, $\dfrac{\partial R}{\partial x} =$ _____, and $\dfrac{\partial Q}{\partial x} =$ _____.

12. Green's Theorem states that if C is a _____, simple _____ curve that bounds a region R in the plane and P and Q have continuous partial derivatives on an open set containing R, then $\int_C P \, dx + Q \, dy =$ _____.

13. If R is a plane region bounded by a piecewise-smooth, simple closed curve C, then the area of R is given by $A =$ _____ $=$ _____ $=$ _____.

14. If $\mathbf{F} = P\mathbf{i} + Q\mathbf{j}$, then Green's Theorem has the vector forms $\oint_C \mathbf{F} \cdot \mathbf{T} \, ds = \oint_C P \, dx + Q \, dy =$ _____ and $\oint_C \mathbf{F} \cdot \mathbf{n} \, ds =$ _____.

15. **a.** A parametric surface S can be represented by the vector equation $\mathbf{r}(u, v) =$ _____, where (u, v) lies in its _____ domain; this vector equation is equivalent to the three _____ _____ $x = x(u, v)$, $y = y(u, v)$, and $z = z(u, v)$.
   b. If S is the graph of the function $z = f(x, y)$, then a vector representation of S is $\mathbf{r}(u, v) =$ _____.

c. If S is a surface obtained by revolving the graph of a nonnegative function $f(x)$, where $a \le x \le b$, about the x-axis, then a vector representation of S is $\mathbf{r}(u, v) =$ _____ with parameter domain $D =$ _____.

16. If a parametric surface S is represented by $\mathbf{r}(u, v) = x(u, v)\mathbf{i} + y(u, v)\mathbf{j} + z(u, v)\mathbf{k}$ with parameter domain D, then its surface area is $A(S) =$ _____.

17. a. If S is defined by $z = f(x, y)$ and the projection of S onto the xy-plane is R, then $\iint_S F(x, y, z)\, dS =$ _____.
   b. If S is defined by $\mathbf{r}(u, v) = x(u, v)\mathbf{i} + y(u, v)\mathbf{j} + z(u, v)\mathbf{k}$ with parameter domain D, then $\iint_S F(x, y, z)\, dS =$ _____.

18. a. The positive orientation of a surface S is the one for which the unit normal vector points _____ from S.
   b. The flux of a vector field $\mathbf{F}$ across an oriented surface S in the direction of the unit normal $\mathbf{n}$ is _____.

REVIEW EXERCISES

In Exercises 1–4, find (a) the divergence and (b) the curl of the vector field $\mathbf{F}$.

1. $\mathbf{F}(x, y, z) = xy^2\mathbf{i} + yz^2\mathbf{j} + zx^2\mathbf{k}$

2. $\mathbf{F}(x, y, z) = xy \cos y\mathbf{i} + y \sin x\mathbf{j} + xz\mathbf{k}$

3. $\mathbf{F}(x, y, z) = e^x \sin y\mathbf{i} + e^x \cos y\mathbf{j} + e^z\mathbf{k}$

4. $\mathbf{F}(x, y, z) = \ln(x^2 + y^2)\mathbf{i} + x \cos y\mathbf{j} + z^2\mathbf{k}$

In Exercises 5–14, evaluate the line integral.

5. $\int_C y\, ds$, where C is the arch of the parabola $y = \sqrt{x}$ from $(1, 1)$ to $(4, 2)$

6. $\int_C (1 - x^2)\, ds$, where C is the upper semicircle centered at the origin and joining $(0, -1)$ to $(0, 1)$

7. $\int_C xy^2\, ds$, where C is the curve given by $\mathbf{r}(t) = \sin t\mathbf{i} + \cos t\mathbf{j} + t\mathbf{k}$, $0 \le t \le \frac{\pi}{2}$

8. $\int_C xyz\, ds$, where C is the line segment from $(1, 1, 0)$ to $(2, 3, 4)$

9. $\int_C x^2y\, dx + x^3y\, dy$, where C is the graph of $y = \sqrt[3]{x}$ from $(1, 1)$ to $(8, 2)$

10. $\int_C xy\, dx - xy^2\, dy$, where C is the quarter-circle from $(0, -1)$ to $(1, 0)$, centered at the origin

11. $\int_C yz\, dx - y \cos x\, dy + y\, dz$, where C is the curve $x = t$, $y = \cos t$, $z = \sin t$, $0 \le t \le \frac{\pi}{2}$

12. $\int_C xe^{-y}\, dx + \cos y\, dy + z^2\, dz$, $C: \mathbf{r}(t) = t\mathbf{i} + t^2\mathbf{j} + t^3\mathbf{k}$, $0 \le t \le 1$

13. $\int_C xy\, dx + e^{-y}\, dy + ze^x\, dz$, where C is the line segment joining $(0, 0, 0)$ to $(1, 1, 2)$

14. $\int_C z\, dx + x\, dy + x^2\, dz$, where C consists of the line segment joining $(0, 0, 0)$ to $(1, 0, 0)$ and the line segment from $(1, 0, 0)$ to $(2, 1, 3)$.

In Exercises 15 and 16, find the work done by the force field $\mathbf{F}$ in moving a particle along the curve C.

15. $\mathbf{F}(x, y, z) = xy\mathbf{i} + (y + z)\mathbf{j} + z^2\mathbf{k}$, where C is the line segment from $(1, 1, 1)$ to $(2, 3, 5)$

16. $\mathbf{F}(x, y, z) = yz\mathbf{i} + zj + x\mathbf{k}$, where C is part of the helix given by $x = 2t$, $y = 2 \sin t$, $z = 2 \cos t$, $0 \le t \le \frac{\pi}{2}$

In Exercises 17 and 18, show that $\mathbf{F}$ is a conservative vector field and find a function f such that $\mathbf{F} = \nabla f$.

17. $\mathbf{F}(x, y) = (4xy + 3y^2)\mathbf{i} + (2x^2 + 6xy)\mathbf{j}$

18. $\mathbf{F}(x, y, z) = (y^2 + 2xz)\mathbf{i} + (2xy - z^2)\mathbf{j} + (x^2 - 2yz)\mathbf{k}$

In Exercises 19 and 20, show that $\mathbf{F}$ is conservative and use this result to evaluate $\int_C \mathbf{F} \cdot \mathbf{T}\, ds$ for the given curve C.

19. $\mathbf{F}(x, y) = (2xy + y^3)\mathbf{i} + (x^2 + 3xy^2)\mathbf{j}$; C is the elliptical path $9x^2 + 25y^2 = 225$ from $(-5, 0)$ to $(0, 3)$ traversed in a counterclockwise direction

20. $\mathbf{F}(x, y, z) = (2xy + yz^2)\mathbf{i} + (x^2 + xz^2)\mathbf{j} + 2xyz\mathbf{k}$; C is the twisted cubic $x = t$, $y = t^2$, $z = t^3$ from $(0, 0, 0)$ to $(1, 1, 1)$

In Exercises 21–24, use Green's Theorem to evaluate the line integral along the positively oriented closed curve C.

21. $\oint_C (y^2 + \sec x)\, dx + (x^2 + y^5)\, dy$, where C is the boundary of the region enclosed by the graphs of $y = 4 - x^2$ and $y = -x + 2$

22. $\oint_C xy \, dx + (x^2 + \sqrt{y}) \, dy$, where C is the boundary of the region enclosed by the graphs of $y = \sqrt{x}$, $y = 0$, and $x = 4$

23. $\oint_C (x^2y + e^x) \, dx + (e^{-y} - xy^2) \, dy$, where C is the circle $x^2 + y^2 = 1$

24. $\oint_C (2y + \cosh x) \, dx + (x - \sinh y) \, dy$, where C is the ellipse $9x^2 + 4y^2 = 36$

In Exercises 25–28, evaluate the surface integral.

25. $\iint_S (y + xz) \, dS$, where S is the part of the plane $2x + 2y + 3z = 6$ in the first octant

26. $\iint_S z \, dS$, where S is the part of the paraboloid $z = 4 - x^2 - y^2$ inside the cylinder $x^2 + y^2 = 1$

27. $\iint_S \mathbf{F} \cdot \mathbf{n} \, dS$, where $\mathbf{F}(x, y, z) = x\mathbf{i} + y\mathbf{j} + z\mathbf{k}$ and S is the part of the paraboloid $y = 1 - x^2 - z^2$ lying to the right of the xz-plane; $\mathbf{n}$ points to the right

28. $\iint_S \mathbf{F} \cdot \mathbf{n} \, dS$, where $\mathbf{F}(x, y, z) = y\mathbf{i} - x\mathbf{j} + z\mathbf{k}$ and S is the part of the paraboloid $z = 5 - x^2 - y^2$ lying above the plane $z = 1$; $\mathbf{n}$ points upward

In Exercises 29 and 30, find the mass of the surface S having the given mass density.

29. S is the part of the plane $x + y + z = 1$ in the first octant; the density at a point P on S is directly proportional to the square of the distance between P and the yz-plane.

30. S is the part of the paraboloid $z = x^2 + y^2$ lying inside the cylinder $x^2 + y^2 = 2$; the density at a point P on S is directly proportional to the distance between P and the xy-plane.

In Exercises 31 and 32, use the Divergence Theorem to find the flux of $\mathbf{F}$ across S; that is, calculate $\iint_S \mathbf{F} \cdot \mathbf{n} \, dS$.

31. $\mathbf{F}(x, y, z) = x\mathbf{i} - y\mathbf{j} + z\mathbf{k}$; S is the surface of the cylinder $x^2 + y^2 = 4$ bounded by the planes $z = 0$ and $z = 3$; $\mathbf{n}$ points outward

32. $\mathbf{F}(x, y, z) = y^2\mathbf{i} + yz^2\mathbf{j} + z^3\mathbf{k}$; S is the unit sphere centered at the origin; $\mathbf{n}$ points outward

In Exercises 33 and 34, use Stokes' Theorem to evaluate $\iint_S$ curl $\mathbf{F} \cdot \mathbf{n} \, dS$.

33. $\mathbf{F}(x, y, z) = (x + y^2 - 2)\mathbf{i} - 2xy\mathbf{j} - (x^2 + yz^2)\mathbf{k}$; S is the part of the paraboloid $z = 4 - x^2 - y^2$ above the xy-plane with an outward normal

34. $\mathbf{F}(x, y, z) = y^2z\mathbf{i} + 2xz\mathbf{j} + \cos z\mathbf{k}$; S is the part of the sphere $x^2 + y^2 + z^2 = 16$ below the plane $z = 2$ with an outward normal

In Exercises 35 and 36, use Stokes' Theorem to evaluate $\oint_C \mathbf{F} \cdot \mathbf{T} \, ds$.

35. $\mathbf{F}(x, y, z) = (2x + y)\mathbf{i} - (3x + z)\mathbf{j} + (y - z)\mathbf{k}$; C is the curve obtained by intersecting the plane $2x + y + z = 6$ with the coordinate planes, oriented clockwise when viewed from the top

36. $\mathbf{F}(x, y, z) = y\mathbf{i} + xz\mathbf{j} + xyz\mathbf{k}$; C is the curve obtained by intersecting the surface $z = x^2y$ with the planes $x = 0$, $x = 1$, $y = 0$, and $y = 1$, oriented counterclockwise when viewed from above

37. Find the work done by the force field $\mathbf{F}(x, y) = 2xy^3\mathbf{i} + 3x^2y^2\mathbf{j}$ when a particle is moved from $(0, 0)$ to $(2, 4)$ along the path shown in the figure.

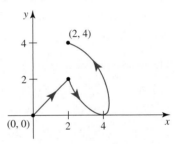

38. Find the work done by the force field $\mathbf{F}(x, y, z) = 2xy\mathbf{i} + (x^2 + 2yz^2)\mathbf{j} + 2y^2z\mathbf{k}$ when a particle is moved from $(2, 0, 0)$ to $(0, 3, 0)$ along the path shown in the figure.

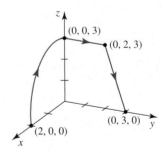

39. Let
$$\mathbf{F}(x, y) = \frac{x - y}{x^2 + y^2}\mathbf{i} + \frac{x + y}{x^2 + y^2}\mathbf{j}$$

Evaluate $\oint_C \mathbf{F} \cdot d\mathbf{r}$, where C is the path shown in the figure.

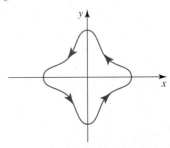

40. Let $\mathbf{F}(x, y, z) = \left(4y - \frac{1}{2}z^2\right)\mathbf{i} + 2xz\mathbf{j} - x^2\mathbf{k}$, and let C be a simple closed curve lying in the plane $2x + 2y + z = 6$ (see the following figure) and oriented in the counterclockwise direction when viewed from above the plane. If $\oint_C \mathbf{F} \cdot d\mathbf{r} = -24$, what is the area of the region enclosed by C?

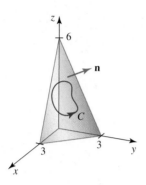

In Exercises 41–48, state whether the statement is true or false. (Assume that all differentiability conditions are met.) Give a reason for your answer.

41. If $\mathbf{F}$ is a vector field, then curl (curl $\mathbf{F}$) is a vector field.

42. If $\mathbf{F}$ and $\mathbf{G}$ are vector fields, then div $(\mathbf{F} \times \mathbf{G})$ is a scalar field.

43. If f is a scalar field, then $\nabla \cdot [\nabla \times (\nabla f)]$ is undefined.

44. If $\mathbf{F}$ is a vector field and f is a scalar field, then $\text{div}[\nabla f \times (\text{curl } \mathbf{F})]$ is undefined.

45. If f has continuous partial derivatives at all points and $\nabla f = \mathbf{0}$, then f is a constant function.

46. If $\nabla \times \mathbf{F} = \mathbf{0}$, then $\mathbf{F}$ is a constant vector field.

47. If div $\mathbf{F} = 0$, then $\iint_S \mathbf{F} \cdot \mathbf{n} \, dS = 0$ for every closed surface S.

48. If $\oint_C \mathbf{F} \cdot d\mathbf{r} = 0$ for every closed path C, then curl $\mathbf{F} = \mathbf{0}$.

CHALLENGE PROBLEMS

1. Find and sketch the domain of the vector-valued function

$$\mathbf{F}(x, y) = \frac{1}{\sqrt{4 - x^2 - 4y^2}}\mathbf{i} + \frac{1}{\sqrt{4x^2 + 4y^2 - 1}}\mathbf{j}$$

2. Find the domain of

$$\mathbf{F}(x, y, z) = \sqrt{|x| - y - 1}\,\mathbf{i}$$

$$+ \ln(1 - |x| - y)\mathbf{j} + \frac{\ln \ln(z - 1)}{\sqrt{z - 3}}\mathbf{k}$$

3. The curve with equation $x^3 + y^3 = 3axy$, where a is a nonzero constant, is called the **folium of Descartes.**

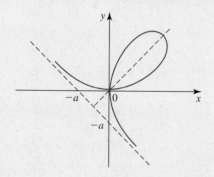

a. Show that a parametric representation of this curve is

$$x = \frac{3at}{1 + t^3} \qquad y = \frac{3at^2}{1 + t^3}$$

Hint: Use the parameter $t = y/x$.

b. Find the area of the region enclosed by the loop of the curve.

4. Let

$$P(x, y) = \frac{x - y}{x^2 + y^2} \qquad \text{and} \qquad Q(x, y) = \frac{x + y}{x^2 + y^2}$$

Evaluate $\oint_C P \, dx + Q \, dy$, where C is the curve shown in the figure.

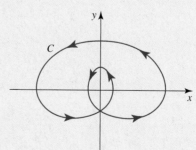

5. Let $\mathbf{F}(x, y, z) = y \cos x\mathbf{i} + (x + \sin x)\mathbf{j} + \cos z\mathbf{k}$, and let C be the curve represented by $\mathbf{r}(t) = (1 + \cos t)\mathbf{i} + (1 + \sin t)\mathbf{j} + (1 - \sin t - \cos t)\mathbf{k}$ for $0 \le t \le 2\pi$. Evaluate $\oint_C \mathbf{F} \cdot d\mathbf{r}$.

6. A differential equation of the form

$$P(x, y)\, dx + Q(x, y)\, dy = 0$$

is called *exact* if $\dfrac{\partial Q}{\partial x} = \dfrac{\partial P}{\partial y}$ for all (x, y). Show that the equation

$$(2y^2 + 6xy - 2)\, dx + (4xy + 3x^2 + 3)\, dy = 0$$

is exact and solve it.

7. Let T be a one-to-one transformation defined by $x = g(u, v)$, $y = h(u, v)$ that maps a region S in the uv-plane onto a region R in the xy-plane. Use Green's Theorem to prove the change of variables formula

$$\iint\limits_{R} dA = \iint\limits_{S} \left| \frac{\partial(x, y)}{\partial(u, v)} \right| du\, dv$$

for the case in which $f(x, y) = 1$. (Compare with Formula (4) in Section 14.8.)

PROOF OF CLAIRAUT'S THEOREM

If $f(x, y)$ and its partial derivatives f_x, f_y, f_{xy}, and f_{yx} are continuous on an open region R, then

$$f_{xy}(x, y) = f_{yx}(x, y)$$

for all (x, y) in R.

PROOF Let (a, b) be any point in R. Fix it, and let h and k be nonzero numbers such that the point $(a + h, b + k)$ also lie in R. Define the function F by

$$F(h, k) = [f(a + h, b + k) - f(a + h, b)] - [f(a, b + k) - f(a, b)] \qquad \textbf{(1)}$$

If we let $g(x) = f(x, b + k) - f(x, b)$, then Equation (1) can be written

$$F(h, k) = g(a + h) - g(a)$$

Applying the Mean Value Theorem to the function g on the interval $[a, a + h]$ (or $[a + h, a]$), we see that there exists a number c lying between a and $a + h$ such that

$$g(a + h) - g(a) = g'(c)h = h[f_x(c, b + k) - f_x(c, b)]$$

If we apply the Mean Value Theorem again to the function f_x on the interval $[b, b + k]$ (or $[b + k, b]$), we see that there exists a number d lying between b and $b + k$ such that

$$f_x(c, b + k) - f_x(c, b) = f_{xy}(c, d)k$$

Therefore,

$$F(h, k) = hk f_{xy}(c, d)$$

If we let $(h, k) \to (0, 0)$, then (c, d) approaches (a, b), and the continuity of f_{xy} implies that

$$\lim_{(h, k) \to (0, 0)} \frac{F(h, k)}{hk} = f_{xy}(a, b)$$

Similarly, if we write

$$F(h, k) = [f(a + h, b + k) - f(a, b + k)] - [f(a + h, b) - f(a, b)]$$

and use the Mean Value Theorem twice, the continuity of f_{yx} at (a, b) implies that

$$\lim_{(h, k) \to (0, 0)} \frac{F(h, k)}{hk} = f_{yx}(a, b)$$

Therefore, $f_{xy}(a, b) = f_{yx}(a, b)$. Finally, since (a, b) is any point in R, we see that this equality holds for any (x, y) in R, and the theorem is proved.

CHAPTER 10

Exercises 10.1 • page 843

1. Parabola (h), vertex $(0, 0)$, focus $(0, -1)$, directrix $y = 1$

3. Parabola (c), vertex $(0, 0)$, focus $(2, 0)$, directrix $x = -2$

5. Ellipse (b), vertices $(\pm 3, 0)$, foci $(\pm \sqrt{5}, 0)$, eccentricity $\frac{\sqrt{5}}{3}$

7. Hyperbola (d), vertices $(\pm 4, 0)$, foci $(\pm 5, 0)$, eccentricity $\frac{5}{4}$

9. Vertex $(0, 0)$,
   focus $\left(0, \frac{1}{8}\right)$,
   directrix $y = -\frac{1}{8}$

11. Vertex $(0, 0)$,
    focus $\left(\frac{1}{8}, 0\right)$,
    directrix $x = -\frac{1}{8}$

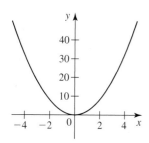

13. Vertex $(0, 0)$,
    focus $\left(\frac{3}{5}, 0\right)$,
    directrix $x = -\frac{3}{5}$

15. Foci $(0, \pm \sqrt{21})$,
    vertices $(0, \pm 5)$

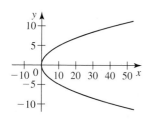

17. Foci $(\pm \sqrt{5}, 0)$,
    vertices $(\pm 3, 0)$

19. Foci $(\pm \sqrt{3}, 0)$,
    vertices $(\pm 2, 0)$

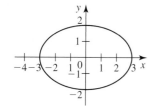
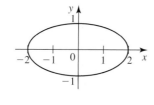

21. Foci $(\pm 13, 0)$,
    vertices $(\pm 5, 0)$,
    asymptotes $y = \pm \frac{12}{5} x$

23. Foci $(\pm \sqrt{2}, 0)$,
    vertices $(\pm 1, 0)$,
    asymptotes $y = \pm x$

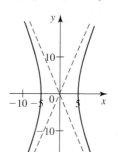
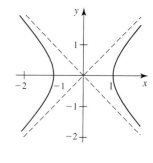

25. Foci $(0, \pm \sqrt{30})$, vertices $(0, \pm 5)$, asymptotes $y = \pm \sqrt{5} x$

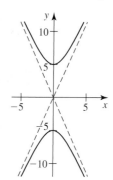

27. $y^2 = 12x$ 29. $y^2 = -10x$

31. $\dfrac{x^2}{9} + \dfrac{y^2}{8} = 1$ 33. $\dfrac{x^2}{8} + \dfrac{y^2}{9} = 1$

35. $\dfrac{x^2}{9} + \dfrac{4y^2}{9} = 1$ 37. $\dfrac{73x^2}{400} + \dfrac{y^2}{25} = 1$

39. $\dfrac{x^2}{9} - \dfrac{y^2}{16} = 1$ 41. $\dfrac{y^2}{21} - \dfrac{x^2}{4} = 1$

43. $\dfrac{x^2}{4} - \dfrac{y^2}{9} = 1$ 45. (b)

47. (c)   49. $(y - 1)^2 = 4(x - 2)$

51. $(y - 2)^2 = -2(x - 2)$ 53. $(x + 3)^2 = -2(y - 2)$

55. $\dfrac{x^2}{9} + \dfrac{(y - 3)^2}{8} = 1$ 57. $\dfrac{x^2}{16} + \dfrac{(y - 2)^2}{15} = 1$

59. $\dfrac{(x-2)^2}{9} + \dfrac{(y-1)^2}{5} = 1$ **61.** $\dfrac{(x-3)^2}{9} - \dfrac{(y-2)^2}{16} = 1$

63. $\dfrac{(x-1)^2}{9} - \dfrac{(y+3)^2}{16} = 1$ **65.** $\dfrac{(y-1)^2}{9} - \dfrac{(x-4)^2}{4} = 1$

67. Vertex $(2, 1)$,
focus $(3, 1)$,
directrix $x = 1$

69. Vertex $(-3, 2)$,
focus $\left(-3, \frac{9}{4}\right)$,
directrix $y = \frac{7}{4}$

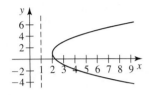

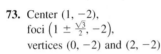

71. Vertex $\left(-1, \frac{1}{2}\right)$,
focus $\left(1, \frac{1}{2}\right)$,
directrix $x = -3$

73. Center $(1, -2)$,
foci $\left(1 \pm \frac{\sqrt{3}}{2}, -2\right)$,
vertices $(0, -2)$ and $(2, -2)$

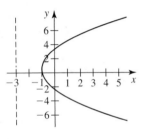

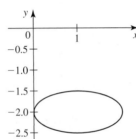

75. Center $(1, -2)$, foci $(1 \pm \sqrt{3}, -2)$, vertices $(-1, -2)$ and $(3, -2)$

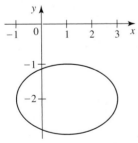

77. Center $\left(\frac{9}{4}, 0\right)$, foci $\left(\frac{9}{4} \pm \frac{\sqrt{105}}{4}, 0\right)$, vertices $\left(\frac{9}{4} \pm \frac{3\sqrt{21}}{4}, 0\right)$

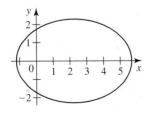

79. Center $(0, -1)$, foci $(\pm\sqrt{7}, -1)$, vertices $(\pm 2, -1)$,
asymptotes $y = \pm\frac{\sqrt{3}}{2}x - 1$

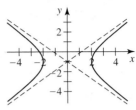

81. Center $(1, 2)$, foci $(1, 2 \pm \sqrt{15})$, vertices $(1, 2 \pm \sqrt{6})$,
asymptotes $y - 2 = \pm\frac{\sqrt{6}}{3}(x - 1)$

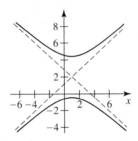

83. Center $(-1, 2)$, foci $(-1 \pm \sqrt{6}, 2)$, vertices $(-1 \pm \sqrt{2}, 2)$,
asymptotes $y - 2 = \pm\sqrt{2}(x + 1)$

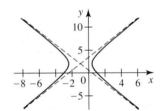

85. $\frac{1}{4}$ ft   **87.** 616 ft   **91.** 6.93 ft   **103.** 23.013

109. True   **111.** True   **113.** False

Exercises 10.2 • page 855

1. a. $x - 2y - 7 = 0$
b.

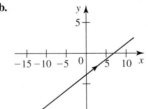

3. a. $y = 9 - x^2, x \geq 0$
b.

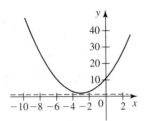

5. a. $y = 2x - 3$, $1 \le x \le 5$
b.

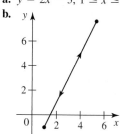

7. a. $x = y^{2/3}$, $-8 \le y \le 8$
b.

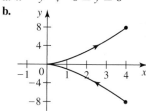

9. a. $x^2 + y^2 = 4$
b.

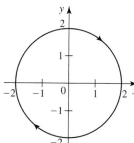

11. a. $\frac{1}{4}x^2 + \frac{1}{9}y^2 = 1$
b.

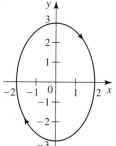

13. a. $\frac{1}{4}(x - 2)^2 + \frac{1}{9}(y + 1)^2 = 1$
b.

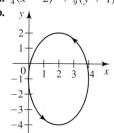

15. a. $y = 2x^2 - 1$, $-1 \le x \le 1$
b.

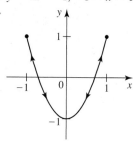

17. a. $x^2 - y^2 = 1$, $x \ge 1$
b.

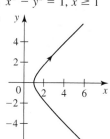

19. a. $y = x^2$, $0 \le x \le 1$
b.

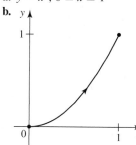

21. a. $y = x^2$, $x < 0$
b.

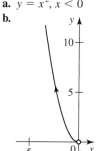

23. a. $y = \frac{1}{4}e^{2x}$
b.

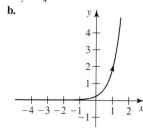

25. a. $x^2 - y^2 = 1$, $x \ge 1$
b.

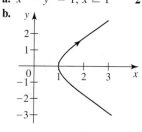

27. a. $y = x^{3/2}$, $0 \le x \le 1$
b.

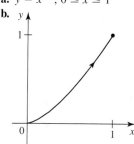

29. As t increases, the particle moves along the parabola $y = \sqrt{x - 1}$ from $(1, 0)$ to $(5, 2)$.

31. The particle starts out at $(2, 2)$ and travels once counterclockwise along the circle of radius 1 centered at $(1, 2)$.

33. The particle starts at $(0, 0)$ and moves to the right along the parabola $y = x^2$ to $(1, 1)$, then back to $(-1, 1)$, then again to $(1, 1)$, and finally back to $(0, 0)$.

35.

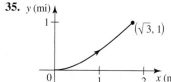

43.

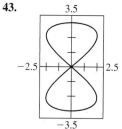

45.

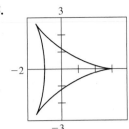

47.

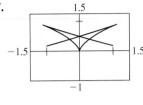

21. $\dfrac{dy}{dx} = \dfrac{1 - \cos\theta}{1 - \sin\theta}$, $\dfrac{d^2y}{dx^2} = \dfrac{\sin\theta + \cos\theta - 1}{(1 - \sin\theta)^3}$

23. $\dfrac{dy}{dx} = \coth t$, $\dfrac{d^2y}{dx^2} = -\dfrac{1}{\sinh^3 t}$

25. Concave downward on $(-\infty, 0)$, concave upward on $(0, \infty)$

27. $dy/dx = -\tan t$, $m = -1$ at $\left(\frac{\sqrt{2}}{4}a, \frac{\sqrt{2}}{4}a\right)$ and $\left(-\frac{\sqrt{2}}{4}a, -\frac{\sqrt{2}}{4}a\right)$, $m = 1$ at $\left(-\frac{\sqrt{2}}{4}a, \frac{\sqrt{2}}{4}a\right)$ and $\left(\frac{\sqrt{2}}{4}a, -\frac{\sqrt{2}}{4}a\right)$

29. Absolute maximum $f(-14) = 8$, absolute minimum $f(94) = -19$

31. $\frac{1}{243}(97^{3/2} - 64)$ **33.** $2\sqrt{5}$ **35.** $\frac{1}{8}a\pi^2$ **37.** $16a$

39. $4\sqrt{2}$ **41.** Approximately 1639 ft

43. $\left(a, -\dfrac{1}{2}\left(\dfrac{eE}{mv_0^2}\right)a^2\right)$

45. a.

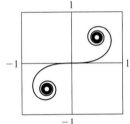

As $t \to \infty$, the curve spirals about and converges to the point $\left(\frac{1}{2}, \frac{1}{2}\right)$. As $t \to -\infty$, the curve spirals about and converges to the point $\left(-\frac{1}{2}, -\frac{1}{2}\right)$.

b. a

47. $3\pi a^2$ **49.** $3\pi a^2/8$ **51.** $32\sqrt{2}/3$

53. $\dfrac{2(247\sqrt{13} + 64)\pi}{1215}$ **55.** $148\pi/5$ **57.** $64\pi/3$

59. $\dfrac{24(\sqrt{2} + 1)\pi}{5}$ **61.** $\pi(e^2 + 2e - 6)$ **63.** $12\pi a^2/5$

65. $4\pi^2 rb$

49.

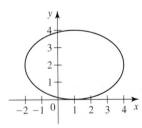

51. False   **53.** True

Exercises 10.3 • page 863

1. $\frac{1}{2}$ **3.** -2 **5.** $-\frac{3}{2}$ **7.** $y = \frac{1}{2}x - \frac{1}{2}$

9. $y = 5x + 2$

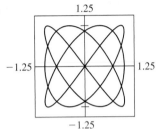

11. $(31, 64)$

13. Horizontal at $(-3, \pm2)$, vertical at $(-4, 0)$

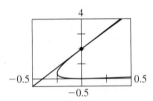

15. Horizontal at $(1, 0)$ and $(1, 4)$, vertical at $(4, 2)$ and $(-2, 2)$

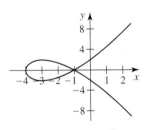

17. $\dfrac{dy}{dx} = t$, $\dfrac{d^2y}{dx^2} = \dfrac{1}{6t}$ **19.** $\dfrac{dy}{dx} = -\dfrac{2}{t^{3/2}}$, $\dfrac{d^2y}{dx^2} = \dfrac{6}{t^2}$

67.

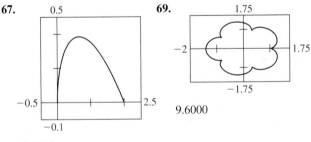

9.6000

2.2469

69.

71. 33.66   **73.** Center $(0, 0)$, radius a

75. $x = \dfrac{3at}{t^3 + 1}$, $y = \dfrac{3at^2}{t^3 + 1}$ **79.** False

Exercises 10.4 • page 877

1.

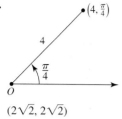

$(2\sqrt{2}, 2\sqrt{2})$

3.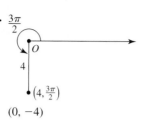

$(0, -4)$

5.

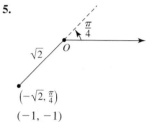

$(-1, -1)$

7.

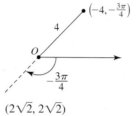

$(2\sqrt{2}, 2\sqrt{2})$

9.

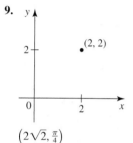

$\left(2\sqrt{2}, \frac{\pi}{4}\right)$

11.

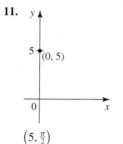

$\left(5, \frac{\pi}{2}\right)$

13.

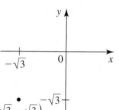

$\left(\sqrt{6}, \frac{5\pi}{4}\right)$

15.

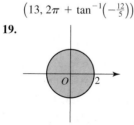

$\left(13, 2\pi + \tan^{-1}\left(-\frac{12}{5}\right)\right)$

17.

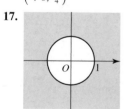

19.

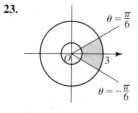

21.

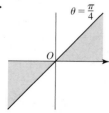

23.

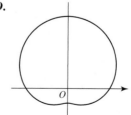

25. $x = 2$ **27.** $2x + 3y = 6$ **29.** $x^2 + y^2 - 4x = 0$

31. $x^2 - 2y - 1 = 0$ **33.** $r = 4 \sec \theta$ **35.** $r = 3$

37. $r^2 = 8 \csc 2\theta$

39.

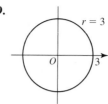

41.

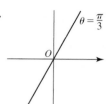

43.

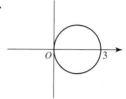

45.

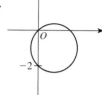

47.

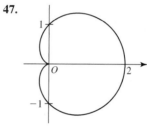

49.

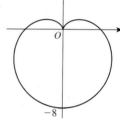

51.

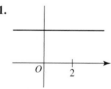

53.

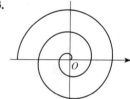

55.

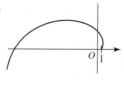

57.

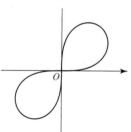

59.

61.

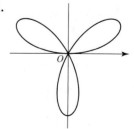

63.

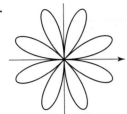

65. $\sqrt{3}/3$ **67.** -1 **69.** π **71.** 0

73. Horizontal at $\left(2\sqrt{2}, \frac{\pi}{4}\right)$ and $\left(-2\sqrt{2}, \frac{3\pi}{4}\right)$, vertical at the pole and $(4, 0)$

75. Horizontal at $(0, 0)$, $(\sin(2\tan^{-1}\sqrt{2}), \tan^{-1}\sqrt{2})$, $(\sin(2\tan^{-1}(-\sqrt{2})), \pi + \tan^{-1}(-\sqrt{2}))$, $(0, \pi)$, $(\sin(2\tan^{-1}\sqrt{2}), \pi + \tan^{-1}\sqrt{2})$, and $(\sin(2\tan^{-1}(-\sqrt{2})), 2\pi + \tan^{-1}(-\sqrt{2}))$; vertical at $\left(\sin\left(2\sin^{-1}\frac{\sqrt{3}}{3}\right), \sin^{-1}\frac{\sqrt{3}}{3}\right)$, $\left(0, \frac{\pi}{2}\right)$, $\left(\sin\left(2\sin^{-1}\left(-\frac{\sqrt{3}}{3}\right)\right), \pi + \sin^{-1}\left(-\frac{\sqrt{3}}{3}\right)\right)$, $\left(\sin\left(2\sin^{-1}\frac{\sqrt{3}}{3}\right), \pi + \sin^{-1}\frac{\sqrt{3}}{3}\right)$, $\left(0, \frac{3\pi}{2}\right)$, and $\left(\sin\left(2\sin^{-1}\left(-\frac{\sqrt{3}}{3}\right)\right), 2\pi + \sin^{-1}\left(-\frac{\sqrt{3}}{3}\right)\right)$

77. Horizontal at $\left(1 + \dfrac{\sqrt{33} - 1}{4}, \cos^{-1}\dfrac{\sqrt{33} - 1}{8}\right)$, $\left(1 - \dfrac{\sqrt{33} + 1}{4}, \cos^{-1}\dfrac{-\sqrt{33} - 1}{8}\right)$, $\left(1 - \dfrac{\sqrt{33} + 1}{4}, 2\pi - \cos^{-1}\left(\dfrac{-1 - \sqrt{33}}{8}\right)\right)$, and $\left(1 + \dfrac{\sqrt{33} - 1}{4}, 2\pi - \cos^{-1}\dfrac{\sqrt{33} - 1}{8}\right)$, vertical at $(3, 0)$, $\left(\frac{1}{2}, \cos^{-1}\left(-\frac{1}{4}\right)\right)$, $(-1, \pi)$, and $\left(\frac{1}{2}, 2\pi - \cos^{-1}\left(-\frac{1}{4}\right)\right)$

81. b. $2\sqrt{3}$

83. a.

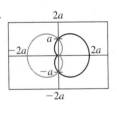

85.

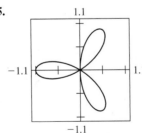

87.

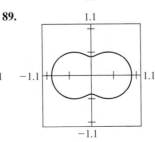

89.

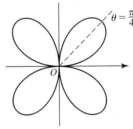

91.

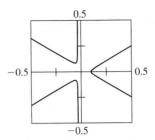

93. False   **95.** False

Exercises 10.5 • page 885

1. a. 4π **b.** 4π **3.** $\frac{1}{6}\pi^3$

5. $\dfrac{e^\pi - 1}{4e^\pi}$ **7.** $\frac{1}{2}$ **9.** $9\pi^3/16$ **11.** $3\pi/4$

13.

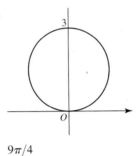

$9\pi/4$

15.

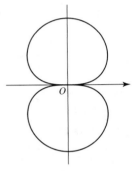

2

17.

2π

19. $\pi/8$ **21.** $\pi/16$ **23.** $\pi - \dfrac{3\sqrt{3}}{2}$ **25.** $\dfrac{\pi - 2}{8}$

27. $\dfrac{3\pi - 1}{4}$ **29.** $\left(1, \frac{\pi}{2}\right)$ and $\left(1, \frac{3\pi}{2}\right)$

31. $\left(2, \frac{\pi}{6}\right)$, $\left(2, \frac{\pi}{3}\right)$, $\left(2, \frac{2\pi}{3}\right)$, $\left(2, \frac{5\pi}{6}\right)$, $\left(2, \frac{7\pi}{6}\right)$, $\left(2, \frac{4\pi}{3}\right)$, $\left(2, \frac{5\pi}{3}\right)$, and $\left(2, \frac{11\pi}{6}\right)$

33. $\left(\frac{\sqrt{3}}{2}, \frac{\pi}{3}\right)$, $\left(-\frac{\sqrt{3}}{2}, \frac{5\pi}{3}\right)$, and the pole   **35.** π

37. $\dfrac{4\pi + 6\sqrt{3}}{3}$ **39.** $\dfrac{4\pi + 9\sqrt{3}}{8}$ **41.** $\dfrac{4\pi - 3\sqrt{3}}{6}$

43. $\dfrac{7\pi - 12\sqrt{3}}{12}$ **45.** $\dfrac{2\pi + 12 - 6\sqrt{3}}{3}$ **47.** 5π

49. $\sqrt{2}(1 - e^{-4\pi})$ **51.** $\dfrac{4\pi - 3\sqrt{3}}{8}$ **53.** $16a/3$

55. 16π **57.** $128\pi/5$ **59.** $2\sqrt{2}\pi$ **61.** 2π

67. a.   **b.** 22.01

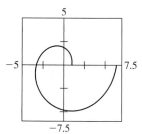

69. a.   1.25   **b.** 5.37

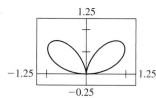

71. a.   **b.** $\left(\frac{21}{20}, 0\right)$

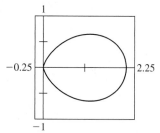

73. a. $r = \dfrac{3\cos\theta\sin\theta}{\cos^3\theta + \sin^3\theta}$, $-\dfrac{\pi}{4} < \theta < \dfrac{3\pi}{4}$

b.   **c.** $\frac{3}{2}$

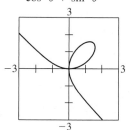

75. True

Exercises 10.6 • page 894

1. $r = \dfrac{2}{1 - \cos\theta}$, parabola

3. $r = \dfrac{2}{2 - \sin\theta}$, ellipse

5. $r = \dfrac{3}{2 + 3\cos\theta}$, hyperbola

7. $r = \dfrac{4}{25 + 10\sin\theta}$, ellipse

9. a. $e = \frac{1}{3}$, $y = 4$
b. Ellipse
c.

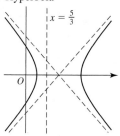

11. a. $e = \frac{3}{2}$, $x = \frac{5}{3}$
b. Hyperbola
c.

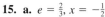

13. a. $e = 1$, $x = \frac{5}{2}$
b. Parabola
c.

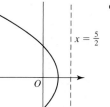

15. a. $e = \frac{2}{3}$, $x = -\frac{1}{2}$
b. Ellipse
c. $x = -\frac{1}{2}$

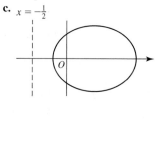

17. a. $e = 1$, $y = -1$
b. Parabola
c.

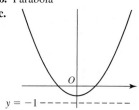

19. a. $e = \frac{1}{2}$, $y = -6$
b. Ellipse
c.

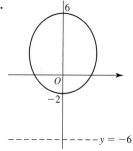

21. $\sqrt{7}/4$ **23.** $\sqrt{2}$ **25.** $\sqrt{10}/3$

33. $r = \dfrac{1.423 \times 10^9}{1 - 0.056\cos\theta}$, perihelion 1.347×10^9 km, aphelion 1.507×10^9 km

35. 0.207

Chapter 10 Concept Review • page 896

1. a. equidistant; point; line; point; focus; line; directrix
b. vertex; focus (or vertex); directrix

3. a. sum; foci; constant
b. foci; major axis; center; minor axis

5. a. difference; foci; constant
b. vertices; transverse; transverse; center; two separate

7. $x = f(t)$, $y = g(t)$; parameter

9. a. $f'(t)$; $g'(t)$; simultaneously zero; endpoints

b. $\int_a^b \sqrt{[f'(t)]^2 + [g'(t)]^2}\, dt = \int_a^b \sqrt{\left(\frac{dx}{dt}\right)^2 + \left(\frac{dy}{dt}\right)^2}\, dt$

11. a. $r\cos\theta$; $r\sin\theta$

b. $x^2 + y^2$; $\dfrac{y}{x}$ $(x \neq 0)$

13. a. $A = \dfrac{1}{2}\int_\alpha^\beta r^2\, d\theta = \dfrac{1}{2}\int_\alpha^\beta [f(\theta)]^2\, d\theta$

b. $A = \dfrac{1}{2}\int_\alpha^\beta \{[f(\theta)]^2 - [g(\theta)]^2\}\, d\theta$

15. $\dfrac{d(P, F)}{d(P, l)} = e$; $0 < e < 1$; $e = 1$; $e > 1$

Chapter 10 Review Exercises • page 897

1. Vertices $(0, \pm 3)$, foci $(0, \pm\sqrt{5})$

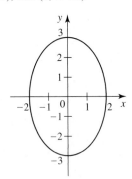

3. Vertices $(\pm 3, 0)$, foci $(\pm\sqrt{10}, 0)$

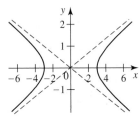

5. Vertices $(0, -1)$ and $(0, -7)$, foci $(0, -4 \pm \sqrt{10})$

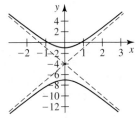

7. $y^2 = -8x$ **9.** $\dfrac{x^2}{49} + \dfrac{y^2}{45} = 1$ **11.** $y^2 - 4x^2 = 9$

15. a. $y = 4 - x$

b.

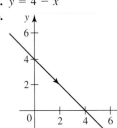

17. a. $(x - 1)^2 + (y - 3)^2 = 4$

b.

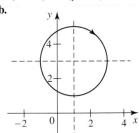

19. $\frac{4}{3}$ **21.** 0   **23.** $\dfrac{dy}{dx} = \dfrac{4(t^2 + 1)}{3t}$, $\dfrac{d^2 y}{dx^2} = \dfrac{4(t^2 - 1)}{9t^4}$

25. Vertical at $\left(\pm\frac{16\sqrt{3}}{9}, \frac{10}{3}\right)$, horizontal at $(0, 2)$

27. $\frac{13}{3}$ **29.** $\sqrt{2}(1 - e^{-\pi/2})$ **31.** 3π

33.

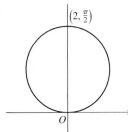

35.

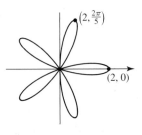

37.

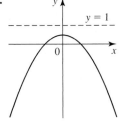

39. -2 **41.** $\left(\frac{1}{2}, \frac{\pi}{6}\right)$, $\left(\frac{1}{2}, \frac{5\pi}{6}\right)$, and the pole   **43.** $9\pi/2$

45. $2(\pi - 2)$ **47.** $\frac{8}{3}[(\pi^2 + 1)^{3/2} - 1]$ **49.** $4\pi^2$

51.

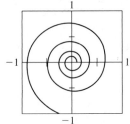

53.

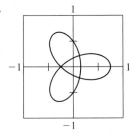

55.

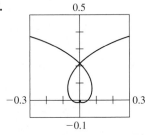

57.

59. $\pi(\pi + 2)$

Chapter 10 Challenge Problems • page 898

1. c.

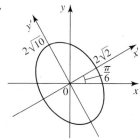

3. $\ln \dfrac{\pi}{2}$ **5. b.** $3a^2/2$

7. a. $r = \dfrac{\sqrt{2}}{2} ae^{(\pi/4)-\theta}$ (for an ant starting at the northeast corner)

  b. a

CHAPTER 11

Exercises 11.1 • page 909

1. a. Scalar **b.** Vector **c.** Scalar **d.** Vector

3. a. No **b.** Yes **c.** No **d.** No **e.** Yes **f.** Yes

9. $\langle -5, 0 \rangle$ **11.** $\langle 3, -4 \rangle$

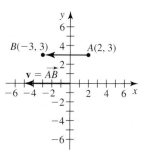 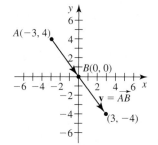

13.

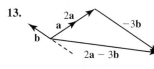

15.

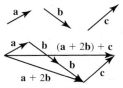

17. $\mathbf{v} = -(\mathbf{a} + \mathbf{b})$ **19.** $\mathbf{v} = 3(\mathbf{b} - \mathbf{a})$

21. $\langle -2, -1 \rangle$ **23.** $\langle -0.3, -0.1 \rangle$

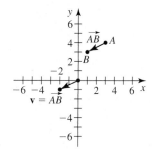 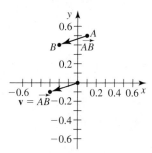

25. $(1, -2)$

27. $2\mathbf{a} = \langle -2, 4 \rangle$, $\mathbf{a} + \mathbf{b} = \langle 2, 3 \rangle$, $\mathbf{a} - \mathbf{b} = \langle -4, 1 \rangle$,
  $2\mathbf{a} + \mathbf{b} = \langle 1, 5 \rangle$, $|2\mathbf{a} + \mathbf{b}| = \sqrt{26}$

29. $2\mathbf{a} = 6\mathbf{i} - 4\mathbf{j}$, $\mathbf{a} + \mathbf{b} = 5\mathbf{i} - 2\mathbf{j}$, $\mathbf{a} - \mathbf{b} = \mathbf{i} - 2\mathbf{j}$,
  $2\mathbf{a} + \mathbf{b} = 8\mathbf{i} - 4\mathbf{j}$, $|2\mathbf{a} + \mathbf{b}| = 4\sqrt{5}$

31. $2\mathbf{a} = \mathbf{i} + 3\mathbf{j}$, $\mathbf{a} + \mathbf{b} = \frac{5}{4}\mathbf{i} + \frac{5}{4}\mathbf{j}$, $\mathbf{a} - \mathbf{b} = -\frac{1}{4}\mathbf{i} + \frac{7}{4}\mathbf{j}$,
  $2\mathbf{a} + \mathbf{b} = \frac{7}{4}\mathbf{i} + \frac{11}{4}\mathbf{j}$, $|2\mathbf{a} + \mathbf{b}| = \frac{\sqrt{170}}{4}$

33. $2\mathbf{a} - 3\mathbf{b} = 4\mathbf{i} + 18\mathbf{j}$, $\frac{1}{2}\mathbf{a} + \frac{1}{3}\mathbf{b} = \mathbf{i} - 2\mathbf{j}$

35. $a = -1.6$, $b = 2.2$ **37.** Parallel   **39.** Not parallel

41. Parallel   **43.** $\frac{8}{3}$

45. a. $\langle \frac{2\sqrt{5}}{5}, \frac{\sqrt{5}}{5} \rangle$ **b.** $\langle -\frac{2\sqrt{5}}{5}, -\frac{\sqrt{5}}{5} \rangle$

47. a. $\langle -\frac{\sqrt{3}}{2}, \frac{1}{2} \rangle$ **b.** $\langle \frac{\sqrt{3}}{2}, -\frac{1}{2} \rangle$

49. $\langle \frac{5\sqrt{2}}{2}, \frac{5\sqrt{2}}{2} \rangle$ **51.** $\sqrt{3}\mathbf{i}$ **53.** $\langle -\frac{27\sqrt{85}}{85}, \frac{6\sqrt{85}}{85} \rangle$

55. $\mathbf{F} = \sqrt{10}\langle \frac{3\sqrt{10}}{10}, \frac{\sqrt{10}}{10} \rangle$, 3, 1   **57.** $\sqrt{39}(\frac{\sqrt{13}}{13}\mathbf{i} + \frac{2\sqrt{39}}{13}\mathbf{j})$, $\sqrt{3}$, 6

59. $2\mathbf{i}$ **61.** $\dfrac{3}{2}\mathbf{i} - \dfrac{3\sqrt{3}}{2}\mathbf{j}$

63. a. $\langle a_1 + a_2, b_1 + b_2 \rangle$; the company produced $a_1 + a_2$
  Model A systems and $b_1 + b_2$ Model B systems

  b. $\langle 1.1(a_1 + a_2), 1.1(b_1 + b_2) \rangle$

65. $400\sqrt{2}\mathbf{i}$ ft/sec, $400\sqrt{2}\mathbf{j}$ ft/sec   **67.** 30 lb

69. $60°$, $\dfrac{\sqrt{3}}{30}$ hr   **71.** 275.4 mph, E 25.8° N

83. False   **85.** True   **87.** False

Exercises 11.2 • page 920

1. **3.**

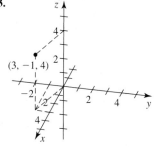

5. $(-3, -2, 4)$

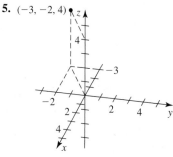

7. $A(2, 5, 5)$, $B(3, -3, -3)$

9.

11.

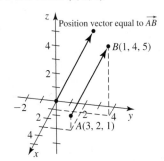

13. The subspace of three-dimensional space that lies on or in front of the plane $x = 3$

15. The subspace of three-dimensional space that lies strictly above the plane $z = 3$

17. a. $A(2, -1, 3)$, $B(-1, 4, 1)$ (in units of 1000 ft)

    b. 6164 ft

19. $\sqrt{21}$, $\sqrt{6}$, $\sqrt{27}$, right   **21.** 3, 2, $\sqrt{5}$, right

23. Collinear   **25.** $(-1, 3, -1)$

27. $(x - 2)^2 + (y - 1)^2 + (z - 3)^2 = 9$

29. $(x - 3)^2 + (y + 1)^2 + (z - 2)^2 = 16$

31. $\left(x - \frac{5}{2}\right)^2 + \left(y + \frac{1}{2}\right)^2 + \left(z - \frac{5}{2}\right)^2 = \frac{35}{4}$

33. $(x + 1)^2 + (y - 2)^2 + (z - 4)^2 = 6$

35. $(1, 2, 3)$, 2   **37.** $(2, -3, 0)$, $\sqrt{13}$ **39.** $\left(\frac{3}{2}, 1, -\frac{1}{2}\right)$, 2

41. All points inside the sphere with radius 2 and center $(0, 0, 0)$

43. All points lying on or between two concentric spheres with radii 1 and 3 and center $(0, 0, 0)$

45. $\langle -2, 2, 4 \rangle$

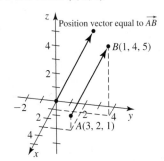

47. $\langle 4, -5, 5 \rangle$

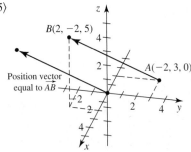

49. $\langle -1, 3, 5 \rangle$

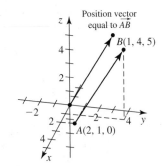

51. $(3, -6, -3)$

53. $\mathbf{a} + \mathbf{b} = \langle 1, 5, -1 \rangle$, $2\mathbf{a} - 3\mathbf{b} = \langle -8, -5, 3 \rangle$, $|3\mathbf{a}| = 3\sqrt{5}$, $|-2\mathbf{b}| = 2\sqrt{14}$, $|\mathbf{a} - \mathbf{b}| = \sqrt{11}$

55. $\mathbf{a} + \mathbf{b} = \langle 1, 6.2, -2.2 \rangle$, $2\mathbf{a} - 3\mathbf{b} = \langle -3, -8.1, 23.6 \rangle$, $|3\mathbf{a}| \approx 11.99$, $|-2\mathbf{b}| \approx 14.02$, $|\mathbf{a} - \mathbf{b}| \approx 9.27$

57. $a = \frac{4}{35}$, $b = \frac{16}{35}$, $c = \frac{2}{5}$

59. Parallel   **61.** Not parallel

63. a. $\langle \frac{1}{3}, \frac{2}{3}, \frac{2}{3} \rangle$ **b.** $\langle -\frac{1}{3}, -\frac{2}{3}, -\frac{2}{3} \rangle$

65. a. $\dfrac{\sqrt{11}}{11}(-\mathbf{i} + 3\mathbf{j} - \mathbf{k})$ **b.** $\dfrac{\sqrt{11}}{11}(\mathbf{i} - 3\mathbf{j} + \mathbf{k})$

67. $\langle \frac{10\sqrt{3}}{3}, \frac{10\sqrt{3}}{3}, \frac{10\sqrt{3}}{3} \rangle$ **69.** $\frac{3\sqrt{5}}{5}\mathbf{i} + \frac{6\sqrt{5}}{5}\mathbf{j}$

71. $\langle \frac{\sqrt{2}}{3}, -\frac{\sqrt{2}}{3}, \frac{4\sqrt{2}}{3} \rangle$ **73.** $5\sqrt{2}\langle -\frac{3\sqrt{2}}{10}, \frac{2\sqrt{2}}{5}, \frac{\sqrt{2}}{2} \rangle$

83. True   **85.** False

Exercises 11.3 • page 932

1. a. Yes **b.** No **c.** No **d.** Yes **e.** Yes **f.** Yes

3. -1 **5.** -4 **7.** 4π **9.** 4   **11.** -75

13. $\langle -24, 48, -12 \rangle$ **15.** 70   **17.** $26.6°$ **19.** $94.7°$

21. $63.1°$ **23.** $\frac{1}{3}$ **25.** Neither   **27.** Neither

29. Orthogonal   **31.** -6

33. $\cos \alpha = \sqrt{14}/14$, $\alpha \approx 74.5°$, $\cos \beta = \sqrt{14}/7$, $\beta \approx 57.7°$, $\cos \gamma = 3\sqrt{14}/14$, $\gamma \approx 36.7°$

35. $\cos \alpha = -\sqrt{35}/35$, $\alpha \approx 99.7°$, $\cos \beta = 3\sqrt{35}/35$, $\beta \approx 59.5°$, $\cos \gamma = \sqrt{35}/7$, $\gamma \approx 32.3°$

37. $\pi/3$ or $2\pi/3$

39. a. $\langle \frac{28}{13}, \frac{42}{13} \rangle$ **b.** $\langle \frac{14}{17}, \frac{56}{17} \rangle$

41. a. $\dfrac{20}{21}\mathbf{i} + \dfrac{10}{21}\mathbf{j} + \dfrac{40}{21}\mathbf{k}$ **b.** $3\mathbf{i} + \mathbf{k}$

43. a. $\langle -\frac{12}{29}, \frac{16}{29}, -\frac{8}{29} \rangle$ **b.** $\langle 0, 4, 0 \rangle$

45. $\mathbf{b} = \langle \frac{7}{5}, \frac{21}{5} \rangle + \langle \frac{3}{5}, -\frac{1}{5} \rangle$

47. $\mathbf{b} = \left(\dfrac{3}{14}\mathbf{i} + \dfrac{3}{7}\mathbf{j} + \dfrac{9}{14}\mathbf{k} \right) + \left(\dfrac{25}{14}\mathbf{i} - \dfrac{10}{7}\mathbf{j} + \dfrac{5}{14}\mathbf{k} \right)$

49. 12   **51.** $54.7°$ **53.** $\theta \approx 67.4°$, $\psi \approx 22.6°$

55. 1039.2 ft-lb

57. 274,955 ft-lb by Tugboat I, 207,846 ft-lb by Tugboat II

61. a.

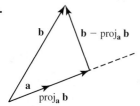

63. c. $\sqrt{13}$ **65.** False   **67.** True

Exercises 11.4 • page 943

1. a. Yes **b.** No **c.** Yes **d.** Yes **e.** Yes **f.** Yes

3. $3\mathbf{i} - 3\mathbf{j} + 2\mathbf{k}$ **5.** $3\mathbf{i} + 5\mathbf{j} + 7\mathbf{k}$ **7.** $-6\mathbf{i} - 7\mathbf{j} + 4\mathbf{k}$

9. $\mathbf{0}$ **11.** $\mathbf{i} + 7\mathbf{j} - 5\mathbf{k}, \ -\mathbf{i} - 7\mathbf{j} + 5\mathbf{k}$

13. $2\mathbf{i} - \mathbf{k}, \ -2\mathbf{i} + \mathbf{k}$ **15.** $\pm\dfrac{\sqrt{26}}{26}(3\mathbf{i} + \mathbf{j} - 4\mathbf{k})$

17. $\sqrt{3}/2$ **19.** $\sqrt{401}/2$ **21.** $-4\mathbf{i} + 6\mathbf{j} + 10\mathbf{k}$

23. $\mathbf{i} - \mathbf{j} + \mathbf{k}$ **25.** 15   **27.** 5   **29.** 21   **31.** $\sqrt{3}$

33. Yes   **35.** Yes   **37.** 32.5 ft-lb

39. $-4.8 \times 10^{-14}\mathbf{j}$ newtons   **53.** $3\sqrt{2}$

55. $a = -\frac{3}{2}, b = -6$ **57.** True   **59.** True   **61.** True

Exercises 11.5 • page 954

1. a. The direction of the required line is the same as the direction of the given line and so can be obtained from the parametric equations of the latter. Use this information with the given point to write down the required equation.
  b. Obtain the vectors $\mathbf{v}_1$ and $\mathbf{v}_2$ from the two given lines. Find the vector $\mathbf{n} = \mathbf{v}_1 \times \mathbf{v}_2$. Then the required line has the direction of $\mathbf{n}$ and contains the given point.
  c. A vector parallel to the required line is $\mathbf{n}$, a vector normal to the plane. It can be obtained from the equation of the plane. Use $\mathbf{n}$ and the given point to write down the required parametric equation.
  d. Obtain $\mathbf{n}_1$ and $\mathbf{n}_2$, the normals to the two planes, from the given equations of the planes. The direction of the required line is given by $\mathbf{v} = \mathbf{n}_1 \times \mathbf{n}_2$. Find a point on the required line by setting one variable, say z, equal to 0 and solving the resulting simultaneous equations in two variables.

3. $x = 1 + 2t, \ y = 3 + 4t, \ z = 2 + 5t$;
$$\frac{x - 1}{2} = \frac{y - 3}{4} = \frac{z - 2}{5}$$

5. $x = 3 + 2t, \ y = -t, \ z = -2 + 3t; \ \dfrac{x - 3}{2} = \dfrac{y}{-1} = \dfrac{z + 2}{3}$

7. $x = 2 - t, \ y = 1 + 2t, \ z = 4 + 3t$;
$$\frac{x - 2}{-1} = \frac{y - 1}{2} = \frac{z - 4}{3}$$

9. $x = -1 + 4t, \ y = -2 + 7t, \ z = -\dfrac{1}{2} - 5t$;
$$\frac{x + 1}{4} = \frac{y + 2}{7} = \frac{z + \frac{1}{2}}{-5}$$

11. $x = 1 + t, \ y = 2 + 2t, \ z = -1 - 3t$;
$$\frac{x - 1}{1} = \frac{y - 2}{2} = \frac{z + 1}{-3}; \ xy\text{-plane: } \left(\tfrac{2}{3}, \tfrac{4}{3}, 0\right),$$
xz-plane: $(0, 0, 2)$, yz-plane: $(0, 0, 2)$

13. Yes   **15.** Intersect at $(5, 4, 5)$

17. Intersect at $(2, 3, 1)$ **19.** Intersect, $49.1°$

21. Skew   **23.** $x + 2y + 4z = 24$

25. $x - 2z = 1$ **27.** $2x + 3y - z = 26$

29. $x - 3z = 8$ **31.** $6x - 4y + 3z = 0$

33. $6x - y - 4z = -5$ **35.** $9x + 11y + 3z = -2$

37. $11x + 10y + 13z = 45$ **39.** Orthogonal

41. Neither, $69.1°$ **43.** $70.5°$

45. $x = \frac{17}{4} - 10t, \ y = 8t, \ z = -\frac{11}{8} + 11t$

47. $x = 2 + 2t, \ y = 3 + 4t, \ z = -1 - 3t$

49. $11x + 5y - 7z = -8$ **51.** $6x - 7y + z = 11$

53. $(5, 0, 1)$ **55.** $4\sqrt{29}/29$ **57.** $2\sqrt{21}/7$

61. $6\sqrt{42}/7$ **65.** $\sqrt{6}/2$ **67.** False

69. False   **71.** False

Exercises 11.6 • page 970

1.

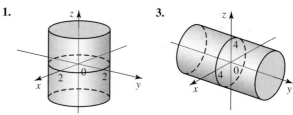

3.

5.

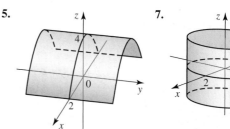

7.

9.

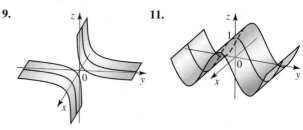

11.

13. (a)   **15.** (f)   **17.** (e)   **19.** (b)

21. $\dfrac{x^2}{1^2} + \dfrac{y^2}{2^2} + \dfrac{z^2}{2^2} = 1$ **23.** $\dfrac{x^2}{2^2} + \dfrac{y^2}{3^2} + \dfrac{z^2}{6^2} = 1$

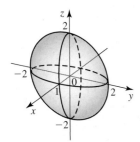   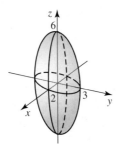

25. $x^2 + y^2 - \dfrac{z^2}{2^2} = 1$ **27.** $\dfrac{x^2}{2^2} + \dfrac{y^2}{1^2} - \dfrac{z^2}{2^2} = 1$

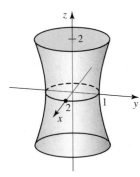   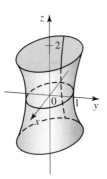

29. $-x^2 - y^2 + z^2 = 1$ **31.** $-\dfrac{x^2}{1^2} + \dfrac{y^2}{2^2} - \dfrac{z^2}{(\sqrt{2})^2} = 1$

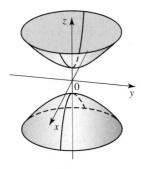   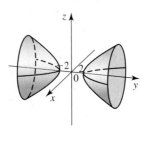

33. $x^2 + y^2 - z^2 = 0$ **35.** $\dfrac{x^2}{1^2} + \dfrac{y^2}{\left(\frac{3}{2}\right)^2} - \dfrac{z^2}{3^2} = 0$

or $\dfrac{x^2}{(2)^2} + \dfrac{y^2}{(3)^2} - \dfrac{z^2}{(6)^2} = 0$

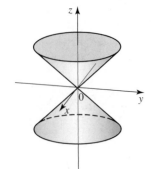   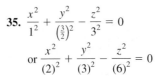

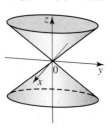

37. $x^2 + y^2 = z$ **39.** $\dfrac{x^2}{3^2} + \dfrac{y^2}{1^2} = \dfrac{z}{3^2}$

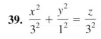

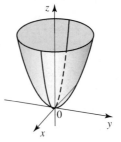   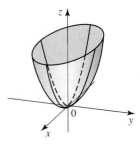

41. $x^2 + y^2 = z - 4$ **43.** $y^2 - x^2 = z$

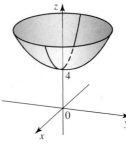   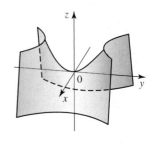

45.   **47.**

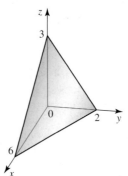   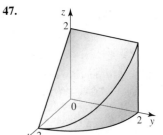

49.

51. $y^2 + z^2 = -12x$, a paraboloid

53. $x + y = 2$, a plane

55.

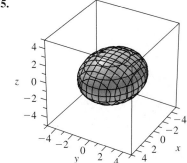

57.

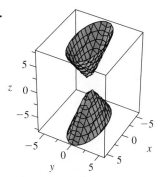

59.

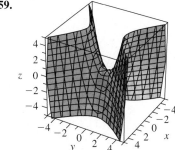

61. False   **63.** True

Exercises 11.7 • page 977

1. $(0, 3, 2)$ **3.** $(1, 1, \sqrt{3})$ **5.** $\left(\frac{3\sqrt{3}}{2}, -\frac{3}{2}, 2\right)$

7. $(2, 0, 3)$ **9.** $\left(2, \frac{\pi}{3}, 5\right)$ **11.** $\left(2, \frac{\pi}{6}, -2\right)$

13. $(0, 0, 5)$ **15.** $(\sqrt{2}, 0, \sqrt{2})$ **17.** $\left(\frac{5\sqrt{6}}{4}, \frac{5\sqrt{2}}{4}, \frac{5\sqrt{2}}{2}\right)$

19. $\left(2, \pi, \frac{\pi}{2}\right)$ **21.** $\left(2, 0, \frac{\pi}{3}\right)$ **23.** $\left(4, \frac{\pi}{2}, \frac{\pi}{3}\right)$

25. $\left(2, \frac{\pi}{4}, \frac{\pi}{2}\right)$ **27.** $\left(4\sqrt{2}, \frac{\pi}{3}, \frac{3\pi}{4}\right)$ **29.** $\left(2\sqrt{13}, \frac{\pi}{6}, \cos^{-1}\left(\frac{3\sqrt{13}}{13}\right)\right)$

31. $(0, 0, 3)$ **33.** $\left(2, \frac{3\pi}{2}, 0\right)$ **35.** $\left(\frac{\sqrt{3}}{2}, \frac{\pi}{4}, \frac{1}{2}\right)$ **37.** $\sqrt{11}$

39. Circular cylinder with radius 2 and central axis the z-axis

41. Sphere with center the origin and radius 2

43. Upper half of a right circular cone with vertex the origin and axis the positive z-axis

45. Paraboloid with vertex $(0, 0, 4)$ and axis the z-axis, opening downward

47. Plane parallel to the xy-plane and three units above it

49. Circular cylinder with radius 2 and central axis the line parallel to the z-axis passing through $(2, 0, 0)$

51. Parabolic cylinder

53. Sphere with radius 4 centered at the origin

55. Plane parallel to the yz-plane passing through $(2, 0, 0)$

57. Two circular cylinders with radii 1 and 2 and axis the z-axis

59. a. $r^2 + z^2 = 4$ **b.** $\rho = 2$

61. a. $r^2 - 2z = 0$ **b.** $\rho \sin^2 \phi - 2 \cos \phi = 0$

63. a. $r(2 \cos \theta + 3 \sin \theta) - 4z = 12$
   b. $\rho(2 \sin \phi \cos \theta + 3 \sin \phi \sin \theta - 4 \cos \phi) = 12$

65. a. $r^2 \cos^2 \theta + z^2 = 4$
   b. $\rho^2(\sin^2 \phi \cos^2 \theta + \cos^2 \phi) = 4$

67.   **69.**

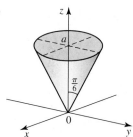

71. a. $(3960, 241.75°, 55.94°)$, $(3960, 2.20°, 41.48°)$
   b. $(-1552.8, -2889.9, 2217.8)$, $(2621.0, 100.7, 2966.8)$
   c. 5663 mi

73. True   **75.** True

Chapter 11 Concept Review • page 978

1. a. direction; magnitude   **b.** arrow; arrow; direction; length
   c. initial; A; terminal; B **d.** direction; magnitude

3. a. $\langle a_1, a_2 \rangle$; a_1; a_2; scalar; $\langle 0, 0 \rangle$
   b. $\langle a_1 + b_1, a_2 + b_2 \rangle$; $\langle ca_1, ca_2 \rangle$

5. a. (h, k, l); r
   b. $\left(\dfrac{x_1 + x_2}{2}, \dfrac{y_1 + y_2}{2}, \dfrac{z_1 + z_2}{2}\right)$

7. a. $a_1 b_1 + a_2 b_2 + a_3 b_3$; scalar
   b. $\sqrt{\mathbf{a} \cdot \mathbf{a}}$
   c. $\dfrac{\mathbf{a} \cdot \mathbf{b}}{|\mathbf{a}||\mathbf{b}|}$

9. a. vector projection; vector component
   b. scalar component
   c. $\left(\dfrac{\mathbf{b} \cdot \mathbf{a}}{|\mathbf{a}|^2}\right)\mathbf{a}$
   d. $\mathbf{F} \cdot \overrightarrow{PQ}$

11. a. $\mathbf{a} \cdot (\mathbf{b} \times \mathbf{c})$ **b.** $|\mathbf{a} \cdot (\mathbf{b} \times \mathbf{c})|$

13. a. $a(x - x_0) + b(y - y_0) + c(z - z_0) = 0$
   b. Plane; $\mathbf{n} = \langle a, b, c \rangle$; normal vectors

Chapter 11 Review Exercises • page 979

1. $\mathbf{i} - 8\mathbf{j} + 9\mathbf{k}$ **3.** $\sqrt{114}$ **5.** $5\mathbf{i} + 7\mathbf{j} - \mathbf{k}$

7. -20 **9.** $12\mathbf{j} + 4\mathbf{k}$

11. $\dfrac{3\sqrt{14}}{14}\mathbf{i} - \dfrac{\sqrt{14}}{7}\mathbf{j} + \dfrac{\sqrt{14}}{14}\mathbf{k}, \ -\dfrac{3\sqrt{14}}{14}\mathbf{i} + \dfrac{\sqrt{14}}{7}\mathbf{j} - \dfrac{\sqrt{14}}{14}\mathbf{k}$

13. $\cos\alpha = \sqrt{6}/6, \ \cos\beta = \sqrt{6}/3, \ \cos\gamma = -\sqrt{6}/6$

15. $-\dfrac{3}{7}\mathbf{i} + \dfrac{3}{14}\mathbf{j} - \dfrac{9}{14}\mathbf{k}$ **17.** 20   **21.** -22 **23.** $70.5°$

25. a. $3\mathbf{i} + 3\mathbf{j} - 4\mathbf{k}$ **b.** $\frac{1}{2}\sqrt{34}$ **27.** 64 J   **29.** $11\mathbf{k}$

31. a. $x = -1 + 3t, \ y = 2 - 3t, \ z = -4 + 7t$

  b. $\dfrac{x+1}{3} = \dfrac{y-2}{-3} = \dfrac{z+4}{7}$

33. a. $x = 1 + 6t, \ y = 2 + t, \ z = 4 - 4t$

  b. $\dfrac{x-1}{6} = \dfrac{y-2}{1} = \dfrac{z-4}{-4}$

35. $2x + 4y - 3z = 3$ **37.** $y = 2$ **39.** $5\sqrt{29}/29$

41. $32.7°$ **43.** $\sqrt{14}/14$ **45.** $5\sqrt{29}/29$

47. All points lying on or inside the (infinite) circular cylinder of radius 2 with axis the z-axis

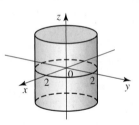

49. All points lying on or inside the prism shown

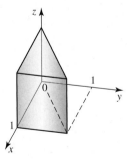

51. A plane perpendicular to the xy-plane

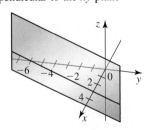

53. A paraboloid

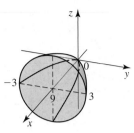

55. An elliptic cone with axis the y-axis

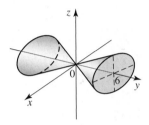

57. A hyperbolic paraboloid.

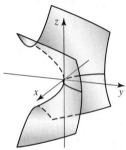

59. $\left(\sqrt{2}, \frac{\pi}{4}, \sqrt{2}\right), \left(2, \frac{\pi}{4}, \frac{\pi}{4}\right)$ **61.** $\left(\frac{\sqrt{6}}{2}, \frac{\sqrt{6}}{2}, 1\right), \left(\sqrt{3}, \frac{\pi}{4}, 1\right)$

63. The half-plane containing the z-axis making an angle of $\pi/3$ with the positive x-axis

65. A right circular cylinder with radius 1 and axis parallel to the z-axis passing through $(0, 1, 0)$

67. a. $r^2 = 2$ **b.** $\rho = \sqrt{2}\csc\phi$

69. a. $r^2 + 2z^2 = 1$ **b.** $\rho^2(1 + \cos^2\phi) = 1$

71. $0 \le r \le z, \ 0 \le \theta \le \frac{\pi}{2}$

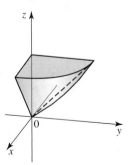

Chapter 11 Challenge Problems • page 981

5. b. $\mathbf{v} = -9\langle 1, 3, 1\rangle - 15\langle 2, -1, 1\rangle + 14\langle 3, 1, 2\rangle$

7. $(1, -2, 8)$ and $(3, 2, 40)$

CHAPTER 12

Exercises 12.1 • page 990

1. $(-\infty, 0) \cup (0, \infty)$ **3.** $(0, 1) \cup (1, \infty)$ **5.** $(0, \infty)$

7. (c)   **9.** (e)   **11.** (b)

13.

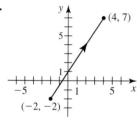

15.

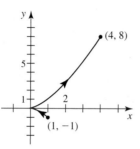

17.

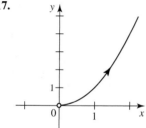

19.

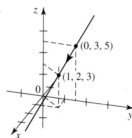

21.

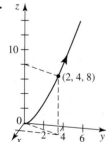

23.

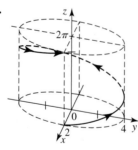

25.

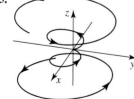

27.

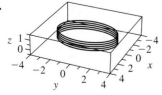

29.

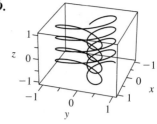

31. b.

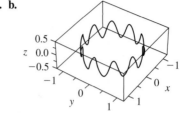

33. $\mathbf{r}(t) = \left\langle \cos t, \sin t, \dfrac{1 - \cos t - \sin t}{2} \right\rangle, \ 0 \le t \le 2\pi$

35. $\mathbf{r}(t) = \left\langle t, \dfrac{2t - 1}{2t - 2}, \dfrac{-2t^2 + 2t - 1}{2t - 2} \right\rangle$

37. $\mathbf{i} + \mathbf{j} - 3\mathbf{k}$ **39.** $\sqrt{2}\mathbf{i} + 4\mathbf{j} + \frac{2}{5}\mathbf{k}$ **41.** $\langle 0, 0, 2\rangle$

43. $[-1, 0)$ and $(0, \infty)$ **45.** $(0, \infty)$

47. $(-\infty, -1)$, $(-1, 1)$, and $(1, 4]$ **49.** 88,000 ft

57. $\mathbf{i} + \frac{1}{2}\mathbf{j}$ **59.** False   **61.** True

Exercises 12.2 • page 998

1. $\mathbf{r}'(t) = \mathbf{i} + 2t\mathbf{j} + 3t^2\mathbf{k}, \ \mathbf{r}''(t) = 2\mathbf{j} + 6t\mathbf{k}$

3. $\mathbf{r}'(t) = \left\langle 2t, \dfrac{t}{\sqrt{t^2 + 1}} \right\rangle, \ \mathbf{r}''(t) = \left\langle 2, \dfrac{1}{(t^2 + 1)^{3/2}} \right\rangle$

5. $\mathbf{r}'(t) = \langle -t \sin t, t \cos t\rangle,$
$\mathbf{r}''(t) = \langle -\sin t - t \cos t, \cos t - t \sin t\rangle$

7. $\mathbf{r}'(t) = (\cos t - \sin t)e^{-t}\mathbf{i} - (\cos t + \sin t)e^{-t}\mathbf{j} + \dfrac{1}{t^2 + 1}\mathbf{k},$

$\mathbf{r}''(t) = -2e^{-t}\cos t\,\mathbf{i} + 2e^{-t}\sin t\,\mathbf{j} - \dfrac{2t}{(t^2 + 1)^2}\mathbf{k}$

9. a. $\mathbf{r}(2) = \sqrt{2}\mathbf{i} - 2\mathbf{j}, \ \mathbf{r}'(2) = \dfrac{\sqrt{2}}{4}\mathbf{i} + \mathbf{j}$

b.

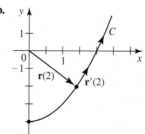

11. a. $\mathbf{r}(\pi/3) = \langle 2, \sqrt{3} \rangle$, $\mathbf{r}'(\pi/3) = \langle -2\sqrt{3}, 1 \rangle$
  b.

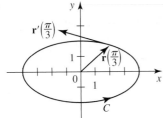

13. a. $\mathbf{r}(1) = 5\mathbf{i} - \mathbf{j}$, $\mathbf{r}'(1) = 3\mathbf{i} - 2\mathbf{j}$
  b.

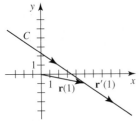

15. a. $\mathbf{r}(\pi/4) = \sqrt{2}\mathbf{i} + 2\mathbf{j}$, $\mathbf{r}'(\pi/4) = \sqrt{2}\mathbf{i} + 4\mathbf{j}$
  b.

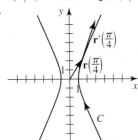

17. $\dfrac{\sqrt{14}}{14}\mathbf{i} + \dfrac{\sqrt{14}}{7}\mathbf{j} + \dfrac{3\sqrt{14}}{14}\mathbf{k}$ **19.** $\dfrac{2\sqrt{31}}{31}\mathbf{i} - \dfrac{3\sqrt{93}}{31}\mathbf{j}$

21. $x = 1 + t, y = 1 + 2t, z = 1 + 3t$

23. $x = 2 + \frac{1}{4}t, y = \frac{1}{3} - \frac{1}{9}t, z = \frac{1}{4} - \frac{1}{8}t$

25. $x = \dfrac{\sqrt{3}\pi}{12} + \left(\dfrac{\sqrt{3}}{2} - \dfrac{\pi}{12} \right)t, y = \dfrac{\pi}{12} + \left(\dfrac{1}{2} + \dfrac{\sqrt{3}\pi}{12} \right)t$,
  $z = \dfrac{\pi}{6}e^{\pi/6} + \left(1 + \dfrac{\pi}{6} \right)e^{\pi/6}t$

27. $\frac{1}{2}t^2\mathbf{i} + \frac{2}{3}t^3\mathbf{j} + 3t\mathbf{k} + \mathbf{C}$

29. $\frac{2}{3}t^{3/2}\mathbf{i} + \ln|t|\mathbf{j} - \frac{2}{5}t^{5/2}\mathbf{k} + \mathbf{C}$

31. $-\frac{1}{2}\cos 2t\mathbf{i} + \frac{1}{2}\sin 2t\mathbf{j} - e^{-t}\mathbf{k} + \mathbf{C}$

33. $(\cos t + t\sin t)\mathbf{i} - \frac{1}{2}\cos t^2\mathbf{j} - \frac{1}{2}e^{t^2} + \mathbf{C}$

35. $(2t + 1)\mathbf{i} + 2t^2\mathbf{j} - (2t^3 - 1)\mathbf{k}$

37. $\mathbf{r}(t) = e^{2t}\mathbf{i} - (3e^{-t} - 2)\mathbf{j} + e^t\mathbf{k}$

39. $\mathbf{r}(t) = \left(\frac{4}{15}t^{5/2} + t + 2 \right)\mathbf{i} - (\ln|\cos t| - 1)\mathbf{j} + (e^t - 2)\mathbf{k}$

55. $-\mathbf{r}'(-t) - \dfrac{1}{t^2}\mathbf{r}'\left(\dfrac{1}{t} \right)$

57. $\mathbf{r}'(t) \cdot [\mathbf{r}'(t) \times \mathbf{r}''(t)] + \mathbf{r}(t) \cdot [\mathbf{r}'(t) \times \mathbf{r}'''(t)]$

63. True   **65.** True

1. $4\sqrt{14}$ **3.** 10π **5.** $\sqrt{3}(e^{2\pi} - 1)$ **7.** e^2

9.

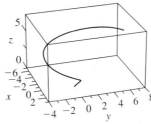

$\pi\sqrt{4\pi^2 + 2} + \ln(\sqrt{2\pi^2 + 1} + \sqrt{2}\pi)$

11. $s(t) = \sqrt{14}t, t \geq 0$;
  $\mathbf{r}(t(s)) = \left(1 + \dfrac{\sqrt{14}}{14}s \right)\mathbf{i} + \left(1 + \dfrac{2\sqrt{14}}{14}s \right)\mathbf{j} + \left(\dfrac{3\sqrt{14}}{14}s \right)\mathbf{k}$,
  $s \geq 0$

13. $s(t) = \sqrt{3}(e^t - 1)$;
  $\mathbf{r}(t(s)) = \left(1 + \dfrac{\sqrt{3}}{3}s \right)\cos\left(\ln\left(1 + \dfrac{\sqrt{3}}{3}s \right) \right)\mathbf{i}$
    $+ \left(1 + \dfrac{\sqrt{3}}{3}s \right)\sin\left(\ln\left(1 + \dfrac{\sqrt{3}}{3}s \right) \right)\mathbf{j}$
    $+ \left(1 + \dfrac{\sqrt{3}}{3}s \right)\mathbf{k}$, $s \geq 0$

15. 0   **17.** $\dfrac{\sqrt{5}}{(1 + 5t^2)^{3/2}}$ **19.** $\frac{1}{4}$ **21.** $\dfrac{6|x|}{(1 + 9x^4)^{3/2}}$

23. $\dfrac{4|\sin 2x|}{(1 + 4\cos^2 2x)^{3/2}}$ **25.** $\dfrac{2|2x^2 - 1|e^{2x^2}}{(e^{2x^2} + 4x^2)^{3/2}}$

27. $\left(-\dfrac{\sqrt{2}}{2}, e^{-1/2} \right), \left(\dfrac{\sqrt{2}}{2}, e^{-1/2} \right)$ **29.** $\left(\ln\dfrac{\sqrt{2}}{2}, \dfrac{\sqrt{2}}{2} \right)$

31. $(-1, -1), (1, 1)$ **33.** (b)   **35.** (d)

37. $\kappa(x) = \dfrac{2|2x^2 - 1|e^{2x^2}}{(e^{2x^2} + 4x^2)^{3/2}}$

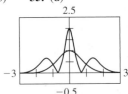

41. 0 for $t \neq 2n\pi$, where n is an integer

43. a. $\dfrac{162}{(81 - 5x^2)^{3/2}}$
  b. $\kappa(3) = \frac{3}{4}$, osculating circle at $(3, 0)$ has equation
    $\left(x - \frac{5}{3} \right)^2 + y^2 = \frac{16}{9}$; $\kappa(0) = \frac{2}{9}$, osculating circle at $(0, 2)$
    has equation $x^2 + \left(y + \frac{5}{2} \right)^2 = \frac{81}{4}$
  c.

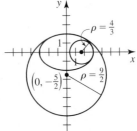

45. a. $\dfrac{dy}{dx} = \tan\dfrac{1}{2}\pi t^2$, $\dfrac{d^2y}{dx^2} = \dfrac{\pi t}{\cos^3\frac{1}{2}\pi t^2}$ **b.** πt

47. $\dfrac{3\sqrt{2}}{4\sqrt{1+\sin\theta}}$

51. $\sqrt{2}\left[\pi\sqrt{1+4\pi^2} + \frac{1}{2}\ln(2\pi + \sqrt{1+4\pi^2})\right]$

53. True   **55.** True   **57.** True

Exercises 12.4 • page 1016

1. $\mathbf{v}(1) = \mathbf{i} - 2\mathbf{j}$, $\mathbf{a}(1) = -2\mathbf{j}$, $|\mathbf{v}(1)| = \sqrt{5}$

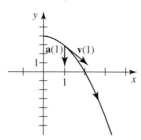

3. $\mathbf{v}\left(\dfrac{\pi}{4}\right) = -\dfrac{\sqrt{2}}{2}\mathbf{i} + \dfrac{3\sqrt{2}}{2}\mathbf{j}$, $\mathbf{a}\left(\dfrac{\pi}{4}\right) = -\dfrac{\sqrt{2}}{2}\mathbf{i} - \dfrac{3\sqrt{2}}{2}\mathbf{j}$,
$\left|\mathbf{v}\left(\dfrac{\pi}{4}\right)\right| = \sqrt{5}$

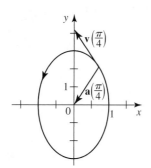

5. $\mathbf{v}(\pi/2) = -\mathbf{i} + \mathbf{k}$, $\mathbf{a}(\pi/2) = -\mathbf{j}$, $|\mathbf{v}(\pi/2)| = \sqrt{2}$

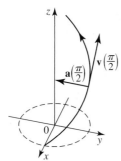

7. $\mathbf{v}(t) = \mathbf{i} + 2t\mathbf{j} + 2t\mathbf{k}$, $\mathbf{a}(t) = 2\mathbf{j} + 2\mathbf{k}$, $|\mathbf{v}(t)| = \sqrt{8t^2+1}$

9. $\mathbf{v}(t) = \mathbf{i} + 2t\mathbf{j} - \dfrac{1}{t^2}\mathbf{k}$, $\mathbf{a}(t) = 2\mathbf{j} + \dfrac{2}{t^3}\mathbf{k}$,
$|\mathbf{v}(t)| = \dfrac{\sqrt{4t^6 + t^4 + 1}}{t^2}$

11. $\mathbf{v}(t) = e^t\langle\cos t - \sin t, \cos t + \sin t, 1\rangle$,
$\mathbf{a}(t) = e^t\langle -2\sin t, 2\cos t, 1\rangle$, $|\mathbf{v}(t)| = \sqrt{3}e^t$

13. $\mathbf{v}(t) = \mathbf{i} + 2\mathbf{j} - 32t\mathbf{k}$, $\mathbf{r}(t) = t\mathbf{i} + 2t\mathbf{j} + (128 - 16t^2)\mathbf{k}$

15. $\mathbf{v}(t) = (t+1)\mathbf{i} - \frac{1}{2}t^2\mathbf{j} + \left(\frac{1}{2}t^2 + t + 1\right)\mathbf{k}$,
$\mathbf{r}(t) = \left(\frac{1}{2}t^2 + t\right)\mathbf{i} + \left(1 - \frac{1}{6}t^3\right)\mathbf{j} + \left(\frac{1}{6}t^3 + \frac{1}{2}t^2 + t + 1\right)\mathbf{k}$

17. $\mathbf{v}(t) = -\sin t\,\mathbf{i} + (\cos t - 1)\mathbf{j} + (t+2)\mathbf{k}$,
$\mathbf{r}(t) = \cos t\,\mathbf{i} + (\sin t - t)\mathbf{j} + \left(\frac{1}{2}t^2 + 2t\right)\mathbf{k}$

23. a. 60,892 ft   **b.** 8789 ft   **c.** 1500 ft/sec

25. a. 61,185 ft   **b.** 8989 ft   **c.** 1504 ft/sec

27. 4.4°

29. a. $-a\omega\sin\omega t\,\mathbf{i} + a\omega\cos\omega t\,\mathbf{j}$
b. $-\omega^2\mathbf{r}(t)$
c. $|\mathbf{v}(t)| = a\omega$, $|\mathbf{a}(t)| = a\omega^2$

31. 74.4 ft/sec $\leq v_0 \leq$ 81.6 ft/sec

35. $\mathbf{r}(t) = \langle 1 + t, 1 + 2t, 1 + 3t\rangle$, $\mathbf{r}(2) = \langle 3, 5, 7\rangle$

37. True

Exercises 12.5 • page 1026

1. $\mathbf{T}(t) = \dfrac{1}{\sqrt{1+16t^2}}\mathbf{i} + \dfrac{4t}{\sqrt{1+16t^2}}\mathbf{j}$,
$\mathbf{N}(t) = -\dfrac{4t}{\sqrt{1+16t^2}}\mathbf{i} + \dfrac{1}{\sqrt{1+16t^2}}\mathbf{j}$

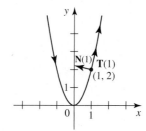

3. $\mathbf{T}(t) = \dfrac{2}{\sqrt{4+9t^2}}\mathbf{i} + \dfrac{3t}{\sqrt{4+9t^2}}\mathbf{j}$,
$\mathbf{N}(t) = -\dfrac{3t}{\sqrt{4+9t^2}}\mathbf{i} + \dfrac{2}{\sqrt{4+9t^2}}\mathbf{j}$

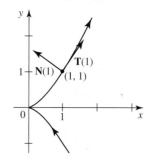

5. $\mathbf{T}(t) = \dfrac{1}{\sqrt{1+4t^2}}\mathbf{j} + \dfrac{2t}{\sqrt{1+4t^2}}\mathbf{k}$,

$\quad\mathbf{N}(t) = -\dfrac{2t}{\sqrt{1+4t^2}}\mathbf{j} + \dfrac{1}{\sqrt{1+4t^2}}\mathbf{k}$

7. $\mathbf{T}(t) = \left\langle \dfrac{2\sqrt{13}}{13}\cos 2t, -\dfrac{2\sqrt{13}}{13}\sin 2t, \dfrac{3\sqrt{13}}{13} \right\rangle$,

$\quad\mathbf{N}(t) = \langle -\sin 2t, -\cos 2t, 0 \rangle$

9. $\mathbf{T}(t) = \left\langle \dfrac{\sqrt{3}}{3}(\cos t - \sin t), \dfrac{\sqrt{3}}{3}(\cos t + \sin t), \dfrac{\sqrt{3}}{3} \right\rangle$,

$\quad\mathbf{N}(t) = \left\langle -\dfrac{\sqrt{2}}{2}(\sin t + \cos t), \dfrac{\sqrt{2}}{2}(\cos t - \sin t), 0 \right\rangle$

11. $a_{\mathbf{T}} = \dfrac{4t}{\sqrt{1+4t^2}}$, $a_{\mathbf{N}} = \dfrac{2}{\sqrt{1+4t^2}}$

13. $a_{\mathbf{T}} = \dfrac{18t^3 + 4t}{\sqrt{9t^4 + 4t^2 + 1}}$, $a_{\mathbf{N}} = \dfrac{2\sqrt{9t^4 + 9t^2 + 1}}{\sqrt{9t^4 + 4t^2 + 1}}$

15. $a_{\mathbf{T}} = 0$, $a_{\mathbf{N}} = 2$

17. $a_{\mathbf{T}} = \sqrt{3}e^t$, $a_{\mathbf{N}} = \sqrt{2}e^t$

19. a.

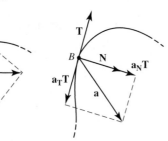

b. Accelerating at A, decelerating at B

21. a. $a_{\mathbf{T}} = -6$, $a_{\mathbf{N}} = 5$ **b.** Decelerating

29. $x - y - \sqrt{2}z = 0$ **31.** $\dfrac{\sqrt{2}}{2}(-\tanh t\mathbf{i} + \mathbf{j} - \operatorname{sech} t\mathbf{k})$

33. $\dfrac{1}{\sqrt{36t^4 + 9t^2 + 4}}(6t^2\mathbf{i} - 3t\mathbf{j} + 2\mathbf{k})$

35. $\frac{1}{2}$ **39.** 2.99×10^{11} m

41. a. $a_{\mathbf{T}} = \dfrac{g(gt - v_0 \sin \alpha)}{\sqrt{(v_0 \cos \alpha)^2 + (v_0 \sin \alpha - gt)^2}}$,

$\quad a_{\mathbf{N}} = \dfrac{gv_0 \cos \alpha}{\sqrt{(v_0 \cos \alpha)^2 + (v_0 \sin \alpha - gt)^2}}$

b. $a_{\mathbf{T}} = 0$, $a_{\mathbf{N}} = g$

45. True   **47.** False

Chapter 12 Concept Review • page 1028

1. a. $\langle f(t), g(t), h(t) \rangle$; real-valued; t; parameter
   b. parameter interval; real numbers

3. a. $\lim_{t\to a} f(t)$; $\lim_{t\to a} g(t)$; $\lim_{t\to a} h(t)$ **b.** $\mathbf{r}(a)$; continuous

5. a. $\mathbf{u}'(t) \cdot \mathbf{v}(t) + \mathbf{u}(t) \cdot \mathbf{v}'(t)$; $\mathbf{u}'(t) \times \mathbf{v}(t) + \mathbf{u}(t) \times \mathbf{v}'(t)$
   b. $\mathbf{u}'(f(t))f'(t)$

7. $\displaystyle\int_a^b \sqrt{[f'(t)]^2 + [g'(t)]^2 + [h'(t)]^2}\, dt$

9. a. $\left|\dfrac{d\mathbf{T}}{ds}\right|$ **b.** $\dfrac{|\mathbf{T}'(t)|}{|\mathbf{r}'(t)|}$ **c.** $\dfrac{|\mathbf{r}'(t) \times \mathbf{r}''(t)|}{|\mathbf{r}'(t)|^3}$ **d.** $\dfrac{|y''|}{[1 + (y')^2]^{3/2}}$

   e. radius of curvature; radius; tangent line; circle of curvature

11. a. $\dfrac{\mathbf{r}'(t)}{|\mathbf{r}'(t)|}$; $\dfrac{\mathbf{T}'(t)}{|\mathbf{T}'(t)|}$ **b.** $\mathbf{T}$; $\mathbf{N}$; v'; κv^2; tangential; normal

   c. $\dfrac{\mathbf{r}'(t) \cdot \mathbf{r}''(t)}{|\mathbf{r}'(t)|}$; $\dfrac{|\mathbf{r}'(t) \times \mathbf{r}''(t)|}{|\mathbf{r}'(t)|}$

Chapter 12 Review Exercises • page 1029

1.

3.

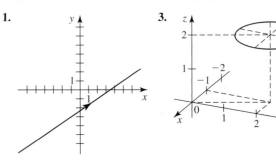

5. $(-1, 0)$ and $(0, 5)$ **7.** $[-1, 1)$ and $(1, 2)$

9. $\mathbf{r}'(t) = \dfrac{1}{2\sqrt{t}}\mathbf{i} + 2t\mathbf{j} - \dfrac{1}{(t+1)^2}\mathbf{k}$,

$\quad\mathbf{r}''(t) = -\dfrac{1}{4t^{3/2}}\mathbf{i} + 2\mathbf{j} + \dfrac{2}{(t+1)^3}\mathbf{k}$

11. $\mathbf{r}'(t) = 2t\mathbf{i} + 2\mathbf{j} + \dfrac{1}{t}\mathbf{k}$, $\mathbf{r}''(t) = 2\mathbf{i} - \dfrac{1}{t^2}\mathbf{k}$

13. $x = 1$, $y = -3 + 2t$, $z = 1$

15. $\frac{2}{3}t^{3/2}\mathbf{i} - \frac{1}{2}e^{-2t}\mathbf{j} + \ln|t + 1|\mathbf{k} + \mathbf{C}$

17. $\left(\dfrac{4}{3}t^{3/2} + 1\right)\mathbf{i} + \left(\dfrac{3}{2\pi}\sin 2\pi t + 2\right)\mathbf{j} + (e^{-t} - 1)\mathbf{k}$

19. $\mathbf{T}(1) = \dfrac{\sqrt{14}}{14}(\mathbf{i} + 2\mathbf{j} + 3\mathbf{k})$,

$\quad\mathbf{N}(1) = \dfrac{\sqrt{266}}{266}(-11\mathbf{i} - 8\mathbf{j} + 9\mathbf{k})$

21. 10   **23.** $\dfrac{2\sqrt{1 + 9t^2 + 9t^4}}{(1 + 4t^2 + 9t^4)^{3/2}}$

25. $\dfrac{4}{(x^2 - 4x + 8)^{3/2}}$, $(2, 1)$

27. $\mathbf{v}(t) = 2\mathbf{i} - 2e^{-2t}\mathbf{j} - \sin t\mathbf{k}$,
   $\mathbf{a}(t) = \mathbf{r}''(t) = 4e^{-2t}\mathbf{j} - \cos t\mathbf{k}$,
   $|\mathbf{v}(t)| = \sqrt{4 + 4e^{-4t} + \sin^2 t}$

29. $\mathbf{v}(t) = \left(\frac{1}{2}t^2 + 2\right)\mathbf{i} + \left(\frac{1}{9}t^3 + 3\right)\mathbf{j} + (3t + 1)\mathbf{k}$,
   $\mathbf{r}(t) = \left(\frac{1}{6}t^3 + 2t\right)\mathbf{i} + \left(\frac{1}{36}t^4 + 3t\right)\mathbf{j} + \left(\frac{3}{2}t^2 + t\right)\mathbf{k}$

31. $a_\mathbf{T} = \dfrac{4t}{\sqrt{1 + 4t^2}}$, $a_\mathbf{N} = \dfrac{2}{\sqrt{1 + 4t^2}}$

33. $a_\mathbf{T} = \dfrac{\sin t \cos t - 8 \sin 2t \cos 2t}{\sqrt{\sin^2 t + 4 \cos^2 2t}}$,

$a_\mathbf{N} = \dfrac{2|2 \sin t \sin 2t + \cos t \cos 2t|}{\sqrt{\sin^2 t + 4 \cos^2 2t}}$

35. a. $20\sqrt{2}\,t\mathbf{i} + (7 + 20\sqrt{2}\,t - 16t^2)\mathbf{j}$ **b.** 56.2 ft

Chapter 12 Challenge Problems • page 1030

3. b. $\dfrac{g(d^2 + h^2)}{2v_0^2}$

5. c. 15000

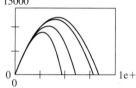

9. a. $|\mathbf{v}(3)| \approx 18.96$ cm/sec, $|\mathbf{a}(3)| \approx 60.54$ cm/sec^2
b. 4π cm/sec^2

CHAPTER 13

Exercises 13.1 • page 1043

Abbreviations: D, domain; R, range.

1. a. 8 **b.** 9 **c.** $4h^2 + 18hk - 4h + 3$

d. $x^2 + 2xh + h^2 + 3xy + 3hy - 2x - 2h + 3$

e. $x^2 + 3xy + 3xk - 2x + 3$

3. a. 6 **b.** $\sqrt{11}$ **c.** $\sqrt{6}\,|t|$ **d.** $\sqrt{6u^2 + 2u + 5}$

e. $\sqrt{15}\,|x|$

5. $D = \{(x, y) \mid -\infty < x < \infty, -\infty < y < \infty\}$,
$R = \{z \mid -\infty < z < \infty\}$

7. $D = \{(u, v) \mid u \neq v\}$, $R = \{z \mid -\infty < z < \infty\}$

9. $D = \{(x, y) \mid x^2 + y^2 \leq 4\}$, $R = \{z \mid 0 \leq z \leq 2\}$

11. $D = \{(x, y, z) \mid x^2 + y^2 + z^2 \leq 9\}$, $R = \{z \mid 0 \leq z \leq 3\}$

13. $D = \{(u, v, w) \mid u \neq \frac{\pi}{2} + n\pi, n$ an integer$\}$,
$R = \{z \mid -\infty < z < \infty\}$

15. $\{(x, y) \mid x \geq 0$ and $y \geq 0\}$ **17.** $\{(u, v) \mid u \neq v, u \neq -v\}$

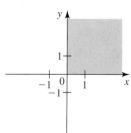
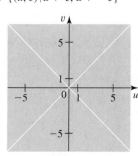

19. $\{(x, y) \mid x > 0, y > 0\}$

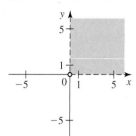

21. $\{(x, y, z) \mid x^2 + y^2 + z^2 \leq 9\}$

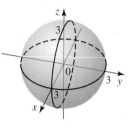

23.

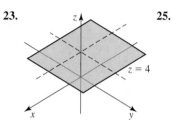

25.

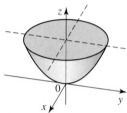

27.

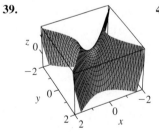

29.

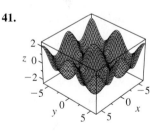

31. a. 200 ft, 400 ft **b.** ascending, ascending **c.** C

33. (c)   **35.** (a)   **37.** (d)

39.

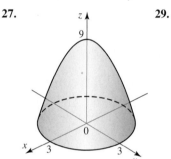

41.

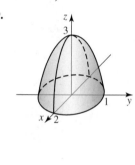

43.

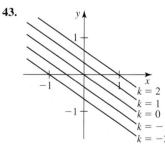

45.

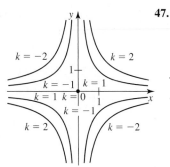

47.

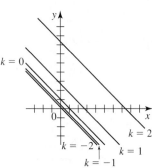

49.

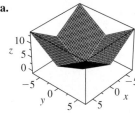

51.

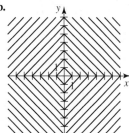

53. A family of parallel planes with normal vector $\langle 2, 4, -3 \rangle$

55. A cone with vertex the origin and axis the z-axis (if $k = 0$), a family of hyperboloids of one sheet with axis the z-axis (if $k > 0$), and a family of hyperboloids of two sheets with axis the z-axis (if $k < 0$)

57. (a)   **59.** (c)   **61.** (e)

63. a.

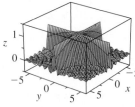

b.

65. a.

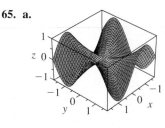

b.

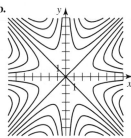

67. $\sqrt{x^2 + y^2} = 5$ **69.** No

71. a.

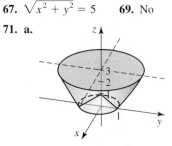

b. For $k = 0$, $x^2 + y^2 = 1$; for $k = \frac{1}{2}$, $x^2 + y^2 = \frac{1}{4}$ or $x^2 + y^2 = \frac{3}{2}$; for $k = 1$, $(0, 0)$ or $x^2 + y^2 = 2$; for $k = 3$, $x^2 + y^2 = 4$

73. $40,000k$ dynes   **77.** $260,552.20, $151,210.04

79. $40.28 g$

83. A family of concentric spheres centered at the origin

85. ≈ 435 Hz   **87.** False   **89.** False

Exercises 13.2 • page 1058

13. 9   **15.** -18 **17.** 3   **19.** -1 **21.** $\frac{1}{4}$ **23.** 0

25. $\frac{11}{3}$ **27.** 0   **29.** 0   **31.** $\{(x, y) \mid 2x + 3y \neq 1\}$

33. $\{(x, y) \mid x \geq 0, |y| \leq x\}$ **35.** $\{(x, y) \mid x \geq 0, y \neq 0\}$

37. $\{(x, y) \mid x^2 + y^2 + z^2 \neq 4\}$ **39.** $\{(x, y, z) \mid yz > 1\}$

41. a. The entire plane   **b.**

43. $(x^2 - xy + y^2) \cos(x^2 - xy + y^2) + \sin(x^2 - xy + y^2)$
The entire plane

45. $\dfrac{2x - y + 2}{2x - y - 1}$; $\{(x, y) \mid 2x - y \neq 1\}$

47. $\cos(x \tan y)$; $\{(x, y) \mid y \neq \frac{\pi}{2} + n\pi, n \text{ an integer}\}$

53. True   **55.** True   **57.** True

Exercises 13.3 • page 1069

1. a. 4, 4

   b. $f_x(2, 1) = 4$ says that the slope of the tangent line to the curve of intersection of the surface $z = x^2 + 2y^2$ and the plane $y = 1$ at the point $(2, 1, 6)$ is 4. $f_y(2, 1) = 4$ says that the slope of the tangent line to the curve of intersection of the surface $z = x^2 + 2y^2$ and the plane $x = 2$ at the point $(2, 1, 6)$ is 4.

   c. $f_x(2, 1) = 4$ says that the rate of change of $f(x, y)$ with respect to x with y fixed at 1 is 4 units per unit change in x. $f_y(2, 1) = 4$ says that the rate of change of $f(x, y)$ with respect to y with x fixed at 2 is 4 units per unit change in y.

3. At P, $\dfrac{\partial f}{\partial x} < 0$ and $\dfrac{\partial f}{\partial y} < 0$. At Q, $\dfrac{\partial f}{\partial x} = \dfrac{\partial f}{\partial y} = 0$. At R, $\dfrac{\partial f}{\partial x} < 0$ and $\dfrac{\partial f}{\partial y} > 0$.

5. $-7.1°$F/in., $-3.8°$F/in.

7. $f_x(x, y) = 4x - 3y$, $f_y(x, y) = -3x + 2y$

9. $\dfrac{\partial z}{\partial x} = \sqrt{y}$, $\dfrac{\partial z}{\partial y} = \dfrac{x}{2\sqrt{y}}$

11. $g_r(r, s) = \dfrac{1}{2\sqrt{r}}$, $g_s(r, s) = 2s$

13. $f_x(x, y) = e^{y/x}\left(1 - \dfrac{y}{x}\right)$, $f_y(x, y) = e^{y/x}$

15. $\dfrac{\partial z}{\partial x} = \dfrac{2x}{1 + (x^2 + y^2)^2}$, $\dfrac{\partial z}{\partial y} = \dfrac{2y}{1 + (x^2 + y^2)^2}$

17. $g_u(u, v) = \dfrac{v(v^3 - u^2)}{(u^2 + v^3)^2}$, $g_v(u, v) = \dfrac{u(u^2 - 2v^3)}{(u^2 + v^3)^2}$

19. $g_x(x, y) = 2x \cosh \dfrac{x}{y} + \dfrac{x^2}{y} \sinh \dfrac{x}{y}$, $g_y(x, y) = -\dfrac{x^3}{y^2} \sinh \dfrac{x}{y}$

21. $f_x(x, y) = y^x \ln y$, $f_y(x, y) = xy^{x-1}$

23. $f_x(x, y) = -xe^{-x}$, $f_y(x, y) = ye^{-y}$

25. $g_x(x, y, z) = \dfrac{\sqrt{xyz}}{2x}$, $g_y(x, y, z) = \dfrac{\sqrt{xyz}}{2y}$, $g_z(x, y, z) = \dfrac{\sqrt{xyz}}{2z}$

27. $\dfrac{\partial u}{\partial x} = e^{y/z}$, $\dfrac{\partial u}{\partial y} = \dfrac{x}{z} e^{y/z}$, $\dfrac{\partial u}{\partial z} = -\dfrac{xy}{z^2} e^{y/z} - 2z$

29. $f_r(r, s, t) = s \ln st$, $f_s(r, s, t) = r \ln st + r$, $f_t(r, s, t) = rs/t$

31. $\dfrac{\partial z}{\partial x} = \dfrac{ye^{-x} - e^y}{e^z}$, $\dfrac{\partial z}{\partial y} = -\dfrac{xe^y + e^{-x}}{e^z}$

33. $\dfrac{\partial z}{\partial x} = -\dfrac{2x(2x^2 + 2z^2 + 1)}{z(3yz^3 + 3x^2yz + 2)}$, $\dfrac{\partial z}{\partial y} = -\dfrac{z^2(x^2 + z^2)}{3yz^3 + 3x^2yz + 2}$

35. $g_{xx}(x, y) = 6xy^2$, $g_{yy} = 2x^3 + 6xy$, $g_{xy} = g_{yx} = 6x^2y + 3y^2$

37. $\dfrac{\partial^2 w}{\partial u^2} = -4 \cos(2u - v) - 4 \sin(2u + v)$,

   $\dfrac{\partial^2 w}{\partial v^2} = -\cos(2u - v) - \sin(2u + v)$,

   $\dfrac{\partial^2 w}{\partial u \, \partial v} = \dfrac{\partial^2 w}{\partial v \, \partial u} = 2 \cos(2u - v) - 2 \sin(2u + v)$

39. $h_{xx}(x, y) = \dfrac{2xy}{(x^2 + y^2)^2}$, $h_{yy}(x, y) = -\dfrac{2xy}{(x^2 + y^2)^2}$,

   $h_{yx}(x, y) = h_{xy}(x, y) = \dfrac{y^2 - x^2}{(x^2 + y^2)^2}$

41. $-8y$ **43.** $-\sin x$ **45.** $4e^x \sin(y + 2z)$

69. $\partial V/\partial T = 0.066512$, $\partial V/\partial P = -0.1596288$

71. $\left(\dfrac{R}{R_1}\right)^2$ **73.** $39 per $1000; $-$975 per 1000 ft^2

75. $-\sqrt{14}kQ/196$ volts per meter

77. $\partial N/\partial x \approx 1.06$, $\partial N/\partial y \approx -2.85$

79. a. $\approx 19.70°$F

   b. $\approx -0.285°$ for each 1 mph increase in wind speed

83. ≈ 0.9872 **87.** $2x + y - z = 2$

89. $0, -\dfrac{4e}{\pi(e^2 + 1)}$ **91.** No   **93.** True   **95.** True

Exercises 13.4 • page 1082

1. a. -0.0386 **b.** -0.04

3. $dz = 6xy^3 \, dx + 9x^2y^2 \, dy$

5. $dz = -\dfrac{2y}{(x - y)^2} \, dx + \dfrac{2x}{(x - y)^2} \, dy$

7. $dz = 12xy^3(2x^2 + 3y^2)^2 \, dx + 3y^2(2x^2 + 9y^2)(2x^2 + 3y^2)^2 \, dy$

9. $dw = 2xye^{x^2-y^2} \, dx + (1 - 2y^2)e^{x^2-y^2} \, dy$

11. $dw = \left[2x \ln(x^2 + y^2) + \dfrac{2x^3}{x^2 + y^2}\right] dx + \dfrac{2x^2y}{x^2 + y^2} \, dy$

13. $dz = 2e^{2x} \cos 3y \, dx - 3e^{2x} \sin 3y \, dy$

15. $dw = (2x + y) \, dx + x \, dy + 2z \, dz$

17. $dw = 2xe^{-yz} \, dx - x^2ze^{-yz} \, dy - x^2ye^{-yz} \, dz$

19. $dw = 2xe^y \, dx + (x^2e^y + \ln z) \, dy + \dfrac{y}{z} \, dz$

21. -0.06 **23.** 0.21 **25.** 1080 in.^{3} **27.** 58.3 ft^2

29. -3.3256 Pa   **31.** 2.73%   **33.** 9%   **35.** 0.3%

37. 0.19 Ω **39.** 7%   **45.** True   **47.** False

Exercises 13.5 • page 1093

1. $4xt - 6yt^2 - 2y$

3. $-2(\cos s + s \cos r)e^{-2t} + (\sin r - r \sin s)(3t^2 - 2)$

5. $2x^2y[3yz - 2xz \sin t + xy(\sin t + t \cos t)]$

7. $\dfrac{-2tz + y^2z - 2tx^2z^3 + \cosh t(y + xy^2 + x^2yz^2)}{y^2(1 + x^2z^2)}$

9. $\dfrac{\partial w}{\partial u} = 6(x^2u + y^2v), \dfrac{\partial w}{\partial v} = 6(x^2v + y^2u)$

11. $\dfrac{\partial w}{\partial u} = e^x\left(\dfrac{2u \cos y}{u^2 + v^2} - \dfrac{\sqrt{uv}\sin y}{2u}\right),$

$\dfrac{\partial w}{\partial v} = e^x\left(\dfrac{2v \cos y}{u^2 + v^2} - \dfrac{\sqrt{uv}\sin y}{2v}\right)$

13. $\dfrac{\partial w}{\partial u} = \dfrac{(\tan^{-1} yz)\sqrt{u}}{2u} - \dfrac{xyv \sin u}{1 + y^2z^2},$

$\dfrac{\partial w}{\partial v} = \dfrac{x(y \cos u - 2ze^{-2v})}{1 + y^2z^2}$

15. $\dfrac{dw}{dt} = \dfrac{\partial w}{\partial r}\dfrac{dr}{dt} + \dfrac{\partial w}{\partial s}\dfrac{ds}{dt} + \dfrac{\partial w}{\partial u}\dfrac{du}{dt} + \dfrac{\partial w}{\partial v}\dfrac{dv}{dt}$

17. $\dfrac{\partial w}{\partial t} = \dfrac{\partial w}{\partial x}\dfrac{\partial x}{\partial t} + \dfrac{\partial w}{\partial y}\dfrac{\partial y}{\partial t} + \dfrac{\partial w}{\partial z}\dfrac{\partial z}{\partial t}$

19. $4x + 2y + xe^t + 2ye^t - 6z^2 \sin 2t$ 21. 0

23. $\dfrac{\partial u}{\partial s} = \dfrac{\csc yz(rt^2 - 2xzst^3 \cot yz - xy \cot yz)}{t^2},$

$\dfrac{\partial u}{\partial t} = \dfrac{sx \csc yz \cot yz (2y - st^3z)}{t^3}$

25. $\dfrac{\partial w}{\partial r} = 4, \dfrac{\partial w}{\partial t} = 2$

27. $\dfrac{\partial u}{\partial x} = \dfrac{1}{4u}, \dfrac{\partial u}{\partial y} = \dfrac{1}{4u}, \dfrac{\partial v}{\partial x} = \dfrac{1}{4v}, \dfrac{\partial v}{\partial y} = -\dfrac{1}{4v}$

29. $\dfrac{3x^2 - 2y}{2x - 3y^2}$ 31. $\dfrac{8x\sqrt{xy} + 3y}{4\sqrt{xy} - 3x}$

33. $\dfrac{\partial z}{\partial x} = \dfrac{2x + y - 2xz}{x^2 - 2yz}, \dfrac{\partial z}{\partial y} = \dfrac{x + z^2}{x^2 - 2yz}$

35. $\dfrac{\partial z}{\partial x} = -\dfrac{ye^y + y^2ze^{xz} + x(2y + x)e^{x/y}}{xy^2e^{xz}},$

$\dfrac{\partial z}{\partial y} = \dfrac{x^3e^{x/y} - xy^2e^y - y^2e^{xz}}{xy^3e^{xz}}$

37. $-\dfrac{x^2 - ay}{y^2 - ax}$ 39. ≈ -13.2 in.^{2}/min

41. -54 mph   45. $\approx -1.53°$F/sec   57. 2   59. 0

63. a. $-\dfrac{f_x^2 f_{yy} - 2f_x f_y f_{xy} + f_y^2 f_{xx}}{f_y^3}$ b. $\dfrac{2xy}{(x - y^2)^3}, \{(x, y)\,|\,x \neq y^2\}$

65. True

Exercises 13.6 • page 1105

1. $6 - \sqrt{3}/2$ 3. 4   5. $5\mathbf{i} + 3\mathbf{j}$

7. $\dfrac{4 - \sqrt{2}\pi}{4}\mathbf{i} + \dfrac{\sqrt{2}}{2}\mathbf{j}$

9. $\mathbf{i} + 2\mathbf{j}$ 11. $-2\sqrt{5}/5$ 13. $\frac{1}{9}$ 15. $7\sqrt{10}/5$

17. $-4\sqrt{13}/13$ 19. $-76\sqrt{3}$ 21. $\frac{1}{6}$

23. $20\sqrt{14}/7$ 25. $-4\sqrt{3}/3$ 27. $\dfrac{\sqrt{6}}{24}(3 - \pi)$

29. $39\sqrt{10}/10$ 31. $\dfrac{12\sqrt{2}}{\sqrt{13\pi^2 + 576}}$ 33. $\mathbf{i} + 6\mathbf{j}, \sqrt{74}/6$

35. $7\mathbf{i} + 8\mathbf{j} + 34\mathbf{k}, 3\sqrt{141}$ 37. $-2\mathbf{i} - \mathbf{j}, \sqrt{5}$

39. $4\mathbf{i} + 2\mathbf{j} - \mathbf{k}, \sqrt{21}/4$ 41. $\mathbf{i} - \mathbf{j}, 440\sqrt{2}$

43. a. Ascending, descending   b. Neither   c. East   d. East

45. a. $\dfrac{100\sqrt{7}}{\pi}\mathbf{i} + \dfrac{500}{\pi}\mathbf{j}$ b.

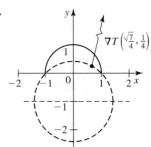

49. a. $x = y^3$ b.

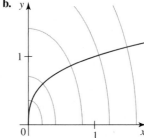

51. 6.25   53. $-3\mathbf{i} + 4\mathbf{j}$

55. a.

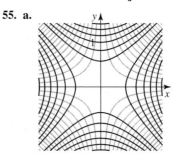

57. f increases most rapidly in a direction along any line emanating from the origin and moving away from it; g increases most rapidly in the x-direction (positive or negative). No.

59. True   61. True

Exercises 13.7 • page 1115

1.

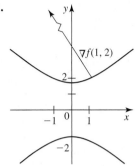

3.

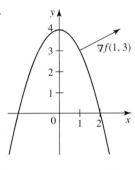

5. $y = \dfrac{\sqrt{3}}{4}x + \dfrac{7}{8},\ y = -\dfrac{4\sqrt{3}}{3}x + 8$

7. $y = -\dfrac{7\sqrt{5}}{5}x + 6,\ y = \dfrac{\sqrt{5}}{7}x - \dfrac{12}{7}$

9.

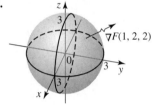

11.

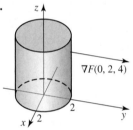

13.

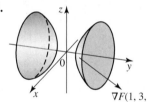

15. $2x + 4y + 9z = 17,\ \dfrac{x-2}{2} = \dfrac{y-1}{4} = \dfrac{z-1}{9}$

17. $x + y + z = 1,\ x - 4 = y + 2 = z + 1$

19. $5x + 4y + 3z = 22,\ \dfrac{x-1}{5} = \dfrac{y-2}{4} = \dfrac{z-3}{3}$

21. $18x - 16y + z = -25,\ \dfrac{x+1}{-18} = \dfrac{y-2}{16} = \dfrac{z-25}{-1}$

23. $x + 3y - 4z = -11,\ x + 2 = \dfrac{y-1}{3} = \dfrac{z-3}{-4}$

25. $x + 2y - z = 0,\ x - 2 = \dfrac{y}{2} = \dfrac{z-2}{-1}$

27. $3y - z = 0,\ x = 3,\ \dfrac{y}{3} = \dfrac{z}{-1}$

29. $x - y + 2z = \dfrac{\pi}{2},\ x - 1 = \dfrac{y-1}{-1} = \dfrac{z-\frac{\pi}{4}}{2}$

31. $x + z = 1,\ \dfrac{x}{1} = \dfrac{z-1}{1},\ y = 3$

35. $\dfrac{xx_0}{a^2} - \dfrac{yy_0}{b^2} - \dfrac{zz_0}{c^2} = 1$ **37.** $(-1, -2, -3)$ and $(1, 2, 3)$

39. $\left(-\dfrac{2\sqrt{5}}{5}, \dfrac{2\sqrt{5}}{5}, -\dfrac{\sqrt{5}}{5}\right)$ and $\left(\dfrac{2\sqrt{5}}{5}, -\dfrac{2\sqrt{5}}{5}, \dfrac{\sqrt{5}}{5}\right)$

45. $x = -\dfrac{\sqrt{2}}{2} + t,\ y = \dfrac{\sqrt{2}}{2} + t,\ z = 1$

47. True   **49.** True

Exercises 13.8 • page 1124

1. Relative minimum $f(1, -2) = -5$

3. Relative maximum $f(2, -1) = 15$

5. Relative minimum $f(0, 0) = 0$

7. Relative minimum $f(5, 6) = -24$

9. Relative minimum $f(0, 0) = 3$, saddle points $(-2, -1, 5)$ and $(2, -1, 5)$

11. Relative minimum $f(0, 0) = 0$, relative maximum $f\left(0, -\dfrac{5}{3}\right) = \dfrac{125}{27}$, saddle points $(-2, -1, 3)$ and $(2, -1, 3)$

13. Relative minimum $f(4, 4) = -12$

15. Relative minimum $f(-1, -2) = 6$

17. Relative maximum $f(0, 0) = 1$

19. Saddle points $(0, 0, 0)$, $(0, \pi, 0)$, and $(0, 2\pi, 0)$

21. None

23. Relative minima at $\left(\dfrac{3}{2}, -\dfrac{3}{2}\right)$ and $\left(\dfrac{3}{2}, \dfrac{3}{2}\right)$, saddle point $(0, 0, 0)$

25. Relative minima $f(-1, 2) = -\dfrac{3}{2}$ and $f(4, -8) = -64$, saddle point $(0, 0, 0)$

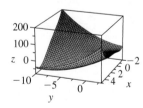

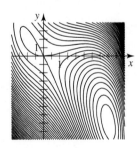

27. Relative minimum $f(-1, -2) = 14$

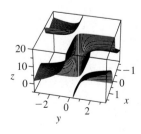

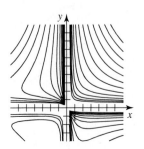

29. Relative minima $f(-1.526, 1) \approx -19.888$ and $f(1.267, 1) \approx -8.620$, saddle point $(0.259, 1, -5.492)$

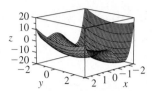

31. Relative maxima $f(\pm 1.225, 1) \approx 0.250$, saddle point $(0, 0.630, -1.528)$

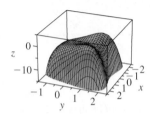

33. Absolute minimum -12, absolute maximum 7

35. Absolute minimum -12, absolute maximum 13

37. Absolute minimum -12, absolute maximum 4

39. Absolute minimum $-\frac{13}{4}$, absolute maximum $\frac{63}{4}$

41. $2\sqrt{6}/3$ **43.** $(2, -3, \pm 1)$, $\sqrt{14}$ **45.** $\frac{500}{3}, \frac{500}{3}, \frac{500}{3}$

47. $2\sqrt{2}$ ft $\times$ $2\sqrt{2}$ ft $\times$ $2\sqrt{2}$ ft

49. $4\sqrt{3}/3 \times 2\sqrt{3} \times 8\sqrt{3}/3$

51. $1 \times \frac{2}{3} \times 2, \frac{4}{3}$ **53.** 2 ft $\times$ 2 ft $\times$ 3 ft   **55.** $\left(\frac{26}{3}, 12\right)$

59. a. $y = 1.59t + 6.69$ **b.** \$19.41 billion

61. a. $y = 3.39x + 23.19$ **b.** \$3.39 billion/yr   **c.** \$60.5 billion

63. b. Maximum 1, minimum -13.5833

65. False   **67.** False

Exercises 13.9 • page 1138

1. Minimum -5, maximum 5

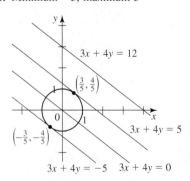

3. Minimum 2

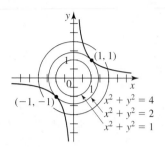

5. Maximum $\frac{3}{2}$ **7.** Minimum $-\frac{1}{4}$, maximum $\frac{1}{4}$

9. Minimum 4, maximum 12   **11.** Minimum $-\frac{1}{2}$

13. Minimum -2, maximum 2

15. Minimum $-2\sqrt{2}$, maximum $2\sqrt{2}$

17. Minimum $2 - 3\sqrt{5}$, maximum $2 + 3\sqrt{5}$

19. Minimum $\frac{1}{2}$, maximum $\frac{3}{2}$ **21.** Minimum $-\frac{4}{3}$, maximum 32

23. $\left(\frac{2}{3}, \frac{4}{3}, \frac{2}{3}\right)$ **25.** $\left(\frac{4}{3}, \frac{5}{3}, -\frac{1}{3}\right)$

27. $(-2^{1/4}, \pm 2^{3/4}, -2^{1/4})$ and $(2^{1/4}, \pm 2^{3/4}, 2^{1/4})$, $2\sqrt[4]{2}$

29. $2\sqrt{2}$ ft $\times$ $2\sqrt{2}$ ft $\times$ $2\sqrt{2}$ ft   **31.** 6 in. $\times$ 6 in. $\times$ 3 in.

33. $\frac{2\sqrt{3}}{3}\, a \times \frac{2\sqrt{3}}{3}\, b \times \frac{2\sqrt{3}}{3}\, c$ **35.** $\frac{a}{3} \times \frac{b}{3} \times \frac{c}{3}, \frac{abc}{27}$

37. $\frac{2}{3}\sqrt[3]{36}$ ft $\times$ $\frac{2}{3}\sqrt[3]{36}$ ft $\times$ $\sqrt[3]{36}$ ft

39. 225 units of labor and 50 units of capital

41. a. $(0, -1)$

b.

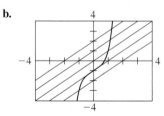

43. $\left(\frac{3}{2}, -\frac{3}{2}, -\frac{7}{2}\right)$ **45. a.** $\dfrac{3\sqrt{2}}{2}\, a$ **47. a.** c **51.** False

Chapter 13 Concept Review • page 1141

1. a. rule; (x, y) **b.** real; real; range

   c. $\{(x, y, z) \mid z = f(x, y), (x, y) \in D\}$

3. $L; L; (a, b)$

5. a. $f(a, b)$ **b.** R

7. a. $\lim\limits_{h \to 0} \dfrac{f(x + h, y) - f(x, y)}{h}$; $y = b$; $(a, b, f(a, b))$;

x; constant; $y = b$

b. y; x

9. a. $f_x\, dx + f_y\, dy$ **b.** dz **c.** Δx; Δy; 0; 0

d. $f_x(a, b)\, \Delta x + f_y(a, b)\, \Delta y + \varepsilon_1\, \Delta x + \varepsilon_2\, \Delta y$; $\varepsilon_1 \to 0$; $\varepsilon_2 \to 0$; $(0, 0)$

11. a. $\dfrac{\partial w}{\partial x}\dfrac{dx}{dt} + \dfrac{\partial w}{\partial y}\dfrac{dy}{dt}$

b. $\dfrac{\partial w}{\partial x}\dfrac{\partial x}{\partial u} + \dfrac{\partial w}{\partial y}\dfrac{\partial y}{\partial u}$

c. $-\dfrac{F_x(x, y)}{F_y(x, y)}$; $F_y(x, y) \neq 0$

d. $-\dfrac{F_x(x, y, z)}{F_z(x, y, z)}$; $-\dfrac{F_y(x, y, z)}{F_z(x, y, z)}$; $F_z(x, y, z) \neq 0$

13. a. $|\nabla f(x, y)|$; $\nabla f(x, y)$ **b.** $-|\nabla f(x, y)|$; $-\nabla f(x, y)$

15. a. Relative maximum **b.** Absolute minimum

c. Does not exist; 0 **d.** Critical point

e. Second Derivative Test

17. a. Constrained **b.** $\lambda \nabla g(x, y)$; multiplier

c. $\lambda \nabla g(x, y)$; $f(x, y)$; critical points; maximum; minimum

Chapter 13 Review Exercises • page 1142

Abbreviations: D, domain; R, range.

1. $D = \{(x, y) \mid 0 < x^2 + y^2 \leq 9\}$

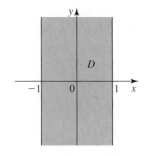

3. $D = \{(x, y) \mid -1 \leq x \leq 1\}$

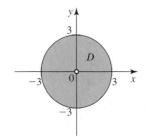

5.

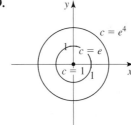

7.

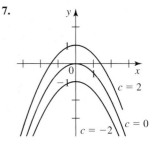

9.

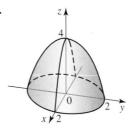

11. $\frac{2}{3}$ **13.** 1 **15.** $\{(x, y) \mid y < x\}$

17. $f_x(x, y) = 4xy - \dfrac{1}{2\sqrt{x}}$, $f_y(x, y) = 2x^2$

19. $f_r(r, s) = (1 - 2r^2)e^{-(r^2 + s^2)}$, $f_s(r, s) = -2rse^{-(r^2 + s^2)}$

21. $f_x(x, y, z) = \dfrac{2x(z^2 - y^2)}{(z^2 - x^2)^2}$, $f_y(x, y, z) = \dfrac{2y}{x^2 - z^2}$,

$f_z(x, y, z) = \dfrac{2z(y^2 - x^2)}{(z^2 - x^2)^2}$

23. $f_{xx}(x, y) = 4(3x^2 - y^3)$, $f_{yy}(x, y) = -12x^2y + 2$,

$f_{xy}(x, y) = f_{yx}(x, y) = -12xy^2$

25. $f_{xx}(x, y, z) = 2yz^3$, $f_{yy}(x, y, z) = 0$, $f_{zz}(x, y, z) = 6x^2yz$,

$f_{xy}(x, y, z) = f_{yx}(x, y, z) = 2xz^3$,

$f_{xz}(x, y, z) = f_{zx}(x, y, z) = 6xyz^2$,

$f_{yz}(x, y, z) = f_{zy}(x, y, z) = 3x^2z^2$

31. $2x \tan^{-1} y^3\, dx + \dfrac{3x^2y^2}{1 + y^6}\, dy$

33. 22.853 **35.** No

37. $4xye^{2t} + \left(\dfrac{1}{2\sqrt{y}} - x^2\right)\sin t$ **39.** $\dfrac{3x^2 - 6xy + 2y^2}{3x^2 - 4xy - 6y^2}$

41. $\dfrac{\sqrt{5}}{5}(\mathbf{i} + 2\mathbf{j})$ **43.** $-11\mathbf{i} - 5\mathbf{j} + 10\mathbf{k}$ **45.** $\frac{127}{5}$

47. $\frac{29}{15}$ **49.** $\frac{5}{4}\mathbf{i} + 8\mathbf{j}$, $\sqrt{1049}/4$

51. $2x + 8y + 9z = 27$, $\dfrac{x - 1}{2} = \dfrac{y - 2}{8} = \dfrac{z - 1}{9}$

53. $9x + 18y - z = 27$, $\dfrac{x - 3}{9} = \dfrac{y - 1}{18} = \dfrac{z - 18}{-1}$

55. a.

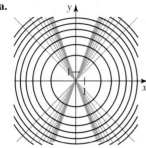

57. Relative minimum $f(6, -7) = -38$

59. Relative minimum $f\left(\frac{3}{2}, \frac{9}{4}\right) = -\frac{27}{16}$, saddle point $(0, 0, 0)$

61. Minimum -11, maximum $\frac{31}{27}$

63. Minimum $-\dfrac{16\sqrt{3}}{9}$, maximum $\dfrac{16\sqrt{3}}{9}$

65. Maximum $\frac{1}{8}$ **69.** $(2, 0, 1)$ **71.** True

Chapter 13 Challenge Problems • page 1144

1. $\{(x, y, z)\,|\,4x^2 + 9y^2 \le 36,\, x^2 - 2x + y^2 > 0,$ and
$4x^2 + 16y + 4y^2 > -15\}$

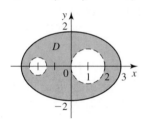

3. a. $D_\mathbf{u}^2 f(x, y) = f_{xx}u_1^2 + 2f_{xy}u_1u_2 + f_{yy}u_2^2$ **b.** $\frac{15}{13}$

7. $\dfrac{\sqrt{14}}{7}(1 - \sqrt{2})$

CHAPTER 14

Exercises 14.1 • page 1155

1. $\frac{29}{2}$ **3.** $\frac{25}{2}$ **5.** $\frac{21}{2}$ **7.** 204   **9.** 15   **11.** 129

13. 24   **15.** 32

17. The wedge bounded above by the cylinder
$z = 4 - x^2$ and below by the triangular base
$R = \{(x, y)\,|\,0 \le y \le x, 0 \le x \le 2\}$

19. $\displaystyle\iint\limits_R (3 - 2x + y)\, dA$

21. 1.28079   **27.** True   **29.** True

Exercises 14.2 • page 1165

1. 5   **3.** $10\sqrt{2}$ **5.** -4 **7.** $\frac{64}{3}$ **9.** $\frac{1}{3}$

11. $\dfrac{2e^6 - 3e^5 + 3e - 2}{6e^3}$ **13.** $\frac{9}{2}$ **15.** $\frac{1}{32}\pi^2(2\sqrt{2} + 1)$

17. $\frac{2}{3}$ **19.** $\frac{32}{5}$ **21.** $\frac{13}{6}$ **23.** $\frac{1}{6}$ **25.** $\frac{93}{10}$

27. $\dfrac{4 - \pi}{8}$ **29.** $\frac{72}{5}$ **31.** $4e^6 - 15e^4 - 1$ **33.** 48

35. 8π **37.** $\frac{64}{3}$ **39.** 4   **41.** 40   **43.** $\frac{3}{35}$

45. $\sqrt{5} + \frac{9}{2}\sin^{-1}\frac{2}{3} + \frac{1}{6}(5\sqrt{5} - 27)$

47.

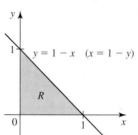

$\displaystyle\int_0^1 \int_0^{1-y} f(x, y)\, dx\, dy$

49.

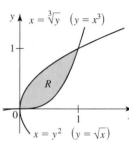

$\displaystyle\int_0^1 \int_{x^3}^{\sqrt{x}} f(x, y)\, dy\, dx$

51.

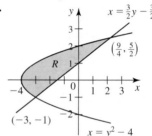

$\displaystyle\int_{-4}^{-3} \int_{-\sqrt{x+4}}^{\sqrt{x+4}} f(x, y)\, dy\, dx + \int_{-3}^{9/4} \int_{(2/3)x+1}^{\sqrt{x+4}} f(x, y)\, dy\, dx$

53.

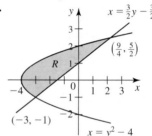

$\displaystyle\int_0^1 \int_{e^y}^{e} f(x, y)\, dx\, dy$

55. $\dfrac{e^4 - 1}{4e^4}$ **57.** $\dfrac{1 - \cos 8}{3}$ **59.** $\frac{4}{3}$

63. $\frac{2}{3}(3\pi + 8)$ slugs   **65.** 2166 people per square mile

67. a.   **b.** 0.550   **c.** 0.062

69. -0.8784 **71.** 0.5610   **73.** True

75. True   **77.** False

Exercises 14.3 • page 1173

1. Rectangular, $\displaystyle\int_0^3 \int_0^{-(2/3)x+2} f(x, y)\, dy\, dx$

3. Polar, $\displaystyle\int_{-\pi/4}^{\pi/4} \int_0^{\sqrt{2}} f(r\cos\theta, r\sin\theta)\, r\, dr\, d\theta$

5.

7.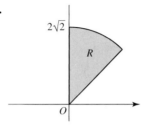

9. 0 **11.** 1 **13.** $\pi/2$ **15.** $\frac{1}{6}$ **17.** 8π

19. $16\pi/3$ **21.** 9π **23.** 16π **25.** $\frac{4}{3}(\sqrt{2} - 1)\pi$

27. $9\pi/4$ **29.** $27\pi/2$ **31.** $\dfrac{2\pi + 3\sqrt{3}}{6} a^2$

33. $8\pi/3$ **35.** $(\pi/2)\ln 2$ **37.** $(\pi/2)(e^4 - 1)$

39. π **41.** $\displaystyle\int_0^{\pi/4} \int_0^2 (r\cos\theta)(r\sin\theta)\, r\, dr\, d\theta$, 1

43. b. $\frac{1}{3}[\sqrt{2} + \ln(\sqrt{2} + 1)]$

45. a.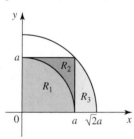

47. $\sqrt{\pi}$ **49.** True

Exercises 14.4 • page 1181

1. $m = 6$, $(\bar{x}, \bar{y}) = \left(\frac{3}{2}, \frac{4}{3}\right)$ **3.** $m = 4$, $(\bar{x}, \bar{y}) = \left(\frac{7}{3}, \frac{1}{3}\right)$

5. $m = \frac{32}{3}$, $(\bar{x}, \bar{y}) = \left(3, \frac{8}{7}\right)$

7. $m = \frac{1}{4}(e^2 + 1)$, $(\bar{x}, \bar{y}) = \left(\dfrac{e^2 - 1}{e^2 + 1}, \dfrac{8(2e^3 + 1)}{27(e^2 + 1)}\right)$

9. $m = \frac{\pi}{4}$, $(\bar{x}, \bar{y}) = \left(\frac{\pi}{2}, \frac{16}{9\pi}\right)$ **11.** $m = \frac{32}{9}$, $(\bar{x}, \bar{y}) = \left(\frac{6}{5}, 0\right)$

13. $\frac{75}{4}$ coulombs **15.** 384.14°F

17. $I_x = \frac{1}{3}\rho a b^3$, $I_y = \frac{1}{3}\rho a^3 b$, $I_0 = \frac{1}{3}\rho a b(a^2 + b^2)$, $\bar{\bar{x}} = \frac{\sqrt{3}}{3}a$, $\bar{\bar{y}} = \frac{\sqrt{3}}{3}b$

19. $I_x = \frac{1}{8}\pi\rho R^4$, $I_y = \frac{1}{8}\pi\rho R^4$, $I_0 = \frac{1}{4}\pi\rho R^4$, $\bar{\bar{x}} = \frac{1}{2}R$, $\bar{\bar{y}} = \frac{1}{2}R$

21. $I_x = 12$, $I_y = 18$, $I_0 = 30$, $\bar{\bar{x}} = \sqrt{3}$, $\bar{\bar{y}} = \sqrt{2}$

23. $I_x = 16$, $I_y = \frac{512}{5}$, $I_0 = \frac{592}{5}$, $\bar{\bar{x}} = \frac{4\sqrt{15}}{5}$, $\bar{\bar{y}} = \frac{\sqrt{6}}{2}$

25. $\ln 2 - \frac{1}{4}$ **27.** True **29.** False

Exercises 14.5 • page 1187

1. $2\sqrt{14}$ **3.** $\frac{1}{3}(3\sqrt{3} - 2\sqrt{2})$

5. $(\pi/6)(37\sqrt{37} - 1)$ **7.** 6π

9. $(2\pi/3)(17\sqrt{17} - 1)$ **11.** $16\pi(2 - \sqrt{2})$

13. $2a^2(\pi - 2)$ **17.** 13.0046 **19.** 13.9783

21. $\displaystyle\int_{-1}^1 \int_{-1}^1 \sqrt{36x^2y^4 + 36x^4y^2 + 1}\, dy\, dx$

23. $\displaystyle\int_0^2 \int_0^x \dfrac{\sqrt{13 + (2x + 3y)^4}}{(2x + 3y)^2}\, dy\, dx$

25. True

Exercises 14.6 • page 1198

1. 18 **3.** 4 **5.** $\frac{3}{8}$ **7.** $-\frac{1}{4}$

9. $\dfrac{16(e - 1)}{3e}$

11. $\displaystyle\int_0^2 \int_0^{4-2x} \int_0^{(12-6x-3y)/4} f(x, y, z)\, dz\, dy\, dx,$

$\displaystyle\int_0^4 \int_0^{(4-y)/2} \int_0^{(12-6x-3y)/4} f(x, y, z)\, dz\, dx\, dy,$

$\displaystyle\int_0^4 \int_0^{(12-3y)/4} \int_0^{(12-3y-4z)/6} f(x, y, z)\, dx\, dz\, dy,$

$\displaystyle\int_0^3 \int_0^{(12-4z)/3} \int_0^{(12-3y-4z)/6} f(x, y, z)\, dx\, dy\, dz,$

$\displaystyle\int_0^2 \int_0^{(6-3x)/2} \int_0^{(12-6x-4z)/3} f(x, y, z)\, dy\, dz\, dx,$

$\displaystyle\int_0^3 \int_0^{(6-2z)/3} \int_0^{(12-6x-4z)/3} f(x, y, z)\, dy\, dx\, dz$

13. $\displaystyle\int_0^1 \int_{-\sqrt{x}}^{\sqrt{x}} \int_0^{1-x} f(x, y, z)\, dz\, dy\, dx,$

$\displaystyle\int_{-1}^1 \int_{y^2}^1 \int_0^{1-x} f(x, y, z)\, dz\, dx\, dy,$

$\displaystyle\int_0^1 \int_0^{1-x} \int_{-\sqrt{x}}^{\sqrt{x}} f(x, y, z)\, dy\, dz\, dx,$

$\displaystyle\int_0^1 \int_0^{1-z} \int_{-\sqrt{x}}^{\sqrt{x}} f(x, y, z)\, dy\, dx\, dz,$

$\displaystyle\int_{-1}^1 \int_0^{1-y^2} \int_{y^2}^{1-z} f(x, y, z)\, dx\, dz\, dy,$

$\displaystyle\int_0^1 \int_{-\sqrt{1-z}}^{\sqrt{1-z}} \int_{y^2}^{1-z} f(x, y, z)\, dx\, dy\, dz$

15. $\frac{1}{24}$ **17.** $\frac{1}{3}$ **19.** $64\pi/3$ **21.** 1

23.

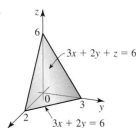

6

25.

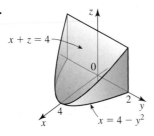

$\frac{128}{5}$

27.

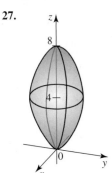

16π

29. 1

31.

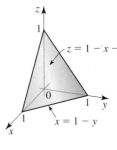

33.

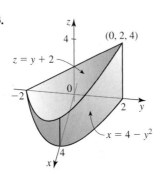

35. $\displaystyle\int_0^6 \int_0^{(6-x)/2} \int_0^{(6-x-2y)/3} f(x, y, z)\, dz\, dy\, dx,$

$\displaystyle\int_0^6 \int_0^{(6-x)/3} \int_0^{(6-x-3z)/2} f(x, y, z)\, dy\, dz\, dx,$

$\displaystyle\int_0^3 \int_0^{6-2y} \int_0^{(6-x-2y)/3} f(x, y, z)\, dz\, dx\, dy,$

$\displaystyle\int_0^3 \int_0^{(6-2y)/3} \int_0^{6-2y-3z} f(x, y, z)\, dx\, dz\, dy,$

$\displaystyle\int_0^2 \int_0^{6-3z} \int_0^{(6-x-3z)/2} f(x, y, z)\, dy\, dx\, dz,$

$\displaystyle\int_0^2 \int_0^{(6-3z)/2} \int_0^{6-2y-3z} f(x, y, z)\, dx\, dy\, dz$

37. $\displaystyle\int_{-1}^1 \int_{-\sqrt{1-x^2}}^{\sqrt{1-x^2}} \int_0^2 f(x, y, z)\, dz\, dy\, dx,$

$\displaystyle\int_{-1}^1 \int_{-\sqrt{1-y^2}}^{\sqrt{1-y^2}} \int_0^2 f(x, y, z)\, dz\, dx\, dy,$

$\displaystyle\int_{-1}^1 \int_0^2 \int_{-\sqrt{1-x^2}}^{\sqrt{1-x^2}} f(x, y, z)\, dy\, dz\, dx,$

$\displaystyle\int_0^2 \int_{-1}^1 \int_{-\sqrt{1-x^2}}^{\sqrt{1-x^2}} f(x, y, z)\, dy\, dx\, dz,$

$\displaystyle\int_{-1}^1 \int_0^2 \int_{-\sqrt{1-y^2}}^{\sqrt{1-y^2}} f(x, y, z)\, dx\, dz\, dy,$

$\displaystyle\int_0^2 \int_{-1}^1 \int_{-\sqrt{1-y^2}}^{\sqrt{1-y^2}} f(x, y, z)\, dx\, dy\, dz$

39. a. 384   **b.** 384   **41.** 0.4439

43. $\left(\frac{2}{5}, \frac{1}{5}, \frac{1}{5}\right)$ **45.** $(2, 0, 0)$

47. $\displaystyle\int_0^1 \int_0^{1-y} \int_0^{\sqrt{1-z^2}} (xy + z^2)\, dx\, dz\, dy$

49. $\displaystyle\int_0^1 \int_0^{(2-y)/2} \int_0^{1-y^2} \sqrt{x^2 + y^2 + z^2}\, dz\, dx\, dy$

51. $I_x = \frac{2}{3}k,\ I_y = \frac{2}{3}k,\ I_z = \frac{2}{3}k$ **53.** $I_x = \frac{1}{180},\ I_y = \frac{1}{90},\ I_z = \frac{1}{90}$

55. 3   **57.** $1/\pi$ **59.** $T = \{(x, y, z)\,|\,2x^2 + 3y^2 + z^2 \le 1\}$

61. True   **63.** True

Exercises 14.7 • page 1207

1.

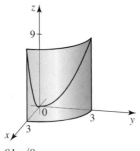

$81\pi/8$

3.

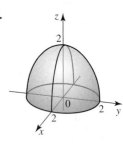

$16\pi/3$

5. $4\pi/3$ **7.** $\frac{64}{15}$ **9.** $512\pi/5$ **11.** $40\pi/3$

13. $(0, 0, 2)$ **15.** $\left(0, 0, \frac{4}{3}\right)$ **17.** $\frac{1}{15}\pi\rho$ **19.** π

21. 0   **23.** 0   **25.** $\pi/3$ **27.** $\sqrt{2}\pi/3$

29. $\left(0, 0, \frac{3}{8}a\right)$ **31.** $\frac{1}{4}ka^4\pi$ **33.** $\frac{48}{5}k\pi$ **35.** $\frac{67}{15}k\pi$

37. $\frac{4}{15}ka^5\pi$ **41.** $7\pi/8$ **43.** $\frac{1}{3}(2 - \sqrt{2})\pi$

45. $108°$F   **47.** True   **49.** True

Exercises 14.8 • page 1217

1.

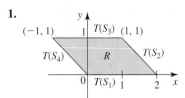

3.

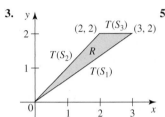

5.

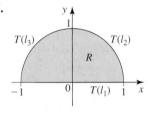

7. $-2(u + 1)$ **9.** $2e^{2u}$ **11.** -4 **13.** 45

15. 0 **17.** 4π **19.** π **21.** $\frac{32}{3}$ **23.** $\dfrac{e^2 - 1}{4e}$

25. $\frac{1}{8}a^2b^2$ **27.** $\frac{4}{3}\pi abc$ **29.** $\frac{1}{64}\pi \rho a^4$

31. $\displaystyle\iiint_R f(x, y, z)\, dV = \iiint_S f(r \cos \theta, r \sin \theta, z)\, r\, dz\, dr\, d\theta$

33. False

Chapter 14 Concept Review • page 1218

1. a. $\displaystyle\sum_{i=1}^{m} \sum_{j=1}^{n} f(x_{ij}^*, y_{ij}^*)\, \Delta A$

b. $\displaystyle\lim_{m,\, n \to \infty} \sum_{i=1}^{m} \sum_{j=1}^{n} f(x_{ij}^*, y_{ij}^*)\, \Delta A;\ (x_{ij}^*, y_{ij}^*)$

c. volume; $z = f(x, y)$

d. $\displaystyle\lim_{m,\, n \to \infty} \sum_{i=1}^{m} \sum_{j=1}^{n} f_D(x_{ij}^*, y_{ij}^*)\, \Delta A;\ f(x, y);\ 0$

3. a. $\displaystyle\int_a^b \int_c^d f(x, y)\, dy\, dx;\ \int_c^d \int_a^b f(x, y)\, dx\, dy$

b. iterated

5. a. $\{(r, \theta)\,|\, a \le r \le b;\ \alpha \le \theta \le \beta\}$

b. $\displaystyle\int_\alpha^\beta \int_a^b f(r \cos \theta, r \sin \theta)\, r\, dr\, d\theta$

c. $\{(r, \theta)\,|\, \alpha \le \theta \le \beta,\ g_1(\theta) \le r \le g_2(\theta)\}$

d. $\displaystyle\int_\alpha^\beta \int_{g_1(\theta)}^{g_2(\theta)} f(r \cos \theta, r \sin \theta)\, r\, dr\, d\theta$

7. a. $\displaystyle\iint_R \sqrt{(f_x)^2 + (f_y)^2 + 1}\, dA$

b. $\displaystyle\iint_R \sqrt{(g_x)^2 + (g_z)^2 + 1}\, dA$

c. $\displaystyle\iint_R \sqrt{(h_y)^2 + (h_z)^2 + 1}\, dA$

9. a. order; $\displaystyle\int_p^q \int_c^d \int_a^b f(x, y, z)\, dx\, dy\, dz$

b. $\displaystyle\iint_R \left[\int_{k_1(x,\, y)}^{k_2(x,\, y)} f(x, y, z)\, dz \right] dA$

11. a. $\displaystyle\int_\alpha^\beta \int_{g_1(\theta)}^{g_2(\theta)} \int_{h_1(r \cos \theta,\, r \sin \theta)}^{h_2(r \cos \theta,\, r \sin \theta)} f(r \cos \theta, r \sin \theta, z)\, r\, dz\, dr\, d\theta$

b. $\displaystyle\int_\alpha^\beta \int_c^d \int_a^b f(\rho \sin \phi \cos \theta, \rho \sin \phi \sin \theta,$
$\rho \cos \phi)\rho^2 \sin \phi\, d\rho\, d\phi\, d\theta$

c. $\displaystyle\int_\alpha^\beta \int_c^d \int_{h_1(\phi,\, \theta)}^{h_2(\phi,\, \theta)} f(\rho \sin \phi \cos \theta,$
$\rho \sin \phi \sin \theta, \rho \cos \phi)\rho^2 \sin \phi\, d\rho\, d\phi\, d\theta$

Chapter 14 Review Exercises • page 1219

1. 18 **3.** $\frac{23}{60}$ **5.** $\dfrac{\pi}{4} - \dfrac{\ln 2}{2}$ **7.** $\frac{8}{15}(5 + \sqrt{2})$

9.

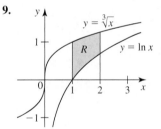

11.

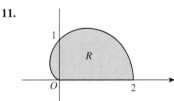

13. $\frac{1}{2}(1 - \cos 1)$ **15.** $\frac{52}{3}$ **17.** $\frac{32}{3}$ **19.** 6 **21.** $\frac{19}{168}$

23. 0 **25.** $\frac{4}{105}$ **27.** $\frac{7}{6}$ **29.** $\pi/2$ **31.** $\dfrac{\pi(e - 1)}{e}$

33. $m = \frac{2}{21}$, $(\bar{x}, \bar{y}) = \left(\frac{21}{32}, \frac{21}{40}\right)$ **35.** $m = \frac{\pi}{6}$, $(\bar{x}, \bar{y}) = \left(\frac{3}{2\pi}, \frac{3}{2\pi}\right)$

37. $I_x = \frac{2}{9}$, $I_y = \frac{4}{45}$, $I_0 = \frac{14}{45}$ **39.** $3\sqrt{14}$ **41.** $16\pi/5$

43. $\displaystyle\int_0^3\int_0^{(6-2x)/3}\int_0^{6-2x-3y} f(x, y, z)\, dz\, dy\, dx,$

$\displaystyle\int_0^2\int_0^{(6-3y)/2}\int_0^{6-2x-3y} f(x, y, z)\, dz\, dx\, dy,$

$\displaystyle\int_0^3\int_0^{6-2x}\int_0^{(6-2x-z)/3} f(x, y, z)\, dy\, dz\, dx,$

$\displaystyle\int_0^6\int_0^{(6-z)/2}\int_0^{(6-2x-z)/3} f(x, y, z)\, dy\, dx\, dz,$

$\displaystyle\int_0^2\int_0^{6-3y}\int_0^{(6-3y-z)/2} f(x, y, z)\, dx\, dz\, dy,$

$\displaystyle\int_0^6\int_0^{(6-z)/3}\int_0^{(6-3y-z)/2} f(x, y, z)\, dx\, dy\, dz$

45. $2 - 16uvw - 4uw$ **47.** $e - 1$

49. True   **51.** False   **53.** True

Chapter 14 Challenge Problems • page 1221

1. a. 10   **7. b.** $\frac{1}{2}(1 - \cos 1)$

CHAPTER 15

Exercises 15.1 • page 1229

1. (b)   **3.** (c)   **5.** (e)

7. 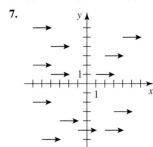   **9.**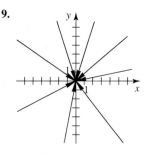

11. 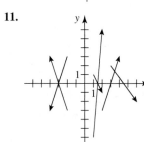   **13.**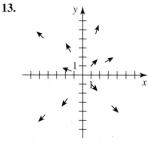

15. 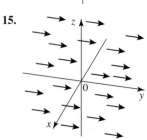   **17.**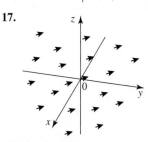

19. (c)   **21.** (a)

23. 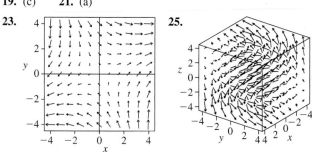   **25.**

27. $2xy\mathbf{i} + (x^2 - 3y^2)\mathbf{j}$ **29.** $yz\mathbf{i} + xz\mathbf{j} + xy\mathbf{k}$

31. $\dfrac{y}{x + z}\mathbf{i} + \ln(x + z)\mathbf{j} + \dfrac{y}{x + z}\mathbf{k}$

33. a. $2\mathbf{i} + 10\mathbf{j} + 4\mathbf{k}$ **b.** $(1.02, 3.1, 2.04)$

37. True   **39.** True

Exercises 15.2 • page 1239

1. a. 0 **b.** 0 **c.** Will not rotate **d.** 0

3. a. Positive **b.** $\dfrac{1}{\sqrt{x^2 + y^2}}$ **c.** Will not rotate **d.** 0

5. a. 0 **b.** 0

7. a. $2x(y^3 + z)$ **b.** $-z^2\mathbf{j} - 3x^2y^2\mathbf{k}$

9. a. $\cos x - x\sin y + \cos z$ **b.** $\cos y\mathbf{k}$

11. a. $1/z$ **b.** 0

13. a. No **b.** No **c.** Yes, a vector field **d.** No

15. a. Yes, a vector field **b.** Yes, a scalar field
  c. No **d.** Yes, a vector field

41. False   **43.** False   **45.** False   **47.** False

Exercises 15.3 • page 1253

1. $\frac{35}{2}$ **3.** $\dfrac{13\sqrt{13} - 8}{54}$ **5.** $\frac{32}{3}$ **7.** 5   **9.** 42

11. $\frac{22}{3}$ **13.** 0   **15.** $\sqrt{6}/2$ **17.** 0

19. $\dfrac{1}{2}e^2 + 4e + \dfrac{1}{e} - \dfrac{9}{2}$ **21.** $\frac{7}{2}$ **23.** $\pi ka, \left(0, \dfrac{2a}{\pi}\right)$

25. $\left(0, \dfrac{a(4 - \pi)}{2(\pi - 2)}\right)$ **27.** $\left(\frac{2}{5}, \frac{2}{5}\right)$

29. a. Negative **b.** -4π

31. $\frac{23}{24}$ **33.** $\frac{1}{2}(e^4 - e + 15)$ **35.** 1   **37.** -720 ft-lb

41. a. (i) $\dfrac{(2\sqrt{5} - \sqrt{2})qQ}{40\pi\varepsilon_0}$

  (ii) $\dfrac{(2\sqrt{5} - \sqrt{2})qQ}{40\pi\varepsilon_0}$

  b. No

43.

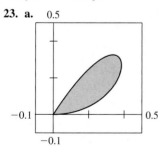

-2π

45. True   **47.** False

Exercises 15.4 • page 1267

1. $2x^2 + 3xy - y^2 + C$ **3.** No   **5.** $y^2 \sin x + 3y + C$

7. No   **9.** $\frac{1}{3}x^3 + y \ln x + \frac{1}{3}y^3 + C$

11. a. $2xy + x + 3y + C$ **b.** 0

13. a. $x^2y^2 + 2xy + C$ **b.** 9

15. a. $\frac{1}{2}x^2 e^{2y} + C$ **b.** $\frac{1}{2}e^2$

17. a. $e^x \sin y + \frac{1}{2}y^2 + C$ **b.** $\frac{1}{2}\pi^2$ **19.** 2   **21.** 10

25. $xyz + C$ **27.** No   **29.** $e^x \cos z + z \cosh y + C$

31. No   **33. a.** $xyz^2 + C$ **b.** 12

35. a. $x \cos y + yz^2 + C$ **b.** $2\pi + 1$ **37.** -1

39. $qkQ\left(\dfrac{\sqrt{14}}{14} - \dfrac{\sqrt{21}}{21}\right)$ **41. c.** No   **43.** False

45. False   **47.** True

Exercises 15.5 • page 1276

1. a. 0   **b.** 0   **3. a.** $\frac{4}{15}$ **b.** $\frac{4}{15}$ **5.** $\frac{1}{3}$ **7.** $\frac{1}{12}$

9. $-\frac{352}{15}$ **11.** $\frac{3}{8}\pi a^2$ **13.** 6π **15.** 6π **17.** $\frac{2}{3}$

19. $\frac{3}{8}\pi a^2$ **21.** $\frac{8}{3}ab$

23. a.

b. $\frac{7}{120}$

25. a.

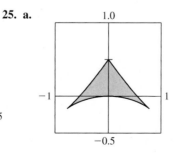

b. $\frac{32}{105}$

27. $288 - 7\pi$ **29.** -12 **31.** $47.5 - \pi$ **33.** 7

37. $\left(0, \frac{18}{5}\right)$ **39.** $\frac{1}{4}\rho a^4 \pi$ **43. c.** No   **45. b.** No   **c.** $\frac{3}{4}$

49. False   **51.** True

Exercises 15.6 • page 1289

1. (b)   **3.** (a)

5. $3x + 2y - 3z = 0$, a plane

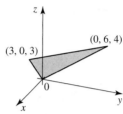

7. $\frac{1}{9}x^2 + \frac{1}{4}y^2 = 1$, $0 \le z \le 2$, a cylinder with an elliptical cross section and axis the z-axis, bounded below by the plane $z = 0$ and above by the plane $z = 2$

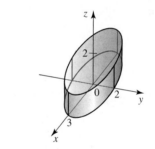

9.   **11.**

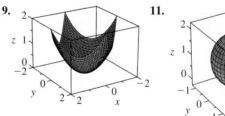

13.

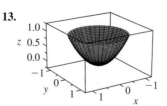

15. $\mathbf{r}(u, v) = (2 + 2u + v)\mathbf{i} + (1 + u - 2v)\mathbf{j} - (3 + u + v)\mathbf{k}$

17. $\mathbf{r}(u, v) = \cos v \cos u\mathbf{i} + \cos v \sin u\mathbf{j} - \sin v\mathbf{k}$ with domain $D = \{(u, v) \,|\, 0 \le u \le 2\pi, 0 \le v \le \pi\}$

19. $\mathbf{r}(u, v) = 2 \cos u\mathbf{i} + 2 \sin u\mathbf{j} + v\mathbf{k}$ with domain $D = \{(u, v) \,|\, 0 \le u \le 2\pi, -1 \le v \le 3\}$

21. $\mathbf{r}(u, v) = v\cos u\mathbf{i} + v\sin u\mathbf{j} + (9 - v^2)\mathbf{k}$ with domain $D = \{(u, v)\,|\,0 \le u \le 2\pi, 0 \le v \le 2\}$

23. $\mathbf{r}(u, v) = u\mathbf{i} + \sqrt{u}\cos v\mathbf{j} + \sqrt{u}\sin v\mathbf{k}$ with domain $D = \{(u, v)\,|\,0 \le u \le 4, 0 \le v \le 2\pi\}$

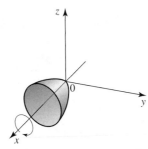

25. $\mathbf{r}(u, v) = (9 - u^2)\cos v\mathbf{i} + u\mathbf{j} + (9 - u^2)\sin v\mathbf{k}$ with domain $D = \{(u, v)\,|\,0 \le u \le 3, 0 \le v \le 2\pi\}$

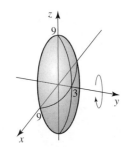

27. $x - y - z = 1$ **29.** $2x + z = -1$
31. $(\ln 2)x - 2z = 0$ **33.** $2\sqrt{3}$ **35.** $4\sqrt{14}\pi$
37. $\dfrac{3\sqrt{2}\pi}{4}$ **39.** $[2\sqrt{2} + \ln(3 + 2\sqrt{2})]\pi$

41. b.
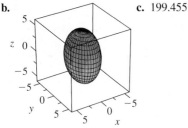
c. 199.455

43. $\sqrt{2}\pi$ **47.** False

Exercises 15.7 • page 1301

1. $5\sqrt{14}$ **3.** $\frac{1}{6}(27 - 5\sqrt{5})$ **5.** $4\sqrt{2}\pi$ **7.** $\sqrt{2}\pi/5$
9. 32   **11.** $\frac{1}{4}[2\sqrt{5} + \ln(\sqrt{5} + 2)]$ **13.** $\frac{8}{3}\pi^2$
15. $18\sqrt{14}k$ **17.** $8k\pi$ **19.** 40π **21.** 18
23. $32\pi/3$ **25.** π **27.** 27π **29.** $\left(0, 0, \frac{a}{2}\right)$
33. $18\sqrt{14}k$ **35.** 42   **37.** -20π

39. a. $\displaystyle\iint_D (-Pg_x + Q - Rg_z)\,dA$

b. 15

41. $\frac{16}{3}\pi kR^3$ **45. b.** $3\sqrt{14}$ **49.** True

Exercises 15.8 • page 1310

5. $\frac{8}{3}$ **7.** $63\pi/2$ **9.** $\frac{5}{24}$ **11.** 216π **13.** 0
15. $8\pi/5$ **17.** 27π **27.** True   **29.** True

Exercises 15.9 • page 1318

5. -8π **7.** 2π **9.** 8π **11.** -36 **13.** -6π
15. $-\frac{1}{2}e^2 + 2e - \frac{3}{2}$ **17.** 2   **23.** 0   **25.** 8π
29. 2π **31.** False

Chapter 15 Concept Review • page 1320

1. vector; vector

3. a. departs, accumulates

b. $\dfrac{\partial P}{\partial x} + \dfrac{\partial Q}{\partial y} + \dfrac{\partial R}{\partial z}$; enters; departs; equals; departs; enters

5. a. $\displaystyle\lim_{n\to\infty} \sum_{k=1}^{n} f(x_k^*, y_k^*)\,\Delta s_k$

b. $\displaystyle\int_a^b f(x(t), y(t))\sqrt{[x'(t)]^2 + [y'(t)]^2}\,dt$

c. $\displaystyle\lim_{n\to\infty} \sum_{k=1}^{n} f(x_k^*, y_k^*)\,\Delta x_k;\ \int_a^b f(x(t), y(t))x'(t)\,dt$

d. $\displaystyle\lim_{n\to\infty} \sum_{k=1}^{n} f(x_k^*, y_k^*)\,\Delta y_k;\ \int_a^b f(x(t), y(t))y'(t)\,dt$

7. a. $\displaystyle\int_a^b \mathbf{F}\cdot\mathbf{T}\,ds = \int_a^b \mathbf{F}(\mathbf{r}(t))\cdot\mathbf{r}'(t)\,dt$

b. work

9. a. closed   **b.** connected; conservative

11. curl $\mathbf{F}$; $\partial Q/\partial z$; $\partial P/\partial z$; $\partial P/\partial y$

13. $\displaystyle\oint_C x\,dy;\ -\oint_C y\,dx;\ \frac{1}{2}\oint_C x\,dy - y\,dx$

15. a. $x(u, v)\mathbf{i} + y(u, v)\mathbf{j} + z(u, v)\mathbf{k}$; parameter; parametric equations

b. $u\mathbf{i} + v\mathbf{j} + f(u, v)\mathbf{k}$

c. $u\mathbf{i} + f(u)\cos v\mathbf{j} + f(u)\sin v\mathbf{k}$; $\{(u, v)\,|\,a \le u \le b, 0 \le v \le 2\pi\}$

17. a. $\displaystyle\iint_R F(x, y, f(x, y))\sqrt{[f_x(x, y)]^2 + [f_y(x, y)]^2 + 1}\,dA$

b. $\displaystyle\iint_D F(\mathbf{r}(u, v))|\mathbf{r}_u \times \mathbf{r}_v|\,dA$

Chapter 15 Review Exercises • page 1321

1. a. $y^2 + z^2 + x^2$ **b.** $-2yz\mathbf{i} - 2xz\mathbf{j} - 2xy\mathbf{k}$

3. a. e^z **b. 0** **5.** $\frac{1}{12}(17\sqrt{17} - 5\sqrt{5})$ **7.** $\sqrt{2}/3$

9. $\dfrac{54{,}229}{110}$ **11.** $\dfrac{3\pi + 10}{12}$ **13.** $\frac{16}{3} - e^{-1}$ **15.** $\frac{109}{2}$

17. $2x^2y + 3xy^2 + C$ **19. 0** **21.** $-\frac{171}{10}$ **23.** $-\pi/2$

25. $\dfrac{9\sqrt{17}}{4}$ **27.** $3\pi/2$ **29.** $\dfrac{\sqrt{3}}{12}k$ **31.** 12π

33. 0 **35. 0** **37.** 256 **39. 0** **41.** True

43. False **45.** True **47.** True

Chapter 15 Challenge Problems • page 1323

1. $\left\{(x, y) \,\middle|\, \dfrac{x^2}{4} + \dfrac{y^2}{1} < 1 \text{ and } x^2 + y^2 > \dfrac{1}{4}\right\}$

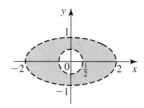

3. b. $\frac{3}{2}a^2$ **5.** π

ALGEBRA

Arithmetic Operations

$$\frac{a+b}{c} = \frac{a}{c} + \frac{b}{c}$$

$$\frac{a}{b} + \frac{c}{d} = \frac{ad+bc}{bd}$$

$$\frac{\left(\dfrac{a}{b}\right)}{\left(\dfrac{c}{d}\right)} = \left(\frac{a}{b}\right)\left(\frac{d}{c}\right) = \frac{ad}{bc}$$

Exponents and Radicals

$$x^m x^n = x^{m+n} \qquad \frac{x^m}{x^n} = x^{m-n} \qquad (x^m)^n = x^{mn}$$

$$x^{-n} = \frac{1}{x^n} \qquad (xy)^n = x^n y^n \qquad \left(\frac{x}{y}\right)^n = \frac{x^n}{y^n}$$

$$x^{n/m} = \sqrt[m]{x^n} \qquad \sqrt[n]{xy} = \sqrt[n]{x}\sqrt[n]{y} \qquad \sqrt[n]{\frac{x}{y}} = \frac{\sqrt[n]{x}}{\sqrt[n]{y}}$$

Factoring

$$x^2 - y^2 = (x-y)(x+y)$$

$$x^3 - y^3 = (x-y)(x^2 + xy + y^2)$$

$$x^3 + y^3 = (x+y)(x^2 - xy + y^2)$$

Binomial Theorem

$$(x+y)^2 = x^2 + 2xy + y^2$$

$$(x-y)^2 = x^2 - 2xy + y^2$$

$$(x+y)^3 = x^3 + 3x^2y + 3xy^2 + y^3$$

$$(x-y)^3 = x^3 - 3x^2y + 3xy^2 - y^3$$

$$(x+y)^n = x^n + nx^{n-1}y + \frac{n(n-1)}{2}x^{n-2}y^2 + \cdots$$

$$+ \binom{n}{k}x^{n-k}y^k + \cdots + nxy^{n-1} + y^n$$

where $\dbinom{n}{k} = \dfrac{n(n-1)\cdots(n-k+1)}{1\cdot 2\cdot 3\cdot\cdots\cdot k}$

Quadratic Formula

If $ax^2 + bx + c = 0$, then $x = \dfrac{-b \pm \sqrt{b^2 - 4ac}}{2a}$.

Inequalities and Absolute Value

If $a < b$ and $b < c$, then $a < c$.

If $a < b$, then $a + c < b + c$.

If $a < b$ and $c > 0$, then $ca < cb$.

If $a < b$ and $c < 0$, then $ca > cb$.

If $a > 0$, then

$|x| = a$ if and only if $x = a$ or $x = -a$

$|x| < a$ if and only if $-a < x < a$

$|x| > a$ if and only if $x > a$ or $x < -a$

GEOMETRY

Geometric Formulas

Formulas for area A, circumference C, and volume V:

Triangle	Circle	Sector of Circle

$A = \frac{1}{2}bh = \frac{1}{2}ab\sin\theta \qquad A = \pi r^2 \qquad \frac{1}{2}r^2\theta$ (θ in radians)

$C = 2\pi r \qquad s = r\theta$

Parallelogram	Trapezoid

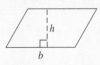

$A = bh \qquad A = \frac{1}{2}(a+b)h$

Sphere	Cylinder	Cone

$V = \frac{4}{3}\pi r^3 \qquad V = \pi r^2 h \qquad V = \frac{1}{3}\pi r^2 h$

$A = 4\pi r^2 \qquad\qquad A = \pi r\sqrt{r^2 + h^2}$

(lateral surface area)

Distance and Midpoint Formulas

Distance between $P_1 = (x_1, y_1)$ and $P_2 = (x_2, y_2)$:
$$d = \sqrt{(x_2 - x_1)^2 + (y_2 - y_1)^2}$$

Midpoint of $\overline{P_1 P_2}$:
$$\left(\frac{x_1 + x_2}{2}, \frac{y_1 + y_2}{2} \right)$$

Lines

Slope of the line through $P_1 = (x_1, y_1)$ and $P_2 = (x_2, y_2)$:
$$m = \frac{y_2 - y_1}{x_2 - x_1}$$

Slope-intercept equation of the line with slope m and y-intercept b:
$$y = mx + b$$

Point-slope equation of the line through $P_1 = (x_1, y_1)$ with slope m:
$$y - y_1 = m(x - x_1)$$

Equation of a Circle

Circle with center (h, k) and radius r:
$$(x - h)^2 + (y - k)^2 = r^2$$

TRIGONOMETRY

Angle Measurement

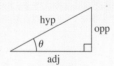

π radians $= 180°$ $\qquad 1° = \dfrac{\pi}{180}$ rad $\qquad 1$ rad $= \dfrac{180°}{\pi}$

Right Triangle Definitions

$$\sin \theta = \frac{\text{opp}}{\text{hyp}} \qquad \cos \theta = \frac{\text{adj}}{\text{hyp}} \qquad \tan \theta = \frac{\text{opp}}{\text{adj}}$$

$$\csc \theta = \frac{\text{hyp}}{\text{opp}} \qquad \sec \theta = \frac{\text{hyp}}{\text{adj}} \qquad \cot \theta = \frac{\text{adj}}{\text{opp}}$$

Trigonometric Functions

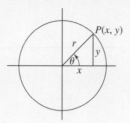

$$\sin \theta = \frac{y}{r} \qquad \cos \theta = \frac{x}{r} \qquad \tan \theta = \frac{y}{x}$$

$$\csc \theta = \frac{r}{y} \qquad \sec \theta = \frac{r}{x} \qquad \cot \theta = \frac{x}{y}$$

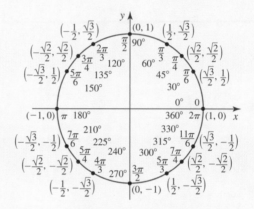

Graphs of Trigonometric Functions

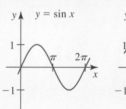
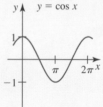

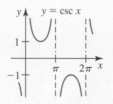
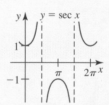

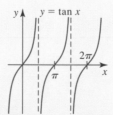
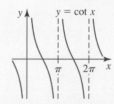